CARBOCYCLE CONSTRUCTION IN TERPENE SYNTHESIS

CARBOCYCLE CONSTRUCTION IN TERPENE SYNTHESIS

Tse-Lok Ho
The NutraSweet Co.

VCH

Tse-Lok Ho
The NutraSweet Company
601 E. Kensington Road
Mt. Prospect, IL 60056

Library of Congress Cataloging-in-Publication Data

Ho, Tse-Lok.
  Carbocycle construction in terpene synthesis.

  Bibliography: p.
  Includes index.
  1. Ring formation (Chemistry)   2. Terpenes.
3. Chemistry, Organic–Synthesis.   I. Title.
QD281.R5H6   1988   547.7'   87-21626
ISBN 0-89573-279-3

©1988 VCH Publishers, Inc.

This work is subject to copyright.

All rights are reserved, whether the whole or part of the material is concerned, specifically those of translation, reprinting, re-use of illustrations, broadcasting, reproduction by photocopying machine or similar means, and storage in data banks.

Registered names, trademarks, etc. used in this book, even when not specifically marked as such, are not to be considered unprotected by law.

Printed in the United States of America.

ISBN 0-89573-279-3 VCH Publishers
ISBN 3-527-26582-1 VCH Verlagsgesellschaft

Distributed in North America by:

VCH Publishers, Inc.
220 East 23rd Street, Suite 909
New York, New York 10010

Distributed Worldwide by:

VCH Verlagsgesellschaft mbH
P.O. Box 1260/1280
D-6940 Weinheim
Federal Republic of Germany

This book is dedicated to

Elias J. Corey

for his seminal contributions to algorithmic exposition and innovations of organic synthesis

Preface

Fundamentally, organic synthesis involves the formation of chain and/or ring structures that may be adorned with various functional groups and heteroatoms. Over the years chemists have focused substantial portions of their efforts on the development of methods for the construction of carbocycles, chiefly because these compounds offer great variety and challenge. Thus, the synthesis of dodecahedrane represents an incandescent achievement of modern organic chemistry. This enchanting molecule of pristine symmetry, which possesses endless interlocking cyclopentane rings, is the hydrocarbon version of a Platonic solid.

Among naturally occurring substances the terpenes constitute a most intriguing family. While biosynthetic schemes depicting assemblage of linear terpenes from acetates via branched five-carbon subunits ("isoprene") are relatively simple, the ways by which ring structures are made from these polyenes seem to know no bounds.

The most spectacular subclass of terpenes is the sesquiterpenes. Nature is able to fashion an undistinguished chain intermediate of 15 carbon atoms into intricate cyclic arrays that span from 3- to 11-membered rings and their combinations, featuring condensed, spiro, and bridged systems.

On the basis of structural variegation, the other terpene groups pale in comparison with the sesquiterpenes. However, their diversity is probably still unsurpassed among natural products, especially when one contemplates the numerous meroterpenoids, which encompass the complex indole alkaloids.

Given such a rich lode of target molecules to test human ingenuity at emulating nature, it is not surprising that terpenes, particularly the sesquiterpenes, have kindled penetrating interests among synthetic chemists for more than three decades. Logically, the cyclic representatives attract the most activity.

In a total synthesis, each and every step is essential. However, the ring formation maneuvers involved in the construction of cyclic molecules must be considered more important than the others, at least in the conceptualization stage. These are the linchpins of synthetic endeavors. This book discusses the most common methods for ring formation and delineates their individual contributions in the terpene area. By comparing the intermediate structures of these compounds, and noting the choice of them as stepping-stones toward the target molecules, the current rubric of synthesis often can be deciphered. It is hoped that this compilation will stimulate further creative thoughts, which will bring forth deeper insights into the strategy and methodology needed to remedy any remaining inadequacies in the field of synthetic chemistry.

In preparing this volume, collection of pertinent material from the chemical literature essentially was restricted to syntheses of naturally occurring terpenes or known precursors. Furthermore, the highly oxygenated members derived from the more barren terpenes having identical skeletons were omitted unless their syntheses offer different annulation strategies. In rare circumstances I have included methods for the preparation of advanced intermediates deemed useful for eventual elaboration into natural products.

It is assumed that the reader possesses a general knowledge of most of the fundamental synthetic protocols and their reaction mechanisms. For space considerations, the introduction to each chapter is usually brief. Many excellent textbooks on synthesis are available in case a review of the subject matter is desired.

I believe that a middle ground should be struck to maximize the value of the present treatment. The mere cataloguing of reactions by showing the reactant(s) and the immediate product(s) has far less heuristic impact than indicating also the final target molecules. On the other hand, detailed exegesis of the whole scheme for each synthesis would have grossly exceeded the scope and intended length of this book. Thus occasional interjections of strategies in various places are appropriate.

I am aware of the inadequacy in not discussing the often interesting and ingenious steps leading to the procyclic intermediates. Again, space limitation dictates this painful omission.

A few words about the mixed treatment of the subject matter are in order. Unlike Chapters 1 to 9, Chapters 10, 11, and 12 are not organized on the basis of theme reactions. However, the merit of the separate grouping is self-evident.

This book owes its birth to the support and indulgence granted me by my family during interminable periods of alienation when I totally lost myself in the writing/drawing process. I must not fail to acknowledge my indebtedness to Honor, Jocelyn and Daphne.

Finally, I wish to thank Professors R. K. Boeckman, Jr., E. J. Corey, M. T. Crimmins, B. Fraser-Reid, R. L. Funk, K. E. Harding, P. Helquist, T. Hudlicky, E. R. Koft, C. Kwasigroch, H-J. Liu, A. P. Marchand, E. A. Mash, G. Majetich, J. A. Marshall, L. A. Paquette, M. C. Pirrung, G. H. Posner, W. Reusch, J. Rigby, B. B. Snider, D. D. Sternbach, D. Ward, and F. E. Ziegler for providing preprints of publications and/or unpublished results.

October, 1987
Mt. Prospect, IL

Notes for the Formulas

While most of the syntheses described in this volume gave rise to racemic products, the formulas indicate the natural isomers whenever possible or convenient. When an equation depicts a common intermediate leading to two or more terpene molecules that happen to possess different absolute configurations at one or more asymmetric centers already present in the intermediate, perforce an enantiomeric modification of one or more of the terpenes must be represented for the sake of consistency (e.g., vetivone **V16,** hinesol **ent-H15**).

Optically active terpenes that have been synthesized, either in the natural or antipodal forms, are so indicated in the text.

In certain formulas an asymmetric center is circled or starred, to denote that an epimeric mixture is produced at that center.

Contents

1	**Introduction**	**1**
2	**Robinson Annulation**	**3**
	2.1 Annulation of Monoterpenes	4
	2.2 Annulation of Methylated Cyclohexanones	14
	2.3 Annulation of 2-Carboalkoxycyclohexanones	24
	2.4 Annulation of 2-Methyl-1,3-cyclohexanedione Wieland–Miescher Ketone and Its Homologs	28
	2.5 Annulation of 1,4-Cyclohexanedione Derivatives	47
	2.6 Annulation of β-Tetralone and Its Analogs	49
	2.7 Annulation of Cyclopentanones	56
	2.8 Miscellaneous	61
3	**Aldol Condensation**	**65**
	3.1 Monocyclic Enones	66
	3.2 Spirocyclic Intermediates	73
	3.3 Bridged Ring Synthesis	77
	3.4 Condensed Rings	91
	3.4.1 Endocyclic Enones	91
	3.4.2 Aldolization Involving Exocyclic Donors	119
	3.4.3 Aldolization Involving Intracyclic Donors	136
	3.5 Twofold Aldol Condensation	137
	3.6 Miscellaneous Aldol Processes	138
4	**Cyclization by Michael–Aldol, Claisen, and Dieckmann Reactions**	**141**
	4.1 Stork Annulation	141
	4.2 Michael and Michael–Aldol Tandem	144
	4.3 Stepwise Michael–Aldol Cyclization	149
	4.4 Wichterle Annulation	160
	4.5 Intramolecular Michael Addition	162
	4.6 Michael–Claisen and Wittig Condensations	174

4.7	Claisen Reactions	179
4.8	Dieckmann and Related Cyclizations	190

5 Intramolecular Alkylation 205

5.1	Six- and Seven-Membered Rings	206
5.2	Five-Membered Rings	215
5.3	Three- and Four-Membered Rings	230
5.4	Medium-Sized Rings	240
5.5	Unconventional Alkylations	247

6 Cationic Cyclization 256

6.1	Acylations		256
	6.1.1	Aromatic Acylations	256
	6.1.2	Nonaromatic Acylations	262
6.2	Alkylations		271
	6.2.1	Alkylation with Aldehydes: Acid-Catalyzed Ene Reactions	271
	6.2.2	Alkylation with Cations: Polyene Cyclizations	277
	6.2.3	Polyene Cyclizations Initiated by Soft Acids	300
	6.2.4	Cyclizations Initiated by Epoxide Opening	309
	6.2.5	Cycloalkylation of Enones	314
	6.2.6	Transannular Cyclizations	326

7 Diels–Alder Reactions 329

7.1	Isoprenes as Diene Units		331
7.2	Piperylene and Its Homologs as Diene		341
7.3	Other Alkyldienes as Diene Units		345
7.4	1,3-Butadiene as Diene Unit		354
7.5	Oxygenated Dienes		360
7.6	Cyclopentadienes		374
7.7	Cyclohexadienes		381
7.8	Cyclohexenone Synthesis		392
7.9	Furans as Dienes		395
7.10	Other Applications of Diels–Alder Adducts		398
7.11	Intramolecular Diels–Alder Reactions		399
	7.11.1	Condensed Rings	399
	7.11.2	Bridged Rings	418
	7.11.3	Reactions Involving Heterocyclic Dienes	428
7.12	Heterodienes		432

8 Other Thermal Processes 433

8.1	[3 + 2]Cycloadditions		433
	8.1.1	Nitrones	433
	8.1.2	Oxyallyl Cations	437
	8.1.3	Trimethylenemethane Complexes	439

	8.2	[4 + 3]Cycloadditions	440
	8.3	[5 + 2]- and [6 + 4]Cycloadditions	441
	8.4	Cycloacylation	445
	8.5	The Ene Reaction	448
	8.6	Cope Rearrangement and Related Processes	458
	8.7	Wittig and Claisen Rearrangements	467
	8.8	Electrocyclization	475
	8.9	Thermal Reorganization of Vinylcyclopropanes	484

9 Radical Cyclization 489

- 9.1 Single Radicals 489
- 9.2 1,1-Diradicals 497
- 9.3 1,3-Diradicals 501
- 9.4 1,n-Coupling 508
- 9.5 Acyloin Condensation 515

10 Synthesis of Small-Ring Compounds 516

- 10.1 Cyclopropanes 516
  - 10.1.1 The Simmons–Smith Reaction 516
  - 10.1.2 Dihalocarbene Addition to Alkenes 517
  - 10.1.3 C-Substituted Carbene and Carbenoid Addition to Alkenes 519
  - 10.1.4 Cyclopropanes from 1,3-Dipolar Adducts 525
  - 10.1.5 Solvolytic Ring Closure 528
  - 10.1.6 Cyclopropanation Using Other 1,1-Dipolar Reagents 530
  - 10.1.7 Miscellaneous Cyclopropanations 531
- 10.2 Cyclobutanes 532
  - 10.2.1 Photoinduced Cycloaddition 532
  - 10.2.2 Thermal Cycloaddition 541
  - 10.2.3 Ring Expansion and Cycloalkylation 544

11 Ring Expansion and Contraction 546

- 11.1 Ring Expansion 546
  - 11.1.1 Expansion of Three- and Four-Membered Rings 546
  - 11.1.2 Expansion of Five- and Six-Membered Rings 563
- 11.2 Ring Contraction 576
- 11.3 Bicyclic Disproportionation 590
  - 11.3.1 Concerted Processes 590
  - 11.3.2 Stepwise Processes 602

12 Transitory Annulation 604

- 12.1 Cyclopropanes 604
  - 12.1.1 Cyclopropanes as Latent Methyl and Functionalized Methylene Groups 604

	12.1.2	One-Carbon Insertion via Cyclopropanation; Ring Expansion	615
	12.1.3	Indirect Alkylation Through Cyclopropanation	619
	12.1.4	Preparation of γ-Functionalized Carbonyl Substances	623
	12.1.5	Ring Contraction via Bicyclo[3.1.0]hexanones	629
	12.1.6	Cyclopropanocycloalkanone Synthons for Stereocontrol	632
	12.1.7	Chain Building Based on Cyclopropanes	634
12.2	Cyclobutanes		635
	12.2.1	Reductive Alkylation of Enones via [2 + 2]Cycloaddition	640
	12.2.2	Additive 1,2-Dialkylation of Alkenes	649
	12.2.3	Alkene Metathesis by a Tandem Photothermal Process	660
	12.2.4	Reductive *gem*-Dialkylation of Carbonyl Compounds	662
12.3	Cyclopentanes		663
	12.3.1	Condensed Systems	663
	12.3.2	Bridged Systems	666
12.4	Cyclohexanes		674
12.5	Heterocycles		697
12.6	Retrograde Diels–Alder Reactions		697

References   **701**

Addendum   **732**

Compound Index   **761**

Chapter 1

Introduction

Alicyclic compounds—those containing one or more rings of carbon atoms—have riveted the attention of innumerable investigators since the dawn of organic chemistry by virtue of their ubiquity in the environment. Preeminent among the alicyclics are the terpenes.

Not only did the formulation of the benzene structure by Kékulé give great impetus to the development of aromatic chemistry, the cyclic nature of the aromatic compounds has had more than a passing influence on the concurrent scrutiny of the more saturated ring substances. It is also true that the most fascinating facets of chemical topology, including stereoelectronic and conformational effects, are ramifications of sustained studies of alicyclic compounds.

The discovery of carbocycles of assorted sizes in various entanglements in natural products has continued to arouse the chemist's desire to duplicate them in the laboratory, besides serving as conclusive structural evidence. In recent years, highly strained unnatural molecules such as tetrahedrane have presented an almost irresistible synthetic challenge to organic chemists. Likewise, topologically remarkable substances (Walba, 1985) including the catenanes (interlocking rings), rotaxanes (threaded rings), knots, and molecular Möbius strips are macrocyclic entities with formidable and tantalizing carbocyclic representatives.

Carbocyclic construction is more critical than the mere adaptation of a C—C or C=C bond formation process. The ease of cyclization falls off in both directions as the number of ring atoms increases or decreases from five and six. This is because strain factors (angular and torsional) discourage formation of the smaller three- and four-membered rings, and the distance factor (entropy) becomes more and more unfavorable as the reactive end groups of a carbon chain are further separated, to the degree that chemists at one time despaired of ever attaining macrocyclic substances.

We are at the threshold of an exciting era that can boast that no reasonable organic molecule is synthetically inaccessible. In reaching this phenomenal level of sophistication, the contribution of carbocyclic construction is paramount. Demonstration of the most important methodologies is often found in the total synthesis of terpenes.

Ring-forming reactions generally do not lend themselves to mechanistic classification, especially regarding their limitations. On the other hand, these

Table 1-1. Selectivity of Ring Closure

Conformation	Cyclization mode	Favored	Disfavored
Tetrahedral	exo	3- to 7-membered	—
Tetrahedral	endo		5- to 6-membered
Trigonal	exo	3- to 7-membered	—
Trigonal	endo	6- to 7-membered	3- to 5-membered
Diagonal	exo	5- to 6-membered	3- to 4-membered
Diagonal	endo	3- to 7-membered	—

cyclizations are subject to stereochemical control about which a set of rules has been deduced (J. E. Baldwin, 1976). These rules provide a useful guide for the selection of proper precursors.

Two types of ring closure are evident. The one designated as *exo* involves bond breaking outside the ring to be created, whereas the *endo* closure connotes a shift of an electron pair toward an atom inside the emerging ring. Realistically, attack on a tetrahedral carbon atom in the endo mode does not lead to a cyclic product.

An ionic cyclization reaction involves backside displacement of a nucleofugal group by an anionic species or a heteroatom with lone-pair electrons. The facility of such an intramolecular S_N2 process depends on the span of the linking chain and hence the size of the ring to be formed. The leaving group could also be a pair of π electrons.

The Baldwin rules (Table 1-1) deal only with the formation of three- to seven-membered rings because restrictions no longer exist for the homologs. In such cases one can always find an active conformation in which the extreme requirement of three-atom collinearity involves the nucleophile, the electrophile, and the leaving group, barring the presence of bulky substituents in the intervening chain, which inhibit such conformation to the extent of exclusivity. Indeed, nonbonding interactions account for much of the inefficiency in achieving closure of 8- to 15-membered rings by conventional methods, in addition to the entropic effects, which bias toward intermolecular reactions. The classical solution to this problem is the employment of a high-dilution technique, which decreases the probability of an encounter between reactive centers from two different molecules. Much better yet hardly general methods involve reactions on a metal surface or template whereby two reactive sites are temporarily immobilized and confined within a short (bonding) distance. The acyloin condensation and the McMurry reaction are distinguished examples.

Chapter 2

Robinson Annulation

The Robinson method for cyclohexenone formation was originally developed in the realm of steroid total synthesis (Rapson, 1935). Essentially, a ketone enolate adds to an α,β-unsaturated ketone, which is generated *in situ* from the corresponding β-oxoalkyl ammonium compound. The Michael adduct (1.5-diketone) is cyclized under the influence of the base present, and the resulting aldol undergoes dehydration.

The major difficulties associated with the Robinson annulation (Gawley, 1976) are as follows: the tendency of β-unsubstituted α,β-enones to polymerize, which actually forced the use of the quaternary ammonium salts; the low yields of cyclohexenone products when the reactions are allowed to run their course; invariant regioselectivity attending annulation of unsymmetrical ketones, such that annulation toward the less substituted α carbon (when desired) cannot be simply diverted; and finally, the frequent intervention of dialkylation.

The usefulness of the structures characterized by the Robinson annulation products undoubtedly has contributed to the tremendous effort to find improved conditions and alternative methods. It has been shown that better yields often result from the stepwise Michael addition/cyclodehydration sequence or the Michael–aldol/dehydration (of ketols) sequence. In certain cases involving acid-stable compounds, the condensations carried out in acid media prove superior.

The enone polymerization problem (cf. Ziegler, 1983) may be circumvented by α-substitution (e.g., silyl group) or replacing the Michael acceptors with iron–enone complexes. On the other hand, activation of the enolates by forming α-hydroxymethylene derivatives or using the corresponding 1,3-diketones helps. Enolates may be generated *in situ* from enol derivatives (e.g., enol silyl ethers). The activation protocol also serves to suppress dialkylation.

Regiochemistry is ensured by blocking techniques. Common blocking groups include α-benzylidene and α-heteromethylene derivatives, and α-dithioacetals.

A regiochemical reversal can be achieved by the use of enamines (Stork, 1963a; Kuehne, 1970*), because (cyclic) enamines tend to localize their double bonds at the less substituted carbon atoms.

---

* When an author is represented by multiple references in a single year, and no letter(s) are given in the text citation, it is to be understood that the citation incorporates all references listed for the year shown. Thus the asterisked item designates Kuehne, 1970a, 1970b.

Other variants of the Robinson annulation include the Wichterle reaction (Wichterle, 1948; Lansbury, 1972), the isoxazole method (Stork, 1967) in which the alkylation step is an S_N2 displacement instead of 1,4-addition. The products require varying degrees of modification to unveil the β-oxoalkyl side chain.

2.1 ANNULATION OF MONOTERPENES

The numerous sesquiterpenes based on a hydronaphthalene skeleton are most amenable to elaboration by the Ronbinson approach. Thus, the majority of the reported syntheses of eudesmane-type compounds involve annulation as a key step.

Because the reaction is under thermodynamic control, 2-substituted cyclohexanones invariably give angularly substituted octalones as major products. A crucial aspect of eudesmane synthesis from disubstituted cyclohexanones such as the menthanones concerns stereoelectronic control in that the formation of the first C—C bond (Michael addition) is favored by axial attack. With 2,5-dialkylcyclohexanones, the major products are those in which the two alkyl groups remain or become trans to each other, regardless of the initial ratio of isomers present in the cyclohexanones.

Thus, (+)-dihydrocarvone gives only about 3 % α-cyperone (**C95**); the major product is a ketol in which the angular methyl group is trans to the isopropenyl group (Howe, 1955).

(+)-α-Cyperone can be converted to (+)-carissone (**C27**) (Pinder, 1961) by epoxidation followed by a redox sequence (LiAlH$_4$; MnO$_2$). Removal of the ketone group of carissone via the dithioketal leads to γ-eudesmol, **E19** (Pinder,

1963). When carissone is submitted to Li–NH₃ reduction and the Bamford–Stevens reaction, α-eudesmol (**E17**) (Humber, 1967) is formed.

C27   **E19**

E17   **S20**

Reduction of the enone system of α-cyperone under thermodynamic control (Na-*n*PrOH), and pyrolysis of the borate derived from the bicyclic alcohol lead to α-selinene (**S20**) (Chetty, 1966).

For rapid stereoselective access to (+)-α-cyperone (Fringuelli, 1969) the use of (—)-3-caranone is profitable. The crucial C—C bond formation is directed away from the β face by the dimethylcyclopropane ring. Another significant feature is the regioselective cleavage of the cyclopropane ring. Apparently a severe steric compression between the angular methyl group and the *syn*-methyl in the cyclopropane ring is the main factor in facilitating the cleavage, and the avoidance of 1,3-diaxial interaction in the product accounts for the selective bond scission.

C95

In a biomimetic approach to cyperolone (**C94**) (Hikino, 1966), an epoxy ketone is crafted from α-cyperone. A circuitous oxidation–reduction sequence is required to accomplish this goal because the β-epoxide cannot be obtained directly.

While the 3β-alcohol promptly assists β-epoxidation of the Δ^4-double bond, the resulting epoxy alcohol undergoes hydride transfer to give the decalolone instead of a ring-contracted hydrindanol. Consequently, reoxidation of the alcohol, rearrangement of the epoxy ketone, and again a tedious differentiation of the oxidation levels at two carbon sites must be carried out.

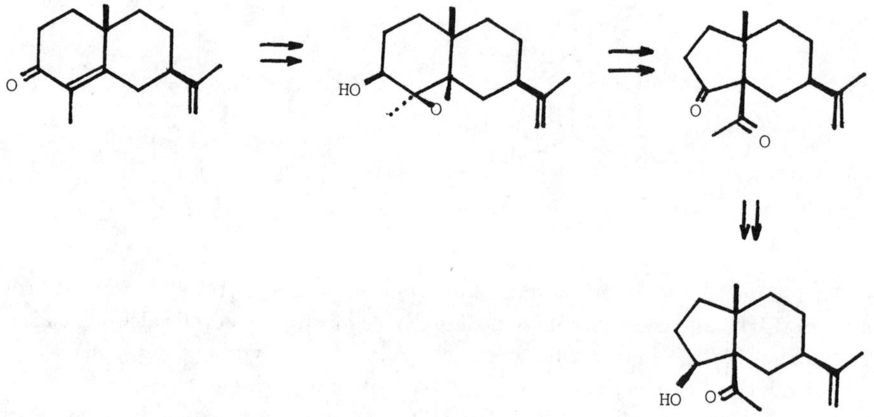

C94

Extensive oxidative readjustment at the A ring of α-cyperone and a net hydration at its isopropenyl group lead to a tetraol. Cleavage of the α-glycol unit followed by reductive cyclization then constitute a biomimetic synthesis of phytuberin (**P28**) (Murai, 1978).

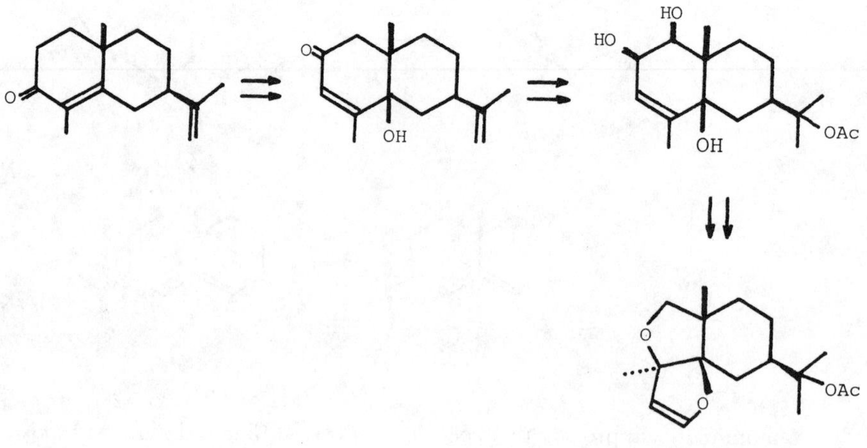

P28

2.1 MONOTERPENES

The major annulation product from dihydrocarvone is useful for synthesizing many other sesquiterpenes, including epi-γ-selinene (**S22**) (Klein, 1970), and α-agarofuran (**A15**) (Barrett, 1967; Asselin, 1968; Marshall, 1968e).

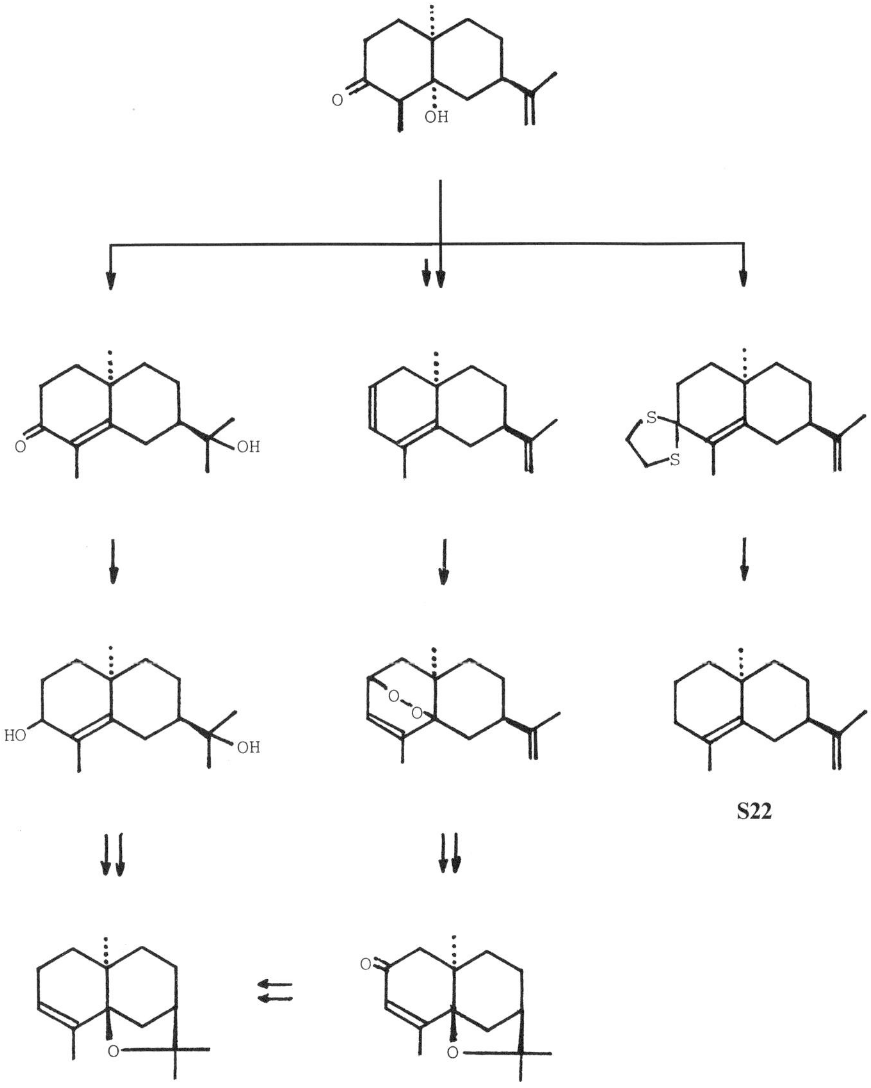

For a synthesis of (+)-occidentalol (**O2**) (Heathcock, 1972), the hydrate of (+)-dihydrocarvone (from hydroxymercuration–demercuration) is employed in the Robinson annulation. The epicarissone intermediate has been used also in an elaboration of evuncifer ether (**E21**) (Baker, 1977), a defensive secretion of the West African soldier termite.

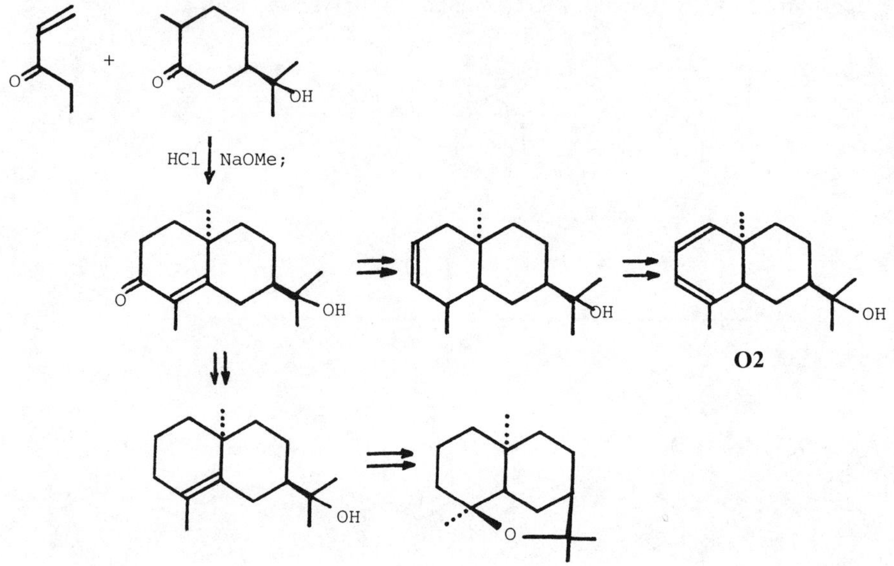

E21

Epi-α-cyperone serves as the starting material for the construction of maaliol (**M1**) (Büchi, 1964b). A cyclopropane ring is formed by connecting C_6 to C_{13}, followed by modification of the A ring via the Δ^3 or $\Delta^{4(12)}$ hydrocarbon.

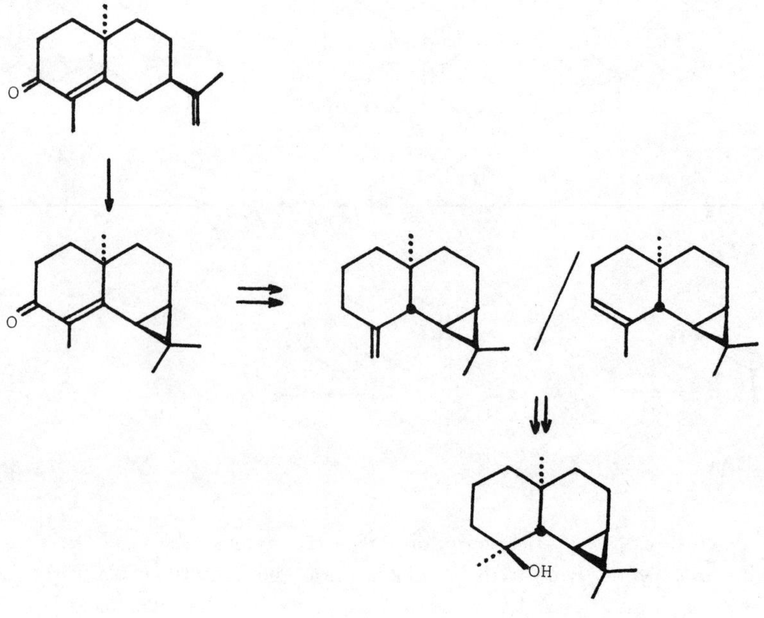

M1

2.1 MONOTERPENES

Less obvious applications of epi-α-cyperone (**C96**) include the syntheses of carotol/daucol (**C29/D9**) (deBroissia, 1972), (—)-ylangocamphor (**Y3**), (+)-sativene (**S11**), and (+)-cyclosativene (**S13**) (Piers, 1973a, 1975c).

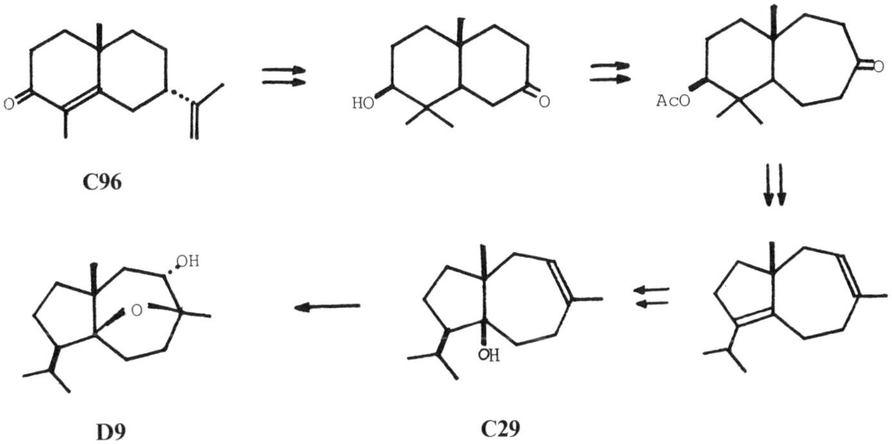

In the former exercise the isopropenyl chain is oxidatively extricated and a systematic B-ring expansion and A-ring contraction are involved. In the latter, the A ring is destroyed and the remnants become the elements of a bridged ring system.

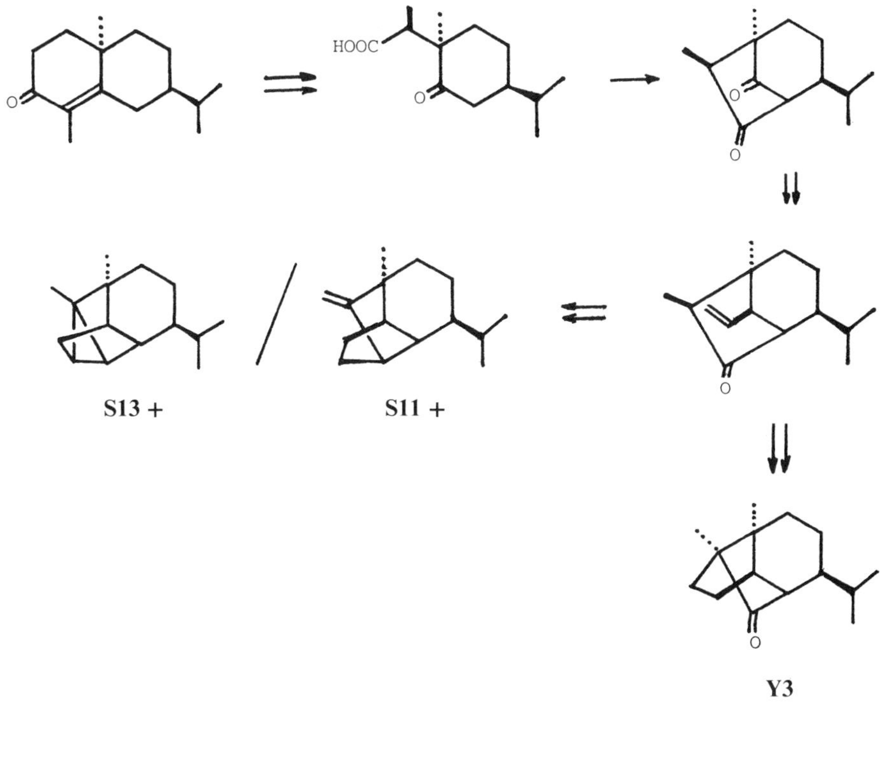

There is little difficulty in preparing (+)-β-cyperone (**C97**) (Roy, 1954) by the Robinson method because it has only one asymmetric center. For acquiring the (+)-isomer (J. P. Kutney, 1980), bicyclic monoterpene ketones such as (—)-thujone must be used. However, additional steps are necessary for cleaving the extra ring.

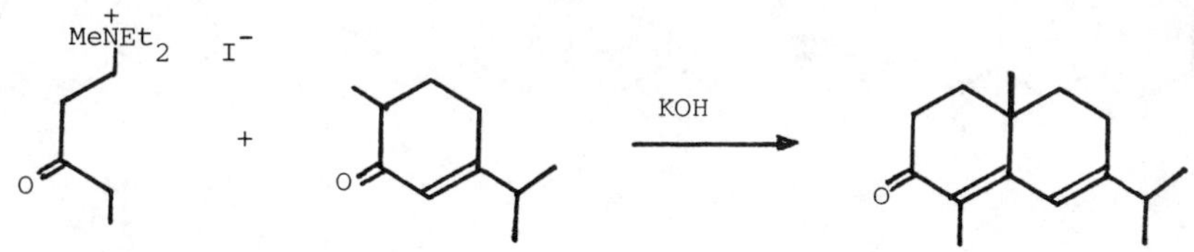

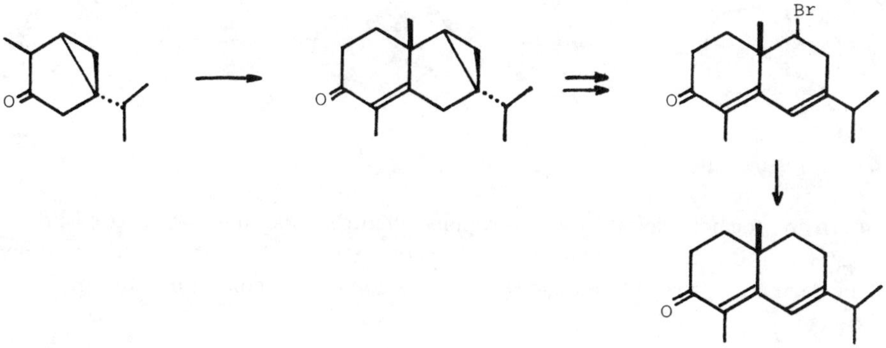

For (+)-β-cyperone → (+)-carissone, see Kutney, 1982b.)

From (—)-dihydrocarvone, (+)-β-eudesmol (**E18**) (Houghton, 1966; Humber, 1967) has been synthesized. The isopropenyl group is the controlling element in establishing the β-configuration of the angular methyl. Cleavage of the double bond at the bicyclic ketol stage allows equilibration of C_7 (axial → equatorial). This appendage also plays a role in ensuring the trans ring juncture when the A ring is elaborated into a methylenecyclohexane moiety.

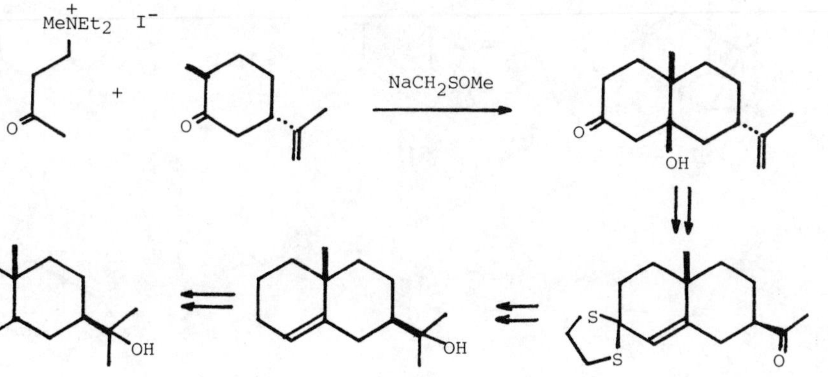

It should be noted that a 3-buten-2-one precursor is the Michael acceptor. The 14-carbon bicyclic ketol has found application in the synthesis of (+)-valeranone (**ent-V2**) (Marshall, 1968) from (—)-dihydrocarvone. The product is the enantiomer of the natural compound.

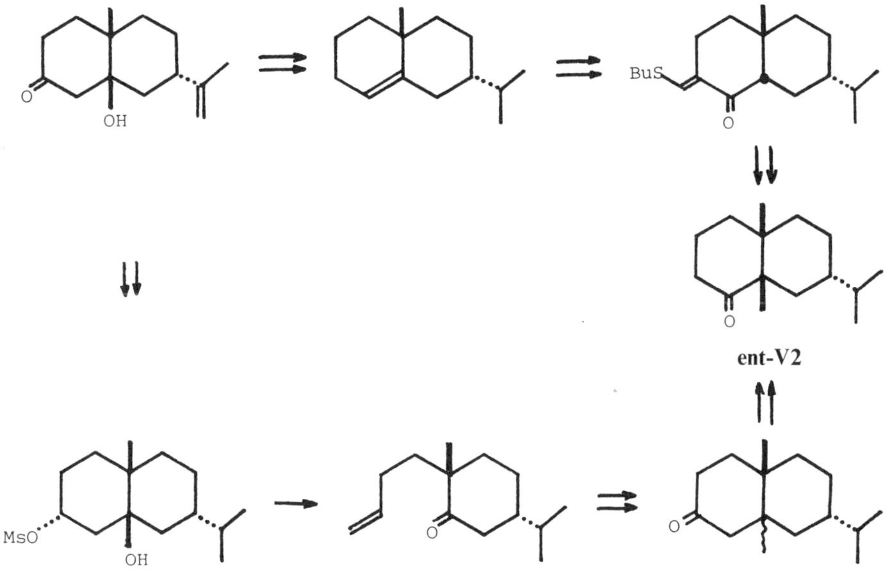

The angular methylation (18 %) of the decalone (blocked on one side with an *n*-butylthiomethylene group) was seriously plagued by competing *O*-methylation (56 %), and less so by the formation of the trans isomer (2 %).

The problem is avoided by way of OH-directed Simmons–Smith reaction and regioselective opening of the cyclopropane ring on treatment with acids. For the latter reaction an oxy function is incorporated in the vinyl ketone precursor (Wenkert, 1967).

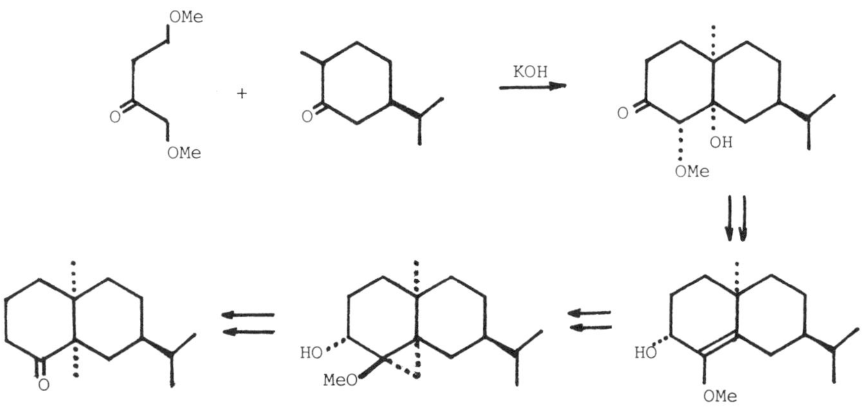

The enantiomer of (+)-axisonitrile (**A55**) has been synthesized from (+)-dihydrocarvone (Caine, 1978). A hydroxyl group is introduced at C_6 of the octalone intermediate via its dienol acetate, and the spirocyclic system is established via a photochemical rearrangement and reductive opening of the cyclopropane ring.

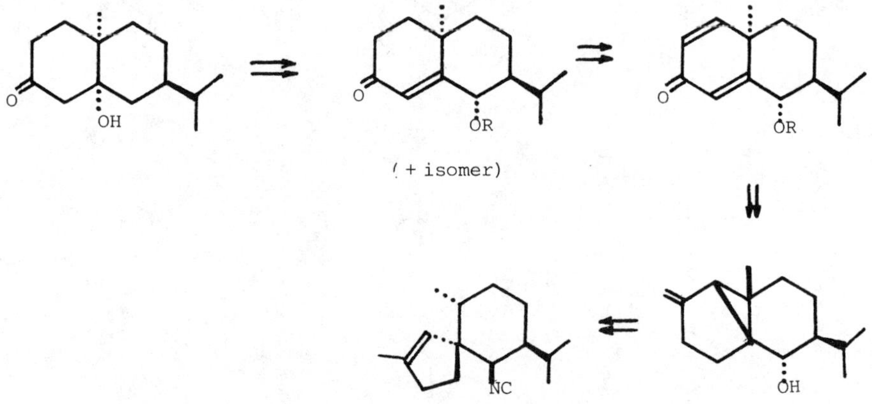

(+ isomer)

ent-A55

3-Phenylthiobut-3-en-2-one is a Michael acceptor with a potential leaving group. Its normal annulation products can therefore be converted into phenols by simple operations. The syntheses of 7-hydroxycalamenene (**C8**) and ferruginol (**F4**) (Takaki, 1984) serve to demonstrate the utility of the synthon.

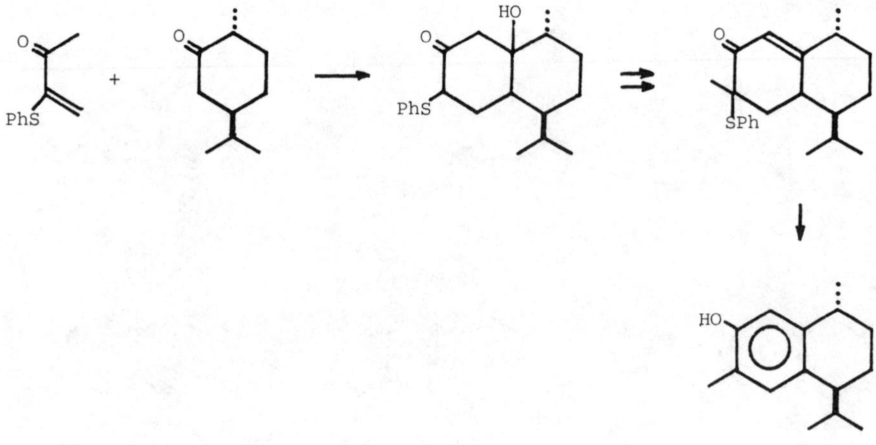

C8

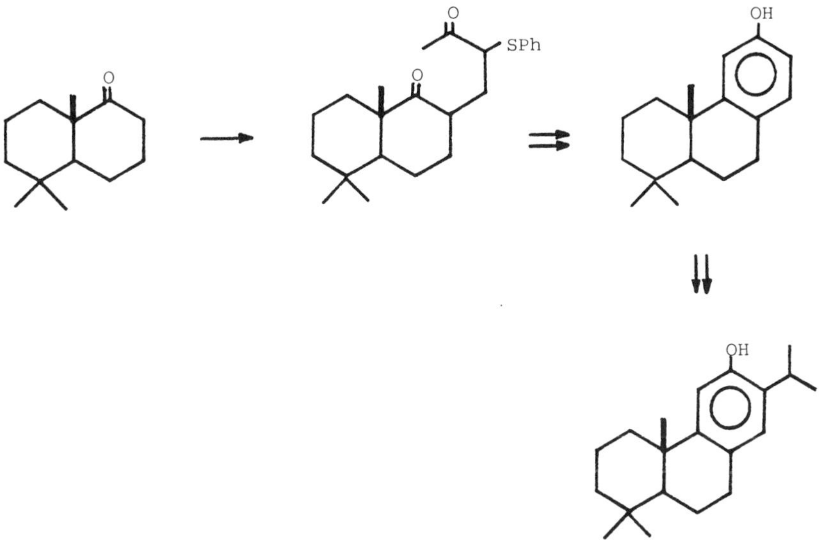

F4

(—)-β-Eudesmol (**ent-E18**) may be obtained from (+)-3-carene (Taticchi, 1969). Steric shielding imposed by the cyclopropane ring of 3-caranone enables annulation on the opposite face of the six-membered ring, thereby ensuring a cis relationship between the angular methyl group and the isopropanol chain. Interestingly, the cleavage of the cyclopropane ring on exposure to hydrogen bromide is regioselective. It reflects a thermodynamic control in which the movement of the quaternary cyclopropane carbon atom away from the angular methyl group is highly favored.

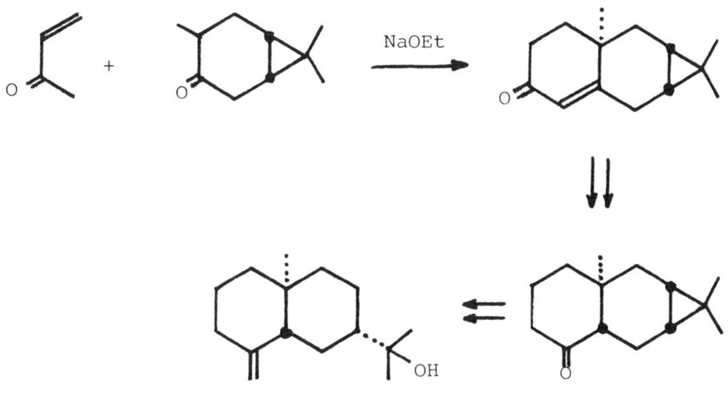

ent-E18

2.2 ANNULATION OF METHYLATED CYCLOHEXANONES

The simplest octalone derived from 2-methylcyclohexanone has been used in a synthesis of ($\pm$)-β-eudesmol (**E18**) (Marshall, 1965c, 1966c).

E18

Protection of the ketone simultaneously deconjugates the double bond, which is hydroborated. The requisite exocyclic methylene group is elaborated via the Wittig reaction on the diketone monoketal, which is equilibrated to establish a trans ring juncture. The ketal is hydrolyzed and the isopropanol chain is attached at the concluding stage of the synthesis.

An advanced intermediate has also been converted into β-selinene (**S21**) (Marshall, 1966c) costol (**C74**), costal (**C71**), and costic acid (**C72**) (Marshall, 1965b, 1966c).

S21	R= Me
C74	R= CH$_2$OH
C71	R= CHO
C72	R= COOH

Alternatively, the octalone can be reductively deconjugated via the dienol benzoate, and the derived halide (β-Cl or β-Br formed by virtue of homoallylic participation, which results in retention of configuration) is homologated by successive treatment with magnesium, carbon dioxide, diazomethane, and methylmagnesium iodide. The β configuration of the side chain is favored (equatorial). The exocyclic methylene group is then erected by the hydroboration–oxidation–Wittig protocol to complete β-eudesmol (**E18**) (Heathcock, 1968). β-Agarofuran (**A16**) is also formally accessible (T. R. Kelly, 1972) from the octalin.

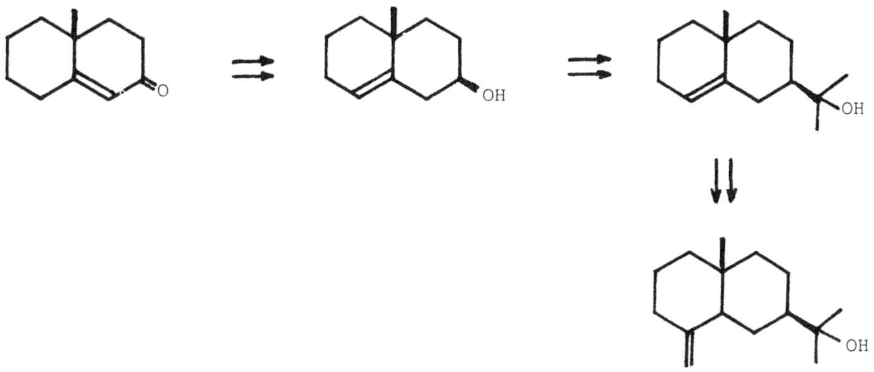

E18

4a-Methyl-1-decalone, readily available from the methyloctalone, is a useful compound for synthesizing β-gorgonene (**G17**) (Boeckman, 1975). Essentially the operation involves moving the enone system one carbon over toward the ring juncture, conjugate addition of three-carbon chain, and methylenation (via the Peterson method).

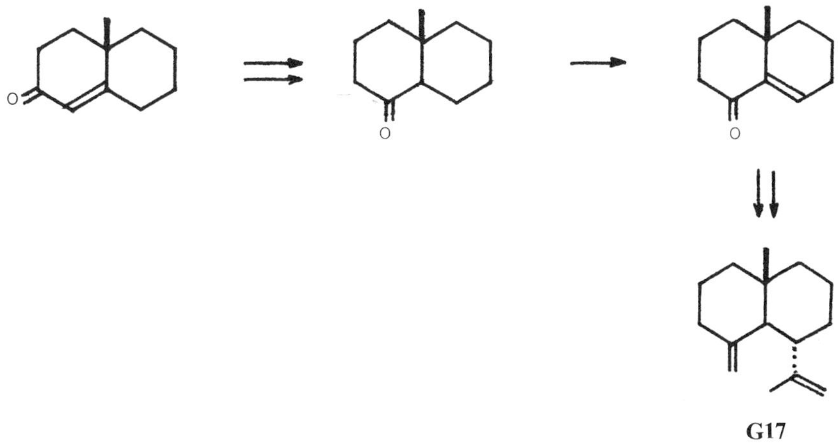

G17

The dimethylation of the octalone shifts the double bond to the 5,6-position. The subsequent removal of the ketone group and reintroduction of another at C_7 constitute the first half of a thujopsene (**T18**) synthesis (Dauben, 1963). The new enone is transformed into widdrol (**W3**) (Enzell, 1962) via a ring expansion process.

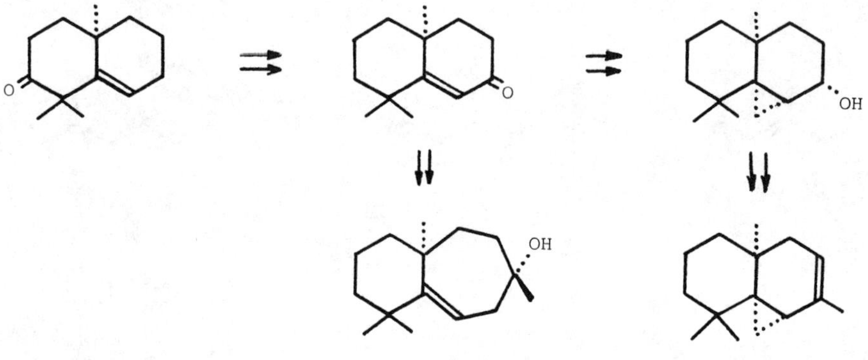

W3   **T18**

The trimethyloctalin has been used to synthesize acolamone (**A3**) and junenol (**J5**) (A. Banerjee, 1980). Interestingly, the equatorial methyl group at C_4 is removed via an oxidation directed by a hydroxyl group to be placed at C_6.

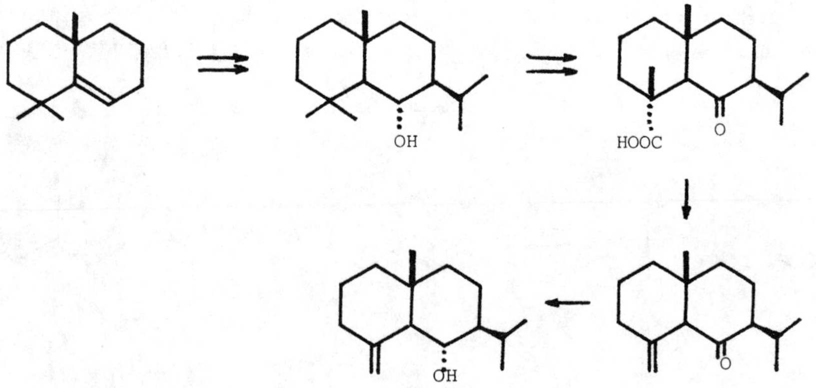

J5   **A3**

The 4-methyl homolog (terpene numbering) of the octalone has been elaborated into tuberiferine (**T34**) (Grieco, 1976b) and temsin (**T10**) (Nishizawa, 1978). The latter compound is an elemanolide, and synthesizing it from the octalone requires cleavage of the A ring. Introduction of the two double bonds is based on selenoxide thermolysis.

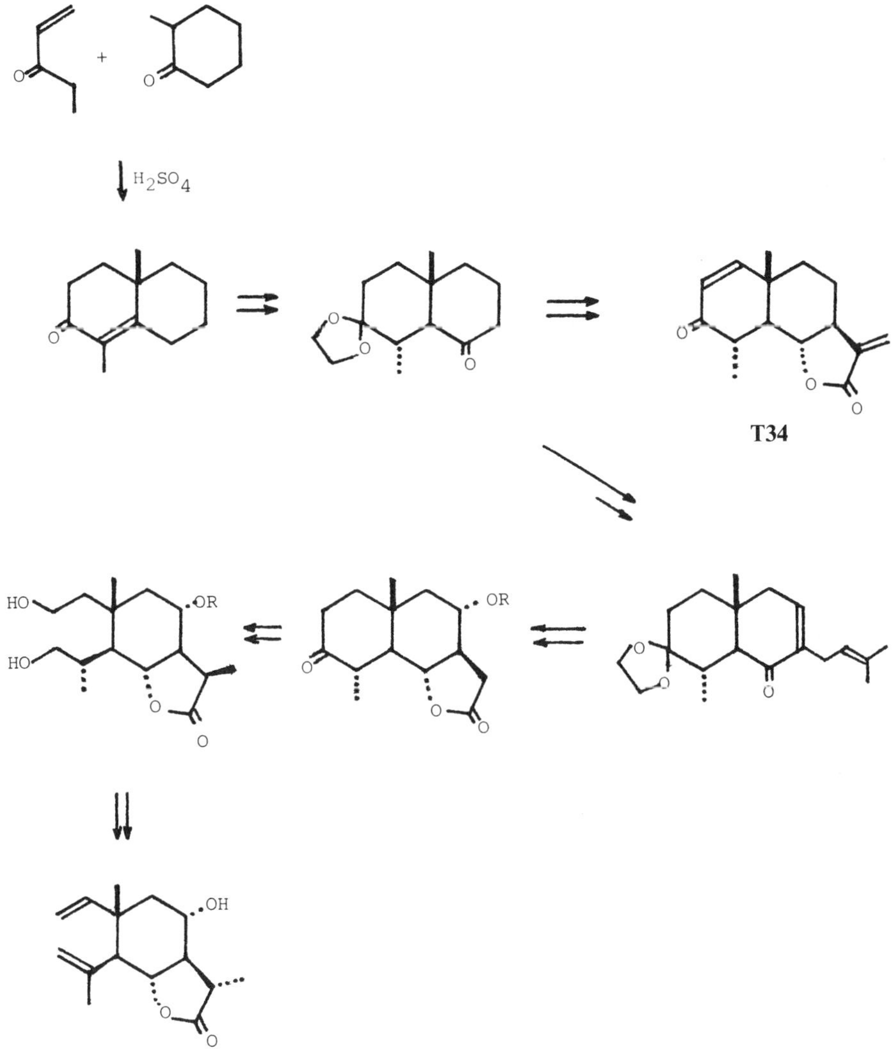

A classical approach to the quassinoids (Heathcock, 1984) involves a C + B + A ring construction scheme using the Robinson annulation method. A high-pressure Michael addition is employed in the attachment of the acetic acid chain to the C ring. Lactonization proves to be difficult; it is effected by exposure of allylic alcohol to pyridinium chlorochromate.

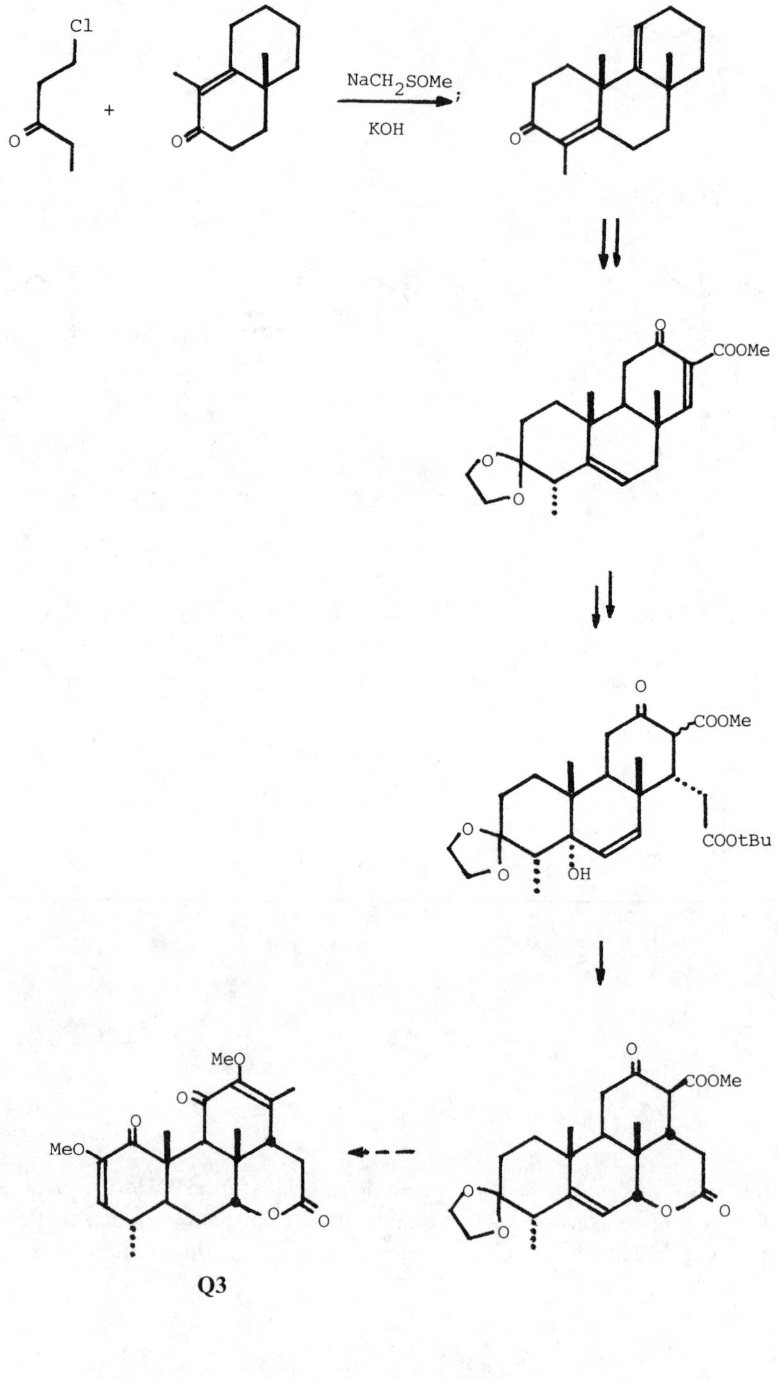

Q3

2.2 METHYLATED CYCLOHEXANONES

Dehydrogenation (with DDQ/TsOH) of the octalone extends the conjugation and permits addition of a cyano group to the far end (C_7). This compound serves as an intermediate for cycloeudesmol (**E20**) (Ando, 1985).

E20

The 6α,10β-dimethyl isomer, obtained from 2,6-dimethylcyclohexanone, serves well as starting material for frullanolide (**F14**) (Still, 1977b). Here the γ-lactone ring is assembled in a stereoselective manner via the Wharton and Claisen rearrangements. The β-stereochemistry is determined in the epoxidation step.

F14

Another frullanolide synthesis (Semmelhack, 1981b) also employs the same octalone. But the ring is cleaved and reconstructed at the same time as the γ-lactone.

The use of the dimethyloctalone in the preparation of geosmin (**G4**) (Marshall, 1968b) and cybullol (**C90**) (W. Ayer, 1976) is readily conceived. The more interesting application involves conversion into and photorearrangement of a cross-conjugated dienone with *in situ* solvolysis of the photochemical product to a spiro[4.5]decenone. The latter compound is a crucial intermediate for α-vetispirene (**V14**) (Caine, 1976a).

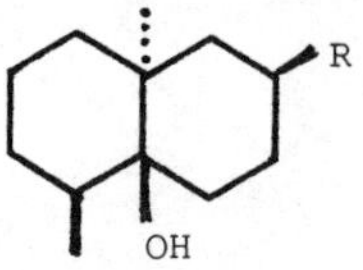

G4	R= H
C90	R= OH

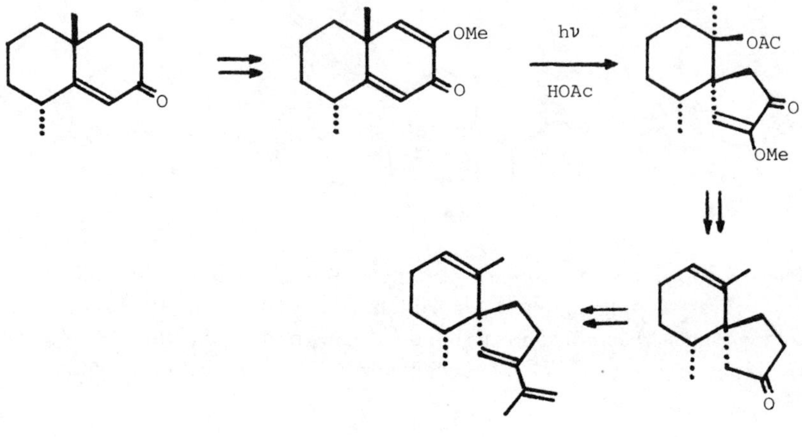

V14

A more straightforward route to the spirovetivanes (Caine, 1976b) involves the 10α-methyl isomer, which is obtained from the dienone by transfer hydrogenation.

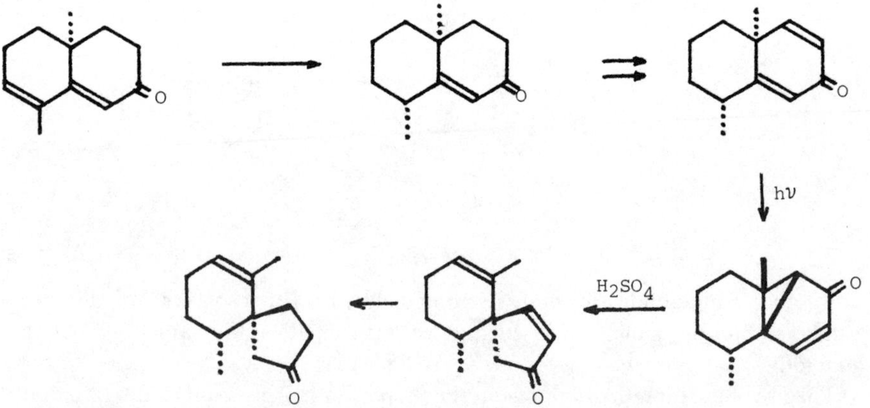

The isomeric hexalone undergoes Michael addition with methylmalonic esters to provide adducts suitable for the synthesis of α-santonin (**S9**), β-santonin (**S10**) (Abe, 1956), and their unnatural isomers.

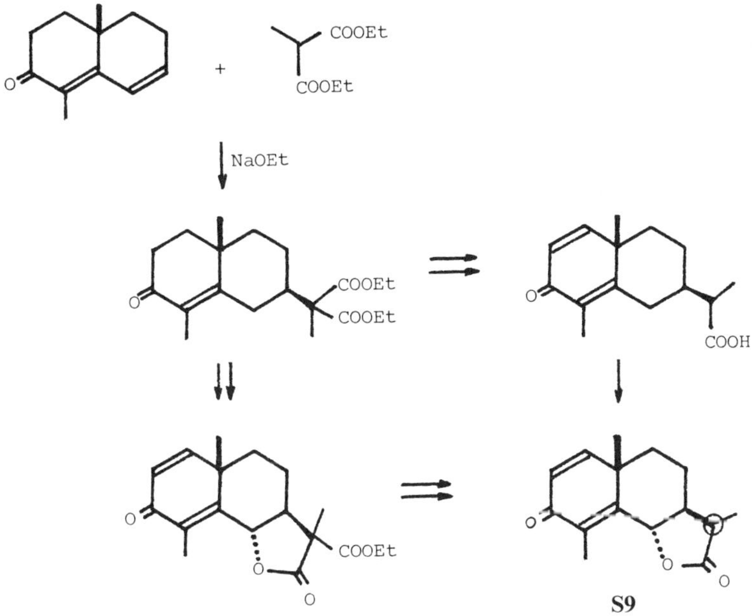

2.2 METHYLATED CYCLOHEXANONES

When the Robinson annulation product from methyl 2-(3-oxo-4-methylcyclohexanyl)propionate is used to elaborate the santonin skeleton, only isomers in which the γ-lactone is cis-fused to the hydronaphthalene skeleton are generated (Abe, 1953).

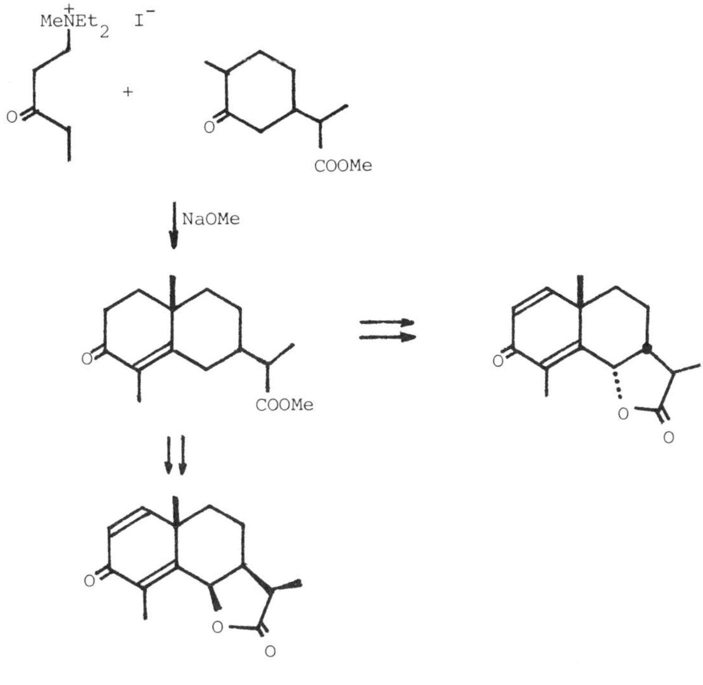

The eremophilane-type sesquiterpenes present a synthetic challenge in terms of the cis stereorelationship of the two vicinal methyl groups, and also the relationship of these methyls with the anglular hydrogen atom. 2,3-Dimethylcyclohexanone undergoes a Robinson annulation with 3-buten-2-one to give a low yield of the product, which consists of a 3:2 mixture of stereoisomers. However, this mixture has been used in a synthesis of aristolone (**A40**) (Berger, 1968). The separation of the isomers was achieved just before the addition of the remaining three-carbon unit.

cis 3
trans 2

A40

Stereoselective establishment of the cis relationship between the vicinal methyl groups of the eremophilane precursors can be accomplished by assigning the cyclohexanone as the B ring. Accordingly, the Michael acceptor is 3-penten-2-one. In the Michael step, the two reactants are marshalled in such a manner that steric repulsion is minimized, while the metal ion associated with the enolate is also coordinated with the enone oxygen to gain maximum electronic stabilization. The result is that the bicyclic product with the desired stereochemistry is favored.

It is not possible to obtain stereoselectively nootkatone (**N11**) from 2-methyl-4-isopropenylcyclohexanone. In fact, 7-epinootkatone (**N11e**) is produced (Odom, 1972) in far greater amounts (9:1). The situation is exactly the same as that encountered in the synthesis of α-cyperone. Axial approach of the enone to the enolate is favored during the first C—C bond formation process.

If C_6 of the cyclohexanone is trigonalized, the reaction becomes stereorandom (Takagi, 1978). In the transition state, the ring is probably quite flat.

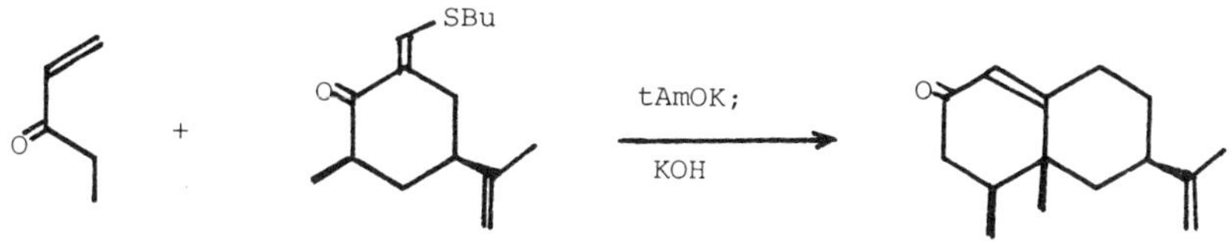

N11e

+

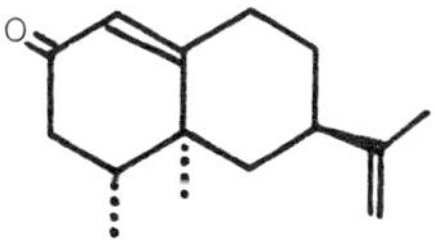

N11

While the isopropenyl group of nootkatone is trans to the methyl group, an opposite relationship exists in eremophilone. For this reason 7-epinootkatone formed directly from a Robinson annulation of 2-methyl-4-isopropenylcyclohexanone is most useful for the functional group transformation into eremophilone (**E13**) (McMurry, 1975).

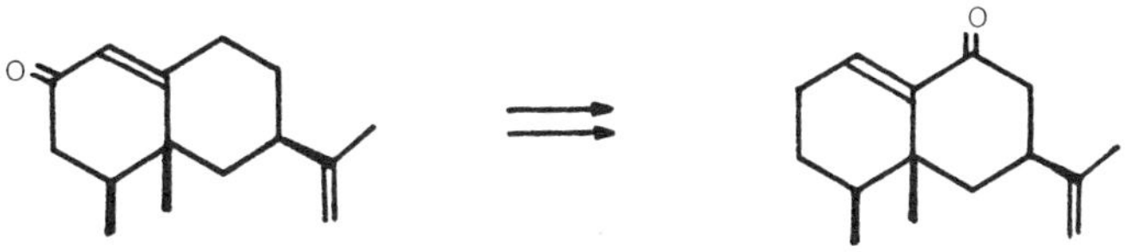

E13

The single example of Robinson annulation on a monoprotected 3-methyl-1,2-cyclohexanedione for terpene synthesis is an access of an optically active synthon for the eudesmanes (Utake, 1986).

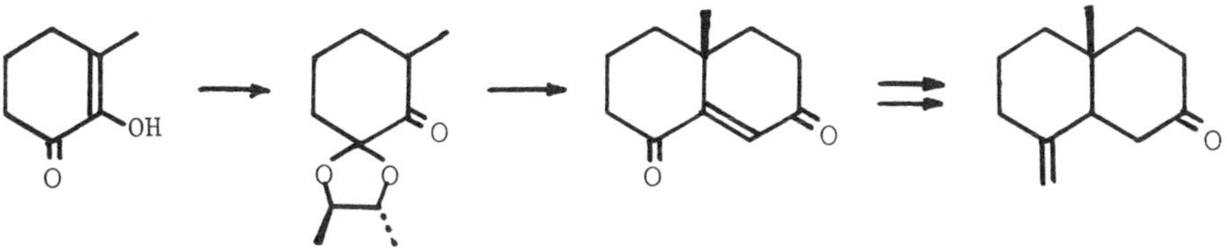

2.3 ANNULATION OF 2-CARBOALKOXYCYCLOHEXANONES

Vernolepin (**V10**) is an antitumor sesquiterpene dilactone that is biosynthesized from a eudesmane precursor by oxidation at the angular methyl and A-ring cleavage. Consequently, a synthesis (Grieco, 1976a, 1977b) vaguely patterned after the central theme of A-ring modification also calls for initial oxygenation of the angular substituent. In the event, submission of 2-carbethoxycyclohexanone to Robinson annulation becomes the first step of the adventure.

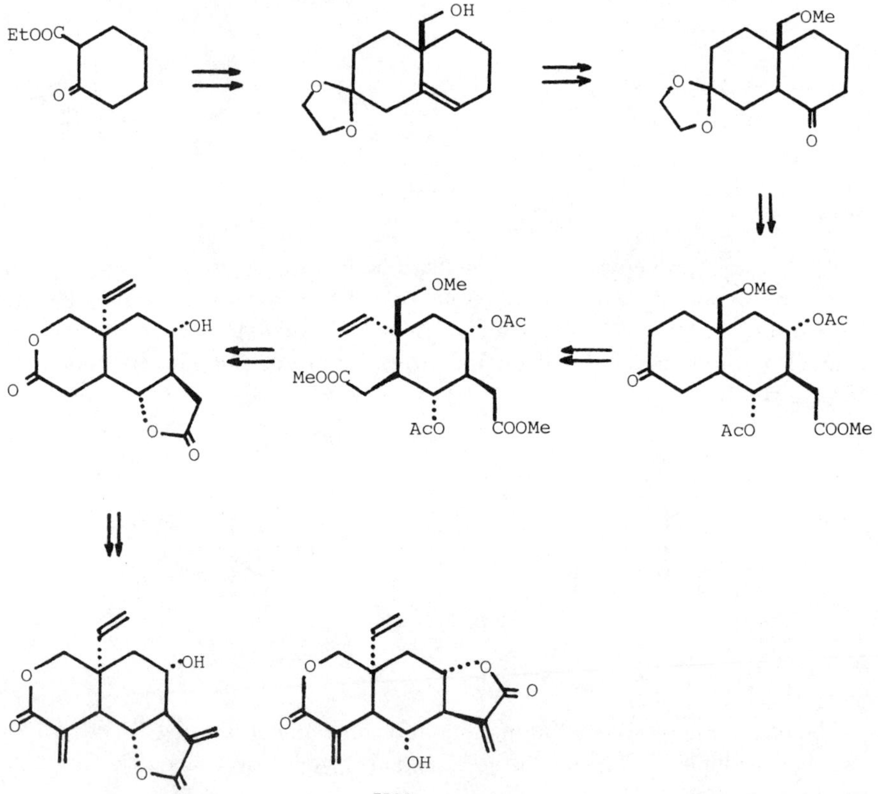

A synthesis of ferruginol (**F4**) using the C + B + A ring construction strategy (Snitman, 1978) depends on two consecutive Robinson annulation processes. An additional ester (carboxyl) group placing at the BC ring juncture not only facilitates the various transformations, it also acts as part of a protecting device for the A-ring enone system while the C ring undergoes modification.

2.3 2-CARBOALKOXYCYCLOHEXANONES

In an extension of this approach to cryptojaponol (**C78**) (Snitman, 1979), the carboxyl group (as lactone) also directs the epoxide cleavage leading to a catechol unit.

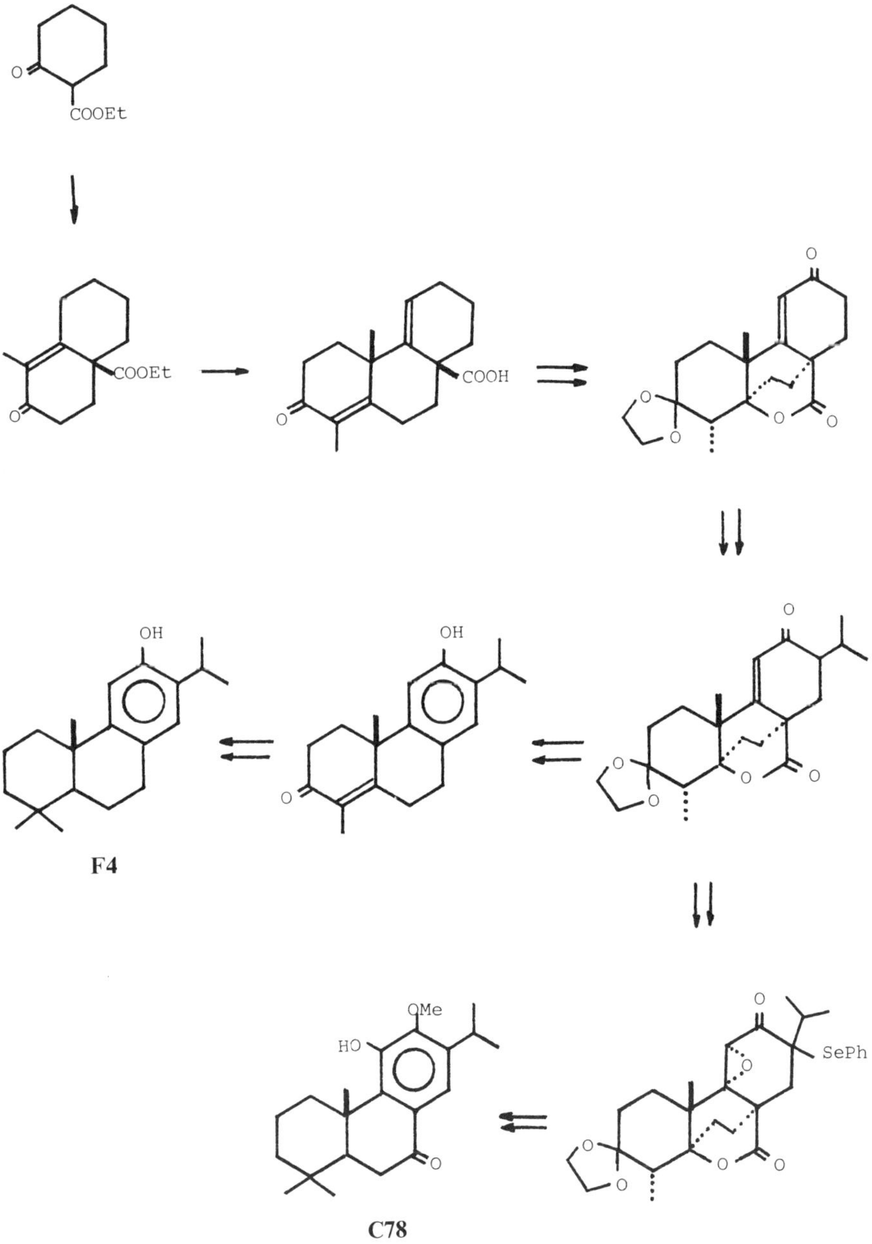

Annulation of a bridged tricyclic ketoester completes the carbocyclic system of veatchine (**V5**) (Masamune, 1964). The piperidine ring is formed later by a photochemical insertion reaction into the axial methyl group at C_4.

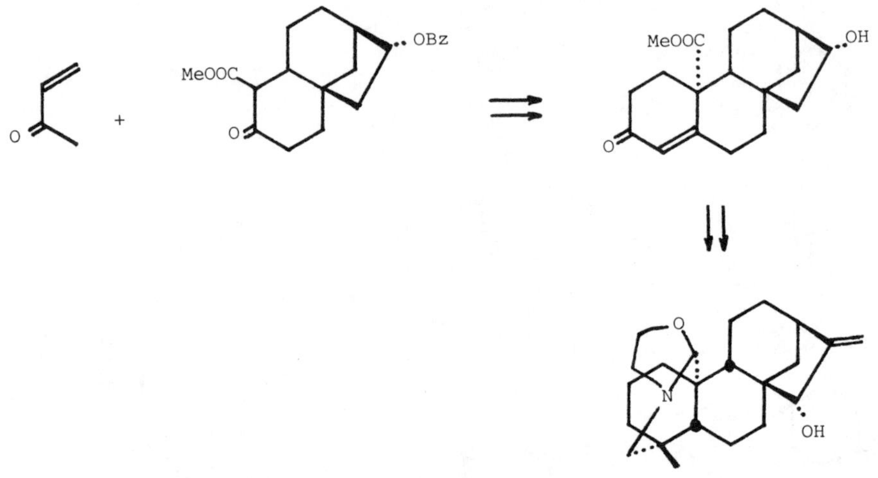

V5

In the synthesis of glutinosone (**G15**) (Murai, 1977), which lacks the angular substituent, the use of a keto diester is still advantageous in the initial annulation step. It serves to maintain and direct the stereochemical buildup at the ring juncture and the C_3, C_4, and C_7 centers (with equatorial substituents). Removal of the extraneous carbethoxy group is by lead tetraacetate oxidation, and unmasking of the ketone group provides a driving force for fixing the double bond between C_1 and C_{10}.

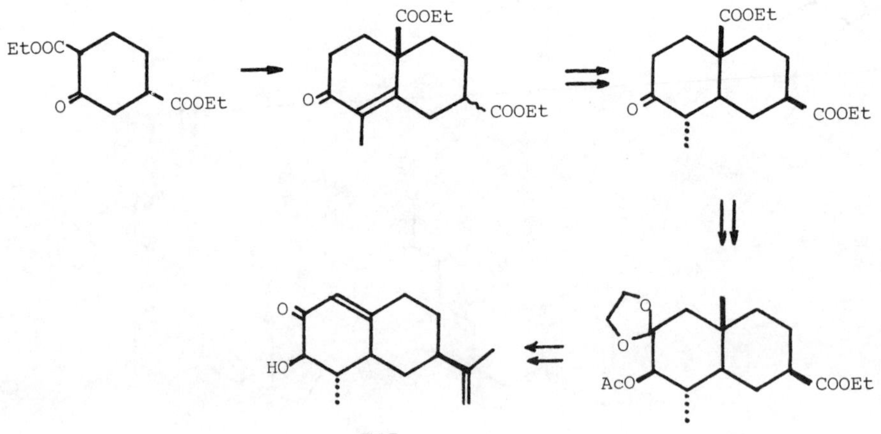

G15

2.3 2-CARBOALKOXYCYCLOHEXANONES

In an earlier effort at solving the *vic/cis*-dimethyl problem concerning nootkatone/isonootkatone (**N11/N12**) synthesis (Marshall, 1970b, 1971c), 2-carbomethoxy-4-alkylidenecyclohexanones were submitted to Robinson annulation. Unfortunately a price had to be paid in additional steps required to reduce the ester function to the methyl group. [*Note*: The ester is unnecesssary, it may be replaced by the methyl group (van der Gen, 1971a, 1971b).]

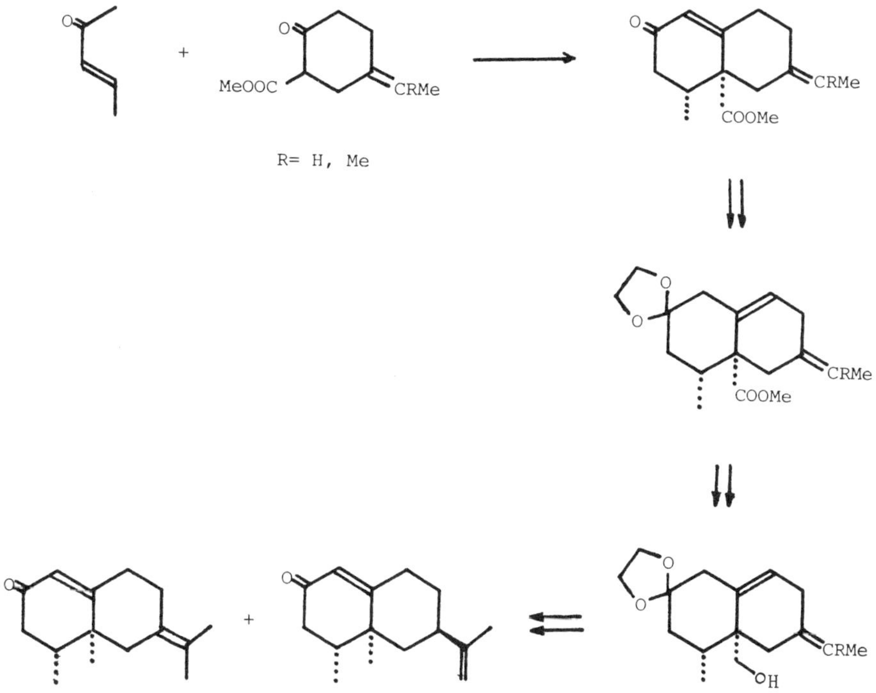

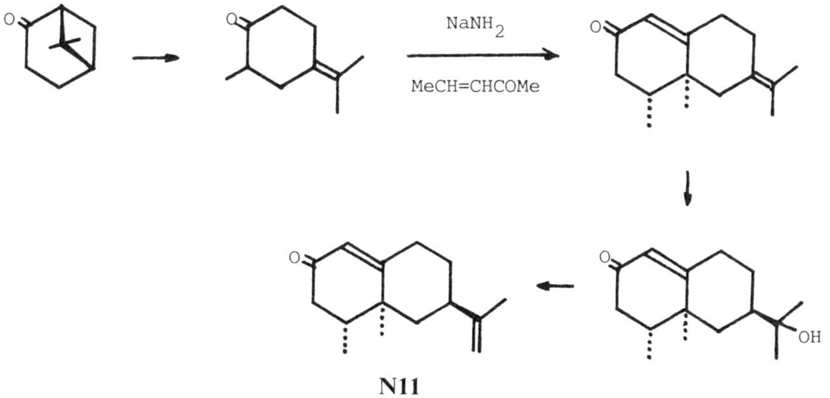

The angular ester group of the octalone was put to good use in an approach to guaiol (**G20**) (Marshall, 1971b). It provides a carbon for expansion of the six-membered ring.

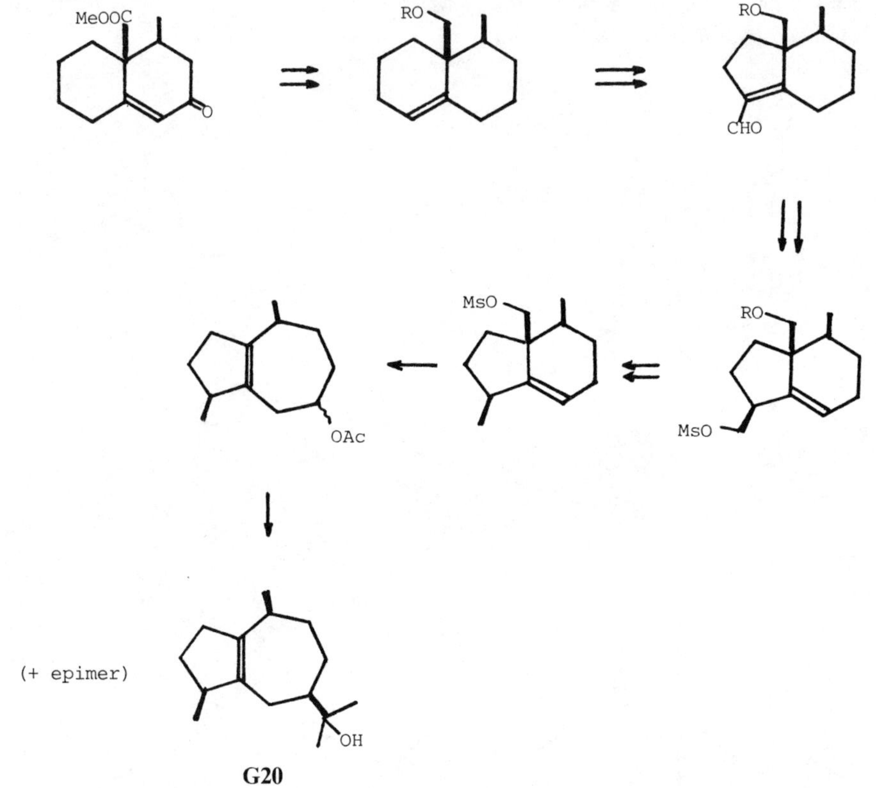

(+ epimer)

G20

2.4 ANNULATION OF 2-METHYL-1,3-CYCLOHEXANEDIONE: WIELAND–MIESCHER KETONE AND ITS HOMOLOGS

The stereochemically uncomplicated lower homolog obtained from Robinson annulation of 2-methyl-1,3-cyclohexanedione is the well-known Wieland–Miescher (W–M) ketone. The diketone has found extensive uses in steroid synthesis. In the sesquiterpene area, its service is equally impressive.

The W–M ketone can be resolved or synthesized by asymmetric induction. The (+)-enantiomer is converted by a route involving selective ketalization, thiomethylation, reductive methylation, Wolff–Kishner reduction, and deketalization into the trimethyldecalone, which is then elaborated into (+)-pallescensin A (**P2**) (Smith, 1984c).

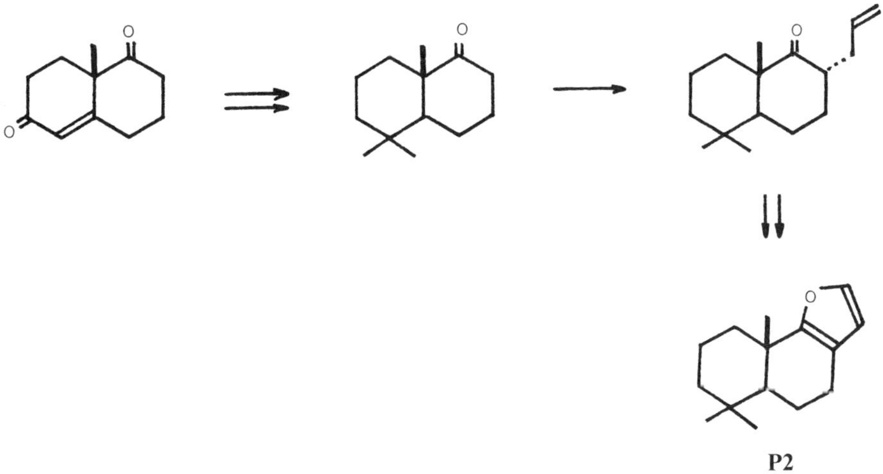

The rearranged drimane sesquiterpene muzigadial has been prepared in racemic form from the W–M ketone (Bosch, 1986). This terpene exhibits potent antifeedant activity against pests such as the African army worms.

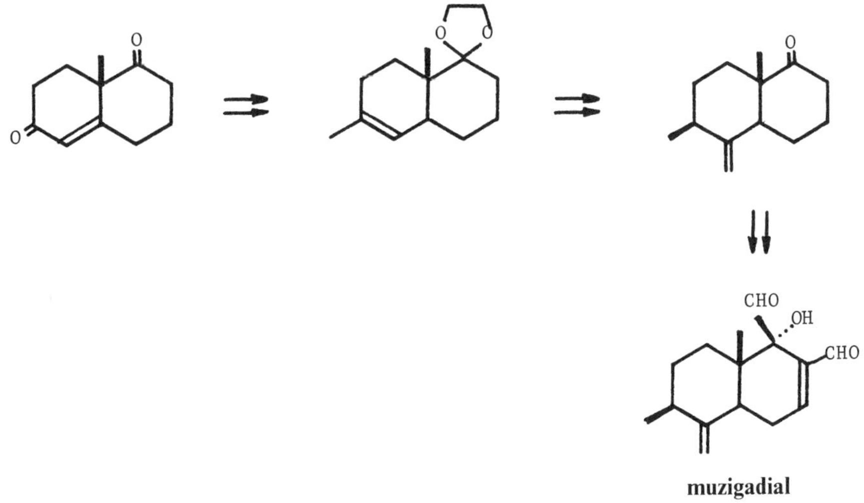

muzigadial

It has been claimed that valeranone (**V2**) (D. K. Banerjee, 1973) may be obtained from the W–M ketone. Thus, the saturated carbonyl is protected by selective reduction and acetylation. Conjugate addition of Grignard reagent

(Cu^{2+} catalyst) affords the second angular methyl substituent. The three-carbon side chain is derived from carbethoxylation and subsequent manipulation.

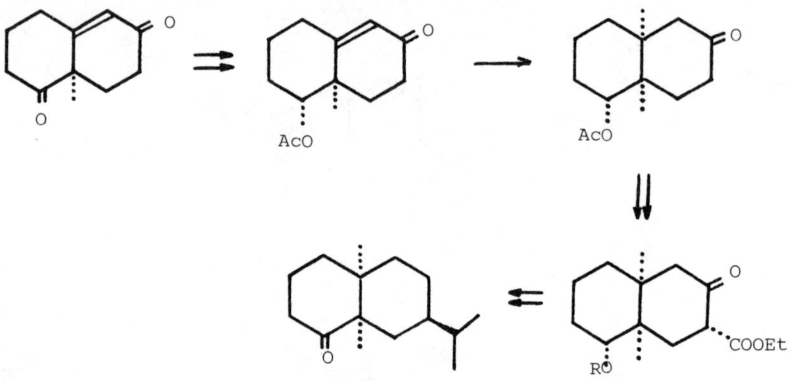

V2

The furanosesquiterpene lindestrene (**L18**) is synthesized (Minato, 1966, 1968b) by a long reaction sequence that commences with the W–M ketone. The furan ring originates at the conjugate ketone, whereas the nonconjugated carbonyl group presents a base for introduction of a double bond. The remaining exocyclic methylene group requires *de novo* oxidation at the allylic position of the starting ketone.

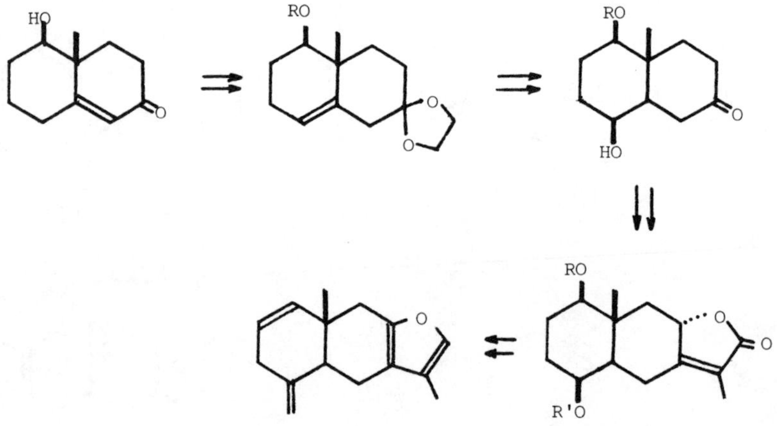

L18

The presence of an equatorial hydroxyl group on the carbon atom adjacent to the angle enables regioselective 1,2-rearrangement, which is translated to a 6:6 → 5:7 ring system mutation. It is natural that certain guaiane sesquiterpenes (Heathcock, 1971; M. Kato, 1970a) have been successfully prepared according to this theme.

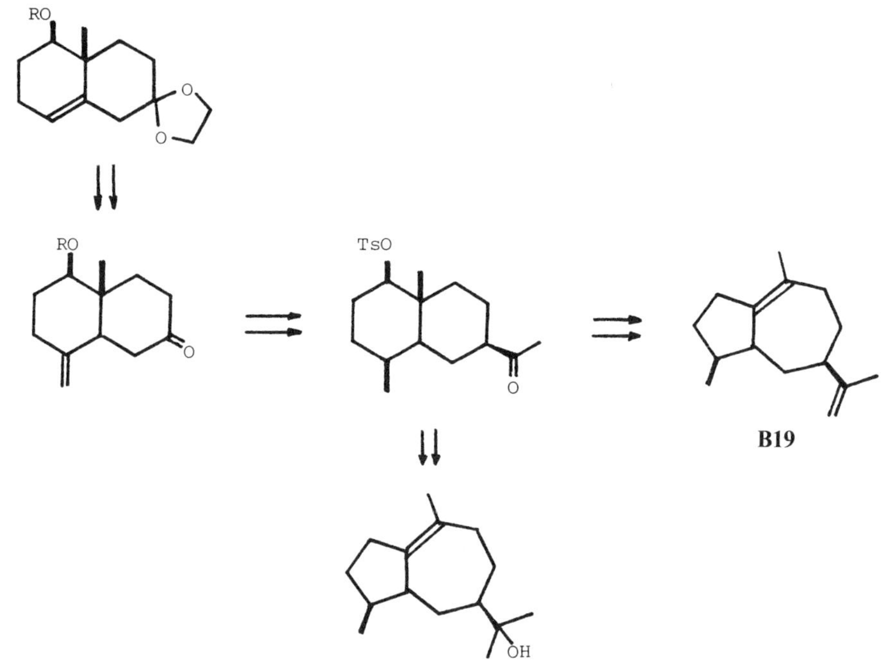

2.4 2-METHYL-1,3-CYCLOHEXANEDIONE

B19

B21

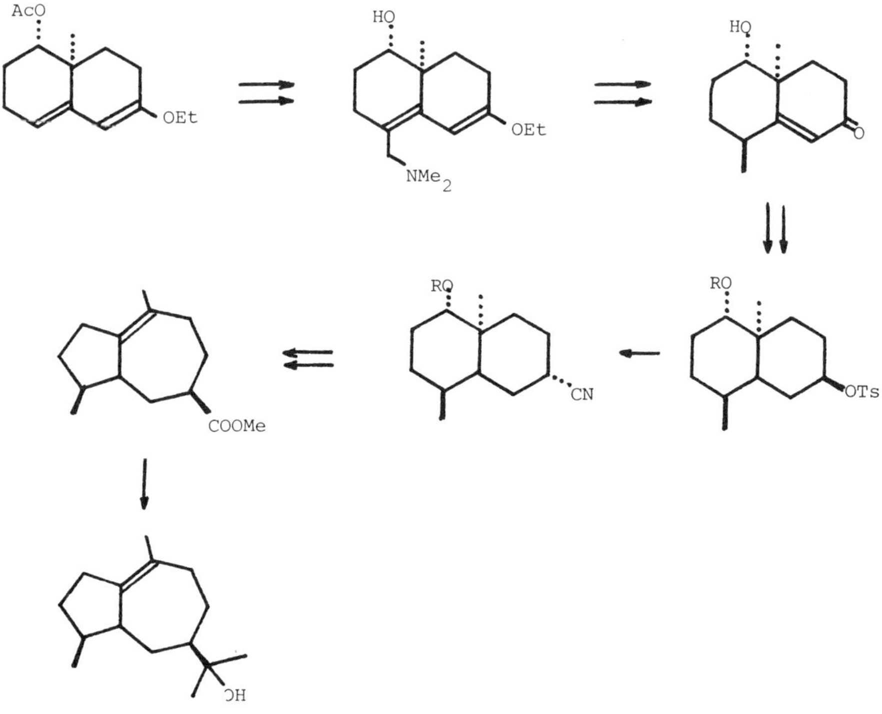

B21

It is interesting to note that in the two syntheses of bulnesol (**B21**), the ring juncture stereochemistry is immaterial, whereas the relative configurations of the angular hydrogen and the secondary methyl group are crucial (as dictated by those present in the sesquiterpene).

Modification of later steps of this work allows the completion of a kessane (**K5**) synthesis (M. Kato, 1970b).

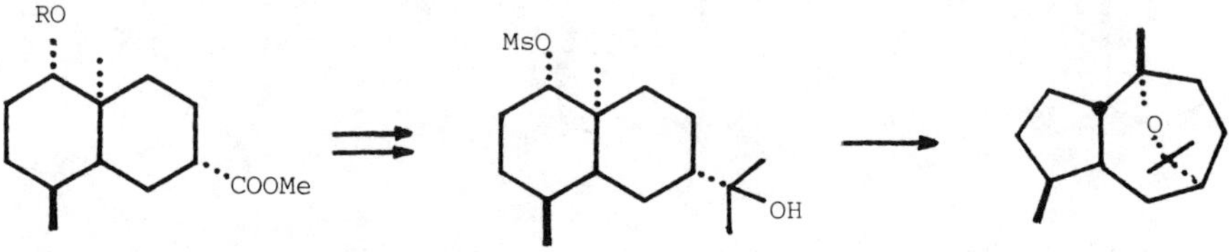

K5

Pseudoguaianolides such as confertin (**C62**) (Heathcock, 1982a) are similarly approachable. Since these compounds contain an angular methyl group and next to it an oxygen function in a cyclopentane ring, the W–M ketone fulfills the structural requirement for the necessary manipulations.

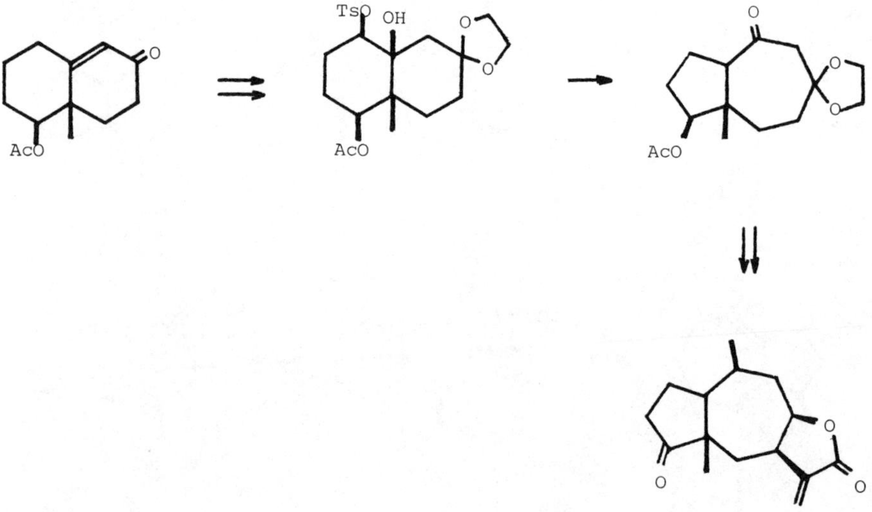

C62

The 6:6 → 5:7 ring transformation may be effected via a fragmentative pathway (biomimetic?). The synthesis of globulol (**G14**) (Marshall, 1974) is exemplary.

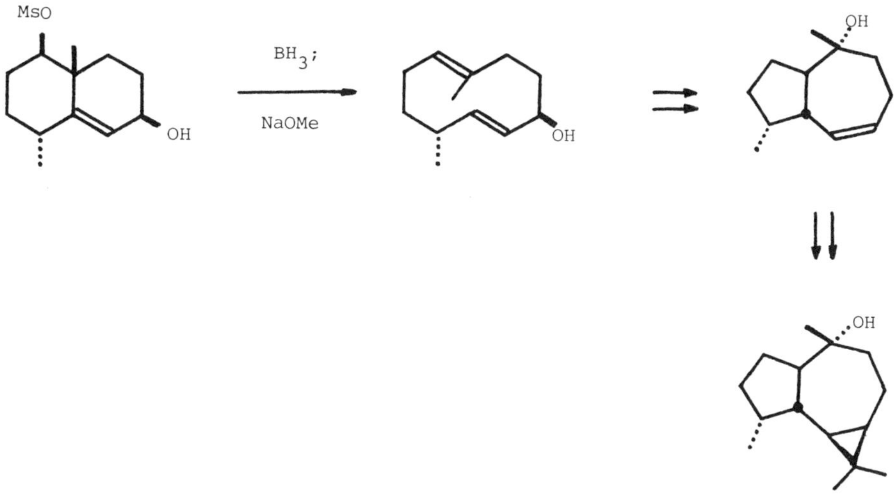

Contrary to an incursion of the intercyclic bond cleavage in the hydronaphthalene skeleton in the synthesis above, a cyclohexane ring is retained in the approach to γ-elemene (**E2**) (M. Kato, 1979) that begins with the W–M ketone.

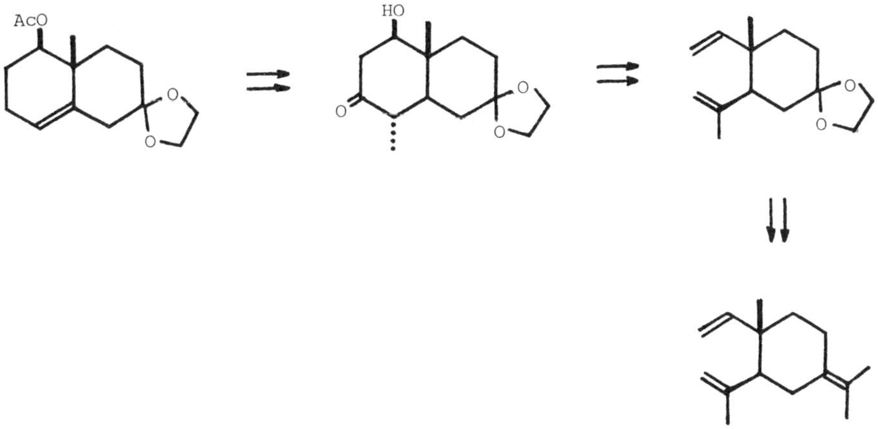

Many tricyclic sesquiterpenes have been synthesized from the W–M ketone by proper adjustment of functional groups and effecting the crucial C—C bond formation. Thus, seychellene (**S29**) (Piers, 1969b) is obtained by adding two methyl groups and a methylene bridge between C_4 and C_8: the angular methyl by conjugate addition, the equatorial secondary methyl out of the carbonyl and the methylene attaching to the α carbon of the original enone system. The roof-shaped *cis*-decalin skeleton forces the approach of reagents from the convex face, and thereby a totally stereoselective synthesis can be achieved.

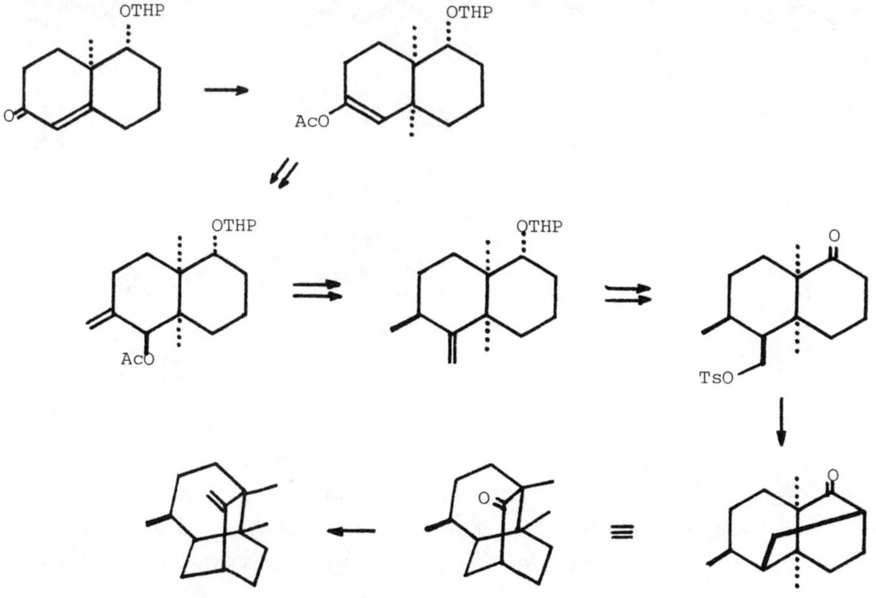

S29

The first total synthesis of longifolene (**L24**) (Corey, 1961, 1964c) involves an intramolecular Michael addition of 6:7-fused enedione as the key step. This enedione is derived from the W–M ketone by selective expansion of the cyclohexenone portion.

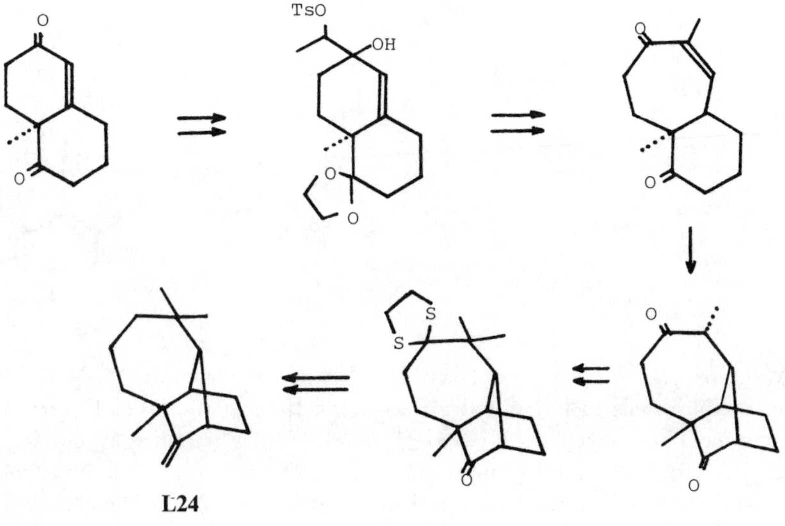

L24

2.4 2-METHYL-1,3-CYCLOHEXANEDIONE

The one weakness of this landmark achievement is inefficient (reversible) cyclization. If the step is rendered irreversible using an alkylation technique, the bridged ketone is formed in high yield. However, a ring expansion step is required (McMurry, 1972).

The idea underlying this improved protocol has been employed in an approach to sativene (**S11**) (McMurry, 1968) (cf. seychellene).

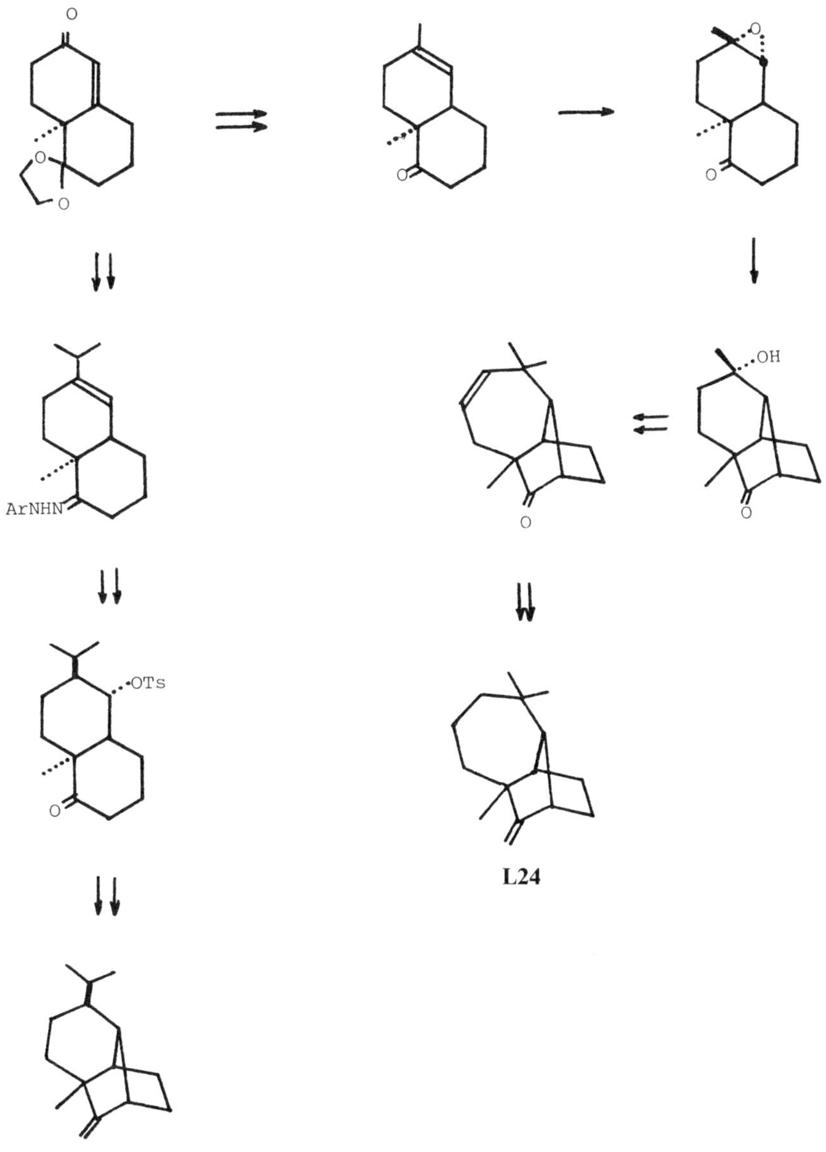

Formation of four-membered ring by the same method provides an entry to such compounds in the ylangane series (Heathcock, 1966) as α-copaene (**C66**) and α-ylangene (**Y1**).

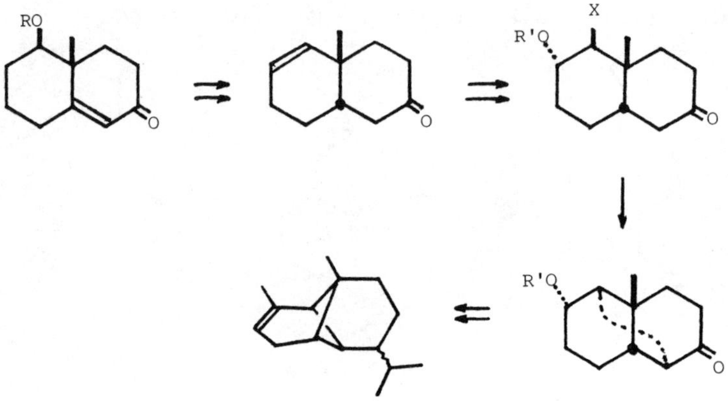

C66
Y1

The longipinenes (**L25**, **L26**) have yielded to synthesis (Miyashita, 1971) only once thus far. The bridged ring system is assembled by a photochemical reaction of a methylenecyclodecadiene. The latter compound is derived from the W–M ketone.

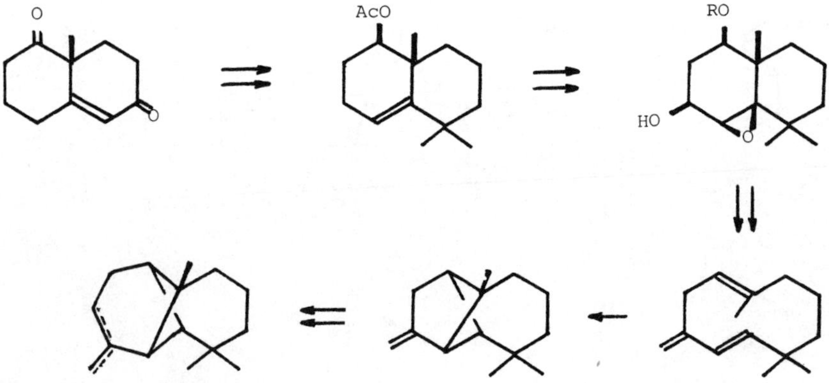

L25
L26

As indicated before (cf. temsin, **T10**), octalones may be elaborated into secoeudesmanes. Additional examples are syntheses of ivangulin (**I20**)

(Grieco, 1977d) and eriolanin (**E14**) (Grieco, 1978), both starting from the W–M ketone.

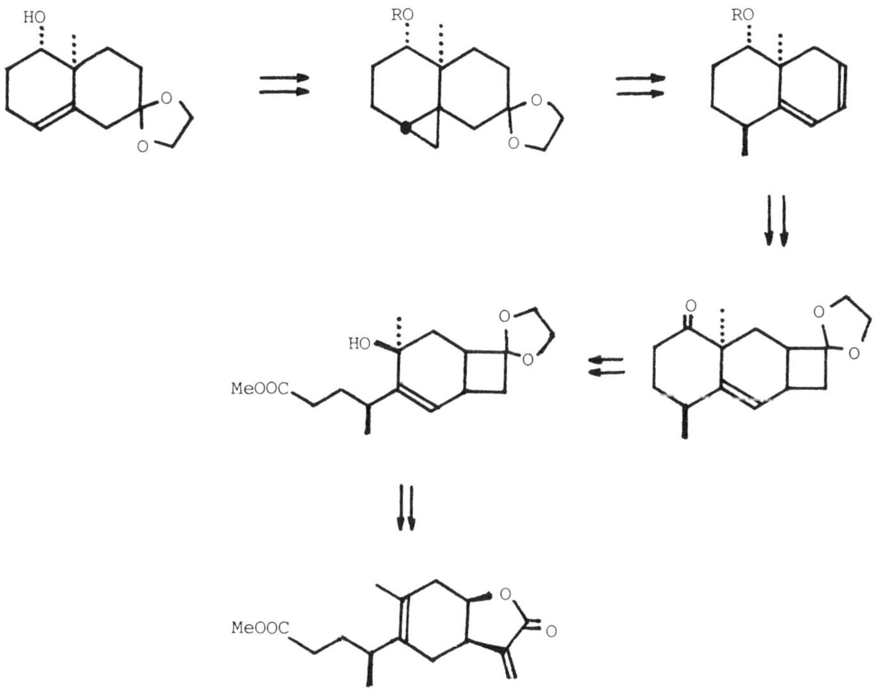

I20

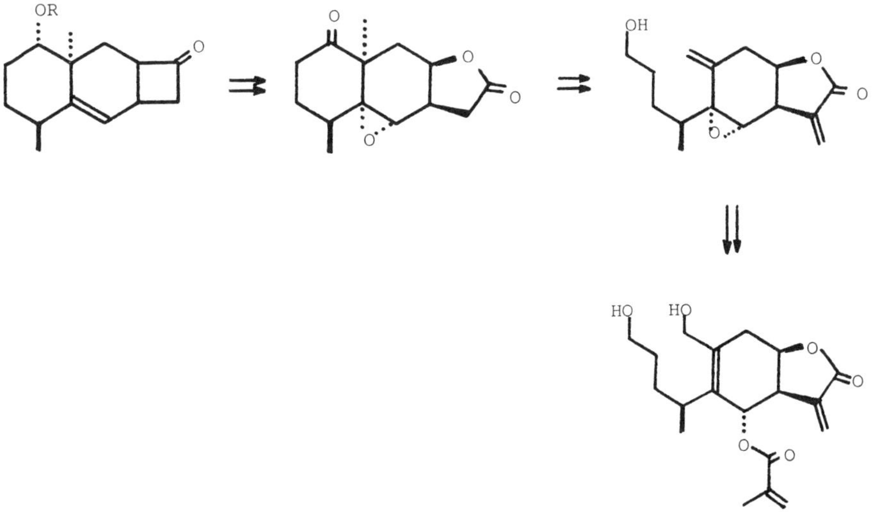

E14

The methyl ether of the 6-methylated W–M ketone has also been elaborated into an isomeric octalone that serves as an AB-ring synthon of quassin (**Q3**) (Grieco, 1980a).

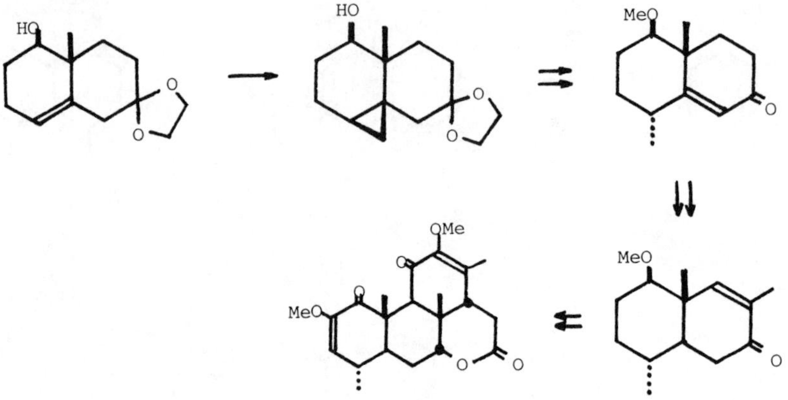

Q3

The 4-methylated W–M ketone proves to be a useful building block for aphidicolin (**A36**) (McMurry, 1979; Trost, 1979a). The stepwise reductive alkylation of which with formaldehyde at C_4 establishes a trans AB-ring juncture and a 4α-hydroxymethyl group. The angular methyl group steers the alkylation away from the β face of the molecule.

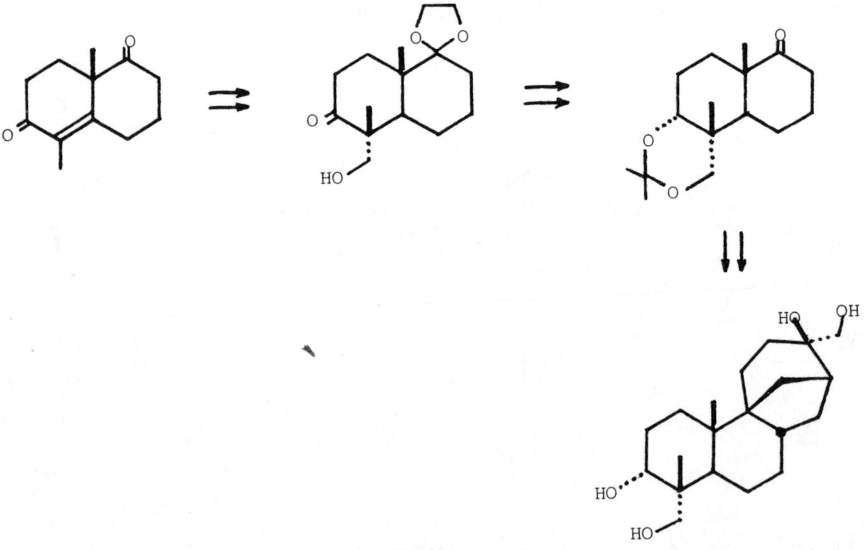

A36

Reductive alkylation of the ketol derived from the same bicyclic ketone represents a convenient method for completing the basic carbon skeleton of avarol (**A54**) (Sarma, 1982), and a marine sponge metabolite, (**M7**) (Sarma, 1985).

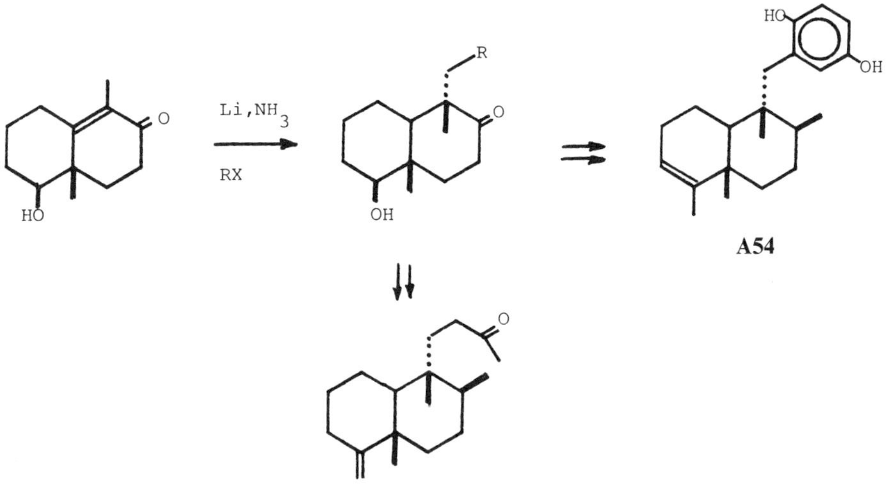

Dolabradiene (**D16**) is an unusual diterpene formed from a tricyclic cation via three 1,2-shifts of methyl and hydrogen. In a retrosynthetic perspective, the AB nucleus may be conceived as derivable from 4-methyl W–M ketone in which the nonconjugate carbonyl is equivalent to the vinylidene substituent, whereas the conjugate ketone provides the foothold for construction of the third ring. The Robinson-type or Michael–aldol annulation is most suitable because a new carbonyl group would appear at C_{13}, where two alkyl substituents must be incorporated. (Y. Kitahara, 1964).

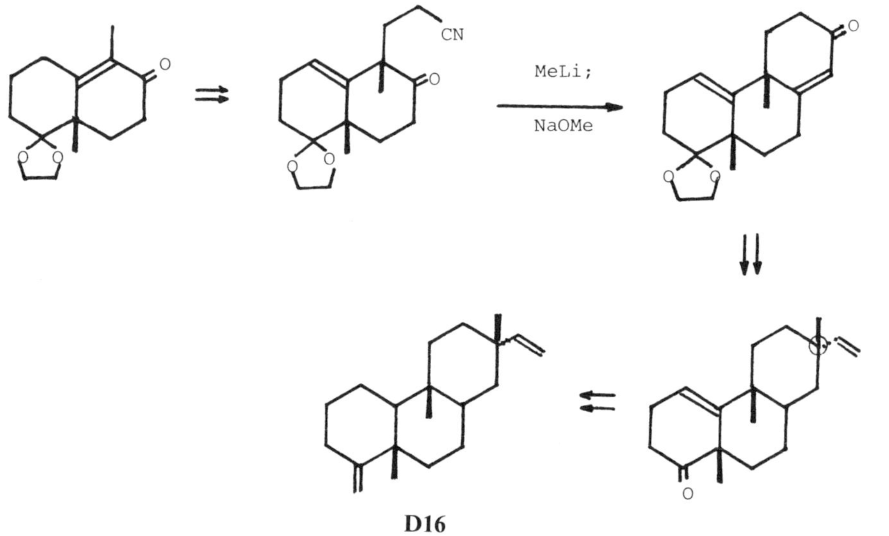

Comparing the bicyclic nucleus of the clerodanes with 4-methyl W–M ketone indicates that the crucial transformations involve alkylation of the enone and the replacement of each carbonyl group with a methyl substituent. Regarding the synthesis of annonene (**A34**) (S. Takahashi, 1979), the alkylation via a Claisen rearrangement of a *trans*-octalin derivative is ideal as a useful exocyclic methylene group is left at the position where a methyl group has to be erected.

A34

Precursors of the more highly oxidized clerodane ajugarin-IV (**A19**) must accommodate an additional 6α-oxygen substituent. This problem is resolved by a methylative 1,3-transposition of the conjugate carbonyl group of the 4-methyl W–M ketone after the ring juncture and the side chain have been established. The various processes are under thermodynamic control; therefore all the newly created asymmetric centers except the angle contain substituents in the equatorial configuration (Kende, 1982).

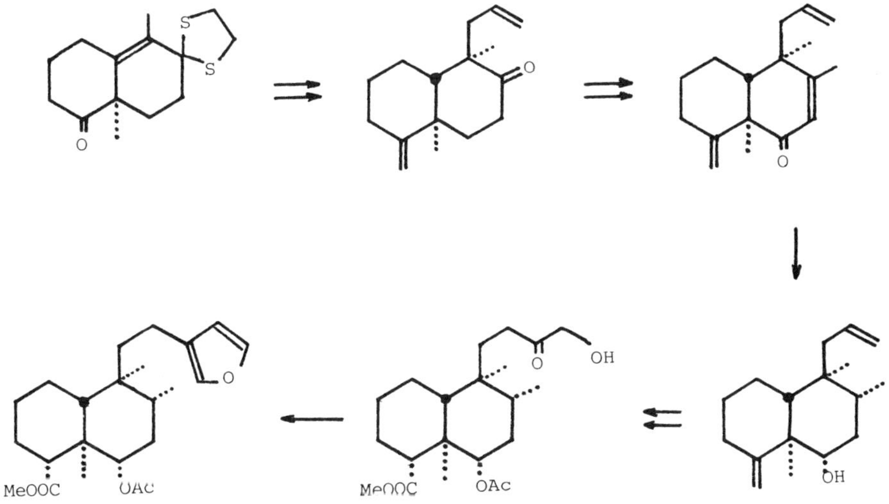

Another clerodane synthon is available from 2-methyl-1,3-cyclohexanedione and 7-hydroxy-1-octen-3-one (ApSimon, 1977).

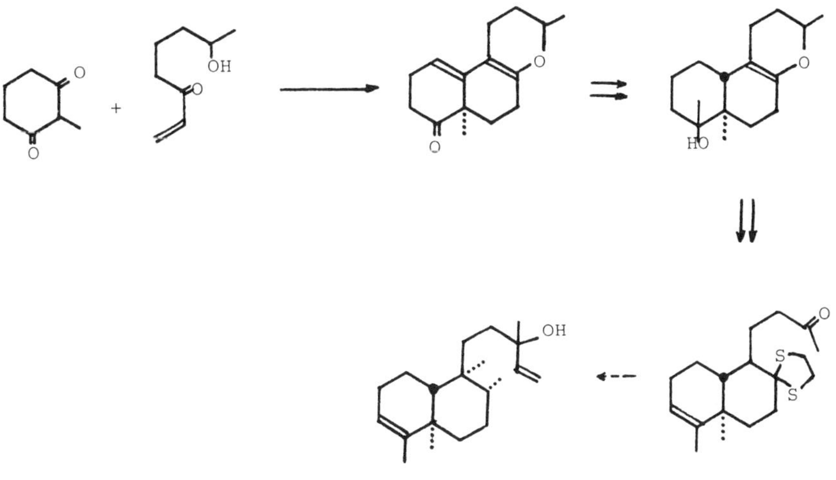

On examining the structure of iresin, one may easily decide that the W–M ketone or its analog is a very attractive synthetic precursor. Elaboration of the enone system to the desired pattern of substitution in the A ring has ample precedent and close analogies in the diterpene area, whereas the nonconjugated carbonyl warrants the addition of the two missing carbon pendants at the adjacent C_8 atom and the ipso position. So far only the diacetate (**I15**) of isoiresin has been synthesized (Pelletier, 1966, 1968a,b). Isoiresin also occurs in nature.

I15

To facilitate A-ring transformations, an ester-appended W–M ketone is prepared. For deprotonation of highly acidic donors, such as β-diketones and β-ketoesters, potassium fluoride is quite sufficient. Thus the Robinson annulation is achievable under these mild conditions.

Methylation of the bicyclic ketoester is stereoselective, occurring trans to the angular methyl group. Hydride reduction of the resulting unconjugated ketoester provides a diol in which the C_3 hydroxyl group is equatorial. The catalytic hydrogenation of the double bond gives a compound with a trans ring juncture.

Since the carbon substituent at the C_9 atom of iresin is equatorial, thermodynamically controlled reactions were implemented. The plan failed to work, however, even if the C_9 configuration could be established without difficulty. Consequently, isoiresin diacetate emerged as the final product.

The bicyclic diketoester synthon can be converted into marrubiin (**M10**) (Mangoni, 1972) and 3β-hydroxynagilactone F (Reuvers, 1986).

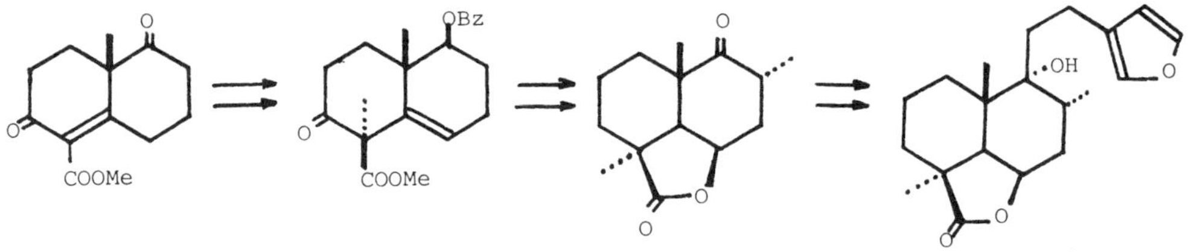

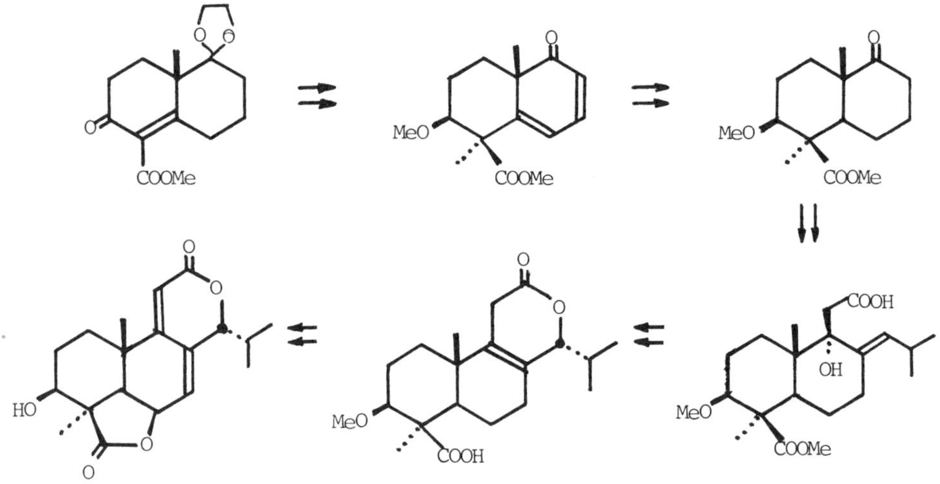

3β-hydroxy nagilactone F

The synthesis of an antifungal metabolite, **LL-Z1271α** (Welch, 1977), can be approached by elaborating the Wieland–Miescher ketone itself through reductive carboxylation and methylation. The carboxyl group is connected to the B ring via activation by the C_9 carbonyl function, which is also the basis of the pyrone ring.

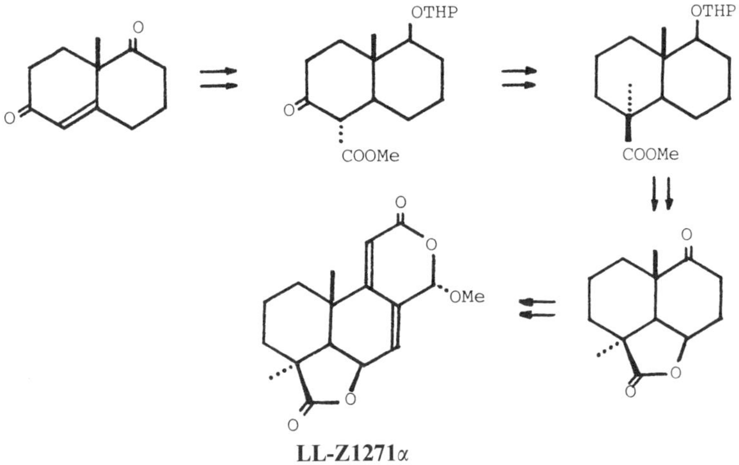

LL-Z1271α

In a synthesis of methyl vinhaticoate (**V17**) (Spencer, 1971), the same reaction sequence is followed. The C ring is then constructed by a Robinson annulation reaction.

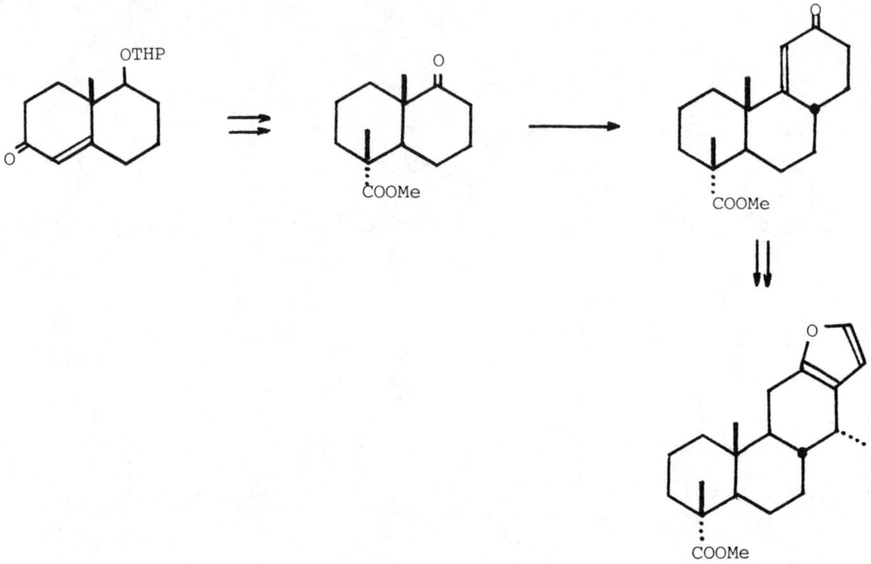

V17

4-Methoxy W–M ketone contains a convenient handle at C_4 for eventual installation of an exocyclic methylene group at that position as demanded in more conventional syntheses of compounds such as colorata-4(13),8-dienolide (**C59**) (deGroot, 1982).

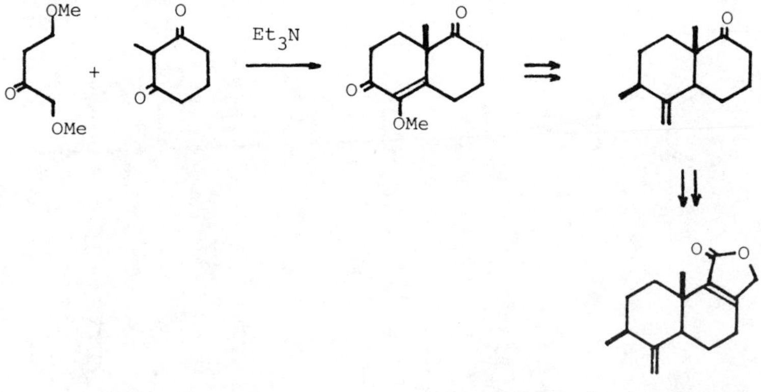

C59

Quite in contrast to the reaction between 2-methylcyclohexanone with 3-penten-2-one, which is highly stereoselective, the annulation of 2-methyl-1,3-

cyclohexanedione (via the pyrrolidinoenamine) gives a stereorandom (1:1) mixture of enones in 27% yield.

The desired isomer for eremophilane synthesis is obtainable by isomer separation only after removal of the conjugated carbonyl group. It has been processed into eremoligenol/eremophilene (**E8/E11**) and, after epimerization of the ester group of the intermediate, valerianol/valencene (**V4/V1**) (Coates, 1970c).

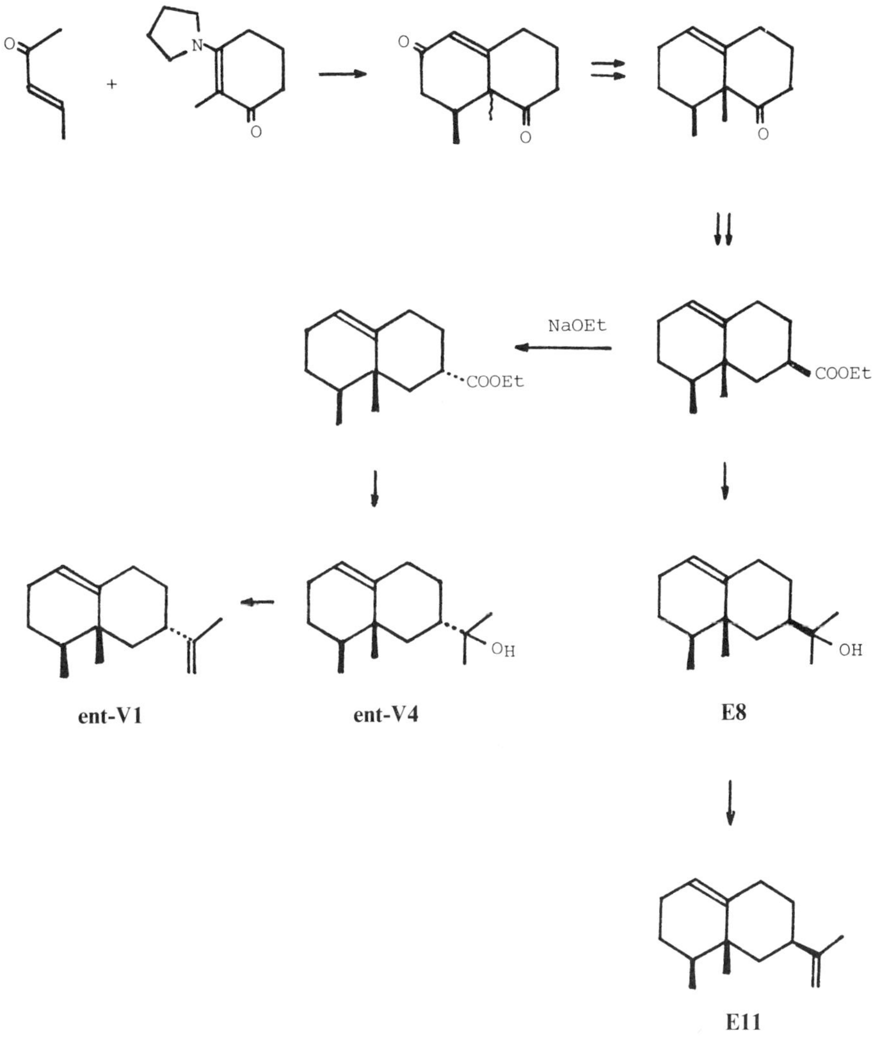

Following a standard reaction sequence (+)-pisiferol has been synthesized (Tamai, 1986). The oxygenated Miescher–Wieland ketone also serves as a precursor of (+)-perrottetianal (Hagiwara, 1987b).

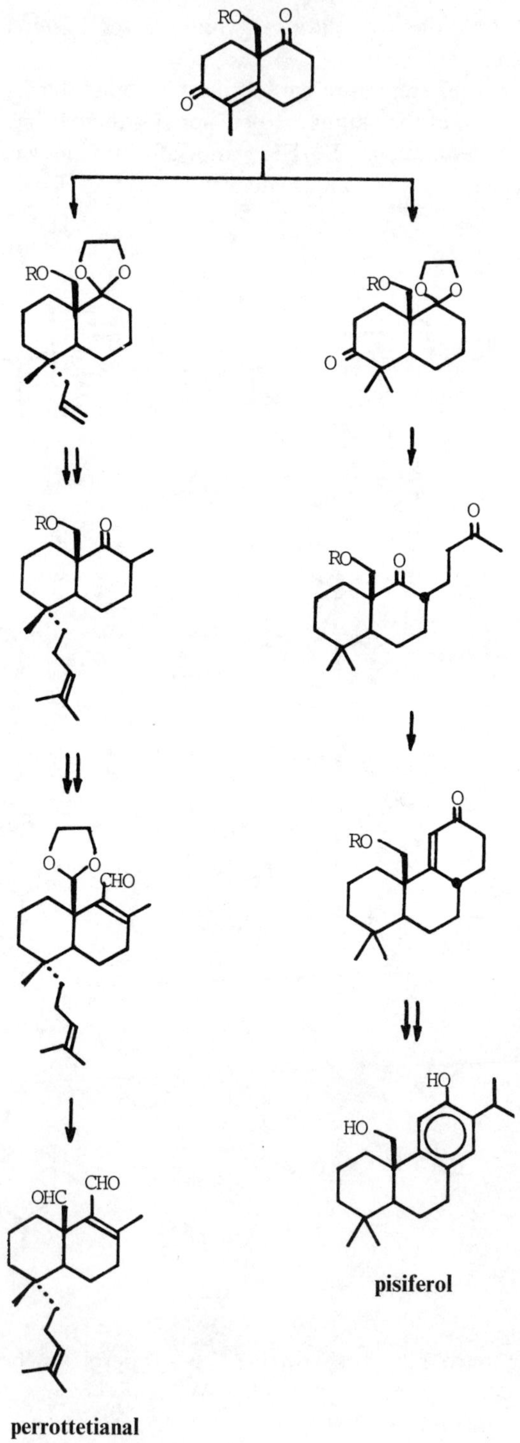

perrottetianal

pisiferol

2.5 ANNULATION OF 1,4-CYCLOHEXANEDIONE DERIVATIVES

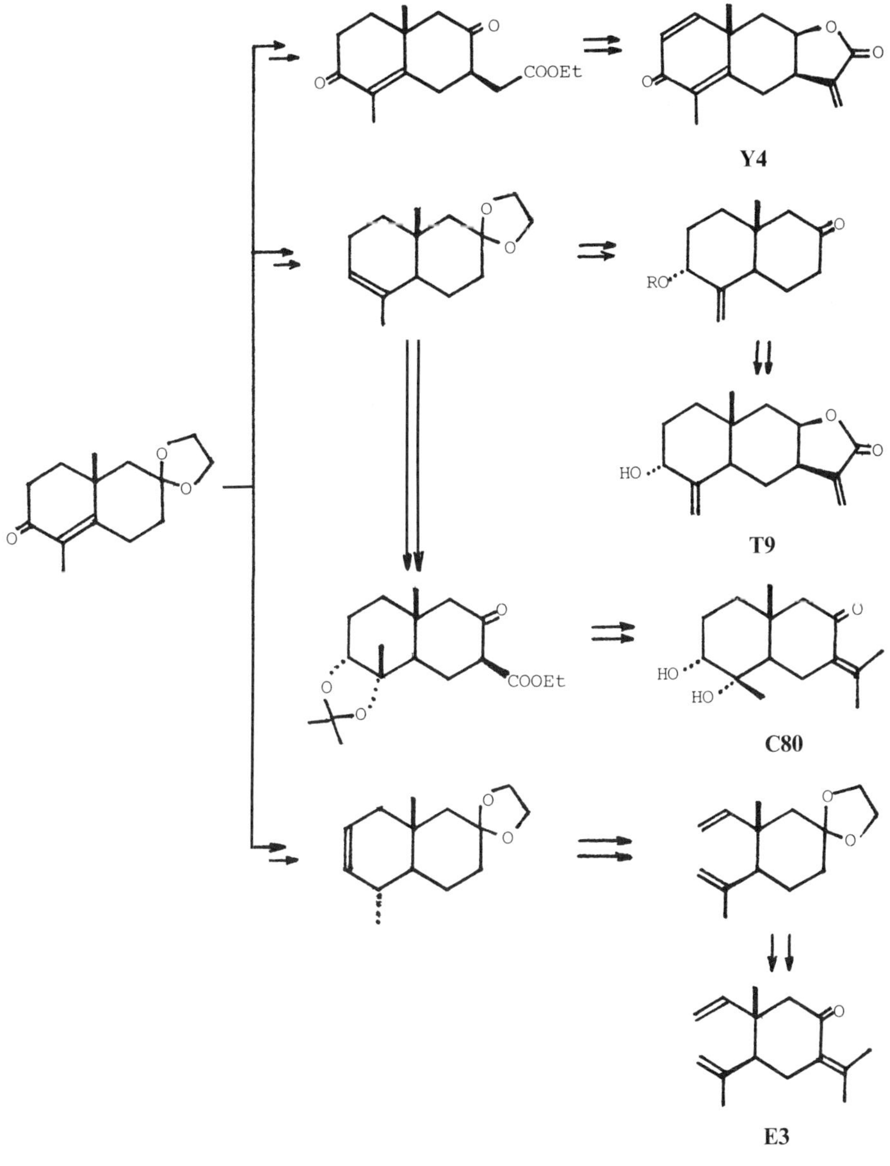

An additional oxygen function at the C_8 of atom the hydronaphthalene system is often required for synthesis of other cyclic sesquiterpenes. The eudesmane cuauhtemone (**C80**) (Goldsmith, 1976), yomogin, (**Y4**) (Caine, 1975), isotelekin (**T9**) (Miller, 1974), the monocyclic ketone β-elemenone (**E3**) (Majetich, 1977), as well as the 4-epimer of aubergenone (R. Kelly, 1978) are available synthetically from the dione (monoketal).

Artemisin (**A46**), or 8α-hydroxy-α-santonin, has been prepared (Nakazaki, 1966) from the corrresponding hydroxyloctalone following the α-santonin scheme. The cross-conjugated ketone is obtained by dehydrogenation with DCDQ, and oxygenation at C_6 is achieved with SeO_2.

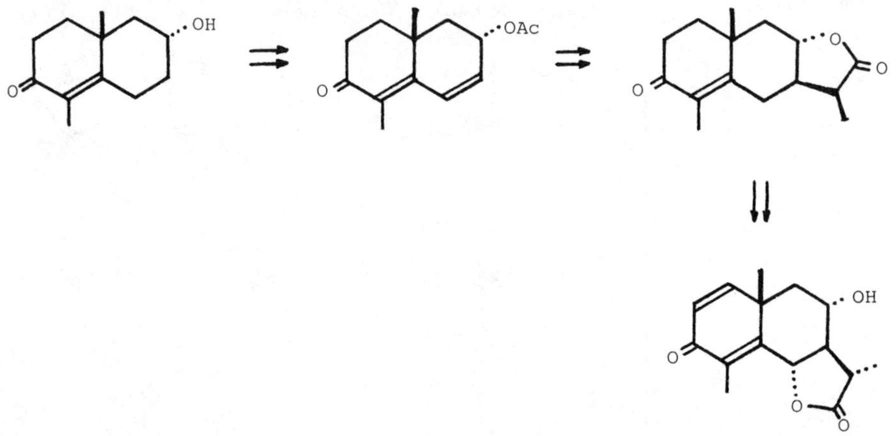

A46

Symmetrization of C_4 of the cyclohexanone in the form of a ketal perhaps represents the best solution to the synthesis of nootkatone (**N11**) (McGuire, 1974), based on the Robinson annulation. An acetyl side chain is attached, which allows the correct orientation to be established and conversion into the isopropenyl group. The disadvantage of this route (for nootkatone) is the wasteful removal of the conjugated carbonyl group and its reincarnation at the same position.

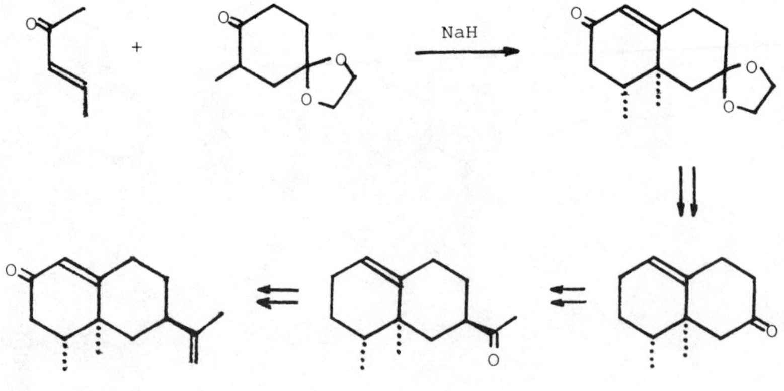

N11

In a synthesis of isopetasol (**P25**) (Bohlmann, 1982a), 3-trimethylsilyl-3-buten-2-one is used in the annulation step.

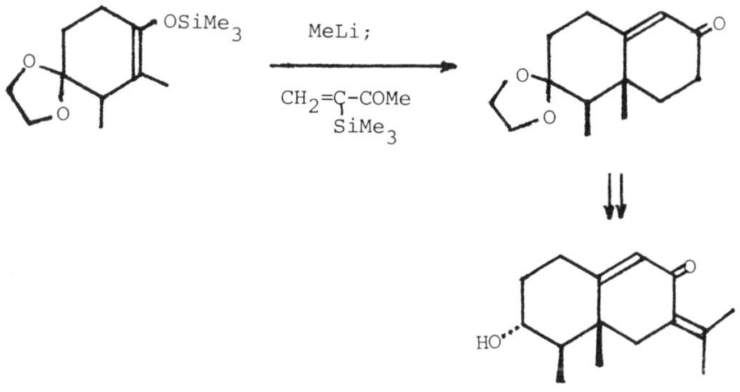

Siccanin (**S33**) (M. Kato, 1981), has an oxygenated angular methyl group. Consequently, a synthesis starting with the corresponding keto ester is most expedient.

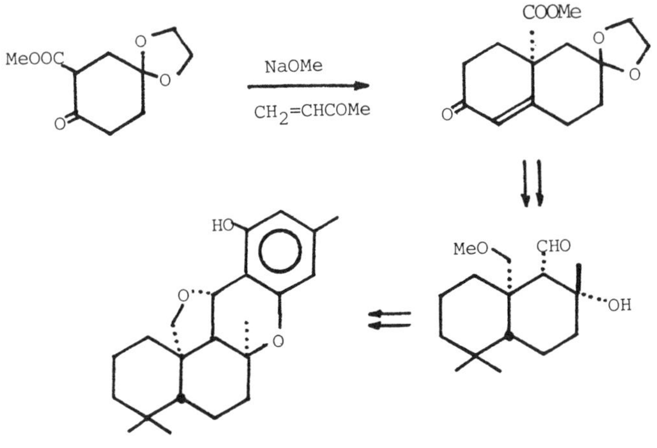

2.6 ANNULATION OF β-TETRALONE AND ITS ANALOGS

Abietic acid (**A1**) is a constituent of many conifer resins. It also arises from acid isomerization of other primary resin acids such as pimaric, isopimaric, levopimaric, and neoabietic acids. A synthesis (Stork, 1962) consisting of the

Robinson annulation of an isopropyltetralone also demonstrates the utility of the enamine alkylation method (i.e., monoalkylation) and the stereoselective introduction of a functionalized chain at C_4, which is degraded to a carboxyl group. Dearomatization of dehydroabietic acid (**A2**) by a Birch-type reduction furnishes the homonuclear diene, which is then isomerized to abietic acid upon exposure to hydrochloric acid. The same diene is transformed into palustric acid (**P6**) by treatment with potassium hydroxide at 210°C (Burgstahler, 1961).

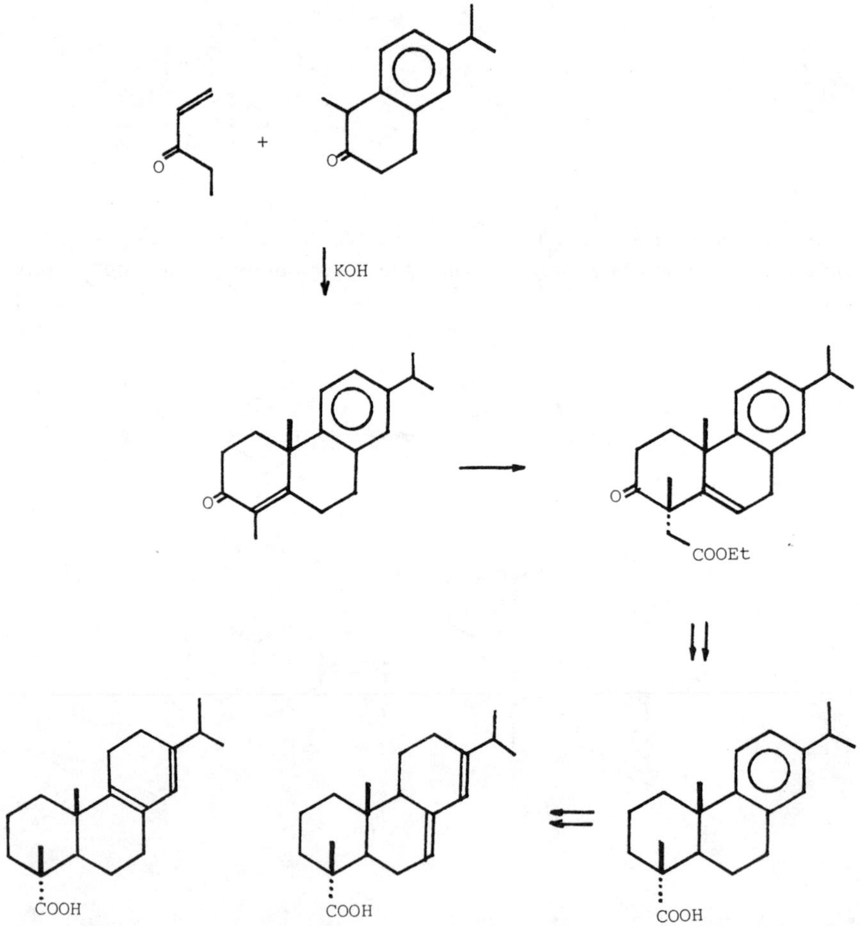

A more direct access to the abietic acid type of substance is by reductive cyanation (with cyanogen chloride) of a tricyclic enone, which is obtained from a Robinson annulation of a tetralone (Kuehne, 1970).

2.6 β-TETRALONE AND ITS ANALOGS

The reductive methylation of β-alkoxy-α,β-unsaturated esters in liquid ammonia forms the key operation of an approach to podocarpic acid (**P43**) and callistrisic acid (**C10**) (Welch, 1977a). The β-ketoester precursor is obtainable by the Robinson annulation of an appropriate tetralone with 3-buten-2-one, followed by reductive carboxylation, and so on, or with methyl 3-oxo-5-methoxypentanoate under acidic conditions (TsOH, then CF_3COOH).

P43 (R= OH, R'= H)
C10 (R= H, R'=iPr)

Various methoxyhydrophenanthrenones derived from the corresponding tetralones are versatile intermediates for terpene synthesis. A classical elaboration of a synthetic relay (degradation product) of phyllocladene (**P27**) from the tricyclic ketone involves an inversion of the C_8 configuration, since alkylation at that site leads to a product with a cis B/C-ring juncture. The inversion is accomplished by ring cleavage and reclosure by a Dieckmann cyclization (Turner, 1966). It should be noted that one of the participatory ester chains comes from the alkylation.

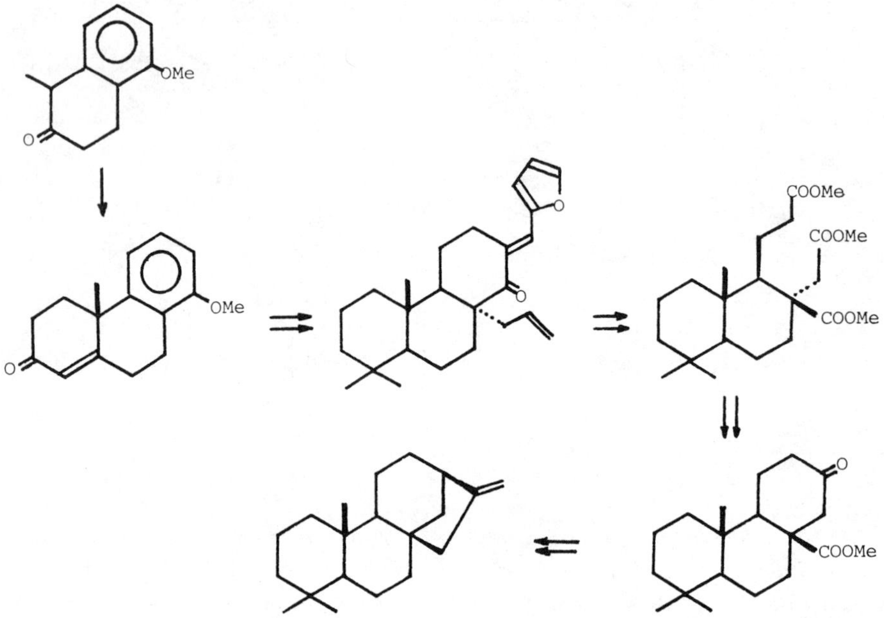

P27

An application of a methoxyphenanthrenone to the synthesis of rimuene (**R4**) (Ireland, 1964) emphasizes the convertibility of the aromatic nucleus to the dimethylated cyclohexane (ring A). The trans–anti stereochemistry of the ring junctures is set up by means of Li–NH$_3$ reduction.

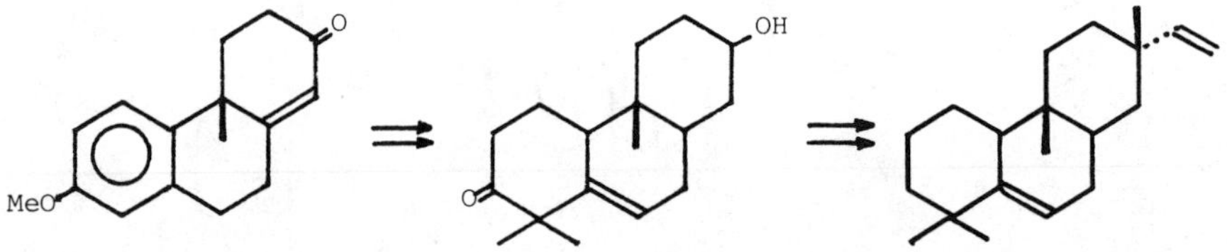

R4

Royleanone (**R9**) routinely qualifies for a synthetic route, which consists of a Robinson annulation of a tetralone (Tachibana, 1975).

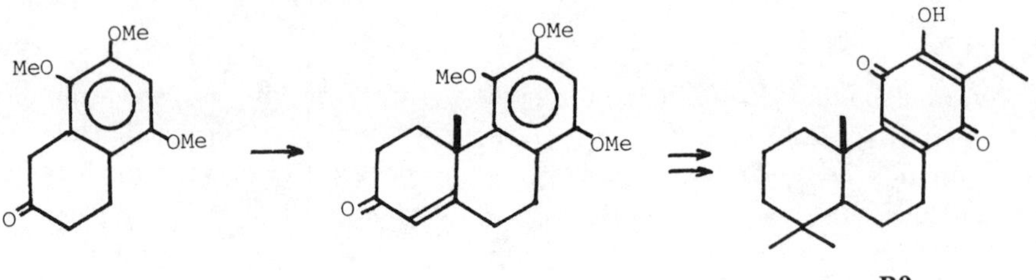

R9

2.6 β-TETRALONE AND ITS ANALOGS

The assembly of alnusenone (**A26**) (Ireland, 1970b) offers a substantial contribution to triterpene synthesis. A convergent route to the pentacyclic framework containing two terminal aromatic nuclei resolves most of the stereochemical problems. A Robinson annulation and intramolecular Friedel–Crafts alkylation are the ring-forming operations. In between these steps, conjugate hydrocyanation serves to establish the 14β-methyl group. The *trans*-decalone system directs the steric course of the Friedel–Crafts reaction. The slight difference of the alkoxy groups (EtO vs. MeO) in the aromatic rings permits selective reduction and the subsequent introduction of the methyl substituents.

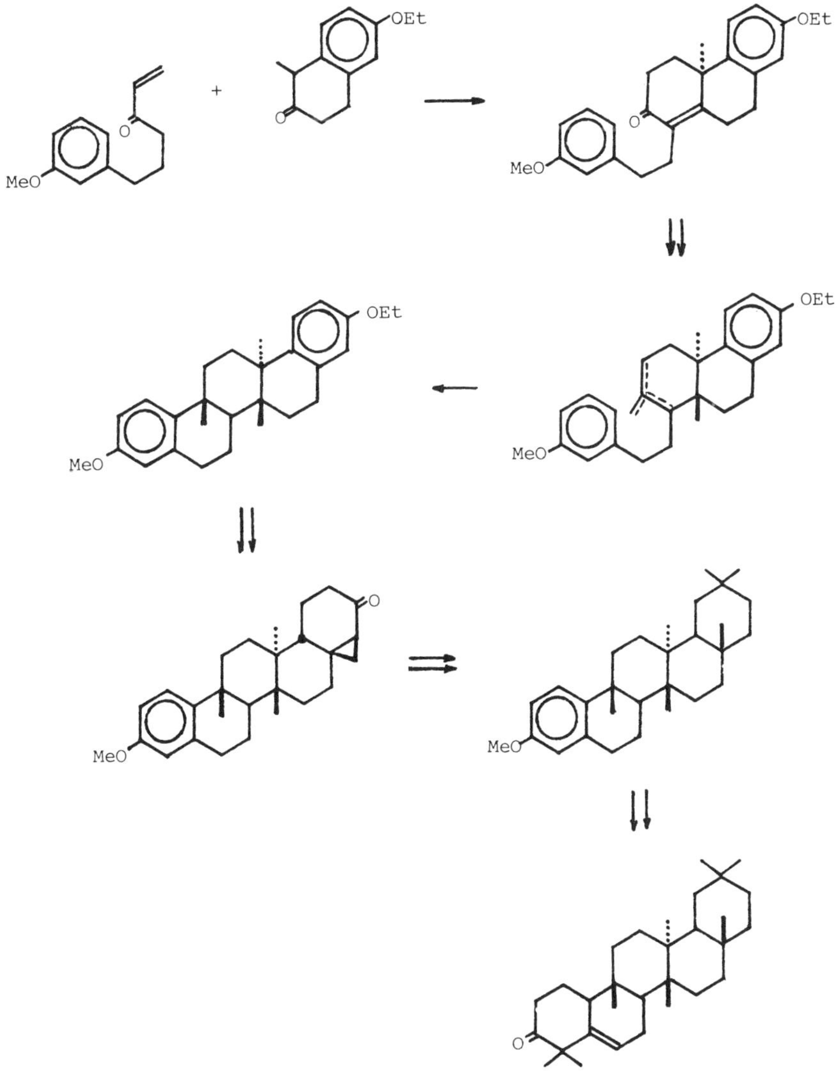

A26

2 ROBINSON ANNULATION

α-Onocerin (**O6**) is a symmetrical tetracyclic triterpene. The most logical approach to this compound entails the coupling of two identical bicyclic precursors. This proposition has been worked out (Stork, 1963b) using Kolbe electrolysis of the C_{15}-keto acid. Subsequent methylenation of the ketone groups is effected.

Several methods for the preparation of the crucial acid precursor have been developed. In general, the Robinson annulation procedure is used in A-ring formation either from a tetralone (Stork, 1963b) or from the Wieland–Miescher ketone (Church, 1962; Danieli, 1967). Note that the olefinic hydroxy acid fails to undergo electrolytic coupling, which otherwise would have given α-onocerin directly.

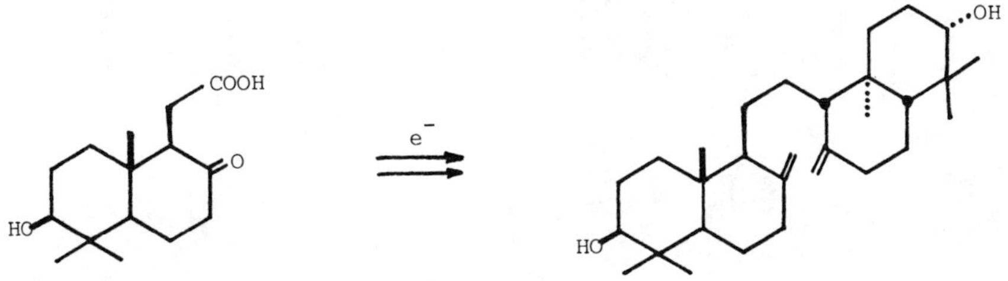

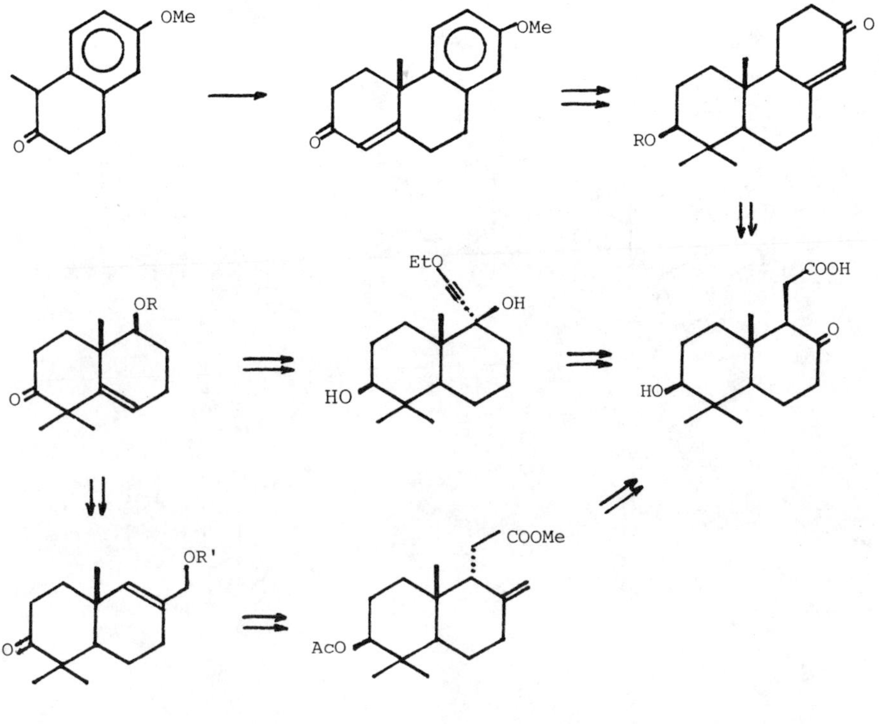

2.6 β-TETRALONE AND ITS ANALOGS

(+)-Podocarpic acid (**P43**) has been synthesized from 1-methyl-2-naphthol via a Robinson annulation (Wenkert, 1958, 1959, 1960), stereoselective methylation of a β-ketoester, and optical resolution at the deoxypodocarpic acid stage. The presence of a quaternary carbon atom in the tricyclic ketone prevents rearomatization of the hydrophenanthrene.

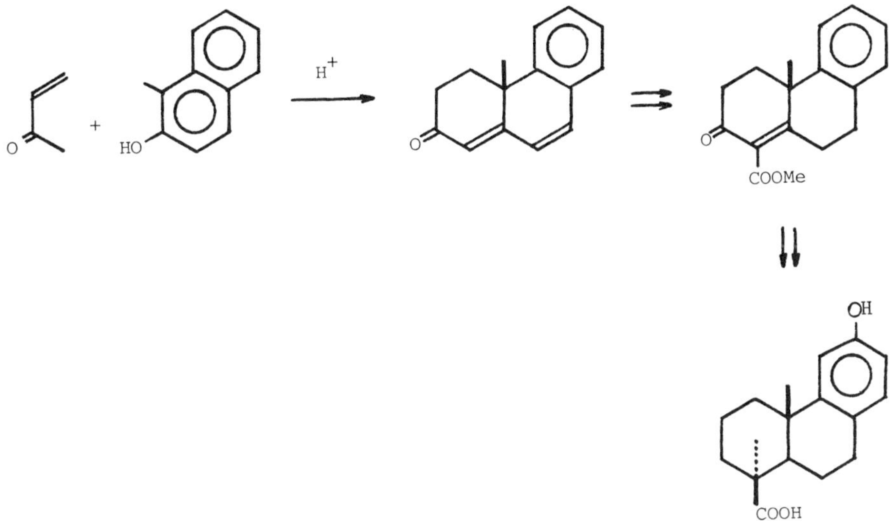

2-Trimethylsilyl-1-penten-3-one is used in the construction of the A ring of stachenone (**S48**) (S. Monti, 1979).

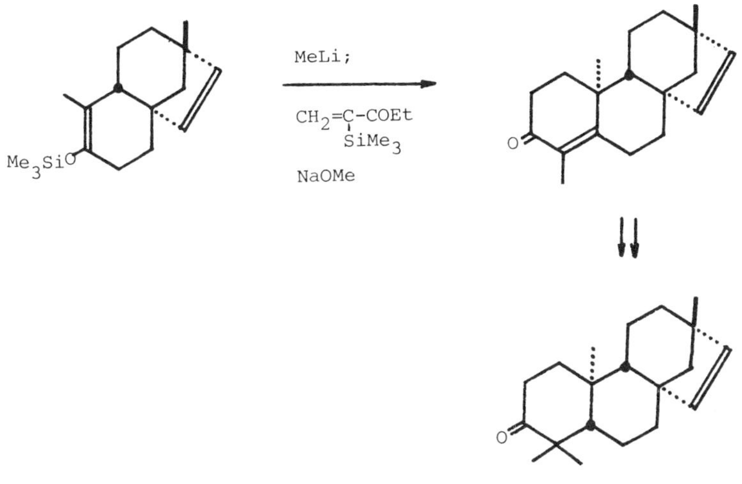

The methoxyhydrophenanthrenone serves as the starting material for bruceantin (**B18**) (Suryawanshi, 1986; F. Kuo, 1987). The synthesis is now at the stage of a pentacyclic intermediate whose ultimate transformation into the terpene molecule should be a relatively uncomplicated matter.

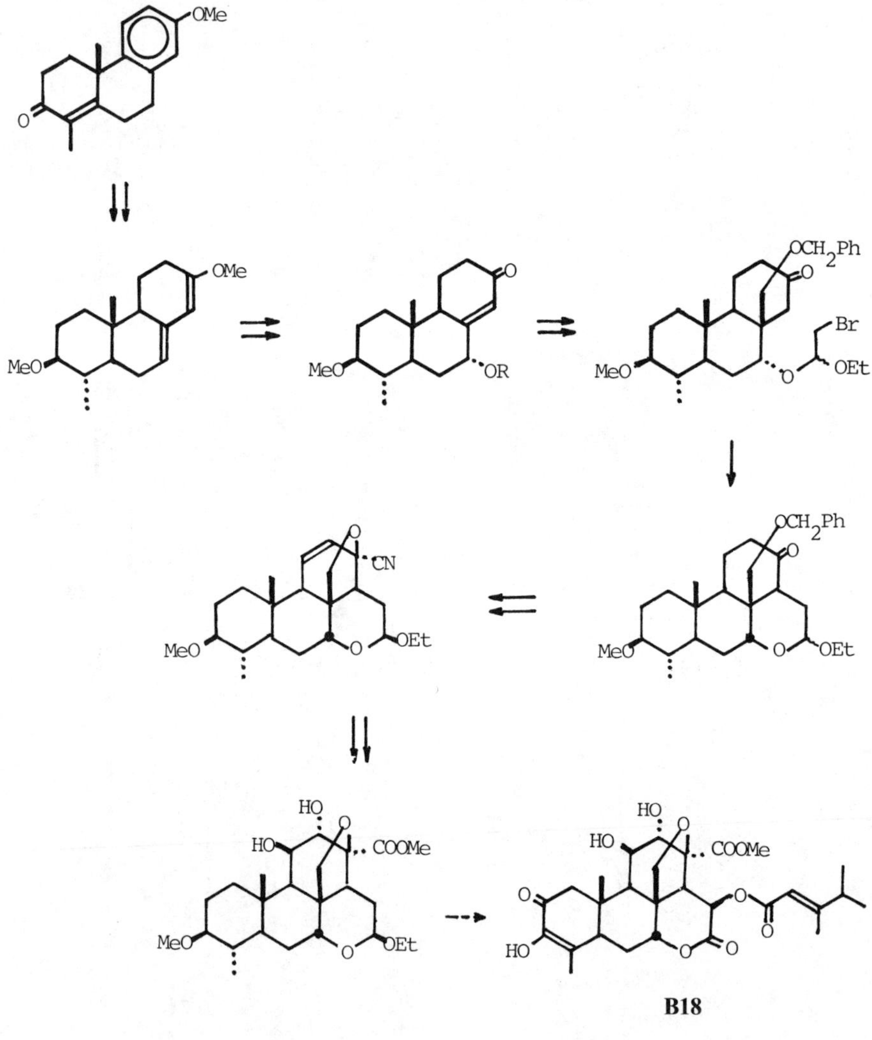

B18

2.7 ANNULATION OF CYCLOPENTANONES

Annulation by an intramolecular ene reaction is an excellent strategy that has been exploited many times in natural product synthesis. It is the basis of the stereoselective formation of the third cyclopentane ring of isocomene (**I14**).

2.7 CYCLOPENTANONES

Because the reaction fails with the bicyclo [3.3.0]octene derivative (for strain reasons), the higher homolog must be employed; in addition, a ring contraction protocol is necessary at a later stage.

The synthesis of isocomene (Oppolzer, 1979) commences with annulation of 2-methylcyclopentanone.

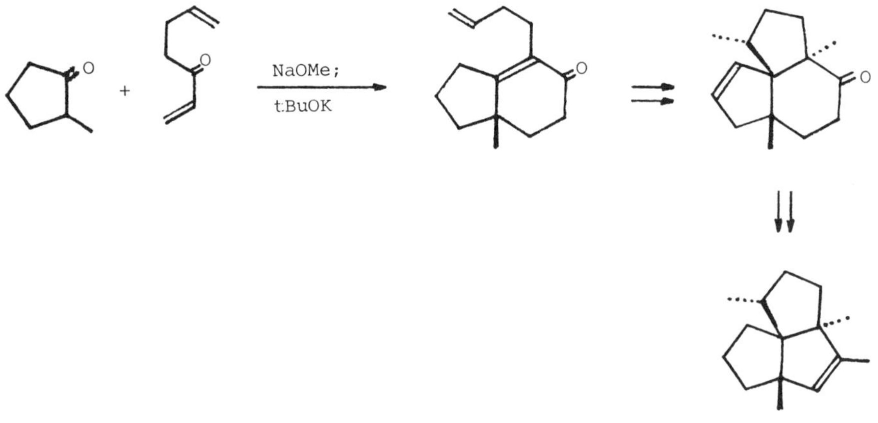

I14

The presence of a hydrindane moiety in gascardic acid (**G1**) has influenced the formulation of a synthetic approach. The work is based on a Robinson annulation and Dieckmann cyclization steps starting from 2-methyl-2-cyclopentenone (Boeckman, 1979).

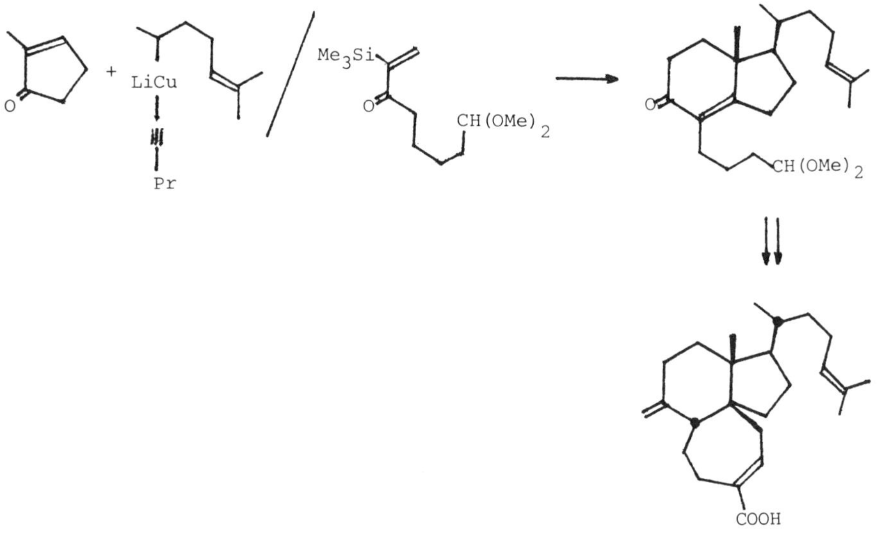

G1

The vicinal dialkylation in a *trans*-selective manner has also been exploited in a synthesis of 8-deoxyanisatin (**A13**, R = H) (Kende, 1985).

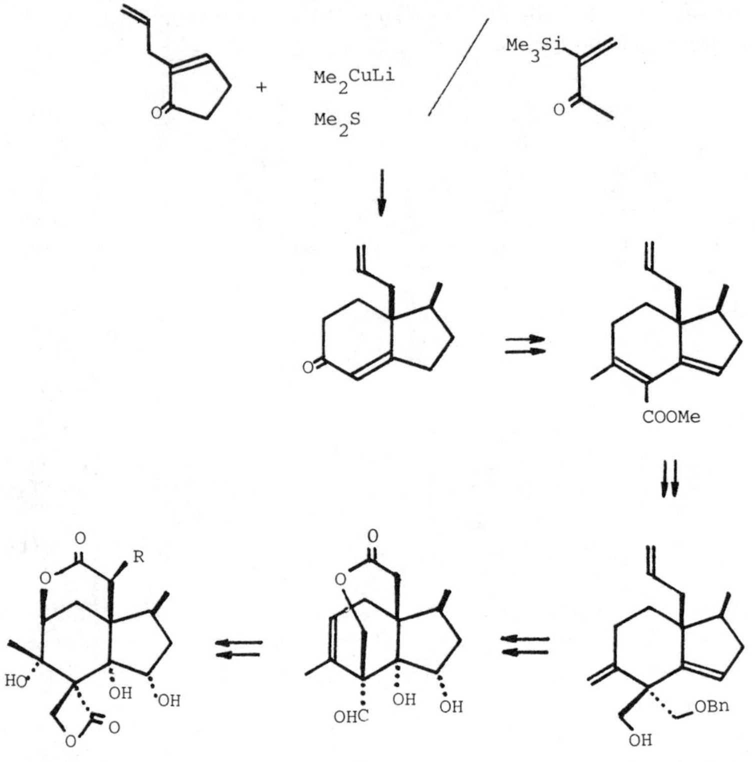

A13 (R=OH)

Prezizaene (**Z2**) is a sesquiterpene in which condensed, spiro, and bridged ring systems are present. A synthesis (Vettel, 1980) starts with building the 5:6-condensed bicyclic moiety from an oxocyclopentanecarboxylic ester obtained from (+)-pulegone. Consequently, the synthetic compound is the optical antipode of the natural terpene.

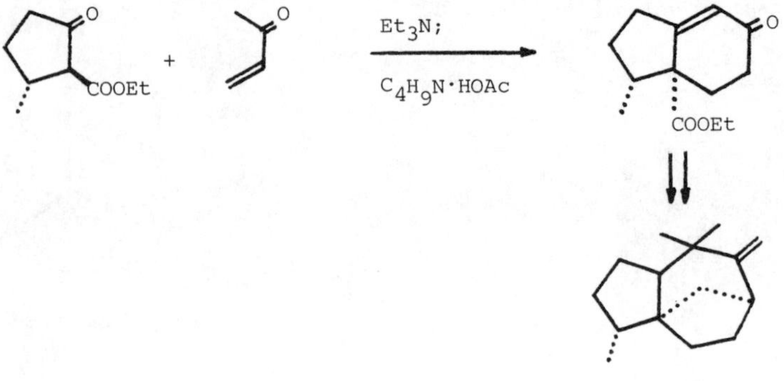

ent-**Z2**

2.7 CYCLOPENTANONES

An interesting synthesis of α-cadinol (**C6**) (Caine, 1977) that starts from preparation of a hydroindenone involves a photochemically induced ring expansion/contraction step. This step corrects the substitution pattern of the two alkyl groups present in the cyclopentanone.

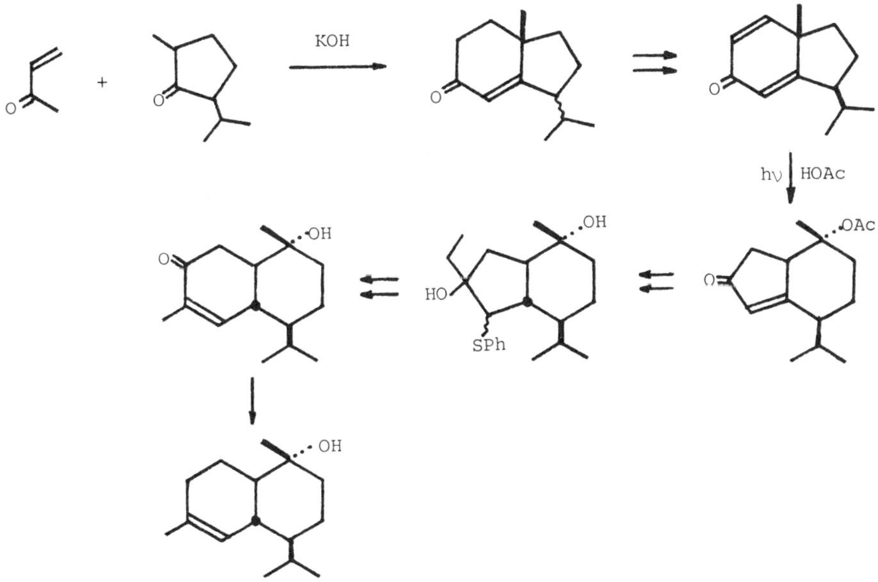

A few hydroindenones prepared by Robinson annulation on cyclopentanone derivatives appear in sesquiterpene work. The lower homolog of the W–M ketone has been used in the construction of the sesquiterpene alkaloid dendrobine (**D11**) (Inubushi, 1972).

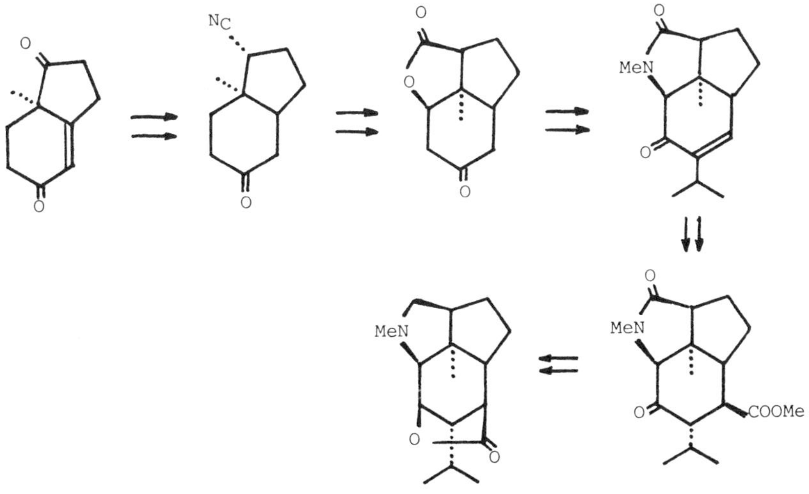

The pyrrolidine ring of dendrobine is assembled by chain extension from the five-membered ketone site and closing at the β'-carbon of the original enone system. The lactone and the isopropyl group are then fashioned systematically from the cyclohexanone moiety.

The ubiquity of a C_5 methyl group and a C_4 oxygen in pseudoguaianolides suggests a profitable enlistment of the lower homologs of the W–M ketone for synthesis. In terms of structural modifications, expansion of the six-membered ring and introduction of the missing functional groups are required. Thus a synthesis of confertin (**C62**) (Marshall, 1976) illustrates the validity of this general concept and also details an elegant plan to meet the specific goal. Needless to say, the preparation of the bicyclic enone involves the Robinson procedure.

C62

Formulation of a synthetic route to verrucarol (**V12**) was instigated by the observation that the relative configuration of two adjacent carbon atoms in the cyclopentane ring of the sesquiterpene is identical to that of the alkoxyhydrindenone related to the W–M ketone. Thus the cleavage of the fused cyclohexenone

V12

portion of the bicyclic precursor, and reelaboration of another cyclohexenol unit from a truncated carbon chain, complete the major undertaking of the synthesis (Schlessinger, 1982).

2.8. MISCELLANEOUS

The last step of a perforenone **A**(**P20**) synthesis (Gonzalez, 1978) consists of a Robinson annulation from a dimethylcycloheptenone. Although the yield is poor (29 %) and the reaction nonstereoselective, this work nevertheless confirms the structure of the marine sesquiterpene.

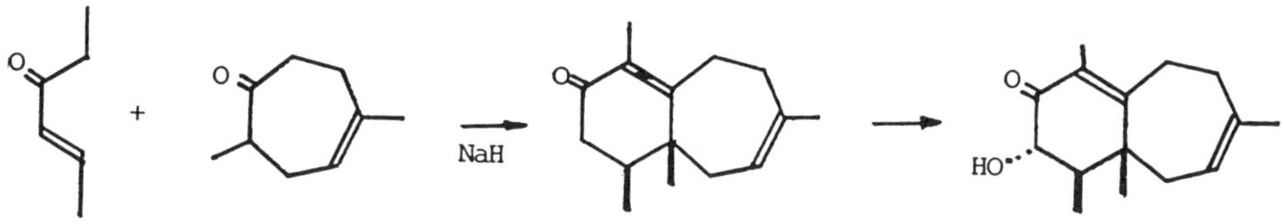

P20

Although aspterric acid (**A47**) contains a 5:7-fused skeleton, its synthesis (Harayama, 1983) commences conveniently with a Robinson annulation. After formation of the ether ring via iodoetherification, *gem*-dimethylation, and introduction of the carboxyl group and the tertiary hydroxyl via the cyanohydrin, the A ring is contracted by treatment of the 3-hydroxy derivative with phosphorus pentachloride.

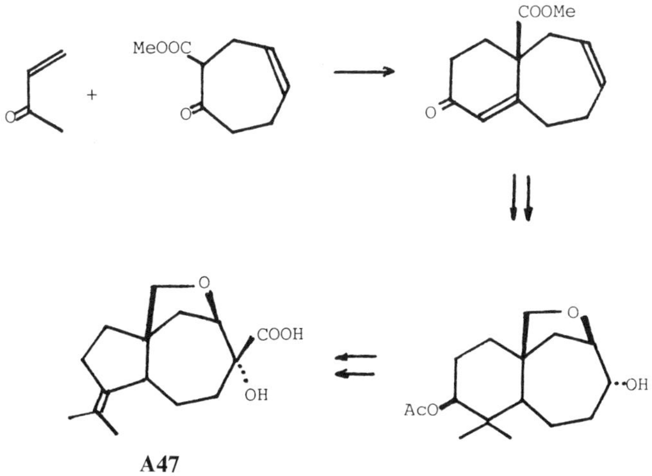

A47

A synthesis of (—)-acorenone (**A10**) (Pesaro, 1978) based on the Robinson annulation to form the spirocyclic system has been reported. In a closely related approach, Stork's enamine method is employed (Lange, 1977, 1978).

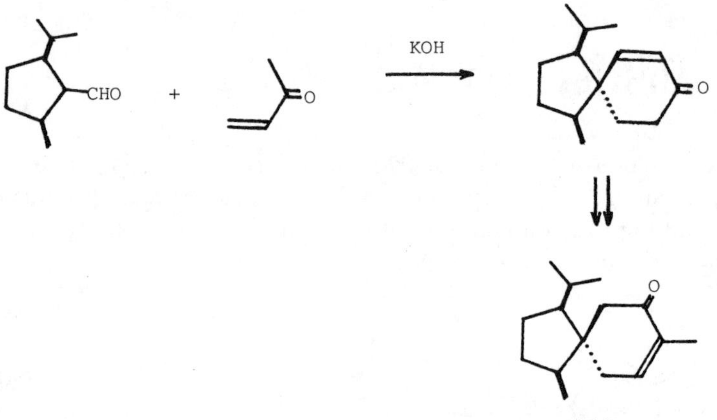

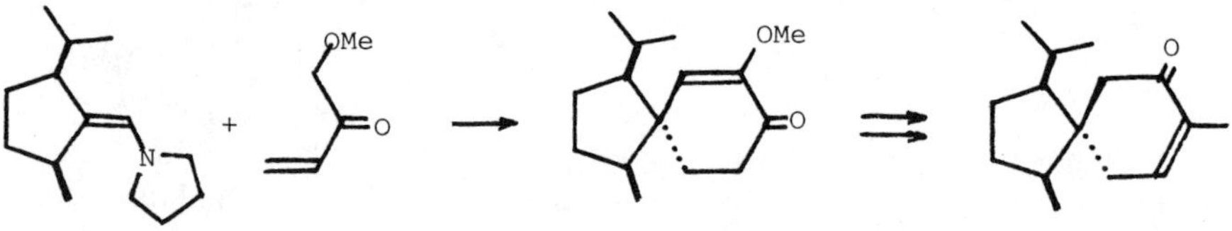

Bicyclic monoterpenes deserve special attention as building blocks for other natural products, especially sesquiterpenes. Stereocontrol is more readily exercised on such networks. For example, (—)-thujone is readily converted to the unnatural (—)-nootkatone (**ent-N11**) (van der Gen, 1971b).

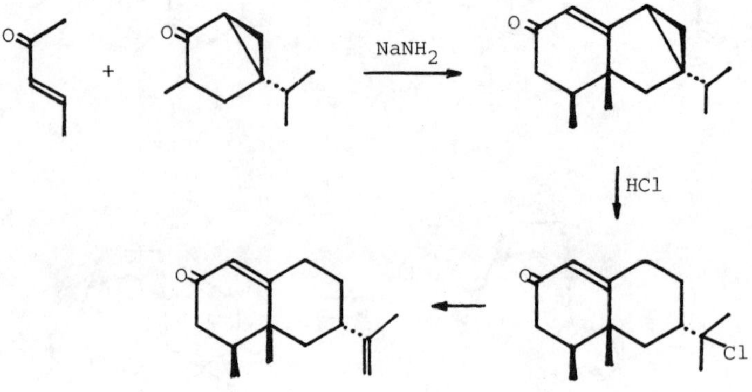

2.8 MISCELLANEOUS

Piperitone (**P34**) and piperitenone (**P33**) are important flavorants as well as being precursors of menthol. Although simple in concept, the route to these compounds via a Michael–aldol tandem is complicated by the alternative aldolization pathways (Walker, 1935, Beereboom, 1966).

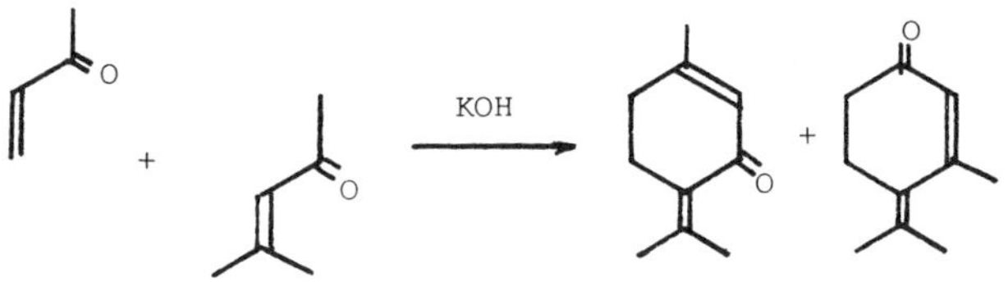

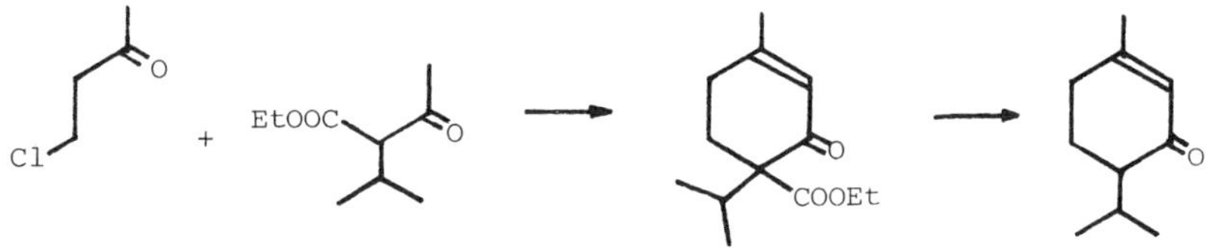

The intramolecular Michael–aldol sequence leading to a bicyclic synthon of retigeranic acid (**R1**) is best induced by zirconium propoxide (Attah-Poku, 1985).

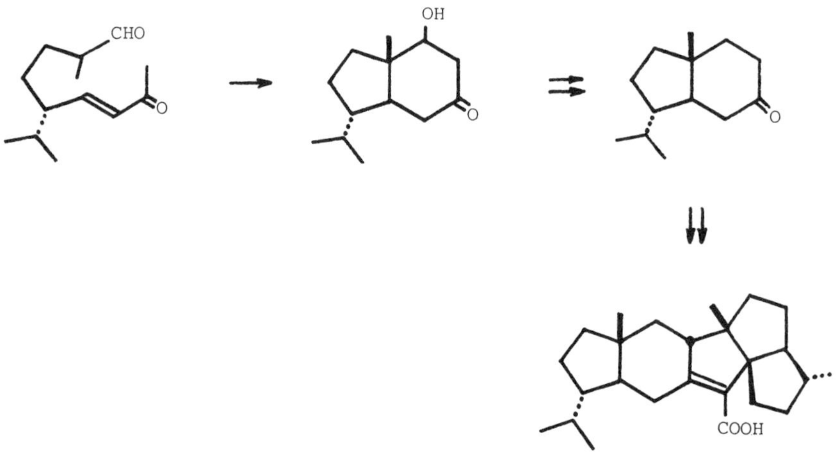

The cyclohexene moiety of Δ^1- and Δ^6-tetrahydrocannabinols (**C16/C17**) may be formed via annulation of a chromanone (Fahrenholtz, 1967).

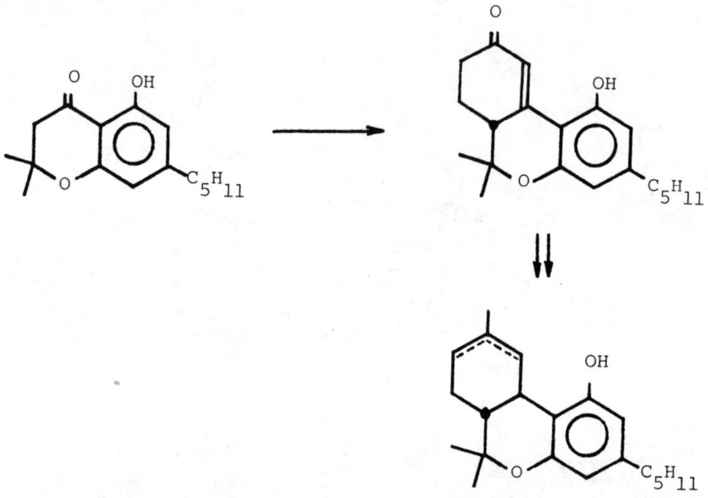

C16, C17

Chapter 3

Aldol Condensation

Aldol condensation is a term generalized from the bimolecular reaction of acetaldehyde, whose product was called "aldol" by Wurtz. Since then the aldol condensation embraces self- and mixed reactions between aldehydes and ketones to afford β-hydroxycarbonyl compounds (β-aldols and β-ketols) (Nielsen, 1968; Mukaiyama, 1982) in the presence of acid or base catalysts.

There are many reactions mechanistically akin to the aldol process. These include the Claisen–Schmidt, Knoevenagel, Doebner, Perkin, Stobbe, and Reformatsky reactions, each characterized by a particular type of reactant. The aldol condensation proper was long beset by side reactions because of its reversibility and the difficulty of achieving specific C—C bond coupling of two different carbonyl substrates. However, recent developments have solved most of the problems, and the aldol condensation has reemerged as one of the most versatile methods for regio- and stereocontrolled C—C bond formation. It has since found spectacular success in the synthesis of complex natural products such as the macrolides.

The nucleophilic enolate anions can be prepared by direct deprotonation of ketones, conjugate addition to α,β-unsaturated ketones, and cleavage of silyl enol ethers with an alkyllithium reagent or fluoride ion. Thus, regioisomeric enolates can be produced as desired.

Masked carbonyl compounds such as Schiff bases, hydrazones, and oximes have been used successfully to generate the enolate equivalents. After their reaction with another carbonyl compound, the nitrogen-containing moiety can be removed and the aldol products recovered.

While lithium enolates are most widely used, their magnesium, zinc, and aluminum counterparts may offer certain advantages in stereochemical control. The same benefits often are easily attainable using vinyloxyboranes as enolate equivalents.

Lewis acid-catalyzed aldol reactions are readily achieved between silyl enol ethers and carbonyl substrates or their acetals/ketals. Titanium tetrachloride, stannic chloride, and boron trifluoride etherate are generally the most effective catalysts.

The aldol process that results in the formation of one C—C bond is suitable for annulation only when it is done intramolecularly.

While intramolecular aldol condensation is embodied in the Robinson annulation, it is also frequently induced from a preformed dicarbonyl compound. In fact, it is imperative to conduct the stepwise method when the cyclic ketol or enone product is not six-membered.

3.1 MONOCYCLIC ENONES

Monocyclic enones prepared by the aldol reaction include the intermediates of cuparene (**C84**) (Kametani, 1979; Casares, 1976; Wenkert, 1978b; Martin, 1980), taylorione (**T8**), and isoshyobunone (**S32i**).

The various cyclopentenones for cuparene synthesis are further methylated according to their substitution patterns. Such an operation may embody conjugate methylation and α-methylation, Simmons–Smith reaction of the corresponding cyclopentenol and subsequent reductive cleavage of the cyclopropane ring, or direct α,α-dimethylation.

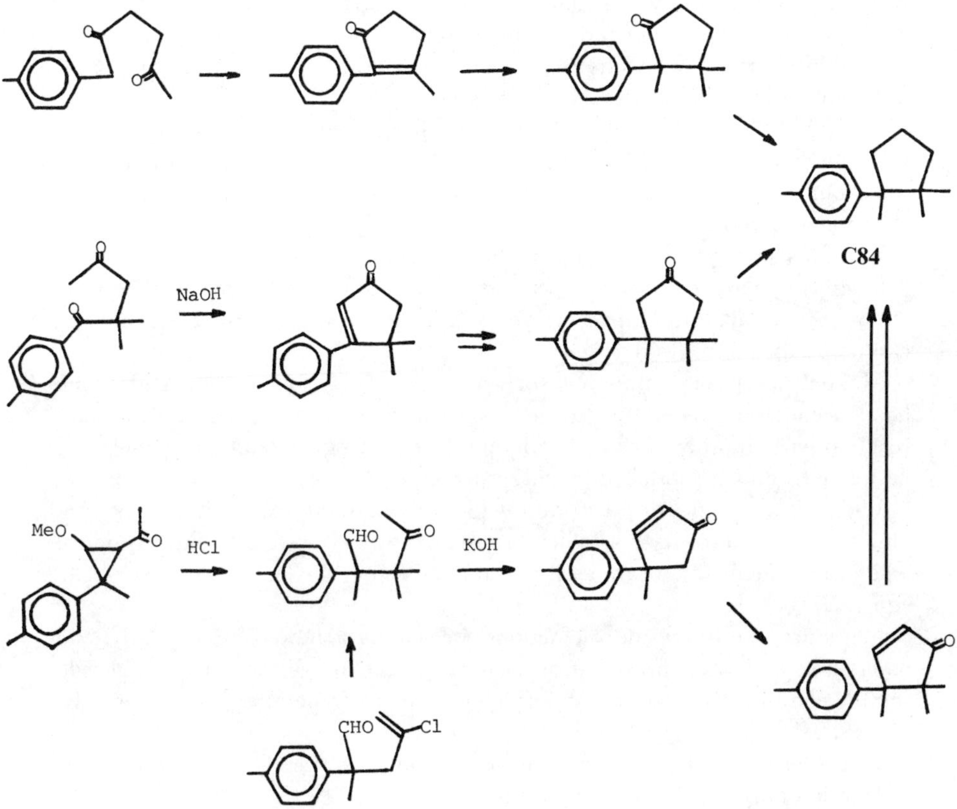

3.1 MONOCYCLIC ENONES

In an access to an optically active cyclopentenone for *ent*-α-cuparenone (**ent-C85**) (Meyers, 1986b) the interesting aspect is a stereoselective *endo*-alkylation of a bicyclooxazolidinone which is definitely roof-shaped.

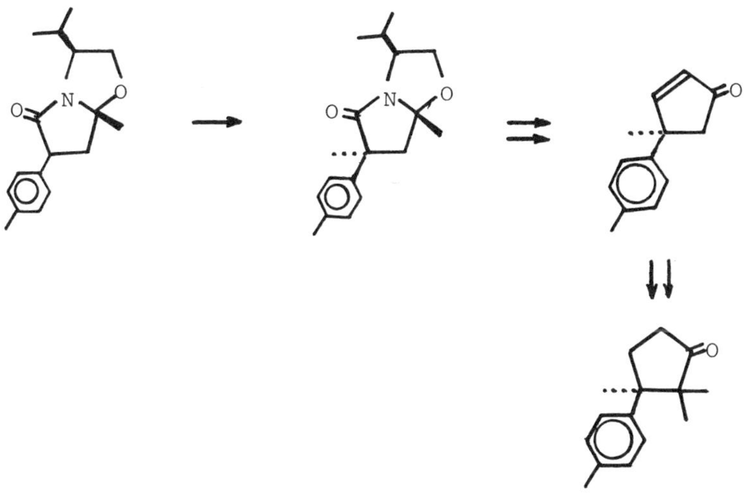

ent-C85

The γ-ketoaldehyde that intervenes in the taylorione synthesis (Nakayama, 1975) is obtainable from (+)-3-carene by cleavage of the cyclohexene ring and selective manipulation of the two side chains. The conversion of the ketoaldehyde into taylorione is straightforward.

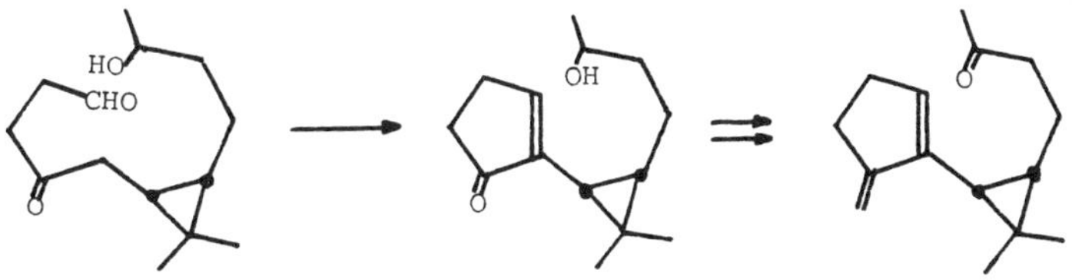

T8

The final step in synthesizing the dehydrofuropelargones (**F19/F20**) combines retroaldol and aldol processes. The ring closure is regioselective, since the alternative aldol product cannot undergo dehydration (Büchi, 1968).

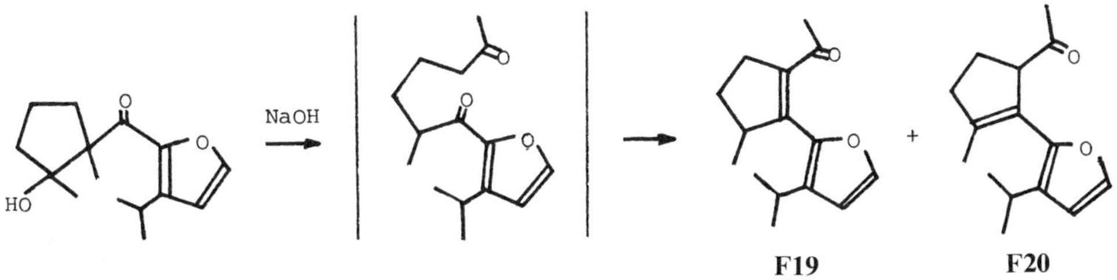

F19   **F20**

(+)-Limonene is the starting material for a synthesis of (—)-daucene (**ent-D8**) (Yamasaki, 1972). A cyclopentenecarboxaldehyde is derived from selective cleavage of the cyclic double bond and aldolization of the resulting ketoaldehyde. To achieve the aldol condensation, piperidinium acetate is used as catalyst; alkaline reagents afford the isomeric methyl ketone. [*Note*: The methyl ketone derived from α-terpineol has been employed in a synthesis of β-vetivone (**V16**: Bozzato, 1974).]

The symmetrical diketodiester derived from oxidative dimerization of ethyl acetoacetate at the γ position readily cyclizes, and the resulting cyclopentenone is a useful precursor of dehydroiridodiol (**I7**) (Kimura, 1982).

Two very similar syntheses of (+)-taonianone (F. Kido, 1986; Huckenstein, 1987) from (—)-carvone involve cleavage of the cyclic double bond and aldolization of the ketoaldehyde with piperidine-acetic acid. Chain elongation of the cyclopentenecarboxaldehyde and coupling with furan-containing reagents follow.

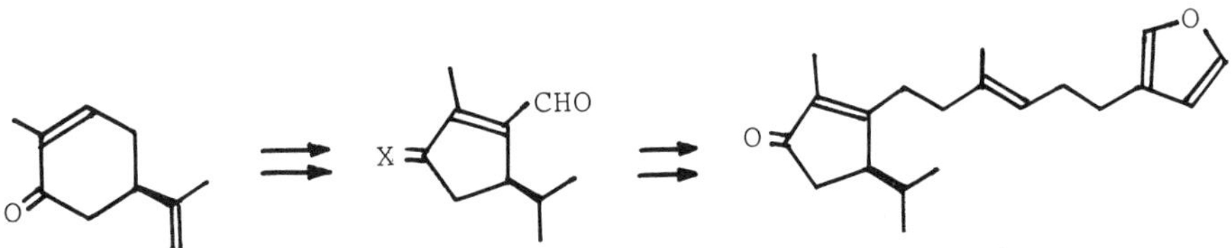

3.1 MONOCYCLIC ENONES

taonianone

In an attempt to utilize the waste stream of an industrial linalool production from pinenes, the plinols have been converted into piperitone (**P34**) (Ho, 1984) by a sequence involving hydrogenation, dehydration, ozonolysis, and aldolization.

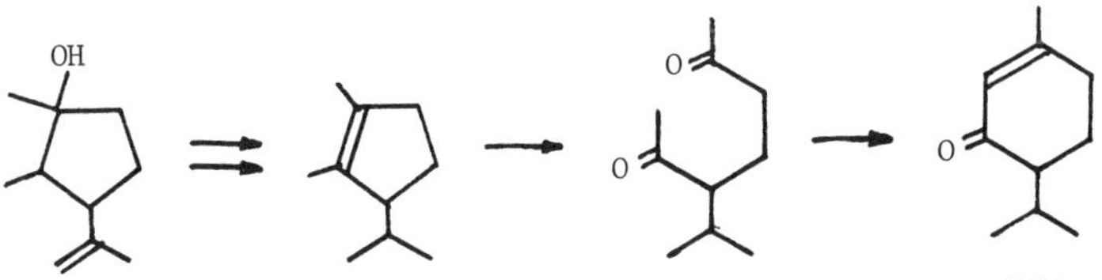

P34

Isoshyobunone (**S32i**) and its 6-epimer can be prepared (Alexandre, 1975, 1977) from a δ-diketone via a 2,3-disubstituted cyclohexenone. However, a monosubstituted cyclohexeneone isomer is also produced from the condensation in large amounts as a result of an alternative mode of aldolization.

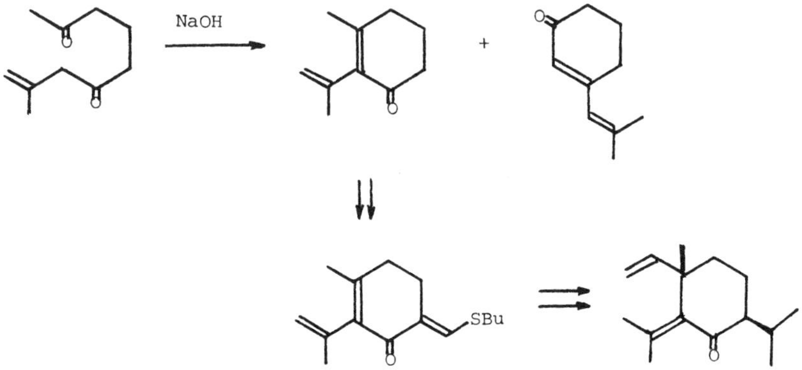

s32i

In syntheses of (—)-khusimone (**K8**) (Liu, 1979, 1982) and (+)-norpatchoulenol (**P13**) (Liu, 1987c) an optically active cyclohexenone in either antipode is used as platform to build the two different tricyclic systems.

It is easy to identify an acyclic ketoaldehyde precursor for carvone (**C31**). The presence of a γ,δ-double bond to the aldehyde function signifies a possible genesis of the compound by a Claisen rearrangement (Vig, 1966).

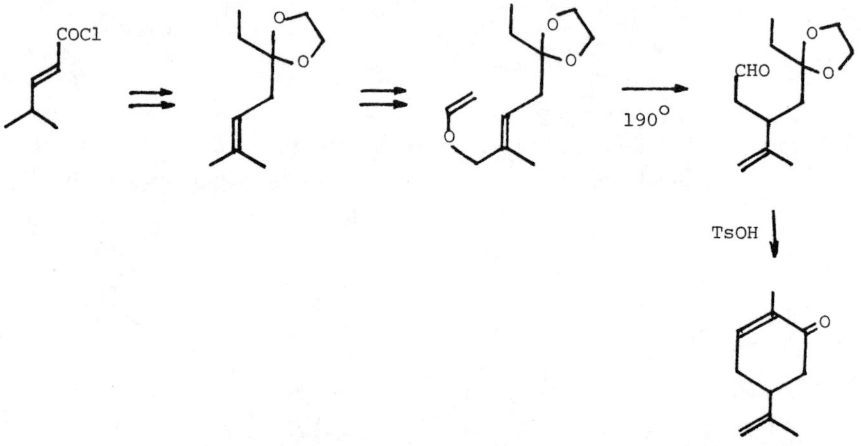

Trisporic acids including 9Z- and 9E-trisporic acid B (**T30/T31**) and congeners such as trisporol B (**T32**) are biosynthetically derived from β-carotene and used by certain fungi in their sexual differentiation. The single cyclohexenone ring present in these molecules automatically indicates an aldol route by which the chemist may attempt synthesis. Indeed, all the reported approaches are based on aldolization, and they differ only in the timing of the cyclization (Edwards, 1971; Isoe, 1971; Secrist, 1977; Prisbylla, 1979; Trost 1983b).

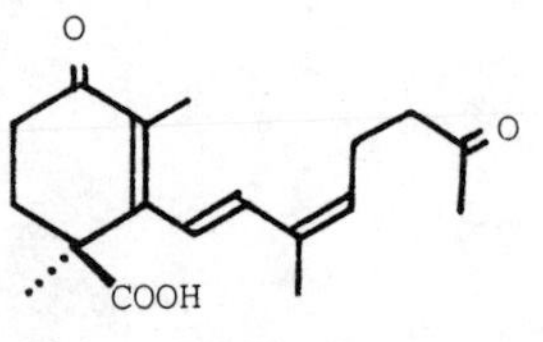

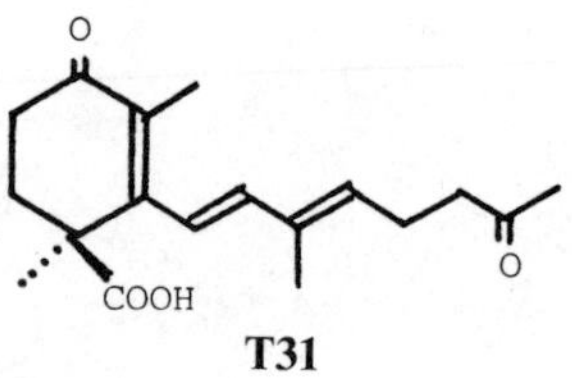

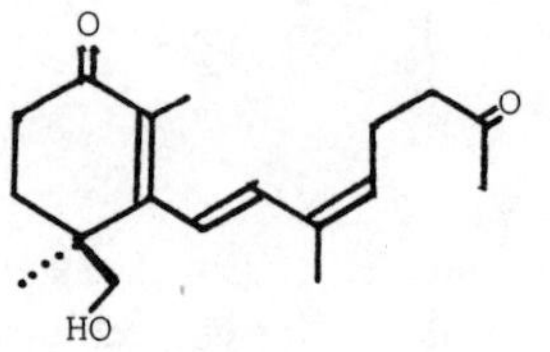

3.1 MONOCYCLIC ENONES

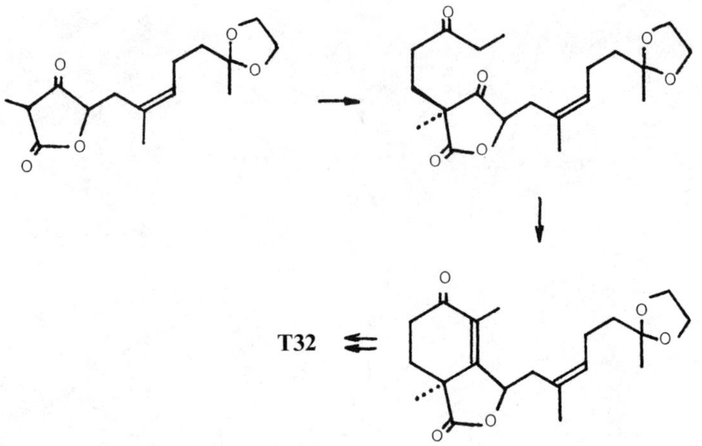

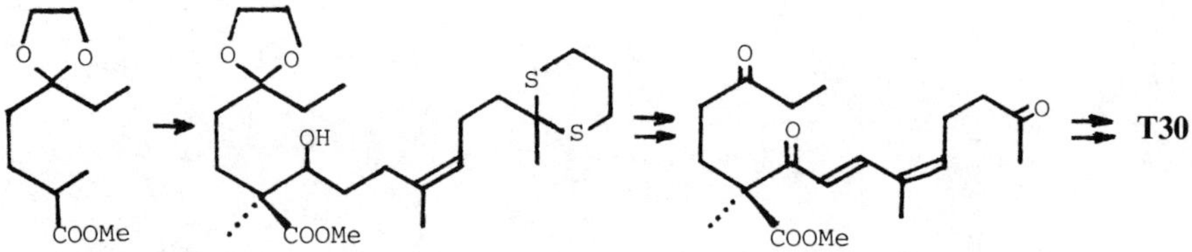

Furanocembranolides such as lophotoxin are challenging targets for synthesis. Assembly of the carbon skeleton has been achieved (A. Kondo, 1987) by an aldol process involving a β-ketoester subunit and α,β-epoxyaldehyde. The epoxide ring is in the best position to react with the ketone function and the overall process accomplishes both macrocyclization and furan formation.

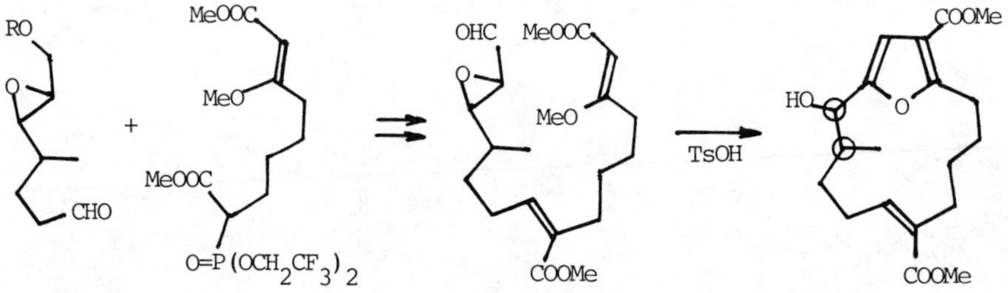

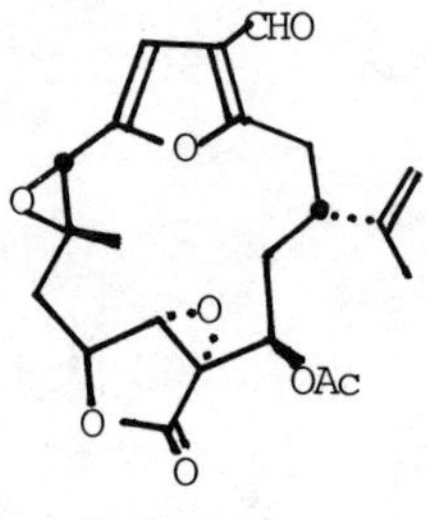

lophotoxin

3.2 SPIROCYCLIC INTERMEDIATES

Acorone (**A11**) and the acorenones (**A10/A10b**) are spirocyclic sesquiterpenes that are obviously approachable by the aldol methodology in forming either five- or six-membered rings. Accordingly, the preparation of the spiro[4.5]decadien-one intermediate (McCrae, 1977; Martin, 1978; Wenkert, 1978b; Ho, 1982) for acorone originates from tetrahydrotolualdehyde by indirect acetonylation.

On the other hand, the presence of an oxygen function only at the six-membered ring of the acorenones indicates the preferred synthetic routes to begin with dialkylcyclopentanes. It is also tactically preferable to use stereodefined cyclopentane compounds to ensure a predictable outcome of the spiroannulation and correct structure of the final product from a rational maneuvering of the intermediates. Thus, spirannulated cyclobutanes are formed and cleaved (Trost, 1975a; S. Baldwin, 1982a). 1,5-Ketoaldehydes are created and condensed intramolecularly to give products requiring transposition of the enone system and methylation to reach the end.

The vetispiranes are another class of spirocyclic sesquiterpenes that has been most actively studied synthetically. The presence of two and sometimes more asymmetric carbon atoms adds a stereochemical burden to the controlled creation of the spirocyclic center.

A10b

A10

A vinylogous aldol condensation is featured in the spirocyclization, which eventually leads to β-vetivone (**V16**) (K. Yamada, 1973). The presence of a carboxyl group and its role in preventing one of the products from reverting to the ketoaldehyde by lactonization is most important in the stereochemical sense. The oxygenation of the α-oriented carbon atom adjacent to the spirocyclic center facilitates the introduction of the three-carbon chain one carbon away.

The secondary methyl group in the six-membered ring is quite effective for inducing asymmetry at the β carbon. It is possible to introduce two carbon chains sequentially and recombine them in a rational manner. Alternatively, the "secoalkylation" technique, which involves the addition of three carbon atoms to a ketone group, thereby converting the latter into a cyclobutanone, and the methodical cleavage of the cyclobutanone to release two carbon chains, is applicable to the synthesis of vetispiranes (Trost, 1975b).

The spiroannulated cyclobutanone derived from 2,6-dimethyl-2-cyclohexenone is activated at the α position in the form of a dithioketal and is subject to attack by a methyl metal reagent. The dithioketal, strained by bond angle deformations and nonbonded interactions, has now acquired the trigger elements for fragmentation to give a masked 1,4-ketoaldehyde. Additional steps leading to the vetispiranes are readily conceived from this point.

The method of preparing 1,4-dicarbonyl compounds by acyl carbenoid addition to enol ethers and subsequent ring opening induced by acids is very appealing. Stereoselectivity is high when the sequence is applied to vetispirane synthesis (Wenkert, 1978b). The only drawback is the competing carbene addition to the endocyclic double bond.

In the approach to (—)-agarospirol (**A17**) pursued by Deighton (1975), the five-membered ring is reformed from a severed methylcyclohexene moiety by the aldol method. In view of the incontrollable regiochemistry, this cyclohexene, upon generation, is contaminated with an undesired isomer.

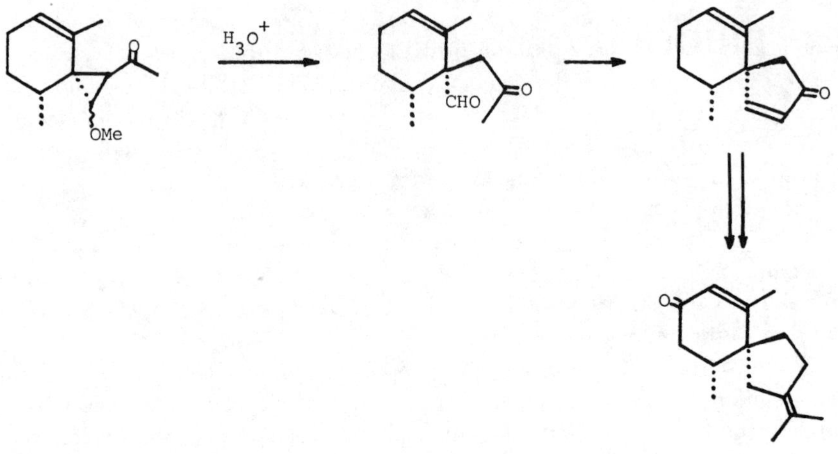

3 ALDOL CONDENSATION

V16

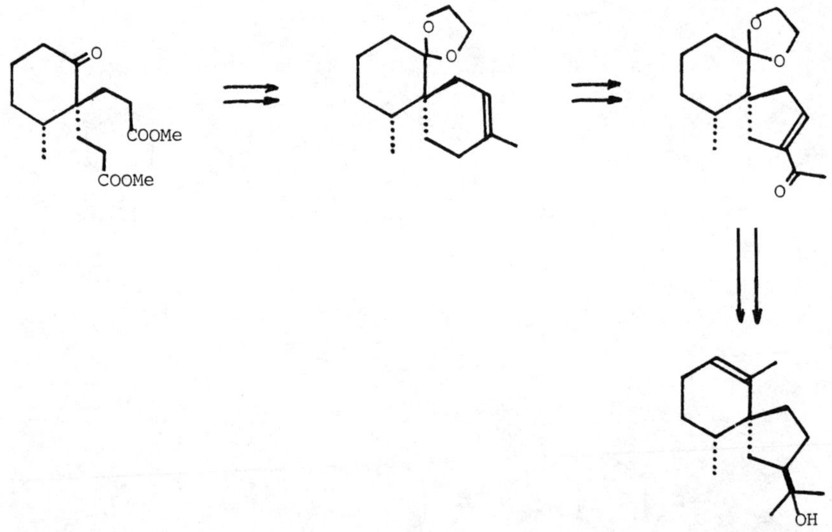

A17

3.3 BRIDGED RING SYNTHESIS

Bridged ring systems occur extensively in sesquiterpenes. Synthesis of these compounds often makes use of the aldol reaction. Thus (—)-helminthosporal (**H9**) has been constructed (Corey, 1963b, 1965b) via a Wichterle reaction on carvomenthone, thereby providing all but one of the carbon atoms. The bicyclic compound is contracted by means of double-bond cleavage and aldolization of the ketoaldehyde.

The presence of a β-hydroxymethylenecyclohexane segment in upial(**U3**) and gymnomitrol (**G21**) suggests an aldol approach, since it is conceivable that this structural feature is accessible from the corresponding hydroxy ketone (e.g., by a Wittig reaction). Furthermore, the separate location of the oxygen functions in two different bridges prevents dehydration and loss of an oxy function under mild conditions.

In the synthesis of upial (Taschner, 1985). (—)-carvone is elaborated into a ketoaldehyde, which is aldolized. The homoallylic alcohol obtained from methylenation of the ketol responds to a redox manipulation to establish the second chiral center. The hydroxyl group then unites with the ester derived from the isopropenyl unit of the carvone to form the γ-lactone. This lactone provides sufficient activation to allow tagging the missing acetaldehyde chain (in a latent form) to the molecule.

Because absolute stereochemistry of upial was unknown at the time, the less expensive (—)-carvone was chosen arbitrarily in this work, resulting in *ent*-upial.

Concerning the synthesis of gymnomitrol, the aldol cyclization is unexpectedly difficult. Although conditions have been found to effect this reaction (Han, 1979), others have adopted a more circuitous protocol (Coates, 1979) or the corresponding Claisen reaction (Welch, 1979).

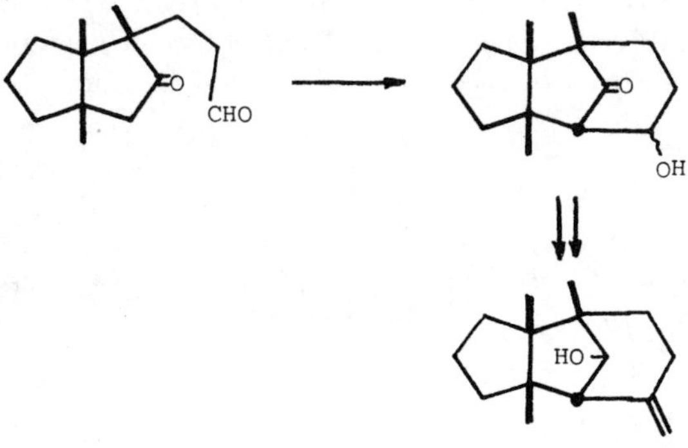

G21

Reluctance to undergo aldolization is also displayed by ketoaldehyde intermediates for trichothecanes. With respect to an approach to trichodermin (**T24**) (Colvin, 1971, 1973), the last resort involving reductive cyclization of an enolactone yielded only 7 % of the bridged ketol; the major product (52 %) is the ketoaldehyde.

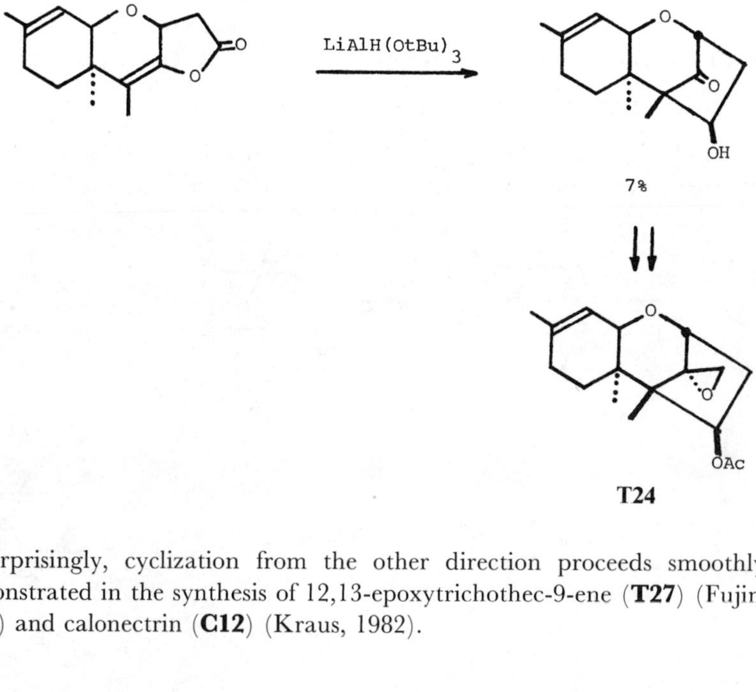

T24

Surprisingly, cyclization from the other direction proceeds smoothly, as demonstrated in the synthesis of 12,13-epoxytrichothec-9-ene (**T27**) (Fujimoto, 1974) and calonectrin (**C12**) (Kraus, 1982).

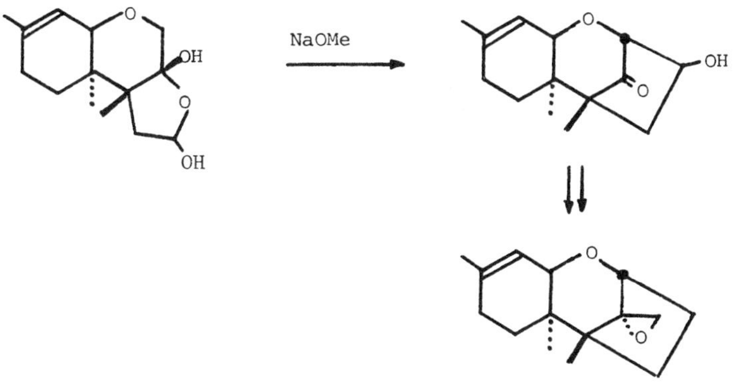

3.3 BRIDGED RING SYNTHESIS

T27

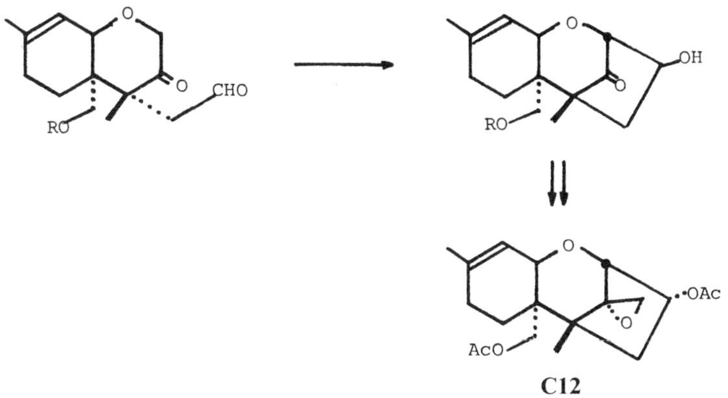

C12

Note: The heterocycle is also assembled by the aldol technique:

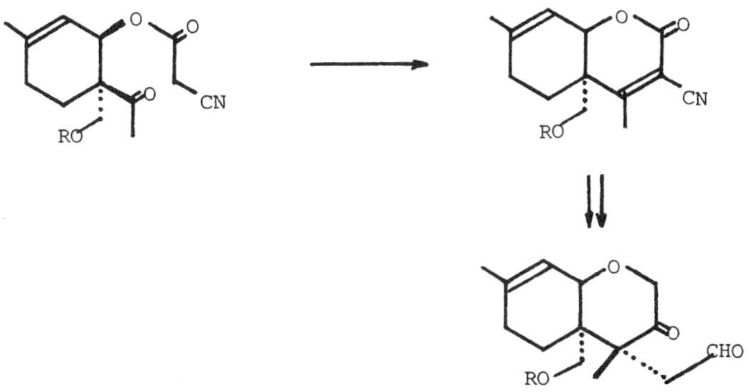

The aldol condensation that follows dibal reduction of enolactone as shown above also figures in the annulation of tetrahydroeucarvone during a synthesis of longicamphor/longicyclene (**L22/L23**) (Welch, 1973a, b, 1974).

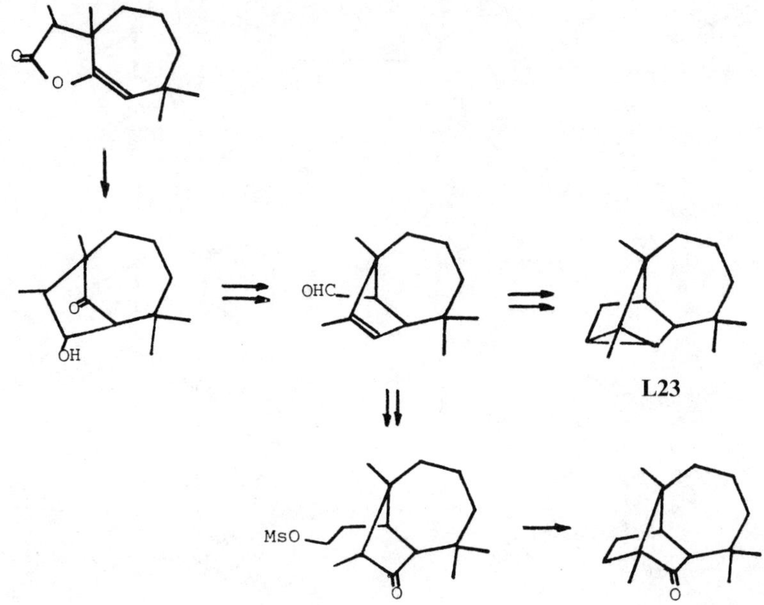

By means of an aldol reaction, a bicyclic precursor of sativene (**S11**) and cyclosativene (**S13**) has been acquired from a ketoaldehyde that originated from (—)-carvone (Hagiwara, 1975).

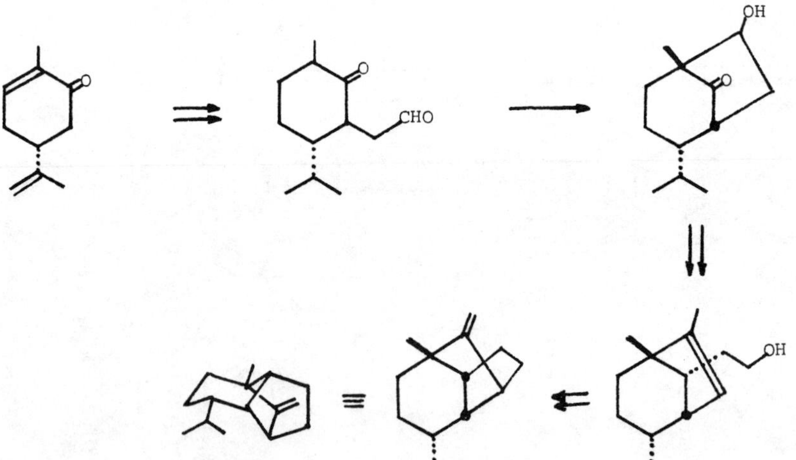

Although there is only one 5:6-fused carbocycle in the skeleton of picrotoxinin (**P29**), the presence of eight contiguous chiral centers in the nine-carbon periphery suffices to test the mettle and acumen of the synthetic chemist.

3.3 BRIDGED RING SYNTHESIS

Stereocontrol offered by bridged ring systems has been seized on in a magnificent picrotoxinin synthesis (Corey, 1979c) from carvone.

Two bridged carbocycles are formed at different stages of this work, both by the aldol method. One of these rings undergoes cleavage later to supply two correctly oriented carboxyl groups destined to become part of the γ-lactones.

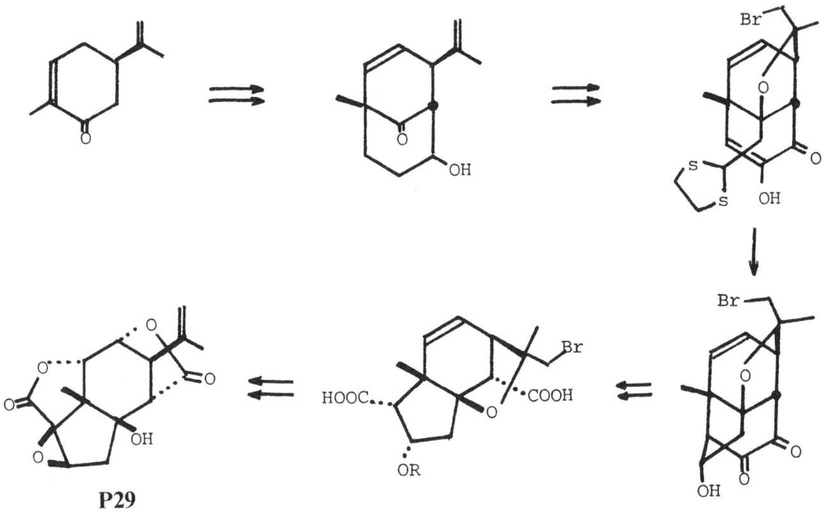

P29

In a synthesis of aphidicolin (**A36**) (Trost, 1979a), a propanal chain is implanted at C_9 of a tricyclic cyclopentanone via axial alkylation. Intramolecular aldol condensation completes the carbon framework.

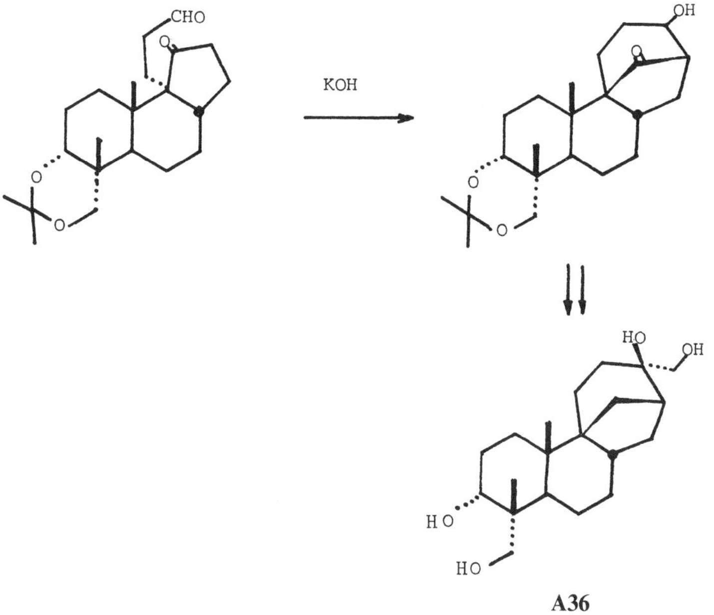

A36

Many tetracarbocyclic diterpenes are characterized by a bicyclo[3.2.1]octane system that shares two carbon atoms with the B ring. Accordingly, the D ring lends itself to bridging by an aldol reaction.

The required ketoaldehyde segment may be constructed by the Claisen rearrangement of an allyl vinyl ether derived from an anisole precursor. The application of this reaction sequence to a synthesis of kaurene (**K4**) (Bell, 1962, 1966b) is illustrative.

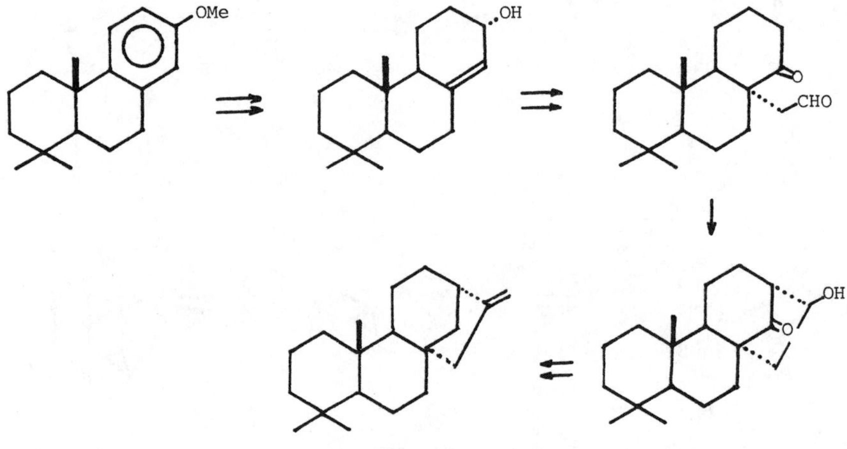

K4

An alternative method for assembling the ketoaldehyde unit involves the use of an isomeric anisole/phenol. In an approach to the diterpene alkaloid veatchine (**V5**) (Guthrie, 1966), the C ring derived from the aromatic precursor is settled in the stage of a cyclohexanone, which is allylated at the angular position. The allyl group is the precursor of an acetaldehyde chain.

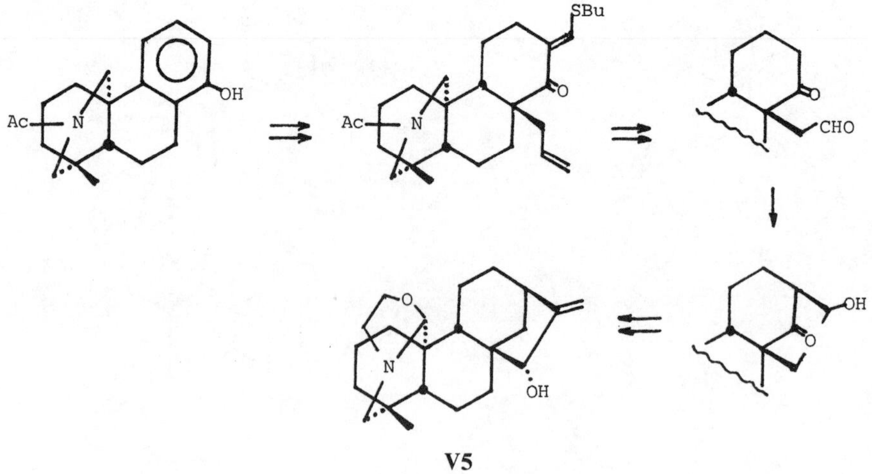

V5

3.3 BRIDGED RING SYNTHESIS

A similar closure of the D ring of grayanotoxin II (**G19**) employs a relay compound (Hamanaka, 1972) that is obtainable from photochemical rearrangement of a cross-conjugated cyclohexadienone (Gasa, 1976).

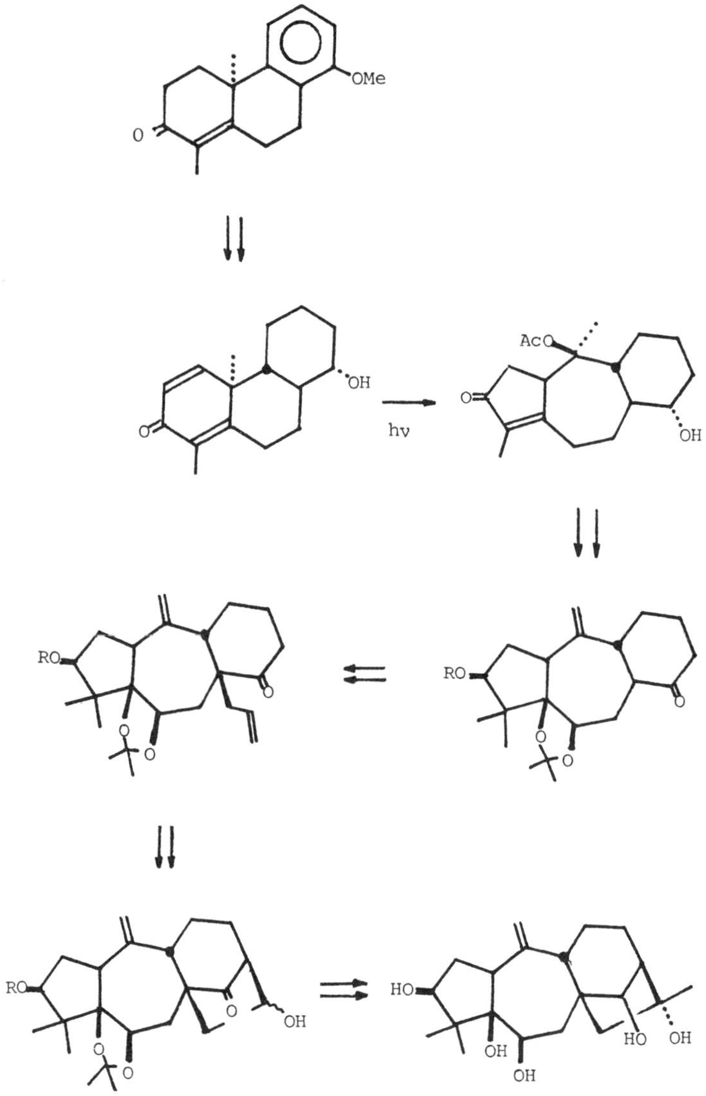

G19

Another synthesis of veatchine (**V5**) (Nagata, 1963a, 1967a) also highlights an aldol process for the construction of the C/D ring system. However, the ring fusion is wrong, and an "inversion" by rearrangement is required.

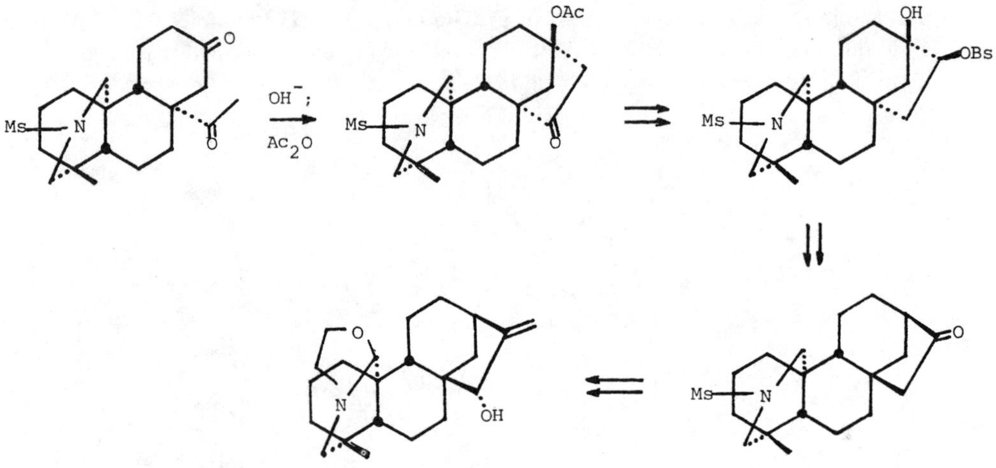

A pivotal tricyclic precursor for gibberellic acid (**G12**) has been resynthesized via a spirocyclic intermediate (Corey, 1979d), which is in turn derived from a cyclohexylideneacetone. The C ring is formed by an intramolecular alkylation protocol, and the D ring is constructed by cleaving the original cyclohexanone ring and effecting an aldol closure.

Regarding the intramolecular alkylation step, stereochemical issue is never a problem. The cis juncture of the hydrindane segment is ensured.

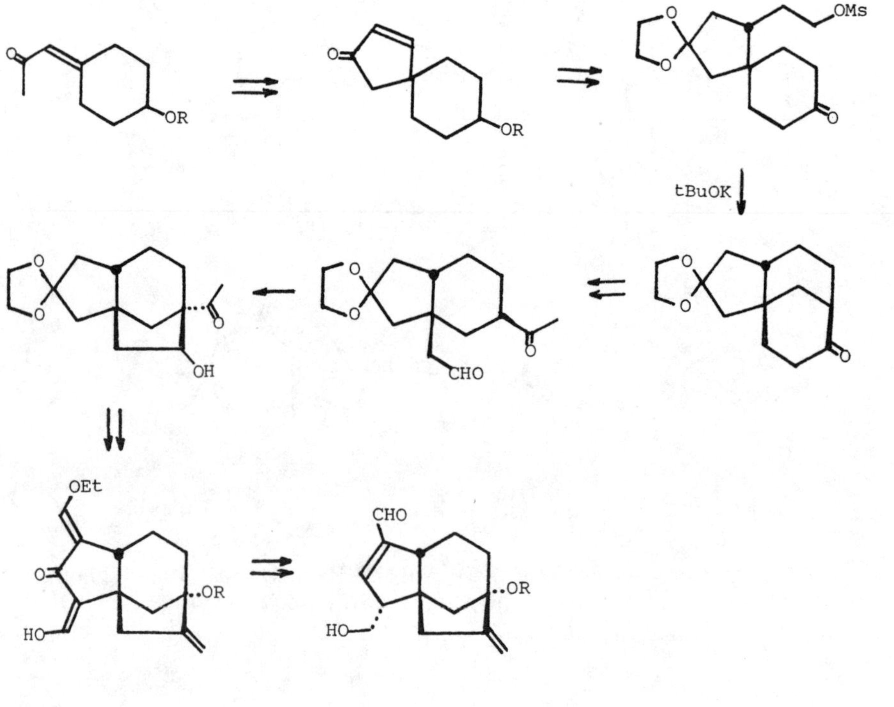

3.3 BRIDGED RING SYNTHESIS

An approach to hibaene (**H11**) (Ireland, 1965; Bell, 1966a) that features D-ring assembly by aldol condensation is less simple. The stereochemistry inherited from the undesired orientation of the acetaldehyde chain requires correction; this is achieved by an allo rearrangement.

H11

Retrosynthetic logic serves to identify an oxygenated C_2 atom of 2-isocyanopupukeanane (**P53**) and thence its synthesis by an aldol route (Corey, 1979a). Location of an additional oxy site at a β carbon and an imaginary disconnection indicate the viability of a *cis*-fused hydrindanone aldehyde as the precursor.

P53

The recognition of a hidden bicyclo[2.2.2]octane subunit in the carbon framework of ishwarane (**I11**) validates a synthetic plan based on aldol cyclization. This approach (R. Kelly, 1971, 1972) was studied at the time when assiduous exploration of the photocycloaddition of allene to enones as an effective synthetic operation was being carried out to accomplish reductive β-alkylation of the enones.

I11

In the kaurene synthesis mentioned above, the ketoaldehyde (monoketal) is actually the minor product of a hydroboration–oxidation sequence. The major compound is well suited to elaboration into atisirene (**A49**) (Bell, 1966b) by an analogous operation involving acid-catalyzed aldolization, etc.

A49

The C/D ring system of stemarin (**S49**) can be assembled by rearrangement from a bicyclo[2.2.2]octane derivative, provided a 12-*anti* leaving group is present. The general method of introducing an acetaldehyde chain at the β position of a cycloalkenone via photocycloaddition with allene followed by aldolization is very useful because it permits an oxygen function at a selected branch of the [2.2.2] skeleton to be retained, and because the stereochemistry of the photocycloaddition is immaterial as long as it is not stereorandom.

The synthesis (R. Kelly, 1980) reveals an α-side approach of allene to the tricyclic enone. Consequently, after degradative ring opening of the fused methylenecyclobutane ring and aldolization of the derived ketoaldehyde, the ketone group is removed. A drawback of this route is that the ketol having the

3.3 BRIDGED RING SYNTHESIS

desired *anti*-hydroxyl configuration is the minor component, the predominant product being favored by hydrogen bonding.

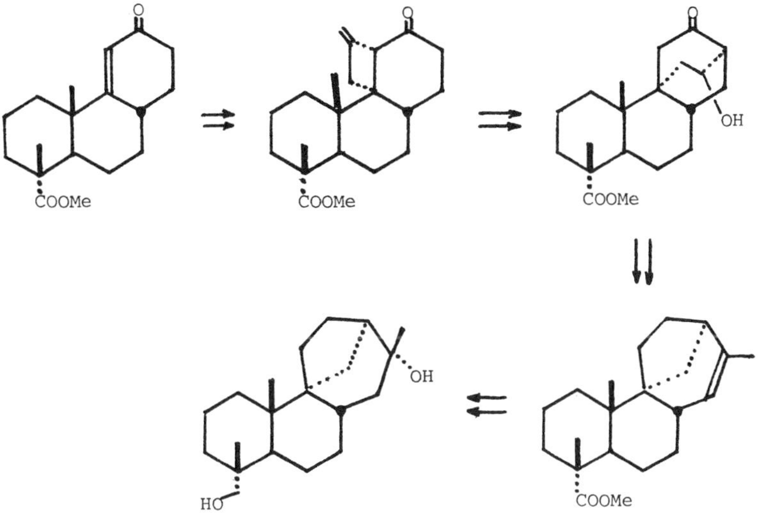

S49

The *syn*-hydroxyl isomer in an analogous series is an ideal precursor of aphidicolin (**A36**) based on the rearrangement protocol (Marini-Bettolo, 1983b). In fact, this route is more efficient than the Diels–Alder approach based on a C-ring diene (van Tamelen, 1983). It is notable that the two addends (allene and maleic anhydride) add from different sides of the tricyclic intermediates.

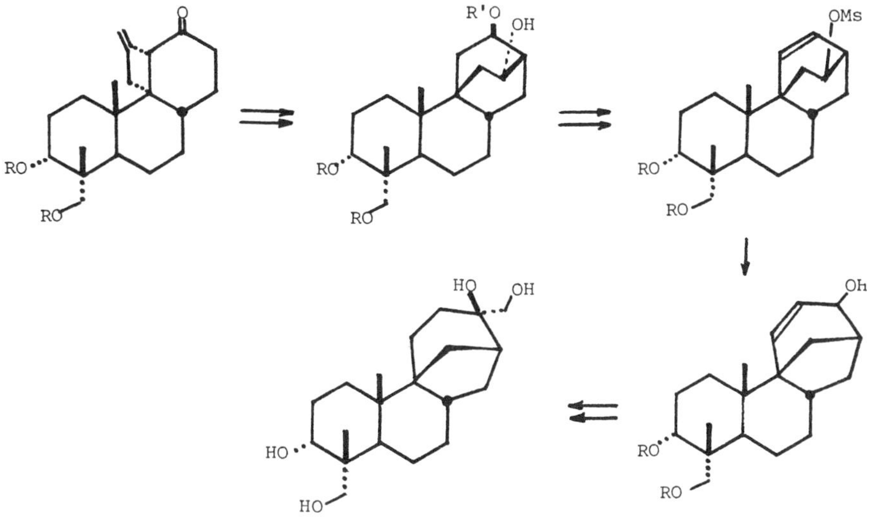

A36

The complex diterpene alkaloid talatisamine (**T1**) has yielded to total synthesis (Wiesner, 1974c). The plan calls for construction of a bicyclo[2.2.2]octanolone and a subsequent ring expansion/contraction maneuver that also involves the B ring. The [2.2.2] system is derived from the photocycloaddition–retroaldol/aldol pathway.

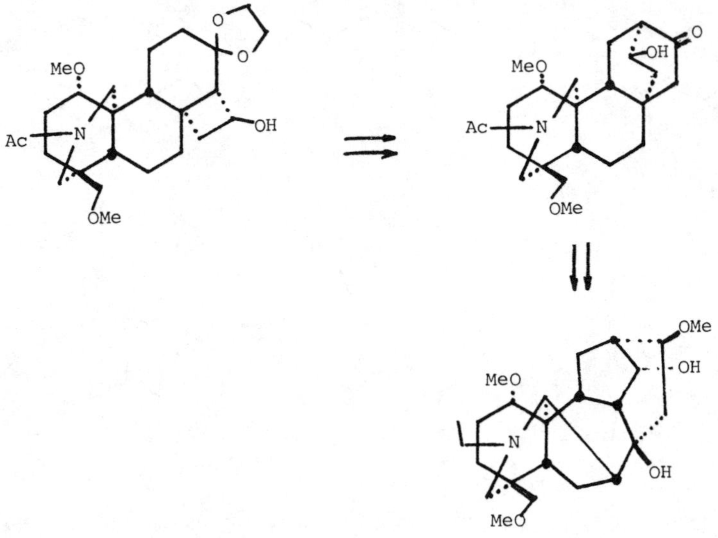

A synthesis of (+)-longifolene (**L24**) from (+)-camphor takes advantage of the chemistry for functionalization of the *syn*-methyl group at C_7. A 7-membered ring is closed by means of the Mukaiyama aldolization method (D. Kuo, 1986).

The methoxylated carbon atom of the tricyclic intermediate is fashioned into the a quaternary center and final unraveling of longifolene is mediated by a Wagner-Meerwein rearrangement.

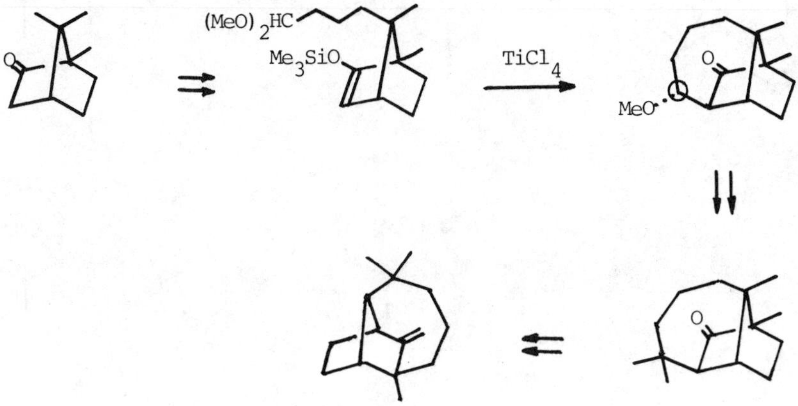

3.3 BRIDGED RING SYNTHESIS

Aldol condensation at the γ position of an α,β-unsaturated ketone has been demonstrated (β-vetivone synthesis: K. Yamada, 1973). Another example pertains to the synthesis of zizaene (**Z1**) (Deljac, 1972). The accumulation of the desired products during the reaction is attributable to the higher stability they possess for being conjugated ketones; the ketols arising from condensation at the α position likely revert to the uncyclized form, which is depleted accordingly.

The total synthesis of napelline (**N3**) (Wiesner, 1974a) requires a strategy somewhat different from that prescribed for veatchine with respect to C/D-ring segment assemblage, chiefly because the C ring contains a hydroxyl group. Aldol condensation at the γ position of the enone system, which is derived from an aromatic ring, accomplishes this goal. The low stereoselectivity accompanying this reaction is due to the presence of a methyl substituent in the acetaldehyde chain.

An exquisite synthetic route (Lombardo, 1980) to gibberellin A_1 and gibberellic acid (**G12**) is based on intramolecular alkylation of an anisole with an α-diazoketone to construct the CD-ring segment. The cyclohexadienone (ex anisole) is then partially reduced and contracted via another diazoketone (Wolff rearrangement). Elaboration of the B ring into a cyclopentenonecarboxylic ester prepares the molecule for the formation of the γ-lactone unit and the A ring by Michael and aldol reactions, respectively.

The crucial stereochemical problem is resolved in the establishment of the C_{10} configuration by attack on the ketone of triallylalane from the same side as the adjacent angular hydrogen. The intramolecular Michael reaction of the derived ester automatically leads to a *cis*-5:5 system, whereas the B-ring ester assumes an equatorial orientation.

The aldol condensation between the lactone and the aldehyde chain affords an epimeric mixture. The 3β-isomer is converted into gibberellin A_1, but the use of the 3α-isomer in acquiring gibberellic acid via the Δ^2-olefin is logistically important.

The A-secoaldehyde derived from methyl gibberellate reverts largely to the latter compound $(3\beta:3\alpha \cong 3:1)$ on treatment with sodium methoxide (Stork, 1979b).

G12

3.4 CONDENSED RINGS

3.4.1 Endocyclic Enones

The Robinson annulation is a special case in which polycyclic enones are prepared by means of the aldol process. When the enone-containing ring is not six-membered, a separate aldol condensation step must be effected. It is not surprising that most fused cyclopentenones are made from 2-(2-oxopropyl)cycloalkanones.

In the first synthesis of cedrol (**C38**) (Stork, 1955, 1961a), its three rings all were built using base-catalyzed reactions. The closure of the second ring involved the aldol method.

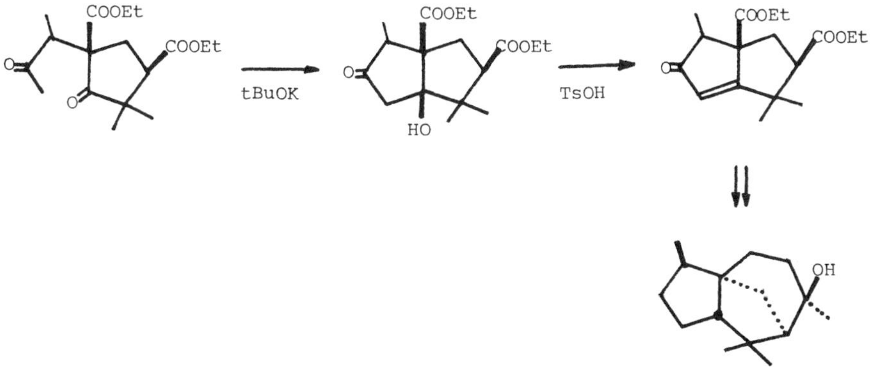

The use of potassium *t*-butoxide is necessary, since the formation of the 5:5-fused ring system by an aldol reaction is much less favorable than the 6:6-fused analog. Further activation of the donor carbon (e.g., by a carboalkoxy or phosphoryl group) is very beneficial (cf. quadrone synthesis: Bornack, 1981; cedrol synthesis: Irie, 1984).

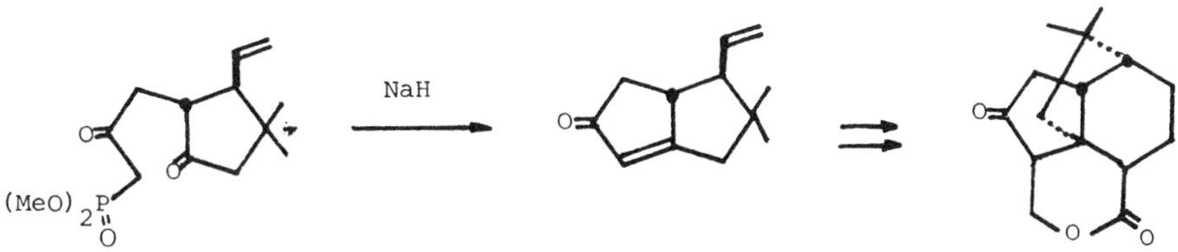

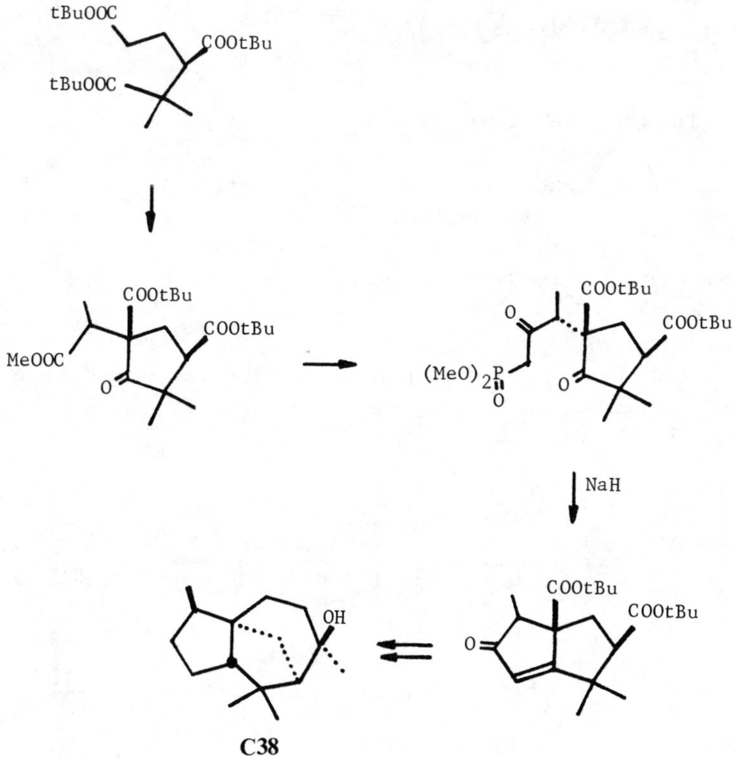

C38

If a β-ketoester chain is involved, the ester group is generally removed after the cyclization. However, this valuable functional group is fully exploited during a preparation of the precursor of pentalenolactone E (**P19**) (Exon, 1981; Paquette, 1981). It is reincarnated as the outer end of a vinylidene group.

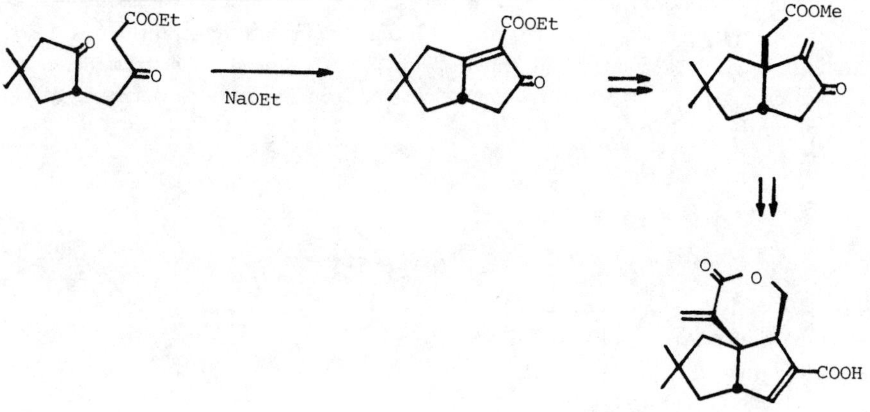

P19

3.4 CONDENSED RINGS

Similar retention of the carbon resource also is featured in a synthesis of $\Delta^{9(12)}$-capnellene-3β,8β,10α-triol (Shibasaki, 1986). The aldol step is apparently highly reversible; its inducement must be accompanied by elimination of the tertiary hydroxyl group as it emerges by derivatization into a siloxy function.

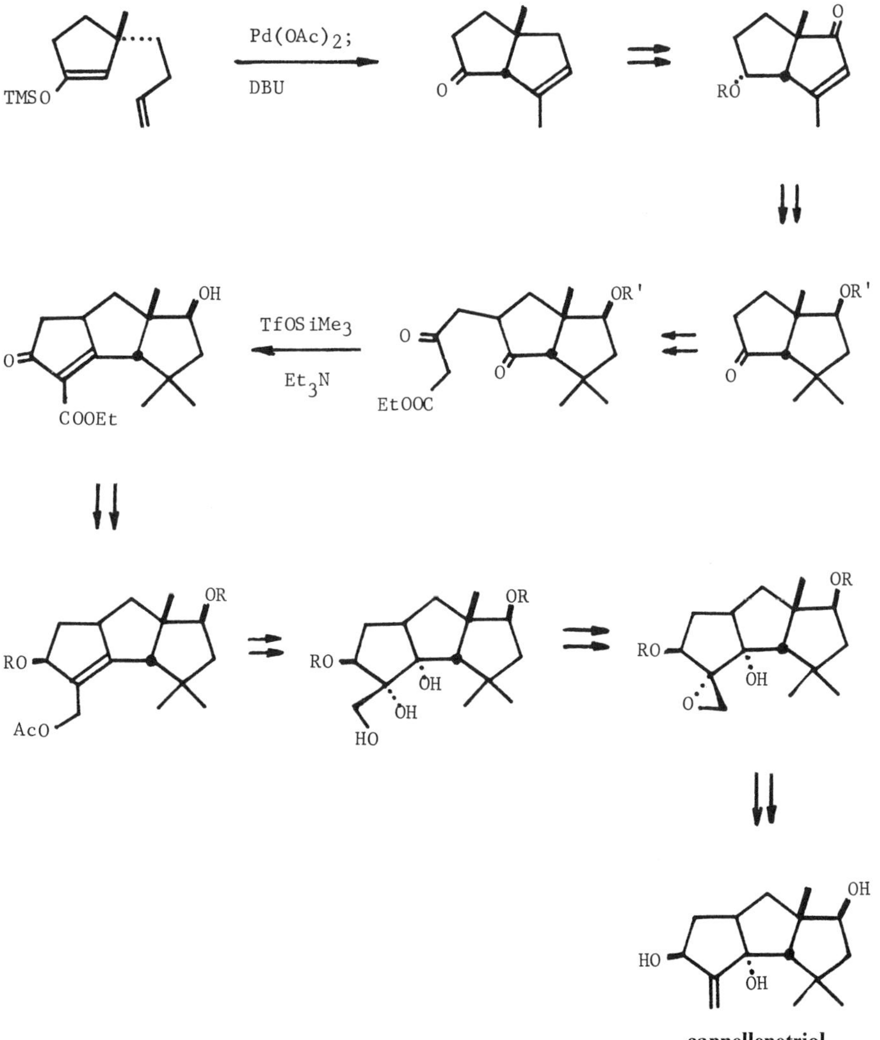

capnellenetriol

Only relatively recently 2-nitro-1-alkenes have been employed as acceptors in the alkylation–aldolization sequence leading to cyclopentenones. It is critical to use silyl enol ethers with Lewis acid catalysts ($SnCl_4$, $TiCl_4$). Direct hydrolysis of the adducts (nitroketones) furnishes 1,4-diketones.

Several simple diquinanes have been prepared by this method and used in terpene synthesis. A route to gymnomitrol (**G21**) (Yoshikoshi, 1985) starts with the generation of a monocyclic diketone (derived from the nitroketone). Similarly, homologs are obtained and applied to the synthesis of isocomene (**I14**) (Paquette, 1979) and velleral (**V6**) (Fex, 1976).

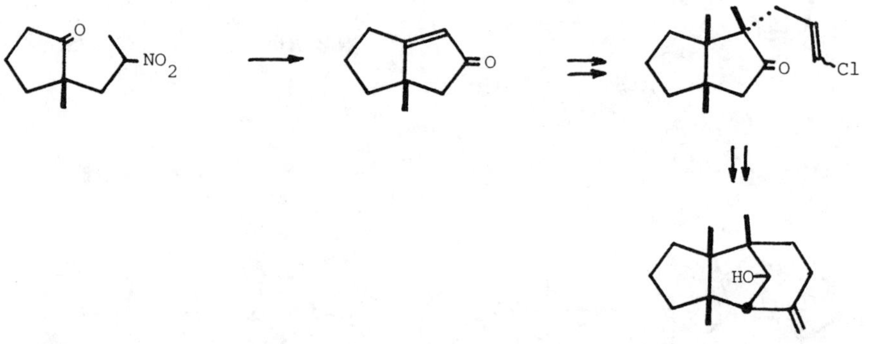

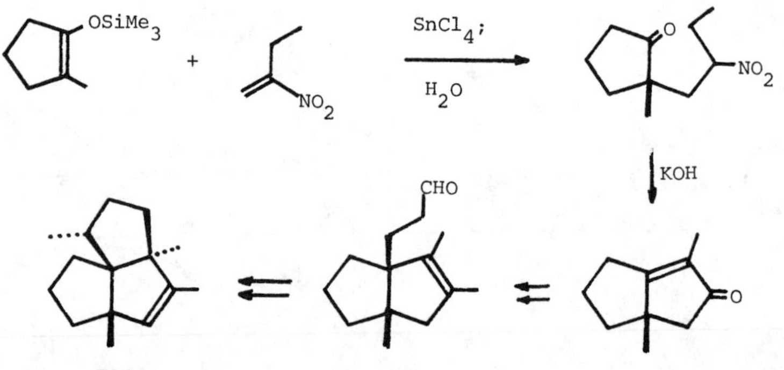

The aldol chemistry is well adapted to construction of linear cis-anti-cis-triquinanes (F. Sakan, 1971; Trost, 1979b, Exon, 1983). In one unfinished hirsutic acid synthesis (Lansbury, 1971a, b) two such processes are incorporated.

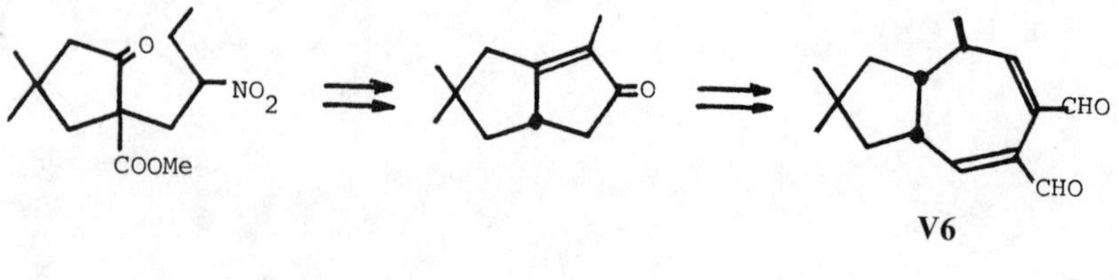

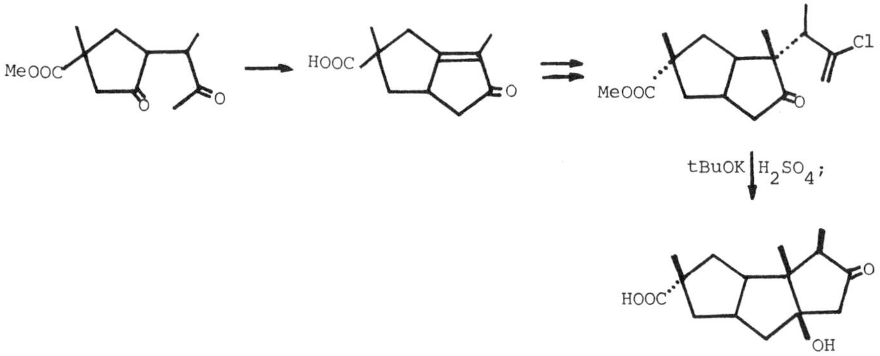

Perhaps it should be emphasized that in applying the aldol cyclization in these approaches, attention must be paid to the order of alkylation steps en route to the closure of the third carbocycle. To attain the correct stereochemistry at the ring juncture characteristic of the hirsutane sesquiterpenes, the carbon chain that participates in the cyclization must be introduced last.

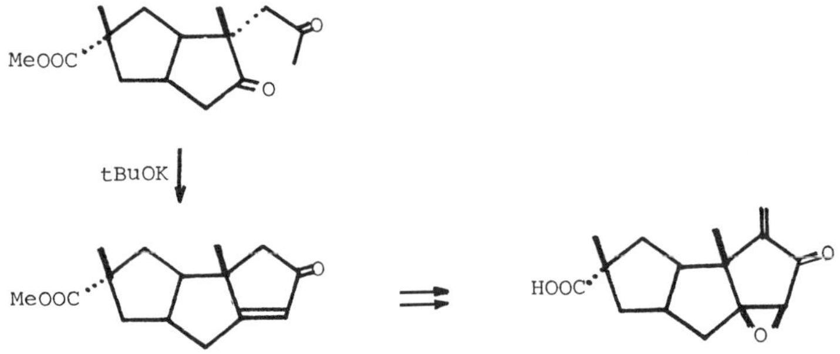

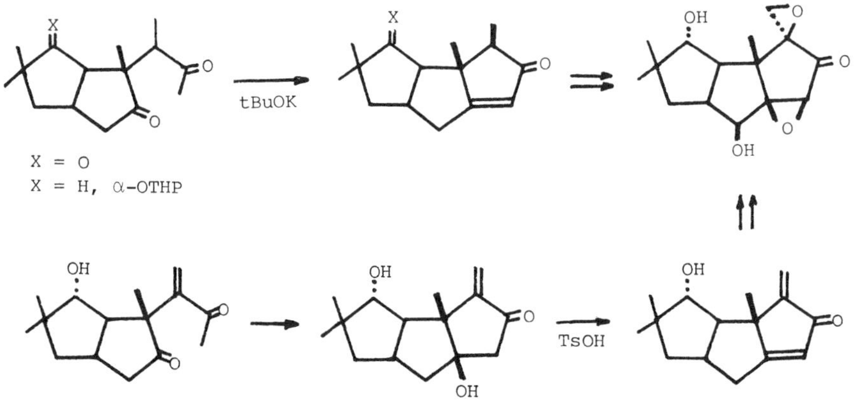

The synthetic approach to hirsutic acid C (**H19**) (Schuda, 1986) from a methanoindenone is conferred with stereocontrol by effective blocking of the endo face of the molecule, allowing establishment of the quaternary carbon center containing the carboxyl function. The etheno bridge is removable completely via oxidation and decarboxylation. A more conventional reaction sequence can then be applied to complete elaboration of the fungal metabolite.

As a footnote to this work, it has been proposed that quadrone (**Q1**) could be advantageously synthesized from a similar substance.

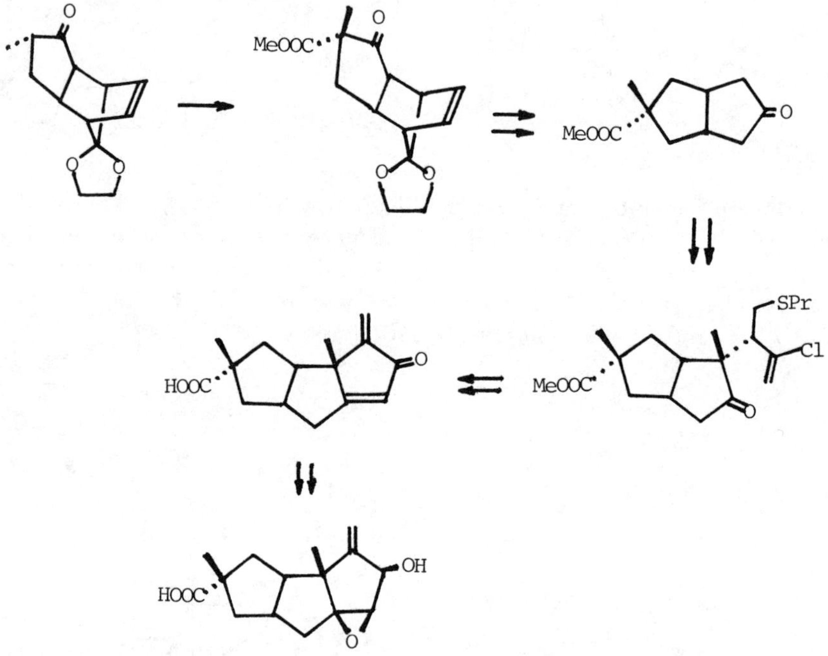

H19

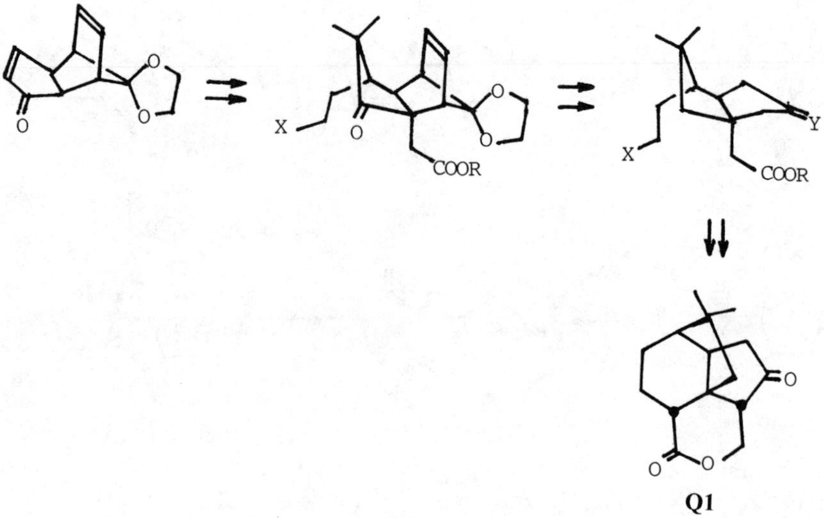

Q1

3.4 CONDENSED RINGS

Quadrone (**Q1**) has a more intriguing carbon framework. It comprises a diquinane subunit bridged by another carbocycle. Because this propylene bridge spans the exo face of the diquinane, and because of the strong bias of such molecules to have exo substituents, the strategy of ring formation from a diquinane derivative is very appealing in the context of quadrone synthesis (Danishefsky, 1980b). In other words, an appropriate precursor containing a nucleophilic and an electrophilic chain is thought to be conducive to ring closure. Such diquinanes are readily available using the aldol reaction.

For an elaboration of the unnatural (—)-terrecyclic acid, the enone acid tautomer of quadrone, from (+)-fenchone via 3,3-dimethyl-4-(2-benzoyloxyethyl)cyclopentanone and a bicyclic enone ester, see Kon (1984).

An intramolecular Diels–Alder route to quadrone (**Q1**) (Dewanckele, 1983; Schlessinger, 1983b) entails the formation of the bridged system and an additional cyclohexene ring in the same step. The strategy implies modification and degradation of the cyclohexene moiety to provide both a carboxyl pendant and an acetone unit. The latter segment is to be condensed with the ketonic bridge to form a cyclopentenone.

In addition to this route, several other syntheses of quadrone (Kende, 1982; Takeda, 1983) identify the tricyclic enone (cyclopentenone) as the key intermediate. On the basis of the anticipated reduction of the double bond to give a less strained *cis*-diquinane moiety, the plans are valid. This reduction is kinetically favorable also because it avoids the spatial interaction of the reagent with one of the tertiary methyl groups.

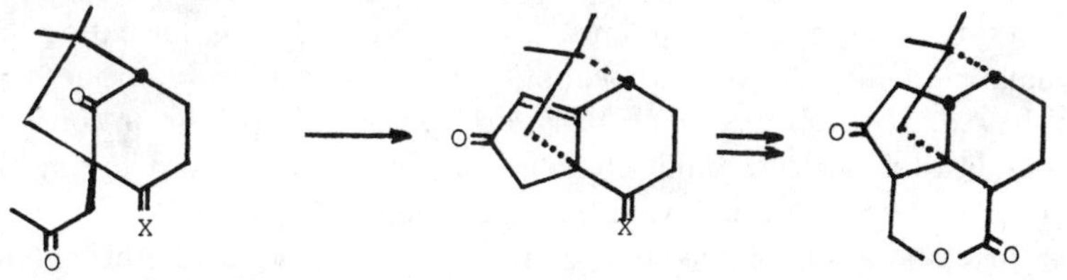

X = CH$_2$
X = α-COOH, β-H
X = α-CH$_2$OMe, β-H

Q1

Based on a hydrophenanthrenone which is formed by mercury(II) induced polyene cyclization the limonoid framework has been assembled via aldol condensation, Simmons–Smith reaction of the 13α-alcohol and reductive cleavage of the cyclopropane after the ketone is restored (Corey, 1987d).

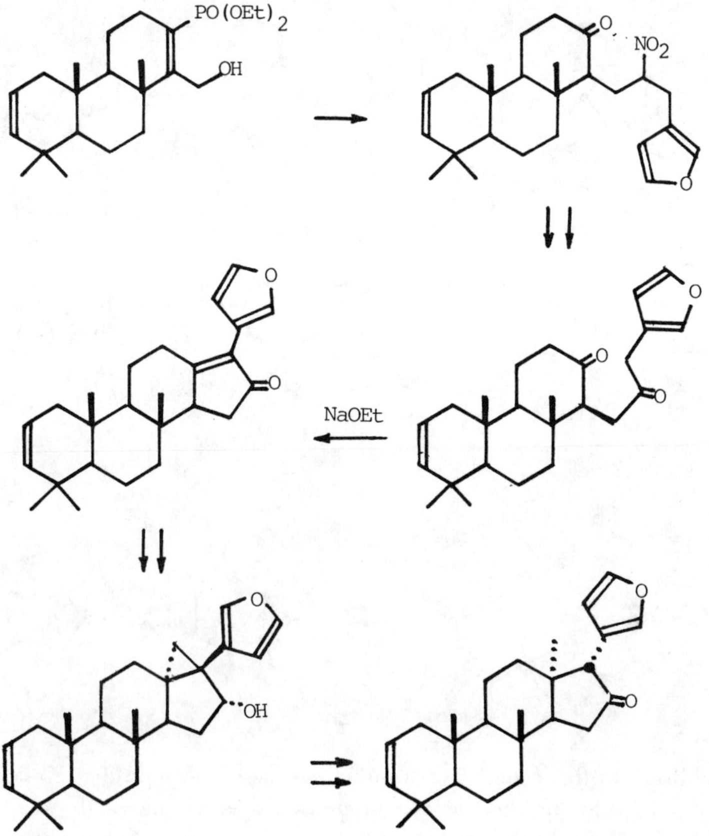

An unwanted aldolization intervenes in the second synthesis of longifolene (**L24**) (McMurry, 1972). However, the retention of both oxygen atoms in a

3.4 CONDENSED RINGS

1,3-relation allows for an easy recovery of the original skeleton.

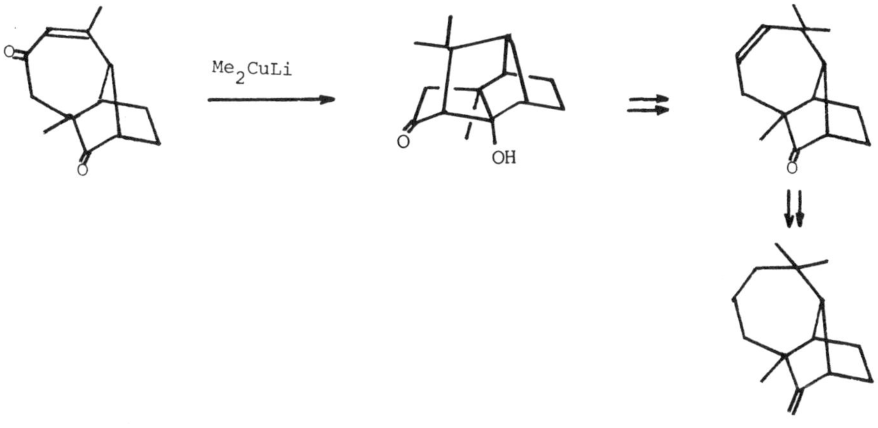

While it is inconsequential to view the aldol condensation leading to the quadrone precursor indicated above as formation of 5:5- or 5:6-fused rings, a 5:6-fused enone is a crucial intermediate for bakkenolide A (**B1**) (Evans, 1973, 1977). The spirocyclic center of this sesquiterpene is most suitably identified with a carbonyl site in synthetic terms.

The ketone also serves as precursor of ligularone (**L13**) (Jacobi, 1981).

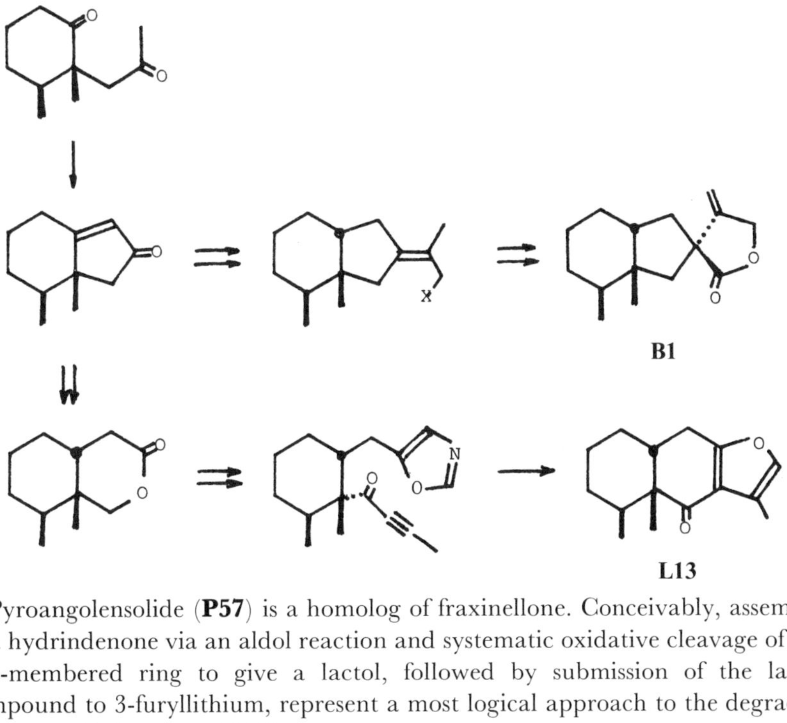

Pyroangolensolide (**P57**) is a homolog of fraxinellone. Conceivably, assembly of a hydrindenone via an aldol reaction and systematic oxidative cleavage of the five-membered ring to give a lactol, followed by submission of the latter compound to 3-furyllithium, represent a most logical approach to the degraded triterpene (Fukuyama, 1973).

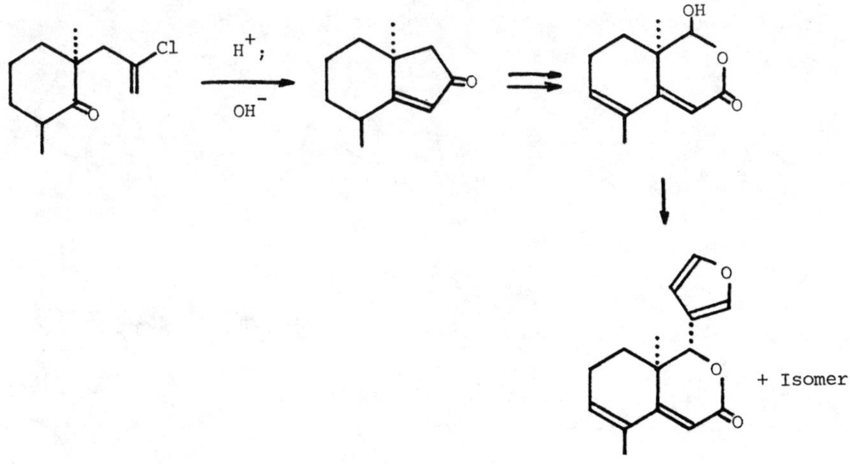

P57

A strategy for aphidicolin (**A36**) synthesis that prescribes D-ring annulation from a tricyclic intermediate equipped with a 9α-alkyl chain must be accommodated by stereoselective installation of the latter structural unit. The situation is partly reminiscent of the approach to the kaurene/phyllocladene-type diterpenes involving a Claisen rearrangement to graft a two-carbon unit onto an angular position. Thus, the chemist targets an allylic alcohol and thence an enone (McMurry, 1979), and the logical reaction leading to such a structure is the aldol process. The configuration of C_8 is safely established in view of the thermodynamically controlled nature of the allylative route for preparation of the diketone precursor.

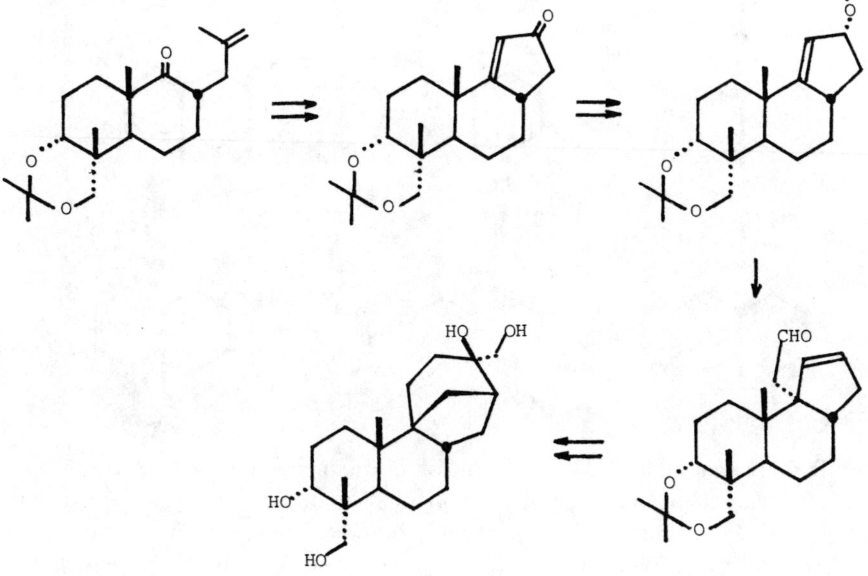

A36

Aldol condensation constitutes the final step of a cyclocolorenone (**C92**) synthesis (Saha, 1986). The diketone is prepared by a reaction sequence consisting of propargylation of a cycloheptenyl silyl ether with a cobalt complex, demetallation, and hydration of the triple bond (with carbonyl participation). Treatment of the diketone with potassium hydroxide in ethanol at room temperature furnishes cyclocolorenone, whereas at reflux, a mixture containing epicolorenone (ring juncture epimer) as the major product ensures. Apparently, the isomer with the active side chain cis to the methyl group undergoes smooth cyclization, but cyclization of the epimeric diketone is disfavored by steric impediment of the cyclopropane ring. At higher temperatures, cyclocolorenone is liable to epimerization under the influence of alkali.

C92

A synthesis of pentalenene (**P16**) (Mehta, 1985a), which follows the general biogenetic idea (but not biomimetic), requires preparation of a 5:8-fused intermediate. This compound is obtainable from a cyclooctenone via alkylation and aldolization.

P16

In an excellent approach to the ABC synthon of ingenol (**I4**) (Rigby, 1985) the [6 + 4]cycloaddition is succeeded by an aldol condensation. Alkylation at the bridgehead provides a latent acetonyl group.

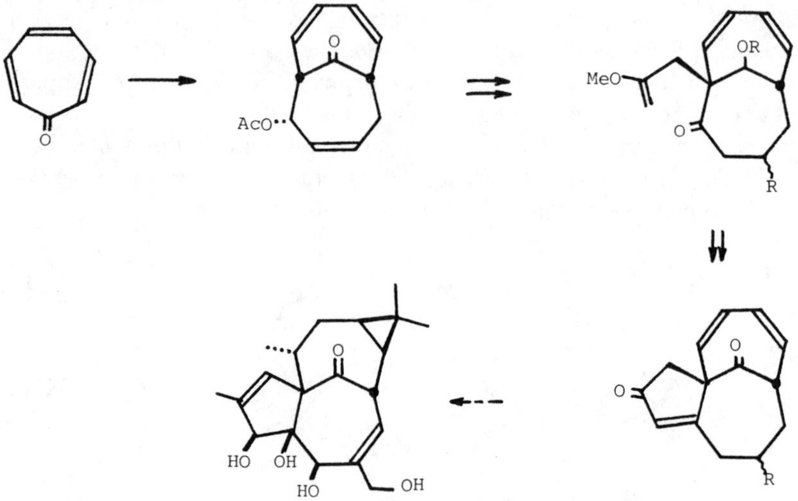

The unsaturated ring of α-bourbonene (**B16**) has been constructed from a diketone that originates from an intramolecular photocycloaddition (Brown, 1968). The question of the cis-anti-cis stereochemistry of the product never arises because the anti relationship corresponds to the more stable exo configuration of the acetyl groups. The apparent regioselectivity leading to the substitution pattern of the bourbonenes is due to the destabilization of the alternative aldol, which experiences strong compression between the angular methyl group with one of the substituents (Me or OH) at the β carbon.

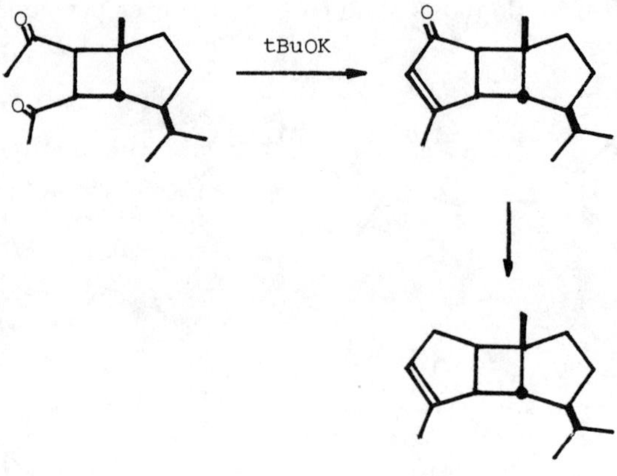

B16

Regiochemistry is not an issue in the synthesis of β-bourbonene (**B17**) (Tomioka, 1982) via an aldol reaction in the formation of the C ring. The ketoaldehyde shown below has only one direction available to react in.

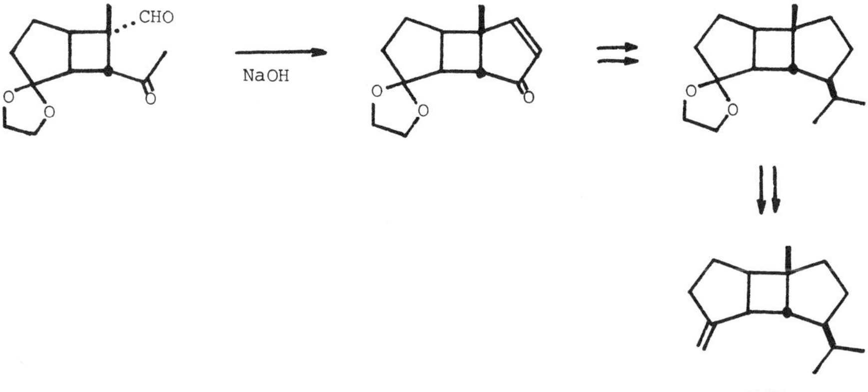

A similar cyclization is involved in the late stage of two synthetic endeavors (K. Stevens, 1981; Oppolzer, 1982b) in $\Delta^{9(12)}$-capnellene (**C21**).

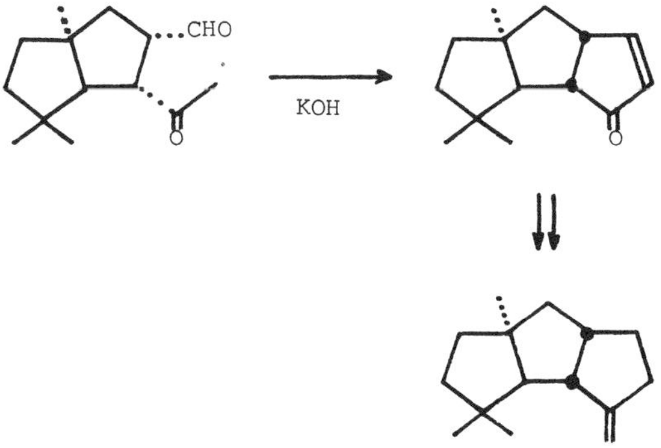

The minor (exo) Diels–Alder adduct of cyclopentadiene and dimethylmaleic anhydride has been converted into albene (**A22**) (J. Baldwin, 1981) by a conventional reaction sequence comprising an aldol step.

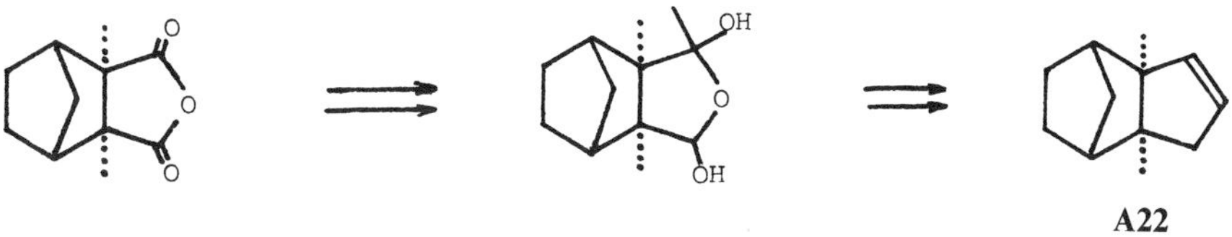

The intramolecular Wittig–Horner reaction is a valuable variant of the aldol process. A facile synthesis of umbellulone (**U2**) and its isomer (Benayache, 1978) is shown below.

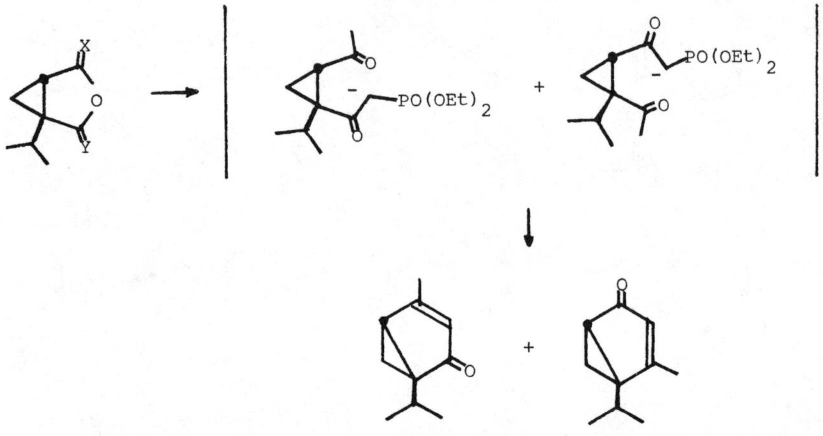

An attractive method for angular triquinane synthesis is by devolution of bridged systems, which are readily formed via Diels–Alder reactions. Cleavage of the etheno bridge of these adducts exposes two carbonyl residues that can be recombined with other nearby functionalities. Thus the angular ring of silphinene (**S34**) (Sternbach, 1985) is closed by aldol reaction following the rupture of the bridged rings.

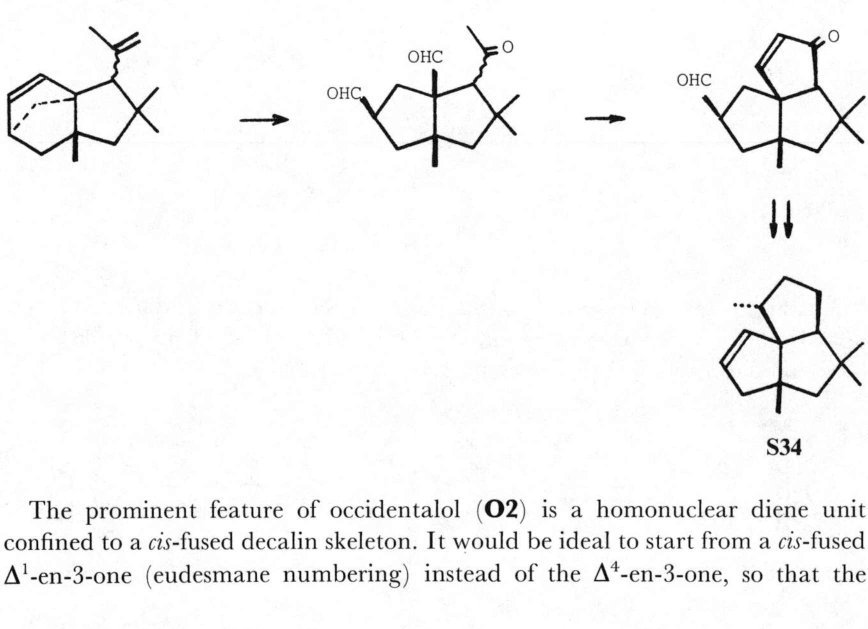

The prominent feature of occidentalol (**O2**) is a homonuclear diene unit confined to a *cis*-fused decalin skeleton. It would be ideal to start from a *cis*-fused Δ^1-en-3-one (eudesmane numbering) instead of the Δ^4-en-3-one, so that the stereochemical issue of the ring juncture no longer mattered. The precursor of the Δ^1-en-3-one is a cyclohexane appended at adjacent carbon atoms with an aldehyde and an acetone chain in a cis relationship. A solution to this problem involves photocycloaddition and unraveling of the adduct (S. Baldwin, 1982b).

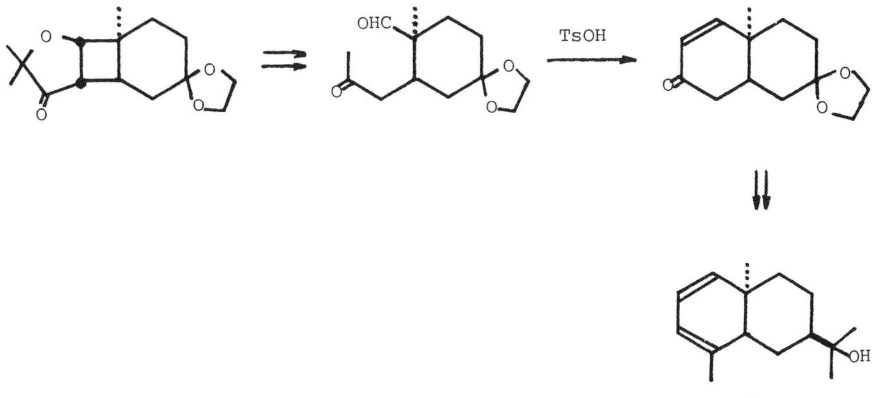

O2

A substituted hydrindanone may be caused to undergo regioselective oxidative cleavage of the five-membered ring (via a β-hydroxysulfide). The resulting ketoaldehyde is condensed again, leading eventually to α-cadinol (**C6**) (Caine, 1977).

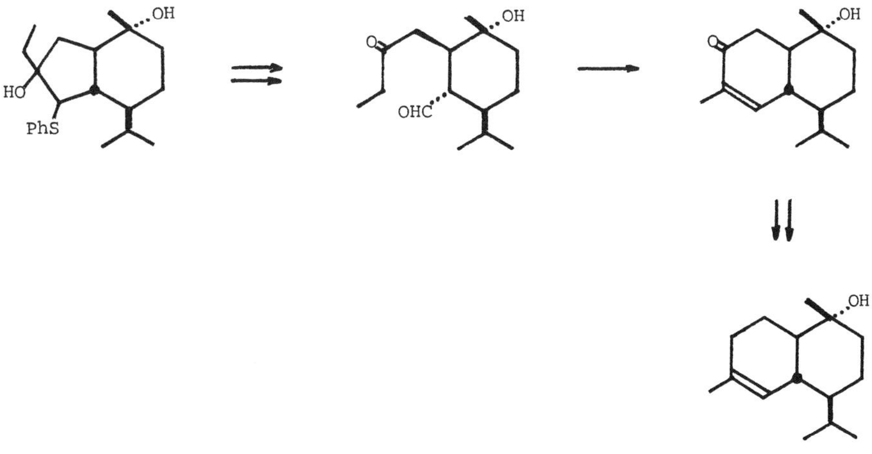

C6

In a synthesis of illudol (**I3**) (Kagawa, 1969; Matsumoto, 1971) a ring expansion operation is required. This is achieved during an α-glycol cleavage. Apparently the strained cyclobutanone is very susceptible to aldol condensation.

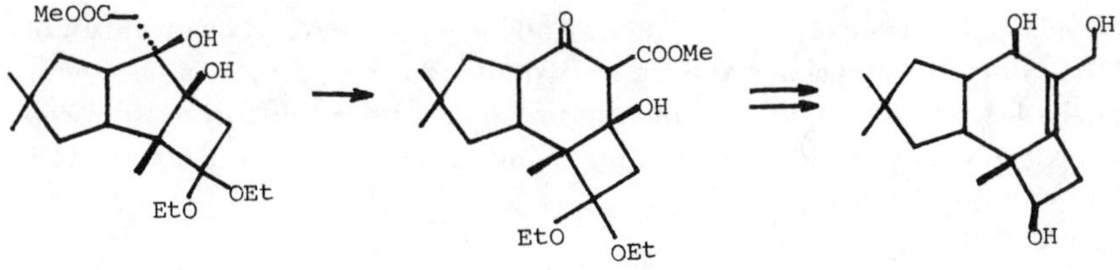

(See also, 6-protoilludene synthesis: Furukawa, 1985.)

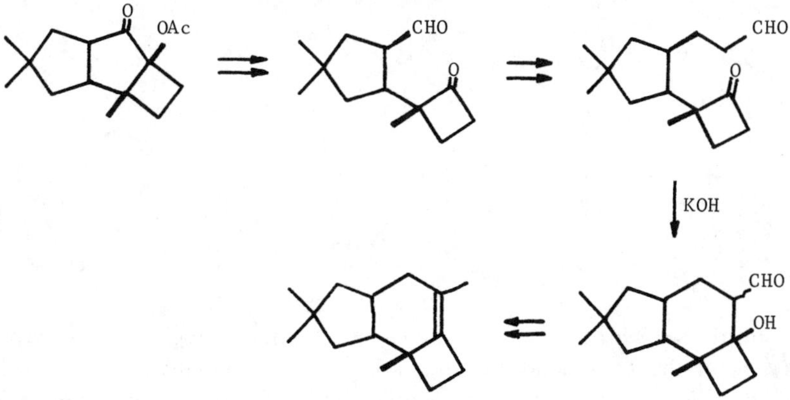

6-protoilludene

Condensation of 2-methyl-5-isopropylcyclopentanone with 1,4-dimethoxy-2-butanone in the presence of potassium ethoxide leads to a dione in low yield. Treatment of the latter compound with potassium hydroxide in ethanol is needed to complete the annulation process and furnish an intermediate of oplopanone (**O7**) (Caine, 1973).

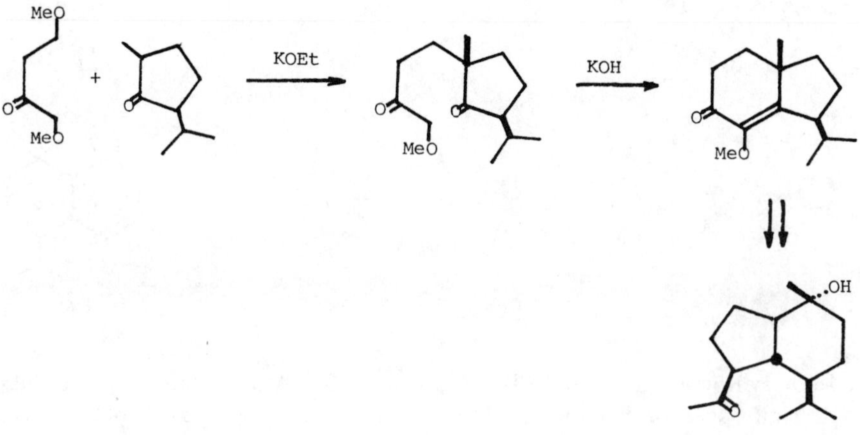

3.4 CONDENSED RINGS

Although the Robinson annulation is convenient, stepwise procedures sometimes offer certain advantages, such as achieving optical induction during the aldol condensation. For example, the $(S)(+)$-isomer of nor-W–M ketone is obtained using (—)-proline as catalyst. This compound has been used to synthesize pinguisone (**P32**) (Bernasconi, 1981), a molecule with a most ususual pattern of methyl substituents.

As pointed out elsewhere, the Robinson annulation is not universally applicable to the synthesis of cyclohexenones. This can be further illustrated by noting that the conventional annulation of camphenilone, whose ketone group is flanked by a quaternary carbon on one side and by a bridgehead on the other, is impossible. According to a synthetic plan for epizizanoic acid (**Z4**) (Kido, 1969) based on 1,2-anionic rearrangement, an annulated camphene is required. Consequently, a 1,5-diketone was prepared by a different method and caused to undergo aldolization.

α-Methylnopinone fails to undergo Robinson annulation with 3-penten-2-one owing to a strong steric compression that develops as the two reactants approach each other. The problem is circumvented by the following manipulations: successively introducing a methylpentenyl chain and a methyl group, unfolding the 1,5-diketone, and conducting an aldol condensation. Cleavage of the four-membered ring accompanies the last operation, which involves treatment with hydrochloric acid. The natural (+)-isomer of nootkatone (**N11**) is produced (Yanami, 1979, 1980).

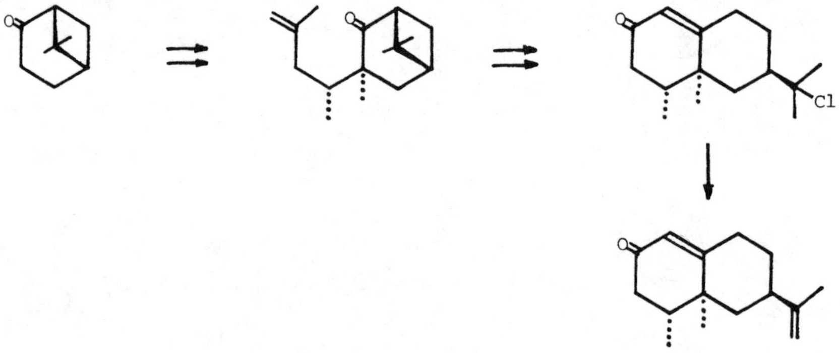

Another impasse stemming from the regiochemistry of the Robinson method is annulation at C_2 and C_3 of the menthan-3-ones. Thus, in a synthesis of arteannuin (Xu, 1986), a cyclohexene predecessor of the peroxyketal ring is constructed via an initial alkylation of the kinetic enolate with a silyl butenone, which is followed by aldolization and other conventional manipulations.

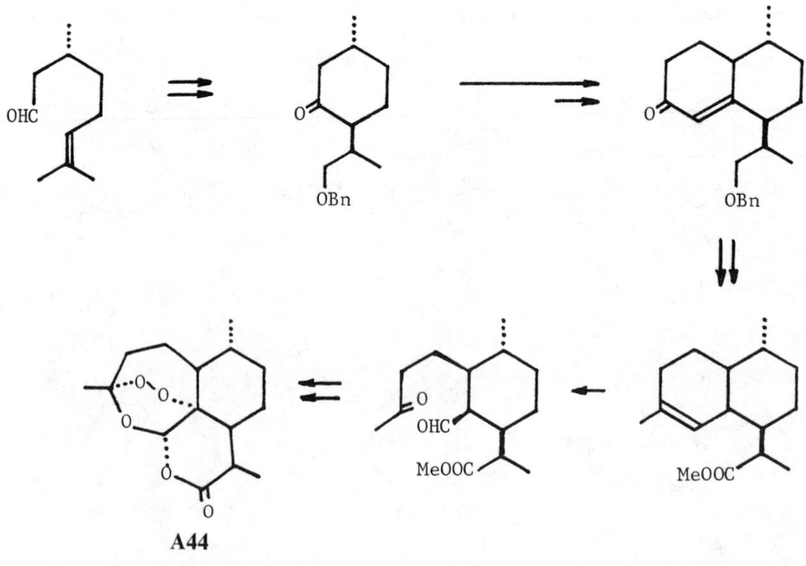

3.4 CONDENSED RINGS

In an interesting route to stachenone (**S48**) (S. Monti, 1979) the CD-ring system evolves from a Diels–Alder adduct. The bridged ketone receives a five-carbon fragment from a Grignard reagent to begin the construction of the B ring, which also carries the angular methyl group. Annulation actually is effected after a [2.2.2] → [3.2.1] skeletal change to expose a 1,5-diketone.

S48

The elegant method developed by Stork as an alternative to Robinson annulation, which consists of alkylation of ketones with isoxazole derivatives to generate latent 1,5-diketones and hence cyclohexenones, has been exploited in a synthesis of ferruginol (**F4**) (Ohashi, 1968). An ideal substitution pattern in the C ring directly obtained after aromatization enables the rapid completion of the synthesis.

Release of the side-chain ketone in this process is by N-ethylation of the heterocycle with Meerwein's reagent ($Et_3O^+BF_4^-$) and base treatment. The base also effects an aldol condensation.

F4

Dehydrofukinone (**F16**) is an isomer of isonootkatone in that the enone moiety is located in the B ring. Stork's annulation also serves ideally in this instance (Ohashi, 1969).

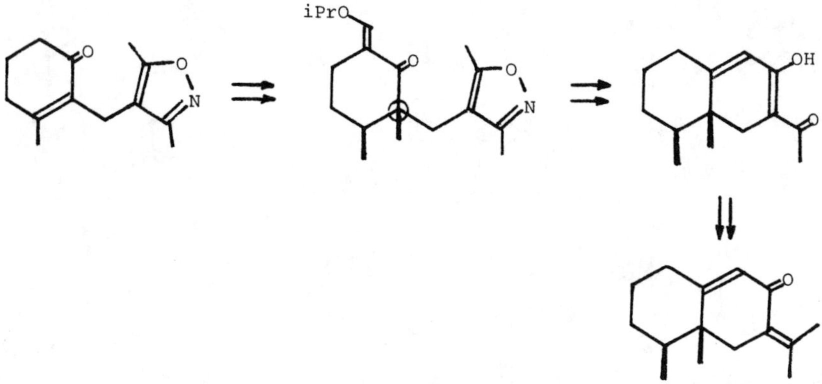

F16

The diketone that undergoes aldolization to afford a bicyclic enone in the synthesis of fukinone (**F15**) (Piers, 1970; Torrence, 1971) is prepared from an enolactone. This method was developed for reconstitution of the A ring of steroids and proved extremely useful in incorporation of radioactive labels at C_4 of such molecules.

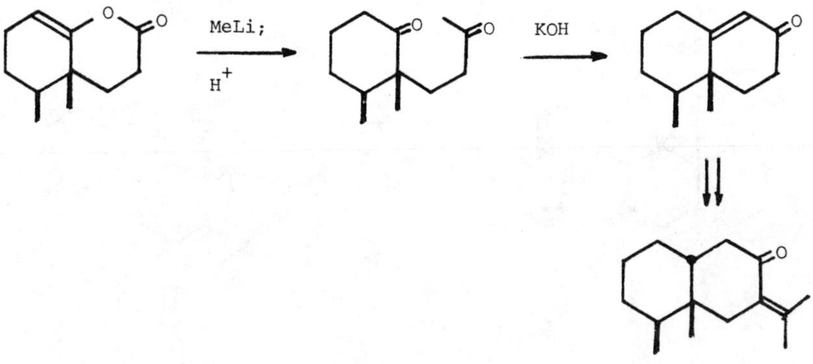

F15

In a synthesis of lupeol (**L28**) (Stork, 1971) both the A and B rings are assembled by stepwise elaboration of an α-allylcyclohexanone unit via the corresponding δ-enolactone. Reaction of the enolactone with a Grignard reagent provides the 1,5-diketone, which is cyclized.

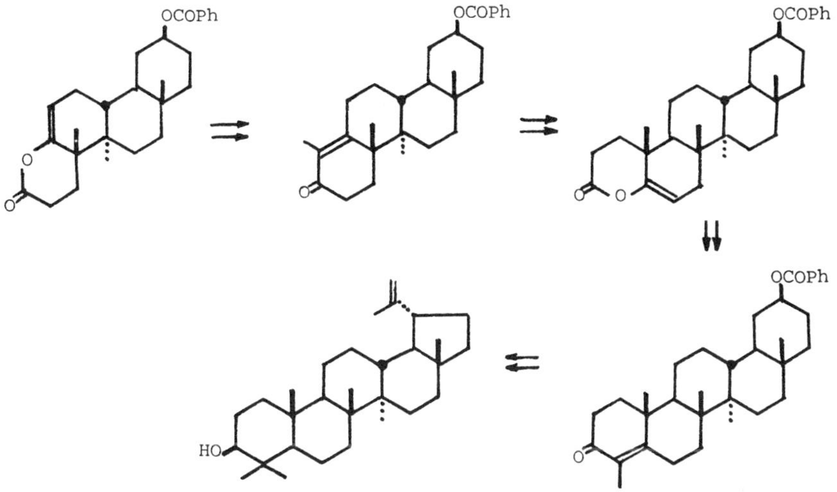

L28

A degradation product of fusidic acid (**F23**) has been synthesized (Dauben, 1972) from a hydrindenone using extensive aldol chemistry. Stereocontrol of the angular methyl groups is based on the principle of least hindrance. It is noteworthy that the first tetracyclic intermediate lacks a methyl group at C_{14}, whereas an extra one is present at C_{13}. Rectification of the substitution pattern is achieved via epoxidation of the cyclopentene double bond and treatment of the resulting epoxide with boron trifluoride etherate.

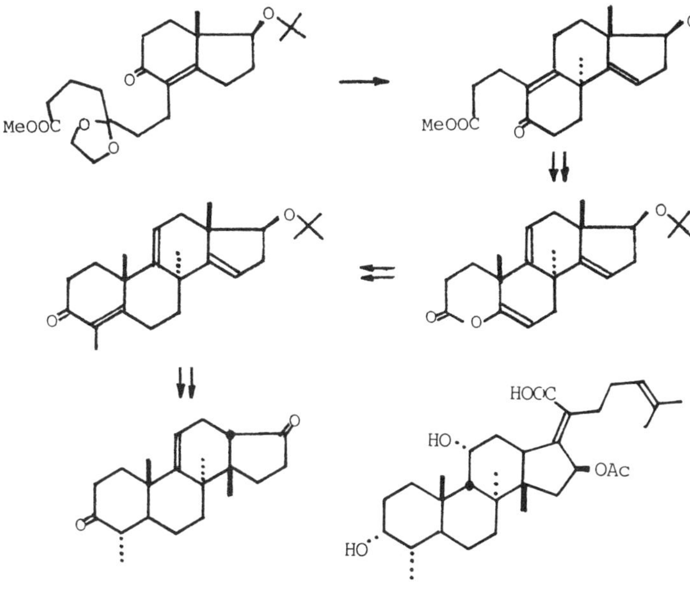

F23

The reaction sequence leading to the formation of the C ring of dolabradiene (**D16**) (Y. Kitahara, 1964) is slightly ususual. While nitriles are less reactive than ketones toward organolithium reagents, the carbonyl group of the octalone apparently survives such treatment. (For steric reasons or because of enolization?) The resulting diketone undergoes cyclization accordingly.

The trimethyl octalone that is the precursor of thujopsene (**T18**) (Amice, 1970) has been obtained by a modified aldol reaction.

More recently, aldol cyclization has found increased employment in the formation of seven-membered rings. Several synthetic studies of hydrazulene sesquiterpenes have made use of this method.

The frequent presence of oxygen functions at C_6 and C_8 of pseudoguaianolides logically identifies the carbonyl precursors at the corresponding positions. Thus a synthesis of helenalin (**H3**) (Ohfune, 1978) proceeds by the acquirement of the 6-keto-8-ol(s), whereas an interesting approach to confertin (**C62**) and damsin (**D5**) (Quallich, 1979), as well as helenalin (**H3**) (M. Roberts, 1979), involves a retroaldol/(Wittig) aldol process, which gives rise to the Δ^9-en-8-one intermediate. This strategy allows for a more flexible assemblage of the two 10-epimeric series of compounds, but it requires functionalization at C_6 (and of course at C_7).

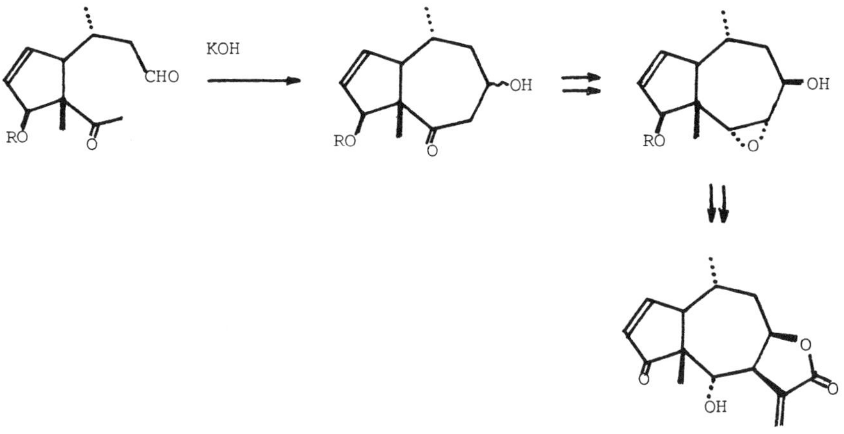

3.4 CONDENSED RINGS

H3

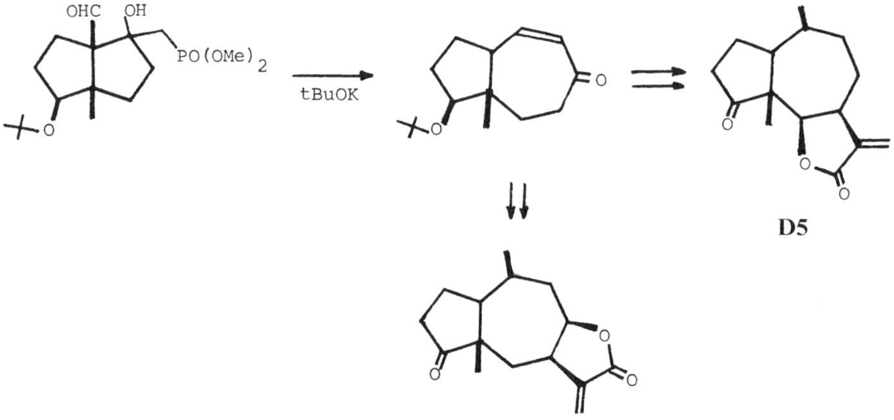

C62

D5

Owing to the accessibility of different ketoaldehydes, other possibilities based on an aldol route to the pseudoguaianolides have been explored. Examples include aromatin (**A43**) (Ziegler, 1981) and a synthon for balduilin (**B2**) (Lansbury, 1985).

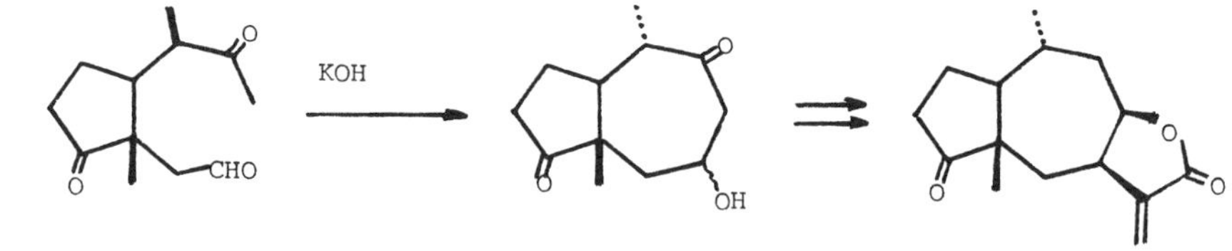

A43

B2

Disconnective analysis indicates possible formation of the bond between C_1 and C_{10} of confertin (**C62**) to give its highly unsaturated precursor from a triketone (Schultz, 1982). The required system is set up by an alkylation of 2-methyl-1,3-cyclopentanedione with a furan derivative. The aldol step cannot be carried out with the alkylation product. However, side reactions are suppressed when the dehydro compound is used.

C62

It appears that hanegokedial (**H1**) is easily derived from a cyclopropanated cycloheptenone via α,β-dialkylation and the Wittig reaction. The required cycloheptenone may be obtained from 3-carene. However, aldol cyclization using traditional reagents is fraught with the dangers of creating a five-membered ring product and/or epimerization of the one-carbon aldehyde chain, rendering the cyclization impossible. The Mukaiyama version of condensation—that is, the reaction of a silyl enol ether with an acetal—is an adequate alternative (Taylor, 1983).

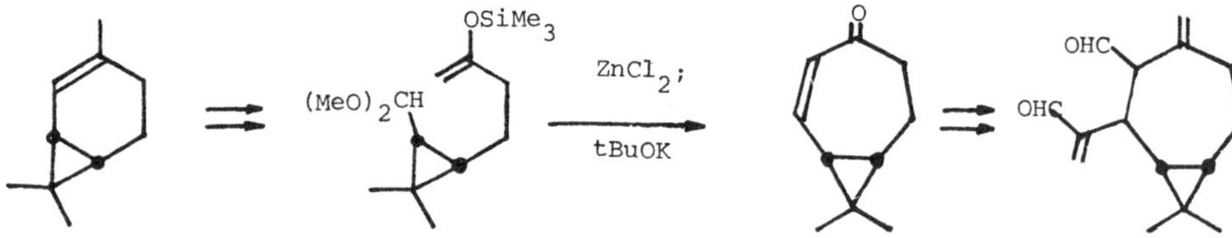

H1

The antitumor diterpene jatrophone (**J3**) contains a 12-membered carbocyclic dienone as well as β-furanone subunits. The critical step of a synthesis is always the macrocyclization, and one logical locale of such a union is between the α and β carbon atoms of the ketone group.

In an actual synthesis (Smith, 1981b), an intramolecular aldol condensation is the designated reaction. The Mukaiyama version is the method of choice; other conditions fail to effect the conversion.

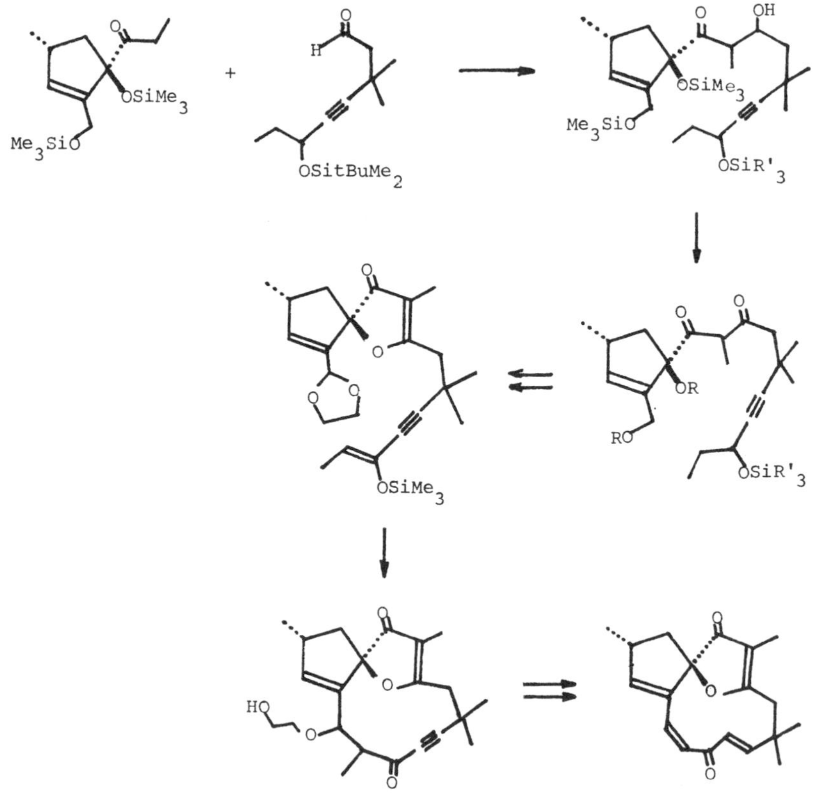

J3

As shown in syntheses of cedrol, quadrone, umbellulone, and certain pseudo-guaianolides, the intramolecular Wittig–Horner–Emmons reaction is an excellent alternative to the classical aldol method. Application of this process to the synthesis of macrocycles has gained considerable attention in recent years, and it serves a key role in a synthesis of asperdiol (Tius, 1986).

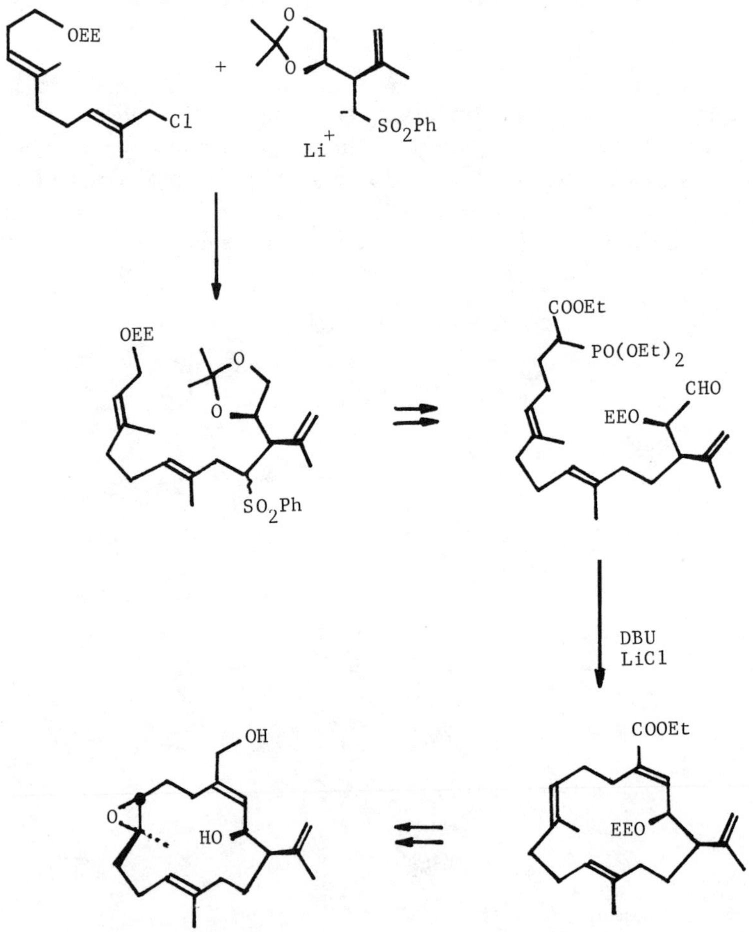

asperdiol

Another recent example is provided by a stereoselective formation of the 11-membered ring of (—)-bertyadionol during its synthesis from (—)-*cis*-chrysanthemic acid (Smith, 1986). Construction of the proper precursor involves rational extension of the two side chains of the monoterpene. A 1,3-dithiane alkylation with a substituted cyclopentenone links up the subunit, which is readily modified into the acceptor moiety. The Wittig–Horner condensation is achieved (28–32 % yield) in a relatively high dilution.

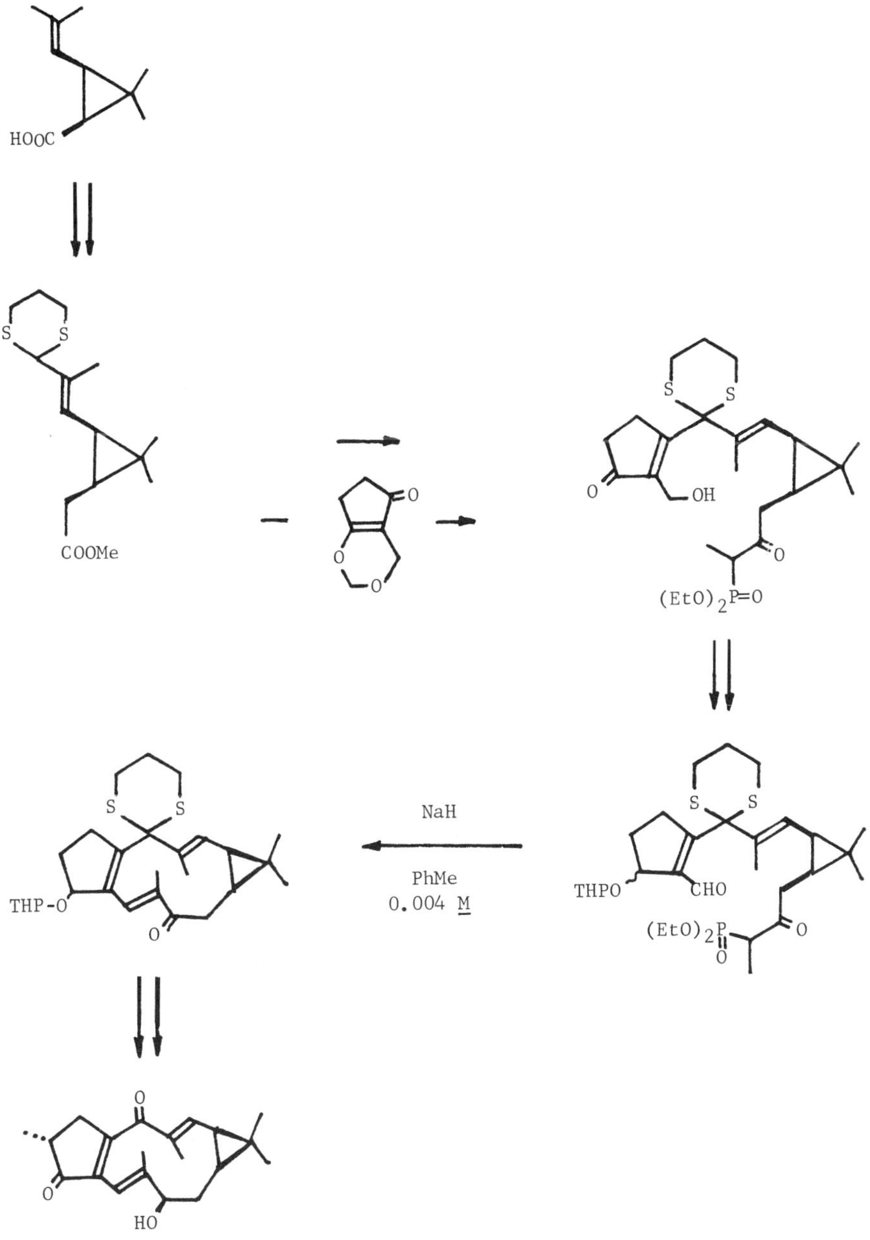

bertyadional

The methyl ester of ceriferic acid-I, a sesterterpene component of an insect wax, has been synthesized (Kodama, 1986a) using a Wittig–Horner–Emmons macrocyclization. The step produces a greater amount (52 %) of the *E*-isomer than the naturally derived ester (24 %).

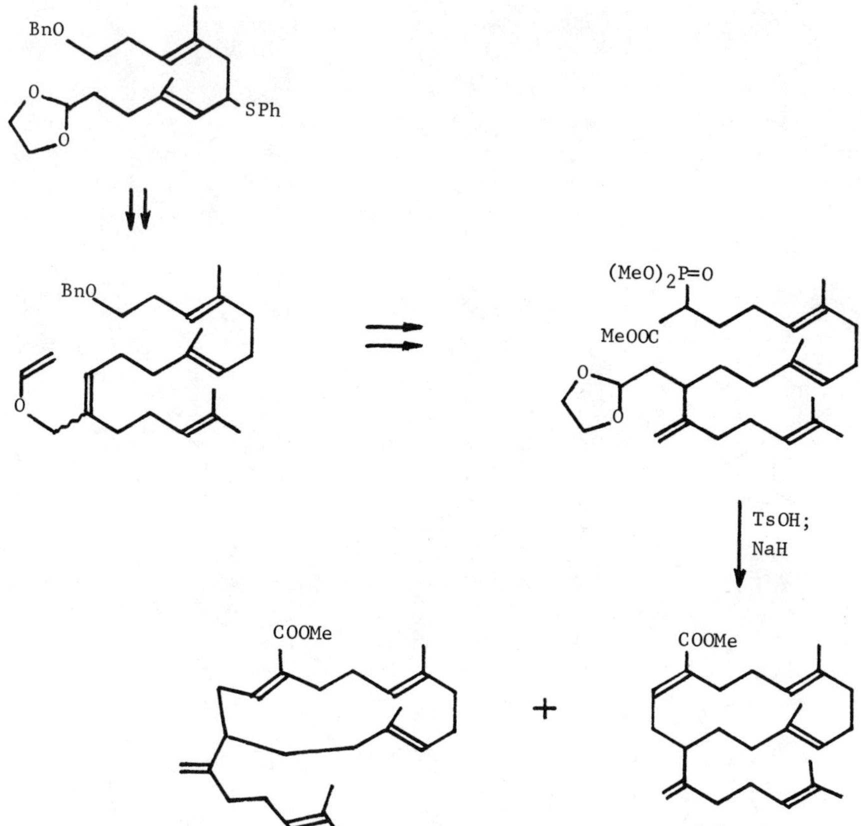

methyl ceriferate-I

A stereoselective assembly of anisomelic acid (Marshall, 1986b) has also appeared in the literature.

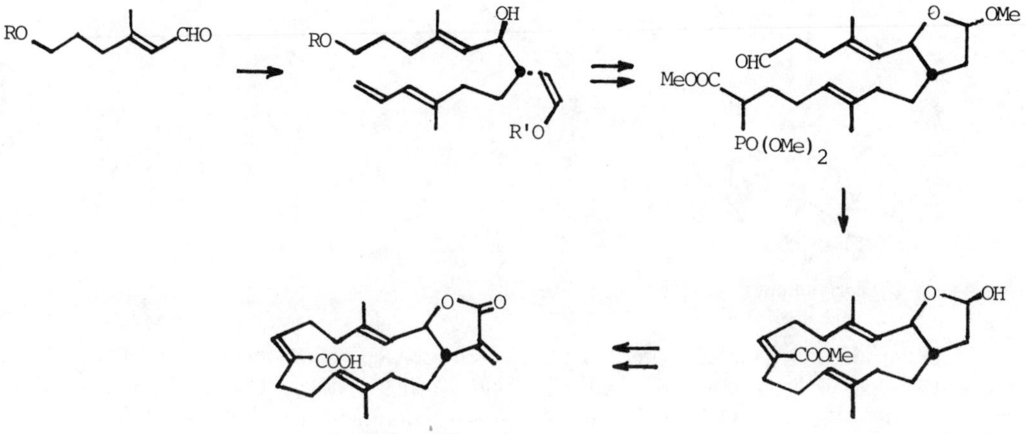

anisomelic acid

3.4.2 Aldolization Involving Exocyclic Donors

When condensation involves exocyclic donors, a carbonyl group is left outside the ring being formed. However, this donor carbonyl can still be a member of a ring structure if it was present originally as such. The intramolecular condensation of a dialdehyde always leads to a product with a formyl pendant. In the final step of a synthesis of vitrenal (**V18**) (Magari, 1982) the aldolization is effective and regiochemically unambiguous.

A very similar strategy has been followed in an independent synthesis of *ent*-vitrenal (ent-**V18**) (M. Kodama, 1986b).

Evidently the norketone of α-vetispirene (**V14**) is formed without complication (Balme, 1985). The alternative reaction pathway would result in a cycloheptenone, which is less favorable.

In the first synthesis of gibberellic acid (**G12**) (Corey, 1978a,b), the BC-ring portion is formed by a Diels–Alder reaction. After the D ring has been completed, the cyclohexene (B ring) is cleaved and reclosed on treatment with dibenzylammonium trifluoroacetate to give an unsaturated aldehyde.

G12

Contraction of a fused cyclohexene obtained from a Diels–Alder reaction to yield a cyclopentanone unit via double-bond cleavage, aldolization, and Beckmann rearrangement constitutes an important phase of a synthesis of $\Delta^{9(12)}$-capnellene (**C21**) (Liu, 1985). The cyclopentanone is converted into the *gem*-dimethylcyclopentane by conventional methods (Wittig and Simmons–Smith reactions and hydrogenation).

C21

Complications arise during preparation of a precursor of dendrobine (**D11**) by the aldol reaction of a dialdehyde derived from the cleavage of the hexalindione (Kende, 1974). The disappointing production of a 1:1 regioisomer mixture contrasts with the selective reaction associated with Woodward's cholesterol synthesis. The latter's success is attributed to the suppression of the donor characteristic of the "lower" acetaldehyde chain by steric interactions with the C_7 methylene group (steroid numbering). A similar situation does not prevail in the present case.

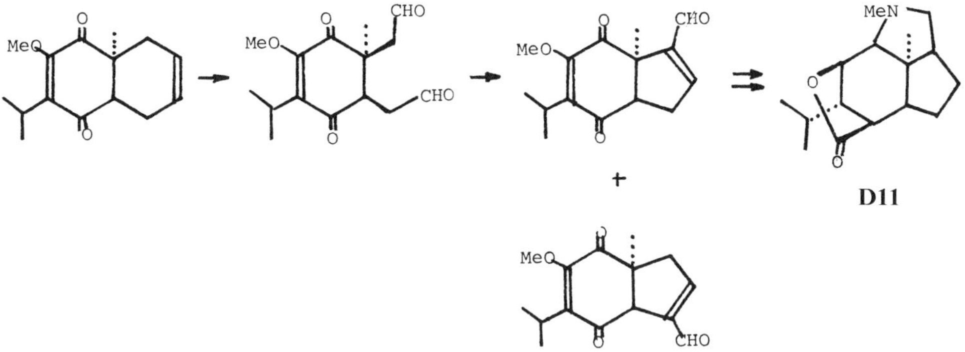

On the other hand, the dialdehyde prepared from (—)-carvone apparently cyclizes in one direction when exposed to titanium tetrachloride/tetraisopropyl titanate. The resulting (+)-sclerosporal (**S16**) (T. Kitahara, 1984) is the enantiomer of the natural product.

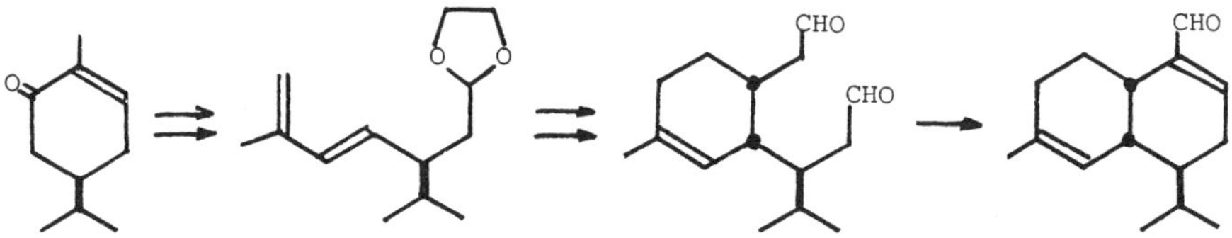

No special attention is necessary in performing the intramolecular condensation of the ketoaldehyde en route to guaiol (**G20**) (Marshall, 1971b, 1972). Here, the desired reaction proceeds according to plan; the alternative mode of cyclization would have given a bridged seven-membered ring product.

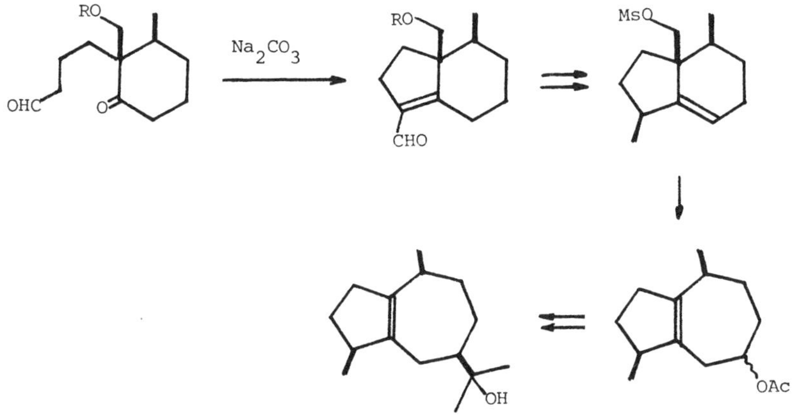

The synthesis of gibberellin A$_{15}$ (**G13**) (Nagata, 1971), one of the least oxygenated of the phytohormones, is somewhat unusual in that the δ-lactone function is disguised as an N-mesylpiperidine in most of the intermediates. Employment of this latent function is a direct consequence of the availability of the tricarbocyclic compounds from a previous synthesis of diterpene alkaloids.

The B ring with its one-carbon appendage is derived from a ring contraction process (cleavage/aldol). Reintroduction of an enone system to the C ring enables the attachment of a chain element for the closure of the bridged skeleton by intramolecular alkylation.

Upon hydrolysis, the diterpene insecticide ryanodine gives 2-pyrrolecarboxylic acid and ryanodol (**R11**). A synthesis of the latter compound (Belanger, 1979) commences with a Diels–Alder reaction that furnishes a 1,4-diketone. An aldol reaction after an *in situ* epimerization locks in a conformation to enable formation of the A ring by another aldol process. The cyclopentanone obtained from the first aldolization is eventually fragmented by a Baeyer–Villiger oxidation.

3.4 CONDENSED RINGS

Still another aldol reaction is used to adjust the bridged framework into the condensed carbocyclic array. This is achieved after cleavage of the tetrasubstituted double bond to the diketone. The condensation proceeds unidirectionally because the alternative pathway leads to a cyclobutanone.

The trans fusion of the newly created rings is favored on steric grounds and by a stabilization of the system with hydrogen bonding.

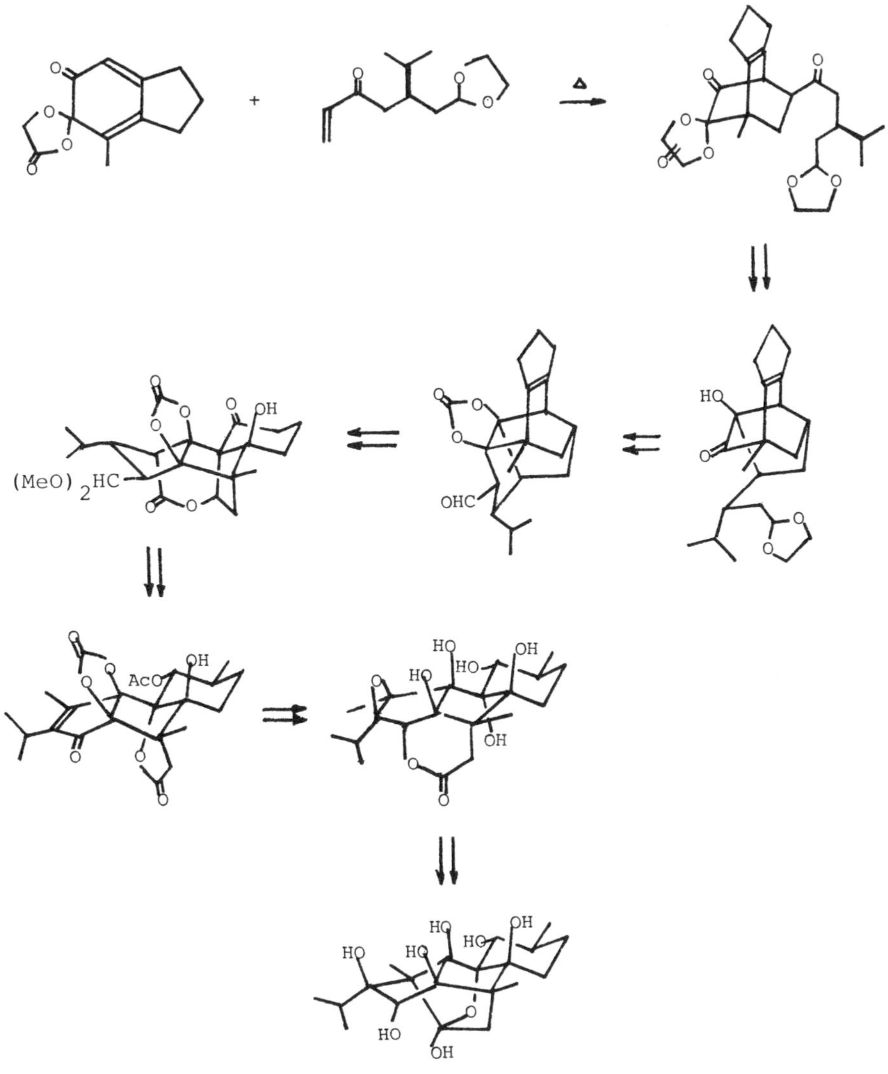

R11

The special oxygenation pattern in the A ring of the diterpene triptolide (**T28**) suggests that an aldol annulation of an aldehyde chain tethered to a β-tetralone would facilitate completion of the tricarbocyclic skeleton (Buckanin, 1980).

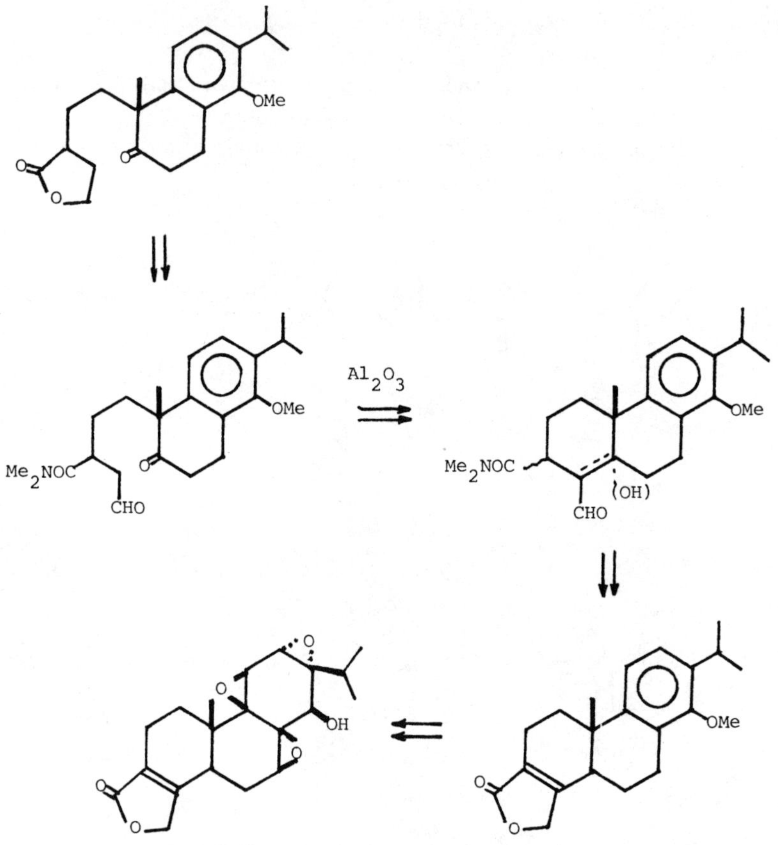

T28

When the carbonyl-containing chain is attached to the β position of a cyclic ketone, the usual product of intramolecular aldolization is the condensed heteronuclear ketol or enone. In these circumstances the side-chain carbonyl acts as the acceptor.

The formation of a cyclopentane on the side of a cycloheptanone is particularly useful for synthesis of pseudoguaianolides that contain oxygen functions at C_4 and C_6, such as parthenin (**P11**) (Heathcock, 1982).

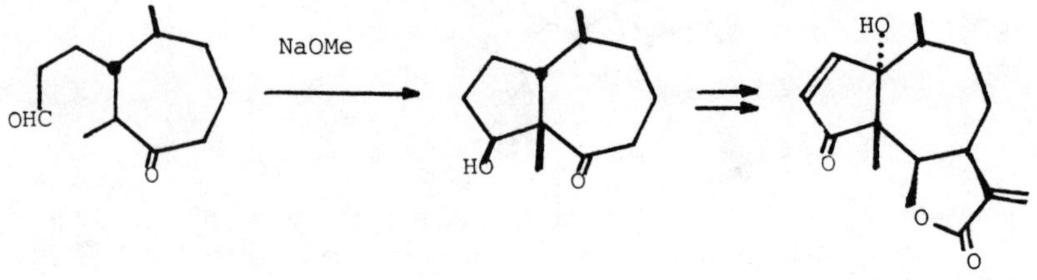

P11

The patchoulenones (mainly 10-epipatchoulenone, (**P14e**) have been obtained from a bridged diketone (Erman, 1971).

P14e

Heteronuclear enones anchoring in a diquinane framework have definite merits in the construction of many recently discovered sesquiterpenes. The efficient copper salt-catalyzed Grignard reaction (Kharasch reaction) of cyclopentenones with a 2-(1,3-dioxolan-2-yl)ethylmagnesium halide makes the preparation of the diquinane enones very reliable. A silphinene (**S34**) synthesis (Leone–Bay, 1982) actually incorporates two sequences of the Kharasch–aldol reactions.

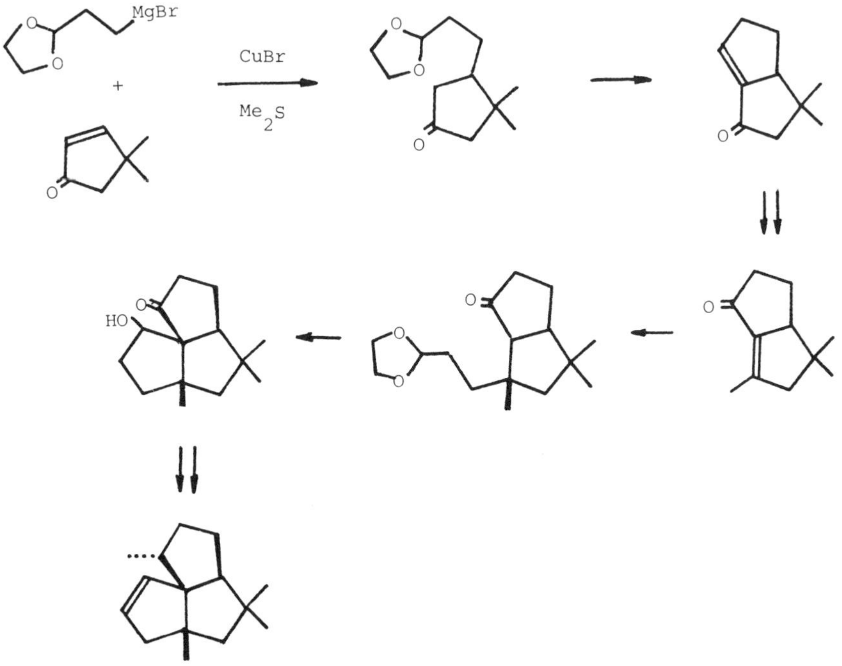

S34

The formation of the third ring by the same process appears in an independent synthesis of silphinene (Tsunoda, 1983). The bicyclic enone arises from a different precursor (by a different method).

The generality of this reaction sequence is further shown in an application to (—)-silphiperfol-6-ene (**S35**) (Paquette, 1984).

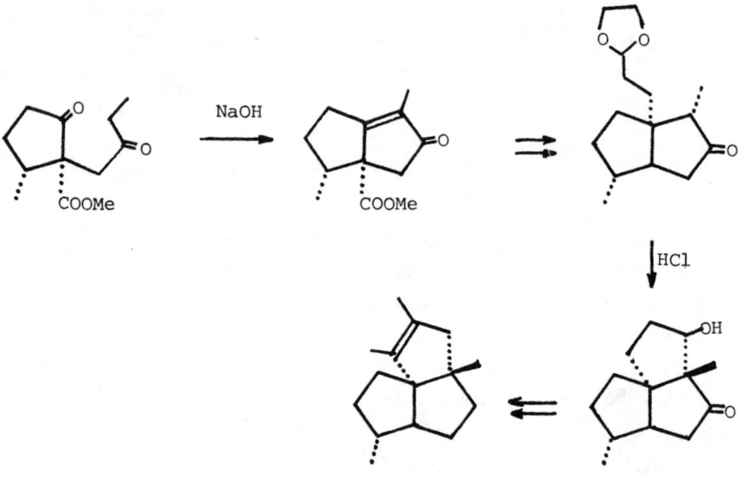

S35

A diquinane enone butenylated at the remaining angle proves to be an excellent intermediate for modhephene (**M12**) (Oppolzer, 1981b). By virtue of an ene reaction, the propellane skeleton is created, thereby solving the stereochemical problem at one stroke.

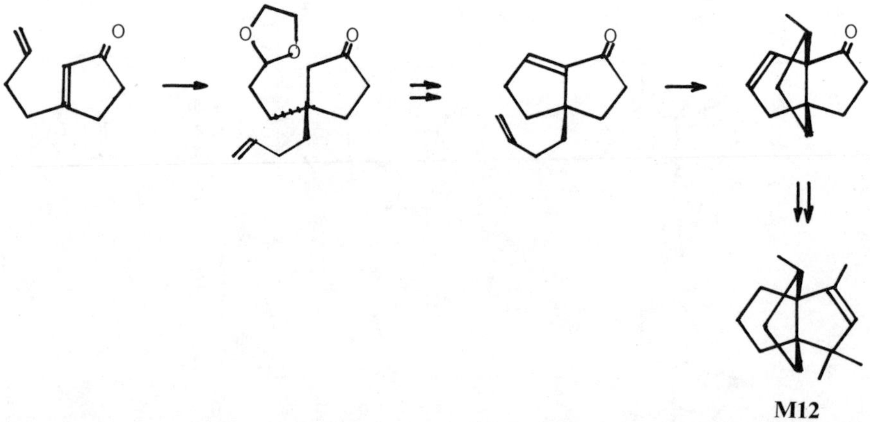

M12

While the preceding examples illustrate the combination of complementary dipolar centers in the reactants, the reaction of 3-alkoxy-2-cyclopentenone dianions with ω-haloalkanals represents a new annulation method. The concentration of negative charges that can be accommodated within the cyclopentadienide framework makes it possible to generate the synthetic equivalent of

2-cyclopentenone-4,5-dicarbanion and hence an expedient synthesis of coriolin (**C70**) (Koreeda, 1983).

C70

The reversible nature of the aldolization causes the β-hydroxy group to assume the exo configuration; thus three stereocenters are established in the first step.

In all but one of the quadrone (**Q1**) syntheses, the two carbon appendends constituting part of the lactone ring are introduced separately. The exceptional case—which includes an aldol reaction, the formation of an unconjugated enone (less strained than the conjugate enone), and cleavage of the double bond (Burke, 1983)—is very refreshing. The acetaldehyde chain of the ketoaldehyde arises from a Claisen rearrangement.

Q1

A synthesis of (+)-pentalenene (**P16**) (Hua, 1986) is interesting in that an optically active allyl sulfoxide is used for kinetic resolution of a bicyclic enone and a (Z)-crotyl sulfoxide to introduce the hydrocarbon elements of the third ring. An aldol condensation accomplishes its closure.

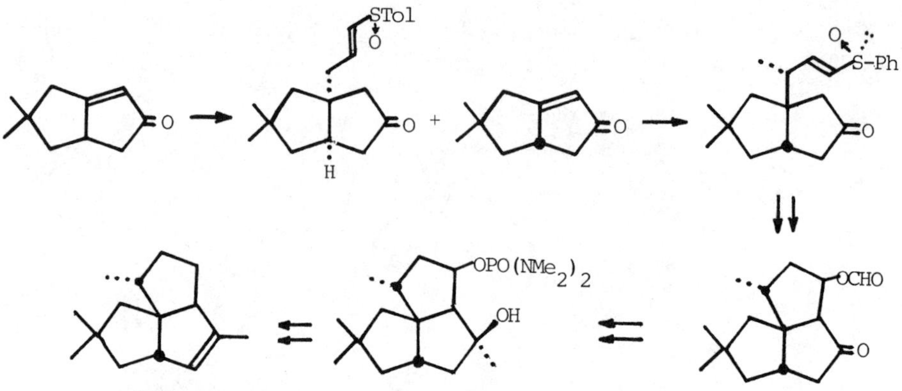

P16

A convergent synthesis of (—)-retigeranic acid A (**R1**) (Paquette, 1987c) from building blocks derived from (—)-limonene and (+)-pulegone via an uncatalyzed Kharasch reaction, aldolization of the ketoaldehyde obtained from the coupled entity, and finally, homologation. The Kharasch reaction gives two diastereomers in a 3:1 ratio. Unfortunately, only the minor product is convertible to the natural molecule. Furthermore, equilibration of the 5:6-ring juncture occurs at the pentacyclic level, leading largely to the undesired cis isomer (20:80).

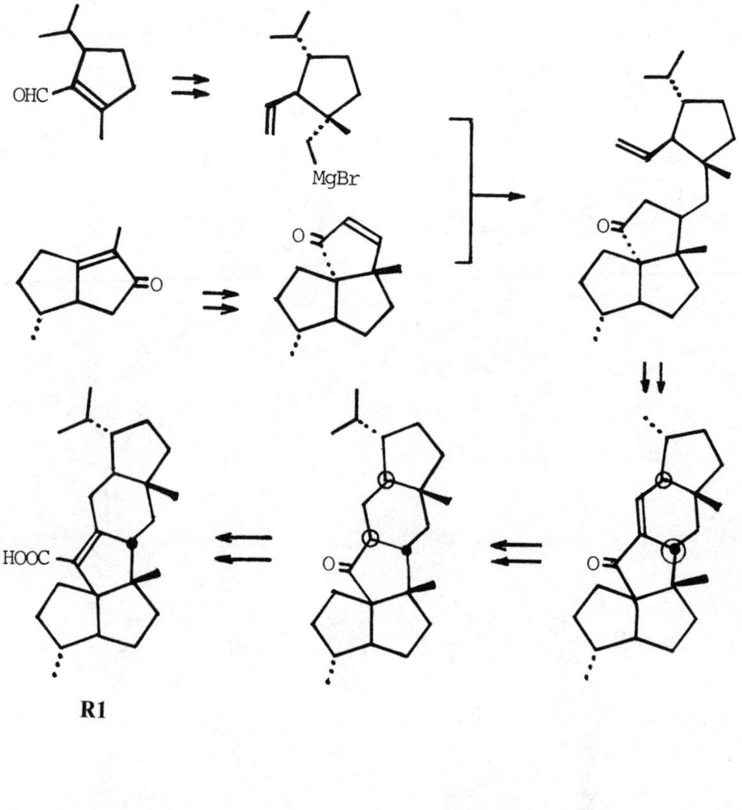

R1

The carbocyclic skeletons of strigol (**S55**) (MacAlpine, 1974, 1976) and alliacol B (**A24**) (Raphael, 1985) have been constructed from cyclopentenones via a Michael–aldol sequence. With respect to alliacol B, the lactonization step gives mainly the epimer, and epoxidation was not successful.

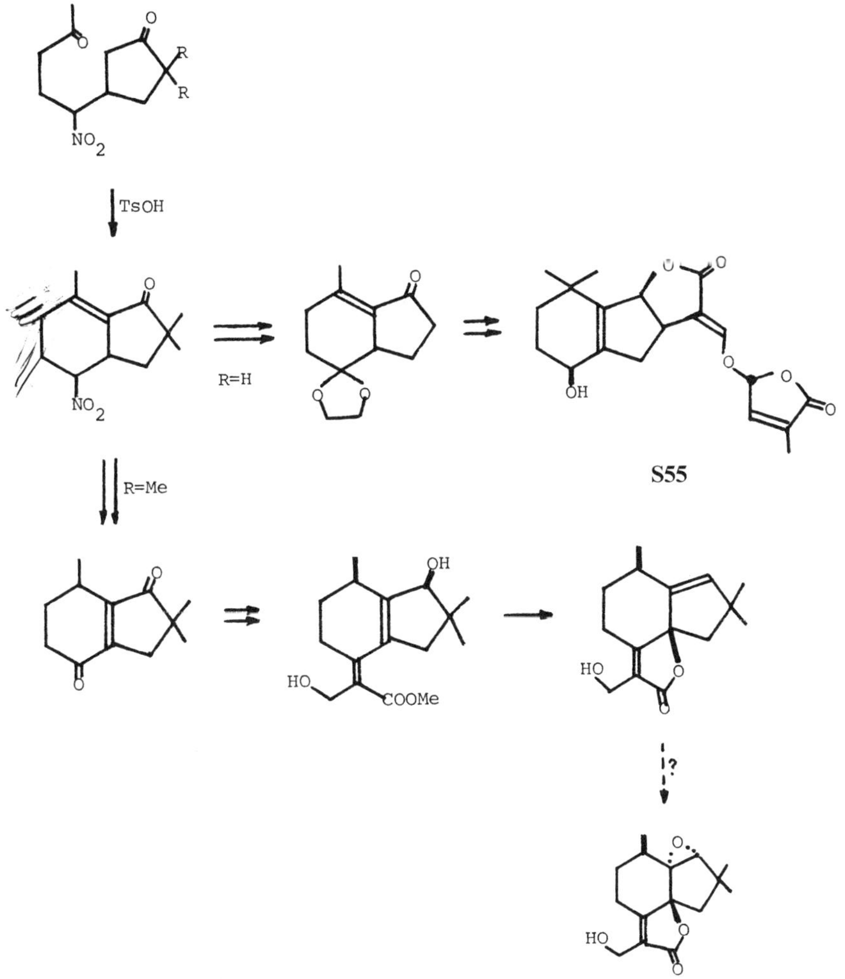

Illudin S (**I1**), a constituent of the poisonous mushroom "jack-o'-lantern," has an engaging structure that is also a challenging synthetic target. A convergent synthesis (Matsumoto, 1971) of this compound is based on building a cyclopentenone and a cyclopropanated chain with multiple functions, and uniting the two fragments by a Michael reaction followed by aldol closure of the six-membered ring.

The sulfinyl group in the "west-side" synthon has an important role not only as an activator, but also in the generation of a carbonyl group at that site by the Pummerer rearrangement.

Laurenene is a diterpene embodying a fenestrane framework. Since its three cyclopentane nuclei and four methyl substituents are the same as in silphinene, synthetic intermediates for the latter compound should be useful for elaboration of laurenene. Indeed, aldol condensation of a properly constructed ketoaldehyde is the key step in three syntheses (Crimmins, 1987a, Paquette, 1988a, Tsunoda, 1987).

Different strategies were formulated for these elaborations. A Claisen rearrangement is used in one route (Tsunoda, 1987) to construct the quaternary carbon atom with appropriate pendants. The second approach (Crimmins, 1987a) involves a high temperature intramolecular photocycloaddition followed by reductive cleavage of the cyclobutane nucleus. In the third synthesis (Paquette,

1988a) a ketone group in the triquinane skeleton becomes the handle on which a cyano group is attached and then a four-carbon chain introduced. All three methods are attended by stereocontrol with regard to the establishment of the quaternary carbon atom.

laurenene

Trihydroxydecipiadiene (**D10**), one of the resinous substances from the bark of an Australian plant, has an unusual skeleton containing a cyclobutane ring fused to two six-membered rings. In contemplating a synthesis of the compound, several salient features have been found to bear directly on the planning (M. L. Greenlee, 1981). The all-*cis* relationship of the methine hydrogens indicates that the secondary methyl group can be generated by hydrogenation of a cyclohexene derivative, thus simplifying the steric problem during creation of the cyclobutane. Since both the quaternary center in the cyclobutane and the endocyclic trigonal carbon atom that carries the hydroxymethyl substituent may be developed from the corresponding ketone function, and the two carbon atoms are β-related, a synthetic pathway based on aldol reaction, and ultimately a ketene cycloaddition, may be formulated.

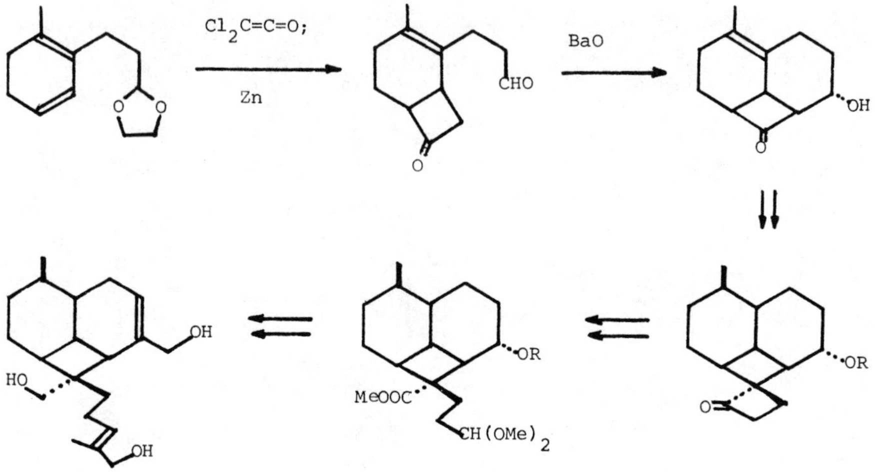

D10

Starting from (—)-cryptone and using the Kharasch reaction followed by Wacker oxidation and aldolization a hydrindenone precursor of (+)-brasilenol (Greene, 1987b) has been obtained. In the presence of Pd-C catalyst under hydrogen, double bond migration occurs in a stereoselective manner. The departing and entering allylic hydrogen atoms are cofacial.

To complete the synthesis it remains at that stage to perform kinetic dimethylation and reduction of the ketone with a hindered complex hydride reagent (LiBHEt$_3$).

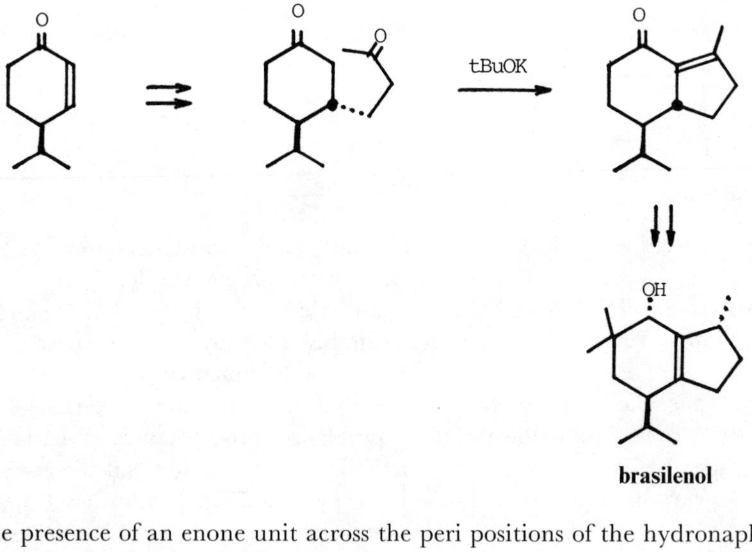

brasilenol

The presence of an enone unit across the peri positions of the hydronaphthalene skeleton of eremophilone (**E13**) is a strong incentive for carrying out synthesis with an aldol reaction in mind. Two possibilities exist, a cyclization

3.4 CONDENSED RINGS

based on precursors with the A ring already in place, or with the B ring already in place. The latter version is the more direct one, although the preparation of the required precursor might or might not be as facile. Aldolization from the A ring necessitates transposition of the functional groups (e.g., via Wharton rearrangement and reoxidation). Both possibilities have been successfully explored (Ziegler, 1974, 1977c; Ficini, 1977).

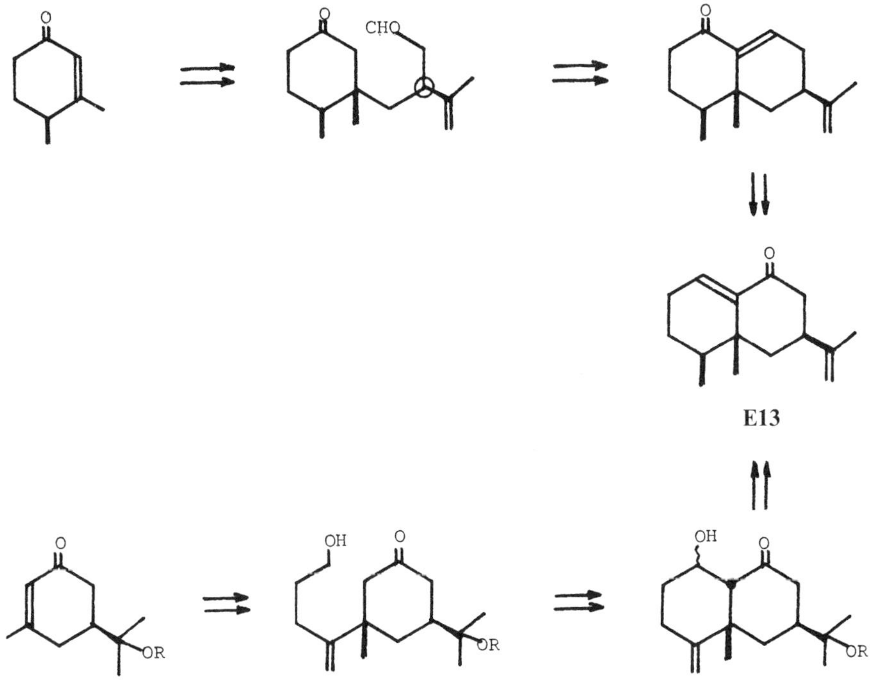

E13

δ-Selinene (**S23**) occurs widely in nature. While the Robinson annulation approach is easily conceived, a more novel option involves the formation of a peri aldol (A. Thomas, 1976).

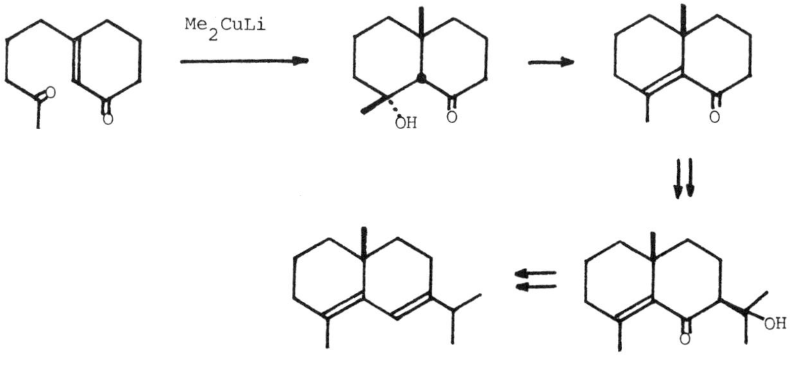

S23

The heteroannular cisoid enone obtained from a 3-(4-oxobutyl)cyclohexanone derivative is readily correlated with the functional groups at the lower portion of ajugarin I. Conjunctive α,β-dialkylation should provide a key intermediate for synthesis of this diterpene (Ley, 1983).

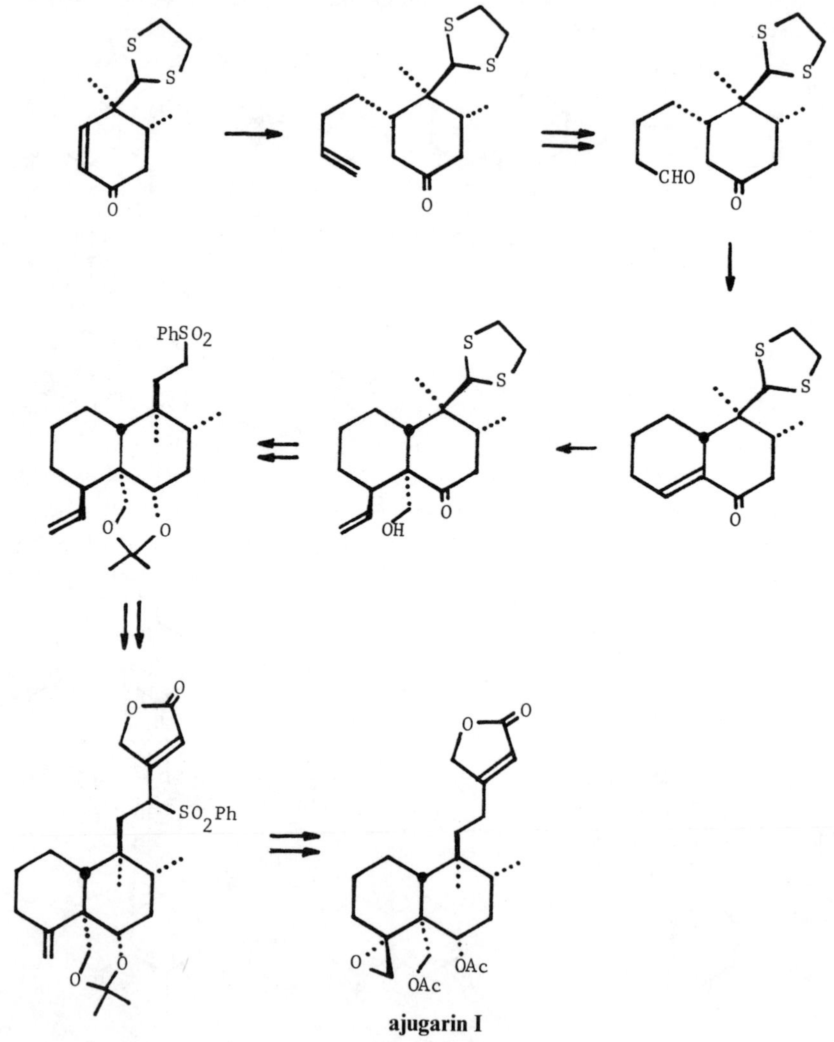

ajugarin I

An ABC + E approach to the triterpene friedelin (**F13**) proves to be relatively simple in terms of pentacyclic construction (ApSimon, 1978). The ABC component is a well-known intermediate for diterpenes. Its alkylation with a tosylate containing the D-ring elements, followed by acid-catalyzed aldol condensation, gives a pentacyclic enone lacking only one angular methyl group, in addition to the requirement of fashioning the aromatic A ring into a dimethylated cyclohexanone. The problem pertaining to the latter aspect has been resolved (Ireland, 1976).

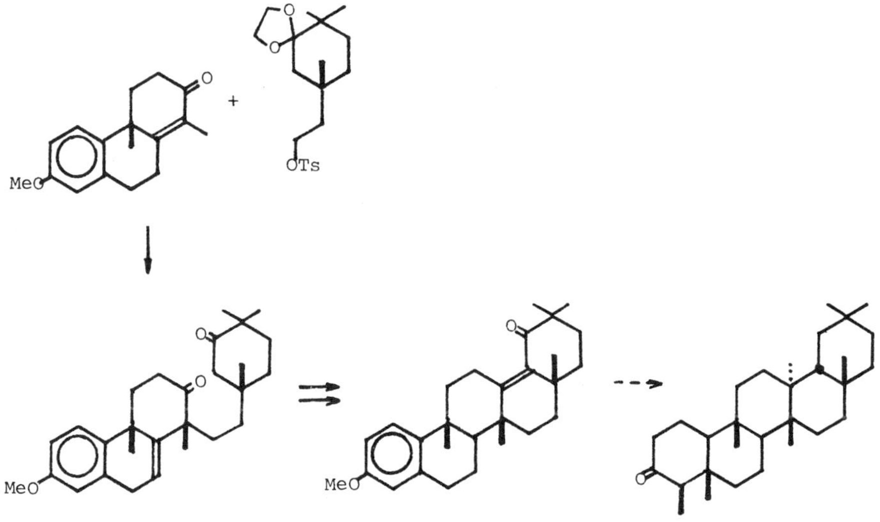

Pleuromutilin (**P39**) has been synthesized in a masterful manner (Gibbons, 1982). A bridged hydrindanone as constructed by a reflexive double Michael reaction is the foundation on which the eight-membered ring evolves. An aldol and modified retroaldol sequence is called upon to accomplish the task. In the latter transformation, an additional driving force must be provided in the form of an electrophilic center created at the β' carbon, which is linked to the same α-carbon atom. Conceivably the enone resulting from the fragmentation will be less prone to revert to the tetracyclic skeleton otherwise favored by steric constraints.

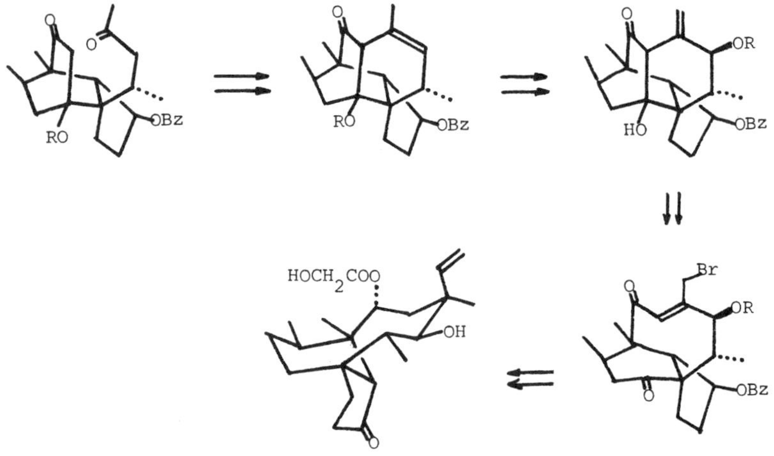

A synthesis of barbatusol (Koft, 1987a) calls for aldol closure of the central ring. In the penultimate stage the trisubstituted double bond is fixed via reduction of the enone tosylhydrazone.

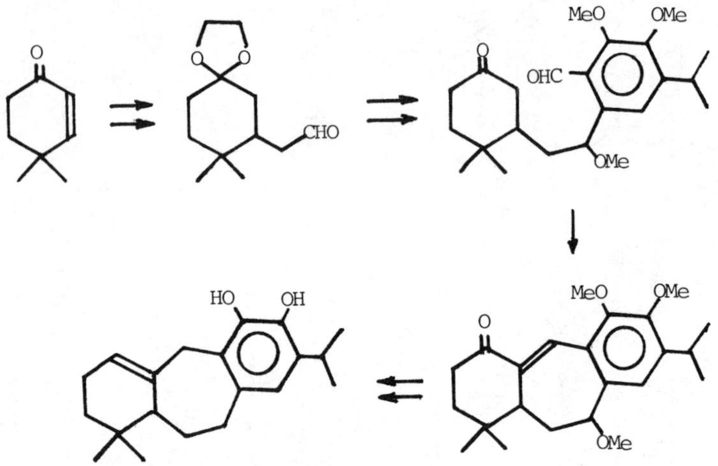

barbatusol

3.4.3 Aldolization Involving Intracyclic Donors

The recyclization of the cyclodecatrione derived from an octalone represents the crucial ingressive step to damsin (**D5**) (Kretchmer, 1976). Methylation of the diketohydroazulene *in situ* provides all but one of the carbon atoms of the target molecule.

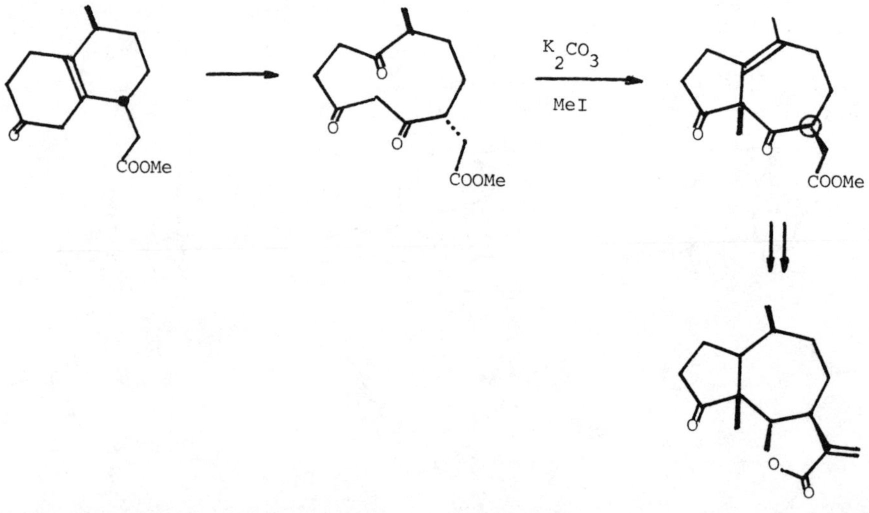

The mesocyclic diketones resulting from the dioxy-Cope rearrangement of a bornanediol undergo aldol cyclization automatically. One of the enones has the same carbon skeleton as patchoulene (**P12**), and the latter's constitution has been proven by the conversion as shown (Leriverend, 1970).

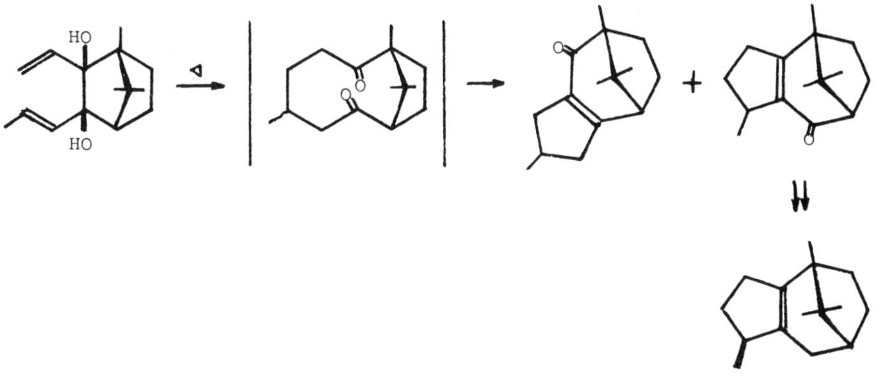

3.5 TWOFOLD ALDOL CONDENSATION

A twofold aldol reaction is encountered in one of the early syntheses of nootkatone (**N11**) (Pesaro, 1968). The ketoaldehyde (mixed with small amounts of its epimer and the O-methyl enol ether) is condensed with methyl acetoacetate to furnish the cross-conjugated dienone ester. The ester group is crucial to the conjugate addition of the methyl group, and also in the correction of its configuration. Thus the cross-conjugated dienone ester is restored, so that hydride reduction results in the production of the desired stereoisomer. The 1,4-addition reactions proceed from the axial and less hindered side.

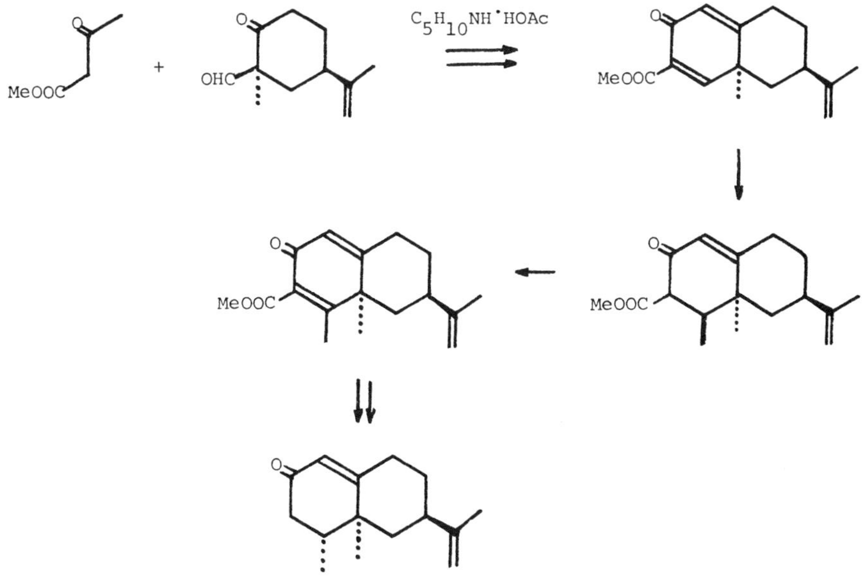

The constitution of isocomene (**I14**) is such that if a diquinane precursor is incorporated with two appropriate angular substituents (i.e., a methyl and a functionalized alkyl group) during its preparation, the synthesis of the terpene is greatly simplified. One of the solutions that has evolved consists of two sets of aldol–Michael reactions (timing unspecified) between dimethyl acetonedicarboxylate and ethyl 4,5-dioxohexanoate (Dauben, 1981).

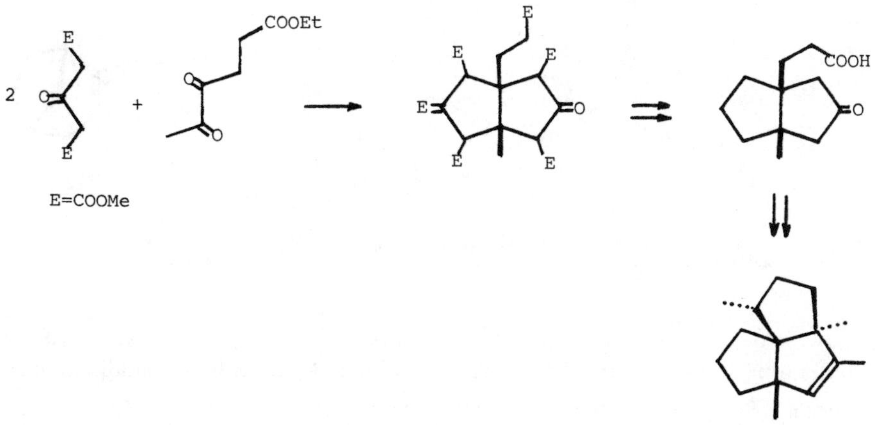

A diketopropellane obtained in one step by a similar process serves as the starting material for modhephene (**M12**) (Wrobel, 1983).

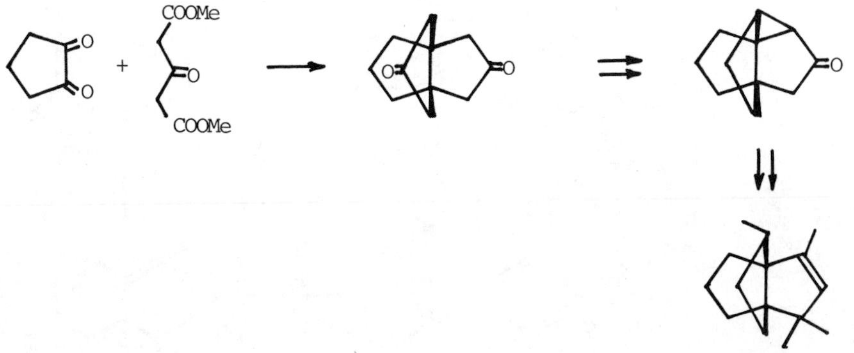

3.6 MISCELLANEOUS ALDOL PROCESSES

In an unfinished synthesis of quadrone (**Q1**) (S. Monti, 1982), an ingenious transformation, whether by design or serendipity, is the cyclization of a 1,5-diketone. By virtue of a pinacolone rearrangement, the initial aldol product mutates into the tricarbocyclic skeleton of quadrone.

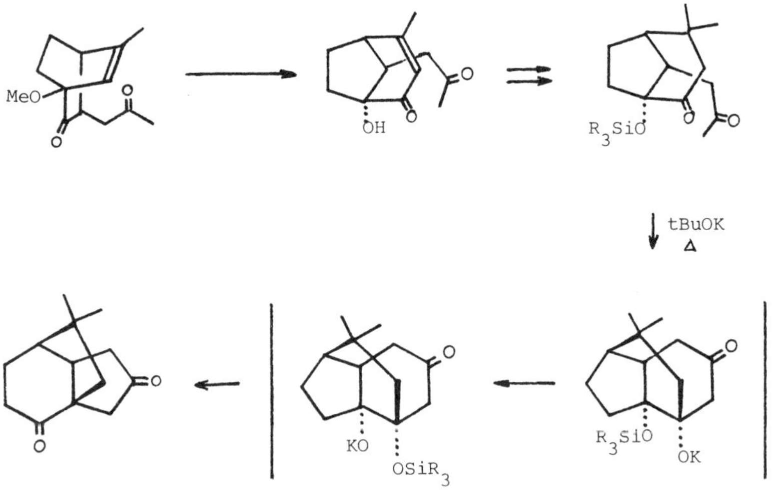

Treatment of a W–M ketone analog with lithium in liquid ammonia affords a cyclopropanol. The cyclization coupled with the opening of the cyclopropanol effects a net structural change of a reductive bond migration. Application of this process to a synthesis of a β-vetivone (**V16**) intermediate has been achieved (Subrahamanian, 1978).

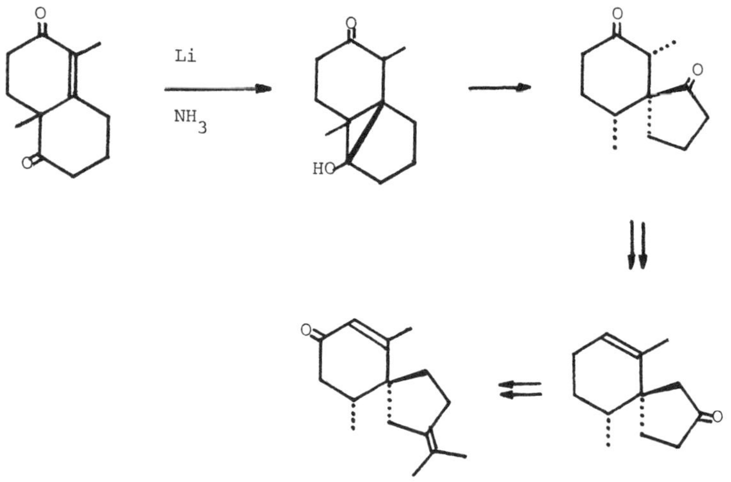

V16

The reaction of α-sulfonyl carbanions with carbonyl compounds is similar to the aldol condensation. Generally, this process suffers more seriously from reversibility. However, the provision of a relieving mechanism such as a leaving group at the β carbon of the carbonyl can drive the condensation forward. This principle is demonstrated in a synthesis of (+)-hinesol (**ent-H15**) (Buddhasukh, 1975; Chass, 1978), antipode of the natural compound.

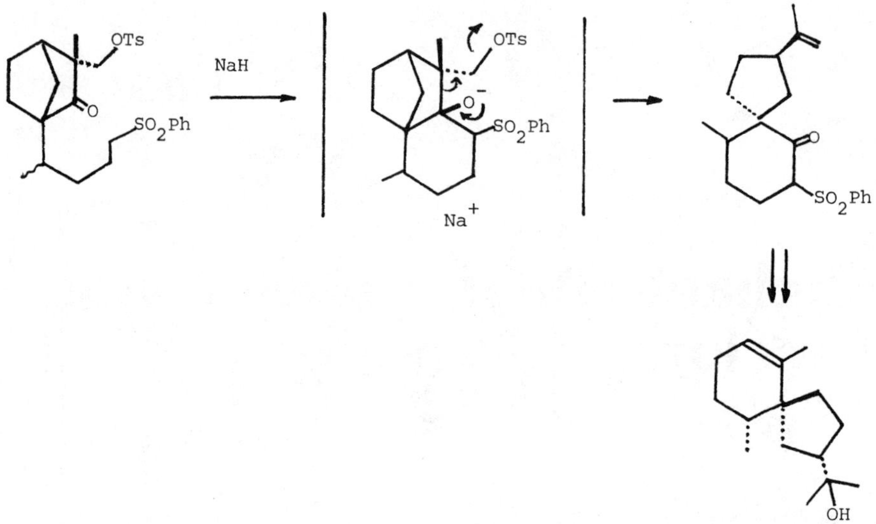

An advanced quassinoid intermediate has been obtained via two annulation sequences involving the Knoevenagel and Claisen reactions (Batt, 1984). In the formation of the B ring under thermodynamic control, the creation of the two new chiral centers is affected by steric factors such that the more bulky substituents (OH, COOEt) are equatorially set. The Claisen reaction serves to assemble the C ring.

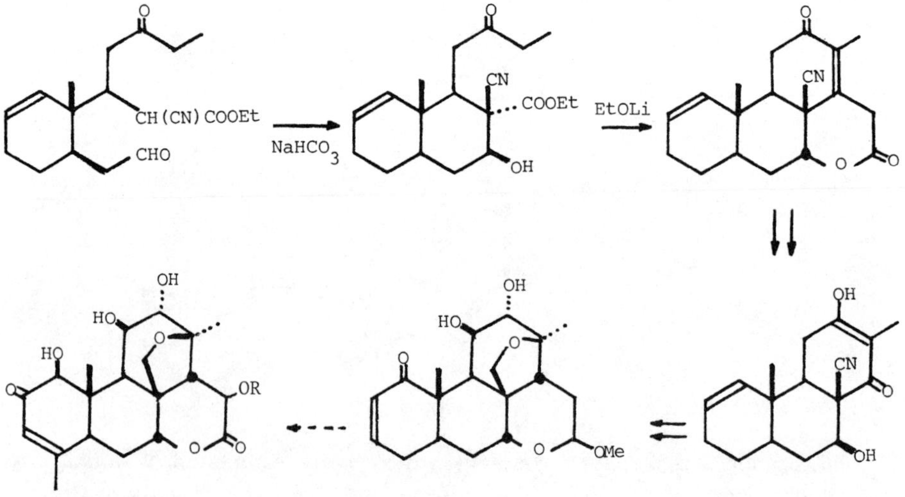

Chapter 4

Cyclization by Michael–Aldol, Claisen, and Dieckmann Reactions

4.1 STORK ANNULATION

A convenient alternative to the Robinson annulation is the enamine condensation with enones developed by Stork (1963a). Enamines are formed when aldehydes or ketones react with a secondary amine with the splitting off a molecule of water. The most widely used secondary amines are pyrrolidine, piperidine, and morpholine.

Enamines of cyclic ketones are readily prepared by refluxing the reactants with or without an acid catalyst (e.g., TsOH) in benzene or toluene under a Dean–Stark trap to remove water continuously. However, aldehydes tend to form the diaminals, and thermal decomposition (distillation) of the latter compounds is required. Naturally, 2 mol-equivalents of the amine must be used to react completely with the aldehyde; the amine is to be recovered after the thermal decomposition.

Enamine alkylation generally occurs at the β-carbon atom. In certain cases (e.g., with allylic halides), N-alkylation takes place, but the ammonium salts undergo a [3.3]sigmatropic rearrangement to give C-allylated carbonyl compounds.

The condensation of an enamine with an enone follows a Michael addition course, the adduct undergoes enamine exchange, and an aldol-type reaction can now take place, leading, upon hydrolysis, to a cyclohexenone. The synthesis of cryptone (**C79**) (Stork 1963a) and α-curcumene (**C87**) (Joshi, 1965, 1968) clearly shows the efficiency of this method.

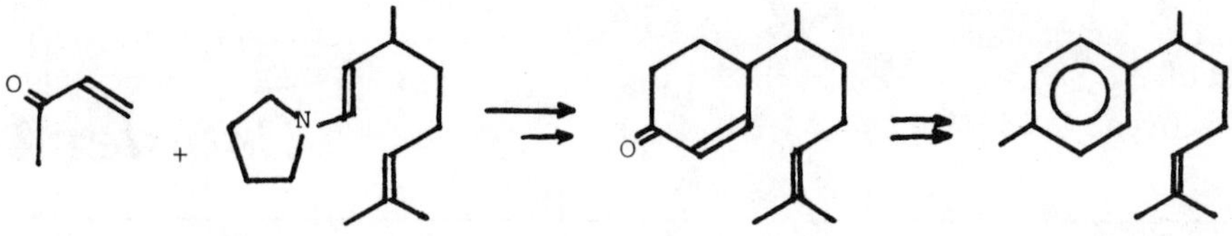

C87

Polycyclic enones are available by the Stork annulation starting from cyclic ketone enamines. It should be emphasized that the product obtained is different from that of a Robinson annulation on an α-alkylcycloalkanone. Two examples of cadinene synthesis are known exploiting the enamine method: cadinene hydrochloride (**C2·HCl**) (Piers, 1975d) and δ-cadinene (**C4**) (Nishimura, 1981).

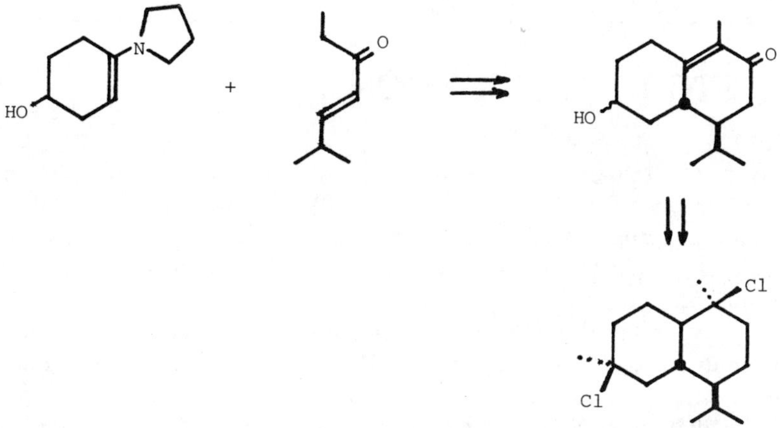

C2·HCl

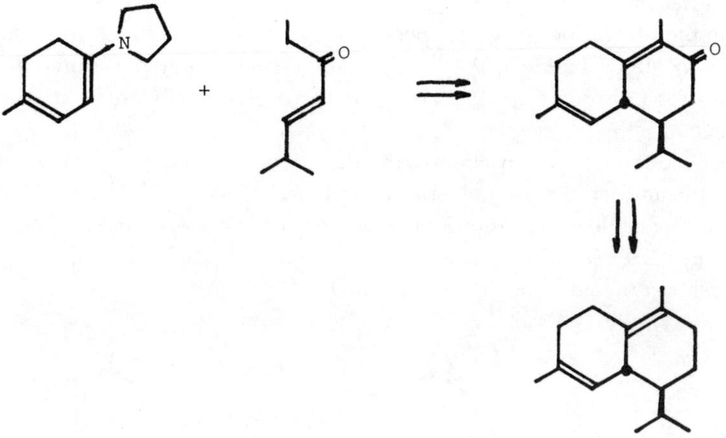

C4

4.1 STORK ANNULATION

The spiroannulation of 3-isopropylidenecyclopentanecarboxaldehyde with 3-buten-2-one by Stork's method constitutes the key step in a synthesis of β-vetivone (**V16**) (Hutchins, 1984).

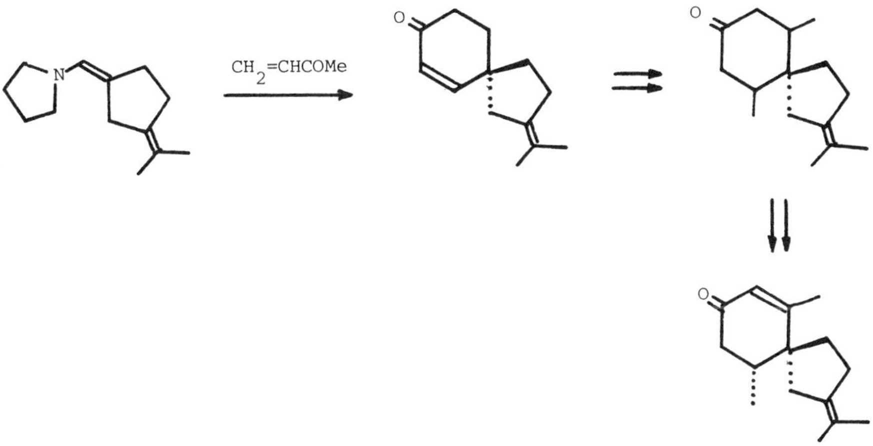

An intramolecular enamine/enal cycloaddition is effected by reaction of (+)-8-oxocitronellal with N-methylaniline. The predominant bicyclic aminodihydropyran product contains all the chiral centers and skeleton of *ent*-nepetalactone (**ent-N5**). The conversion of this cycloadduct to the terpene is straightforward (Schreiber, 1986).

Ketene dithioacetals may also be used as the ene component in the cycloaddition under Lewis acid catalysis. (+)-Nepetalactone (**N5**) is available from the appropriate intermediate (Denmark, 1986).

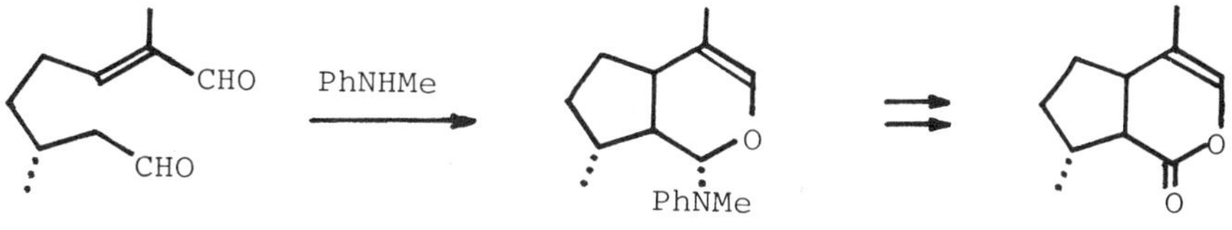

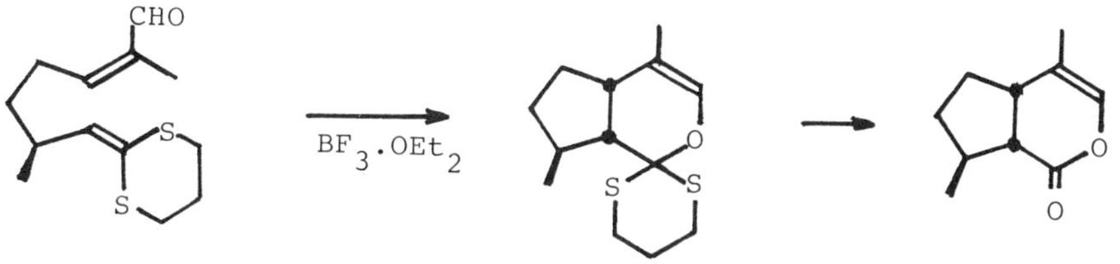

4.2 MICHAEL AND MICHAEL–ALDOL TANDEM

Synthesis of (+)-norpatchoulenol (**P13**) (Liu, 1987c) from (+)-campholenic acid illustrates a discernment of molecular architectural relationship. The optical antipode of the cyclohexenoneacetic ester intermediate has been employed previously in an approach to (—)-khusimone (**K8**) (Liu, 1979, 1982). In the norpatchoulenol synthesis the ester chain is extended into an enone system which is allowed to partake a Michael reaction to form the bicyclo[2.2.2]octanone framework.

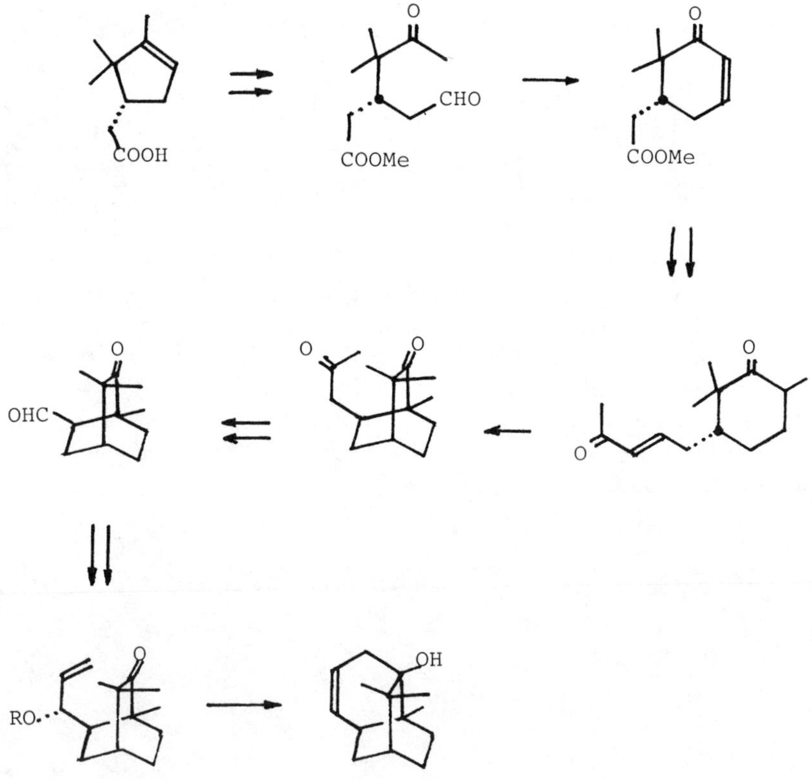

Asymmetric induction by a sulfoxide substituent in a Michael acceptor is instrumental to an enantioselective synthesis of aphidicolin (**A36**) (Holton, 1987). Another Michael reaction is employed in closing the B-ring. The octalone intermediate contains the complete set of skeletal carbon atoms and functionality to form the remaining bridged system by known methods.

4.2 MICHAEL—ALDOL TANDEM

A36

While the prototype of the Michael–aldol reaction sequence is the Robinson annulation, the intramolecular condensation of a cyclic enone with a side chain containing a carbonyl group could lead to a vastly different ring skeleton.

The tricyclic skeleton of seychellene (**S29**) and patchouli alcohol (**P15**) has been assembled in one step from a cyclohexenone substituted at C_4 with an appropriate aldehyde chain (K. Yamada, 1979). The intermediate is convertible also to cycloseychellene (**S30**).

The stereoselective formation of the desired epimer (about the secondary methyl group) is due to its origin from the least encumbered transition state. Generation of the unwanted epimer must involve a transition state that suffers 1,3-diaxial Me–Me interaction.

Replacing the secondary methyl group by a cyano function and effecting the double cyclization of the cyclohexenone aldehyde leads to a useful intermediate for the elaboration of norpatchoulenol (**P13**) (Niwa, 1984a). The cyano group is hydrolyzed, and subsequent oxidative decarboxylation installs the double bond.

4.2 MICHAEL—ALDOL TANDEM

The patchouli alcohol synthesis shows the feasibility of achieving hydroxylation at a bridgehead of the tricyclic framework. This experience enables a synthetic design that incorporates the same general ring formation scheme for (—)-picrotoxinin (**P29**) and (+)-coriamyrtin (**C69**) (Niwa, 19984b).

In this synthesis a shorter chain is required to form an isotwistane system. The bridge that corresponds to the original aldehydic carbon is dissociated from the cyclopentane ring by means of oxidation maneuvers (including α-hydroxylation).

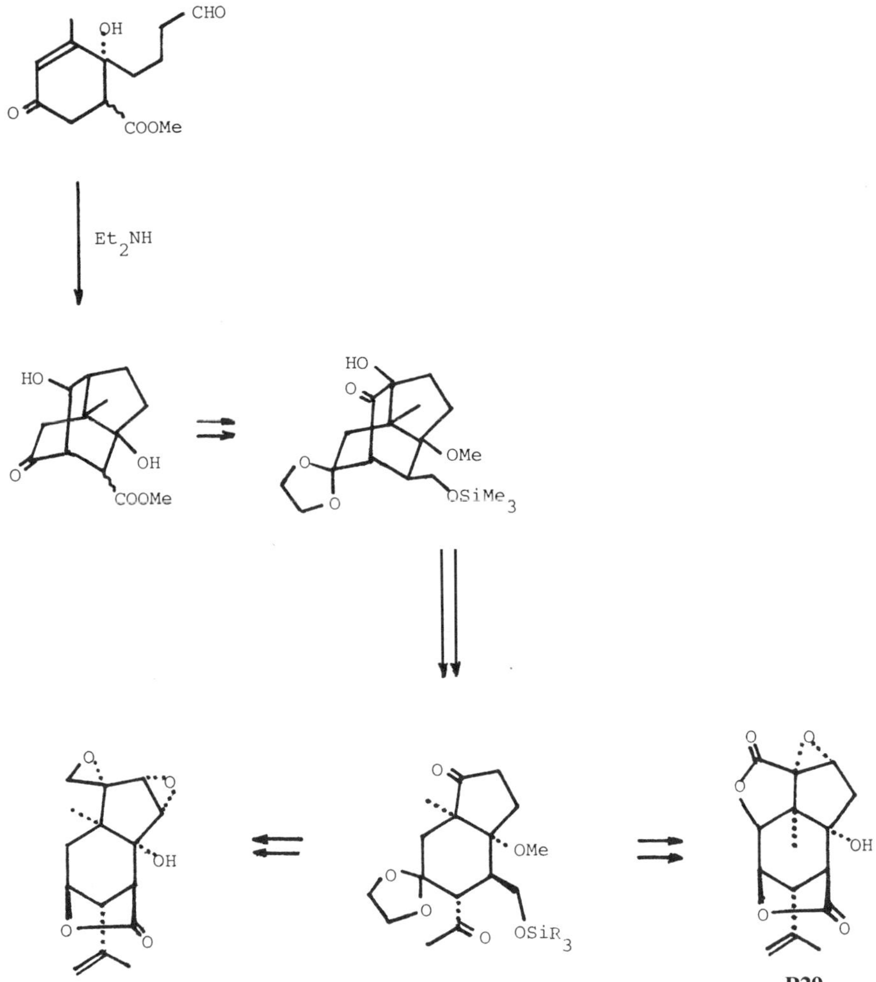

In another dendrobine (**D11**) synthesis (K. Yamada, 1972), the annulation of the cyclopentane residue onto an existing cyclohexenone via intramolecular Michael addition also goes overboard. The aldol process that follows the initial

Michael cyclization gives a bowl-shaped tricyclic ketol. (This type of transformation was first observed by W. S. Johnson during a study of steriod synthesis.) The ketol is useful in the investigation of synthetic compounds.

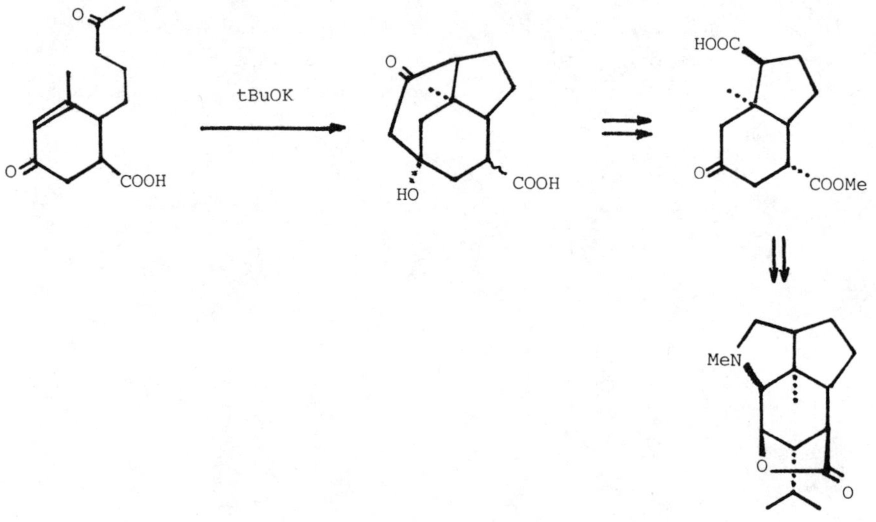

The rapid assemblage of Coates's ketone intermediate for synthesis of zizaene (**Z1**) from a substituted cyclopentanone is also based on the Michael–aldol tandem (Alexakis, 1978).

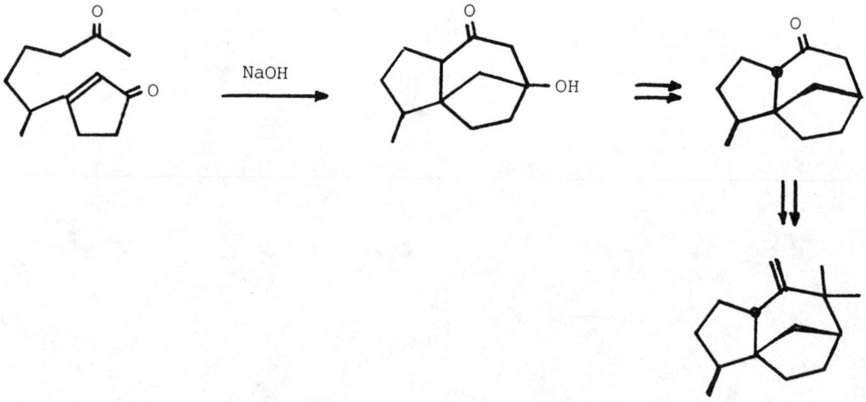

An alternative (to the Robinson method) octalone synthesis involves the preparation of an α-methylenecyclohexanone and condensation of the latter compound with an acetoacetic ester. Admittedly this reaction sequence is far more laborious; however, it has been applied to a synthesis of (+)-epizonarene (**Z5**) from (—)-menthone (Belavadi, 1976).

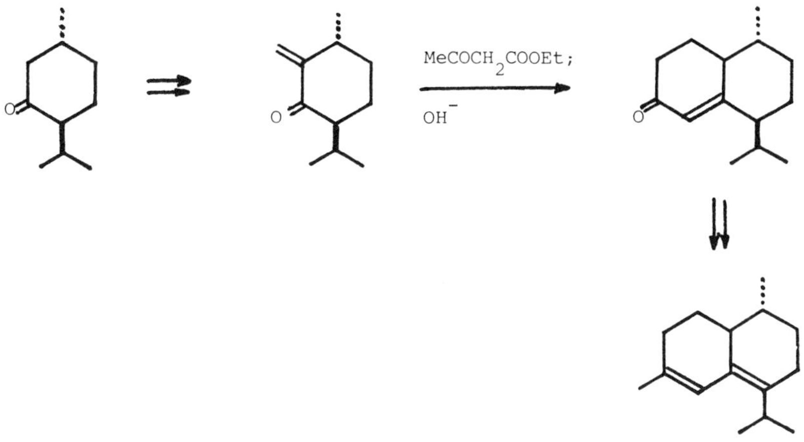

The Michael–aldol tandem for annulation proves valuable in the preparation of a bicyclic precursor for (—)-methyl kolavenate (Iio, 1987). The enone is readily accessible by conjunctive alkylation and elimination of the α-hydroxymethylene ketone.

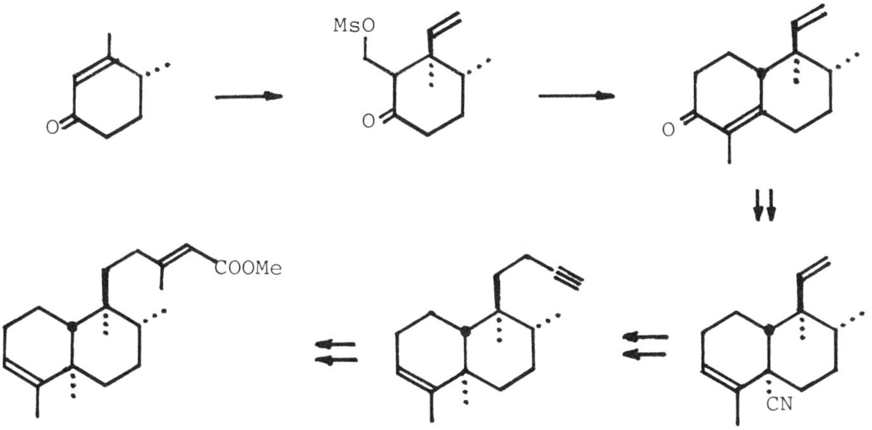

4.3. STEPWISE MICHAEL–ALDOL CYCLIZATION

The separation of the Robinson annulation into two distinct operations is occasionally profitable in terms of product yields. At other times it is just impossible to carry out the annulation in one step, when the cyclization of the 1,5-diketone requires acid catalysis.

In the synthesis of α-cyperone (**C95**) (Caine, 1974) from (—)-2-carone, the base-catalyzed reaction stops at the Michael addition stage. Subsequent treatment of the diketone with HCl accomplishes the aldol condensation (and dehydration) as well as the opening of the cyclopropane ring; therefore the number of steps remains the same as was probably planned.

The bicyclic 2-carone undergoes alkylation from the α face for steric reasons.

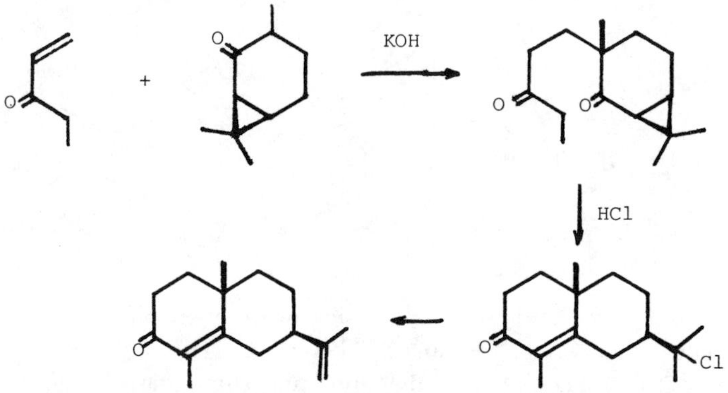

C95

The reaction of 2,3-dimethyl-1,4-cyclohexanedione with 3-buten-2-one in the presence of sodium ethoxide gives a monocyclic triketone and two bridged ketones. These bridged diketones are converted mainly into the octalindione required for the synthesis of isopetasol (**P25**) and warburgiadione (**W2**) (Yamakawa, 1974, 1979).

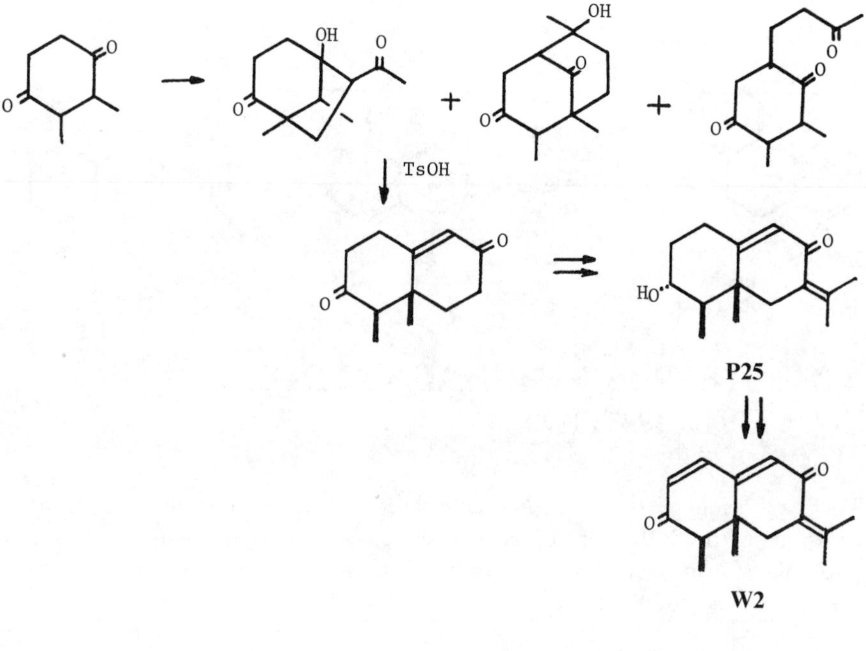

P25

W2

4.3 STEPWISE MICHAEL—ALDOL CYCLIZATION

As discussed in Chapter 2, activation of the donor ketone is beneficial and sometimes mandatory when conducting the Robinson type of annulation. To synthesize (+)-calamenene (**C7**) (Ladwa, 1968) from (—)-menthone, alkylation at the methylene site necessitates the introduction of a formyl (or related) substituent at that carbon atom. The octalone is then transformed into the alkylated tetralin by standard procedures.

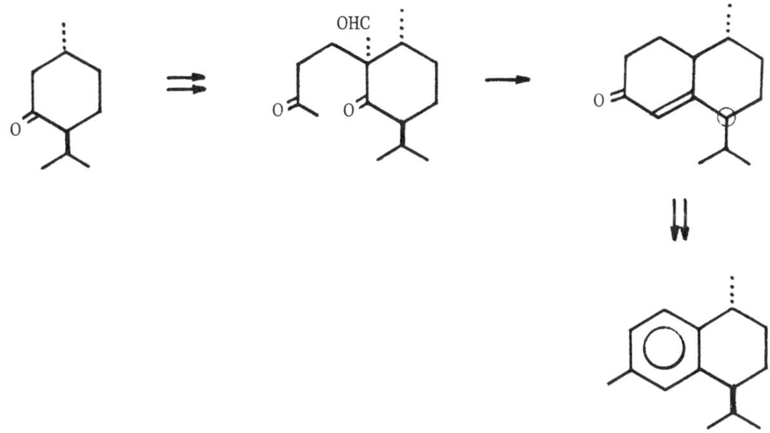

The α-decalone also needs activation during a synthesis of taxodione (**T7**) (A. K. Banerjee, 1983). This annulation sequence is modeled after a synthesis of methyl deisopropyldehydroabietate (**A2′**) (Spencer, 1968).

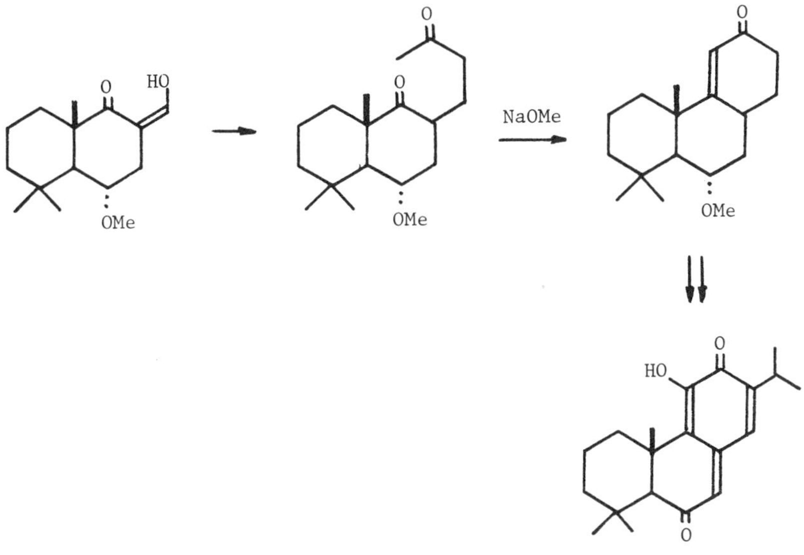

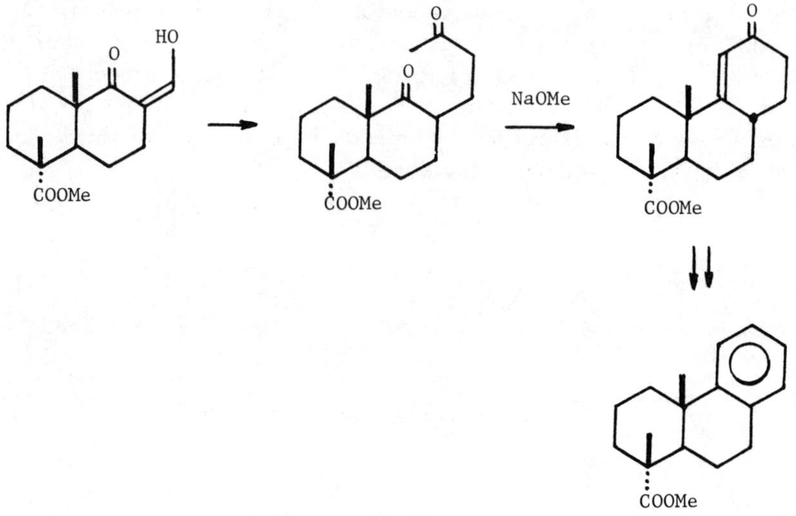

Methyl 2-oxo-6,6-dimethylcyclohexanecarboxylate is unreactive toward 3-buten-2-one under the conventional Robinson annulation conditions. However, a Michael adduct is obtained by conducting the reaction in tetrahydrofuran in the presence of a catalytic amount of potassium methoxide. Cyclization is then effected using methanolic potassium carbonate. The octalone is processed into spiniferin I (**S45**) (Marshall, 1983), while converting the activating ester group into a sulfonyloxymethylene and thence the methylene bridge present in the terpene molecule.

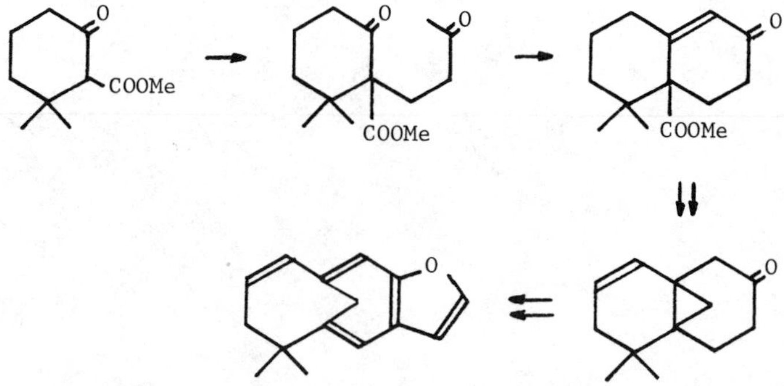

One of the advantages of employing the formyl group for activation during the Michael reaction is its simple disengagement upon treatment with aqueous bases of the adducts. On the other hand, an approach to anguidine (**A32**) (Brooks, 1983) involves the deliberate retention and participation of such a functional group in the annulation.

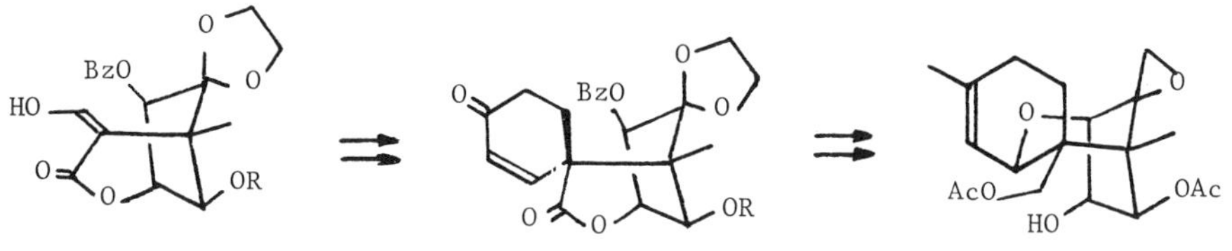

4.3 STEPWISE MICHAEL–ALDOL CYCLIZATION

A32

The optically active γ-hydroxymethyl-γ-butyrolactone is a valuable synthon for many natural products. Its use has been extended to the preparation of an enantiomer of a known intermediate of calonectrin (**C12**) via a cyclohexenone (Tomioka, 1987). The lactone moiety provides three oxygen functions which are fully exploited.

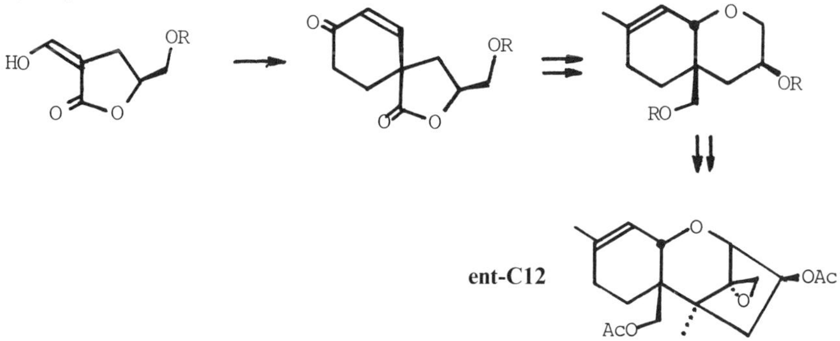

ent-C12

Similarly, the syntheses of aphidicolin (**A36**) (Corey, 1980a) and stemodin (**S50**) (Corey, 1980b) feature a spiroannulation of formyldecalones, which is followed by intramolecular alkylation. The aphidicolane and the stemodane skeletons differ in the stereorelationship pertaining to the CD rings.

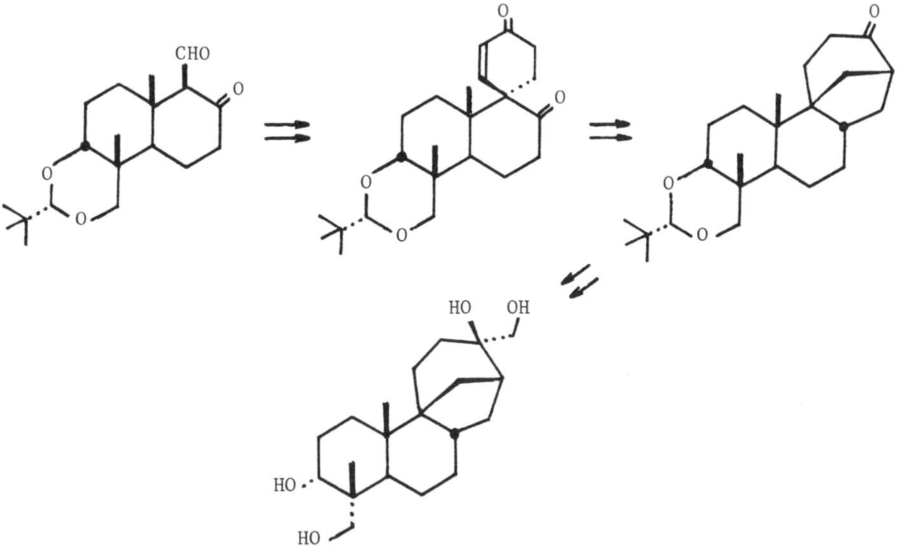

A36

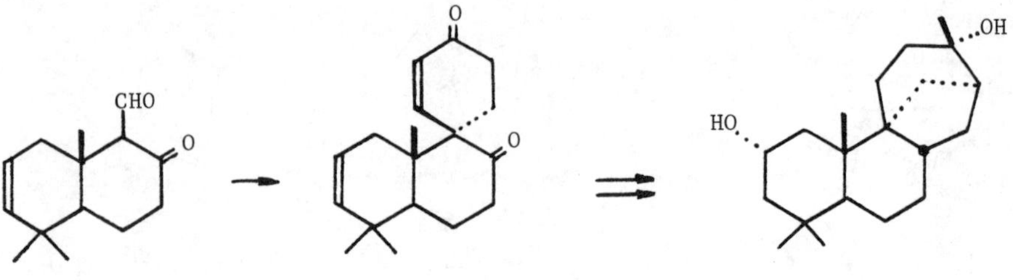

The A + B + C approach to tricyclic diterpenes offers flexibility in the variation of the C-ring substituents. Essentially the decalone is prepared and elaborated into an 8-formyl-Δ^8-en-7-one as a prelude to C-ring annulation by a Michael–aldol reaction sequence. The required decalones are easily obtained by the Robinson method or via cyanoketones and 1,5-diketones. The synthesis of sugiol (**S56**), ferruginol (**F4**), and nimbiol (**N10**) (Meyer, 1975) shows the incorporation of the formyl carbon in the final products.

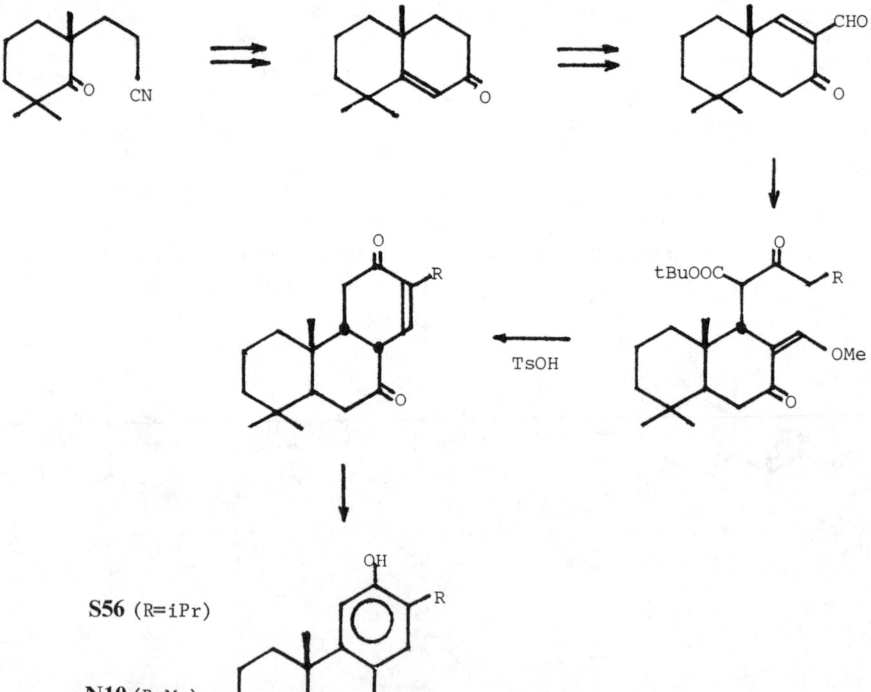

For synthesis of dehydroabietic acid (**A2**) (Meyer, 1977), an ester-bearing *trans*-decalone is similarly processed. After aromatization of the C-ring, both oxygen functions are removed by hydrogenolysis.

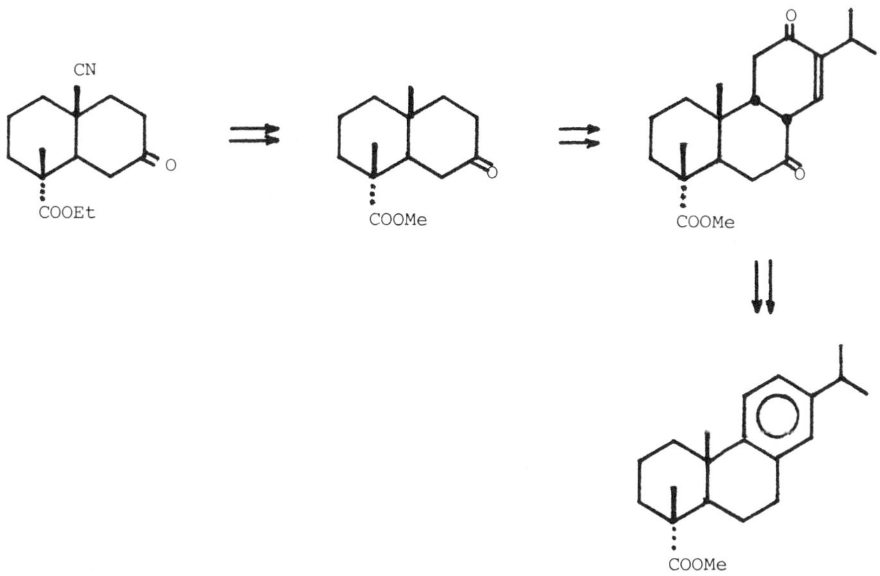

A2E

An isomeric decalone is required for the synthesis of carnosic acid (**C28**) (in the form of dimethyl ether) (Meyer, 1976), the ester group being at the angular position. The additional oxygen function at C_{11} is to be introduced after aromatization via the diazo coupling protocol. An improved method involves the use of an α-sulfinyl ketone as the Michael donor to initiate C-ring formation. Pummerer rearrangement of the Michael adduct exposes the C_{11} oxygen, and a second Michael addition furnishes the tricyclic framework.

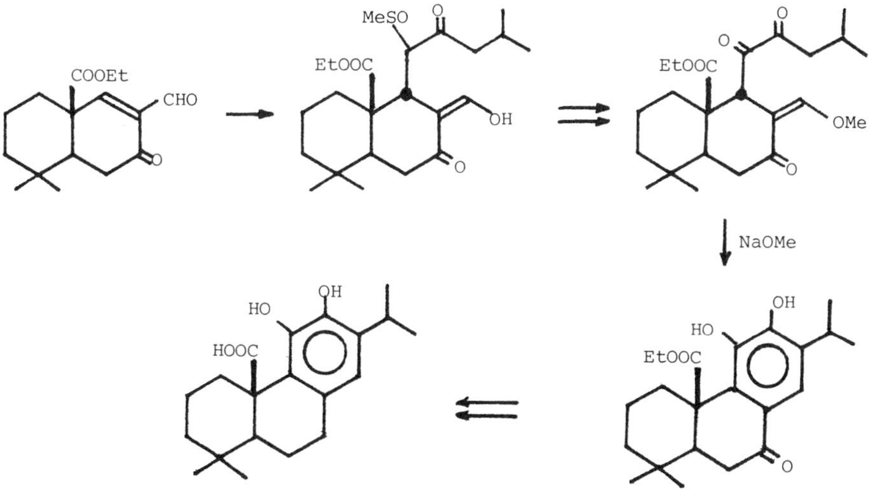

C28

The Michael–aldol reaction sequence for annulation may be modified by reversing the roles of the reactants. The cyclic ketone component may act as the acceptor by means of α-methylation, which is usually achieved by dehydration (indirectly) of the formaldehyde adduct.

The dicarbocyclic skeleton of both *cis*- and *trans*-clerodanes is available from a common intermediate (Tokoroyama, 1983). The A ring is formed by a Michael–aldol reaction sequence, and the ring juncture is erected by proper selection of the one-carbon addends.

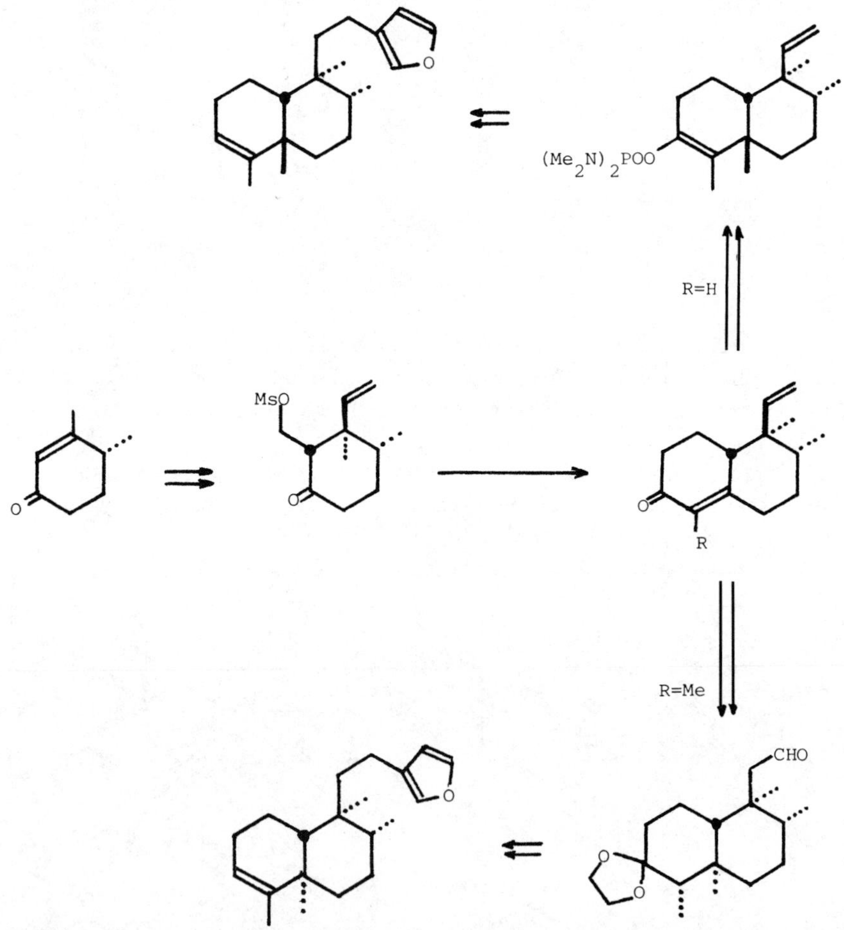

A34

An important intermediate for tetracyclic diterpene synthesis is $\Delta^{8(14)}$-podocarpen-13-one (**P42**). While this substance is easily prepared from a tetralone or by degradation of manool, an alternative method is also available, involving the reaction of methyl acetoacetate with the α-methylenedecalone. The resulting diketoester undergoes cyclization readily (Skeean, 1976).

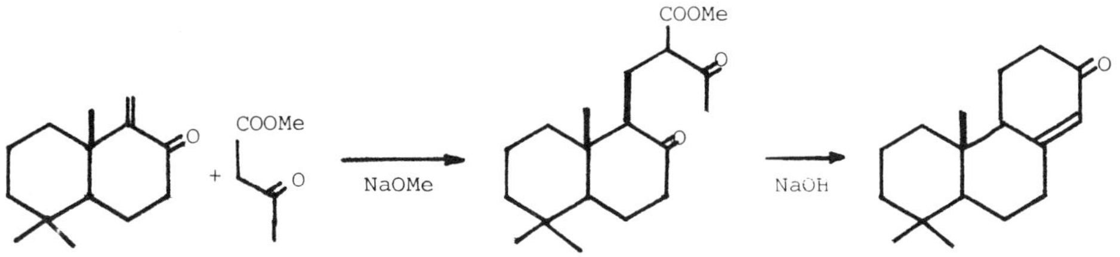

P42

The α-methyleneoctalone is quite readily obtained from β-ionone, and it has been converted into the podocarpenone in the same fashion. By means of an intramolecular Wittig–Horner–Emmons reaction, jolkinolide E (**J4**) is produced (Katsumura, 1982).

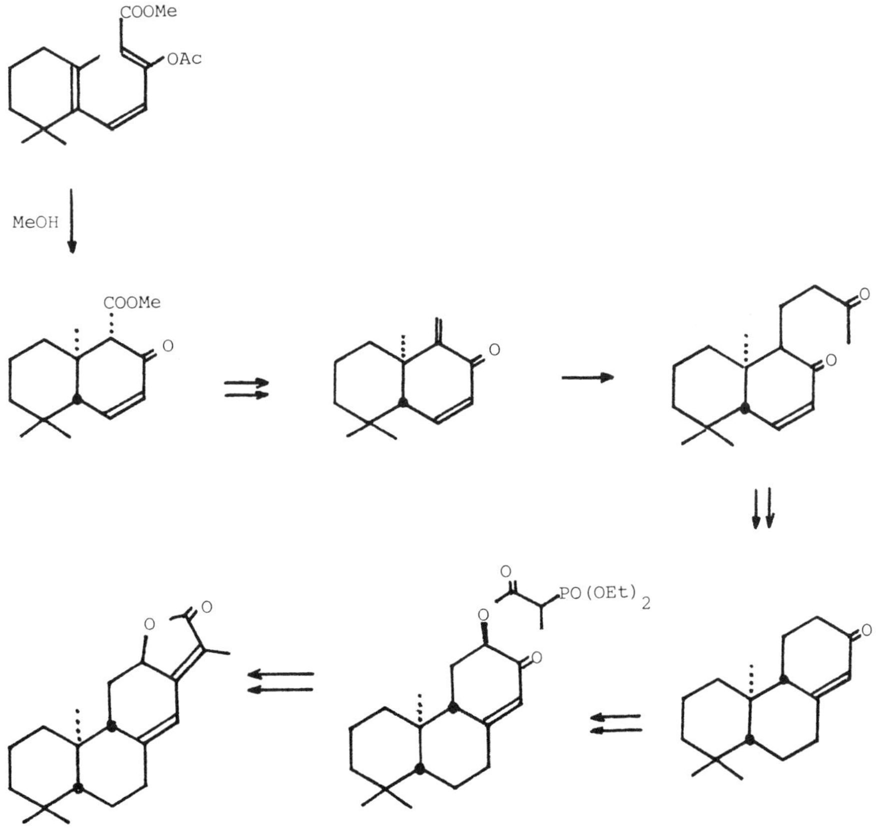

J4

A delightful synthesis of quadrone (**Q1**) (Burke, 1982) delineates an unusual Michael–aldol sequence involving three different carbonyl groups. Two carbocycles are formed in succession.

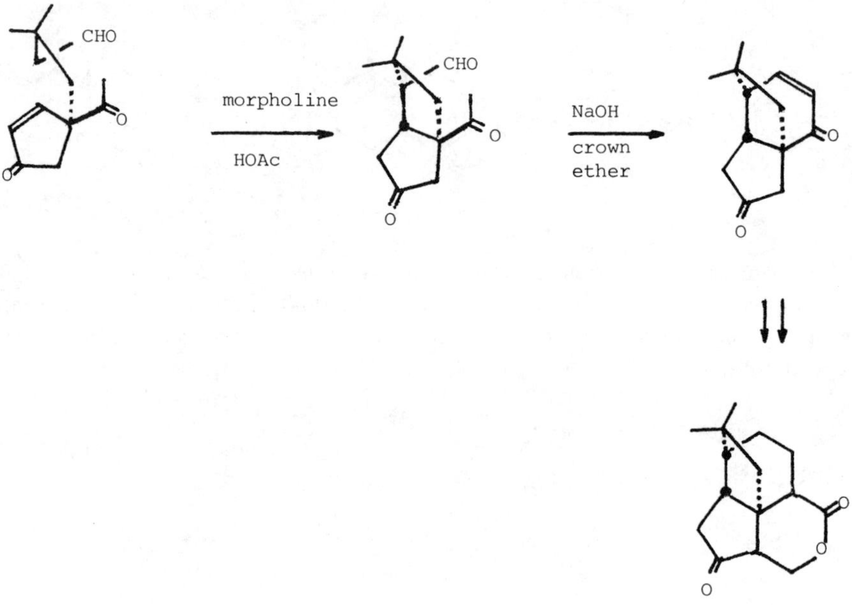

Q1

The construction of a bicyclic enedione synthon for coriolin (**C70**) (Danishefsky, 1980c,d) has been delegated to a two-step Michael–aldol process.

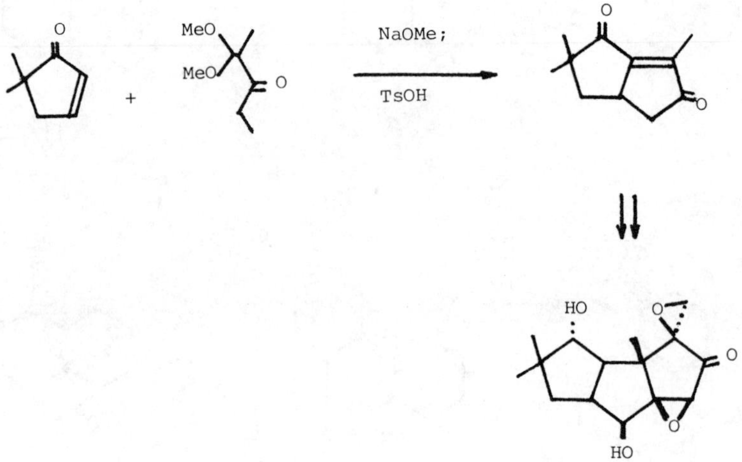

C70

4.3 STEPWISE MICHAEL–ALDOL CYCLIZATION

A search for new Michael acceptors containing significant structural features that may facilitate and shorten the synthesis of complex natural products has resulted in the development of two butenolides. They supply not only four carbon atoms to the incipient carbocyclic ring but also an intact γ-lactone moiety, which is present (perhaps with minor modifications in its alkyl substituents) in many sesquiterpenes. Their application in the synthesis of dihydrocallistrisin (**C11**) (Schultz, 1976; Godfrey, 1979) frullanolide (**F14**) (Kido, 1979a, b), paniculide A (**P9**) (Kido, 1981b), and furoventalene (**F21**) (Kido, 1982) illustrates the efficient 1,6-addition and the aldol condensation with an allylically activated donor atom.

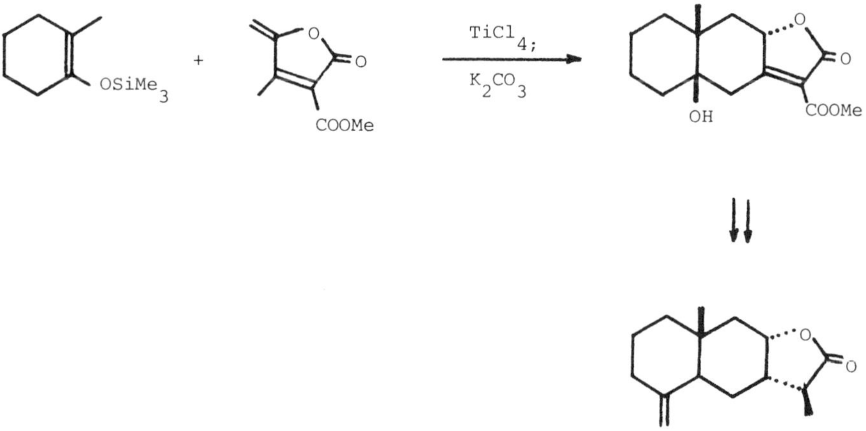

C11

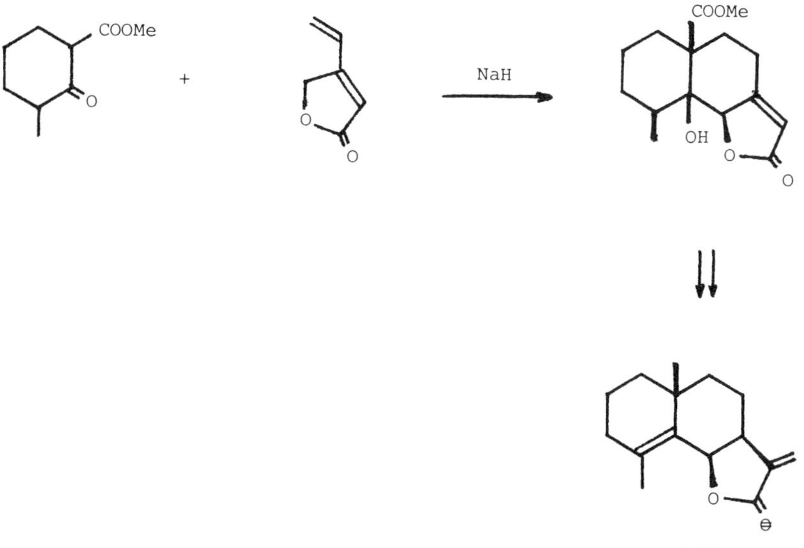

F14

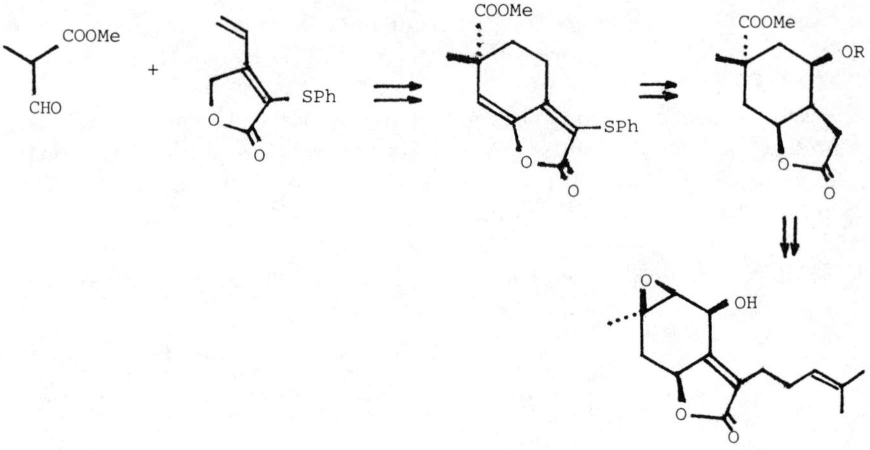

P9

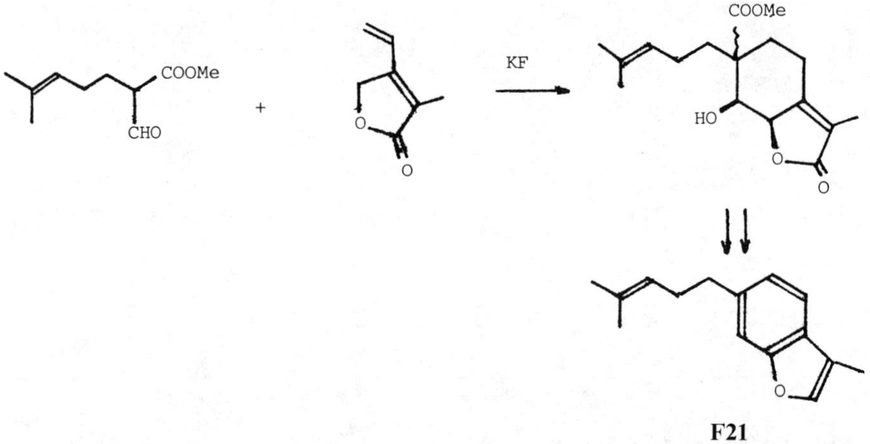

F21

The Michael condensation between a silyl enol ether and a conjugated carbonyl compound should find more widespread application (Huffman, 1985), since it complements the sometimes troublesome annulation protocol.

4.4 WICHTERLE ANNULATION

While the Robinson annulation of cyclic ketones most often gives rise to condensed ring systems, the Wichterle process proper (Wichterle, 1948) consists of alkylating ketone enolates with 1,3-dichloro-2-butene and treatment of the products with an acid such as concentrated sulfuric acid. The acid treatment releases a ketone group from the chloroalkene chain, and the diketones undergo aldolization in a sense favoring bridged ring systems. Nowadays, the bridged-ring formation process is considered to be the Wichterle reaction, regardless the origin of the diketones.

4.4 WICHTERLE ANNULATION

In this spirit, it seems proper to include the synthesis of helminthosporal (**H9**) (Corey, 1963b, 1965b), seychellene/cycloseychellene (**S29/S30**) (Welch, 1984), bulnesol (**B21**) (Marshall, 1968d, 1969b), and guaiol (**G20**) (Buchanan, 1971, 1973) under this topic.

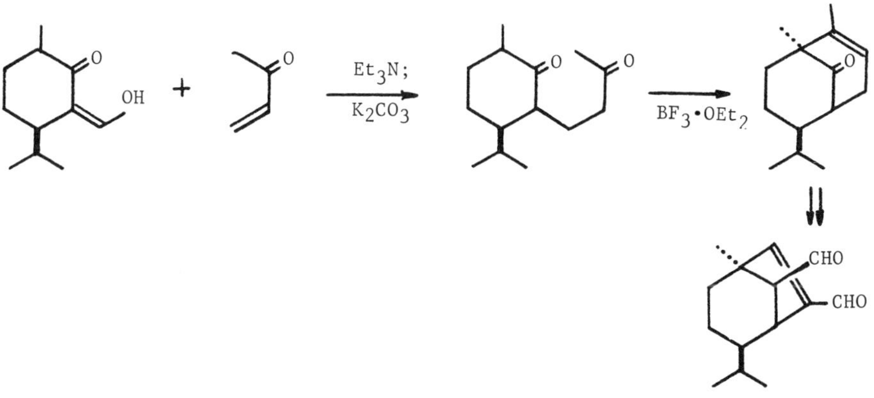

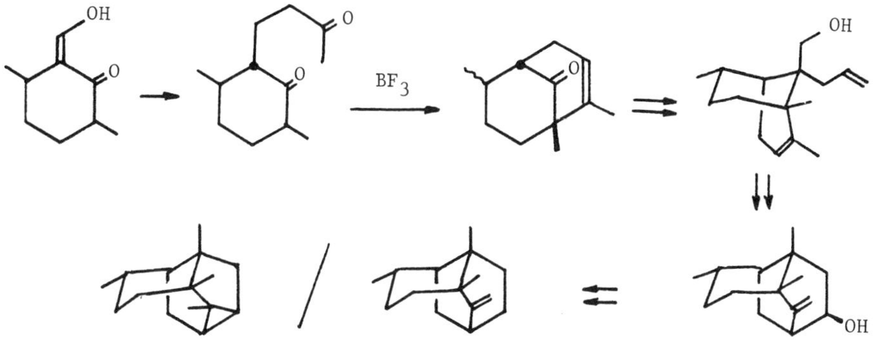

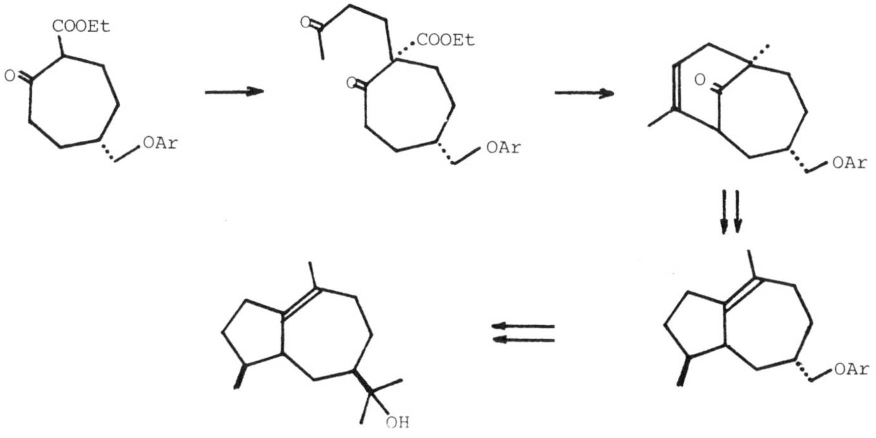

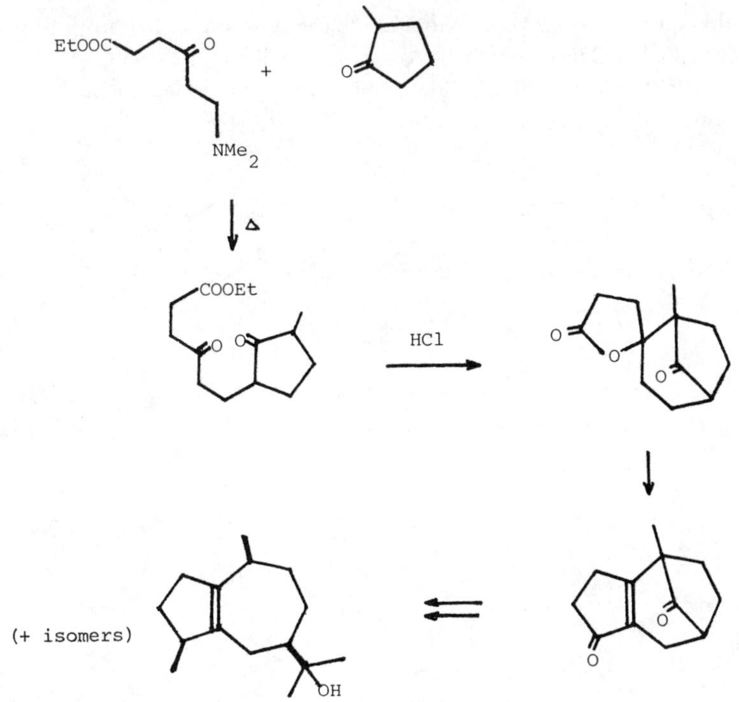

G20

It should be emphasized that certain structural characteristics may suppress the Wicherle reaction and favor the Robinson annulation under acidic conditions. A case in point is the formation of β-cyperone (**C97**) (Gammill, 1976) from the cyclohexenone. Apparently the stability of the doubly conjugated hexalone drives the reaction toward the result observed.

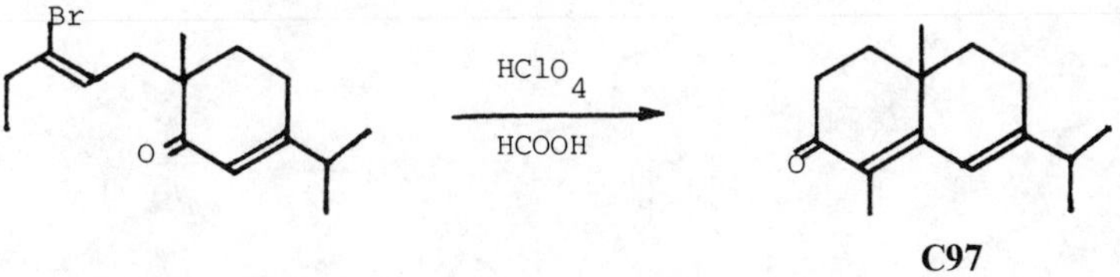

C97

4.5 INTRAMOLECULAR MICHAEL ADDITION

Intramolecular Michael addition could lead to ring structures. Among sesquiterpene syntheses that have employed this process, the best-known example is perhaps that of longifolene (**L24**) (Corey, 1961, 1964c). Despite the low yield (reversible reaction!), the formation of the bridged tricyclic diketone from a homooctalindione ensures the success of the endeavor.

L24

The titanium(IV)-catalyzed addition of siloxyalkenes to enones to form 1,5-diketones constitutes a key step of a short synthesis of seychellene (**S29**) (Jung, 1981).

S29

In most synthetic approaches to cedrene (**C37**), the *gem*-dimethyl group is present in the early intermediates. This feature limits many options. The replacement of this isopropylidene subunit with a carbonyl group has enabled an application of the Michael reaction to close the six-membered ring (Horton, 1983, 1984).

C37

In a stereoselective preparation of the crucial tricyclic intermediate for gibberellic acid (**G12**), an intramolecular Michael addition leads to a hydrindanedione in which the angular substituent is readily elaborated into a propargyl group for eventual closure of the D ring (Stork, 1979b).

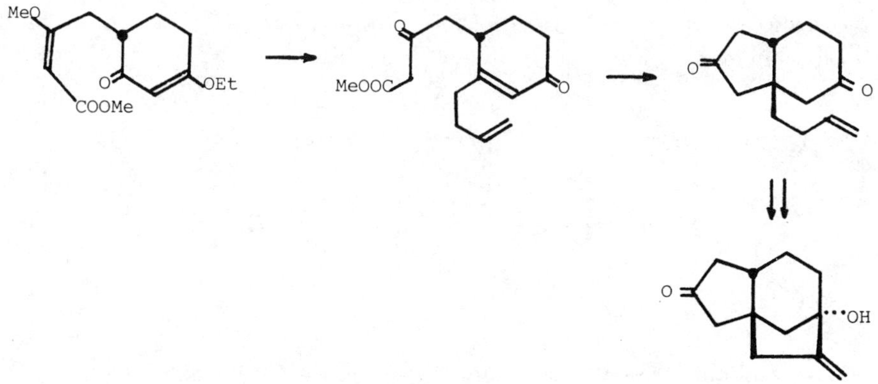

By virtue of a Michael reaction, a bridged ring system has been built to help establishing all the asymmetric centers at the ring junctures and the most inaccessible carboxylated carbon atom of hirsutic acid C (**H19**) (Trost, 1979).

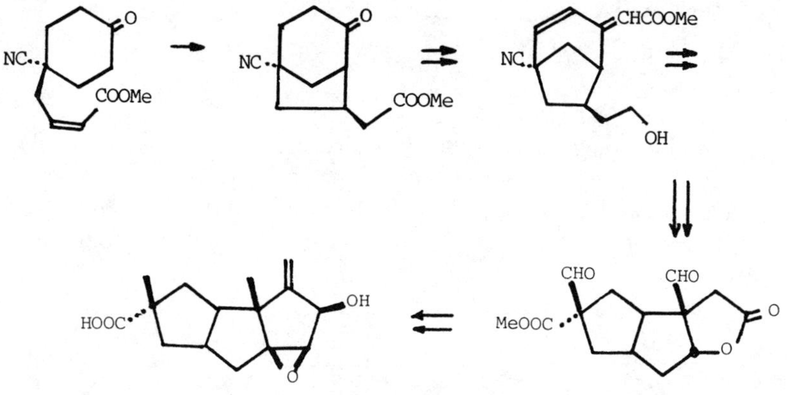

H19

Based on a biogenetic consideration of the iridodials (**I5**) that predicates their origin from 10-oxocitronellal, a synthesis has been carried out by acid treatment of the monoketal derivative (Clark, 1959). The ring formation step is an intramolecular Michael reaction. A similar synthesis of the dolichodials (**D18**) has also been realized (Cavill, 1964), although the major product (trans, trans isomer) does not occur in nature.

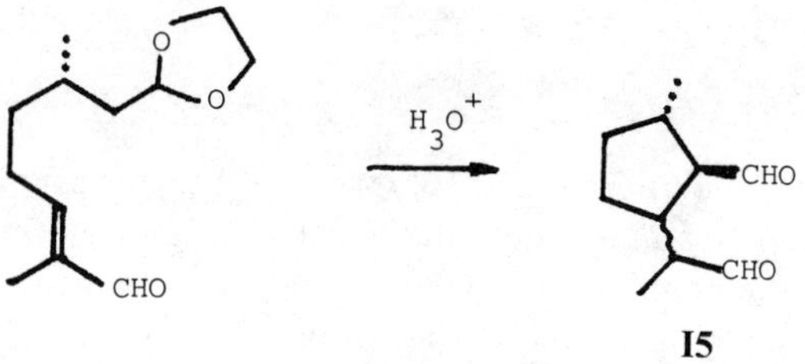

I5

4.5 INTRAMOLECULAR MICHAEL ADDITION

Cyclization of ω-oxocitral is successful only with dilute alkali. A mixture of chrysomelidial (**C54**) and dehydroiridodial (**I6**) is produced (Bellesia, 1986).

The mold metabolite pleuromutilin (**P39**) possesses an intriguing structure that may be considered to be a hydrindanone bridged by a cyclooctane unit. Not only does the assemblage of such a framework provide an enormous challenge, the eight stereocenters must be meticulously addressed during any synthetic assault.

An elegant approach (Gibbons, 1982) to this diterpene molecule is based on the Michael and aldol processes and a modified retroaldol reaction. Thus the hydrindanone precursor, which contains an additional bridge and side chain for elaboration of the cyclooctane subunit, is created via a reflexive double Michael condensation initiated by the kinetic enolate of a 3-alkoxy-2-cyclohexenone. Four stereocenters are established in this step.

Equally engaging is the formation of the atisirane skeleton in an AD → ABCD fashion using a reflexive Michael reaction (Ihara, 1985). The stereoselectivity of the process is endowed by a metal-chelated transition state from which a product with three relevant asymmetric carbon atoms at the ring junctures in a correct relative configuration is generated. While the ester pendant is crucial for the double annulation, it is subsequently removed. The resulting atisiran-15-one is an apparent precursor of *ent*-isoatisirene (**ent-A50**), the optical antipode of a natural diterpene.

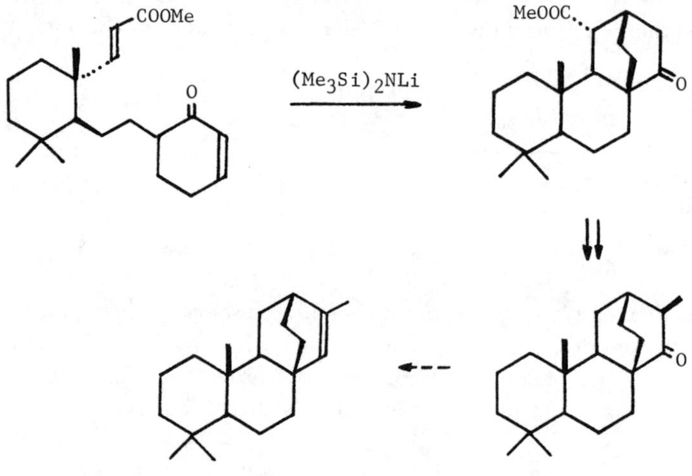

The reflexive Michael cyclization is the key step for a synthesis of pentalenic acid (**P17**) (Ihara, 1987a). The major tricyclic diketone product contains four chiral centers corresponding to the terpene molecule. It requires only a ring contraction and redox maneuvers to reach the target.

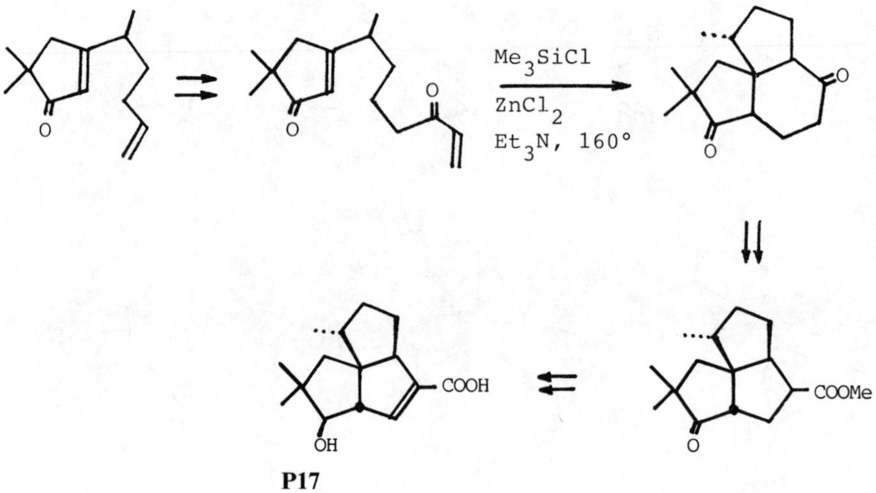

Reflexive Michael addition of 3-alkoxy-2-cyclohexenone with ethyl sorbate sets up a bicyclo[2.2.2]octanone system in which another Michael closure is later induced during a synthesis of sanadaol (Nagaoka, 1987). Ethyl sorbate actually contributes three carbon atoms to the skeleton of the diterpene; the propenyl group is to be degraded, leaving behind a carbon center modifiable to an electrofugal state. Subsequent fragmentation unveils the desired carbon framework, and thereafter only cosmetic trimming of functionalities suffice to complete the synthesis.

An interesting aspect of the approach concerns the Michael closure. Stereoselective formation of the tricyclic intermediate (6:1 at −20°C) is due to the absence of steric and electrostatic interaction between the enolate and the ester group in the transition state leading to the desired product.

sanadaol

The Lewis acid-catalyzed reaction of 2-silyloxy-1,3-butadienes with 3-buten-2-one is a formal Diels–Alder reaction, yet evidence indicates a reflexive Michael pathway. This process has been used in a synthesis of ε-cadinene (**C5**) (Hagiwara, 1987a).

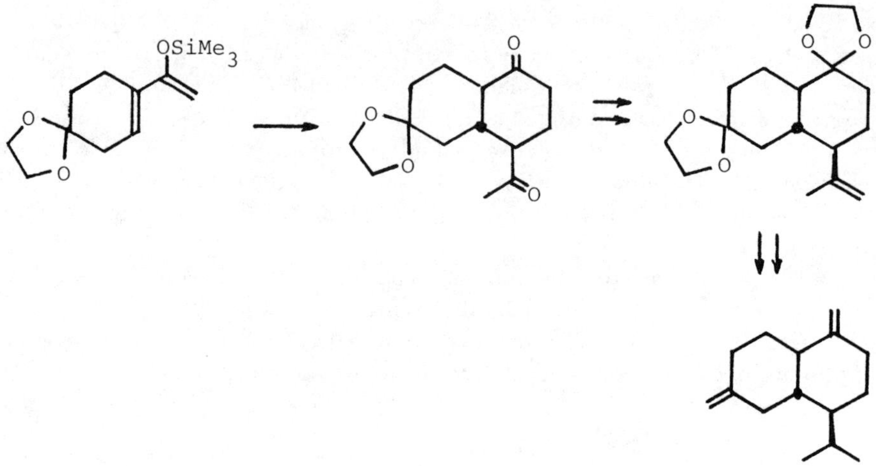

The double Michael addition of dimethyl acetonedicarboxylate to 2,3-dimethyl-4-methylene-2-cyclohexeneone conveniently affords a highly functionalized decalindione diester. The mandatory cis ring juncture produced by this process on account of the intramolecular nature of the second addition step favors the cis arrangement of the vicinal methyl groups. Consequently, furanoeremophilane (**E9**) is easily realized from this intermediate (Irie, 1978).

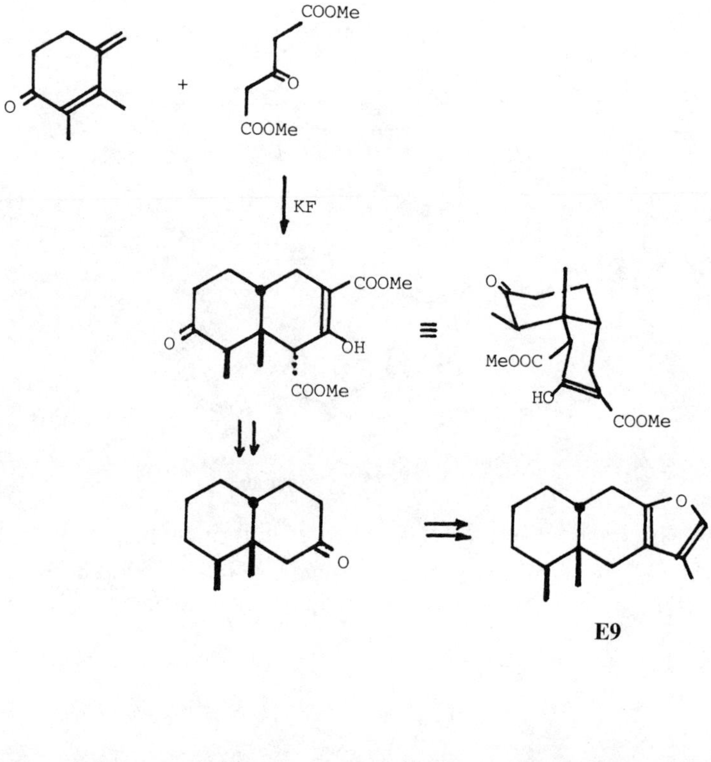

4.5 INTRAMOLECULAR MICHAEL ADDITION

The formation of three C—C bonds in one chemical operation is an exciting prospect. Ishwarane (**I11**) seems to be the unique molecule that enjoys the high honor of being synthesized by two triple-coupling methods. One of these routes is highlighted by a tandem reflexive Michael addition followed by an internal alkylation reaction. The first reaction must be initiated by a kinetic enolate ion (Hagiwara, 1979).

I11

A Lewis acid-catalyzed triple Michael reaction between 1,2-dimethyl-3-trimethylsiloxy-1,3-cyclohexadiene and 1,4-pentadien-3-one furnishes the diketone intermediate of seychellene (**S29**) (Hagiwara, 1985).

S29

Another efficient bimolecular condensation that makes two rings in one operation has great potential for synthesizing aflavinine (**A14**) in a few steps (Danishefsky, 1984). While noraflavinine has been obtained, the presence of a secondary methyl group in the cyclizing chain forces a conformational change that results in a product with a trans juncture between the two new rings. Some device to maintain a chair conformation of the initially formed cyclohexane is imperative if this strategy for acquiring aflavinine is to be deployed successfully. The key processes consist of double Michael and Reformatsky reactions.

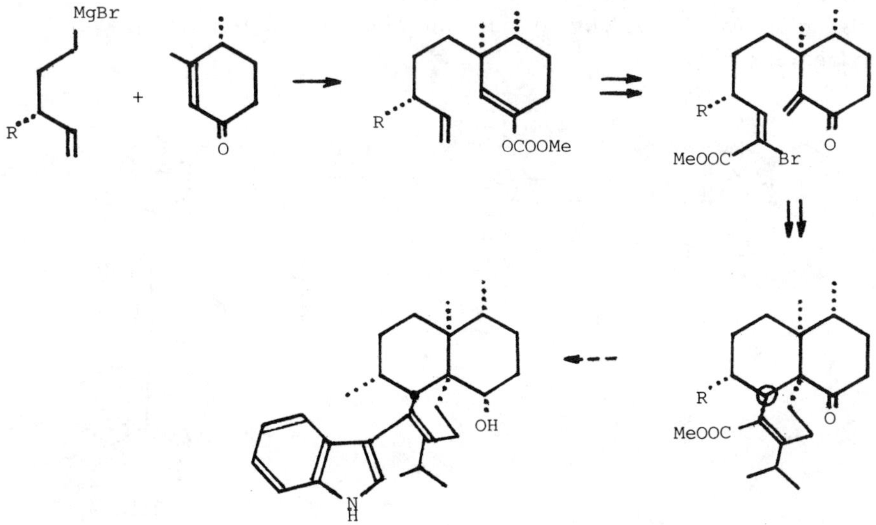

A14

R=H cis
R=Me trans

The reaction of α,β-unsaturated carbonyl compounds with ylides to yield cyclopropane derivatives actually involves a tandem Michael and cycloalkylation process. For the synthesis of chrysanthemic acid (**C51**), ylide reagents include ethyl dimethylsulfuranylideneacetate (Rama Rao, 1984), diphenylsulfonium isopropylide (Corey, 1967), and isopropylidenetriphenylphosphorane (Sevrin, 1976; DeVos, 1976, 1983; Mulzer, 1983).

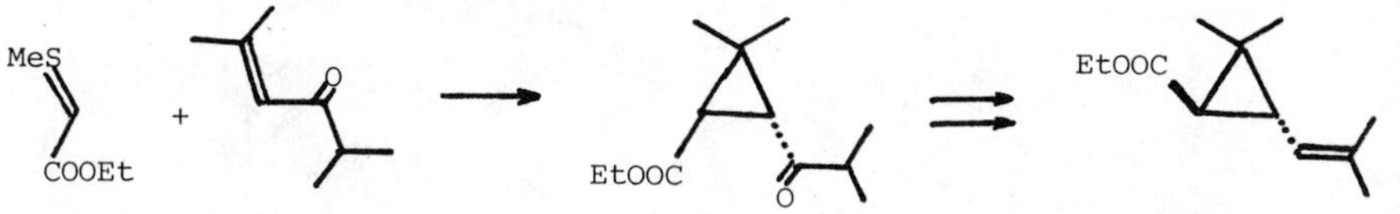

C51E′

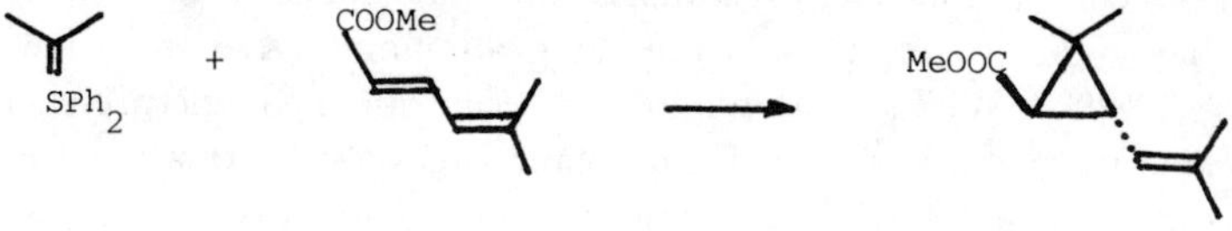

C51E

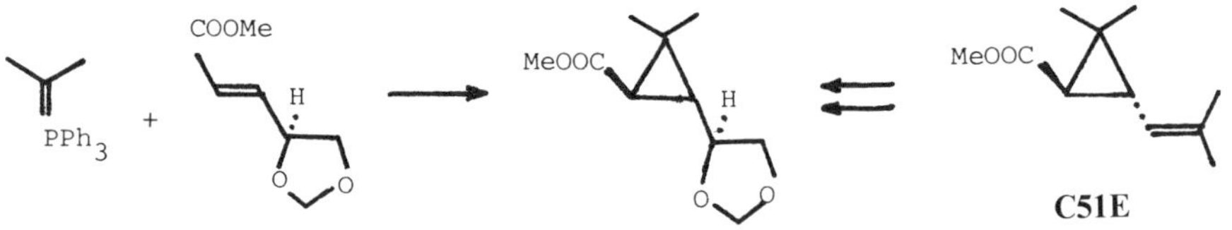

4.5 INTRAMOLECULAR MICHAEL ADDITION

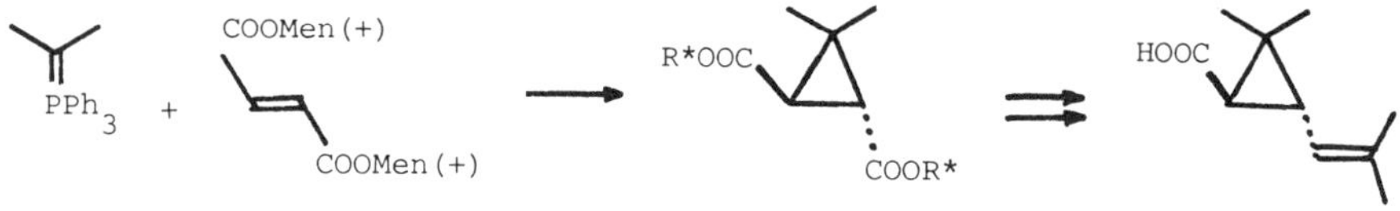

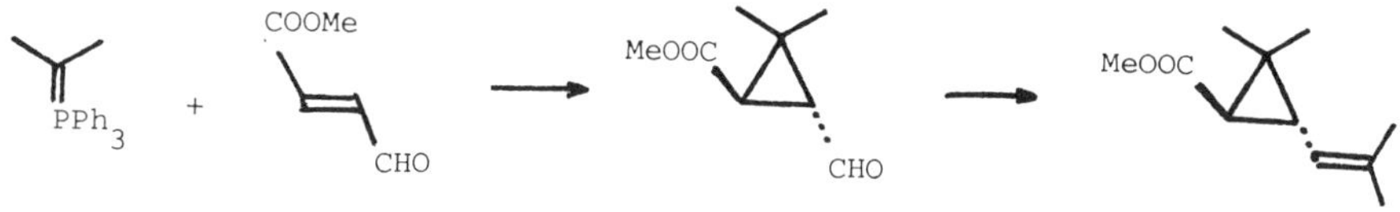

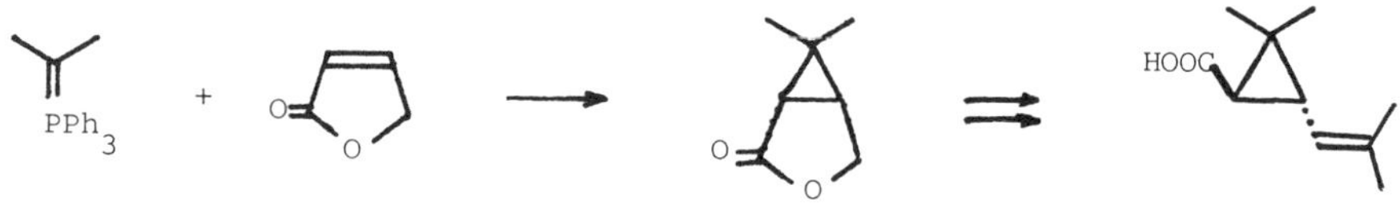

The sulfonyl and nitro groups acidify the hydrogen atoms at the α carbon and under certain conditions they act as leaving groups also. Consequently, cyclopropane rings may be formed by the Michael–cycloalkylation sequence, as demonstrated in some simple routes to the chrysanthemic acid-type substances (M. Julia, 1967; Martel, 1967; Babler, 1985; Krief, 1985).

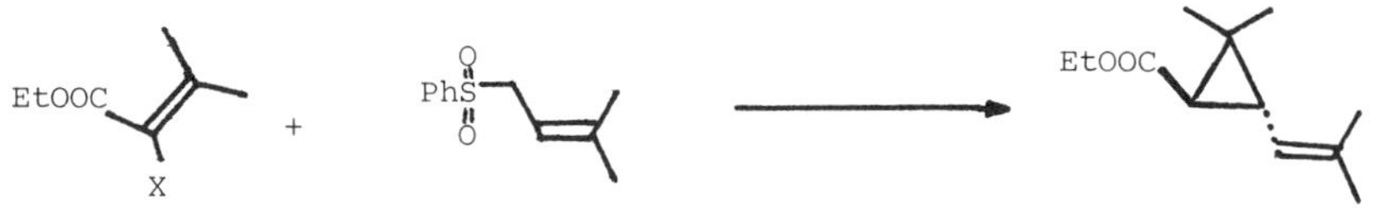

X= H; COOEt

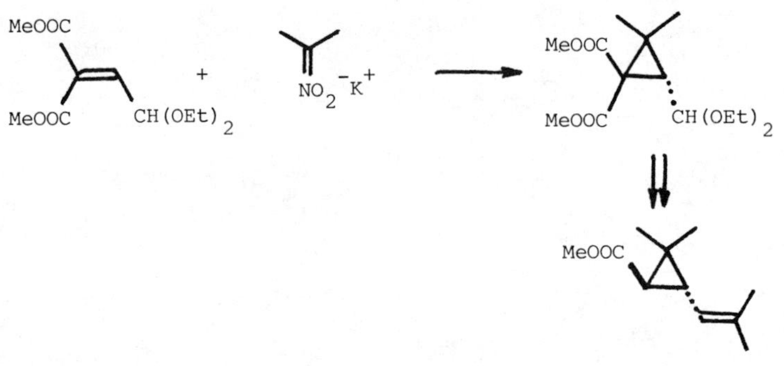

C51E

The elusive biointermediate between farnesol and squalene, called presqualene alcohol (**P48**), has been synthesized also by a base-catalyzed condensation of E,E-farnesyl phenylsulfone with ethyl E,E-farnesate, followed by hydride reduction (R. V. M. Campbell, 1971). The cyclopropanation step involves a Michael addition and 1,3-elimination.

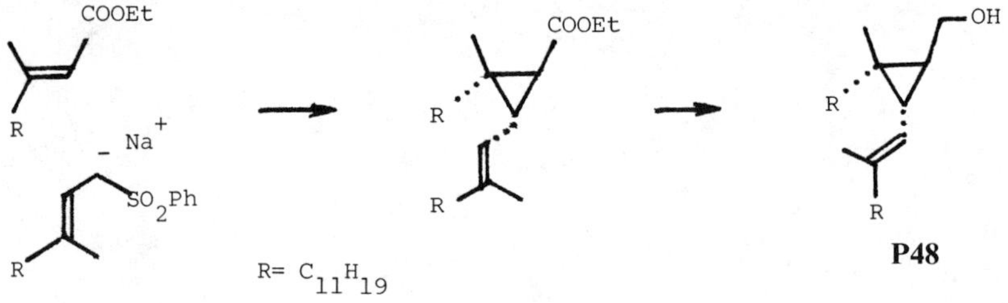

P48

Another variant of the Michael–cycloalkylation process operates when the Michael acceptor contains a leaving group at the γ position (DeVos, 1979). The cyanide ion, acting here as the donor, not only adds to the β carbon of the alkylidenemalonate faithfully, but also mediates a decarbomethoxylation.

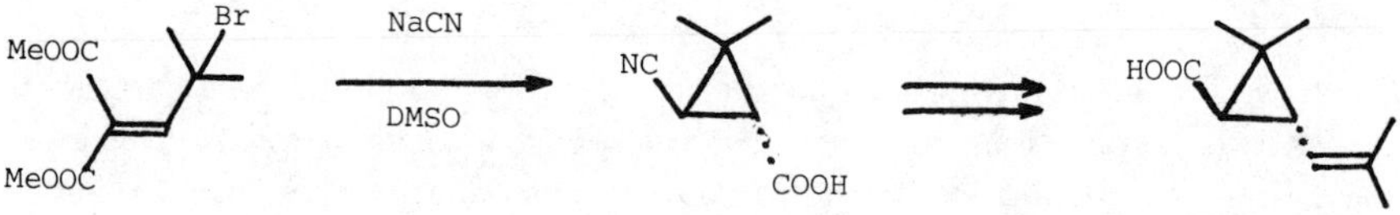

C51

When the Michael acceptor also contains a potential leaving group at the α position, cyclopropanation can be achieved in one step. For the preparation of a chrysanthemic acid precursor, see Babler (1981).

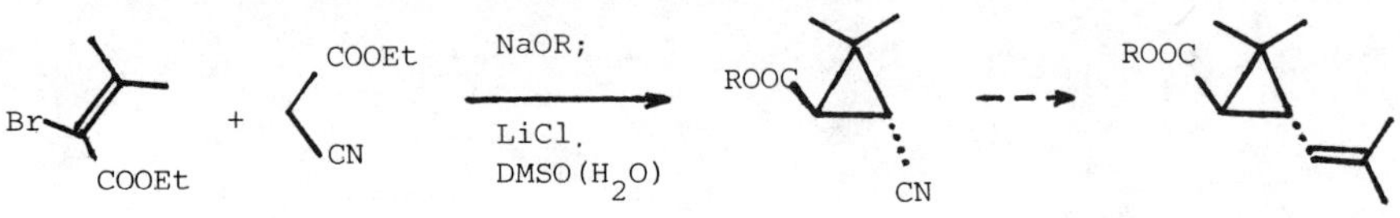

C51

4.5 INTRAMOLECULAR MICHAEL ADDITION

The reaction between an enolactone and an ynamine is intriguing. A dipolar intermediate is formed by the attack of the ynamine on the carbonyl group of the lactone, and the collapse of this intermediate leads to a β-enaminocycloalkenone. The utility of this reaction in synthesizing acoradiene III (**A6**) (Ficini, 1981) has been reported.

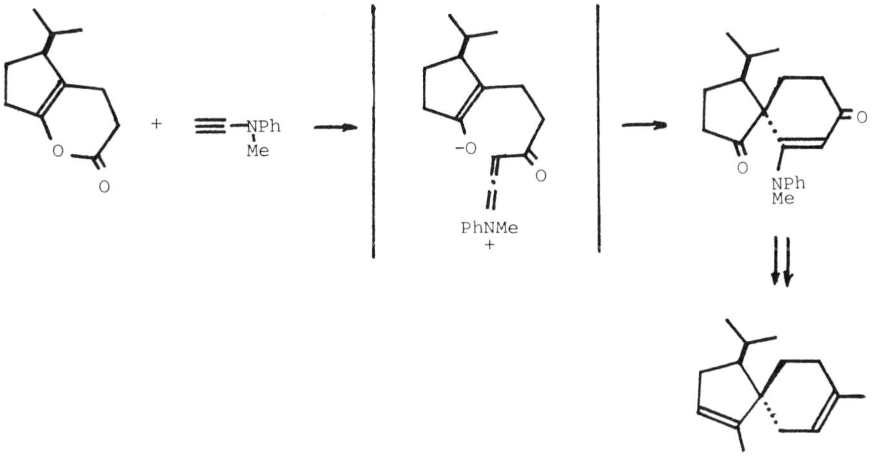

A bicyclo[4.3.1]decenone that is expected to be converted into vernolepin (**V10**) (Harding, 1981) has been prepared starting from an acylation–Michael addition tandem between an enamine and a γ-oxygenated crotonyl chloride. This approach is interesting in that three of the five asymmetric centers around the cyclohexane ring are readily controlled.

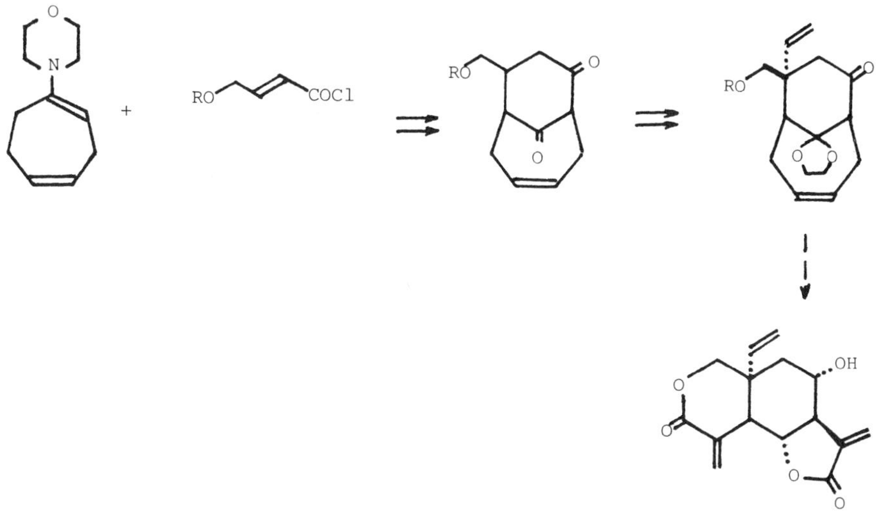

Aromatic compounds become much more electrophilic on complexation with transition metal species. Remarkably, the addition of carbanions to arene tricarbonylchromium complexes shows a regioselectivity in favor of forming the new C—C bond at the meta carbon atom to an existing alkoxy substituent. The net result of this reaction (after proper demetalation) is the same as achieving an internal Michael addition to the highly unstable, and hence mostly unavailable, 4-unsubstituted 2,5-cyclohexadienones. The scope of this reaction has been extended to the synthesis of spirosesquiterpenes such as acorenone (**A10**) and acorenone B (**A10b**) (Semmelhack, 1980b).

4.6 MICHAEL–CLAISEN AND WITTIG CONDENSATIONS

The combination of a Michael addition with another base-catalyzed reaction leading to cyclic products has frequently been exploited. Pertinent to the present discussion is the Michael–Claisen route to 1,3-cyclohexanediones. Since such compounds are easily transformed into cyclohexenones via the monoalkyl enol ethers, bilobanone is a prime candidate for use of the method (Escalone, 1980). The more difficult part of the synthesis is the preparation of the furanated enone.

bilobanone

The tandem Michael–Claisen condensations of α,β-unsaturated ketones with 3-(phenylthio)-1-(trimethylsiloxy)-1-methoxy-1,3-butadiene is a versatile method of annulation because the adducts are adorned with functionalities to allow modification and introduction of substituents almost at will. The successful synthesis of aristolone (**A40**) and fukinone (**F15**) (Prasad, 1987) from *trans*- and *cis*-octalindiones, respectively, shows its potential. The synthesis of fukinone is aided by easy access of the *cis*-fused diketone but somewhat marrred by the stereorandom hydrogenation of the trisubstituted double bond of a later intermediate. Introduction of the isopropylidene unit is greatly facilitated by the presence of the conjugated double bond.

A40

The application of the Wittig reaction to ring formation has been studied in recent years. A particularly interesting version involves the reaction of an enone with a vinylphosphorane, which results in a 1,3-cyclohexadiene. A facile synthesis of occidol (**O3**) (Dauben, 1973) presents itself.

A new preparation of ethyl safranate used in a synthesis of β-damascenone (**D2**) (Büchi, 1977) embodies a slightly different condensation.

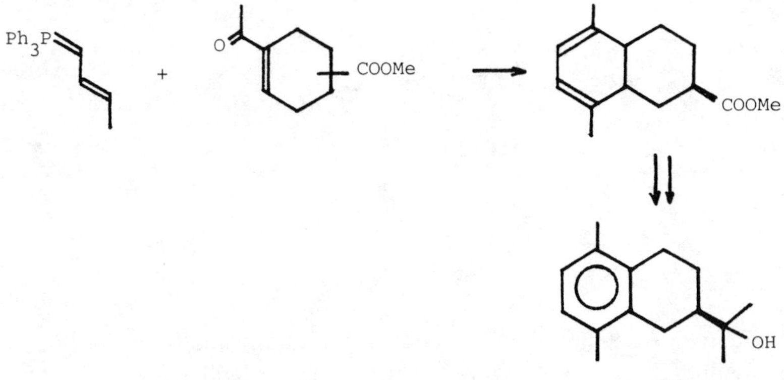

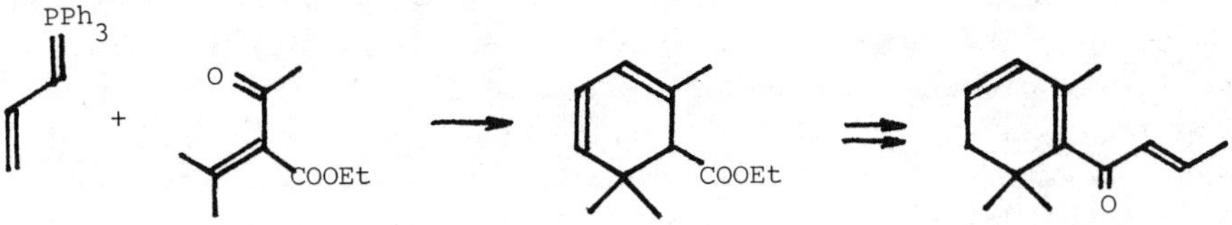

Reaction of vinylphosphonium salts with nucleophiles furnish Wittig reagents that may be trapped by properly situated carbonyl groups intramolecularly to give cyclic products. Thus, functionalized diquinanes have been generated as synthons for chrysomelidial (**C54**), loganin (**L20**), and hirsutene (**H17**), (Hewson, 1985).

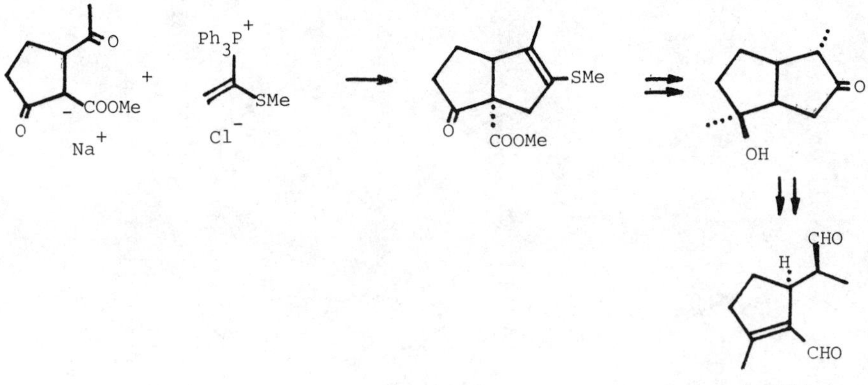

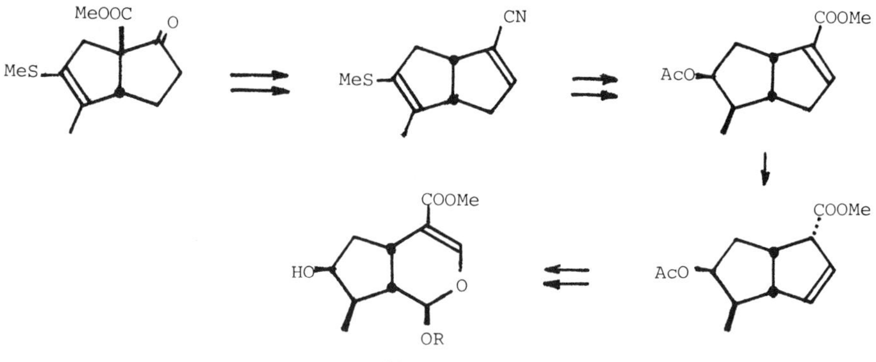

4.6 MICHAEL–CLAISEN AND WITTIG CONDENSATIONS

L20

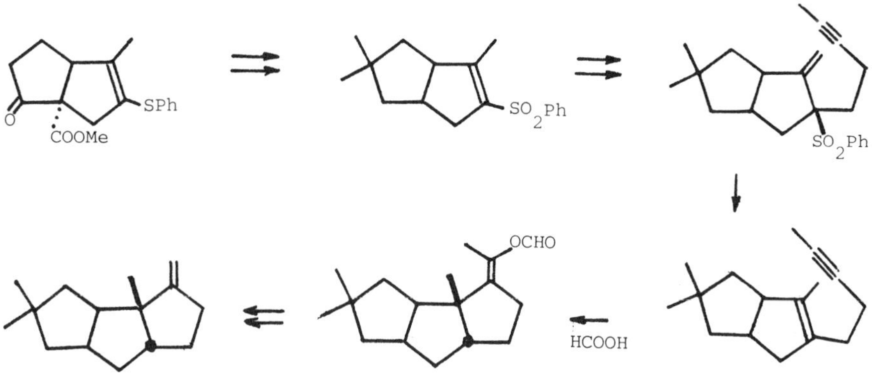

17

Cyclopentannulation has been effected by condensation of a nascent γ-oxo ester α-anion with a vinylphosphonium salt. The nucleophilic species can be obtained from β-(silyloxy)cyclopropyl esters.

Synthesis of pentalenolactone E methyl ester (**P19**) (Marino, 1987) demonstrates the promising reaction. All proper functionalities are incorporated in this step.

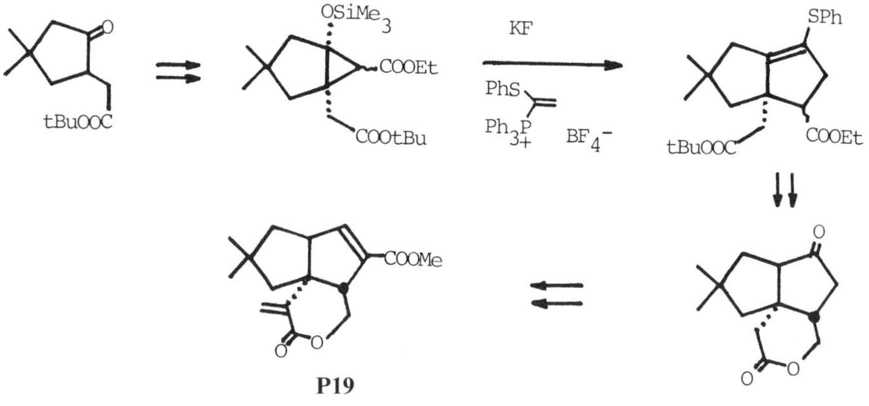

P19

The annulation step in a synthesis of spirovetivanes including hinesol (**H15**), α-vetispirene (**V14**), β-vetispirene (**V15**), β-vetivone (**V16**), and anhydro-β-rotunol (**R8**) (Dauben, 1975b, 1977b) can be classified mechanistically as a homo-Michael addition/Wittig combination. The scavenger of the Wittig reagent is the formyl group.

R8 **V16** **V15** **V14** **ent-H15**

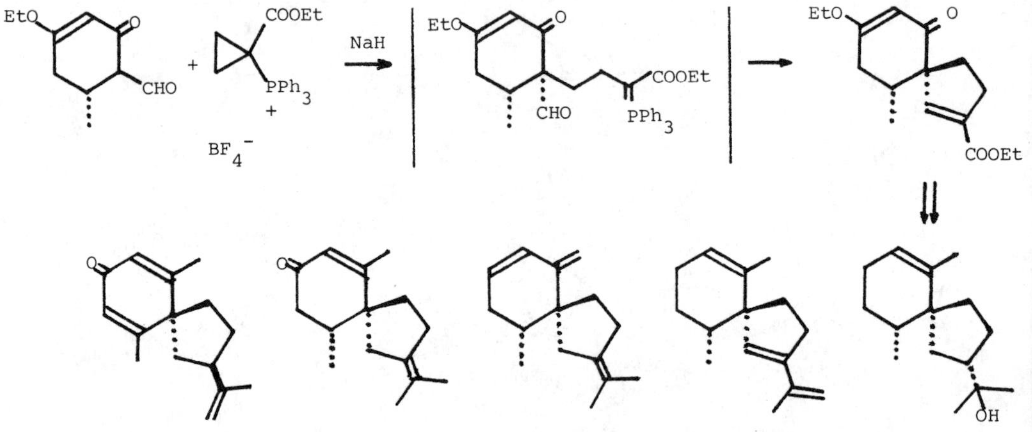

In a model study aimed toward synthesis of the ophiobolins (Dauben, 1977a), annulation of a five-membered ring onto an existing 5:8-fused bicyclic compound takes advantage of the β-ketoester function that evolves from a ring expansion operation (enamine + acetylenedicarboxylate) and subsequent hydrogenation.

C44

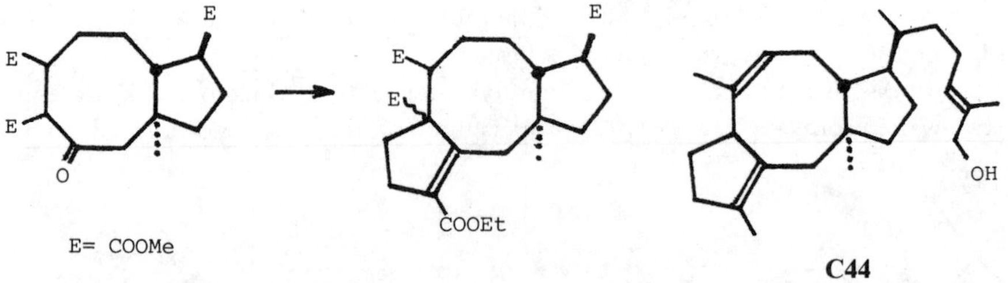

A one-step preparation of the spirocyclic precursor of acorone (**A11**) (Altenbach, 1979) from 4-methyl-3-cyclohexene-1-carboxaldehyde comprises a tandem alkylation–Wittig pathway. Unfortunately, the yield is only 14 %.

A11

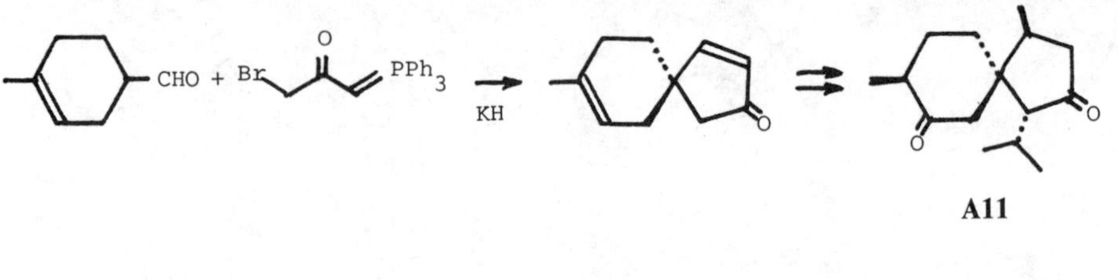

4.7 CLAISEN REACTIONS

Although the Claisen condensation generally denotes the β-ketoester formation process (Hauser, 1942) by self- or cross-reaction between two ester molecules that may be identical or different, Claisen was also the first chemist to effect the synthesis of a β-diketone (from acetophenone and ethyl benzoate). Therefore we should also designate the β-diketone synthesis as a Claisen reaction.

The Claisen reaction is closely related to the aldol process. The difference is the oxidation state of the acceptor carbonyl carbon. When it is desirable to retain two oxygenated carbon atoms or when there exist other restrictions in acquiring the aldol precursors, the Claisen reaction is the prime alternative method.

The monocarbocyclic antitumor sesquiterpene lactone vernolepin (**V10**) contains two oxygenated carbon atoms in a cyclohexane ring. The 1,3-relationship of these carbon atoms is most easily correlated with an aliphatic ketoester by means of a Claisen reaction. 5-Carbethoxy-5-vinyl-1,3-cyclohexanedione, despite its simplicity, contains all the strategic oxygen atoms present in the complex terpene molecule and has been successfully elaborated into the latter (Kieczykowski, 1978).

The analogous 5-methyl-5-vinyl-1,3-cyclohexanedione is the foundation of curzerenone (**C88**) (Miyashita, 1981). The missing isopropenyl group and the furan ring are to be attached to the six-membered ring at the activated positions to complete a synthesis of the elemane sesquiterpene.

While the symmetrical 1,3-cyclohexanediones are exploited to great advantage as shown in the two preceding examples, it must be emphasized that when differentiation of the two oxygenated carbon atoms is desired, there is a more serious drawback associated with the Claisen reaction than with the aldol process. When the product does not possess a plane of symmetry, the possibility of obtaining a mixture of regioisomeric enols is real. In the light of this tendency, the approach to the drimanes (Wenkert, 1964) based on a Claisen reaction to elaborate the B ring is fortunate in that the major enol (84 %) is transformed into the Δ^8-7-one from which the side chain at C_9 is then introduced by hydrocyanation. Drimenin (**D20**), the first natural member of this series so reachable, has been further converted into isodrimenin (**D21**), confertifolin (**C61**), drimenol (**D22**), and others. It is noteworthy that drimanes having a chiral C_9 uniformly possess a β-oriented carbon chain. The equatorial configuration naturally facilitates synthesis of these terpenes.

4.7 CLAISEN REACTIONS

Perhaps it should be stated more clearly that in the enolalkylation or protection of a β-diketone, the less hindered carbonyl group reacts preferably in the enol form. This trend is also evident in 4-isopropyl-1,3-cyclohexanedione, which is a precursor of cryptone (**C79**) (Mukherji, 1963).

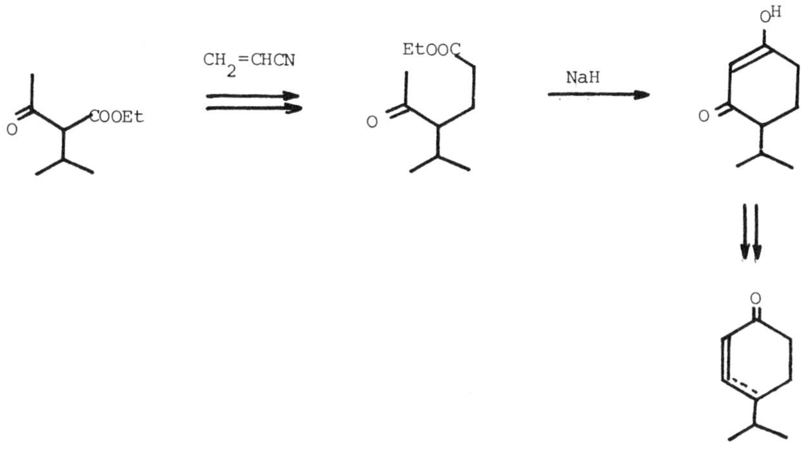

C79

The ready availability of a 5:5-fused γ-lactone from (+)-pulegone opens a way to (—)-acoradiene (**ent-A4**) (Solas, 1983). The synthesis hinges on a stereoselective alkylation from the convex face of the lactone and a Claisen reaction of a ketoester.

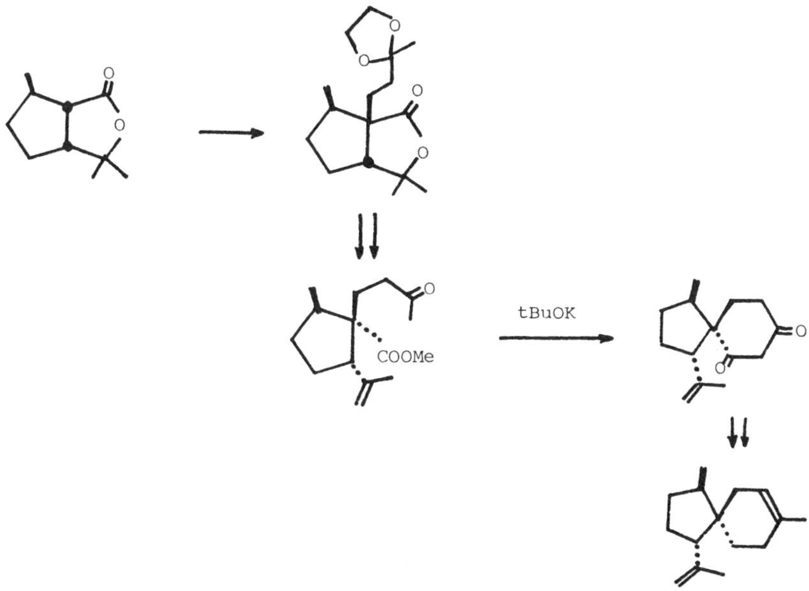

ent-A4

The difficulty in bridging the third ring of gymnomitrol (**G21**) by the rather obvious aldol approach has been circumvented by one group of chemists (Welch, 1979) by changing the cyclization procedure to a Claisen reaction. The two ketone groups of the product are easy to differentiate because only one of them can undergo enolization.

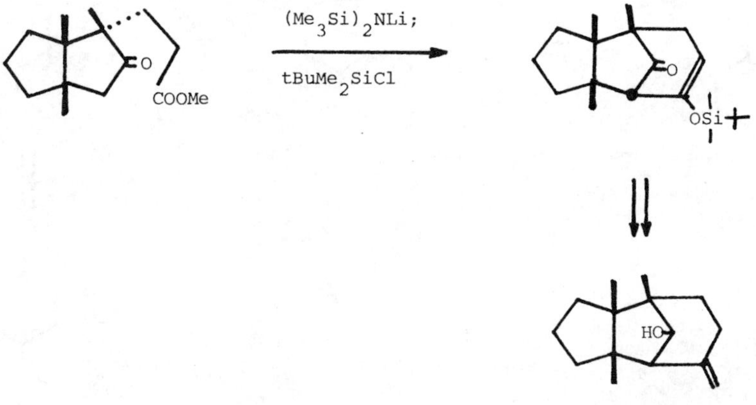

G21

In one of the routes to copacamphor (**C65**), ylangocamphor (**Y3**), and their corresponding alcohols (Piers 1971a, 1975a), the Claisen reaction is called into service. The rigid bicyclo[3.2.1]octanedione intermediates are readily maneuvered into the target sesquiterpenes.

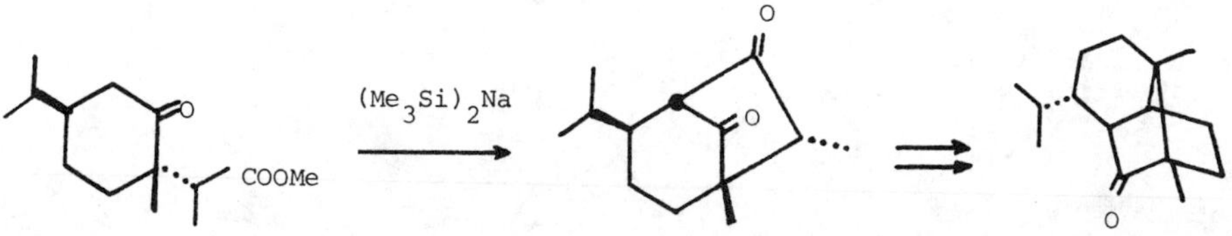

C65

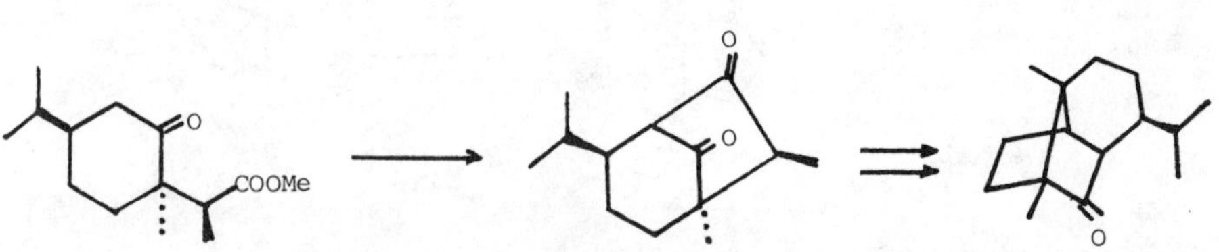

Y3

The diketone of culmorin (**C83**) is available synthetically from tetrahydroeucarvone (B. Roberts, 1969). The bridged ring is formed first using the

4.7 CLAISEN REACTIONS

Claisen reaction, and the remaining two-carbon unit is then spanned between the carbonylated bridges by conventional procedures.

C83k

In the first synthesis of cedrol (**C38**) (Stork, 1955, 1961a) a 5:5-fused ring system is constructed to supply functionalized chains for the eventual bridging of the six-membered ring. The advantage of this strategy lies in the steric disposition of the ring juncture of the bicyclic precursor. Since the configuration of the ester group is fixed, the epimerizability of the ketone chain has no effect on the cyclization. Furthermore, the active exo configuration of the ketone chain is favorably oriented for steric reasons.

The operation after the Claisen step is blessed by the subtle steric factor that favors enolization of the more exposed carbonyl group. Complex metal hydride reduction of the diketone results in the secondary alcohol directly as the main product.

C38

The trapping of the ketone enolate derived from an enone via conjugate addition by a sterically accessible ester group in the same molecule represents an expedient method for the synthesis of angularly substituted peri-dioxobicyclo-[*m.n.*0]alkanes. The value of this procedure can be judged by its gainly application to the eudesmanolides such as isoalantolactone (**A21**) and dihydrocallistrisin (**C11**) (Tada, 1982). In contrast to most other syntheses of these substances, the present approach obviates an alkylation step for the installation of the lactone unit. Also important to this route is the presence of carbonyl group(s) adjacent to the ring juncture, which permits rectification of the stereochemistry.

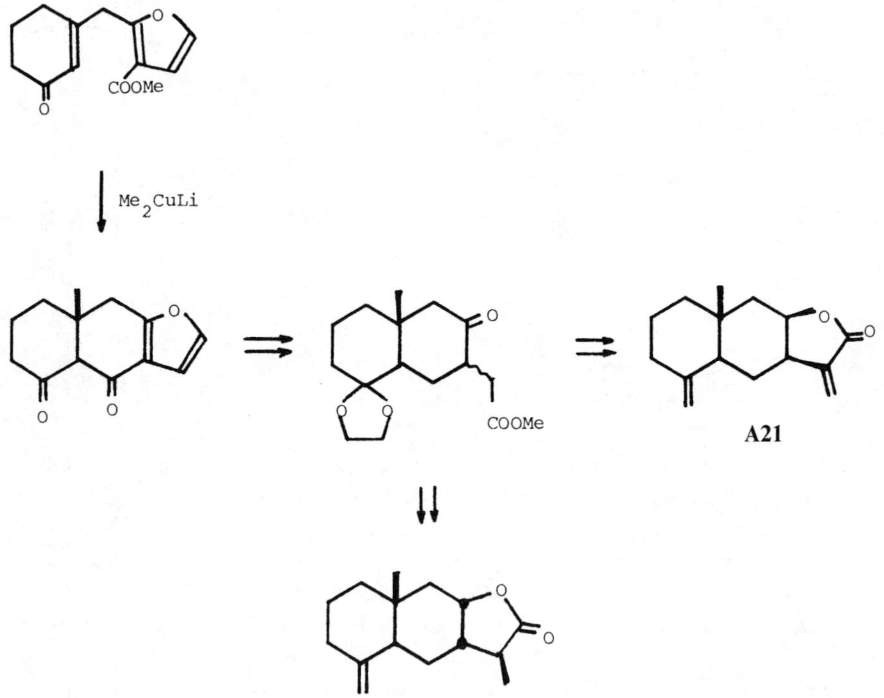

A remarkable process whereby a diquinane is constructed containing a vast majority of the carbon atoms of pentalenolactone (**P18**) embodies an allylic Claisen reaction (Parsons, 1980). At the annulation stage the nonangular tetrahedral carbon atom is subject to equilibration, which results in the desired diastereoisomer only.

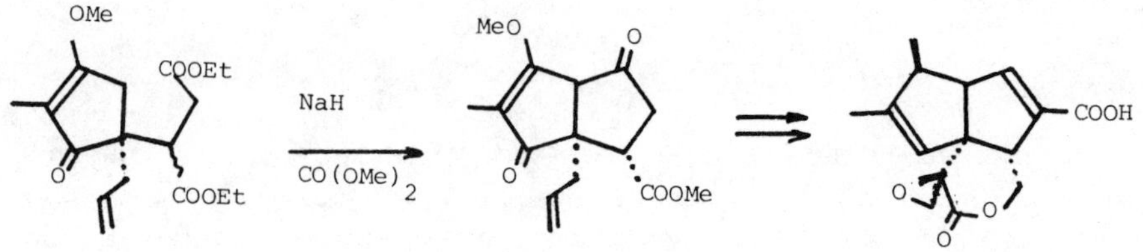

4.7 CLAISEN REACTIONS

The presence of a peripheral oxygen substituent in pentalenic acid (**P17**), and particularly its positioning in relation to the α,β-unsaturated acid unit, influences a synthetic design concerning the union of the two functionalized rings. The possibility of forming the spiro system via an intramolecular photocycloaddition–fragmentation protocol further suggests the Claisen reaction as the follow-up reaction. Since the head-to-head cycloaddition of an alkene to a cyclopentenone is favored by an electron-withdrawing group at the alkenic terminus, an α,β-unsaturated ester tethered by a three-carbon chain to the enone moiety fulfills the optimal structural requirement. The strained cyclobutane bond flanked on each side by a carbonyl group of the photoadduct is expected to undergo reductive cleavage selectively to unravel the spirocyclic subtarget.

Pentalenic acid has indeed been synthesized (Crimmins, 1984) according to this plan. To facilitate photocycloaddition and subsequent cyclobutane scission, the cyclopentenone portion is further polarized by an additional ester group at the α position in the actual synthesis.

Like the aldol condensation, the Claisen reaction is occasionally performed under acid catalysis. Examples from terpene synthesis include isocomene (**I14**) (Dauben, 1981), ligularone (**L13**), and furanoeremophilan-14,6α-olide (**E10**) (Tada, 1980).

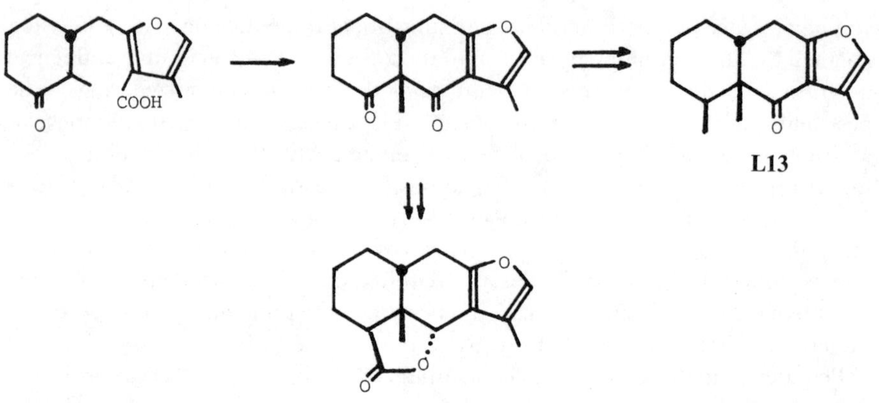

L13

E10

A synthesis of (+)-phyllocladene (**P27**) from (—)-abietic acid (Shimagaki, 1975) entails an acid-catalyzed Claisen reaction to form the D ring. The crucial stereochemical problem concerns with the configuration of the acetic acid side chain, which is resolved during hydrogenation of the aromatic ring. Hydrogen delivery occurs from the opposite side of the angular methyl group.

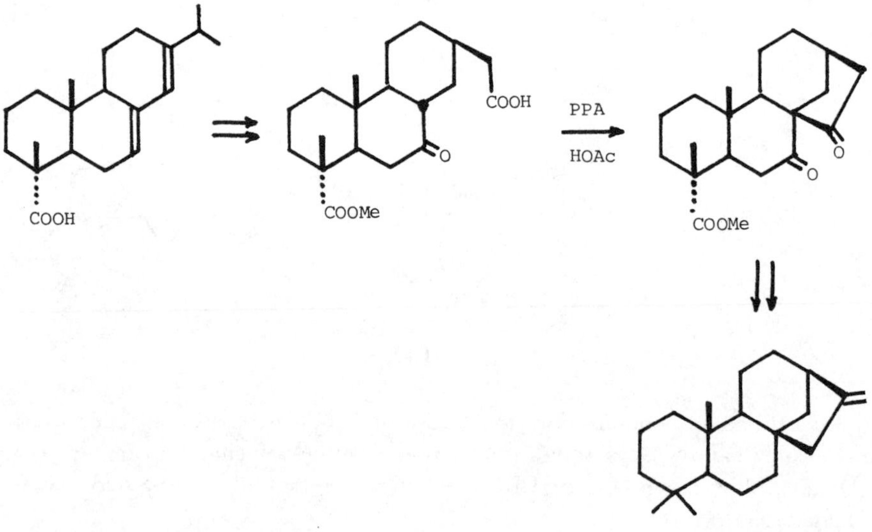

P27

In an approach to the Garrya alkaloids such as veatchine (**V5**) (Guthrie, 1966), one method for creating the bridged piperidine ring is to modify a cyclopentene, which is in turn obtainable from a ketone. A β-diketone precursor of the latter compound is the thermodynamic product of a Claisen reaction. The alternative bridged diketone can be isomerized to the useful material.

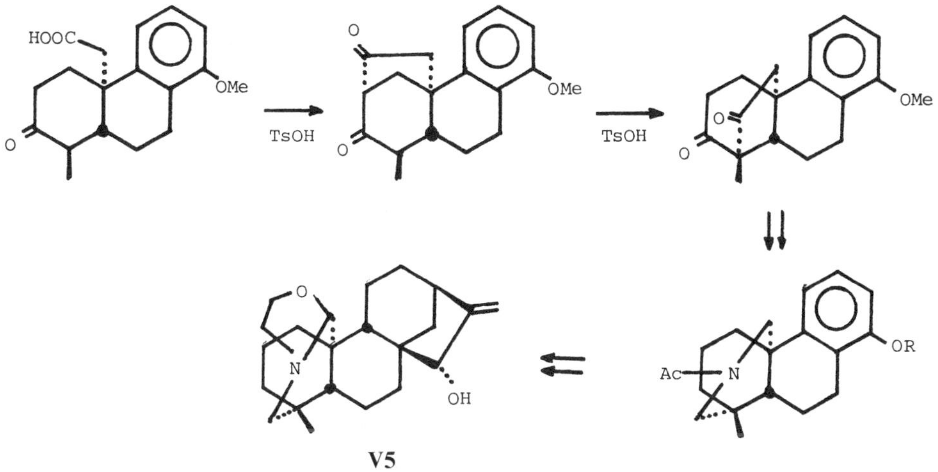

A photochemical version of the Claisen reaction is used as the key step in the conversion of pimaradiene into hibaene (**H11**) (Wenkert, 1973a). Thus the enolactone is isomerized into a bridged diketone upon exposure to light (254 nm) and then is reduced and acetylated. Pyrolytic elimination of one acetic acid molecule introduces the double bond, whereupon the inactive acetate group on the single carbon bridge is removed.

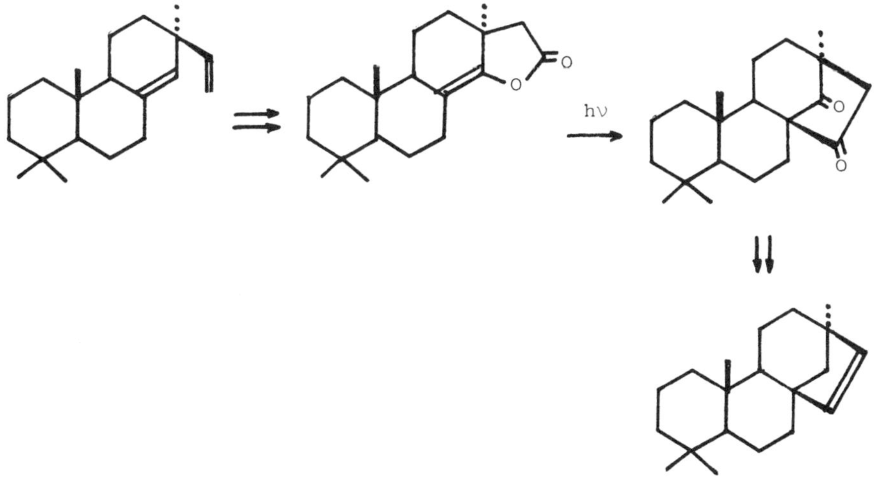

The condensation of an ester enolate with an aldehyde or ketone is perhaps more appropriately considered to be a Claisen reaction. Although in modern synthesis this condensation is superseded by the Reformatsky reaction (Rathke, 1975) or the Wittig reaction (Maercker, 1965) when an α,β-unsaturated ester product is desired, it is somewhat revived by the observation that acetic ester enolates can be generated using strong amide bases such as lithium diisopropylamide (LDA). However, the application of this new procedure is still quite limited.

Because of its proximity to a ketone group, the ester anion acquired by conjugate addition of isopropenylmagnesium bromide to the acrylic ester chain of a synthetic intermediate of coriamyrtin (**C69**) attacks the ketone and effectively closes the cyclohexane ring (K. Tanaka, 1982). Completion of the synthesis involves an inversion of configurations of two stereocenters and other routine transformations.

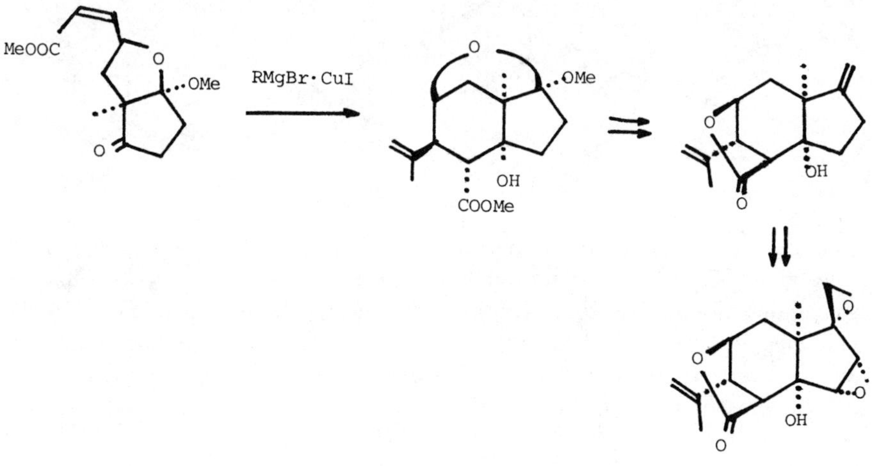

C69

Highly pertinent to the Claisen reaction is the reaction of an ester with an amide group under the influence of soda lime at high temperature. This method is incorporated into a synthesis of deoxynupharidine (**N14**) (Kaneko, 1959; Kotake, 1960).

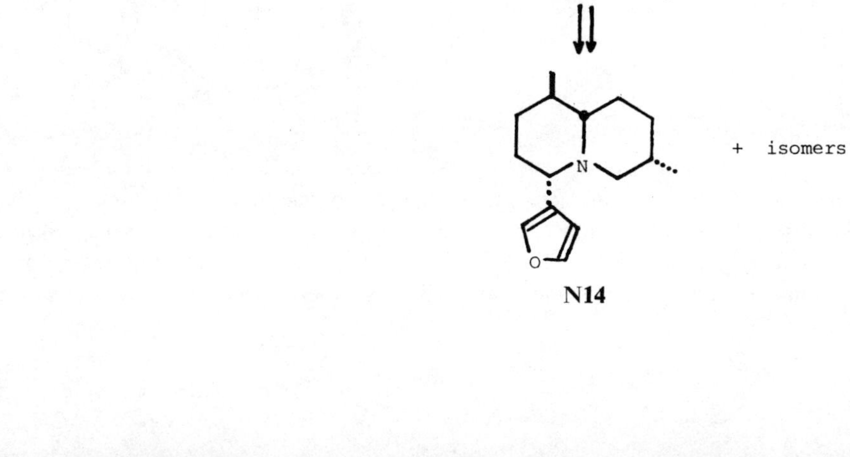

+ isomers

N14

An intramolecular N → C acyl transfer reaction constitutes the key step in a synthesis of isocaryophyllene (**C33**) and caryophyllene (**C32**) (Ohtsuka, 1984). The uncommon carbocyclization takes advantage of conformational restrictions that bring the reactive centers close together. It is noteworthy that in the synthesis of isocaryophyllene the ring juncture between the cyclobutane and the macrocycle is cis, whereas in the synthesis of caryophyllene, the *trans*-fused isomer is employed.

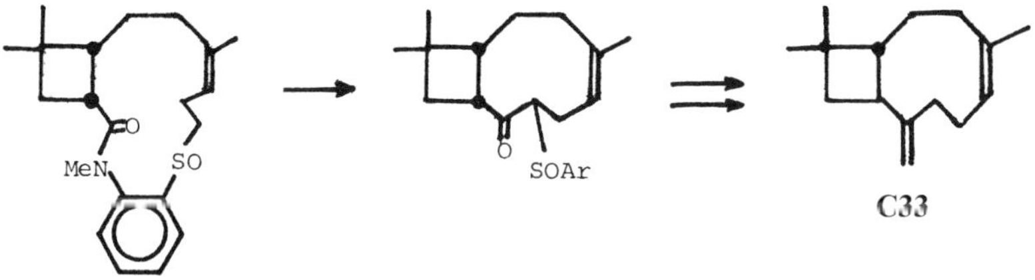

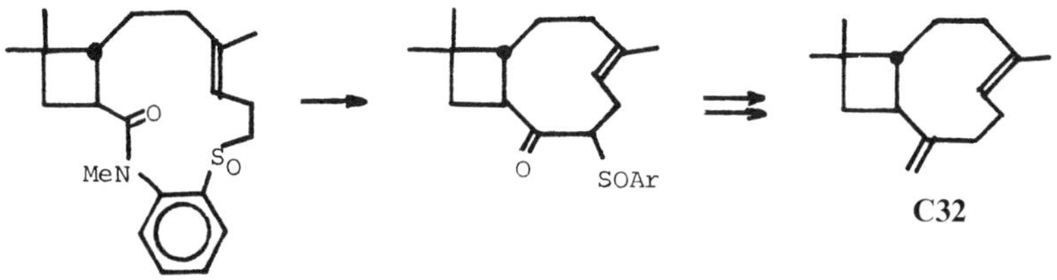

During an approach to eriolanin (**E14**) (M. Roberts, 1981), a substituted cyclohexenone is stereoselectively constructed via a bicyclo[2.2.2]octane intermediate by a retrograde Claisen reaction.

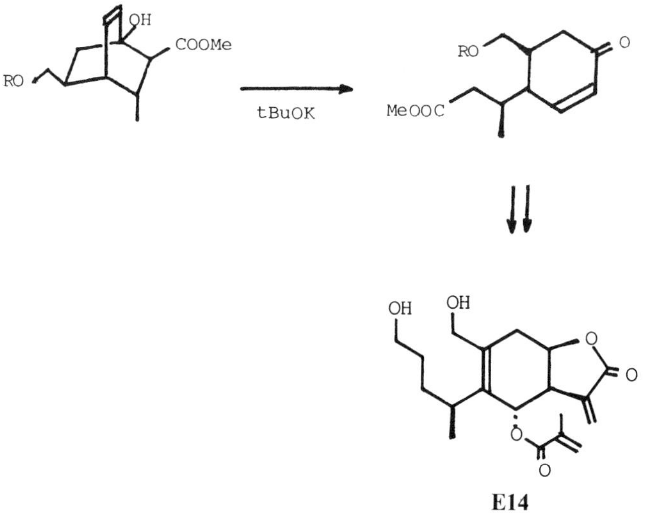

4.8 DIECKMANN AND RELATED CYCLIZATIONS

The internal Claisen condensation of an alkanedioic ester is known as the Dieckmann cyclization (Schaefer, 1967). This reaction remains an important method for making common-sized ring compounds because of its relative ease of operation and generally reliable yields, and the availability of starting materials. Accordingly, the Dieckmann cyclization lends itself well to the preparation of terpene intermediates.

An early work concerned with the synthesis of cuparene (**C84**) (Parker, 1962) consists of a 6 → 5 ring contraction step. The execution in fact involves the cleavage of a cyclohexanone and the recyclization of the diester by base treatment.

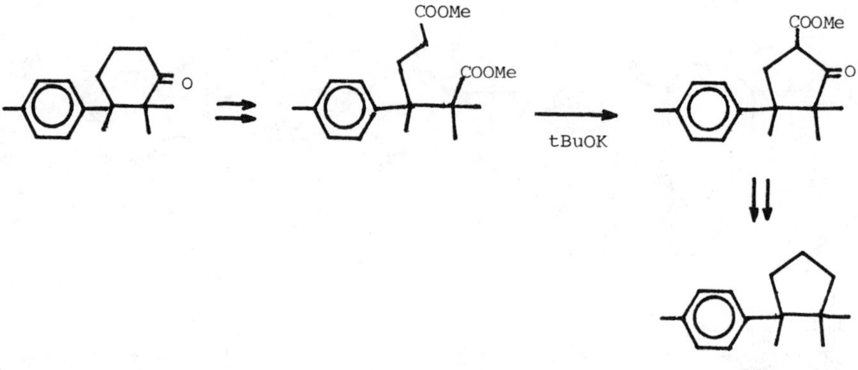

The first cyclic intermediate for systematic buildup toward cedrol (**C38**) (Stork, 1955, 1961a) is obtained by the Dieckmann reaction. On this foundation the A ring is annexed, and finally the six-membered ring is closed.

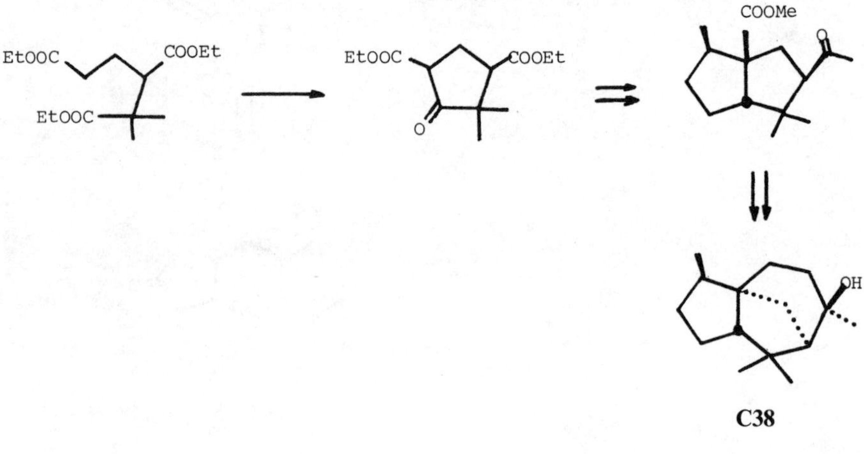

A practical synthesis of strigol (**S55**), the potent germination stimulant for the parasitic *Striga* plants, is based on an annulation of a β-cyclogeranic ester derivative (Brooks, 1985). The cyclopentane ring is closed by a Dieckmann reaction.

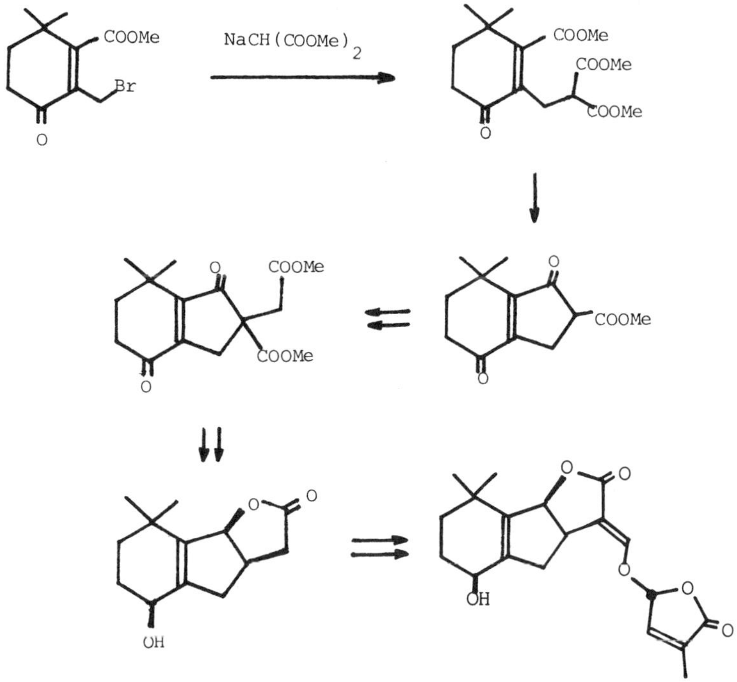

The first synthesis of camphor (**C15**) (Komppa, 1903), which firmly corroborated the constitution of the terpene, begins with a ring-forming reaction between diethyl 3,3-dimethylglutarate and diethyl oxalate. The symmetrical cyclopentanedione diester is monomethylated and then reduced to afford camphoric acid. Homologation of the less hindered carboxyl function and distillation of the derived calcium homocamphorate yield camphor.

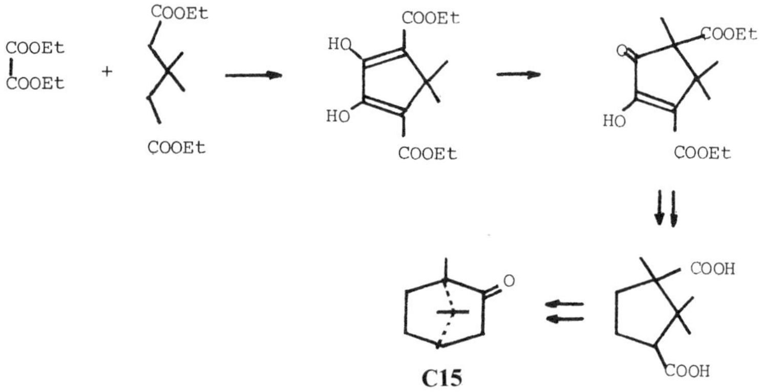

Although far less in use nowadays, the pyrolysis of metal dicarboxylates can be a convenient method for making cyclopentanones (and cyclohexanones). A second route to camphoric acid (Perkin, 1904) involves heating a tricarboxylic acid sodium salt to give, after acidification, 2,2-dimethyl-3-oxocyclopentane-carboxylic acid, which on Grignard reaction furnishes α-campholactone. Three more steps convert the lactone into camphoric acid.

C15

The cycloalkanone formation alternative (to Dieckmann cyclization) also figures in a reconstitution of phyllocladene (**P27**) from a C_{18} tricyclic ketoester. The ketone group is homologated to establish an acetic acid chain, and the barium salt of the diacid is pyrolyzed. (Turner, 1966).

P27

Trichodiene (**T26**) is the simplest representative of the trichothecanes. The presence of a methylenecyclopentane subunit in trichodiene inevitably invites approaches involving a Dieckmann condensation and concluding by a Wittig reaction (Welch, 1976a, b).

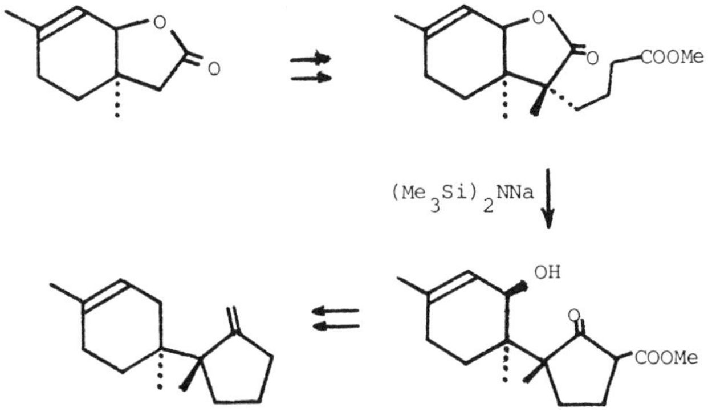

4.8 DIECKMANN AND RELATED CYCLIZATIONS

T26

The use of lactone as one of the ester equivalents in the Dieckmann condensation is not new. But in the synthesis of trichodiene, as shown, it affords assurance of having the correct stereochemistry at the quaternary carbon atoms.

An earlier report on the use of an ester–lactone for Dieckmann cyclization features the construction of a hydroxycyclopentanone intermediate for the virtuosic synthesis of caryophyllene (**C32**) and isocaryophyllene (**C33**) (Corey, 1963a, 1964b).

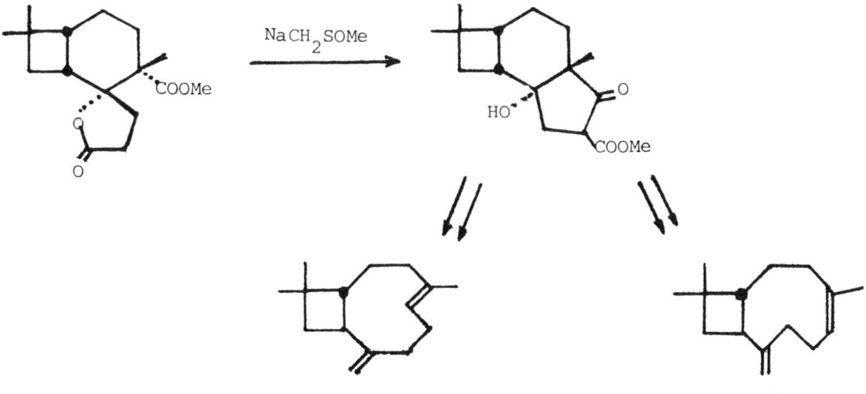

C32   **C33**

Several six-membered ketone intermediates used in terpene synthesis are readily available by means of the Dieckmann condensation. Two of these are involved in the synthesis of isonootkatone (**N12**) (Marshall, 1967) and yomogin (**Y4**) (Caine, 1975), respectively.

N12

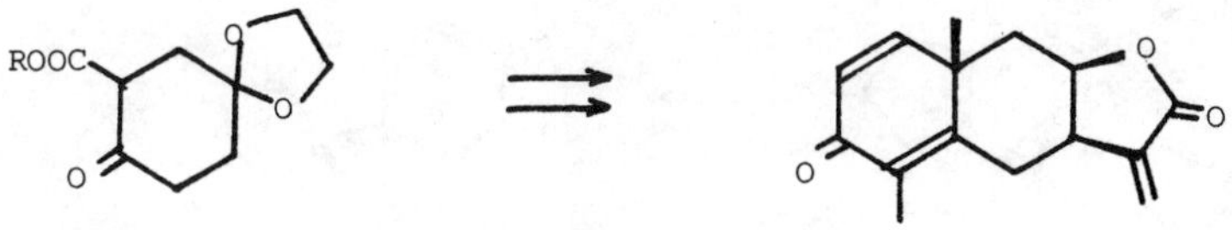

Y4

In an approach to nootkatone (**N11**) (Marshall, 1970b) analogous to that used for isonootkatone, the monocyclic ketoester is obtained while subjecting dimethyl 4-oxoheptanedioate to Wittig reaction conditions.

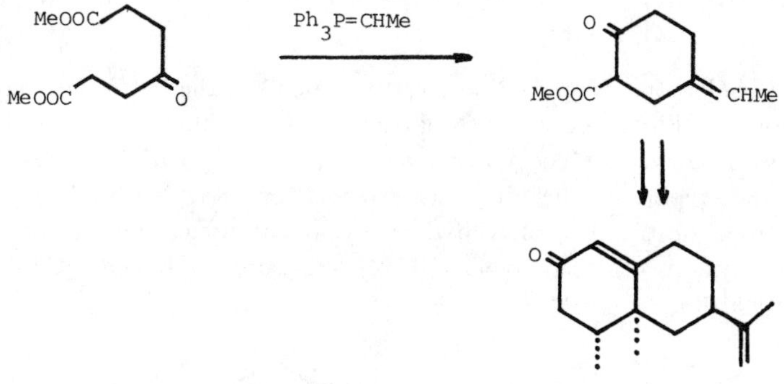

N11

Isobisabolene (**B11**) is a 4-substituted methylenecyclohexane. In a synthesis (Vig, 1969) of this material, the ring closure was conducted in a straightforward manner.

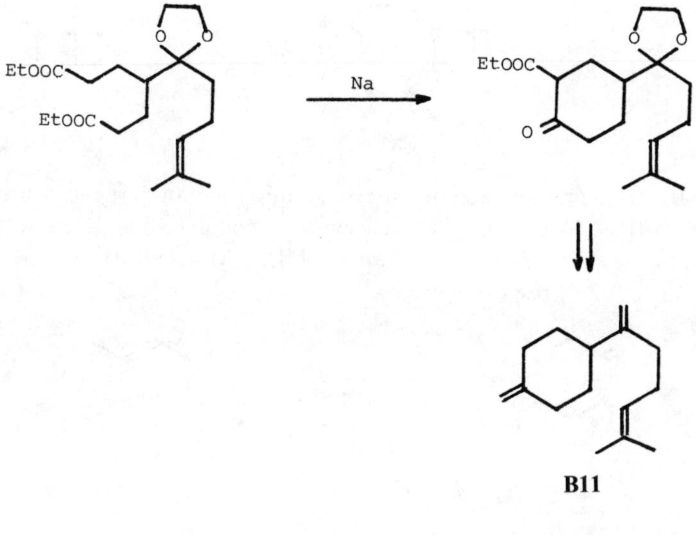

B11

A spirocyclic diketone is the intermediate en route to (—)-agarospirol (**A17**) (Deighton, 1975). The diketone is derived from the ketodiester by the Dieckmann cyclization.

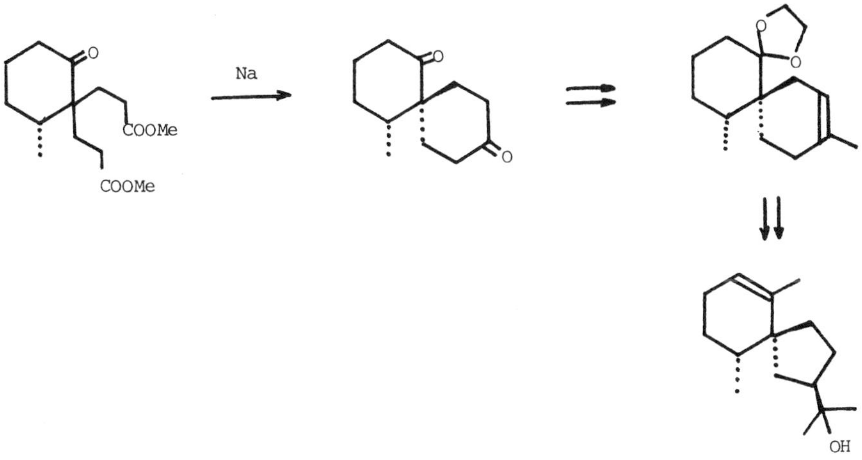

An *in situ* Dieckmann condensation effected by conjugate addition of divinylcuprate reagent to dimethyl 2-hexenedioate gives a valuable intermediate for mitsugashiwalactone (Nugent, 1986).

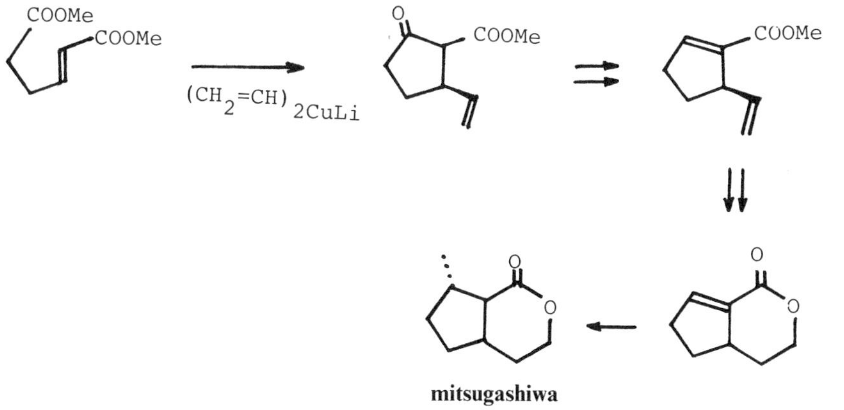

mitsugashiwa lactone

The cembrane derivative obtained by a macrocycloacylation can be elaborated into 3α-acetoxy-15β-hydroxy-7,16-secotrinervita-7,11-diene (**S17**) via functional group manipulation and a Dieckmann cyclization (T. Kato, 1987).

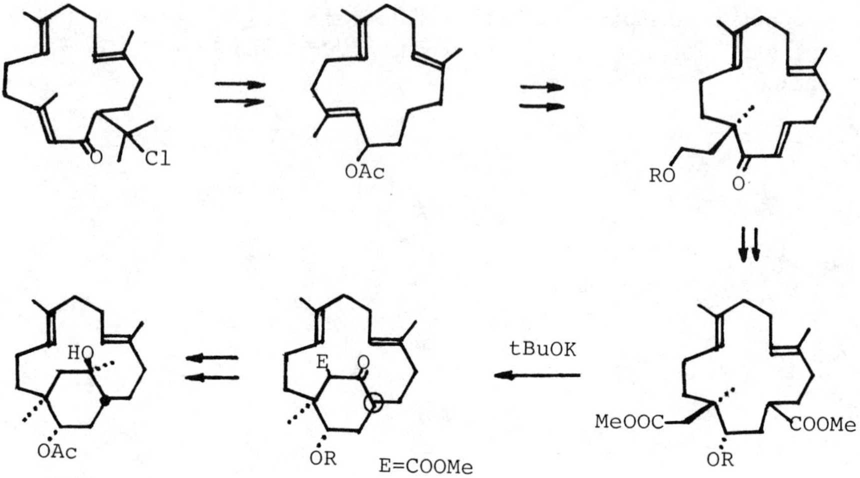

S17

α,α'-Dialkylation of succinic esters has received some attention in organic synthesis recently. Annulation is readily achieved when an α,ω-dihaloalkane is used as alkylating agent. (See, for example, synthesis of modhephene (**M12**), Wilkening, 1984). Further modification of the substrate structure allows the preparation of cyclic ketones via a Dieckmann condensation.

A synthesis of vetiselinene (Garratt, 1986) is based on such annulation. Removal of the extraneous ester group, conversion of the remaining one into an angular methyl substituent leave a carbon skeleton suitable for elaboration into the sesquiterpene. Thus the peri carbonyl is the exocyclic methylene equivalent and the double bond provides a site for an indirect alkylation (via a ketone).

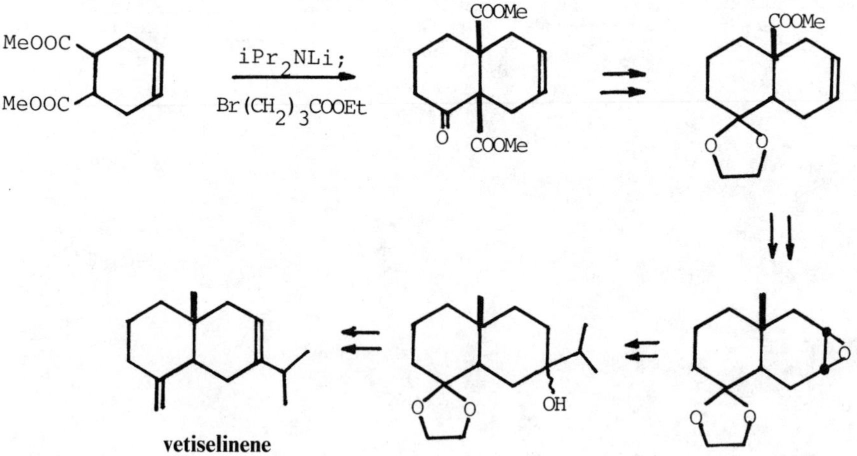

vetiselinene

A total synthesis of bilobalide (Corey, 1987e) is distinguished by many new and useful reactions. Among them are a formal Michael–Dieckmann condensation tandem with a propiolic ester as acceptor to form a cyclopentenone, deoxygenation of a highly functionalized epoxide with triethylsilane, and the reduction of oxalic ester with tributyltin hydride.

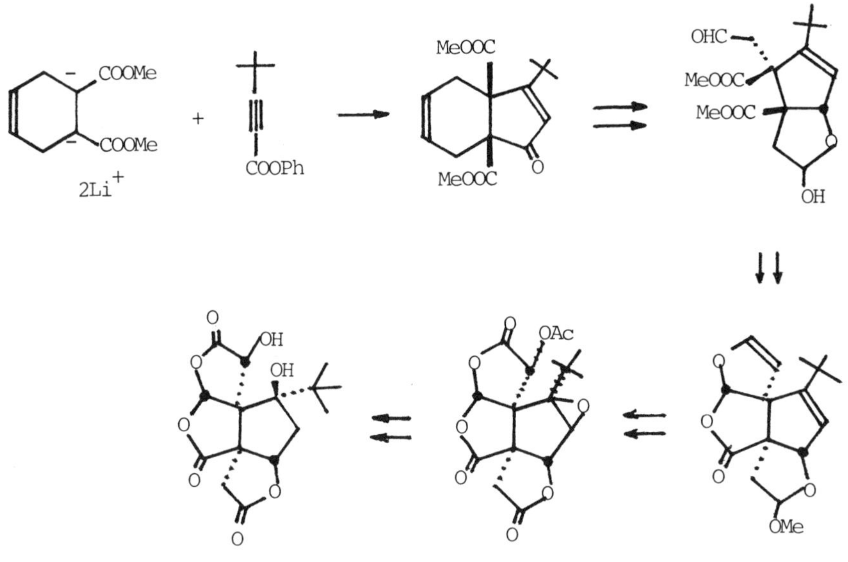

bilobalide

Alkylation of α-lithiated tertiary amides derived from optically active amino alcohols is a most powerful method for the rational establishment of two adjacent chiral centers in a carbon chain. This method has been licensed to effect diastereoselective Michael addition. For synthesis of (+)-dehydroiridodiol (**I7**) and (−)-isodehydroiridodiol (**I7'**) (Yamaguchi, 1986), the Michael process also induces *in situ* Dieckmann condensation.

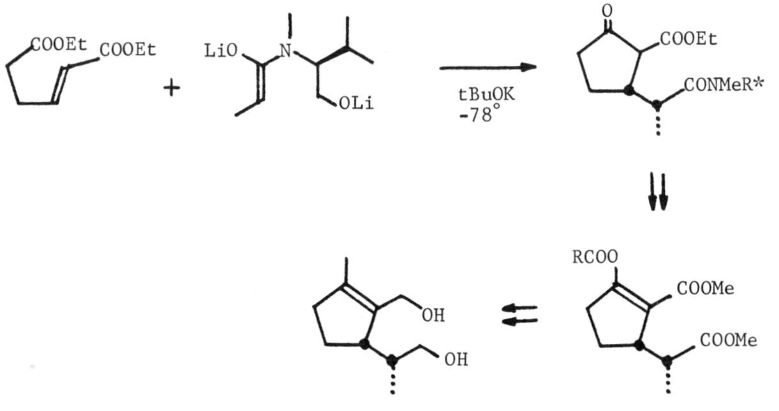

I7

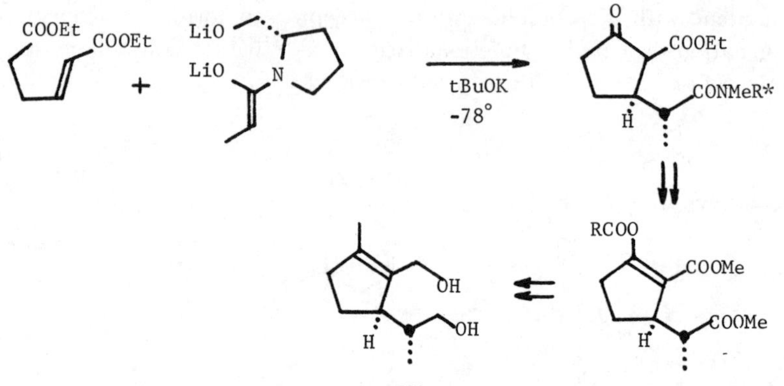

Khusitene (**K9**) is supposedly synthesized by two different ring-forming procedures: the Diels–Alder and the Dieckmann methods (Vig, 1974).

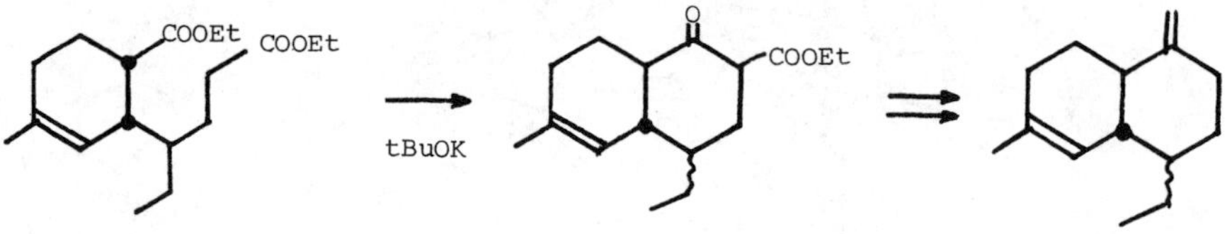

The tricyclic aldehyde intermediate for a kaurene synthesis is not directly suitable for phyllocladene (**P27**) because of the α-configuration of the carbon chain at C_8. However, inversion at this center has been achieved via cleavage of the existing double bond and a Dieckmann reclosure of the C ring (Church, 1960, 1966). The initial reaction is apparently a methanolysis of the anhydride at the least hindered carboxyl group to generate a diester.

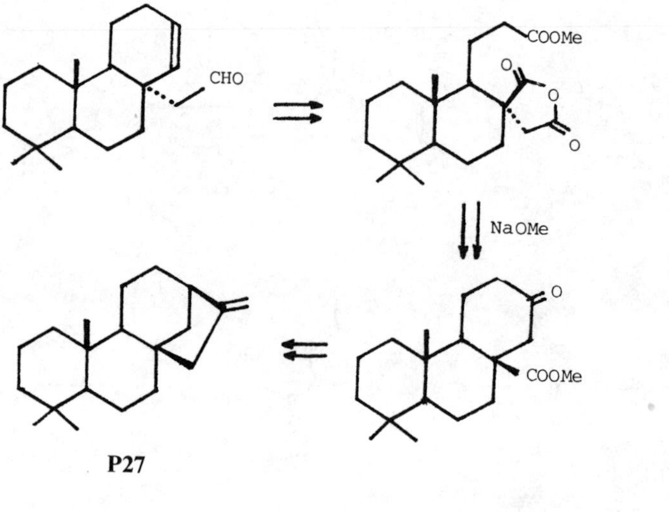

4.8 DIECKMANN AND RELATED CYCLIZATIONS

While Dieckmann cyclization is most useful in the synthesis of cyclopentanones and cyclohexanones, its application in a synthesis of gascardic acid (**G1**) (Boeckman, 1979) demonstrates the particular advantage of being able to retain the one-carbon pendant.

The cyclization of a hydrindane diester is expected to give the correct substitution pattern in the cycloheptane ring on account of the preferred enolization toward the least hindered methylene group. Of course the two ester chains must be properly introduced. The acetate chain at the ring juncture evolves stereoselectively from an allylic alcohol, and the α-oriented butanoate chain arises from hydroboration of the transposed double bond. The crucial stereochemistry at this point was actually uncertain before completion of this synthesis.

G1

The presence of an intact furan ring in gnididione (**G16**) indicates a perfect platform from which the 7- and 5-membered rings are constructed. Dieckmann cyclization of a furan-2,3-diester accomplishes the first task. Monoalkylation of the dianion of the diketoester provides all the carbon fragments for the sesquiterpene. Retention of the ester function ensures regioselectivity of the alkylation. Aldol condensation is the step for completing the A-ring (Dell, 1987).

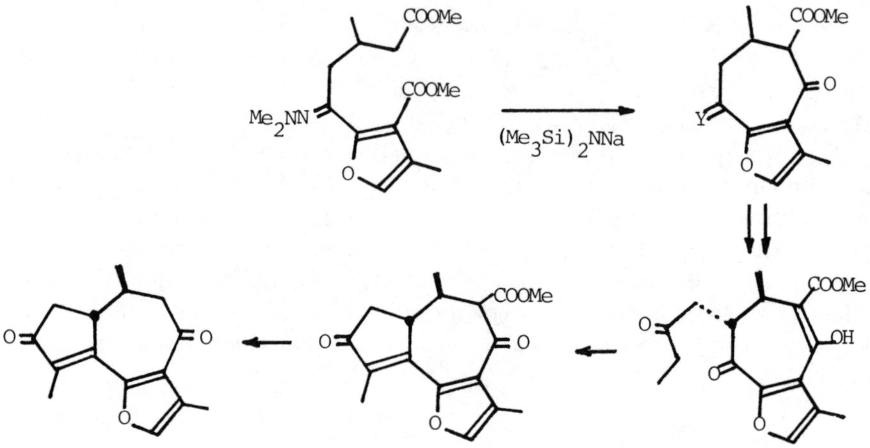

G16

A synthesis of Δ^1- and Δ^6-tetrahydrocannabinols (**C16** and **C17**) (Fahrenholtz, 1967) starting from a von Pechmann reaction of olivetol includes formation of a cyclohexanone by a formal Dieckmann condensation. The carbanion generated at the vinylic methyl group is stabilized by coordination of its countercation with the phenolic oxygen atom.

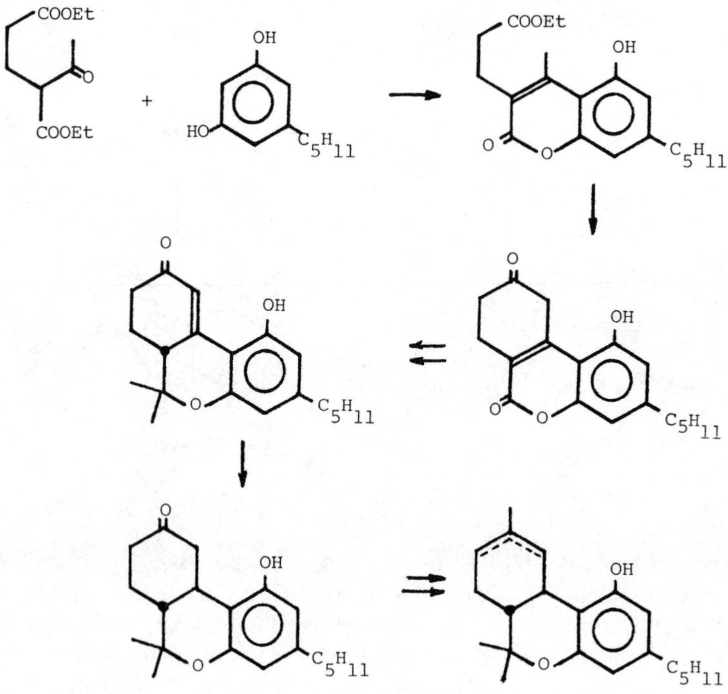

C16/C17

4.8 DIECKMANN AND RELATED CYCLIZATIONS

The intramolecular reaction of an α-sulfonyl carbanion with an ester is analogous to the Dieckmann cyclization. A synthesis of cyclosativene (**S13**) (Heissler, 1980) employs this process to close the nonbridged cyclohexane ring.

Tandem alkylation–Dieckmann condensations are a valuable annulation protocol. The version involving sulfone group as activator has the advantage of its easy removal, as demonstrated in a synthesis of maturone (Ghera, 1986). Thus, a substituted naphthalene is rapidly assembled and the attached oxygen functions modified to give the natural product.

maturone

An advanced intermediate for artecalin (**A45**) and tuberiferine (**T34**) has been assembled from a carbohydrate derivative (Georges, 1985b). The A ring is also formed by the dehydroalkoxylation of a sulfone carboxylate, and the B ring is constructed by a conventional Claisen condensation. While only one chiral center of the carbohydrate precursor would be retained in the terpene targets (at

C_6), asymmetry at the other sites is induced during chain building by an S_N2 displacement (for C_7), methylation (at C_{10}) by virtue of a 5:6-ring fusion, and by inversion (of 5β-H) after the lactone has been opened and the primary alcohol oxidized (Georges, 1985c).

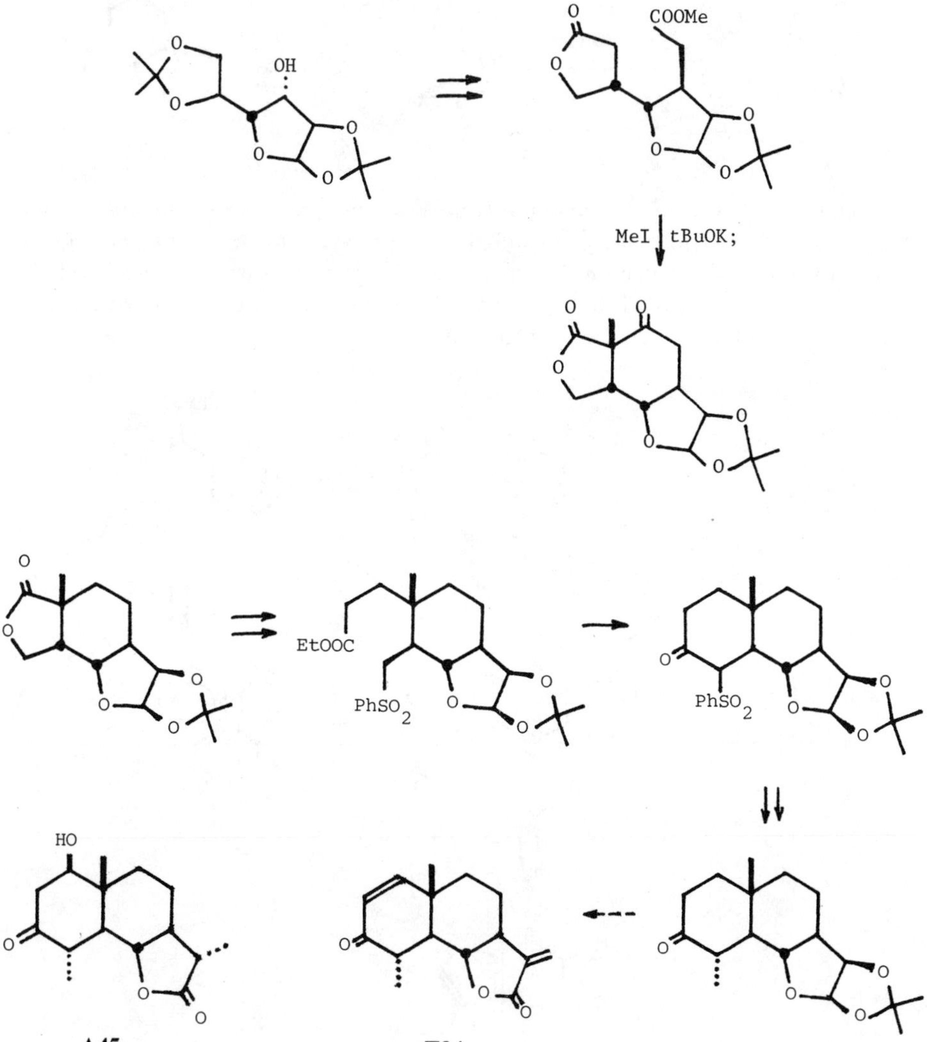

A45   **T34**

A synthesis of (+)-confertin (**C62**) (Quinkert, 1987) from an optically active cyclopropane-1,1-dicarboxylic ester involves ring expansion and inversion of the chiral center. The cyclopentanone product arises from a homo-Michael addition followed by Dieckmann condensation. Ene reaction serves to close the 7-membered ring.

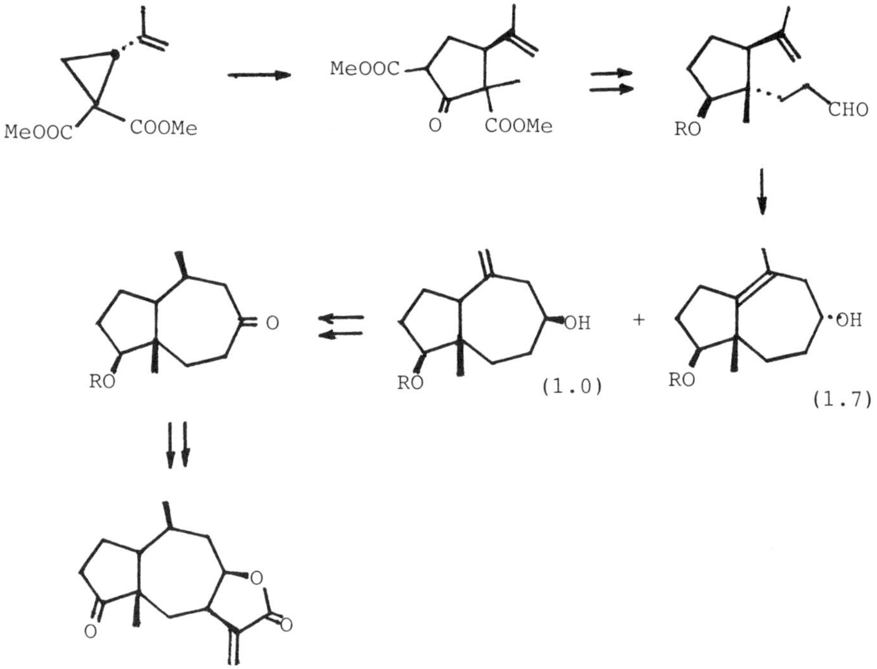

C62

The Thorpe–Ziegler reaction (Schaeffer, 1967) is rarely encountered in natural product synthesis nowadays. However, in providing the D-ring elements of stemodin (**S50**) (Piers, 1982a, 1985a) a chain elongation is needed, and since a dinitrile is the immediate product, an attempt to cyclize it directly is most logical.

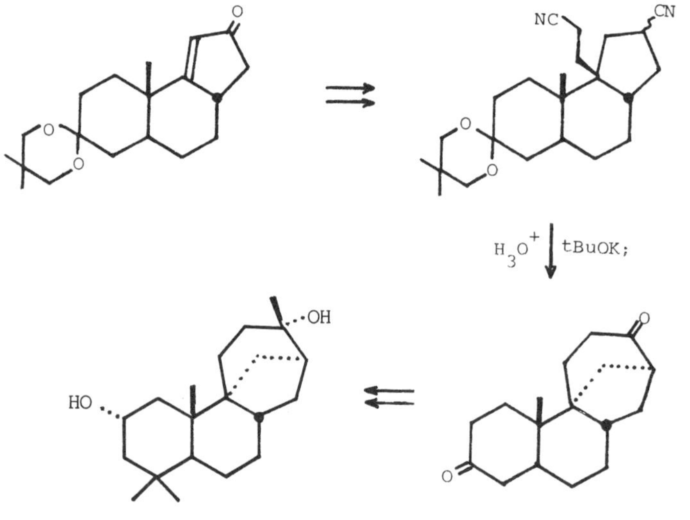

S50

Chapter 5

Intramolecular Alkylation

Alkylation reactions constitute the backbone of synthetic operation. Broadly speaking, the Michael reaction (hence the Robinson annulation) and even the aldol condensation may be categorized as alkylation reactions. On the other hand, chemists commonly use "alkylation" to refer only to the process of C—C bond formation by means of S_N2 displacement of alkyl halides, sulfonates, and epoxides with enolate species. The extensively practiced aromatic alkylation under acidic conditions is usually specified as the Friedel–Crafts alkylation.

Carbonyl compounds must first undergo enolization by submitting an α proton to a base. Most ketones have pK_a values of about 20, therefore sodium hydride and alkali metal alkoxides and amides are adequate bases. A most popular base in current use is lithium diisopropylamide, because it is strong, nonnucleophilic, and soluble in most organic solvents. Often, by using this base at low temperatures, it is possible to generate kinetic enolates and accomplish alkylation without causing equilibration and/or polyalkylation. Equilibration is a vexatious problem attending alkylation reactions promoted by alkoxides, unless the ketone is activated on one side. The chemist is also frequently forced to use blocking groups to achieve regioselective alkylation.

An "indirect" method of enolate formation involving the preparation (and purification) of enol derivatives such as trimethylsilyl enol ether and enol acetates has proved to be profitable. These derivatives are cleaved just prior to the addition of the alkylating agents.

Solvents play an important role in alkylation. The degree of solvation of the enolate affects the structure of the latter species, which could be a solvent-separated or contact ion pair, in O-metalated or C-metalated form. Accordingly, their rates of reaction and the extent of the competing O-alkylation change. The hardness of the alkylation agent also contributes to the divergence of the alkylation pathway.

α,β-Unsaturated ketones generally undergo alkylation at the α position, provided the γ carbon atom is not fully substituted. Enolization leading to a conjugate dienolate is much more favorable. If the α position already carries a substituent, the alkylation results in α,α-disubstituted β,γ-unsaturated ketones.

Enolates may be generated regioselectively via reduction of an enone system with lithium in liquid ammonia. Alkylation may then be accomplished *in situ* (Stork, 1965). This method is very powerful because the stereochemistry at the β carbon of the reduced ketones usually can be predicted (thermodynamic control).

It has been shown that internal alkylation at the β carbon occurs in the presence of an appropriate alkylating agent. Steric constraints may impose a configuration different from that normally expected at the β carbon.

Conjugate addition to enones by copper salt-catalyzed Grignard reagents (Kharasch reaction) or with lithium diorganocuprates gives saturated ketones after hydrolysis. The enolate intermediates may react with an alkyl halide, in effect achieving α,β-dialkylation.

β-Alkoxy-α,β-unsaturated ketones can be deprotonated kinetically at the α' carbon and alkylated at that position (Stork, 1973a: β-vetivone synthesis).

In cyclic systems the stereochemistry of alkylation follows the stereoelectronically preferred pathway unless it is prohibited sterically. Axial alkylation of cyclohexanone enolates is generally observed.

Alkylation procedures that have significant implications in ring formation include the first step of the Wichterle annulation (i.e., alkylation with 1,3-dichloro-2-butene) and haloketal annulation (Hajos, 1967). These procedures were developed to remedy the deficiency of the Michael reaction involving 1-alken-3-ones under aprotic conditions.

The β-haloketals are formed from the enones by their admixture with a diol (e.g., ethylene glycol) in the presence of a hydrogen halide. Other haloketals can be made accordingly, and an annulation study (Stork, 1973b) based on this property has been made. A fascinating cation effect on the steric course of the formation of 9-cyano-2,7-decalindione diethylene ketal has been noted: the cis product is formed when M = K, and the trans product, when M = Li.

3-Alkyl-4-chloromethylisoxazoles are latent 3-oxoalkyl chlorides. Their high reactivity toward ketone enolates makes them promising substitutes for most enone electrophiles (Stork, 1967). The isoxazole portion undergoes reductive cleavage, and treatment of the ring-opened species with base causes consecutive retro-Claisen reaction and aldol condensation in excellent yields. The reaction sequence often gives better results than the Robinson annulation.

Alkylation of ketones and aldehydes via their enamines (Stork, 1963) or metalloimines gives excellent control of monoalkylation and regioselectivity. With unsymmetrical ketones, enamines are generally formed toward the less substituted α carbon. The alkylation amounts to a kinetically controlled reaction even though the enamines are formed thermodynamically. In cycloalkanone enamines, an allylic ($A^{1,3}$) strain dictates the conformation of the existing substituents in the ring, hence the stereochemistry of the incoming alkyl group. However, the enamine alkylation does not lend itself well to annulation operations those involving enones and acrylyl chlorides (Stork reaction), for the obvious reason that enamine formation is incompatible with the presence of a leaving group. (It is essential that intramolecular alkylation be initiated by deprotonation of ketones with a hard base, to avoid direct displacement of the halide, sulfonate, etc.)

Since only rarely do 2-metallo-1,3-dithianes find application in ring formation, presumably due to the nucleophilic bases used in the proton abstraction step, this very important synthetic method of ketone synthesis by the umpolung principle (Seebach, 1969, 1979) is not discussed. Suffice it to identify the 1,3-dithiane anion as the forebear of the numerous acylanion equivalents; thus its development has turned a new leaf in the synthesis album.

As an annulation method, the intramolecular alkylation protocol has undeniable value because the removal of extraneous functional groups can be kept to a minimum (e.g., a double bond or hydroxy group in aldol). However, structural exactitude, especially the stereochemistry of the precursors, can hardly be compromised. Furthermore, once the cyclic structures have been formed, it is usually impossible to make changes unless contingent allocation of other functionalities has been planned. Generally, well-designed substrates undergo intramolecular alkylation with high efficiency.

Cycloalkylation works well in the construction of three-, four-, five-, and six-membered rings in either condensed, spiro, or bridged systems. The donor group is usually an enolizable carbonyl, and the leaving group can be a halide, sulfonate ion, or epoxide ring. With epoxy ketones, the products are the ketols.

5.1 SIX- AND SEVEN-MEMBERED RINGS

As expected, intramolecular alkylation represents an excellent method for making common rings. Two such steps are interspersed in the middle and late stages of a synthesis of ishwarone (**I12**) (Piers, 1977a, 1980). The preparation of the octalone starts from 3,4-dimethyl-2-cyclohexenone via a conjugate addition of a vinyl group, which also establishes the *cis*-dimethyl unit. A chain elongation via an acetylene derivative follows. With regard to the annulation step, products from alternative modes of cyclization (e.g., allylic displacement) have not been isolated.

Several other condensed-ring systems are made in connection with synthesis of ishwarone (**I12**) (Cory, 1979), sinularene (**S37**) (Collins, (1979), copacamphor/ylangocamphor (**C65/Y3**) (Hodgson, 1972; Eck, 1974), seychellene (**S29**) (Mirrington, 1972, Jung, 1978, Spitzner, 1978), and damsin (**D5**) (Grieco, 1977c).

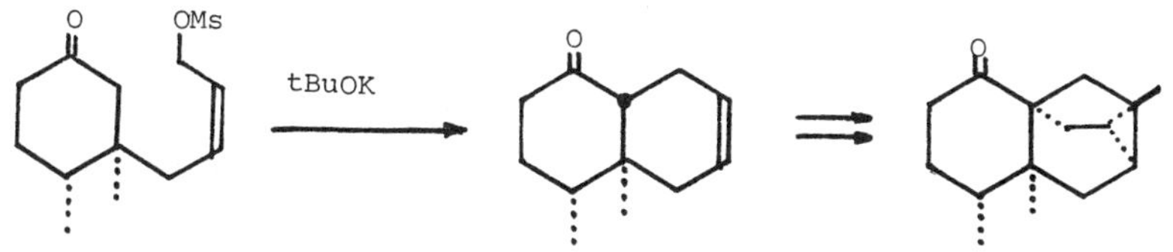

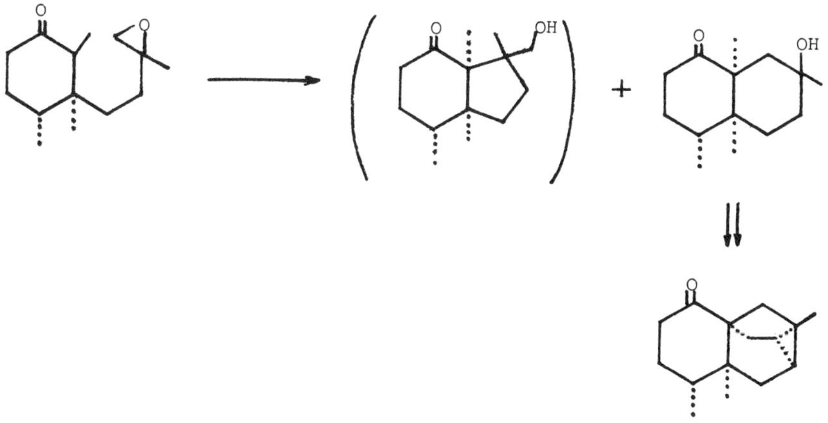

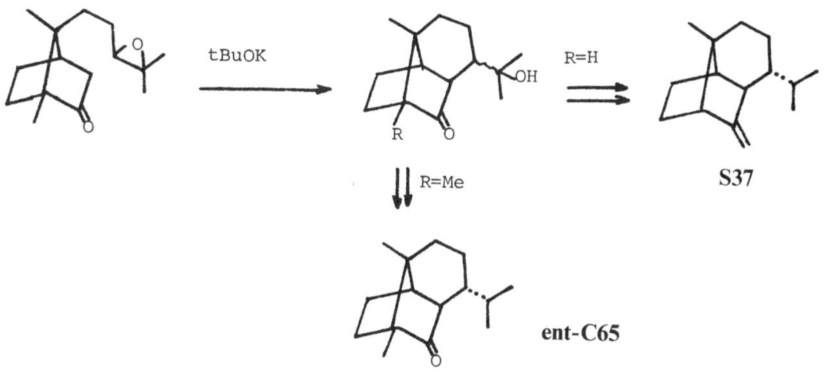

An α-decalone prepared by intramolecular alkylation is the keystone for synthesis of arteannuin B (**A44**) (Lansbury, 1986). Two of the chiral centers in this ketone which cannot be equilibrated emerge from stereoselective addition of the carbon chain to 4-methyl-2-cyclohexenone.

Another important feature of the synthesis concerns the regioselective formation of a trisubstituted double bond for establishment of the γ-lactone system via a γ-hydroxy crotonitrile unit. The hydroxyl group is a stereodirector during reduction of the double bond by complex metal hydride reagent.

The unusual sequiterpenes isolated from a marine sponge, represented by axamide-1, are particularly amenable to synthesis (Piers, 1986c) using the conjunctive alkylation method. The ketone group is the pivot to direct this annulation and to enable introduction of the functionalized side chain. This latter operation necessarily gives rise to a mixture of diastereomers.

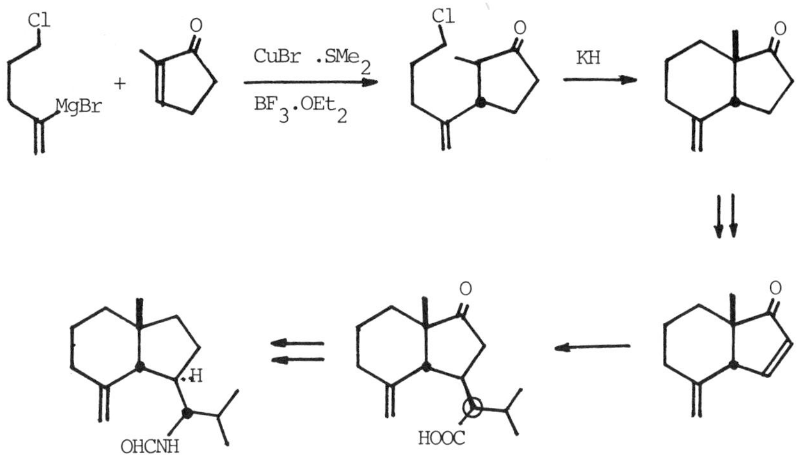

axamide-1

The basic strategy for preparation of the methylenedecalin system during synthesis of palauolide (Piers, 1987) is the same.

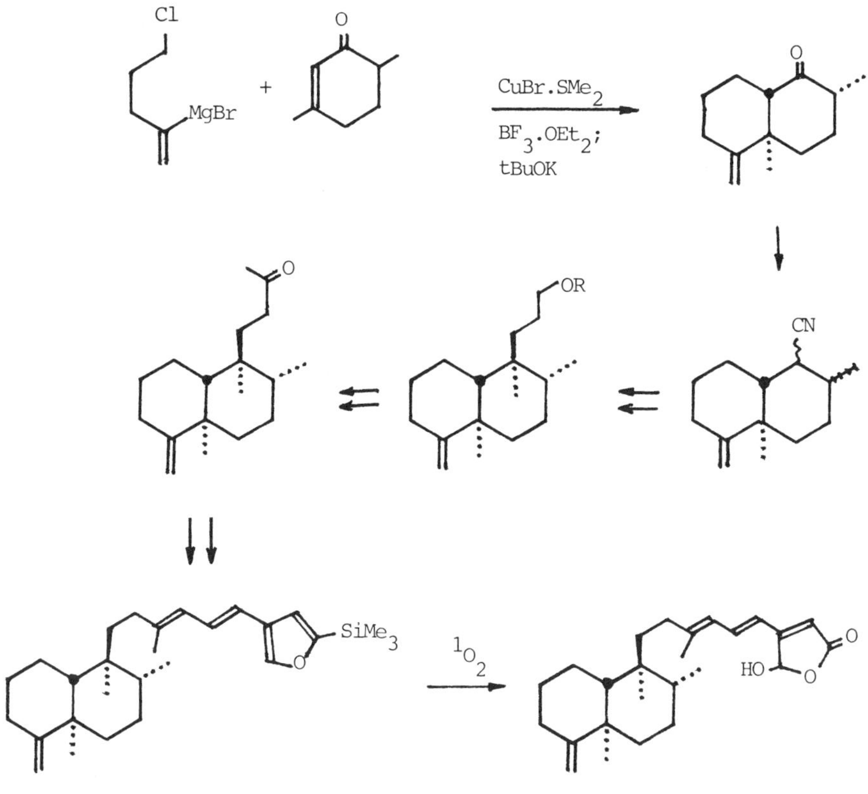

palauolide

Bridged systems are quire amenable to construction by the intramolecular alkylation procedure. The various skeletons of seychellene (**S29**) (Piers, 1969b), 9-isocyanopupukeanane (**P54**) (Corey, 1979a), and quadrone (**Q1**) (Danishefsky, 1980b, Bornack, 1981) have been fashioned according to the following bridging schemes.

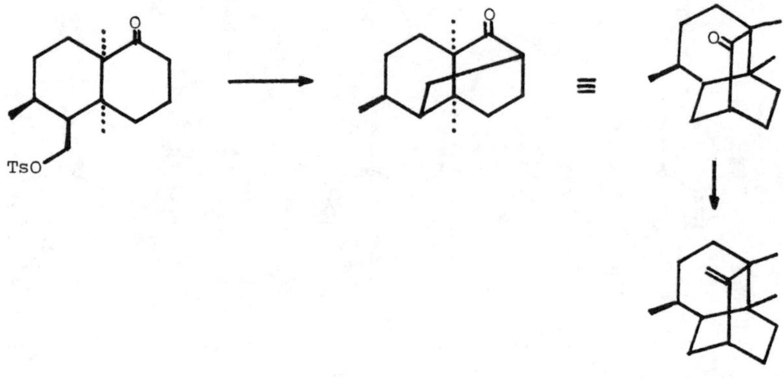

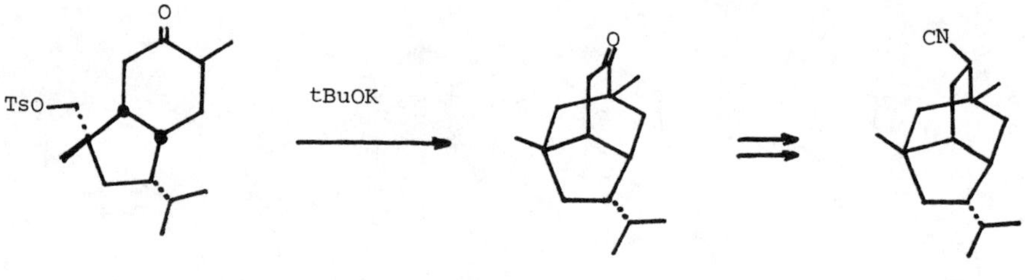

This cyclization intervenes also in the acquisition of natural (—)-quadrone (**Q1**) (Liu, 1987) from (—)-α-campholenic acid.

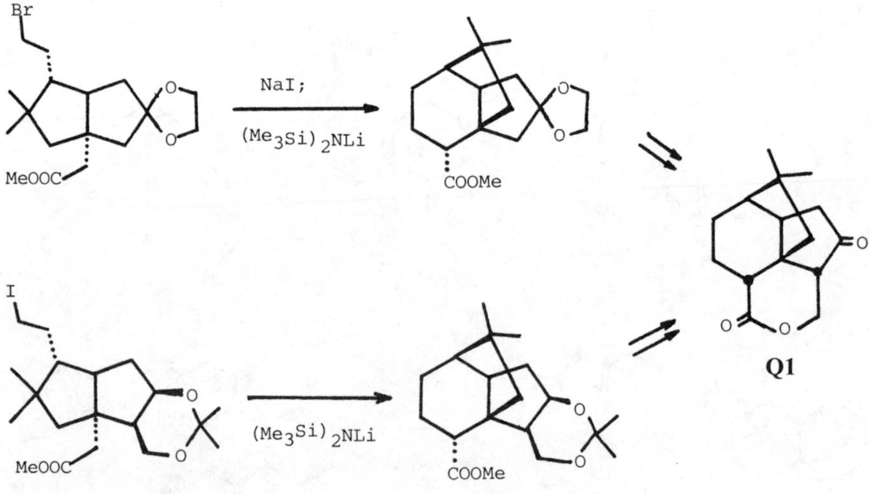

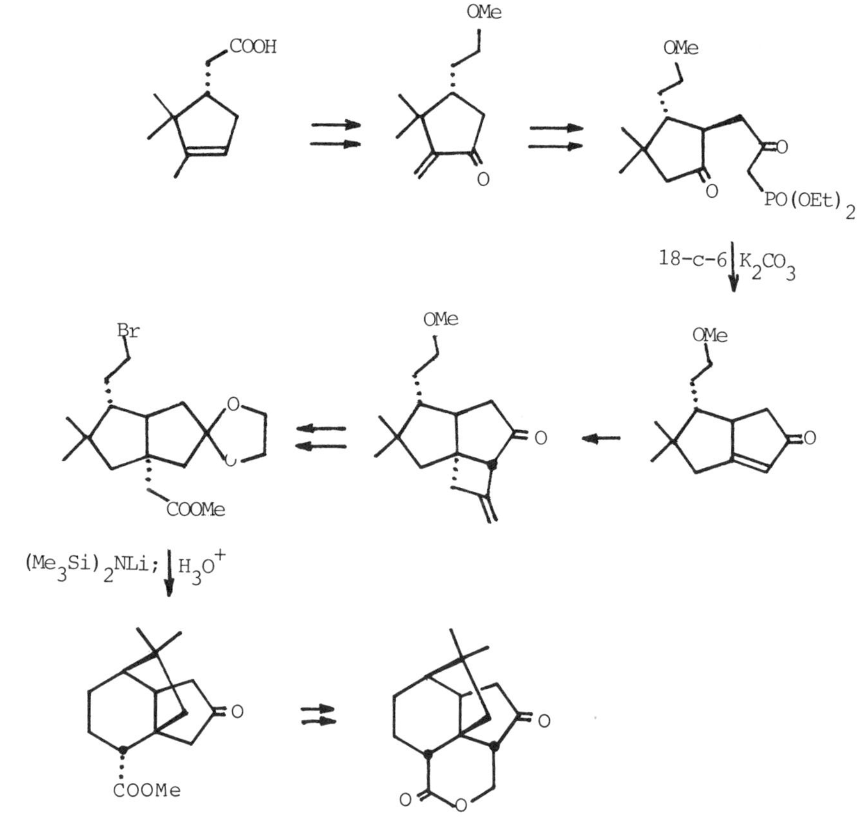

Q1

Intramolecular alkylation is the key step in a synthesis of geijerone (**G2**) (Kim, 1987). The interesting feature is the stereoselectivity originating from avoidance of an alternative transition state which suffers an $A^{1,3}$ strain.

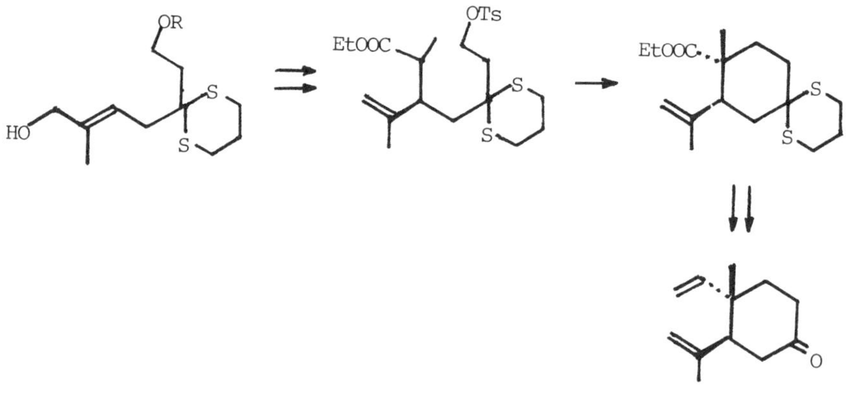

G2

The high efficiency (95% yield) of the α,α'-dialkylation of 2,4,4-trimethylcyclohexanone enolate with 3,4-bischloromethylfuran renders the synthesis of pyrovellerolactone (**V8**) (Froborg, 1975) via ring contraction of the bridged ketone particularly facile.

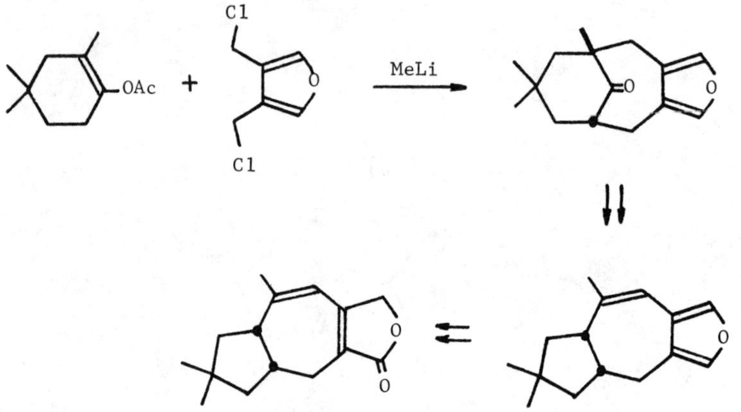

The synthesis of gibberellic acid (**G12**) embodying an intramolecular Diels–Alder reaction for constituting the A ring is predicated on proper selection of a tricyclic precursor. This crucial intermediate can be assembled from a spirocyclic compound via cycloalkylation, D-ring contraction, and so on (Corey, 1979d).

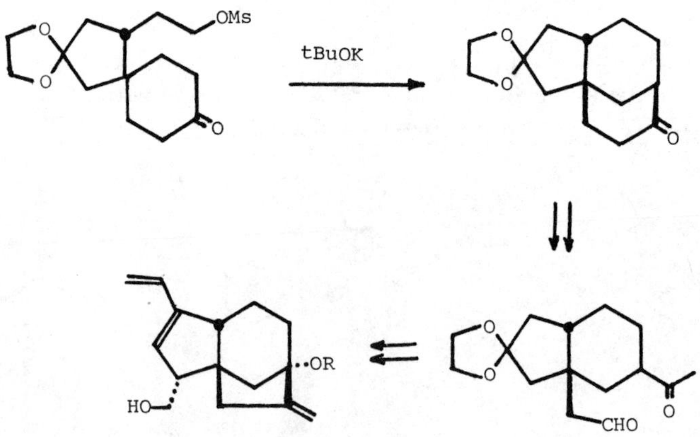

α,α-Dialkylation of 3R-methyl-δ-valerolactone is the key step of an intriguing synthesis of (—)-β-vetivone (**V16**) (Posner, 1987). The spirocyclic lactone can then be transformed into a δ,ε-unsaturated aldehyde which undergoes an ene reaction.

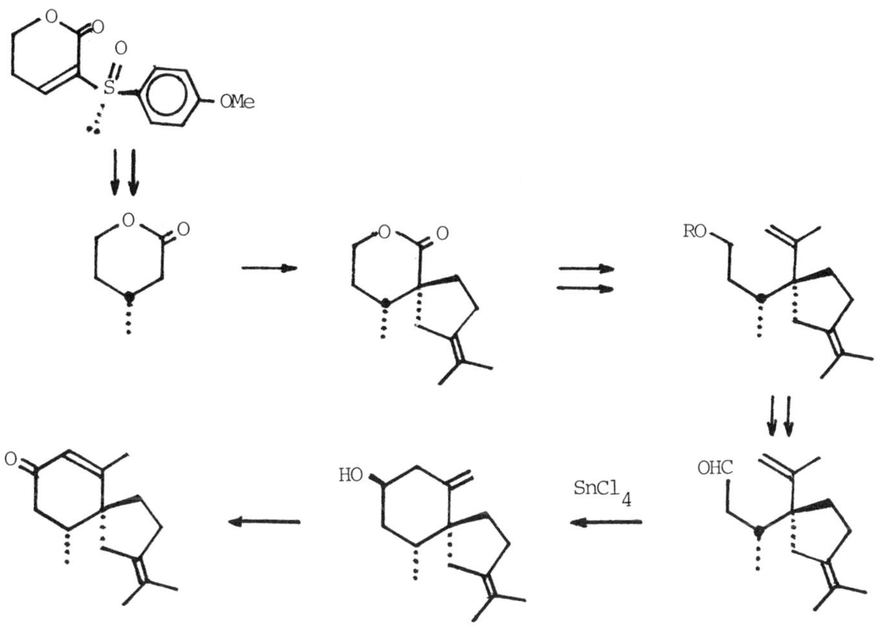

The stereoselective Michael reaction of α,β-unsaturated esters has a significant implication in the elaboration of vetispiranes (Yamaguchi, 1987). The Michael adduct is easily converted into an iodoketone which undergoes intramolecular alkylation.

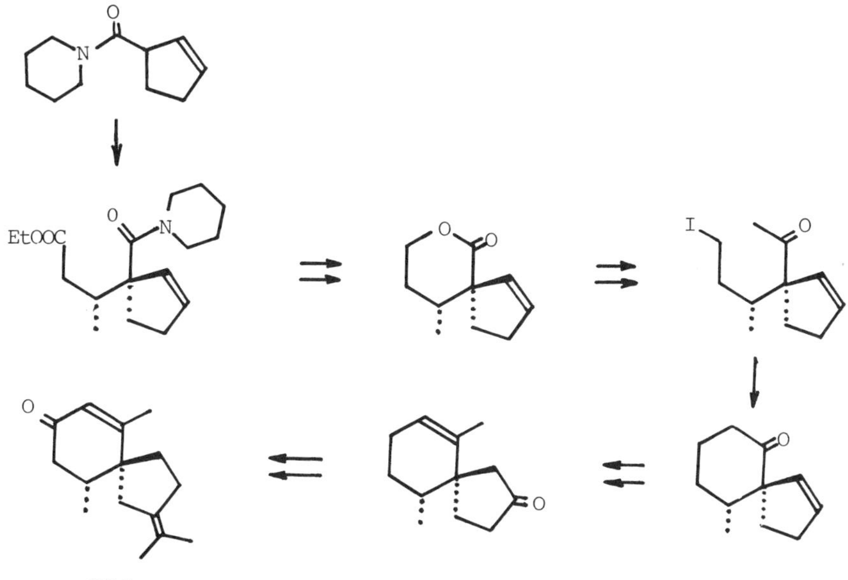

V16

An intermediate for veatchine synthesis has been converted into atisine (**A48**) (Nagata, 1963b, 1967b). In the process, reopening of the two-carbon bridge is required. This manipulation also involves a 1,2-transposition of a functional group for elaboration of the bicyclo[2.2.2]octane subunit by cycloalkylation.

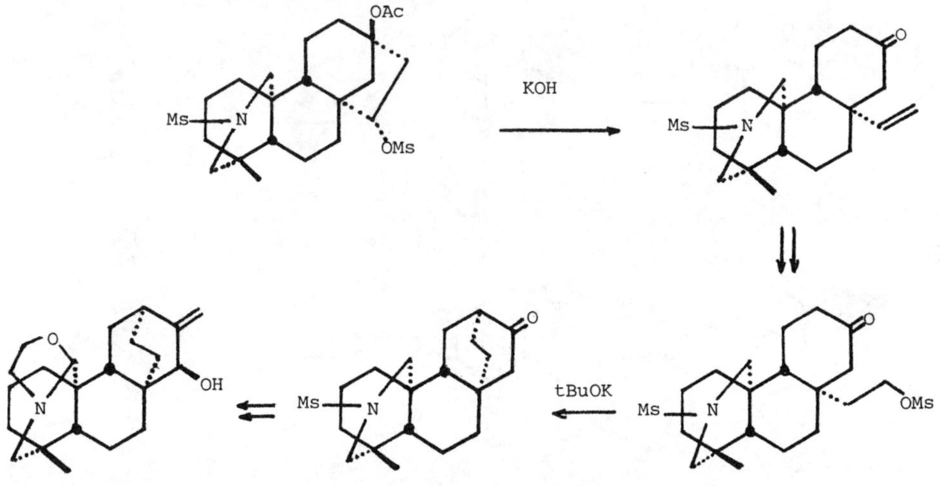

An apparent precursor of damsinic acid (**D6**) may be constructed by an intramolecular alkylation of an α-alkoxy nitrile anion, an acyl anion equivalent (Ziegler, 1980). Perhaps the more interesting aspect of this approach is the way in which a substituted cyclopentane is prepared. Tandem Cope–Claisen rearrangements, both proceeding via a chair transition state, serve to establish three stereocenters also.

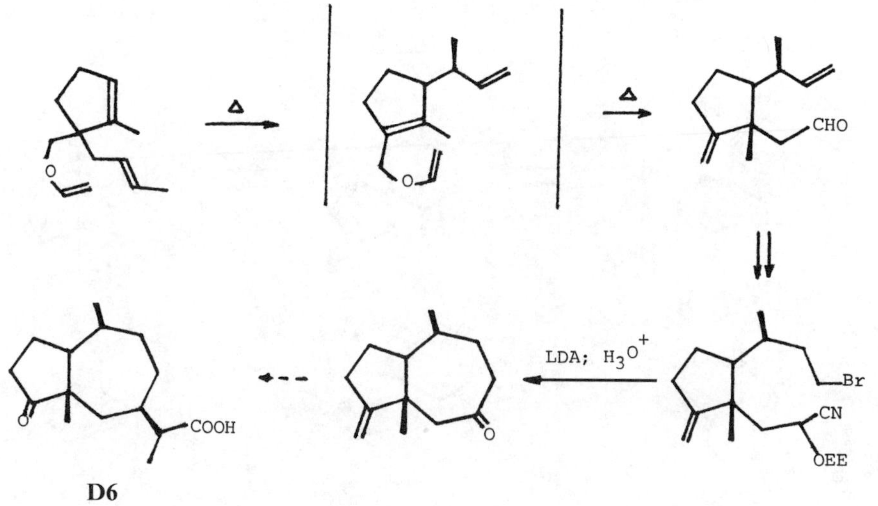

5.2 FIVE-MEMBERED RINGS

Formation of a five-membered ring as part of a bridged system by the intramolecular alkylation route again enjoys great popularity. In fact, the khusimone (**K8**) framework has been closed from two directions based on hydrindane derivatives (Liu, 1979, 1982; Piers, 1979a; Oppolzer, 1982c).

Ishwarone (**I12**) has also been synthesized via a similar cyclopentanation (Piers, 1977a). The bridged nature of the ring being formed simplifies the stereochemical issues slightly.

In addition to enolate and sulfone carbanions, phenolates are effective nucleophiles. Thus a route to hinesol (**H15**) (Marshall, 1969a) consists of an internal displacement reaction to set up a tricyclic dieneone. The secondary methyl group can then be introduced stereoselectively via a conjugate addition, and the central ring is functionalized and cleaved to provide a spirocyclic compound containing proper substituents for the completion of the synthesis.

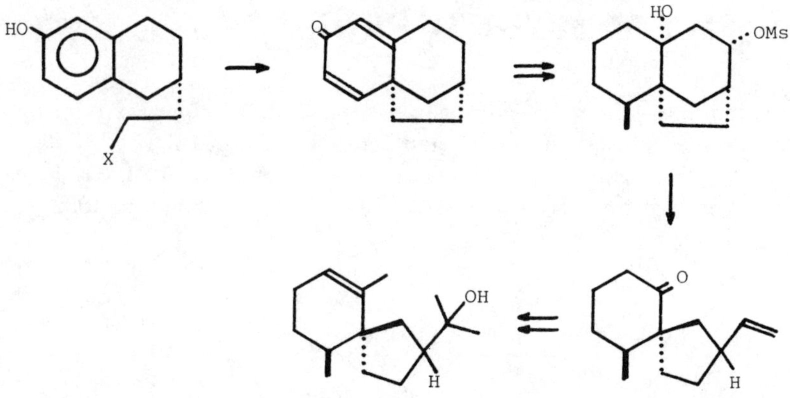

H15

Another route to β-vetivone (**V16**) (Uneyama, 1977; Torii, 1978) consists of intramolecular displacement of a tosylate by a phenolate ion at its para position. Conjugate addition of a methyl group to the resulting cross-conjugated dienone no longer enjoys the steric bias imposed by the substituent on a neighboring carbon atom. Consequently, diastereomers emerge.

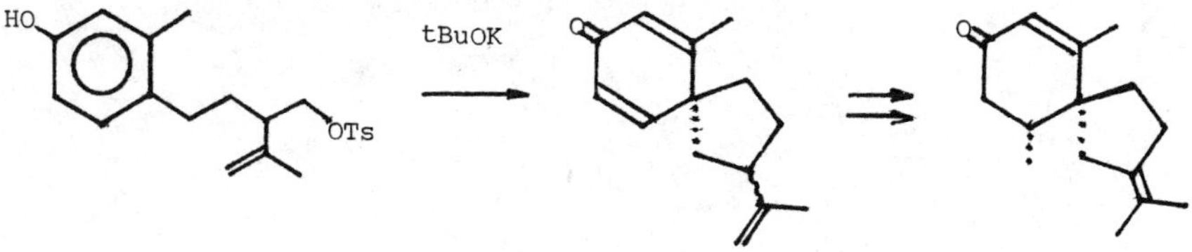

V16

A synthesis of anhydro-β-rotunol (**R8**) (Marx, 1987) based on intramolecular alkylation of phenolate anion involves an interesting step of transmesylation of the precursor. A phenyl mesylate is transformed into a quinone methide and coupled with the allyl carbanion.

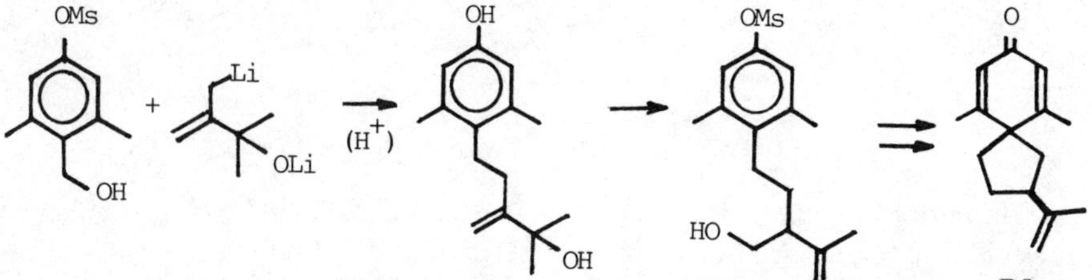

R8

Cedrene/cedrol (**C37/C38**) is probably derived from acoradienes in nature. Synthetic pathways that emulate the final steps of the biosynthesis have been devised and executed (Corey, 1969e; Crandall, 1969). The formation of a spirocyclic dienone can be approached from two different directions.

One of the cross-conjugated spirodienone intermediates resulting from the Ar_1^--5 participatory alkylation can be elaborated into β-acorenol (**A9**) and β-acoradiene (**A5**) (Iwata, 1985a).

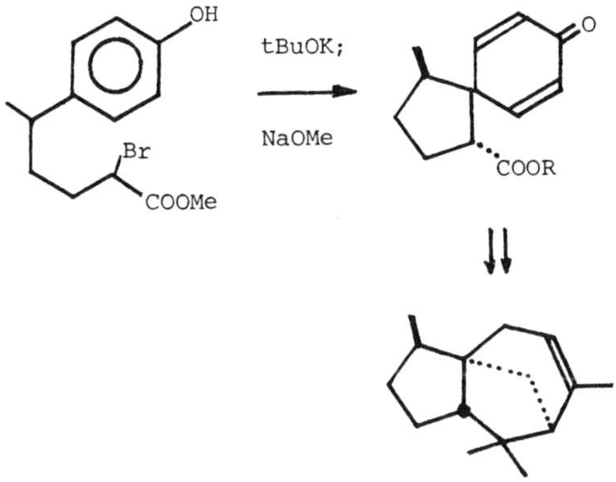

C37

The BC + D + A approach to kaurene (**K4**) (Masamune, 1964) is expedient for it dispenses with an additional functional group required for conventional bridging operation. At least one step for the functional group removal is saved. The D ring is closed by cycloalkylation of a phenolate ion.

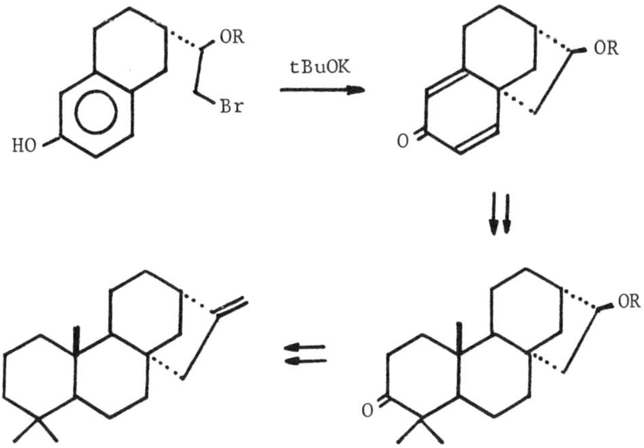

K4

A synthesis of the 15-deoxy analog of effusin (Kenny, 1986) is patterned, at the late stages, the biogenesis of the B-secokaurene. The bridged system is assembled by phenolic alkylation, and the A-ring by α-alkylation of a δ-lactone.

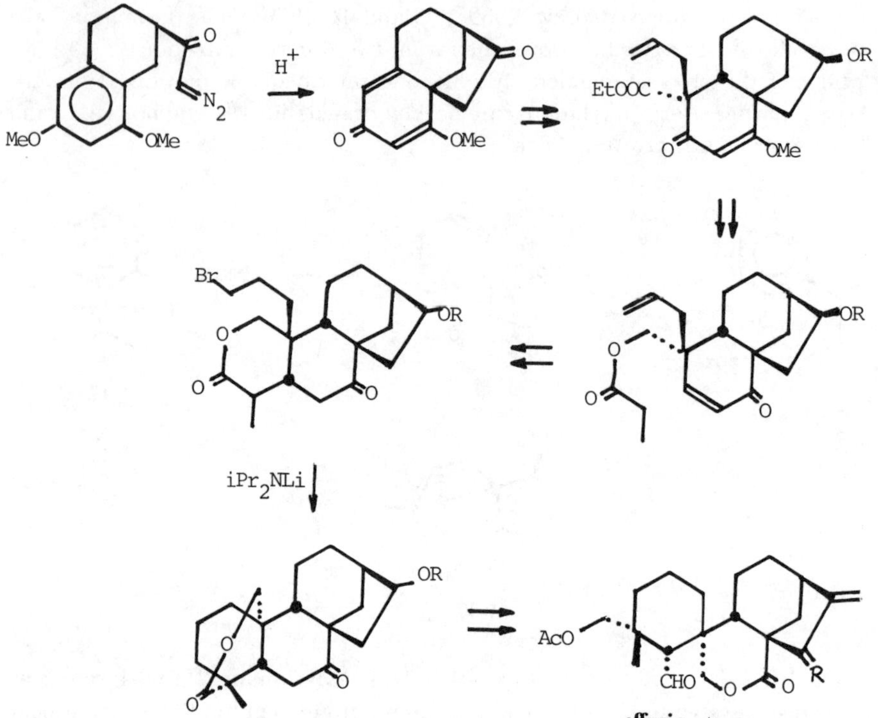

effusin (R=O)

The acid-catalyzed displacement of a diazonium ion (S_EAr) is a complementary to the anionic process described above.

The attachment of a cyclopentane ring to an existing bicyclo[3.2.1]octane unit is an effective strategy for building the entire skeleton of sativene (**S11**) (Yanagiya, 1979), copacamphor (**C65**), and ylangocamphor (**Y3**) (Piers, 1971, 1975a). In practice, the substrates are prepared from menthane derivatives.

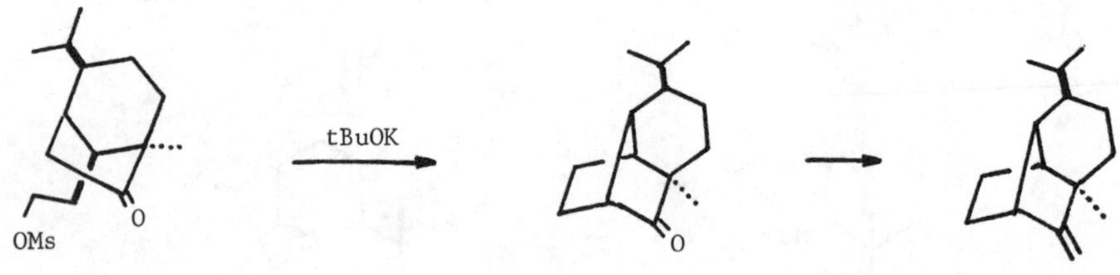

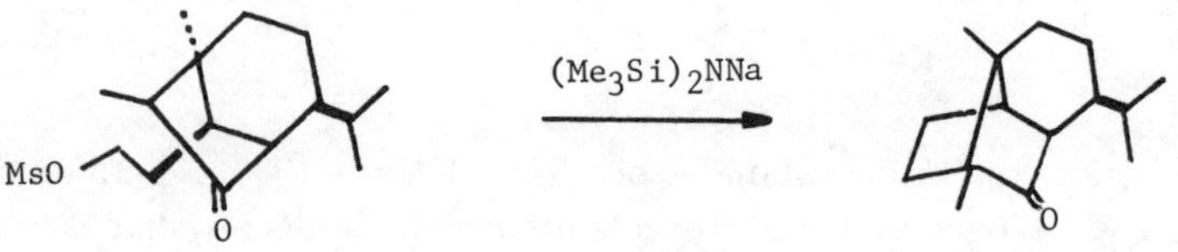

A common entry into the ring system of sativene (**S11**) (McMurry, 1968), copacamphene (**C63**) (McMurry, 1971), and longifolene (**L24**) (McMurry, 1972) is gained by an alkylative maneuver on substituted *cis*-α-decalones. The angular position adjacent to the ketone group is blocked; therefore there is no danger of epimerization. The corrresponding *trans*-fused isomers cannot be induced to form the extra intercyclic bond.

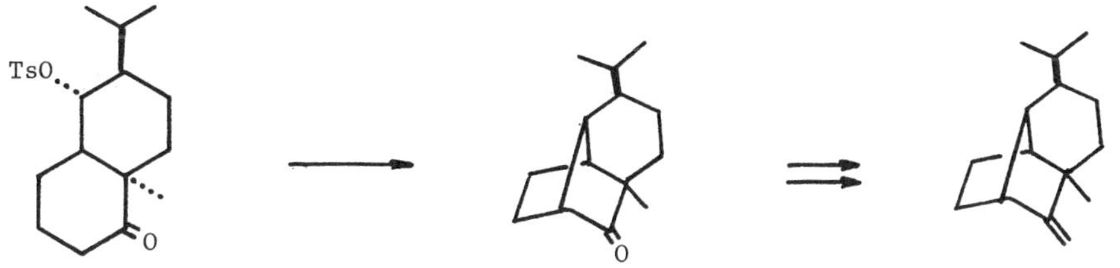

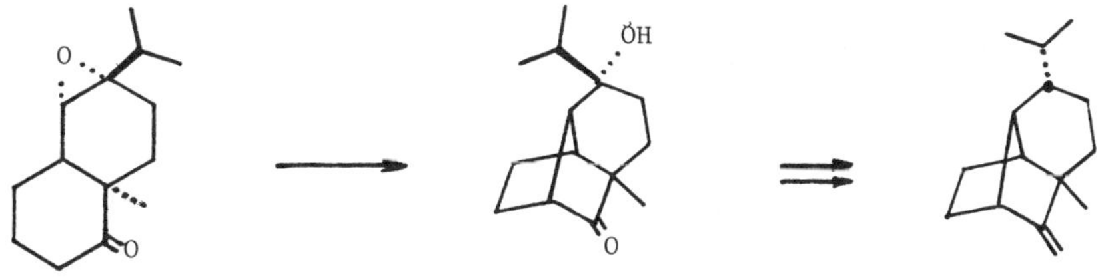

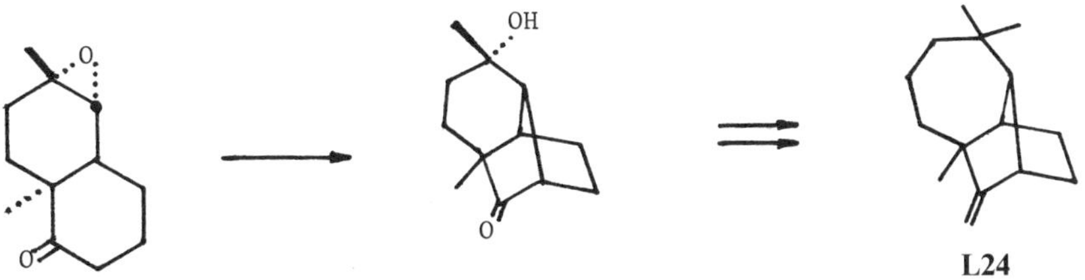

A synthesis of aphidicolin (**A36**) (Corey, 1980a) features intramolecular alkylation of a spirocyclic keto tosylate to close the C ring. The regioselectivity of this reaction shows a delicate dependence on the enolization procedure. Cyclization in the desired manner is achieved at low temperatures using a hindered base (-120 to $-130°$C, tBu$_2$NLi). On the other hand, exclusive bond formation from the α' carbon of the ketone results under thermodynamically controlled conditions (0°C, NaOMe, MeOH).

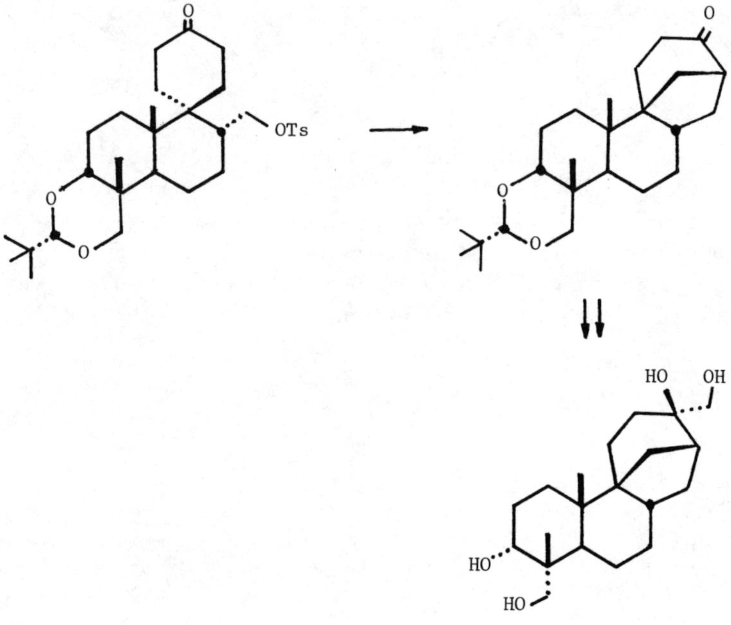

A36

A similar alkylation to construct the ring skeleton of stemodin (**S50**) has also been reported (Corey, 1980b). Here the task is much simpler because the other α position is rendered nonenolizable by a double bond.

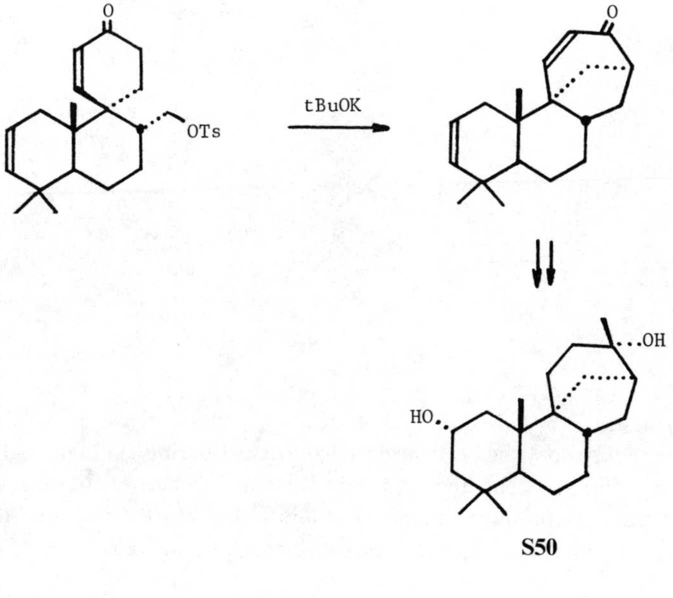

S50

5.2 FIVE-MEMBERED RINGS

The reaction sequence involving photocycloaddition of an acetylene equivalent to 3,4,4-trimethyl-2-cyclopentenone and subsequent opening of the small ring to provide a functionalized ethyl chain at the carbon is a prelude to closure of the bicyclo[2.2.1]heptane nucleus. Necessarily, the alkylation leads to epicamphor (Boust, 1974). However, the ready intercoversion of epicamphor and camphor (**C15**) means that the well-known terpene is also accesible.

C15

It is interesting to note that transformation of the cyclopentadiene–maleic anhydride adduct to forsythide dimethyl ester (**F10**) (Furuichi, 1974) entails cleavage of both carbocycles. A main purpose of the bicyclic template is the stereoselective creation of a new cyclopentane nucleus by a cycloalkylation route. The first ring cleavage is accomplished by a Norrish type I photochemical fission, leaving a double bond in the second ring, which is the site of another cleavage.

F10

Several synthetic pathways to a tetracyclic aromatic intermediate for diterpene alkaloids such as veatchine (**V5**) focus on the formation of a bridged ketone with which insertion of an amino group is then called for. Intramolecular alkylation of the 3-ketone is a most appropriate method (Grafen, 1968).

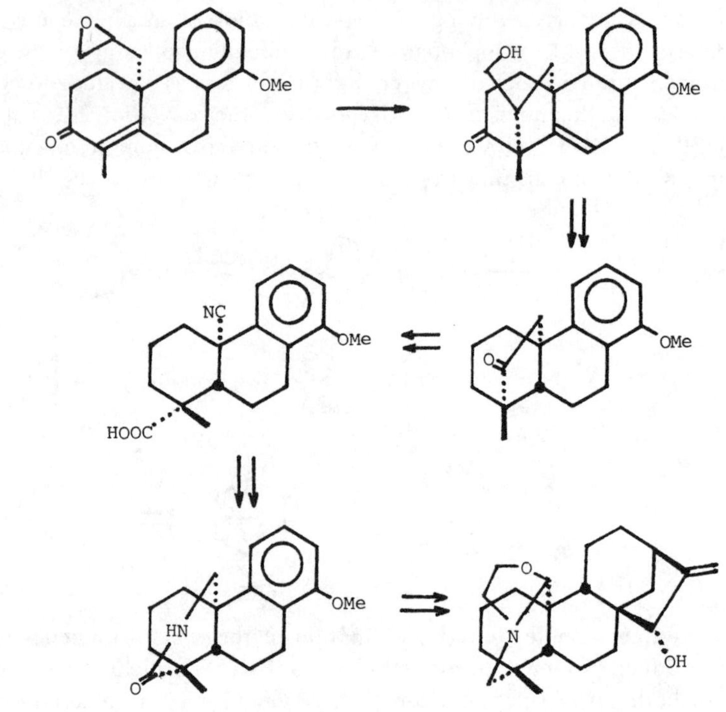

The construction of the D ring in a synthesis of gibberellin A_{15} (**G13**) (Nagata, 1971) entails the internal displacement of a tosylate by an enamine. The aldehyde group is destined to become the vinylidene fragment.

The cycloalkylation is unusual. The strong inclination of the aldehyde to form enamine and the propinquity of the reactive centers are the critical factors favoring the observed events.

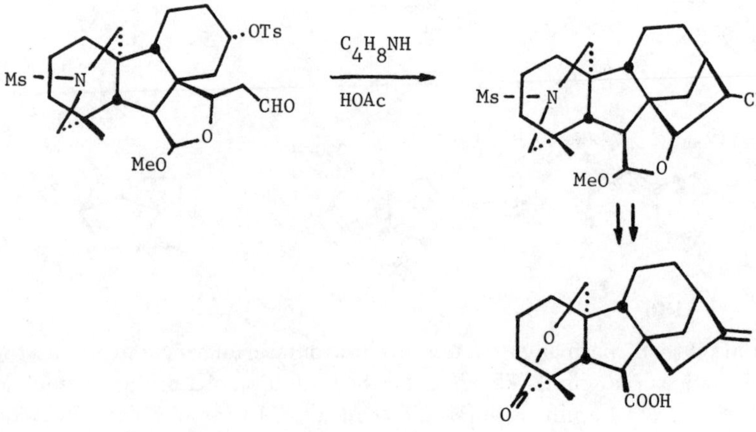

G13

5.2 FIVE-MEMBERED RINGS

In trichothecane synthesis, two reaction pathways for bridging the cyclopentane within a hydrobenzopyran framework by intramolecular aldol condensation have been studied. The third possibility, in which the ring closure involves two carbon chains, is more difficult in stereochemical terms because one more asymmetric center must be established prior to the C—C bond formation. However, this approach has certain redeeming features: namely, since the substituents on the pyran periphery may be assembled from a sugar molecule, optically active compounds are obtainable.

The synthesis of an advanced intermediate suitable for elaboration into verrucarol (**V12**) and maybe some other trichothecanes is initiated from D-glucose. The cyclohexene ring is attached via a cationic cyclization. Although the homoallylic ether is produced, an accidentally found acid-catalyzed isomerization of a fully functionalized compound already possessing the complete skeleton of the terpene to a separable mixture provides a solution to a most critical problem (Tsang, 1985).

The cyclopentane ring is formed by an intramolecular displacement of a sulfonate by an α-amido carbanion. The exo configuration of the amide group is desirable because it can be more easily degraded into a hydroxy group without resorting to configuration inversion.

V12

Probably due to the many available alternative methods, cycloalkylation is much less often used in synthesizing the condensed-ring components of terpenes. However, expedient formation of a few fused cyclopentanes has been witnessed in connection with the synthesis of oplopanone (**O7**) (Taber, 1978), hirsutene (**H17**) (Magnus, 1982), and modhephene (**M12**) (Wilkening, 1984). In the last example, an α,α'-dialkylation of dimethyl succinate proves to be an excellent protocol for securing 1,2-cyclopentanedicarboxylate. The formation of the second ring follows a very similar path.

In an elegant synthesis of lupeol (**L28**) (Stork, 1971) the E ring is masked as a cyclohexanol at the start for tactical reasons. A hydrophenanthrenone derived from 6-methoxy-α-tetralone is used because of its facile accessibility by the Robinson annulation. On reaching the pentacyclic intermediate, the E ring is modified via a secoester tosylate and reclosed to a five-membered carbocycle upon treatment with a strong base. The newly created asymmetric center assumes the natural configuration in which the side chain is pseudoequatorial.

L28

Several methods are available for preparation of the BCD-tricyclic intermediate of gibberellic acid. One method involves cycloalkylation of a cyclohexanecarbonitrile with subsequent transformation of the angular cyano substituent into a propargyl group (Stork, 1979a). The hydrindane product is contaminated by 5–8% of the trans isomer.

By virtue of symmetry, the homologous series does not require selective protection of one of the carbonyl groups before the reductive cyclization of the D ring. However, a ring contraction maneuver must be performed at the tricyclic level.

The characteristic arrangement of the ring system and the various substituents in the pentalenane and isocomane sesquiterpenes tend to elicit annulation sequences based on diquinane enones (e.g., via conjugate addition and subsequent cyclization). The stereochemistry at the ring juncture is self-regulated, since the existing angular substituent (hydrogen or others) imposes exo addition of external reagents. The alternative attack is strongly disfavored, and not only on steric grounds: the enormous strain of the *trans*-fused product would be felt in the transition state of its formation.

The many syntheses based on this concept differ somewhat in the choice of the functionalized chain that must accommodate the secondary methyl group or the introduction thereof. In one route to pentalenene (**P16**) (Piers, 1984a), an alkenylcopper reagent is employed, and the ring is created immediately following the attachment of the chlorobutenyl chain by an intramolecular alkylation reaction.

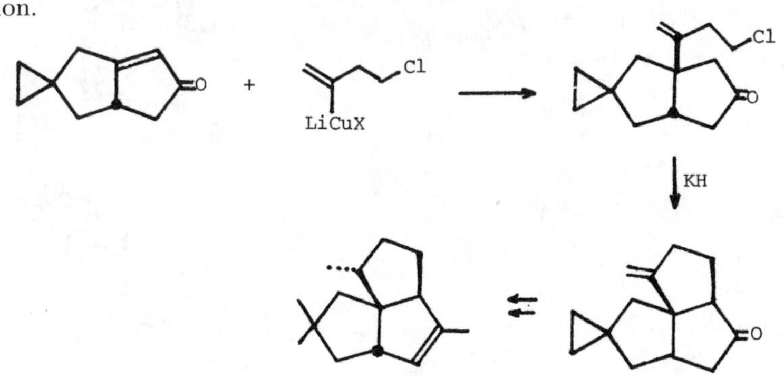

The general method also finds an application to synthesis of anhydrooplopanone (**O9**), oplopanone (**O7**), and 8-epioplopanone (**O8**) (Piers, 1986b) (see Sec. 5.1 also). The ethylidene group is transformed into the acetyl chain by hydroboration and oxidation.

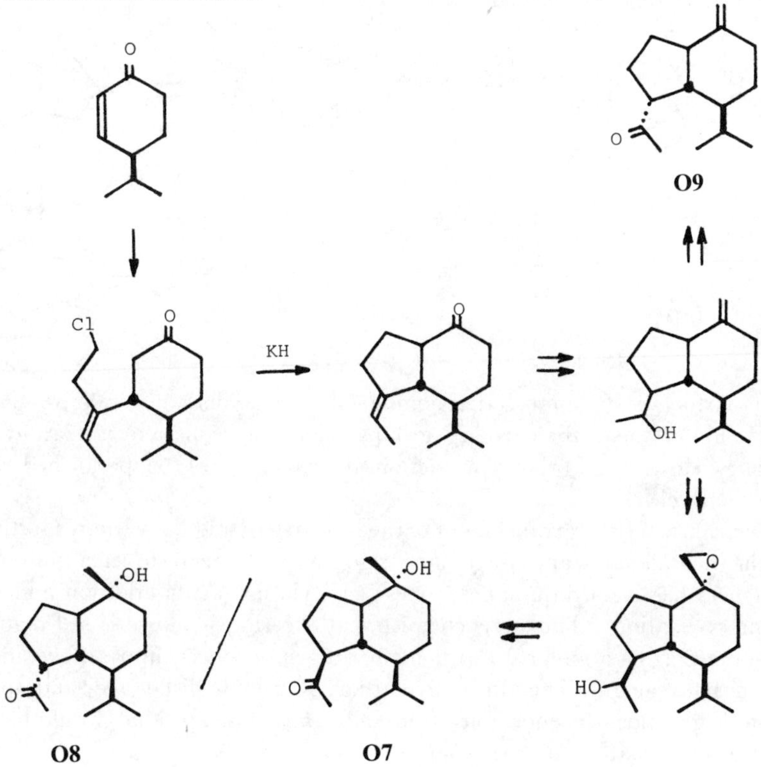

5.2 FIVE-MEMBERED RINGS

The vitality of the 1,3-conjunctive alkylation as an annulation protocol is further shown in a distinguished approach to $\Delta^{9(12)}$-capnellene (**C21**) (Piers, 1984b). The same conjunctive reagent is used on two occasions. The advantage of this scheme lies in the direct incorporation of the vinylidene group, which is present in the sesquiterpene. Furthermore, this unit can be easily transformed into the *gem*-dimethyl substituents.

The tactical priority of the synthesis is therefore assigned to the construction of the diquinane nucleus containing all the methyl groups. It should be noted that the route is very efficient for a nonconvergent approach because the ketone function is employed to the fullest extent without recourse to protection, transposition, and other nonproductive operations.

It is never redundant to stress the economic appeal of reiterative annulation techniques. A prepared reagent that can serve more than once in a synthesis is always pleasing.

C21

The linear triquinane sesquiterpene (+)-hirsutene (**H17**) has been prepared from S-(—)-allyl p-toyl sulfoxide. Thus, conjugate addition of the sulfinyl carbanion to 2-methylcyclopentenone is followed by reduction, cyclization of the vinyl sulfide, conversion to a monoprotected bicyclooctenedione, and repeating the addition–cyclization sequence (with different reagents and procedures) (Hua, 1985).

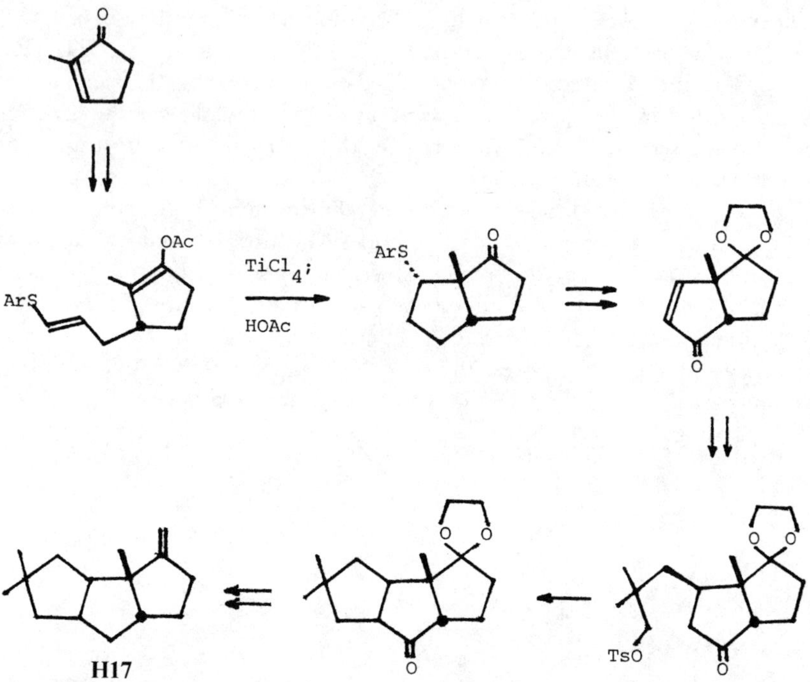

H17

Double alkylation of malonates using α,ω-dihaloalkanes to afford certain cyclic compounds is a well-established method. Extension of the scheme to monocarboxylic esters became possible only when suitable amide bases were found. An excellent application of the new procedure is seen in the cyclopentanation leading to a precursor of bakkenolide A (**B1**) (Greene, 1985a). The observed 3:1 stereoselectivity in favor of exo ester in the bicyclic product is the result of stereocontrol by the existing tertiary methyl group.

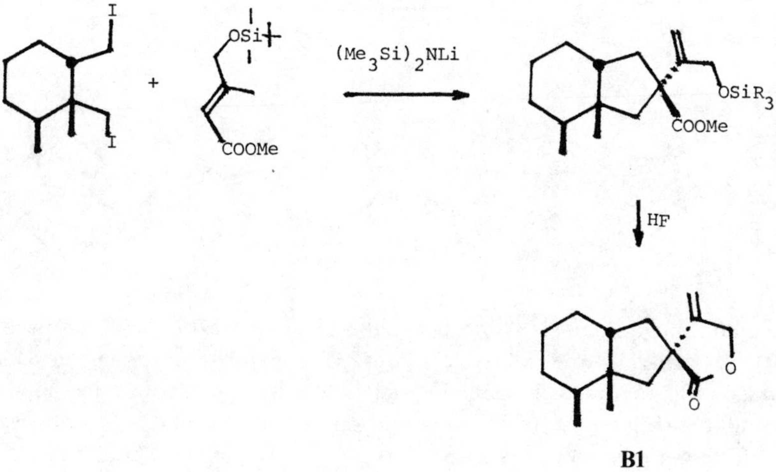

B1

5.2 FIVE-MEMBERED RINGS

Transannular cyclization occurs easily in medium rings. Thus, the treatment of 1,3-cyclooctadiene monoxide with alkali metal amides furnishes a bicyclo[3.3.0]octenol, which is conceivably a fitting precursor of many triquinane sesquiterpenes, such as coriolin (**C70**) (Shibasaki, 1980).

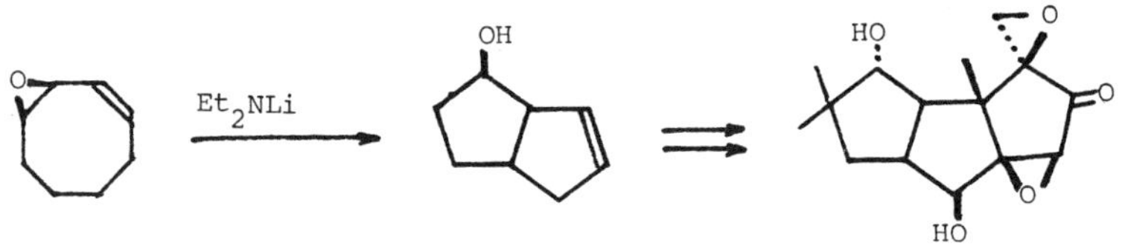

C70

The marine algal diterpene udoteatrial (**U1**), containing a cyclopentane ring flanked by two lactol ethers, is another synthetic target that can be related to the bicyclooctenol (Whitesell, 1983).

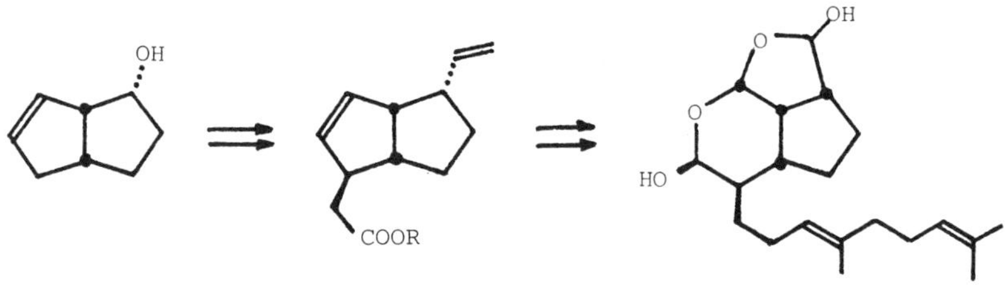

U1

Allamcin has been constructed from the bicyclo[3.3.0]octenone (Parkes, 1986). The additional five-carbon lactone fragment is attached via homologation of the carbonyl group and reaction of the derived α-acetoxyaldehyde with a 2-butenoic acid synthon.

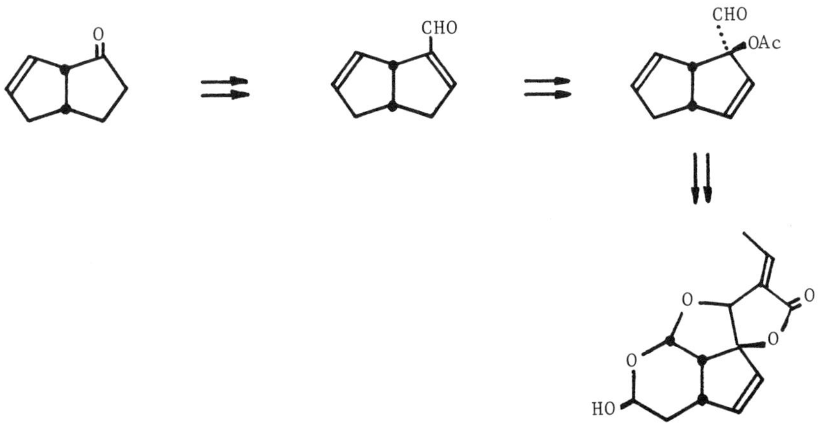

allamcin

Vetispiranes as a family have elicited the most intense synthetic inquiry among all the sesquiterpenes. In terms of brevity and stereocontrol, the magnificent method of twofold alkylation (Stork, 1973a) for a synthesis of β-vetivone (**V16**) deserves special accolade. This work firmly establishes a method for the kinetic enolization of 3-alkoxy-2-cyclohexenones.

Stereocontrol is important in the second step, the reaction of the nonallylic chloride. The secondary methyl group in the cyclohexenone moiety is solely responsible for the observed 1,2-asymmetric induction.

A modified procedure involving alkylation of a β-ketoester is of preparative value. A base is not needed in the cyclization step because the enolate can be generated from demethoxycarbonylation with lithium chloride (Eilerman, 1981).

5.3 THREE- AND FOUR-MEMBERED RINGS

The seemingly limited application of the internal alkylation method to the construction of cyclobutanes in the terpene domain is due to the relatively meager number of such compounds containing a four-membered ring. It is also probable that the efficient photocycloaddition method has totally eclipsed others.

5.3 THREE- AND FOUR-MEMBERED RINGS

Copaene/ylangene (**C66/Y1**) and α-*trans*-bergamotene (**B5**) represent two structural types that have yielded to alkylative annulation. The two syntheses of copaene/ylangene differ more in the manner of assembling the bicyclic intermediates. Thus the development of a Diels–Alder route to *cis*-fused homonuclear hexalone facilitates the preparation of the required precursor (Corey, 1973b), whereas a more traditional approach from the Wieland–Miescher ketone (Heathcock, 1966, 1967) gives rise to a tricyclic alkoxyketone from which the skeletal ethereal carbon atom is destined to become an unsaturation site.

The indicated planar structures of the two tricyclic intermediates are less than immediately correlative. On closer examination, the spatial relationships between the ketone and the alkoxy group and that between the ketone and the trigonal carbon are identical (i.e., 1,4-related).

The synthesis of α-*trans*-bergamotene (**B5**) (Larsen, 1977) and α-pinene (**P30**) (S. Monti, 1978) involving intramolecular displacement of a mesylate group represents a splendid exercise in design. The alkylation step leads to a tricyclic intermediate that is ready for a directed fragmentation to unveil the characteristic framework. The versatility of the keone function should be noted. It is an activator in the cycloalkylation step and an acceptor to the amide anion in the Haller–Bauer type of reaction.

B5

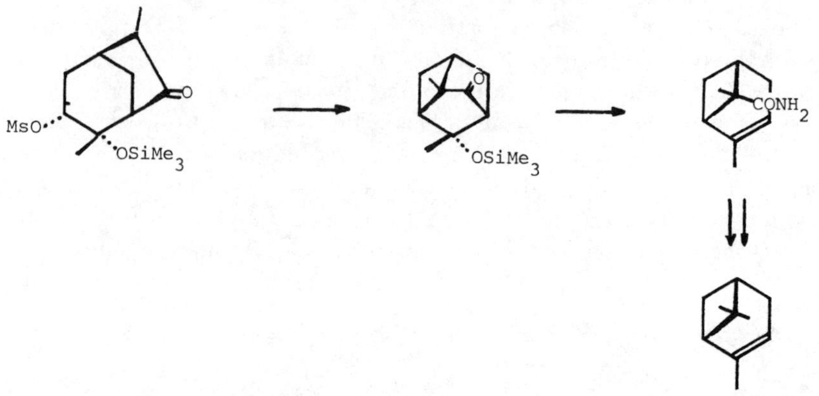

Another synthesis of the pinenes (**P30/P31**) (M. Thomas, 1973) hinges on a cycloalkylation within the framework of a cyclohexanone. Blocking of the less hindered α-methylene group is needed to achieve regioselectivity.

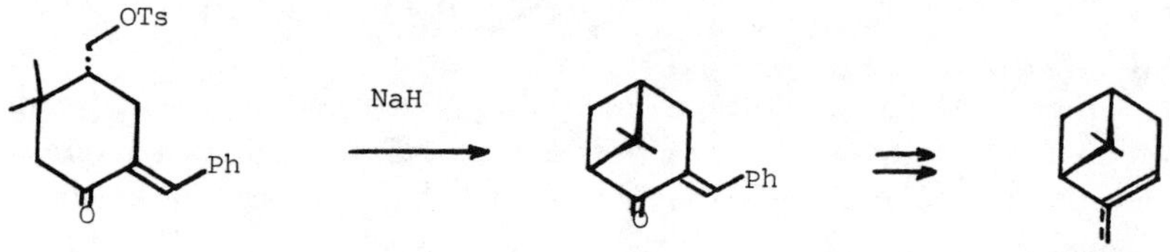

Two syntheses of grandisol (**G18**) embody cyclobutylation. The one involving epoxide opening by an α-cyanosubstituted carbanion (Stork, 1974) shows a high degree of stereoselectivity, which arises from steric origin. The other method entails alkylation of an ester enolate (Babler, 1975).

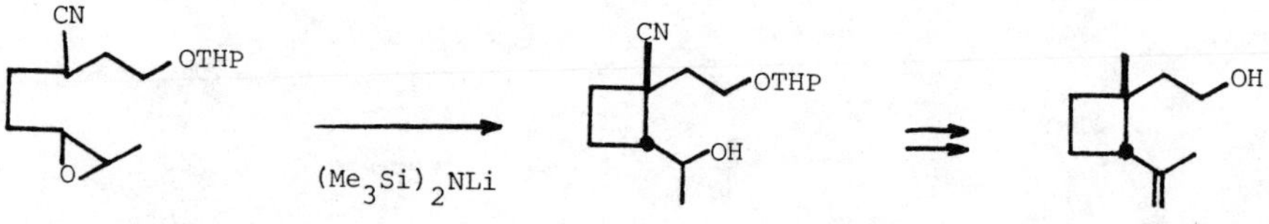

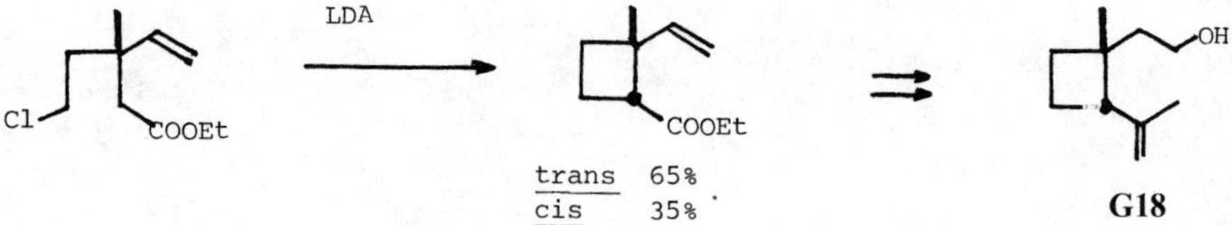

(+)-Lineatin (**L19**) has been assembled from D-ribonolactone (Kandil, 1985) by a process featuring internal cyclization of a cyanomesylate.

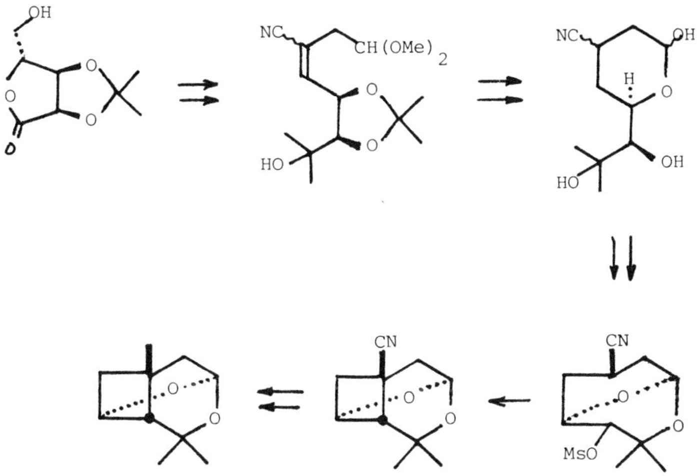

L19

Cyclopropanation via 1,3-elimination is an integral operation in a number of terpene syntheses. Bicyclic derivatives are formed by base treatment of the proper γ-halo- or γ-sulfonyloxy-carbonyl compounds, and these are employed in the synthesis of chaminic acid (**C49**) from dimethyl 5-hydroxyisophthalate (Gensler, 1973), cycloartenol (**C91**) from lanosterol involving functionalization at C_{19} by photolysis of an 11β-nitrito derivative in the presence of iodine, and cyclization of the iodoketone (Barton, 1969) and ishwarane (**I11**) from a tricyclic tosyloxy-ketone (R. B. Kelly, 1972).

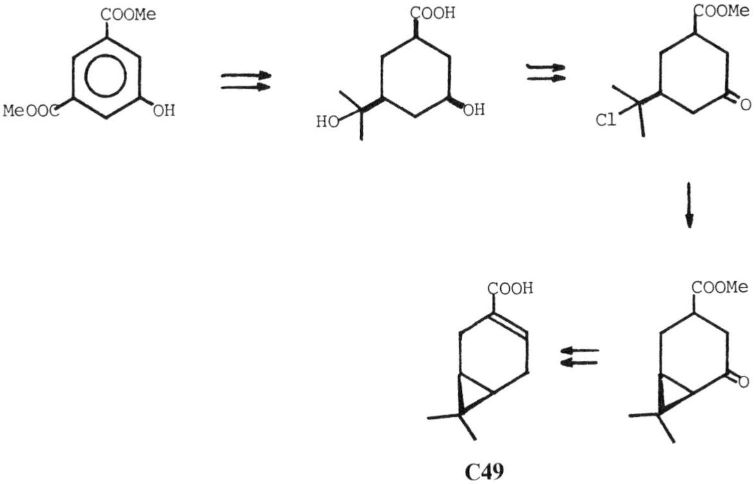

C49

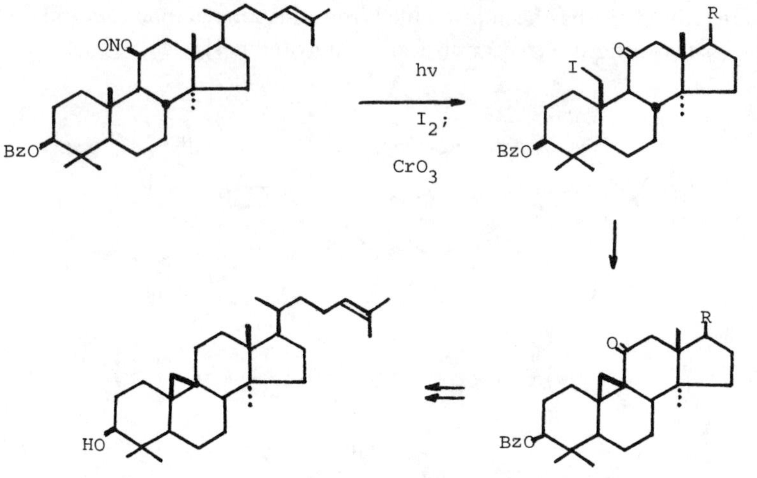

A formal 1,3-dehydration of γ-hydroxyketones occurs when certain structural requirements are met. Sabina ketone (**S1′**) has been obtained in such a manner (Alexandre, 1972).

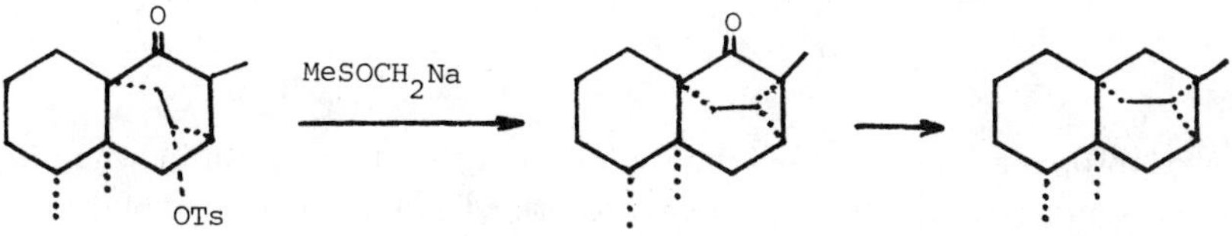

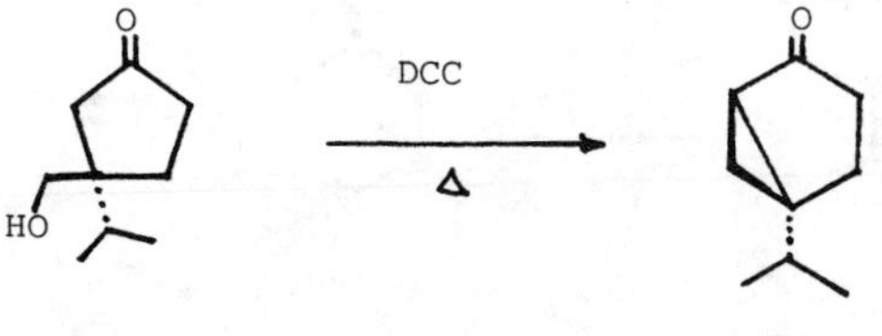

Cyano and ester groups are also capable of activating the α carbon to effect cyclopropanation, although stronger bases for deprotonation are usually required.

Marasmic acid (**M6**) contains an oxapropellane moiety in which a cyclopropane ring is astride a tetrahydrophthalaldic acid (in lactol form). Two successful syntheses (W. J. Greenlee, 1976; Boeckman, 1980) follow the intramolecular Diels–Alder strategy to provide the two carbocycles. Useful pendant groups for the elaboration of the remaining molecular architecture are also incorporated. The most important step is a 1,3-elimination.

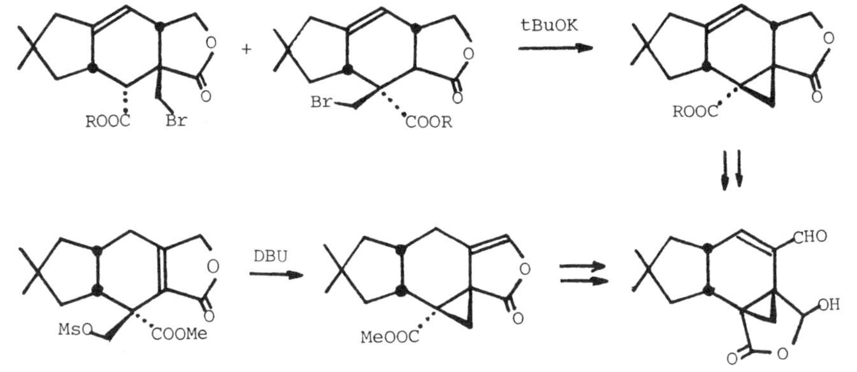

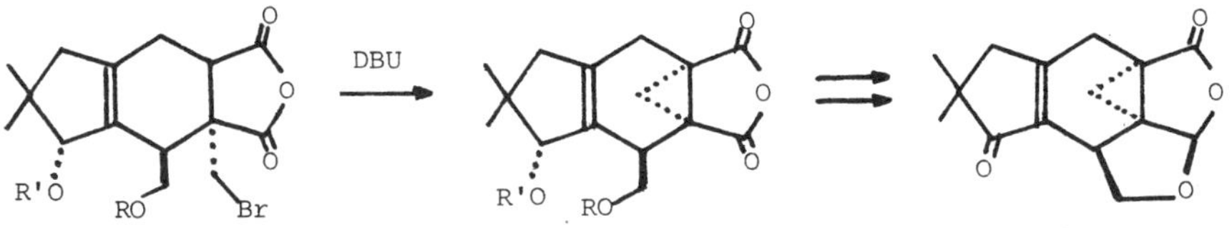

In structural terms, sterepolide (**S51**) differs from marasmic acid only in a transposed cyclopropane ring, and the oxidation level at various skeletal carbon atoms. The convenience of the Diels–Alder approach (Trost, 1985a) followed by a 1,3-elimination process to this molecule has been admirably demonstrated.

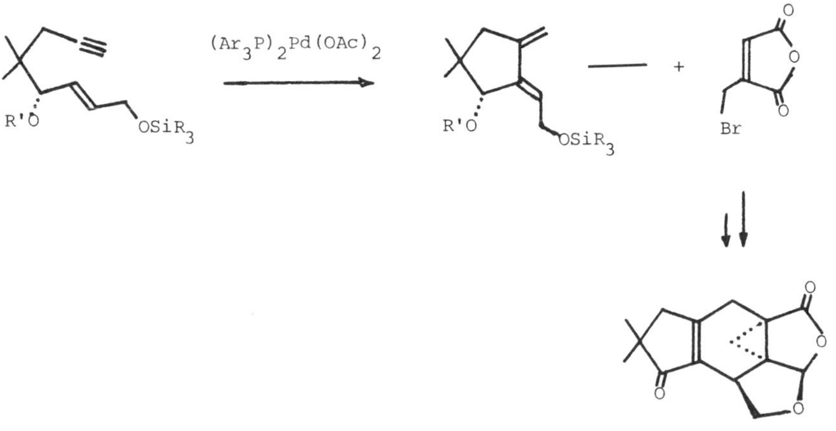

A key operation in a synthesis of cycloeudesmol (**E20**) (Ando, 1985) is the segmentation of the B-ring of an octalin epoxide by intramolecular alkylation. The required configuration of the epoxide ring is established with the aid of an allylic hydroxyl group.

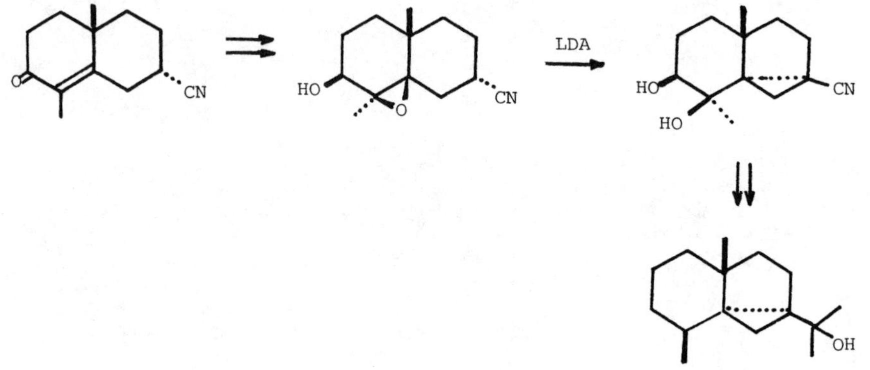

E20

The key operation is also used in a synthesis of (1*R*,3*S*)-*cis*-chrysanthemic acid (**C51c**) (Buisson, 1984). The optically active alcohol intermediate is obtained by microbiological reduction of the symmetrical diketone.

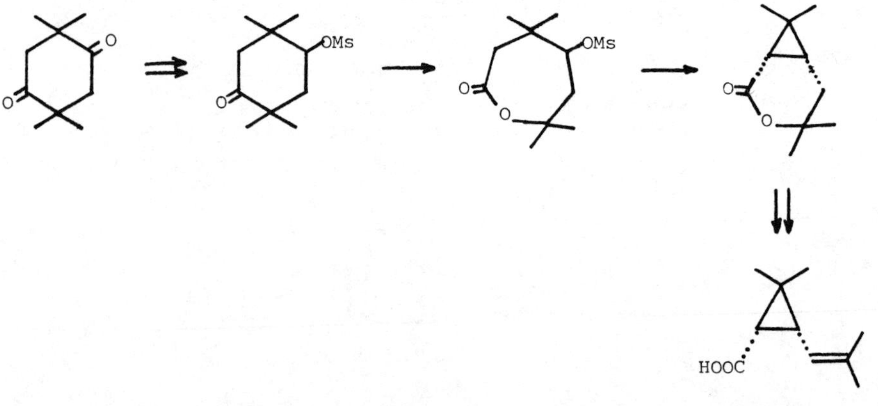

C51c

Many approaches to *trans*-chrysanthemic acid (**C51**) are based on cyclization of γ-chloroesters (M. Julia, 1964; Takeda, 1977; Garbers, 1978).

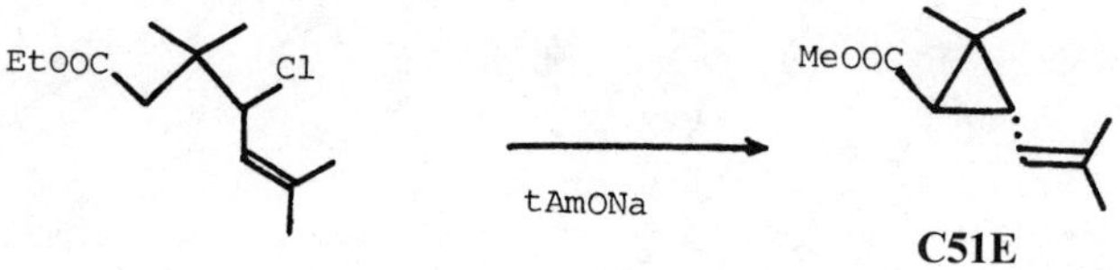

C51E

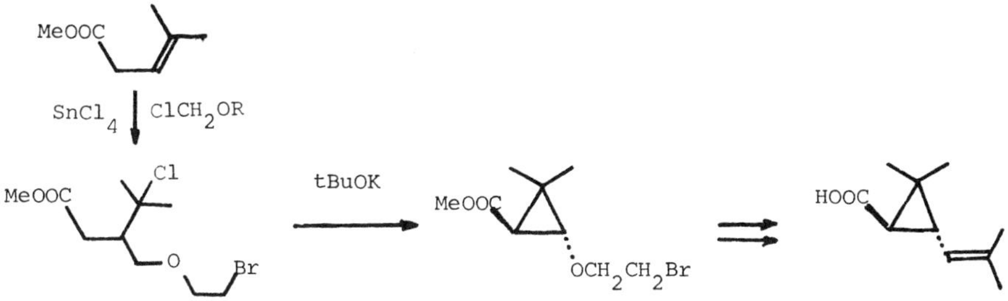

C51

C51E′

γ,δ-Epoxycarboxylic derivatives, which are readily available via homologation of allylic alcohols by the Claisen rearrangement and epoxidation, undergo cycloalkylation to give cyclopropyl compounds. Application of this sequence to the synthesis of chrysanthemic acid is on record (Babler, 1976; Matsuo, 1976; Majewski, 1984).

C51A

C51E′

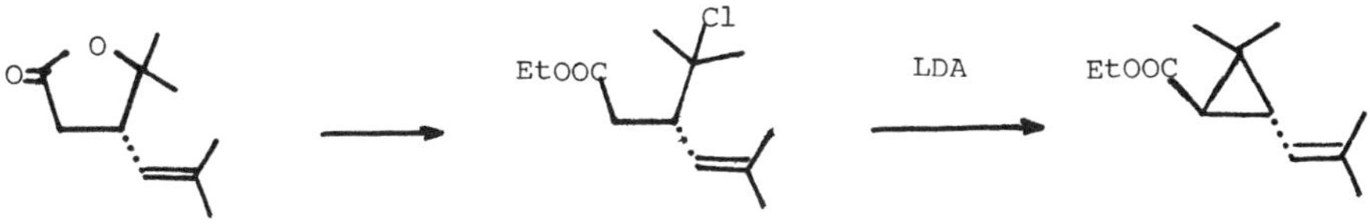

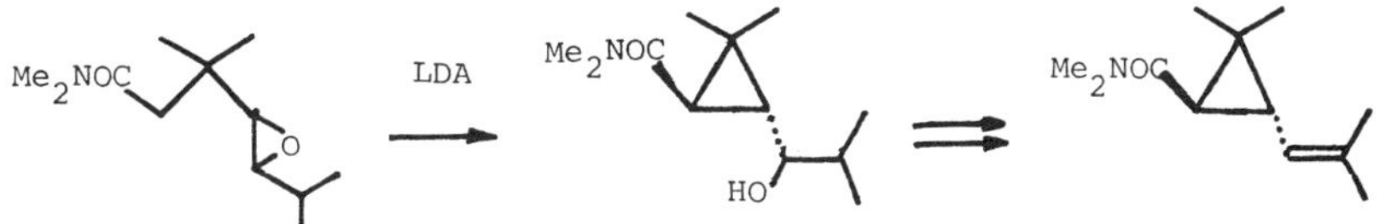

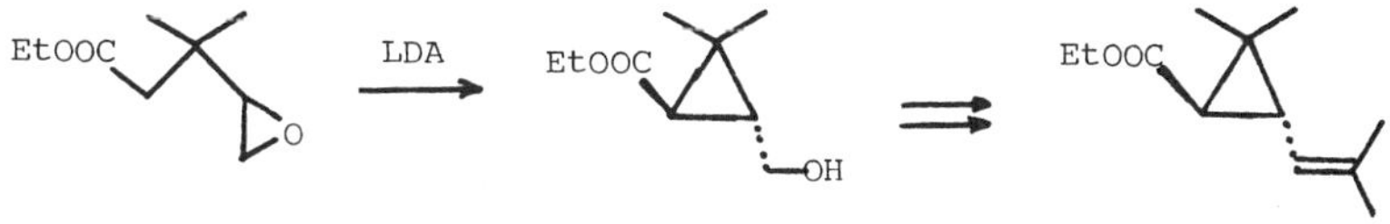

Phenylthiolate anion appears to be an adequate leaving group in the internal displacement by an allylic carbanion. The cyclization of C_{10} alcohol to form chrysanthemol (**C52**) as a cis/trans mixture (2:8) is biomimetic (Babin, 1981).

trans 80
cis 20

C52

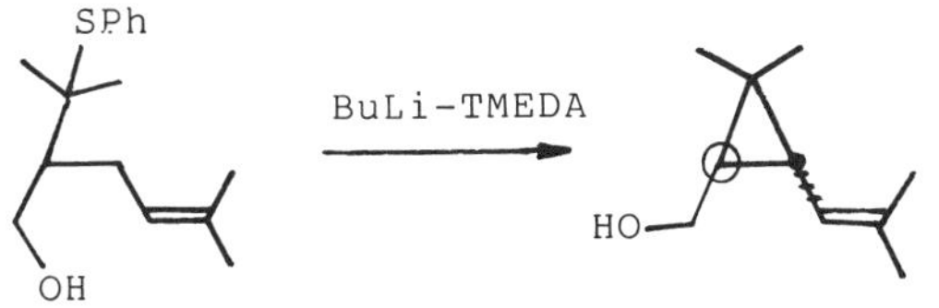

The intramolecular S_N2' displacement of an acyloxy group is also effective in cyclopropane formation. While both cis and trans isomers are produced in about equal quantities (Ficini, 1983; cf. Genet, 1980), a stereoselective annulation is observed with the cyanosulfone analog (Genet, 1981) that enables a facile acquisition of *cis*-chrysanthemic acid (**C51c**).

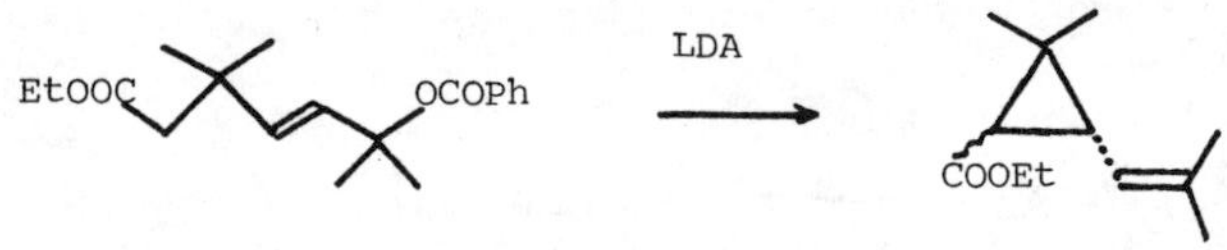

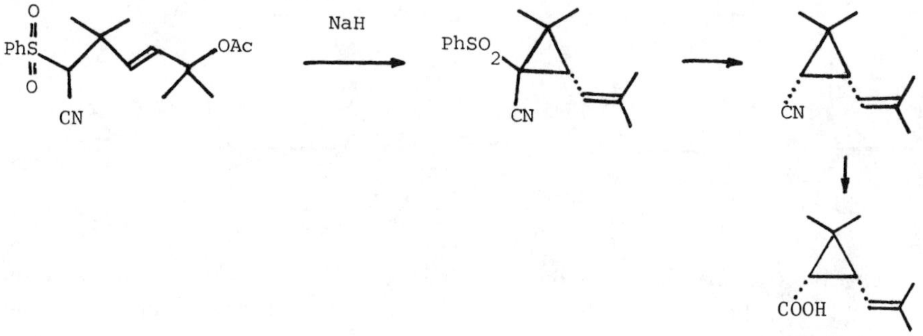

The ability to achieve a long-range displacement of an η bromine atom from a γ,δ; δ,ζ-unsaturated ester to provide a hemicaronic aldehydic acid is remarkable (Kondo, 1976).

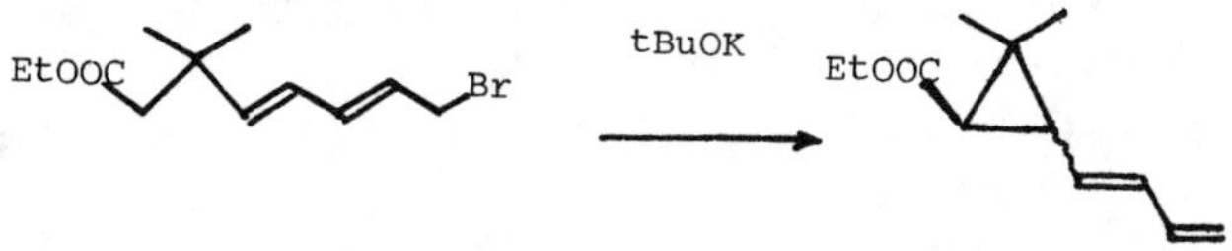

From carvone, a chloroisopropyl lactone is available by treatment with hydrogen chloride and modification of the enone system. Cyclization of the lactone gives dihydrochrysanthemolactone, a well-known precursor of *trans*-chrysanthemic acid (**C51**) (Torii, 1983). Alternatively, methyldihydrocarvone can be elaborated into the same compound (Ho, 1982b).

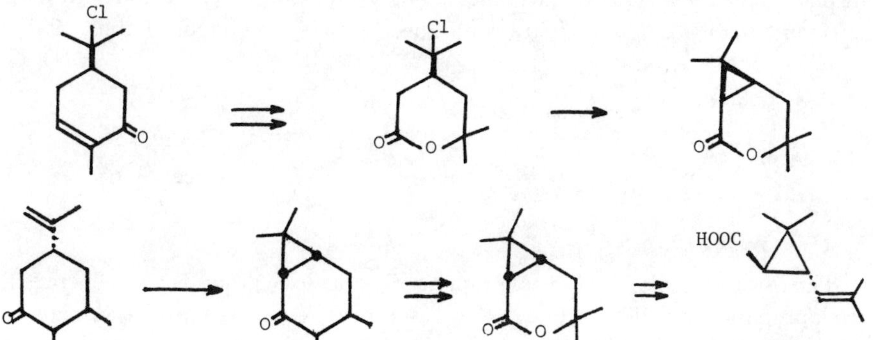

5.3 THREE- AND FOUR-MEMBERED RINGS

Cyclopropanation across the γ and ε positions of conjugated carbonyl compounds does not present any difficulties. Several terpene intermediates including sirenin (**S38**) (Garbers, 1975), (—)-aromadendrene (**ent-A41**), (Büchi, 1966, 1969), sativenediol (**S12**) (McMurry, 1976), and maaliol (**M1**) (Bates, 1960) are available in this way.

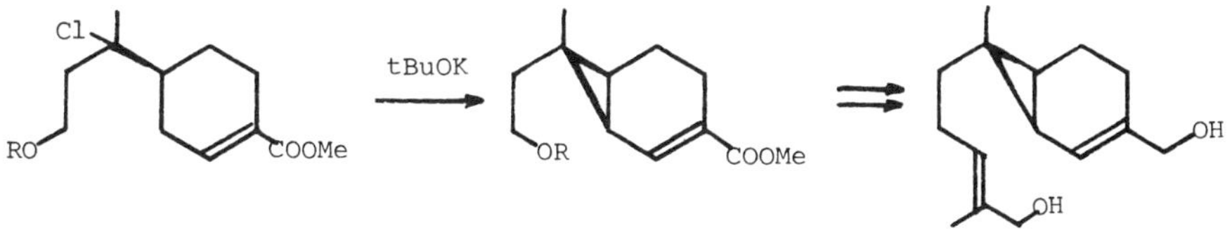

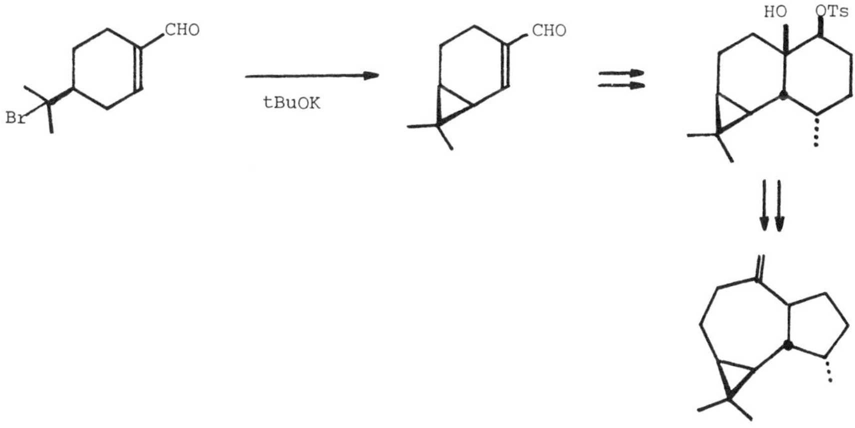

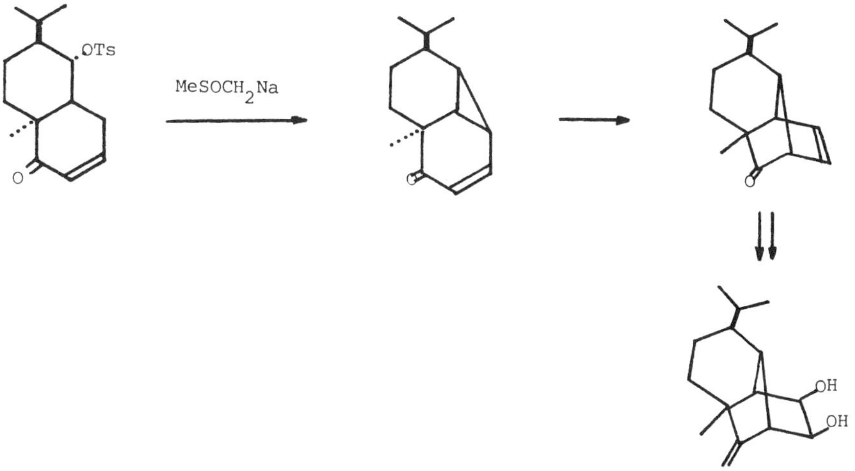

A tautomeric shift (electrocycloreversion) of the norcaradiene to cycloheptatriene is virtually automatic and complete except for compounds substituted heavily with electron-withdrawing groups. This knowledge enables synthesis of cycloheptatrienes via fusion of a cyclopropane to a six-membered ring and modification of the latter moiety into a homonuclear diene unit.

The remarkable sponge metabolite called spiniferin I (**S45**) is a bridged cycloheptatriene. Indeed the most viable synthetic plan (Marshall, 1983) embodies an internal reductive alkylation.

5.4 MEDIUM-SIZED RINGS

On account of unfavorable entropy and nonbonding interactions in the transition states leading to mesocycles, medium-sized ring compounds have eluded attempts at synthesis for many years. Usually, the presence of multitrigonal atoms in the carbon chain to be cyclized restricts the number of possible conformations and therefore facilitates the closure of medium-sized rings containing several multiple bonds, while simultaneously creating certain pitfalls.

5.4 MEDIUM-SIZED RINGS

The germacrene skeleton is now attainable via intramolecular opening of an epoxide by an α-sulfenyl carbanion, as shown in the synthesis of hedycaryol (**H2**) (Kodama, 1976) and obscuronatin (**O1**) (Kodama, 1984). Other macrocyclic terpenes that have yielded to an analogous approach include cembrene A (**C40**) (Kodama, 1975), 3Z-cembrene A (**C41**) (K. Shimada, 1981), and cubitene (**C82**) (Kodama, 1982). [Dotted lines in the formulas indicate bonds that are formed by this process.]

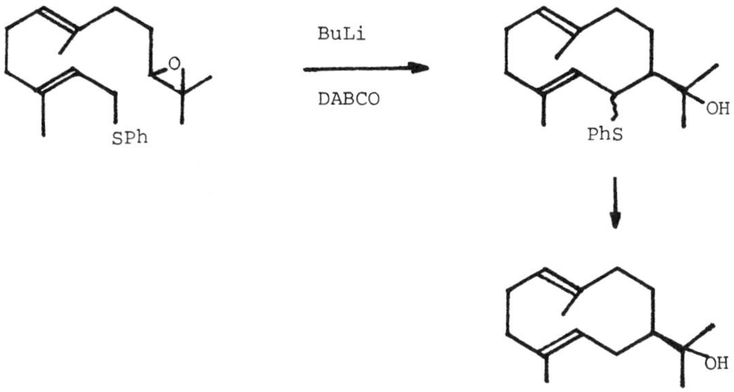

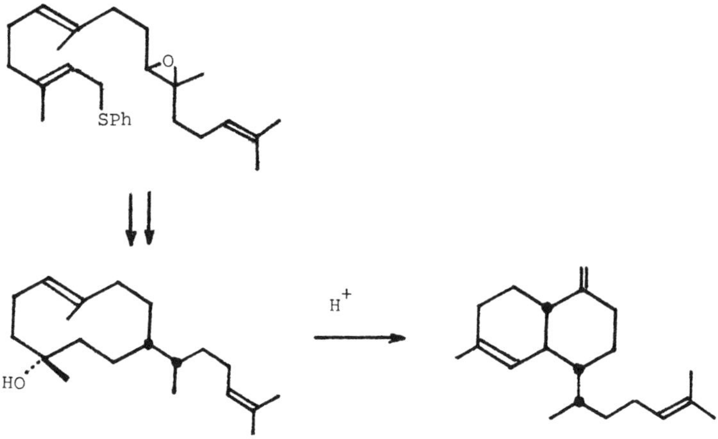

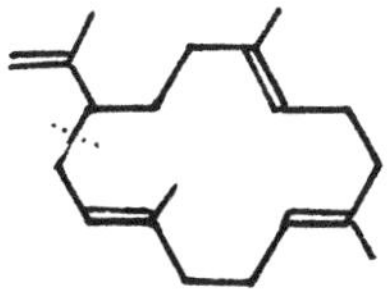

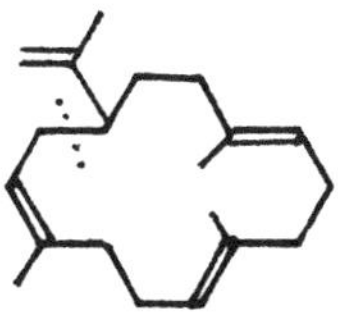

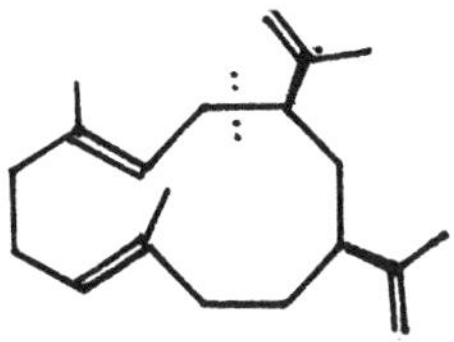

The intramolecular alkylation of an α-phenylthioacrylate to provide a cyclodecene is the pivotal operation in a synthesis of (—)-periplanone B (**P22**) (T. Kitahara, 1986) from (+)-limonene.

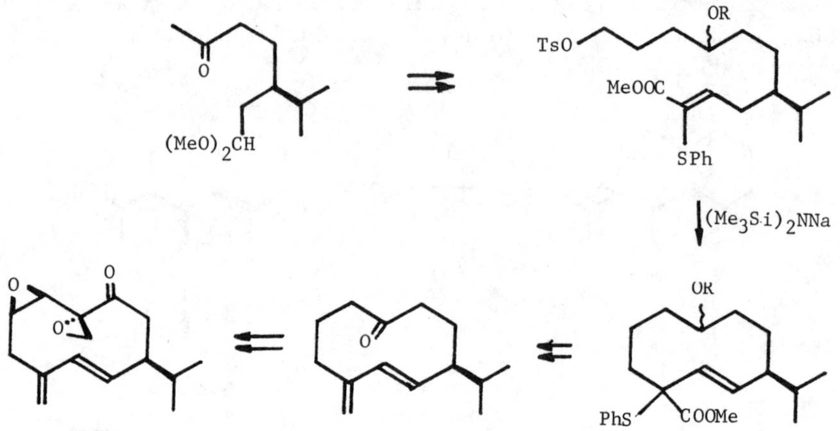

Synthetically, α-sulfonyl carbanions are easier to generate and equally effective in undergoing alkylation in comparison with their sulfenyl counterparts. Among the uses of such anions is the formation of 14-membered trienes characteristic of the cembranes (cf. neocembrene, (**C42**: Takayanagi, 1978).

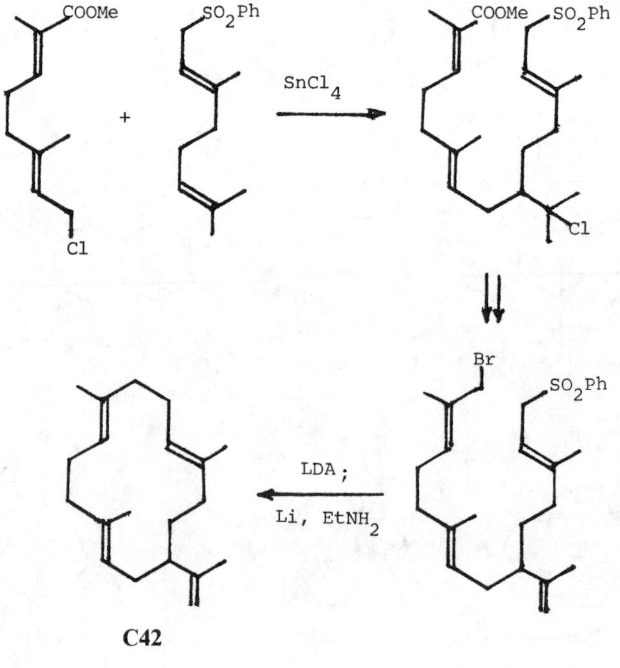

A synthesis of asperdiol entails closure of the medium-sized ring by an internal alkylation of an ω-halosulfone with $(Me_3Si)_2NK$/18-crown-6 (Marshall, 1986b). The acyclic intermediate is assembled by a Wittig–Horner–Emmons condensation. Interestingly, an independently conceived approach to the same diterpene (Tius, 1986) employs these two key processes in reverse order. In other words, the sulfone alkylation serves to join the building block derived from geraniol and a synthon containing the two asymmetric carbon atoms; the cyclization is achieved by a Wittig-type reaction.

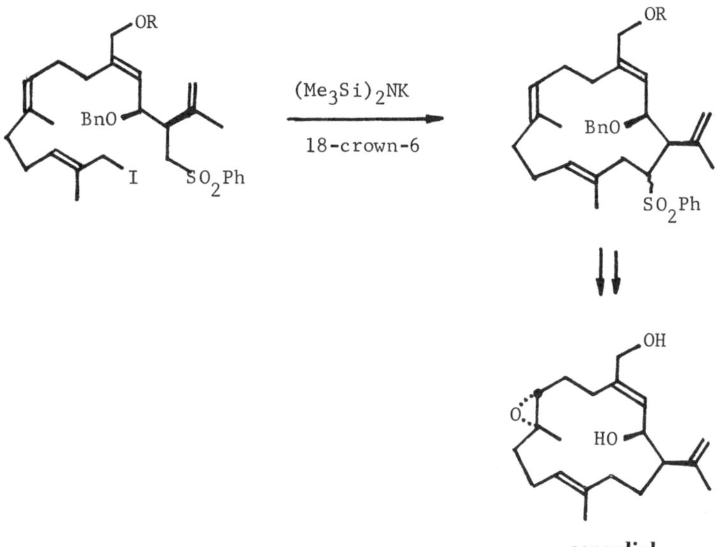

asperdiol

Another method applicable to the synthesis of these mesocyclic polyenes is alkylation of nitriles. Protected cyanohydrins are particularly useful in delivering cyclic ketones: germacrone (**G9**) and acoragermacrone (**G10**) (Takahashi, 1983a, 1983b) and periplanone B (**P22**) (Takahashi, 1986a).

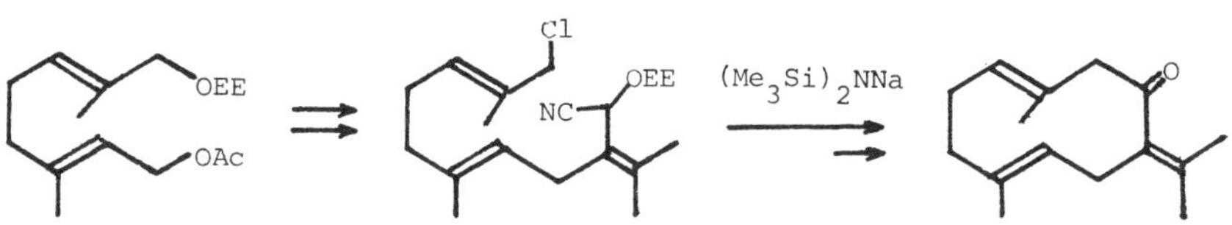

G9

G10

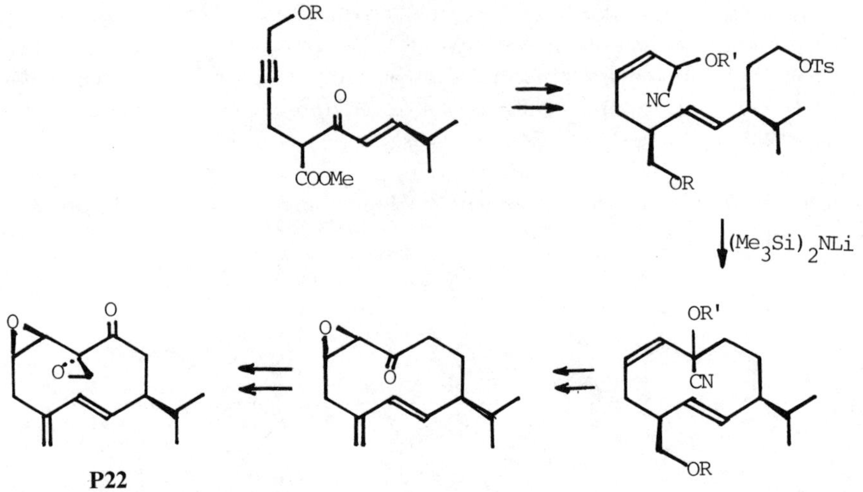

P22

The preparation of the secotaxane skeleton (cf. verticillol, **V13**) has been realized (T. Kato, 1978).

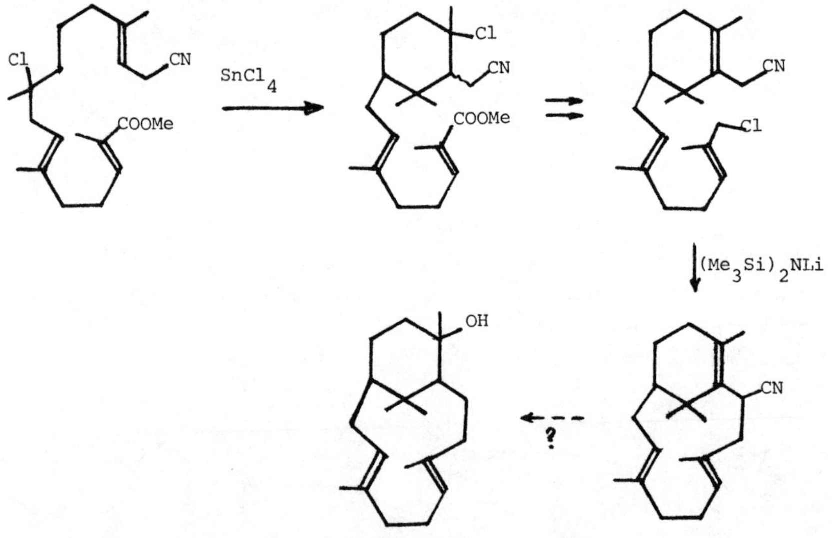

V13

Allylic acetates are rendered electrophilic on treatment with palladium(0) species to form η^3-allyl complexes that react with stabilized carbanions. The template effect of the organometallic compounds has a dramatic influence on the efficiency of ring closure. In this context, a synthesis of the hop constituent humulene (**H22**) (Y. Kitagawa, 1977) and germacrene A (**G5**) (Itoh, 1978) based on the 11-membered β-ketoester is distinguished by its conciseness.

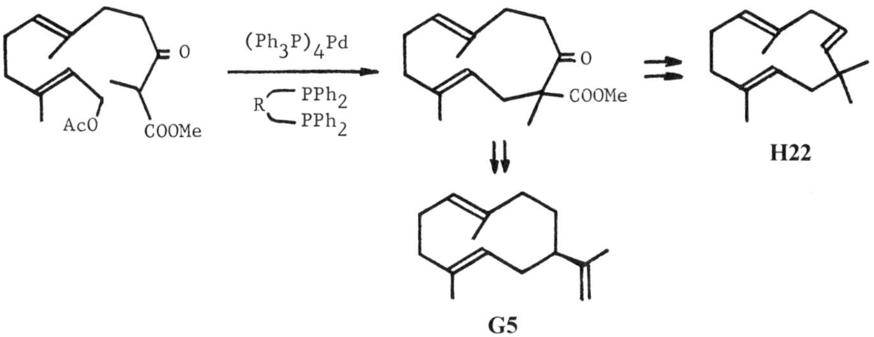

5.4 MEDIUM-SIZED RINGS 245

G5

H22

The macrocyclization of sulfone ester initiated by η^3-allylpalladium species is also featured in a total synthesis of isolobophytolide (Marshall, 1986a).

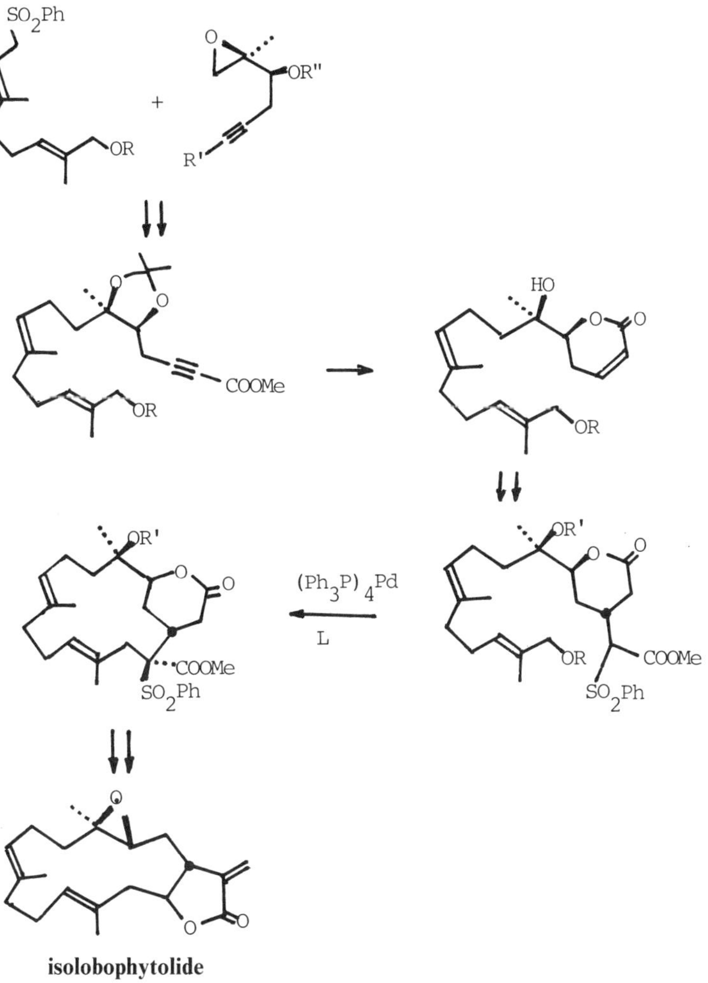

isolobophytolide

A 14-membered cembranolide precursor is readily available by macrocyclization of an ω-(α-alkoxy allylstannyl) aldehyde (Marshall, 1987a): the syn isomer being generated predominantly (ratio 7:1). The corresponding acetylenic ketone undergoes β-methylation with a cuprate reagent and equilibration to the more stable all-E triene. Conventional steps culminated in the *L. michaelae* cembranolide.

L. michaelae cembranolide

Incipient vinylic anions generated from vinylboranes also react with η^3-allylpalladium species. Consequently, an even more direct approach to humulene (**H22**) (Miyaura, 1984) is realizable.

It may be proper to consider the coupling of allylic halides as a reductive alkylation process. The reaction mediated by nickel carbonyl proceeds via a bis(η^3-allyl)nickel complex (Semmelhack, 1972). Successful exploitation of an intramolecular version of this coupling has enabled synthesis of many cyclic 1,5-dienes including humulene (**H22**) (Corey, 1964a).

In the actual pursuit of this intriguing terpene, the E,Z,E-isomer of the dibromoundecatriene is the preferred substrate, although an isomerization of the central double bond $(Z \to E)$ with light and diphenyl disulfide adds an extra step to the endeavor.

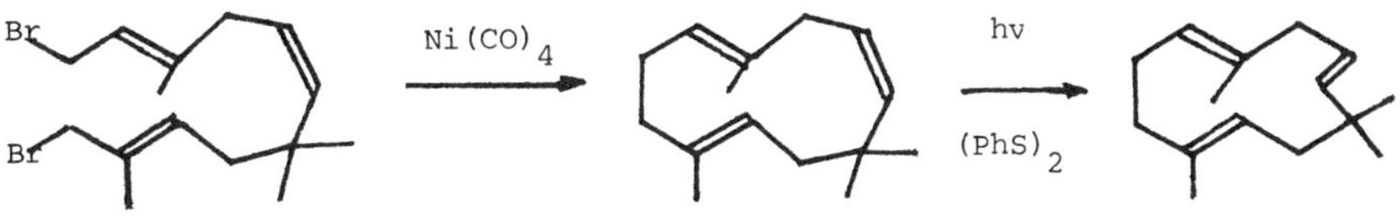

H22

Syntheses of cembrene (**C39**) (Dauben, 1974, 1975a) and casbene (**C34**) (Crombie, 1976) are keyed on the same coupling method.

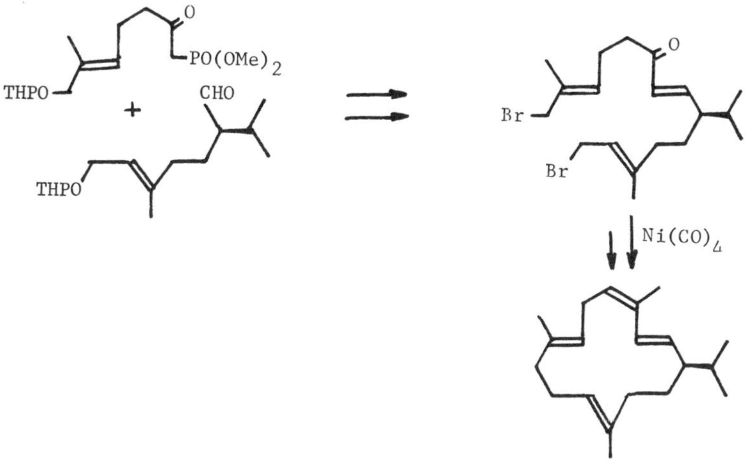

C39

C34

5.5 UNCONVENTIONAL ALKYLATIONS

In earlier sections alkylation of carbonyl compounds invariably involves bond formation at the α-carbon atom, with a carbon bearing the leaving group. The alternative mode of bond formation, which identifies the carbonyl group as an acceptor, is traditionally much more difficult to perform intramolecularly.

Intramolecular Grignard and related reactions are generally impractical because of poor selectivity and other complications. On the other hand, the treatement of a halide or an activated ether with sodium metal can lead to the desired reaction when a carbonyl group is present nearby. Regardless of the actual mechanism (ketyl intermediate for coupling?), cyclization is the basis of several efforts toward synthesis of patchouli alcohol (**P15**) and norpatchoulenol (**P13**) (Danishefsky, 1968; Teisseire, 1974; Bertrand, 1980; Liu, 1987).

The discovery of a homoallylic alkoxide fragmentation permits unraveling of trichodiene (**T26**) norketone from a tricyclic skeleton (Snowden, 1984). The homoallylic alcohol intermediate used in this synthesis is obtainable from a bromoketone.

Pyrolysis of potassium 2-alkenylcyclobutanolates creates carbonyl compounds with an allyl anion attaching to the β position. Such compounds collapse automatically to give cyclohexenols (Cohen, 1983). A synthesis of (—)-β-selinene (**ent-S21**) helps demonstrate the utility of this method.

5.5 UNCONVENTIONAL ALKYLATIONS

ent-S21

An interesting cyclization of a bromoketone en route to a quassin (**Q3**) precursor (Stojanac, 1979) is mediated by zinc metal. The new ring is designated as a latent δ-lactone, which is to be established after cleaving the glycol, converting the ketol to a cyclopentanone, and subjecting the latter to a Baeyer–Villiger reaction.

The cyclization is ususual and may actually involve a β-elimination to generate an allylic osmate salt, which then undergoes annulation. The proximity of the reacting centers accounts for the success of this reaction.

Q3

A unique spiroannulation embodies the addition of a carbon nucleophile to a fulvene moiety and interception of the resulting cyclopentadienide ion with a carbonyl group in the side chain. Proper selection of the reagent (Me$_2$CuLi) and chain length permits the elaboration of a vetispirane intermediate (Büchi, 1976). Critical to this brilliant effort is the chemoselectivity of organocopper reagents and the curious absence of vicinal alkylation to give hydrazulene products.

V16

The attack on carbonyl groups by η^3-allylnickel species results in homoallylic alcohols. When the central atom of the allyl residue carries a bromine atom, further carbonylation at that site occurs and an α-methylene-γ-butyrolactone can be obtained in one step. This efficient transformation is included in a stereoselective synthesis of frullanolide (**F14**) (Semmelhack, 1981a).

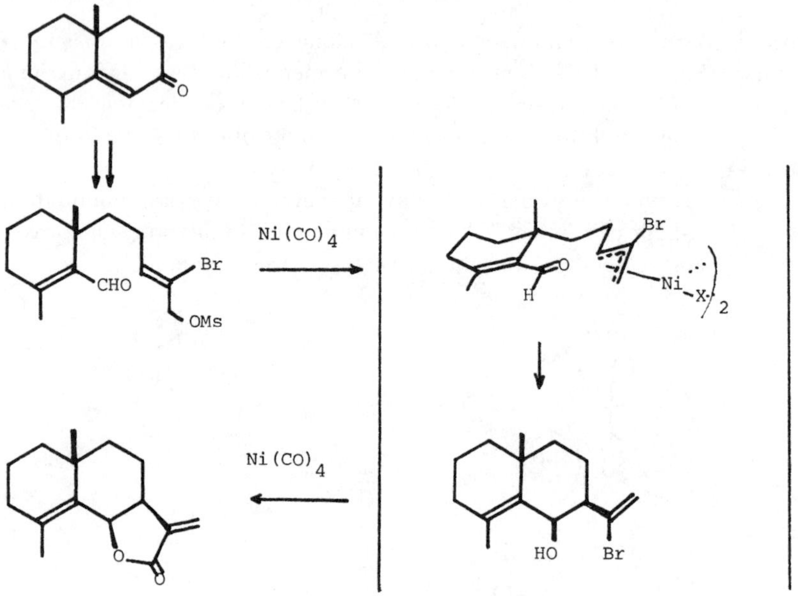

F14

The *in situ* carbonylation may be omitted if the central carbon atom of the allylic system is already installed with an ester function. In this case nickel(0) reagents other than the very toxic nickel carbonyl may be used. A synthesis of confertin (**C62**) (Semmelhack, 1978) employs a coupling of an allylic sulfonium salt with an aldehyde.

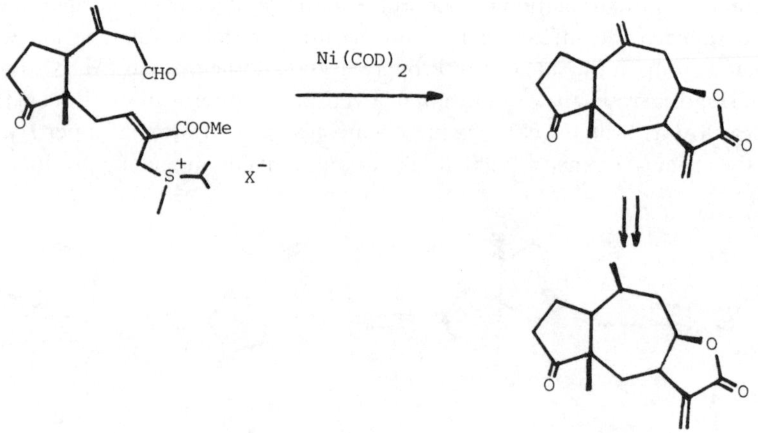

C62

5.5 UNCONVENTIONAL ALKYLATIONS

Allylchromium species are formed by reaction of allylic bromides with low-valent chromium reagents such as $CrCl_2$. Since the latter reagents do not attack carbonyl groups readily, intramolecular allylation may be accomplished. This process not only provides a partial solution to a synthetic condition that incapacitates the Grignard reaction, it is particularly valuable in assembling terpenes such as asperdiol (Still, 1983a), which contain a homoallylic alcohol system, and dihydrocostunolide (**C76**) (Shibuya, 1986), in which the corresponding unsaturation is oxidatively masked.

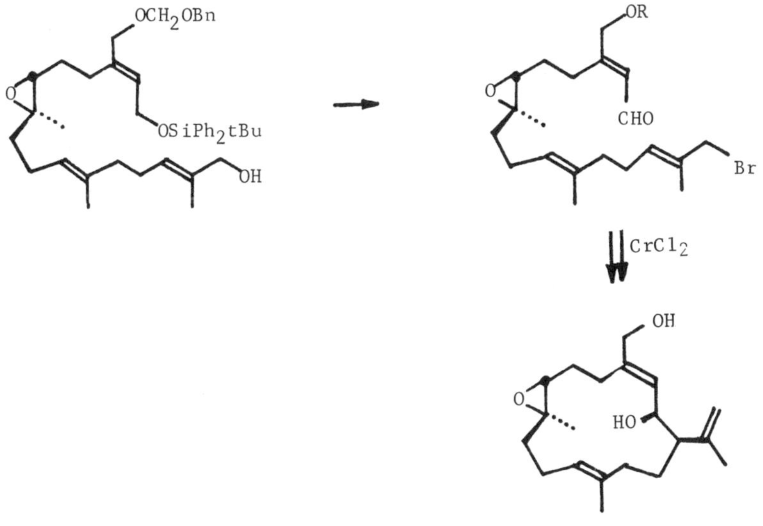

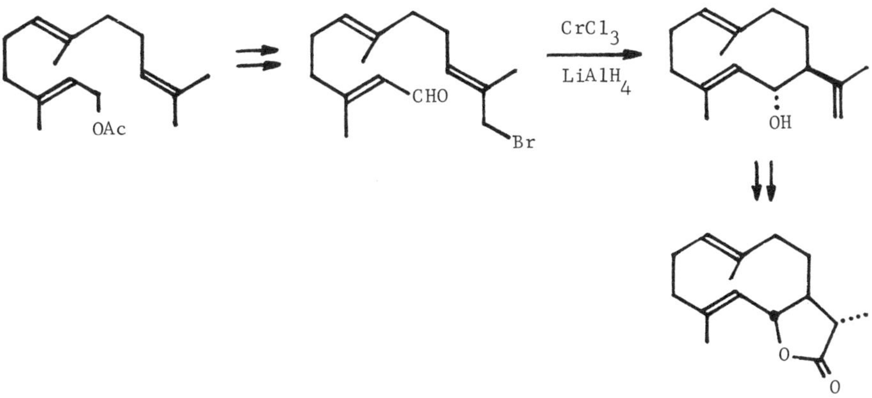

Incipient allyl carbanions can be generated from allylsilanes on treatment with fluoride ions. A cyclopentanation based on carbonyl trapping of a such carbanion is part of a synthesis of coriolin (**C70**) (Trost, 1980), which is distinguished by an unusual direction of ring growth.

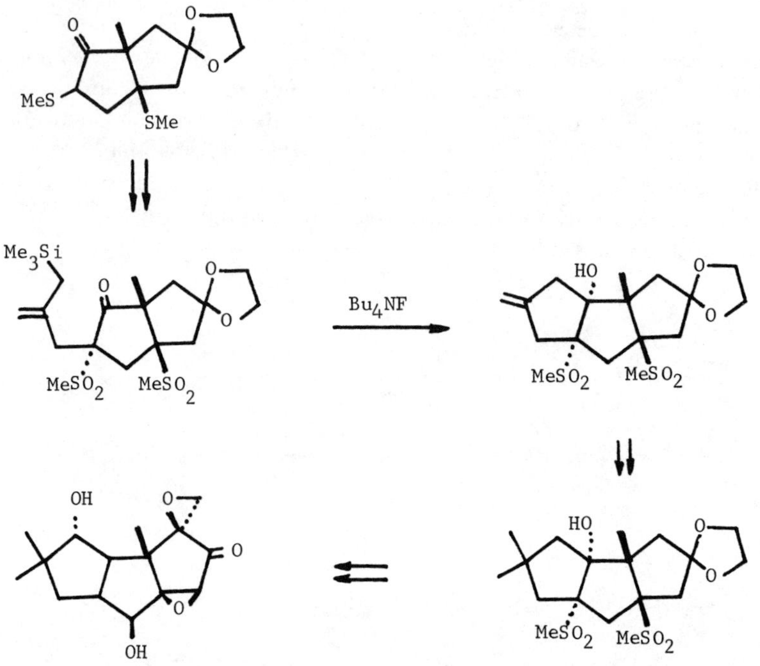

C70

An intramolecular Reformatsky-like reaction has been conducted in an attempt to synthesize arteannuin B (**A44**) (Goldberg, 1980). Unfortunately, only its stereoisomers are obtainable.

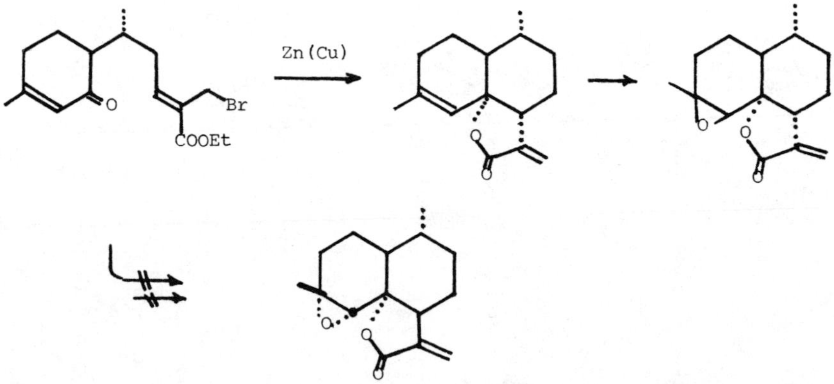

A44

The pivotal tricyclic ketone intermediate for gibberellic acid (**G12**) (Corey, 1979a,b) is independently available via reductive cyclization of a propargyl diketone monoethyleneketal with potassium in liquid ammonia (Stork, 1979), and by chemoselective coupling of an angular 2-bromopropenyl chain with the cyclohexanone moiety of the bicyclic diketone (Corey, 1982b; cf. Grootaert,

1982). The selectivity has been attributed to the lower strain of the cyclized product vis-à-vis that derived from coupling with the cyclopentanone.

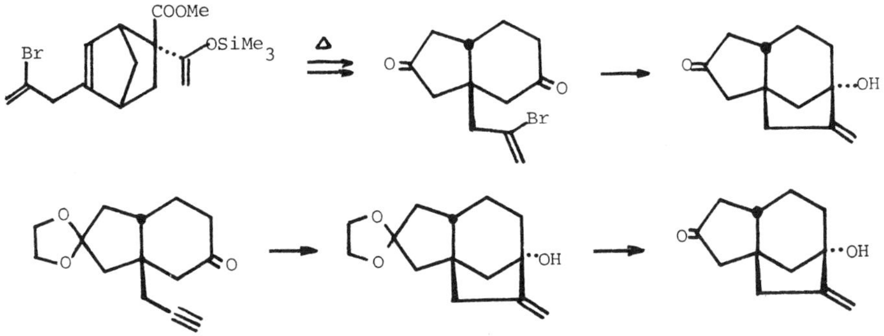

Coupling of a vinyl triflate with a vinylstannane leads to conjugated diene. The intramolecular version is a valuable annulation procedure in which diene units such as that present in (14S)-dolasta-1(15),7,9-trien-14-ol (**D17**) can be regioselectively created. In a synthesis of this terpene (Piers, 1986a) the allylic alcohol segment arises from a Grignard reaction.

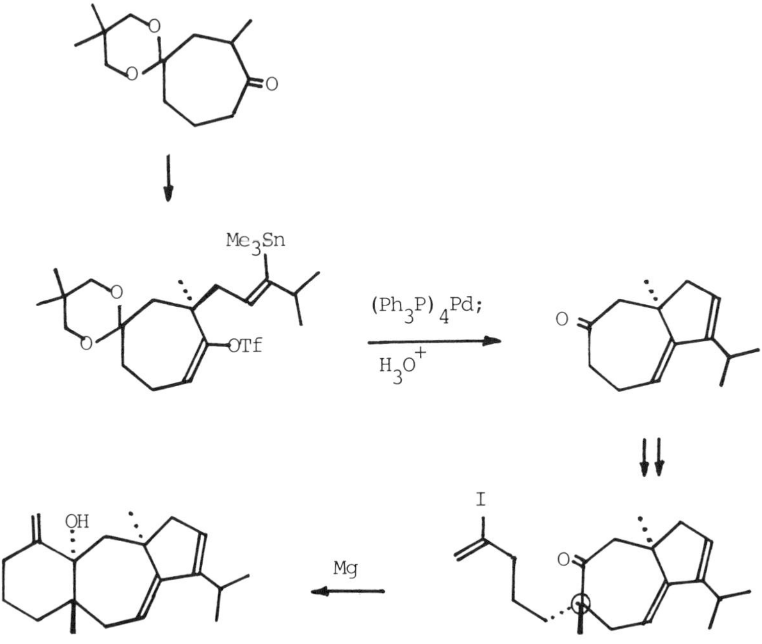

Application of a relatively new carbonylative alkylation method involving displacement of a tosylate with disodium tetracarbonylferrate and intramolecularly trapping the acylferrate intermediate with an alkene to induce the formation of a bicyclo[3.2.1]octanone moiety highlights a synthesis of aphidicolin

(**A36**) (McMurry, 1979). The remarkable feature of this reaction is the preferential formation of a six-membered (vs. five-membered) ketone.

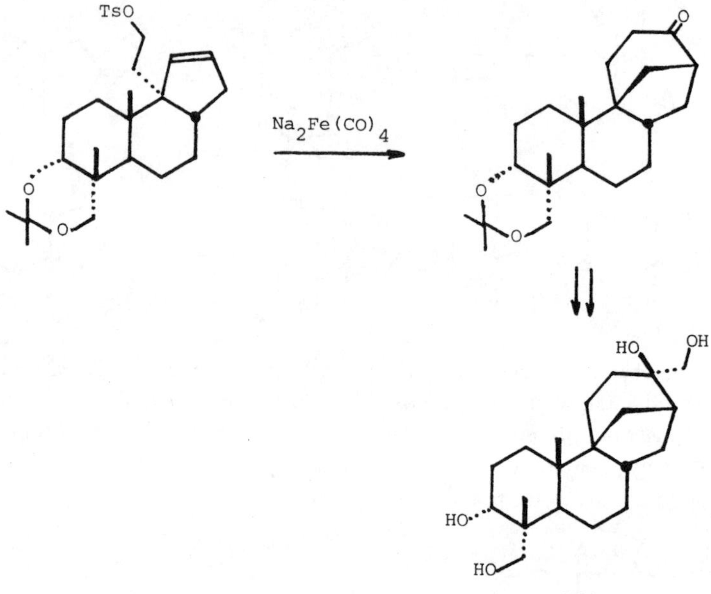

A36

The intramolecular interaction of an enol(ate) species with an alkene linkage via *in situ* activation of the latter represents a relatively new development in C—C bond formation techniques. *N*-Phenylselenylphthalimide is a soft electrophile that readily forms selenonium ions with alkenes. The neutralization of a selenonium ion by an appropriate carbon nucleophile in the same molecule results in a new ring structure. The selenyl group is removable by reduction.

This reaction sequence has been successfully applied to a synthesis of hirsutene (**H17**) (Ley, 1982). Significantly, only the cis-anti-cis product is found.

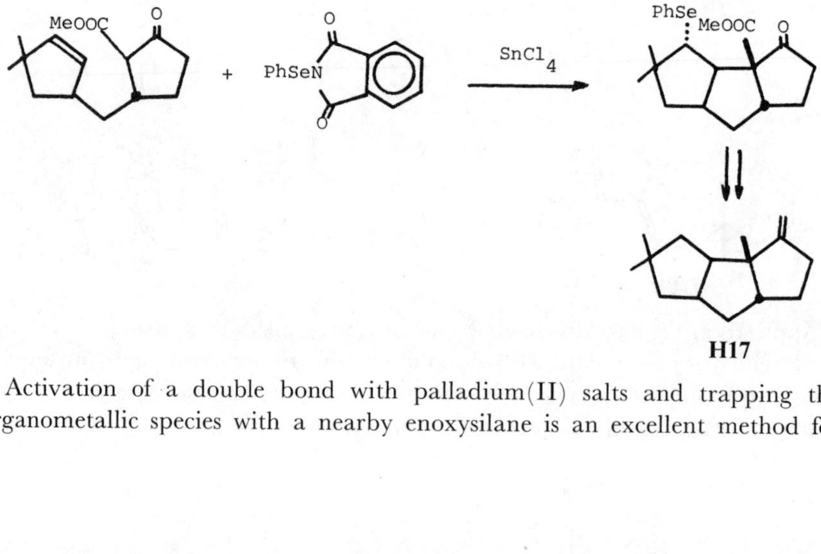

H17

Activation of a double bond with palladium(II) salts and trapping the organometallic species with a nearby enoxysilane is an excellent method for

5.5 UNCONVENTIONAL ALKYLATIONS

generating β,γ-unsaturated ketones. A synthesis of quadrone (**Q1**) (Kende, 1982) shows an efficiency seldom expected of a more classical approach.

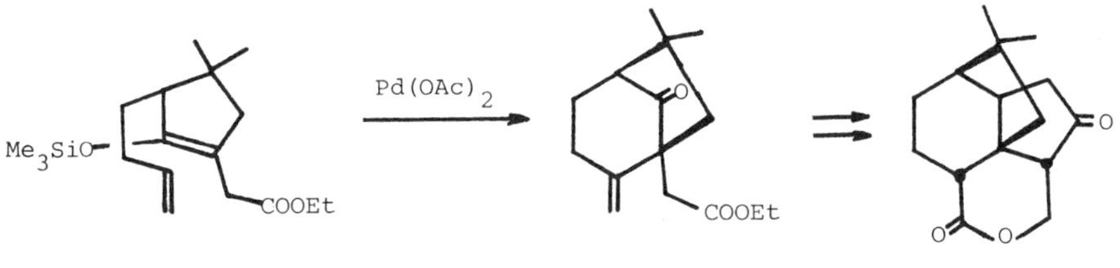

This reaction also provides a convenient entry to the bicyclic enone intermediate of $\Delta^{9(12)}$-capnellene-$3\beta,8\beta,10\alpha$-triol (Shibasaki, 1986). The deconjugated endocyclic enone is obtainable via equilibration of the intial mixture of double bond isomers.

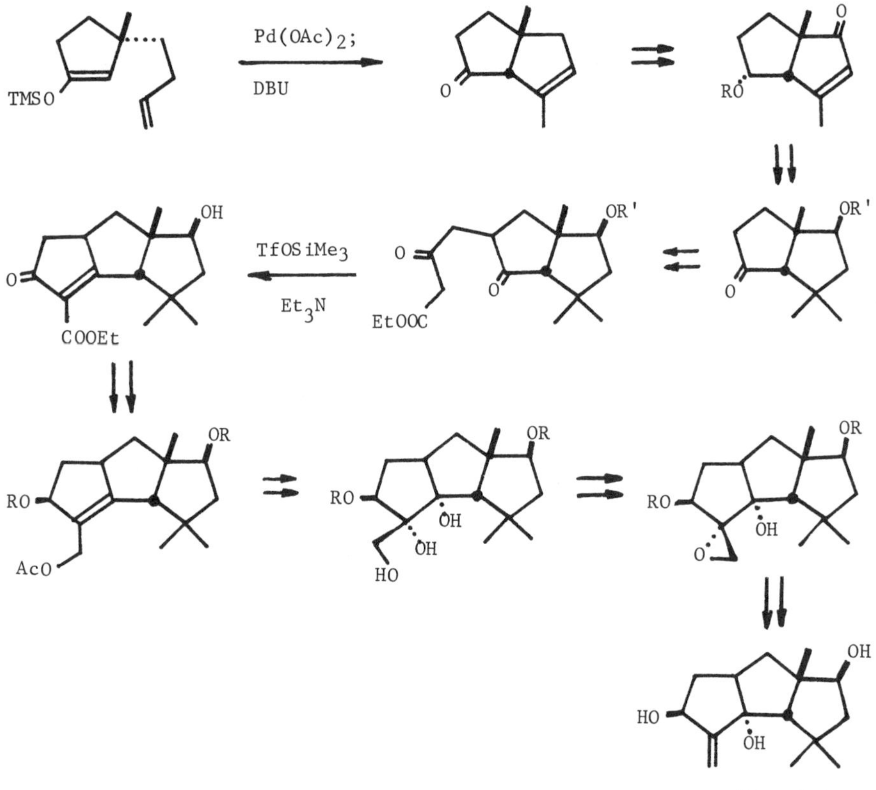

capnellenetriol

A related reaction is the palladium(II)-catalyzed cyclization of enynes, which supplies the better half of the carbon constituents of stereopolide (**S51**) (Trost, 1985a).

Chapter 6

Cationic Cyclizations

Intramolecular reaction of a cationic center or incipient cation with a multiple bond results in cyclic products, which can be unsaturated ketones, alcohols, or alkenes, depending on the nature of the substrates. As will be seen later, saturated cycloalkanones may also be formed from compounds in which the nucleophilic multiple bond is substituted with a heteroatom. The cyclizations are mediated by acidic reagents, including protonic and Lewis acids.

6.1 ACYLATIONS

Intramolecular acylation usually denotes a reaction between an activated carboxylic acid derivative such as acyl halide with an unsaturated C—C bond. Acylium ion intermediates are often implicated. This reaction is useful for making five-, six-, and seven-membered ketones, although compounds of larger ring size may be formed under high dilution conditions and/or with certain templates. The donor group also includes aromatic nuclei (Friedel–Crafts acylations).

6.1.1 Aromatic Acylations

The Friedel–Crafts acylation is commonly effected with a carboxylic acid in the presence of hydrogen fluoride, sulfuric acid, polyphosphoric acid, or trifluoroacetic anhydride. Alternatively, an acid chloride or acid anhydride is used in conjunction with a Lewis acid catalyst such as aluminum chloride, boron trifluoride (etherate), tin (IV) chloride, and zinc chloride. The direct cyclization of carboxylic acids is suitable mostly for intramolecular acylation.

With respect to sesquiterpene synthesis, indanones and α-tetralones are generally made for their obvious structural similarity to the molecular skeletons of the targets. These encompass pterosin E (**P49**) (Nambudiry, 1974), hypolepin B (**H24**) (Y. Hayashi, 1972), emmotin H (**E5**), (Reddy, 1979), rishitinol (**R5**) (Katsui, 1971), occidol (**O3**) (Ho, 1971; Reddy, 1980), β-eudesmol (**E18**) (Huffman, 1971), mansonone D (**M4**) (Viswanatha, 1974a, b), lacinilene C methyl ether (**L2**) (McCormick, 1978), cacalol (**C1**) (Inouye, 1975; Yuste, 1976; Huffman, 1979), and pallescensin E (**P3**) (Baker, 1981).

6.1 ACYLATIONS

P49

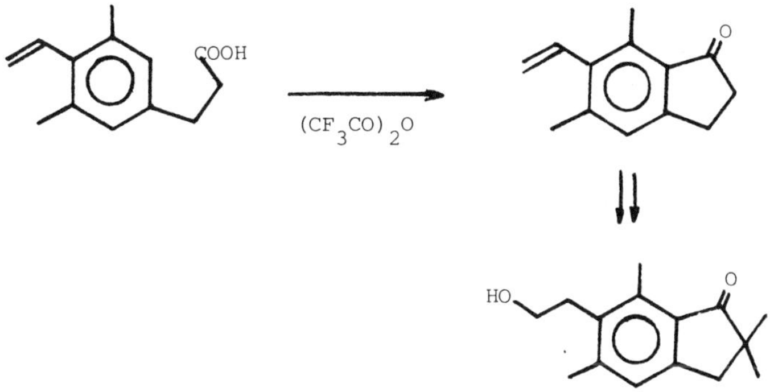

H24

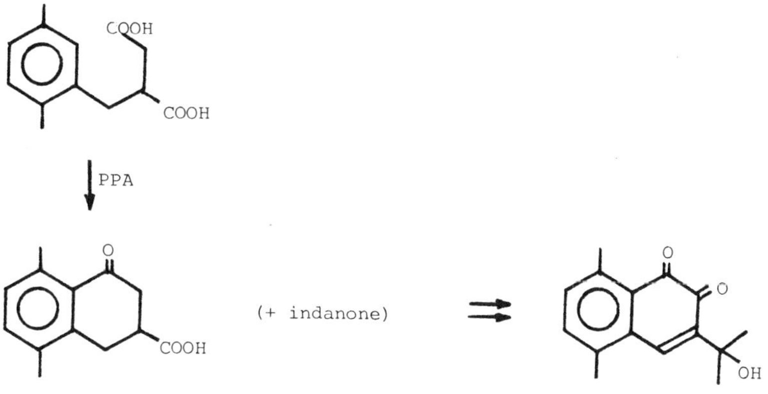

E5

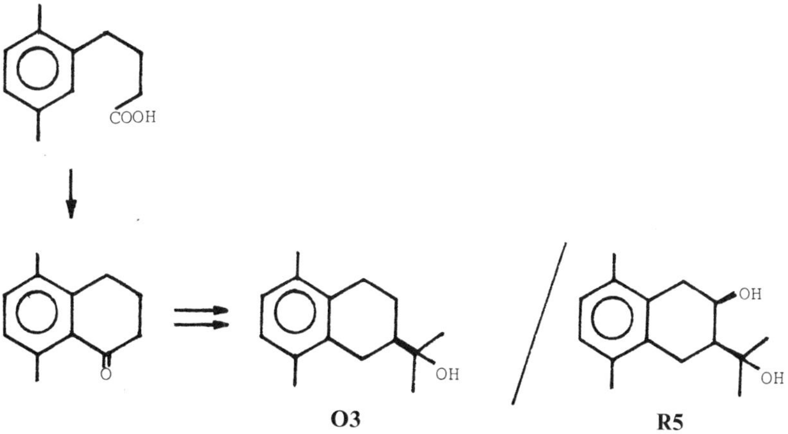

O3   **R5**

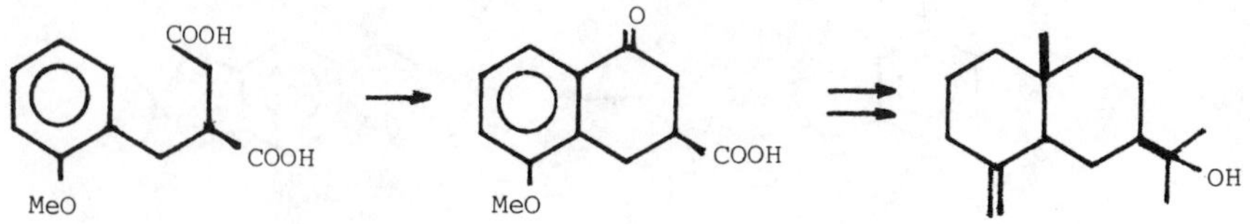
COOH
COOH
MeO
O
COOH
MeO
OH

E18

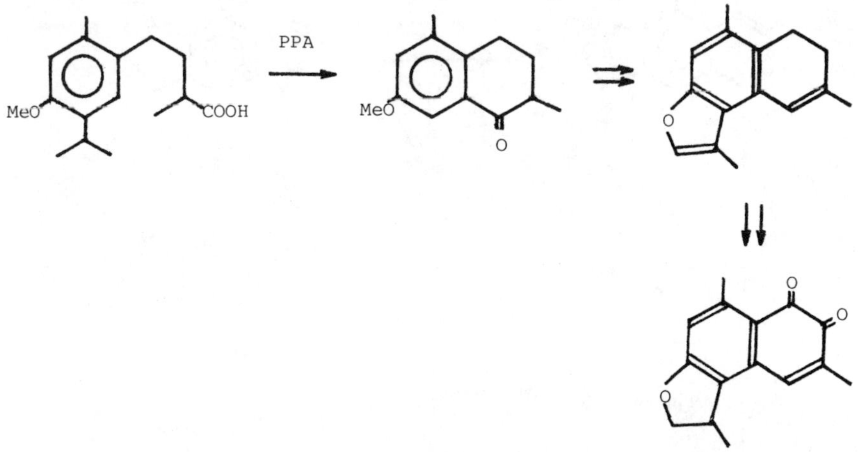
PPA
MeO
COOH
MeO
O
O
O
O
O

M4

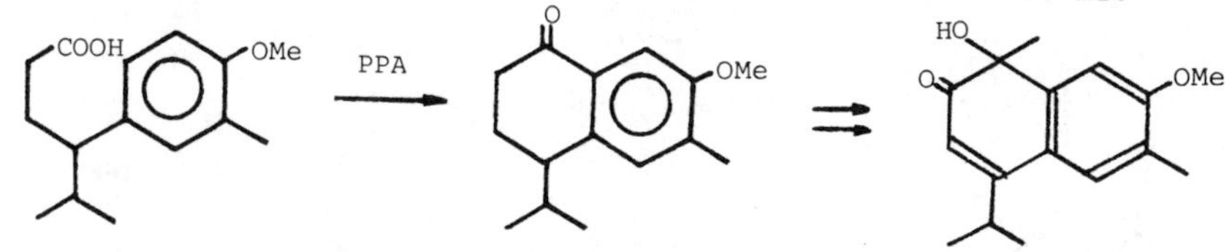
COOH
OMe
PPA
O
OMe
HO
O
OMe

L2

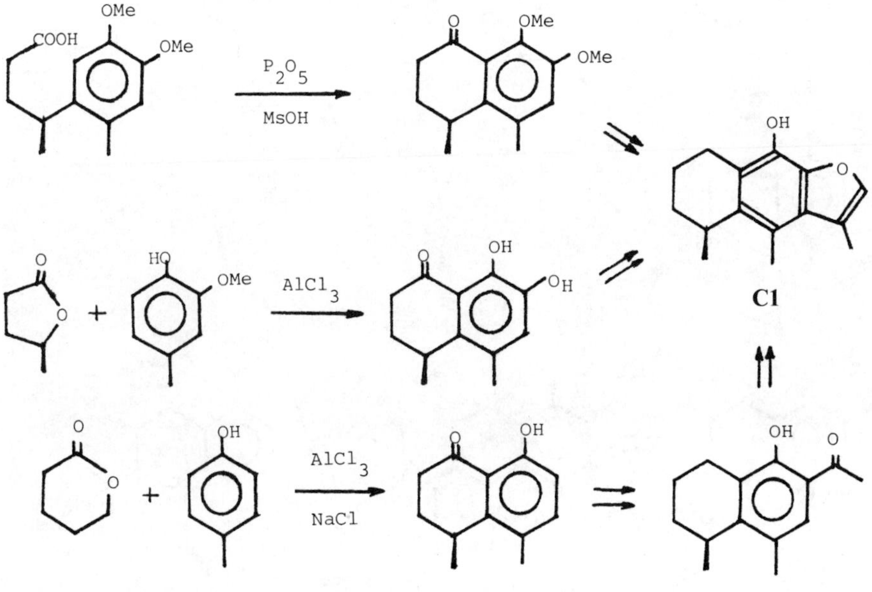
OMe
COOH
OMe
P2O5
MsOH
O
OMe
OMe
OH
O
O
HO
OMe
O
AlCl3
O
OH
OH
O
OH
O
AlCl3
NaCl
O
OH
OH
O

C1

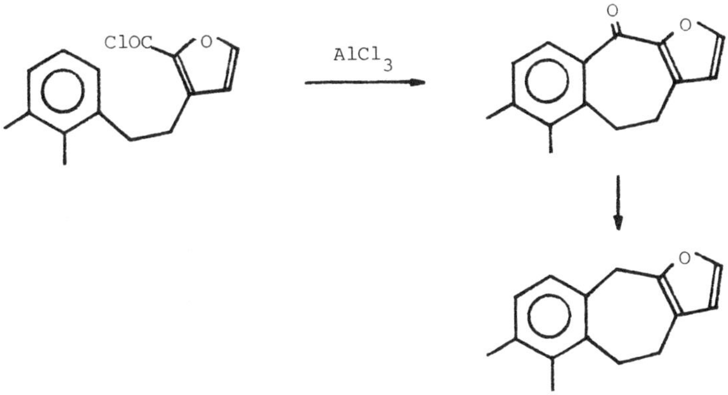

In the synthesis of mansonone D, the acylation is accompanied by a deisopropylation, which is not uncommon under the Friedel–Crafts reaction conditions.

Reductive methylation of an α-tetralone allows entry into functionalized intermediates suitable for further elaboration into atractylon and lindestrene (**L18**) (Honan, 1985). Naturally, the α-tetralone is obtainable from a Friedel–Crafts acylation.

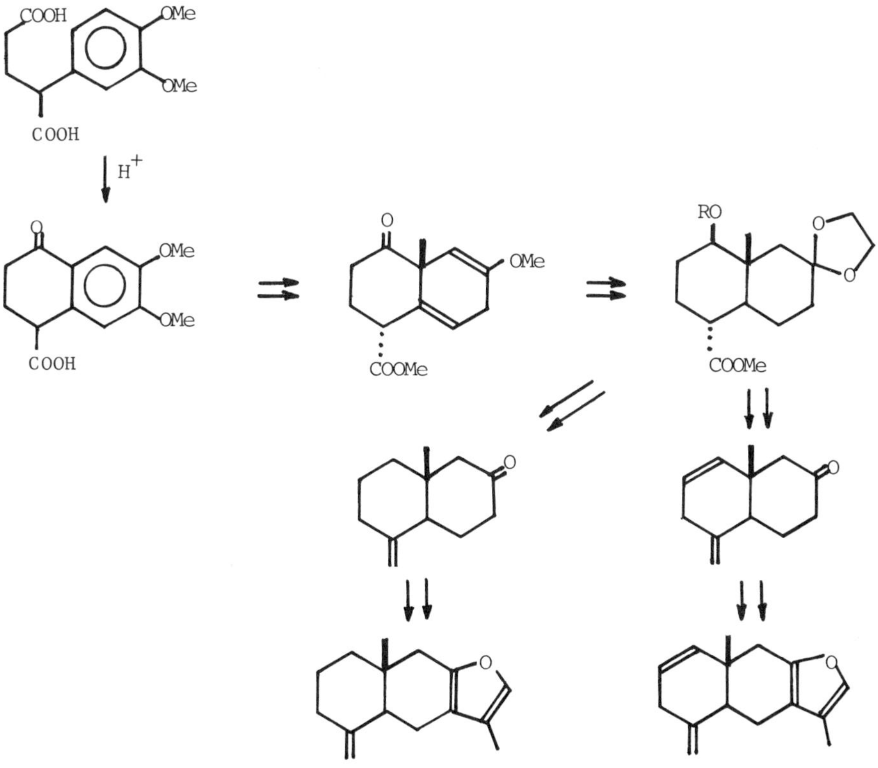

Brasilenol has been acquired from a dihydroindenone through Birch reduction and further adjustment of the peripheral substituents (Greene, 1986).

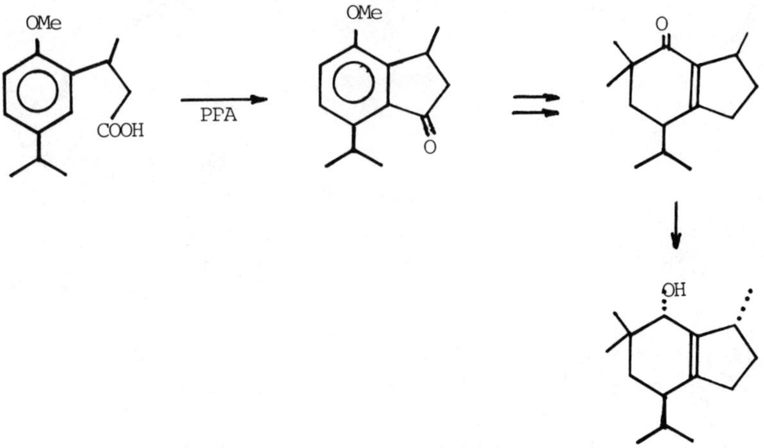

brasilenol

An AC + B approach to the diterpene resin acids (Ireland, 1966) involves formation of a *trans*-β-decalone (obtained from a Robinson annulation) and its selective cleavage to release two acetic acid chains. One of these chains participates in an intramolecular acylation to complete the B ring. A synthetic route to dehydroabietic acid (**A2**) is shown in the following equation.

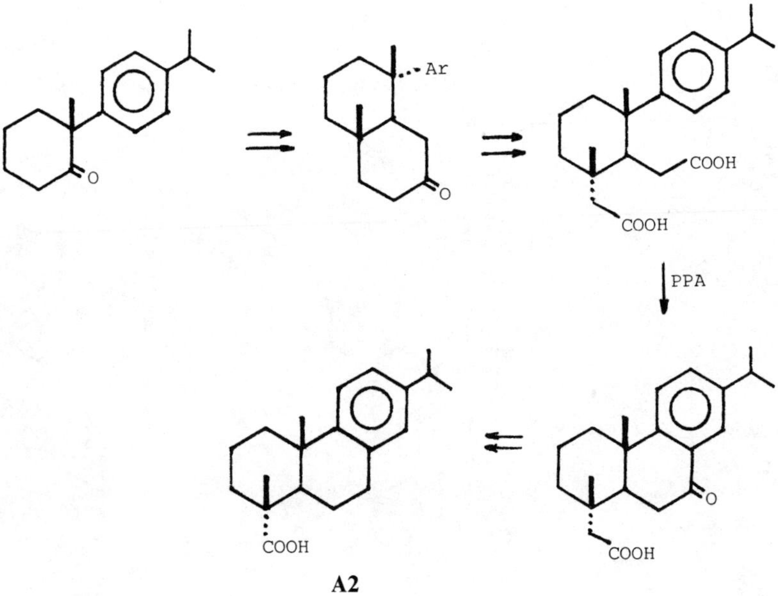

A2

6.1 ACYLATIONS

The order of the alkylation of the cyclohexanone in furnishing the elements of the B ring and the C_4 substituents correlates with the two different series of resin acids. This holds true also in an improved route consisting of a hydrindenone intermediate, obviating a degradative sequence to install the C_4 substituents. A synthesis of podocarpic acid (**P43**) (Giarrusso, 1968) is demonstrated.

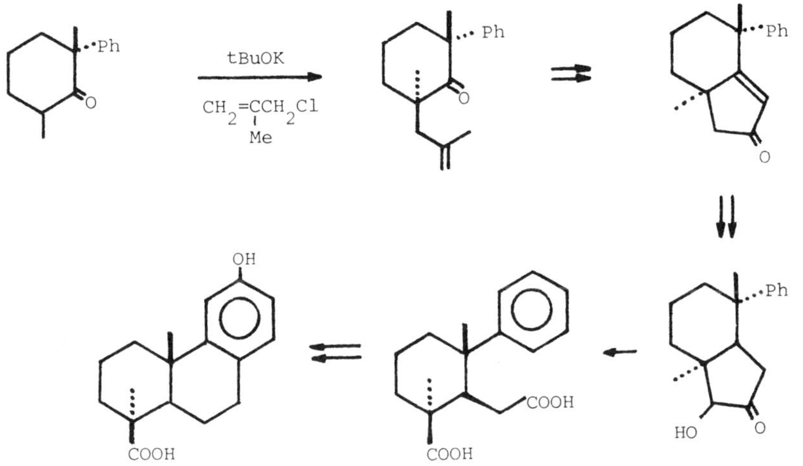

It should be noted that an ester may sometimes be used as an acylating agent, as shown in the preparation of the indanone for a synthesis of 9-isocyanopupukeanane (**P54**) (Corey, 1979a).

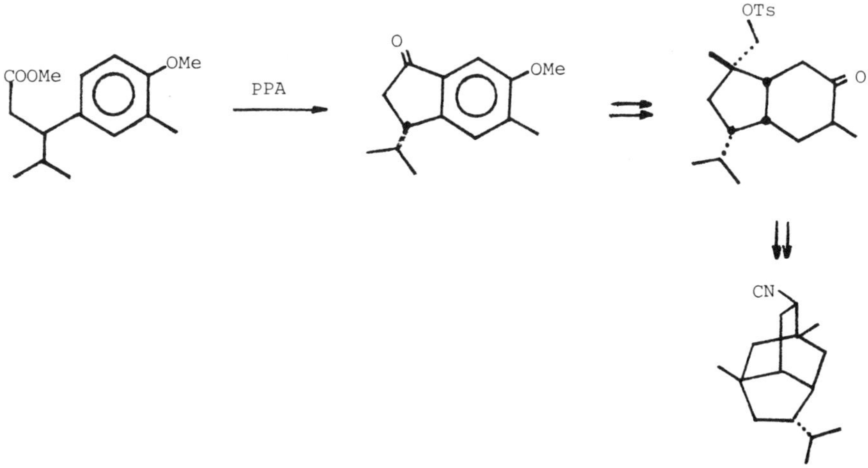

6.1.2 Nonaromatic Acylations

Many menthenones are conveniently accessible from acyclic acids (Kuhn, 1953): for example, pulegone (**P51**) from citronellic acid, piperitenone (**P33**) from levandulic acid, and piperitone (**P34**) from dihydrolevandulic acid. Geometrical or positional isomerization of one of the double bonds occurs in many cases.

P51

P33/P34

A convenient route to (—)-curcuphenol acetate (Murali, 1987) from *R*-citronellal is via intramolecular acylation of the dienoic acid obtained from a Wittig–Horner–Emmons reaction.

curcuphenol acetate

Hypacrone is a secoprotoilludane in which two carbocycles are connected by a double bond. It is easy to conceive that a certain aldol-type reaction is suitable for the preparation of hypacrone (**H23**). Indeed, an approach based on this concept has reached its destination (Y. Hayashi, 1975); the cyclopentenone synthon is available via a Friedel–Crafts acylation.

H23

6.1 ACYLATIONS

The general applicability of the intramolecular acylation method is better illustrated by its appearance in the more complex environment—for example, in the syntheses of pentalenolactone (**P18**) (Danishefsky, 1978, 1979) and dactylol (**D1**) (Paquette, 1985b).

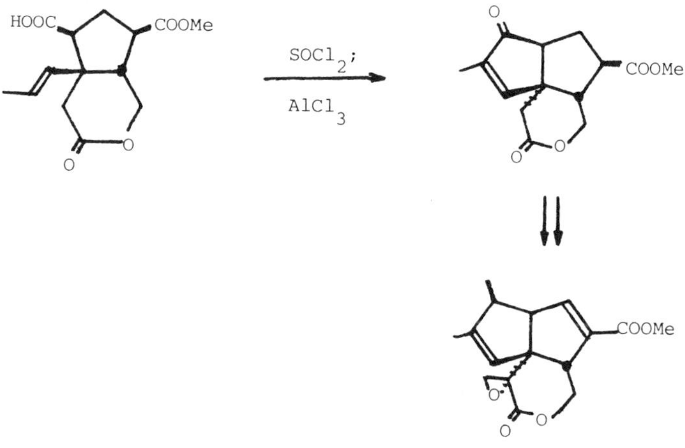

P18

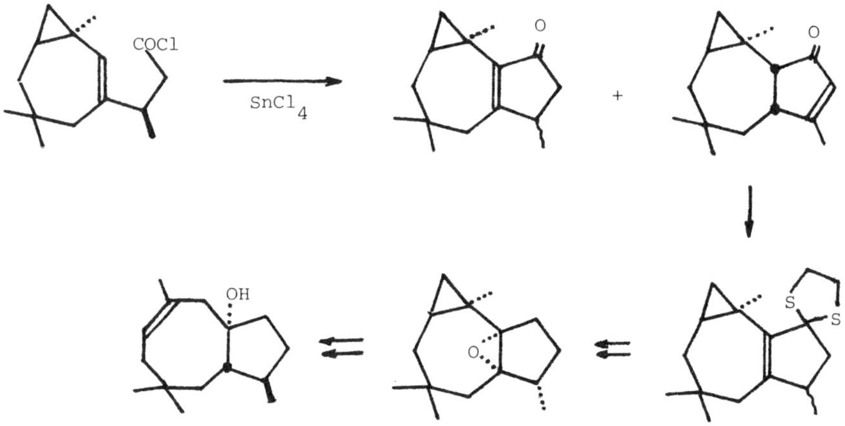

D1

A synthesis of africanol (Paquette, 1986a) hinges on the conformational bias exhibited by a trimethylbicyclo[5.1.0]octanone. Thus, the acetyl derivative undergoes stereoselective reduction and chirality transfer by a Claisen rearrangement establishes the configuration of the tertiary carbon atom which will become a ring juncture. A cyclopentenone is formed by Friedel–Crafts reaction. Introduction of the angular hydroxyl group is somewhat complicated by nonstereoselective formation of the epoxide.

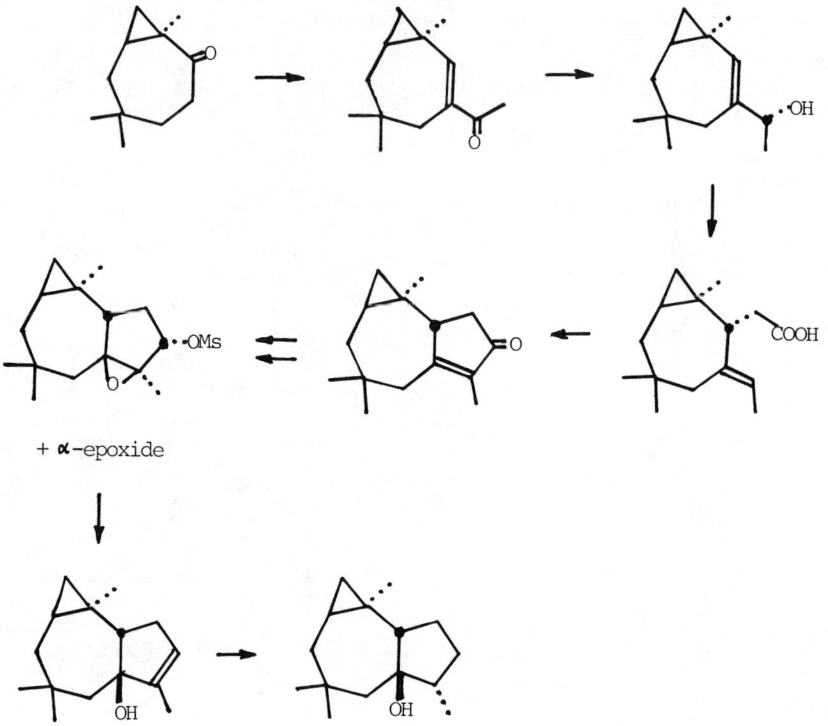

africanol

A diquinane diester gives rise to a lactone on reaction with methyllithium. Treatment of this lactone with phosphorus pentoxide in methanesulfonic acid furnishes a precursor of modhephene (**M12**) (Wilkening, 1984). The regioselective attack of methyllithium on the less hindered ester group, and then the more reactive ketone, is expected.

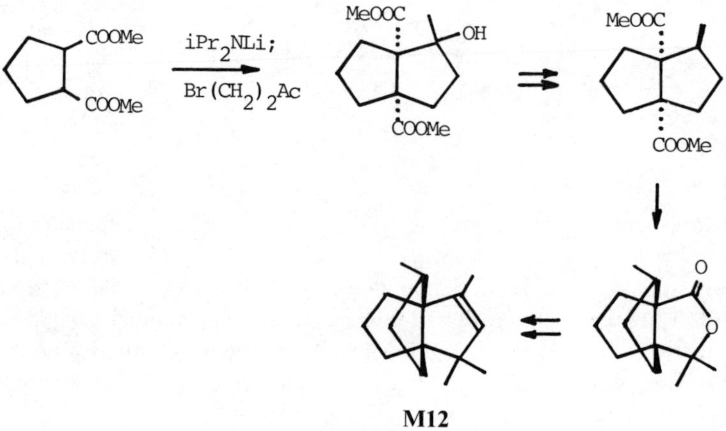

M12

6.1 ACYLATIONS

When the double bond is tetrasubstituted, a conjugate cycloalkenone product cannot be formed, and the cycloacylation is more reluctant. Often, the expelled chloride ion from the acyl chloride returns to neutralize the positive charge, thereby leading to a β-chloroketone. The first attempt at a synthesis of hirsutene (**H17**) (Nozoe, 1976) encountered this problem.

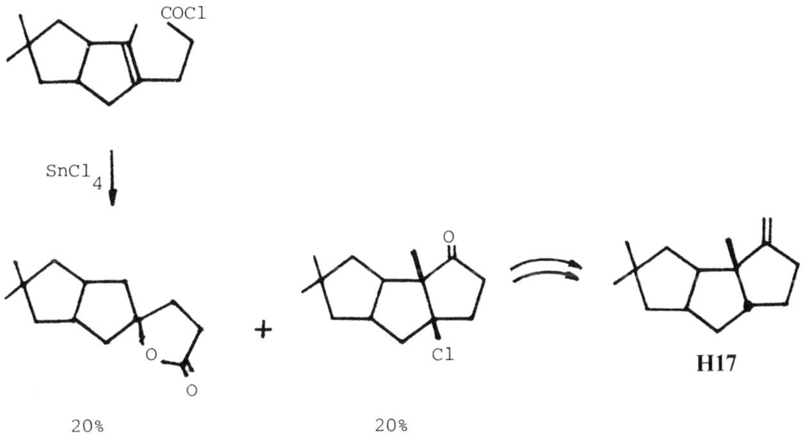

Little difficulty arises in effecting intramolecular acylation that results in spirocyclic enones. Thus, the synthesis of compounds related to β-vetivone (**V16**) (Bozzato, 1974) may be approached in that manner. Even under the influence of Lewis acids, the rearrangement of the spirocyclic products into condensed-ring isomers can be avoided.

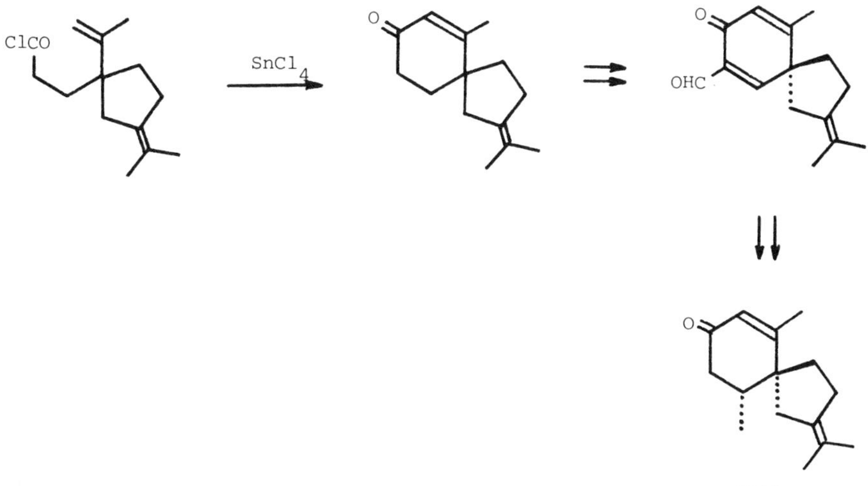

A spirocyclic dienone is a pivotal intermediate in a synthesis of quadrone (**Q1**) (Burke, 1982). It has been obtained by a silicon-directed intramolecular acylation (Burke, 1981). Note that the silicon atom stabilizes a positive charge at a β atom and is electrofugal.

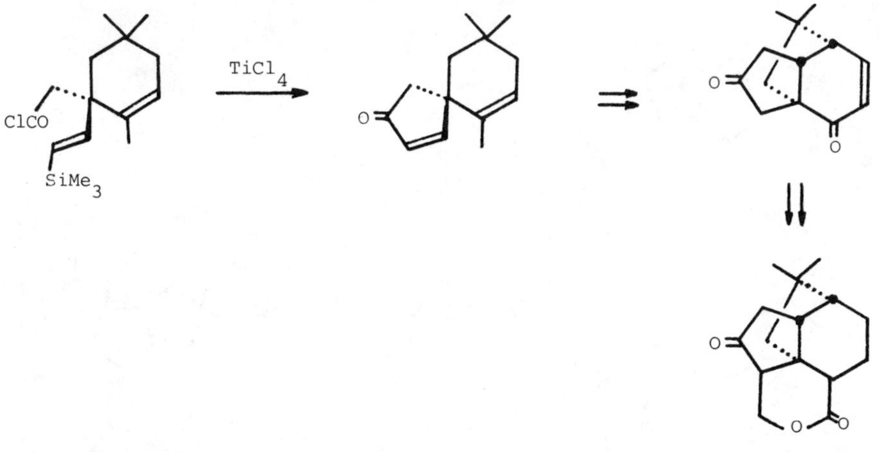

An earlier paragraph describes an approach to β-eudesmol (**E18**) involving the preparation of a methoxytetralone through reductive manipulations. The same bicyclic intermediate may be secured by an acylation route (Carlson, 1972).

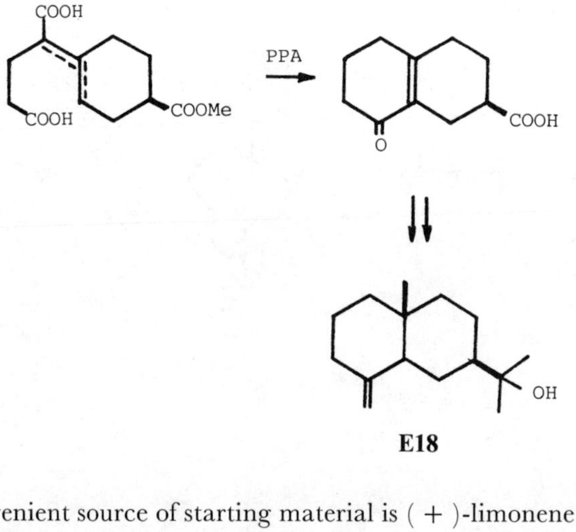

A more convenient source of starting material is (+)-limonene. The Friedel–Crafts acylation of its monohydrochloride with 3-butenoyl chloride leads to a bicyclic enone (contaminated with about the same amount of an isomer). This enone is convertible to β-selinene (**S21**) (MacKenzie, 1979).

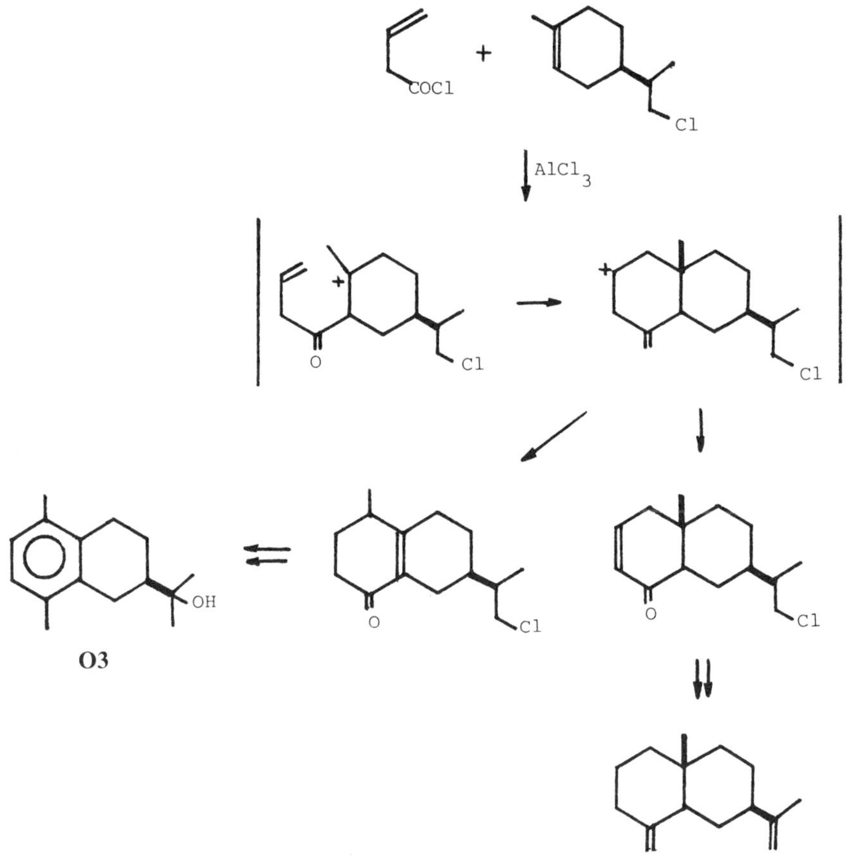

The isomer is formed by successive hydride and methyl shifts of the annulated intermediate before deprotonation occurs. It is useful for another synthesis of (+)-occidol (**O3**) (MacKenzie, 1979).

In the fortuitous synthesis (that involves skeletal rearrangements) of patchouli alcohol (**P15**) (Büchi, 1964a), an intramolecular acylation was prescribed to gain access to the relay compound β-patchoulene. Previous to that time it was thought that the alcohol and the hydrocarbon possessed the same carbon framework, since chemists were unaware of the intervention of a rearrangement during degradation of the alcohol and a backward rearrangement during the reconstitution of the latter compound from β-patchoulene.

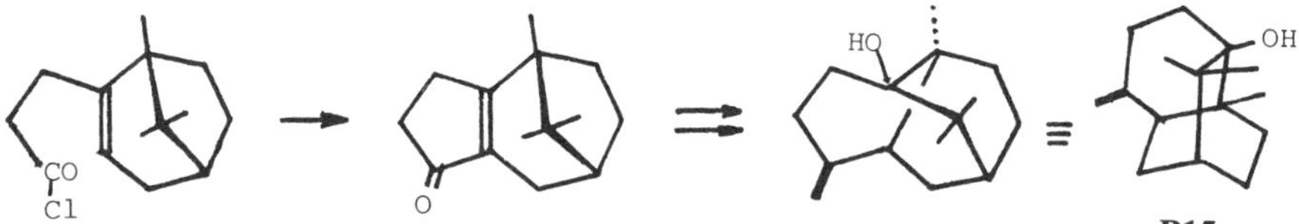

6 CATIONIC CYCLIZATIONS

3-Methylbicyclo[3.2.1]oct-3-en-6-one is an excellent intermediate for terpene synthesis. It has been elaborated into α-santalene (**S3**) (S. Monti, 1978), α-*trans*-bergamotene (**B5**) (Larsen 1977), and *threo*-juvabione (**J9**) (Larsen, 1979). In the juvabione work the cyclopentanone formation serves to introduce the secondary methyl group in the correct relative configuration only. The original cyclohexene nucleus is restored immediately after the new asymmetric center has been established.

Macrocyclic acylation is demonstrated in the synthesis of neocembrene (**C42**) (Y. Kitahara, 1976), asperdiol (Aoki, 1983), and ceriferol (**C43**) (Fujiwara, 1984). An accrued benefit from this approach to ceriferol is the aid provided by the enone carbonyl in realizing a regioselective functionalization of the nearby vinylic methyl group (via reduction, epoxidation, etc.) and the configurational change of the double bond.

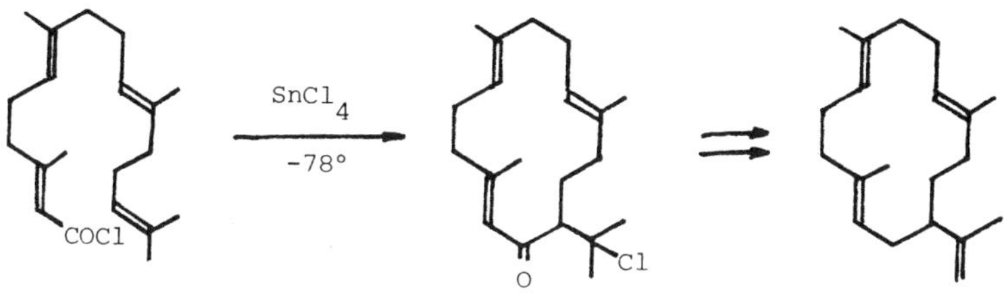

C42

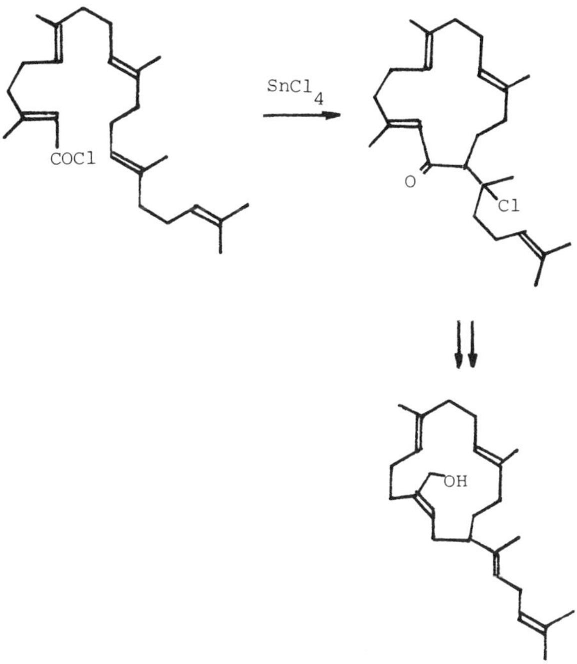

C43

While it is not an acylation, the acid-catalyzed reaction of an α-diazoketone with a π bond is a very efficient method for C—C bond formation (Smith, 1981). This reaction has been incorporated into synthetic schemes of many diterpenes containing a bridged CD ring system. For example, in a hydrofluorene approach to gibberellic acid (**G12**) (Hook, 1984), the linear tricyclic portion is constructed by alkylation of a cyclohexenone enolate equivalent in the form of a dihydroaromatic anion with a benzyl iodide (for the A ring) and Friedel–Crafts alkylation of the product. Cyanohydrination of the cyclohexenone carbonyl provides a foothold for a chain that eventually grows into the diazoketone and the D-ring elements.

This route is convenient in that three rings are assembled by simple steps. The BC + D + A route to gibberellin A_{15} (**G13**) (Lombardo, 1980) also involves D-ring formation in the same fashion.

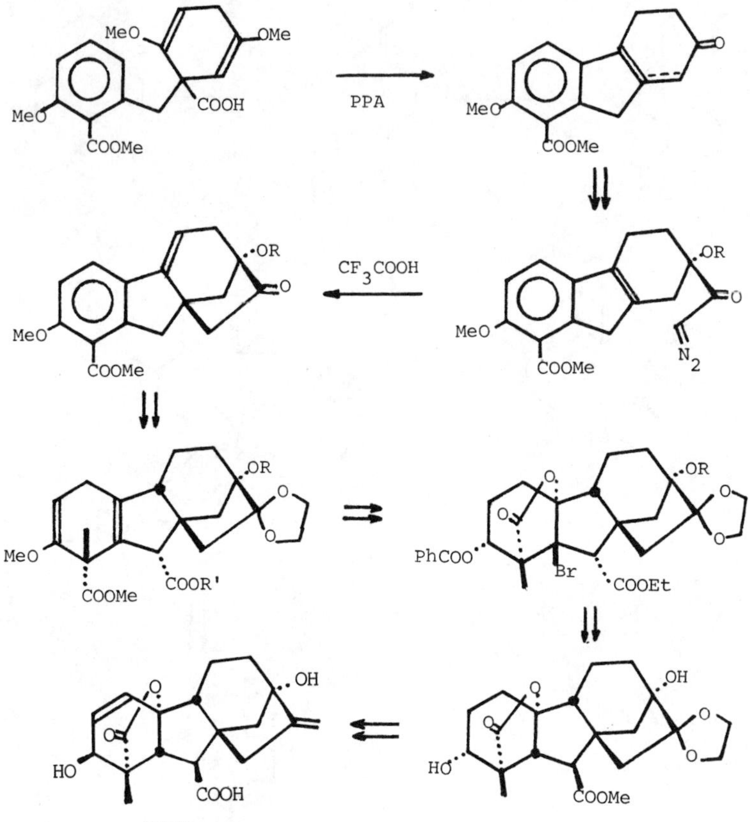

G12

A simpler case of bridging is found in a synthesis of patchoulenone (**P14**) (Erman, 1971). Unfortunately, the 10-epimer is the major product, as a result of preferential hydrogenation of the trisubstituted double bond from the side opposite the isopropylidene bridge.

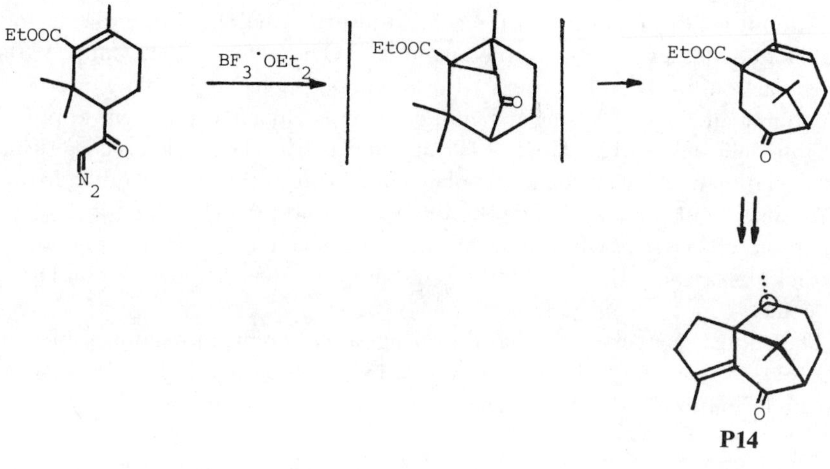

P14

The efficient cyclization of unsaturated α-diazoketones under the influence of a Lewis acid to afford cyclobutanones was somewhat unexpected. Nevertheless, this property has been developed into a straightforward access to filifolone (**F6**) (Hudlicky, 1980).

6.2 ALKYLATIONS

6.2.1 Alkylation with Aldehydes: Acid-Catalyzed Ene Reactions

Intramolecular interaction of oxyfunctional groups with unsaturated moieties to form cyclic products in the same molecules is not restricted to acyl halides. Examples showing the attack of aldehydes by a nearby double bond include the formation of key intermediates in the synthesis of β-vetivone (**V16**) (McCurry, 1973a,b), isocomene (**I14**) (Paquette, 1979), pentalenene (**P16**) (Annis, 1982), sativenediol (**S12**) (Piers, 1976, 1977c), bulnesol (**B21**) (Andersen, 1973a), guaiol (**G20**) (Andersen, 1973b), kessanol (**K7**) (Andersen, 1977), and junenol (**J5**) (Schwartz, 1972).

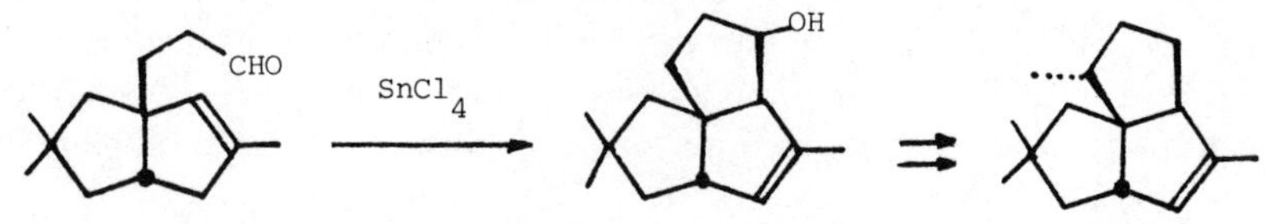

P16

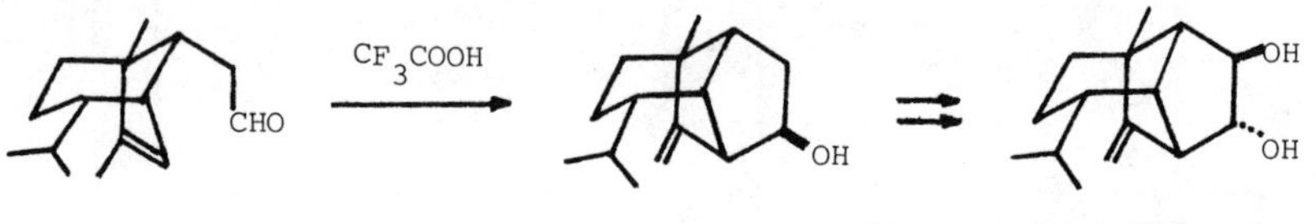

S12

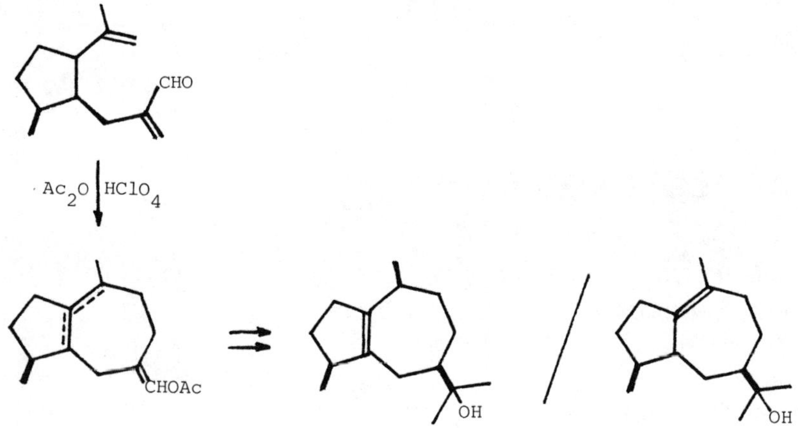

Ac$_2$O | HClO$_4$

G20 / **B21**

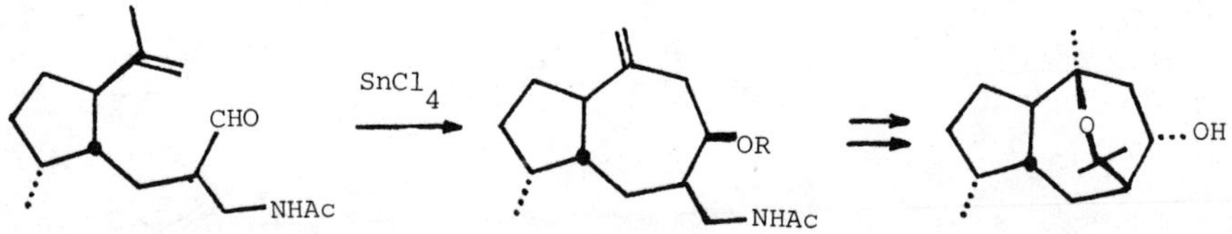

K7

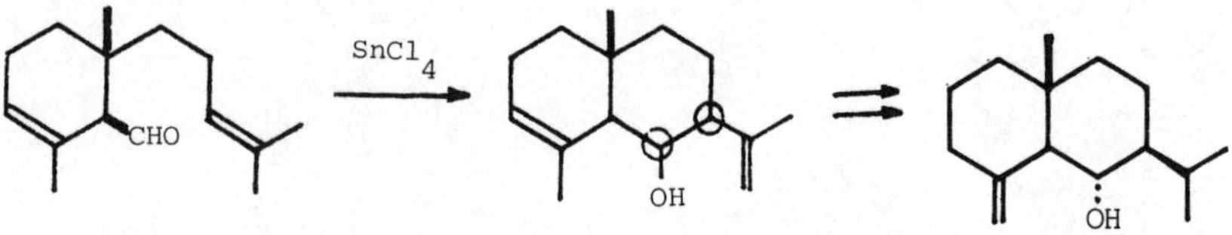

J5

This process is generally regarded as an ene reaction (see Chapter 8). It has also been employed to great advantage in the formation of the A ring of

(+)-ivalin (**I19**) (Tomioka, 1984). This eudesmanolide contains an exocyclic methylene group at C_4 and an α-hydroxyl function at C_2, which can be connected by the ene reaction despite incorrect configuration of the hydroxyl group thus created.

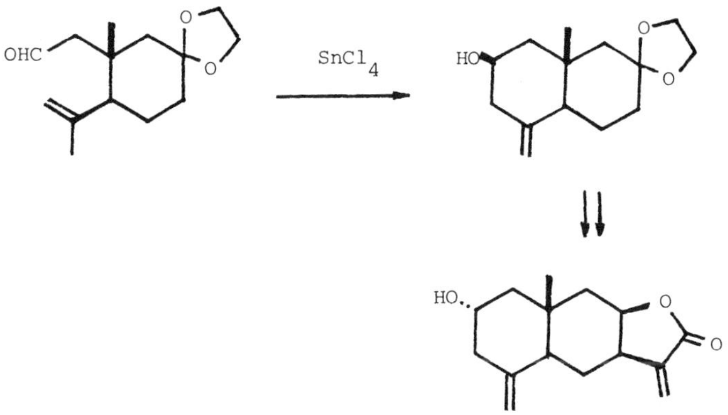

The requirement for hydroxyl inversion does not detract from the synthetic effort. Rather straightforward steps performed on the homoallylic alcohol intermediate lead to the antileukemic sesquiterpene.

Cyclopentanation on a cycloheptadiene derivative by means of the ene reaction represents a refreshing way for the synthesis of guaianolides such as dehydrocostus lactone (**C77**) and estafiatin (**E15**) (Rigby, 1984). The cis ring juncture is ensured by this annulation, and a double bond is left in the position where the γ-lactone eventually will be installed.

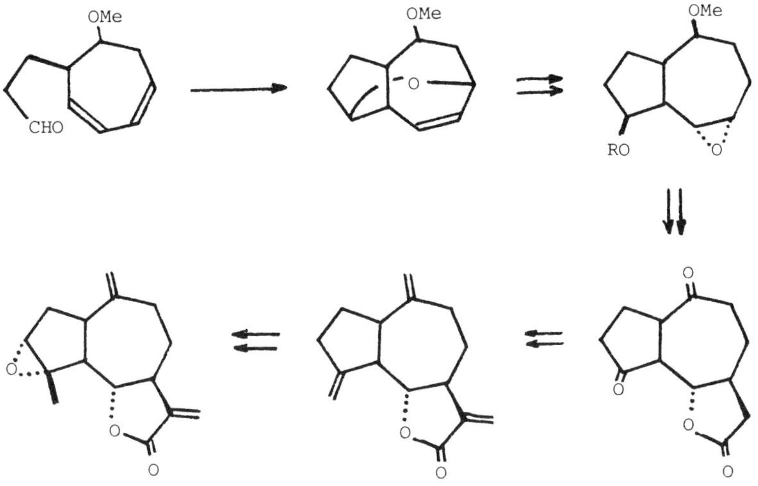

Very facile cyclization occurs when the aldehyde group and a double bond are in an enforced propinquity. Thus the hydroxyseychellene is directly isolated from the oxidative cleavage of a bicyclic α-glycol under neutral conditions (Welch, 1984). Accordingly, deoxygenation of the alcohol leads to seychellene (**S29**).

Sometimes a derivative of the aldehyde group such as thioacetal monoxide may be used in lieu of the carbonyl compound. In one α-santonin (**S9**) synthesis (Marshall, 1978), the carbocyclic framework is assembled from such a derivative. It is possible that *in situ* hydrolysis to the aldehyde precedes cyclization. Since the thioacetal monoxide is acquired by a chain-lengthening process, not by deliberate protection of an existing aldehyde group, a step is actually saved by subjecting the derivative directly to acid treatment. (For a use of the corresponding aldehyde in a synthesis of the diterpene dictyolene, **D14**, see Marshall, 1978.)

Conjugated carbonyl compounds apparently do not react directly with an alkene to afford the allylic alcohol. A Michael-type reaction emerges, and the reaction course has been well adapted to a synthesis of occidentalol (**O2**) (Marshall, 1977).

The devolution of a bridged framework into a condensed one by a similar process is an extremely valuable transformation. It has a tremendous impact in the design of synthesis toward nootkatone (**N11**) (Dastur, 1973, 1974) and glutinosone (**G15**) (Murai, 1980).

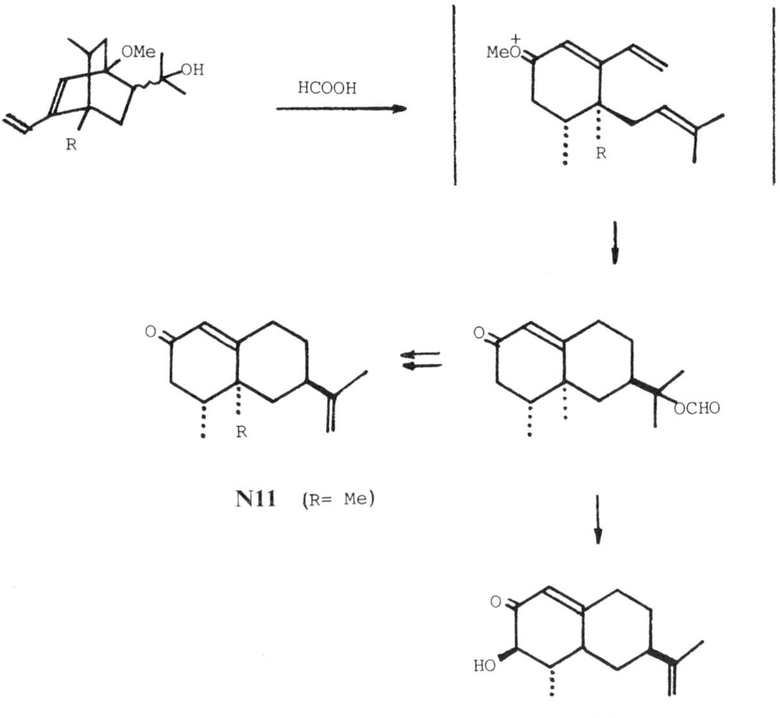

6.2 ALKYLATIONS

N11 (R= Me)

G15

Despite its rare occurrence, the reaction between a ketone and a double bond materializes when 2,5-dimethyl-2-(4-methyl-3-penten-1-yl)-4-cycloheptenone is exposed to tin(IV) chloride. The oxetane product undergoes rearrangement to give acoratriene (**A7**) (Demole, 1971) under the influence of a Lewis acid/reducing agent.

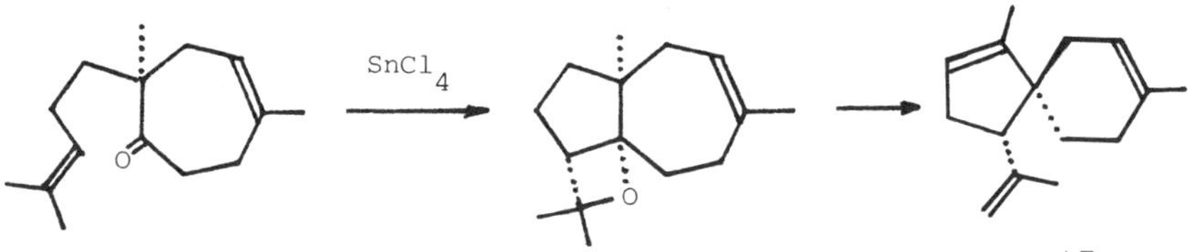

A7

The well-known Friedel–Crafts alkylation of arenes with ketones renders excellent service in the formation of the D ring during a synthesis of germanicol (**G11**) (Ireland, 1970a). In this approach the E ring appears in a latent aromatic form that links to a tricyclic moiety. Preparation of the latter component is by a reaction sequence dominated by the Robinson annulation method.

G11

6.2.2 Alkylation with Cations: Polyene Cyclizations

A carbocation (usually tertiary or allylic) generated in the presence of a double bond three to six atoms away is often intercepted by the latter. The source of the cation can be an alchol, an epoxide ring, or another double bond: all ionize on treatment with protonic or Lewis acids.

The pioneering contribution in the formation of cyclic terpenes must be credited to Ruzicka and Capato (1925), who treated nerolidol with acetic anhydride to obtain a hydrocarbon mixture containing bisabolene.

Apparently the allopatric cation (i.e., the relocated cation generated through interaction of the initial ion with a double bond) loses a proton to yield the unsaturated hydrocarbon. On the other hand, this allopatric cation may be neutralized by a solvent molecule, and bisabolol (**B12**) may be obtained under proper conditions. Thus bisabolol is formed in a larger amount by electrolysis of nerolidol in neutral solvent (Uneyama, 1984).

B12

Annulative trapping of an allylic cation has been used quite extensively in terpene synthesis, and probably it parallels the biogenetic pathways of some of those terpene molecules. Among the spirocyclic members, the skeletons of acoratriene (**A7**) (Naegeli, 1972), acorenone B (**A10b**) (Wolf, 1975), β-vetivone (**V16**) (McCurry, 1973b; Murai, 1981), and β-chamigrene (**C47**) (Kanno, 1967) have been assembled via π-bond participation.

A7

A10b

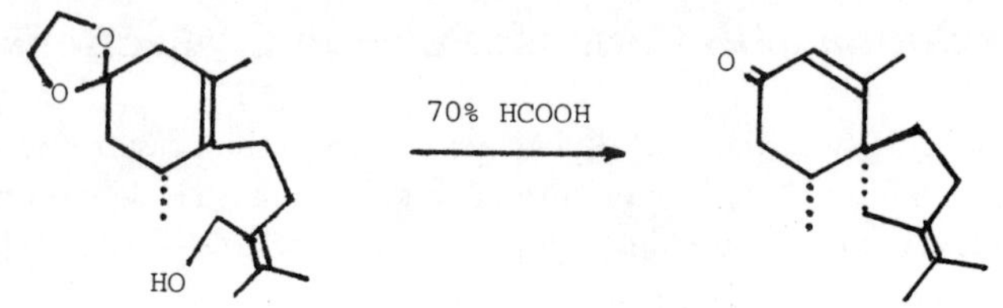

A desilylative cyclization of allylic alcohol derivatives with retention of the double-bond geometry is instrumental in the precise synthesis of *E*-γ-bisabolene (Corey, 1986a). For access to *Z*-γ-bisabolene (Corey, 1986b), the ring formation is induced by a boron-to-carbon alkyl migration of a boranate. Judging from the reverse steric course, a different mechanism must be operating for this cyclization.

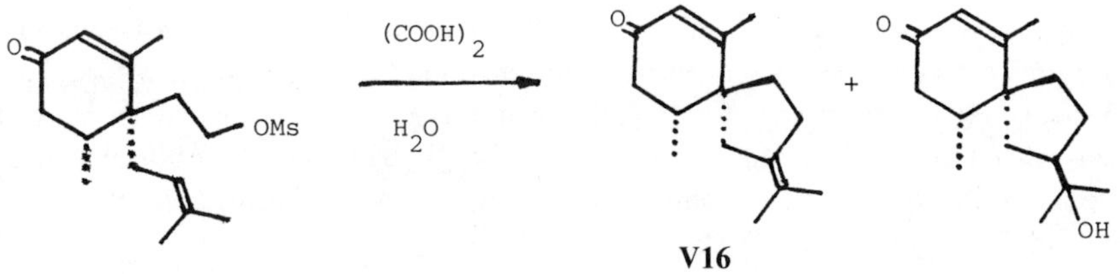

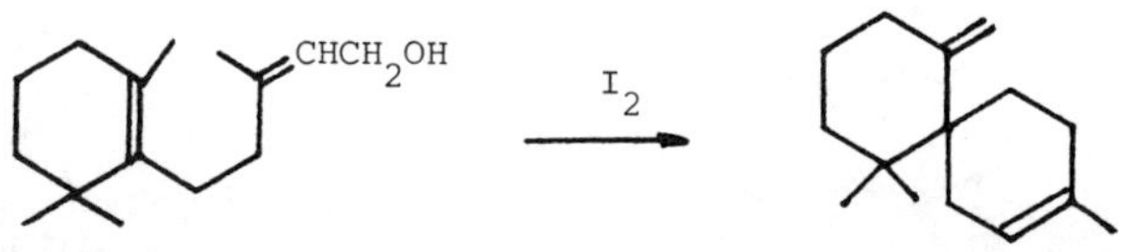

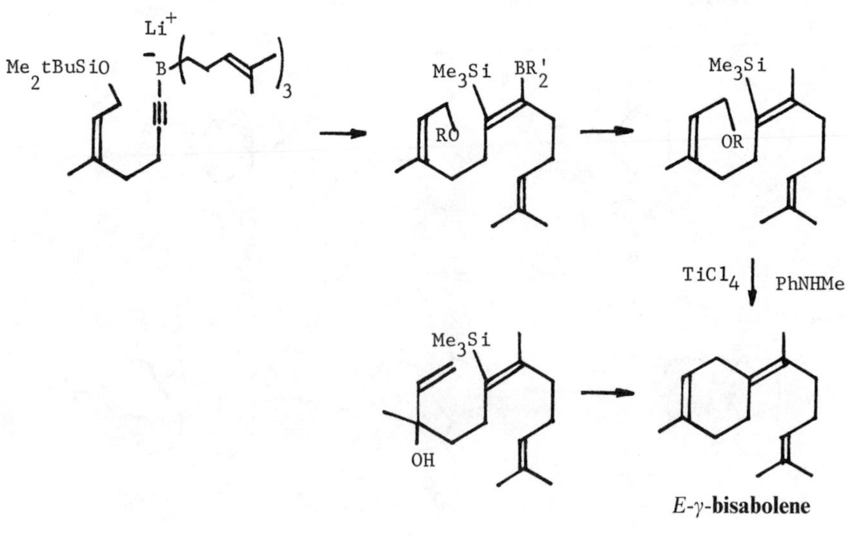

E-γ-**bisabolene**

Z-γ-**bisabolene**

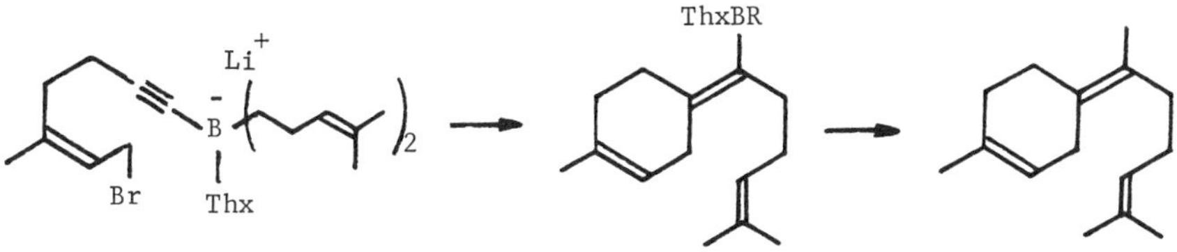

Condensed-ring systems are formed with equal facility from the appropriate cyclic allylic or tertiary alcohols. In the synthesis of (+)-valeranone (**ent-V2**) (Marshall, 1968a), fukinone (**F15**) (Marshall, 1971a), eremophilenolide (**E12**) (Pennanen, 1980), γ-eudesmol (**E19**) (Marshall, 1966b), and alantolactone (**A20**) (Marshall, 1965a, 1966a), the fundamental frameworks are constructed according to the following general scheme.

ent-V2

F15 **E12**

E19 **A20**

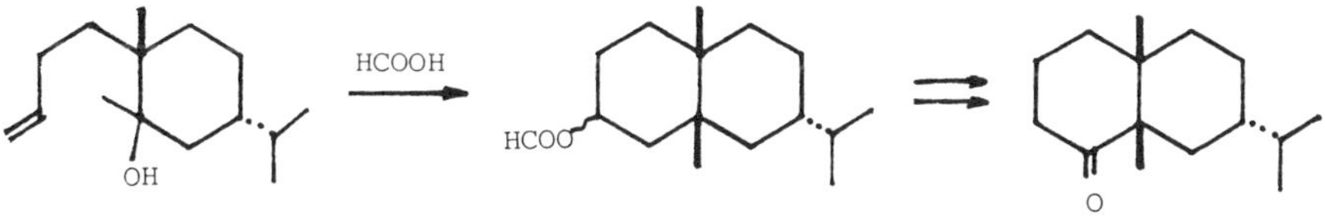

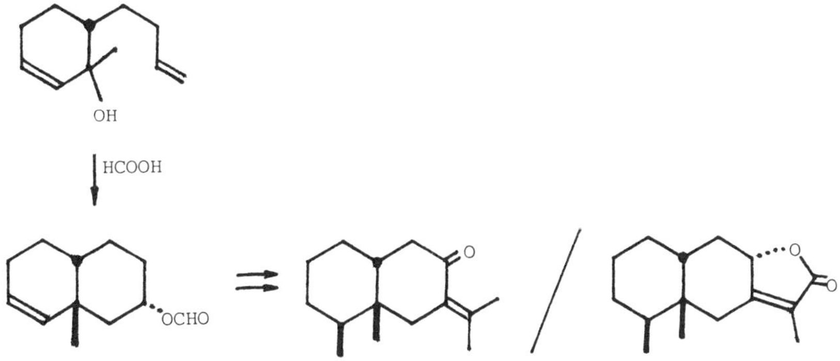

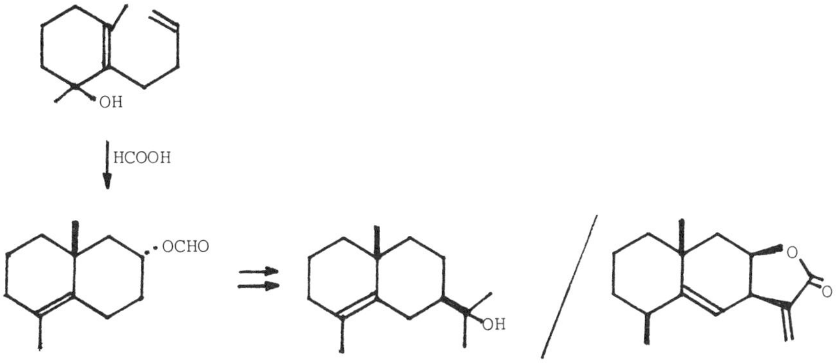

In a hirsutene (**H17**) synthesis, cationic cyclization proceeds from protonation of a tetrasubstituted double bond (Hewson, 1985). For thermodynamic reasons, the product has a *cis-anti-cis* skeleton.

H17

A biogenetic-type synthesis of rosenonolactone (**R6**) from methyl isocupressate involves constitution of the C ring by cationic cyclization (nonstereoselective) and a Nametkin shift (McCreadie, 1969). As expected, the donor–acceptor relation changes when the oxidation state of the terminal carbon atom is altered.

R6

Another monocyclization process locking the pentacyclic oleanane framework is by treatment of a ditertiary alcohol with acid (Ghera, 1964). The diol is obtained from a stepwise union of two isomeric trimethyldecalones with an acetylene molecule followed by reduction, Serini reaction, and a Grignard reaction.

One of the hydrocarbon products, 18α-olean-12-ene (**04**) is then converted to β-amyrin (**A30**) (Ghera, 1964; Barltrop, 1962) through a series of oxidation maneuvers including inversion of C_{18}, oximation at C_1 via photolysis of the 11α-nitrite, and oxygen function transposition to C_3.

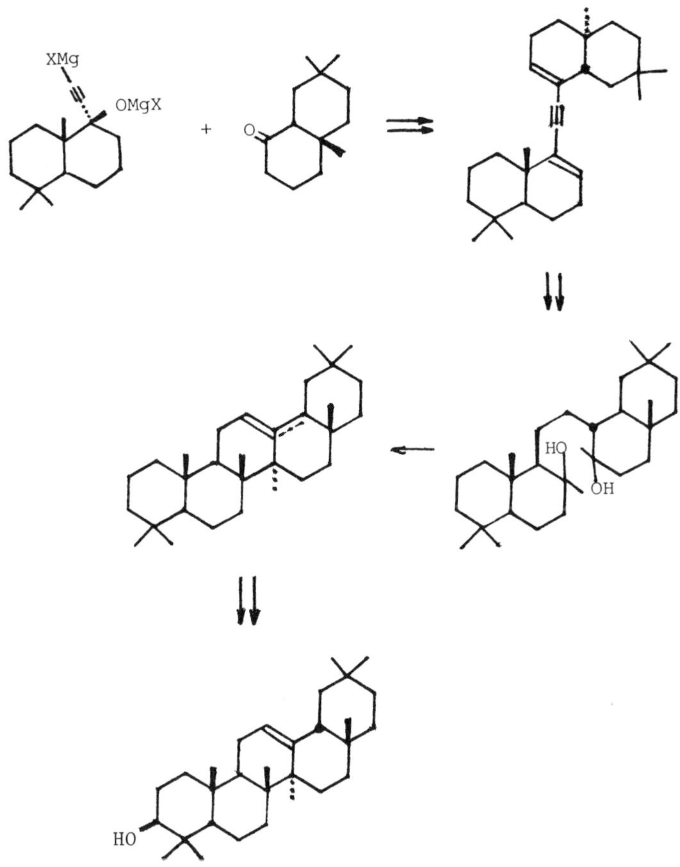

A30

Generally, heterosubstituted alkenes are better cation terminators, as attested by the successful syntheses of cedrol (**C38**) (Corey, 1969e; cf. Lansbury, 1974), campherenone (**C14**) (Hodgson, 1971), α-santalene (**S3**) (Hodgson, 1971), β-cuparenone (**C86**) (Lansbury, 1969), and (—)-albene (**A22**) (Kreiser, 1979).

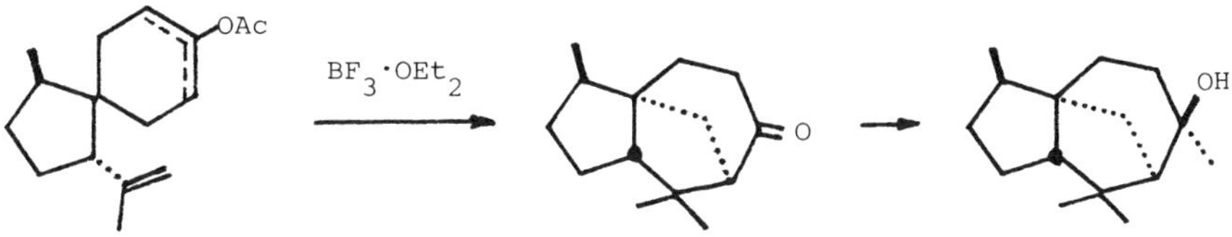

A38

A37

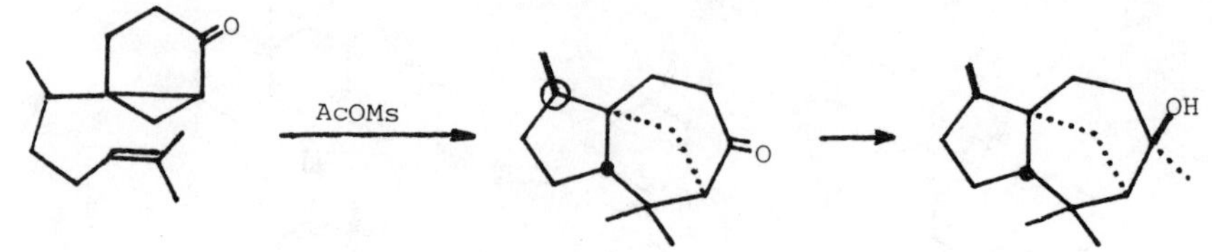

The reaction course of the cyclization of the tertiary alcohol derived from (+)-camphenilone was totally unexpected. Only because of an unusual Nametkin shift of an *endo*-methyl group preceding the cyclopentanation is albene accessible.

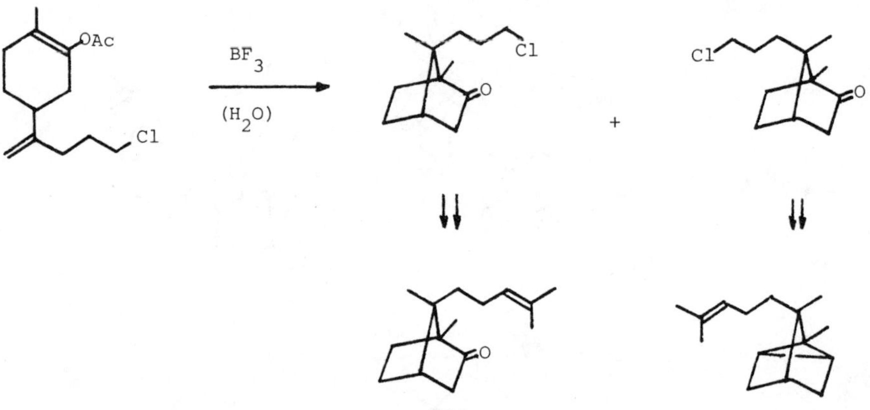

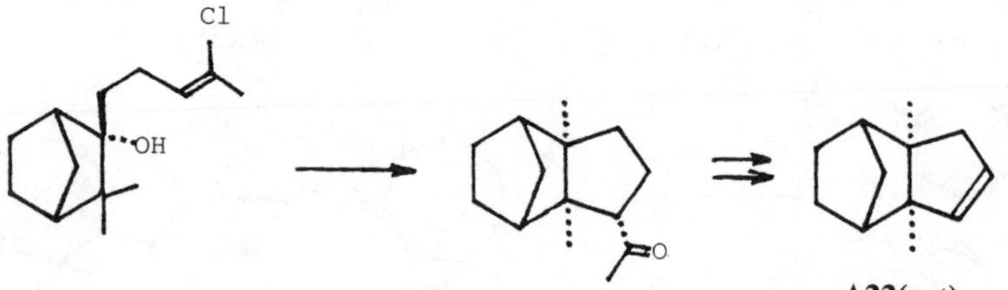

An optically active tetramethyl α-decalone which serves as convenient precursor for many sesquiterpenes and diterpenes has been obtained from (—)-carvone via kinetic alkylation and cyclization of the Lansbury type (Gesson, 1986). Judicious execution of the alkylation sequence may lead to both cis and trans ketones.

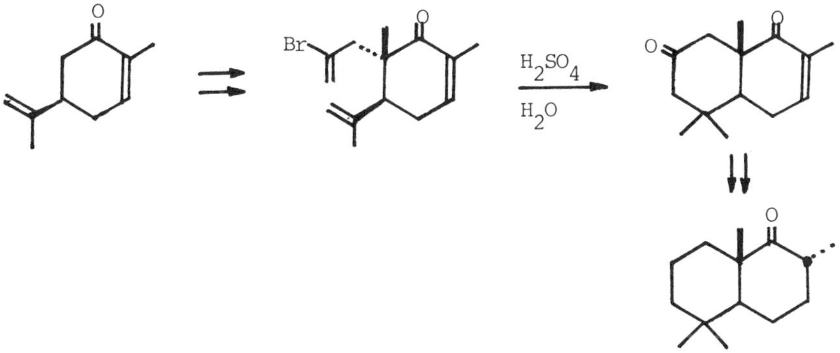

A preparation of an advanced intermediate for verrucarol (**V12**) has been accomplished starting from D-glucose. The cyclohexene ring is made via a cationic cyclization in which the donor is a β-ketoester (Tsang, 1985).

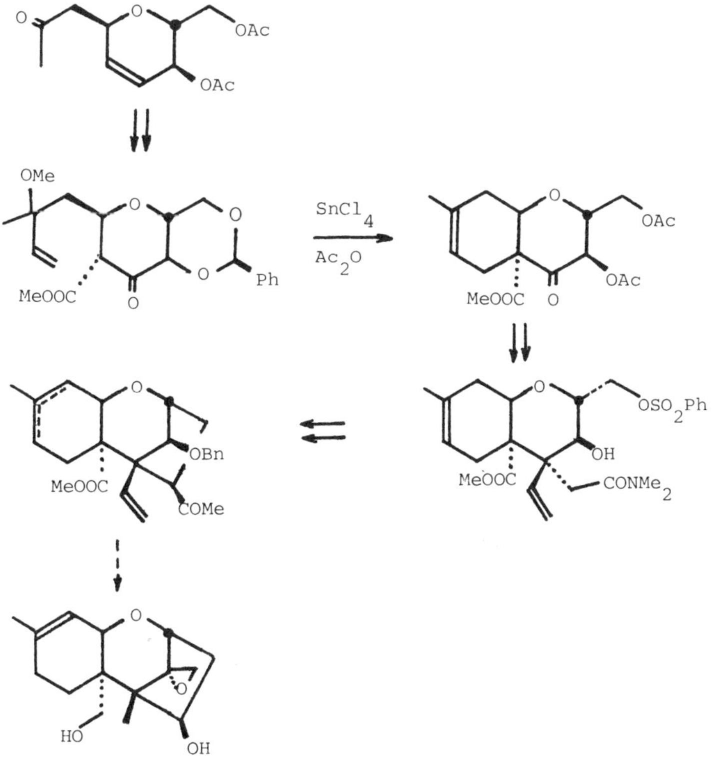

V12

Allyl cations stabilized by heterosubstituents are also capable of C—C bond formation with alkenes. Benzoannulation of ketones based on this reaction is exemplified by a synthesis of the (+)-calamenenes (**C7**) from (—)-menthone (Dieter, 1985).

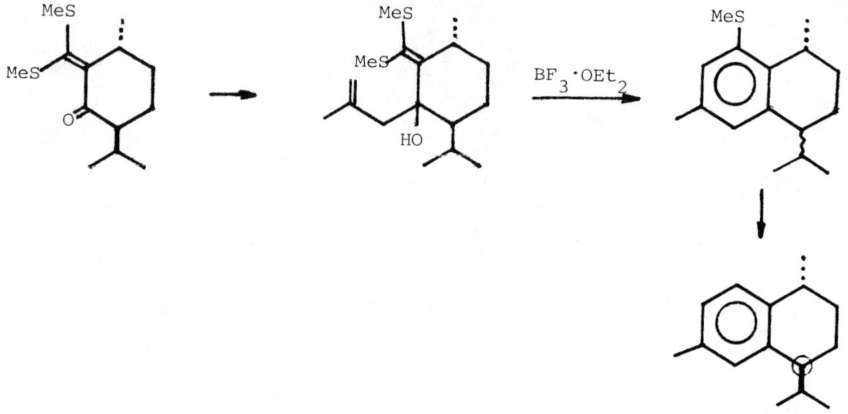

Although rarer applications have been witnessed, acetylene participation in cyclization can be as effective as the alkene. Thus, for elaboration of shionone (**S31**) (Ireland, 1975b) and friedelin (**F13**) (Ireland, 1976) from their respective A-ring aromatic precursors, an excellent protocol consists of Birch reduction, fragmentation of the α,β-epoxyketone on treatment with tosylhydrazine, conversion of the resulting acetylenic ketone to the tertiary alcohol, and reclosure of the A ring via solvolysis, which provides an enol ester and thereby facilitates introduction of the secondary methyl group via a Simmons–Smith reaction.

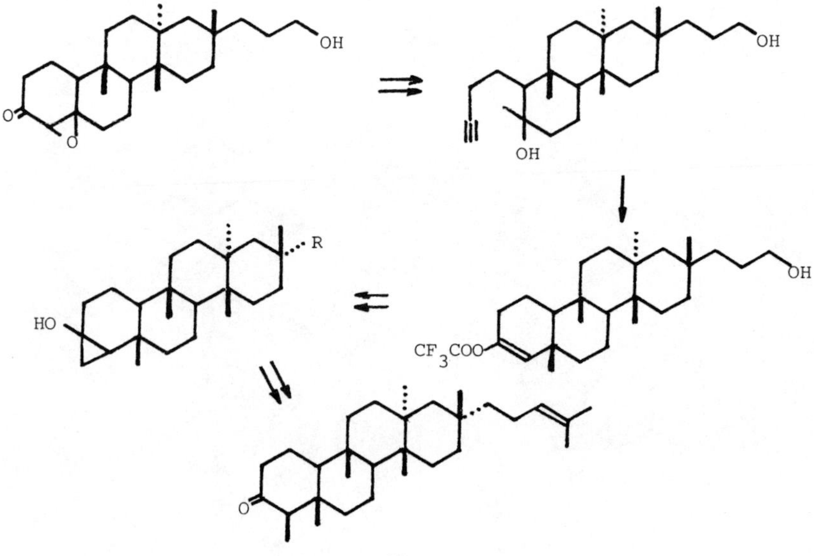

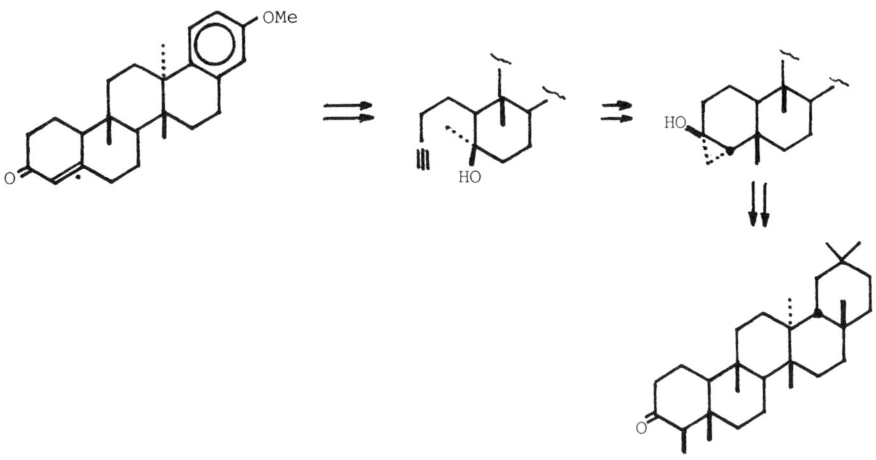

F13

Concerning the stereochemistry of the A/B ring juncture that emerges during solvolysis, equatorial attack of the alkyne side chain on the cation is coerced by the axial methyl group at C_9.

Entry into the hydrazulenes such as daucene (**D8**) (Yamasaki, 1972), aromaticin (**A42**) (Lansbury, 1979, 1980), and damsinic acid (**D6**) (Lansbury, 1978) may be gained by forming the seven-membered ring from the cyclopentane nucleus. The ionizing allylic alcohol may be placed in the side chain or directly associated with the cyclopentane moiety. The cyclization of germacrene-type substances is biogenetically significant. The stereoselective elaboration of globulol (**G14**) (Marshall, 1974) is patterned after this pathway.

D8

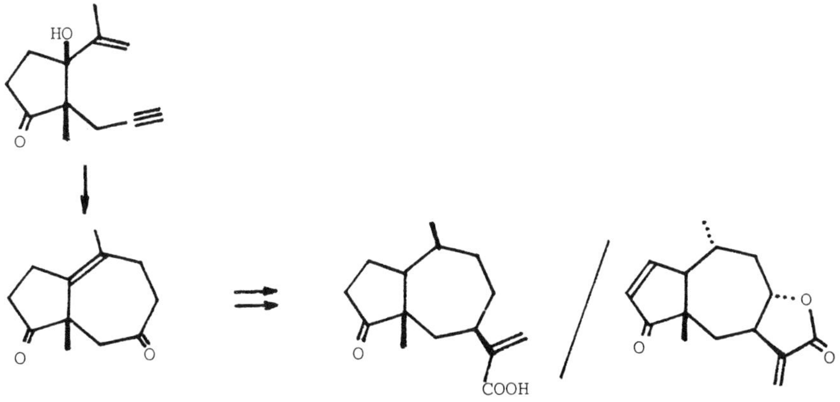

D6   **A42**

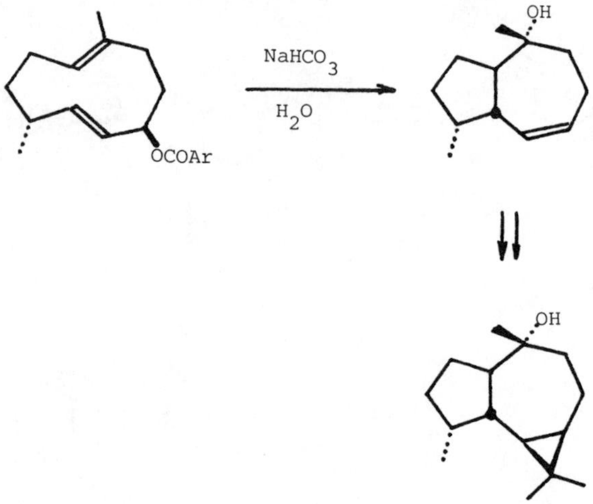

G14

A synthesis of karahanaenone (**K1**) (Hashimoto, 1977) from 6-trimethylsiloxy-neryl acetate shows a high degree of chemoselectivity. It is also interesting to note the facile formation of a seven-membered ring en route to nezukone (**N8**) (T. Saito, 1979).

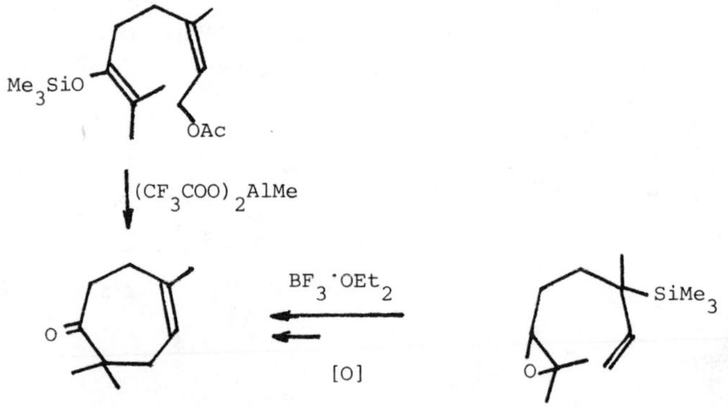

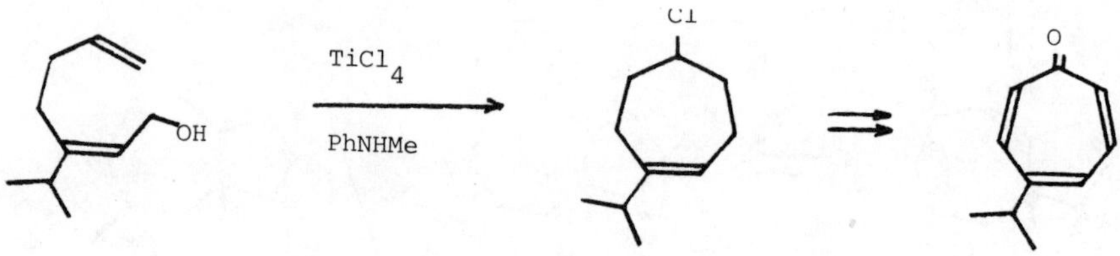

6.2 ALKYLATIONS

Aromatic alkylation leading to condensed-ring systems is quite well known. For example, the *gem*-dimethylated A ring of tanshinone II (**T2**) is closed by acid treatment of the tertiary alcohol. Interestingly, the B ring is also derived from a Friedel–Crafts reaction (acylation), the aromatization of which is accompanied by loss of a methoxy group (Tateishi, 1971).

T2

In the early synthesis of podocarpic acid (**P43**) (King, 1956) and ferruginol (**F4**) (King, 1957), an interesting feature is the formation of the hydrophenanthrene framework mainly with a trans ring fusion.

R=H, X=COOEt
R=OMe, X=H

P43 X=COOH, R'=H
F4 X=Me, R'=iPr

This approach has also been extended to a synthesis of totarol (**T21**) (Barltrop, 1958). This necessitates the implementation of an indirect isopropylation involving a redox maneuver of the aromatic nucleus, which is punctuated by alkylation of the enone intermediate.

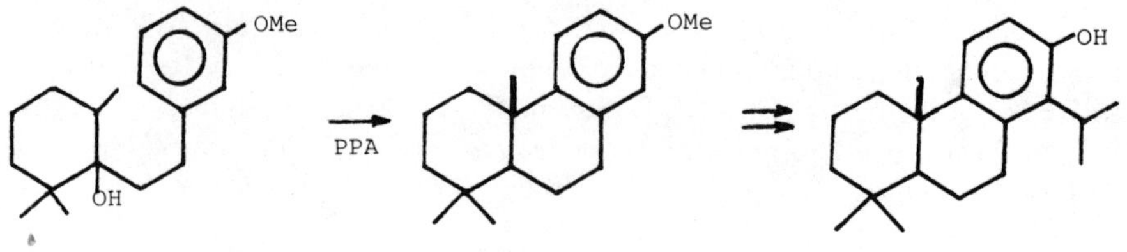

The same tricyclic ether has been employed in the elaboration of tetracyclic diterpenes. Thus, the C ring is fashioned into a cyclohexenol to which an acetaldehyde chain is then introduced at the nearest ring juncture for later cyclization. By proper manipulations kaurene (**K4**) (Bell, 1962, 1966b), hibaene (**H11**) (Ireland, 1965; Bell, 1966a), and phyllocladene (**P27**) (Church, 1960, 1966) have been acquired.

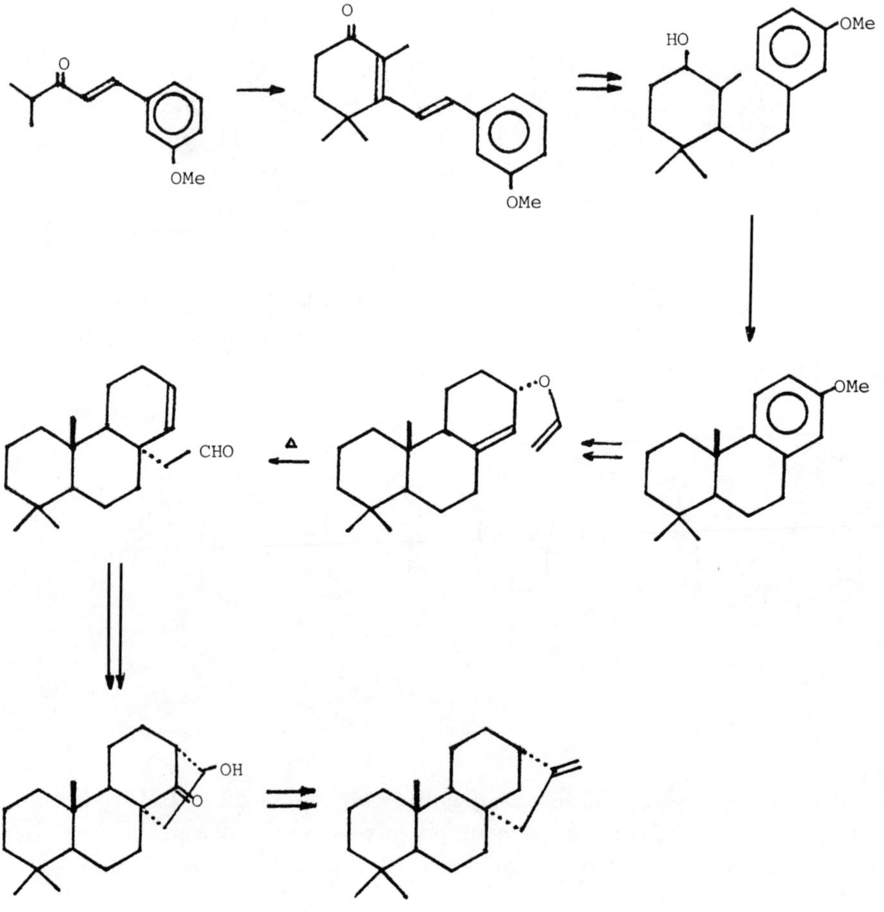

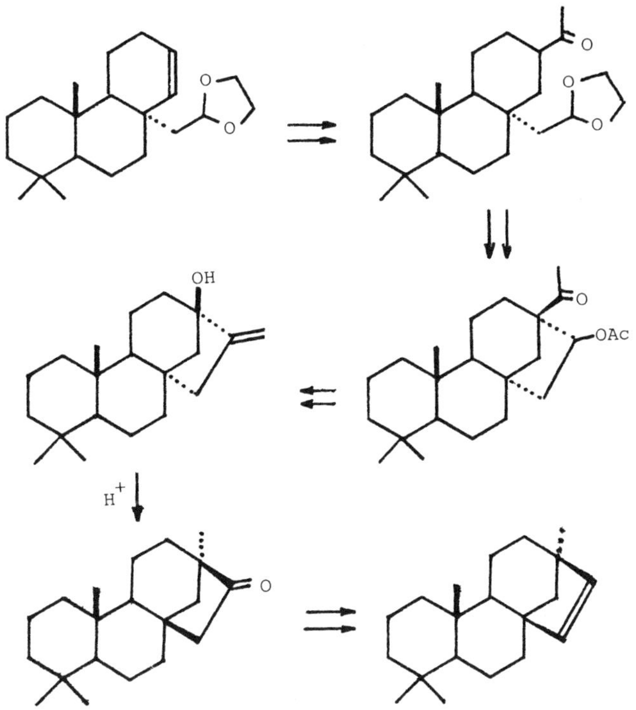

H11

In the hibaene work, a stereoinversion of the CD-ring segment is modeled after the allogibberic-to-gibberic acid rearrangement.

The stereochemistry of the A/B ring juncture and the C_4 configuration of podocarpic acid (**P43**) can be controlled during its synthesis by a cationic cyclization route (Kanjilal, 1981). Enforced backside attack of the bridged lactone by the aromatic moiety can give rise only to a compound of the podocarpic acid series.

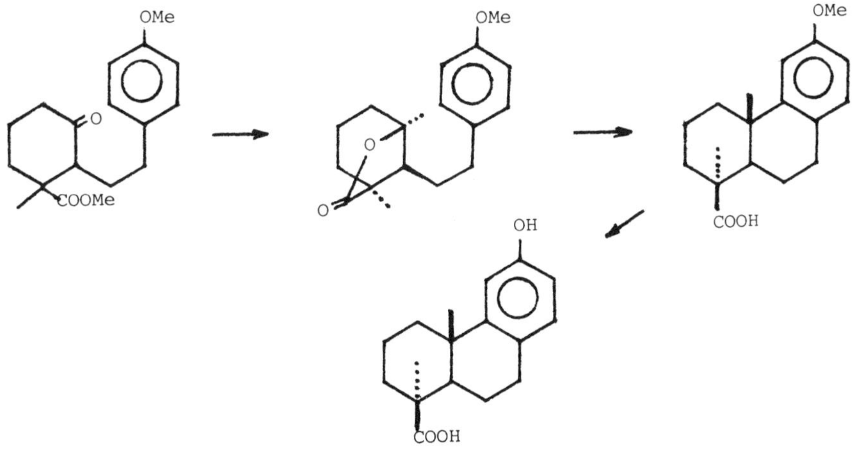

P43

The remarkably stereoselective cationic cyclization of polyenes to afford many polycyclic terpenes in nature has received attention since the beginning of this century. The greatest impact is evident in the Stork–Eschenmoser postulate that asserts a concerted, stereoelectronically favored interaction among neighboring double bonds along the folded carbon chains as a prerequisite for generating the unique terpene structures.

The biomimetic polyene cyclization investigated by W. S. Johnson (Bartlett, 1984), which culminated in stereocontrolled (six or more asymmetric centers) production of the steroid skeleton from acyclic compounds, must be counted as one of the monumental achievements of organic chemistry. It is important to create the initial cation regioselectively at the terminal double bond or its equivalent, and to use conditions mild enough to avoid random protonation at other positions.

The synthesis of fichtelite (**F5**) (W. Johnson, 1966) represents one of the initial successes of biomimetic cationic cyclization of polyenes beyond the bicyclic level. Three contiguous chiral centers are formed in the solvolysis of the cyclohexenol precursor.

F5

The cyclization of a trienol anchored at the two ends with an anisyl group and a cyclohexenol ring furnishes a pentacyclic substance containing six contiguous chiral centers, five of which corresponding to those in alnusenone (**A26**) (Ireland, 1975a). While synthetic correlation of the compound with a known intermediate has failed, the fruition of a slightly modified approach (Ireland, 1975d) is most gratifying.

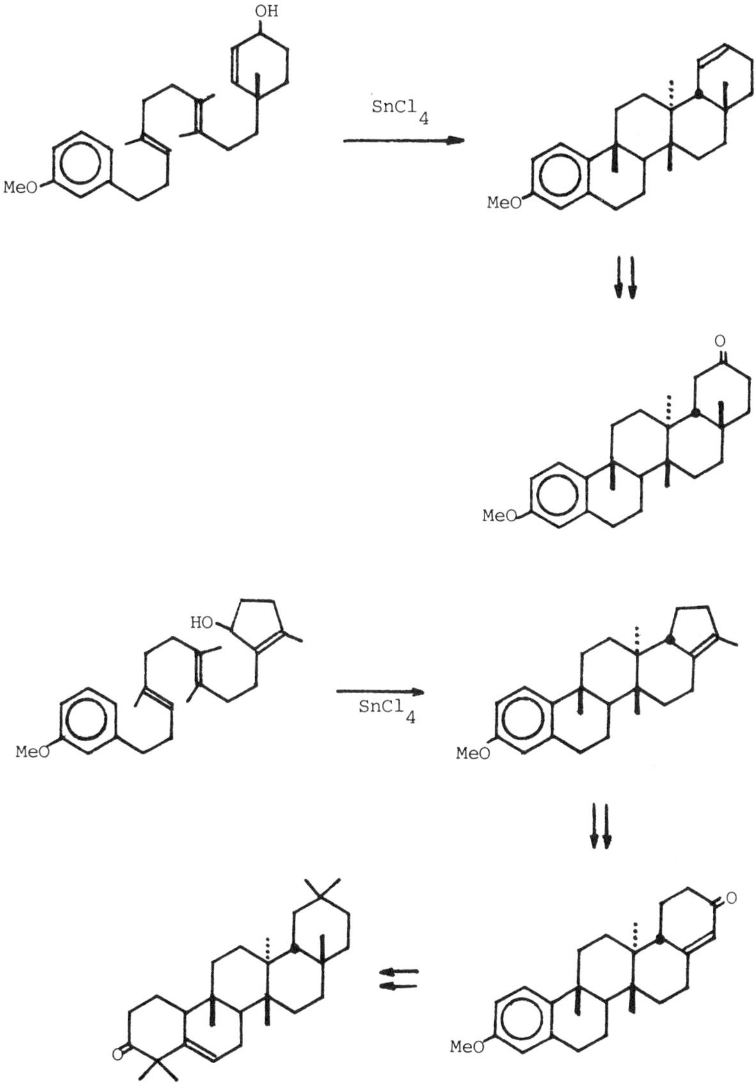

A26

The allylic cation that is not confined to a ring may also initiate cyclization. Thus, a route to the tricyclic resin acids has been devised (Bryson, 1982). The ring size effect (six- vs. eight-membered) dictates the formation of hydrophenanthrene derivatives exclusively. While the E-geometry of the central double bond bequeathes to the products a trans ring juncture, the C_4 configuration is largely determined by the preferential equatorial orientation of the larger alkyl substituent on the cationic center during cyclization (selectivity, 69:31). Consequently, abietane-type products predominate.

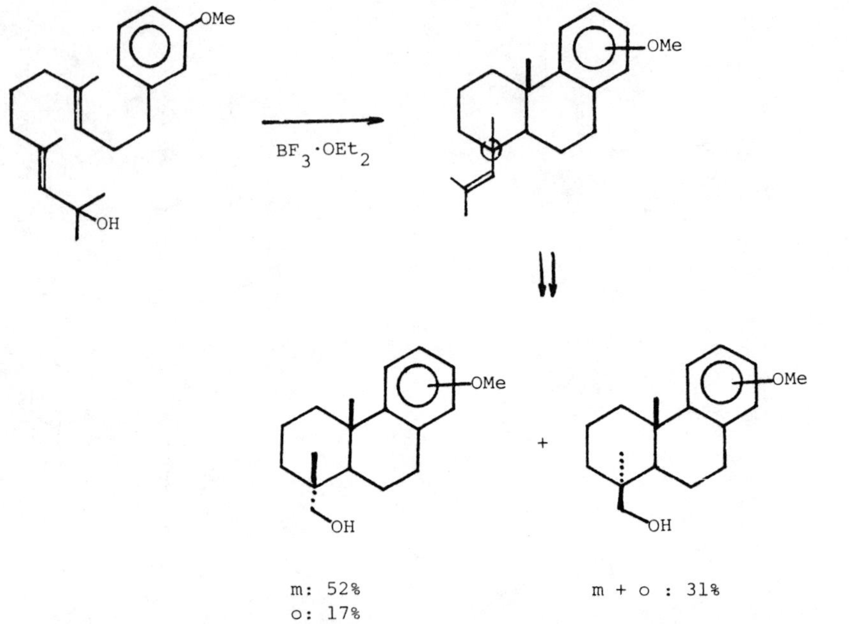

m: 52%
o: 17%

m + o : 31%

An annulation method that enables the assemblage of polycyclic substances containing a trimethyl trans-decalone unit commonly found in diterpenes and triterpenes is available. The initiator is an alcohol that ionizes to afford a symmetrical allylic cation. The symmetry is essential to avoiding generation of isomeric products. An example bearing on this strategem is a synthesis of taxodione (**T7**) (W. Johnson, 1982).

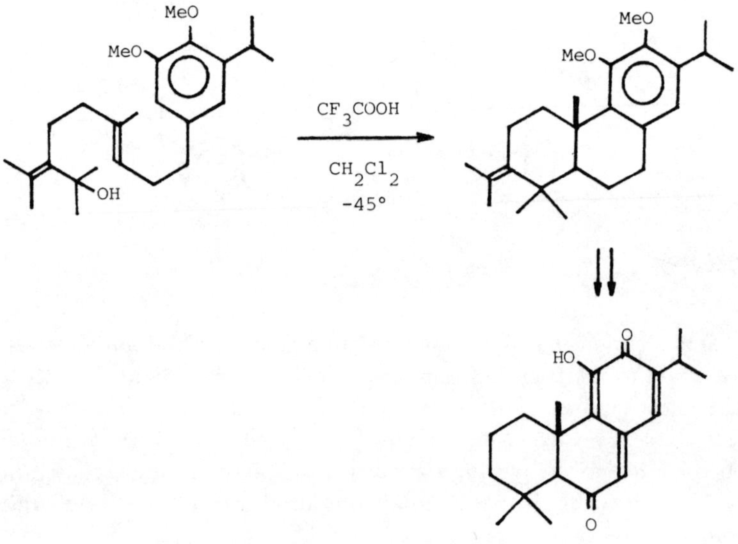

T7

Based on the same line of thought, a bicyclic precursor for serratenediol (**S25**) has been constructed (Prestwich, 1974).

S25

On account of the smaller number of rings, the structural prerequisites for the preparation by polyene cyclization of sesquiterpenes or their intermediates are less stringent. However, much impressive work has accumulated in this area. Thus, acyclic or monocyclic dienes or trienes have yielded the precursors of boll weevil pheromone components (**BW1**, **BW2**) (Janes, 1973, Bedoukian, 1975), dihydroactinidiolide (**A13**) (Torii, 1976), α-damascone (**D3**) (Nakatani, 1974; Schulte-Elte, 1975), β-damascenone (**D2**) (Torii, 1979), deoxytrisporone (**T33**) (Uneyama, 1976), junenol (**J5**) (Schwartz, 1972), drimenol (**D22**) (Caliezi, 1949), albicanyl acetate (**A23**) (Armstrong, 1982), confertifolin (**C61**) Akita, 1979), isodrimenin (**D21**) (Akita 1979), cinnamolide (**C56**) (Suzuki, 1970; Yanagawa, 1970), polygodial (**P44**) (T. Kato, 1971), pallescensin 1 (**P1**) (Tius, 1982), pallescensin A (**P2**) (Nasipuri, 1979; Matsumoto, 1983), and puupehenone (**P56**) (Trammell, 1978).

BW1

BW2

6.2 ALKYLATIONS

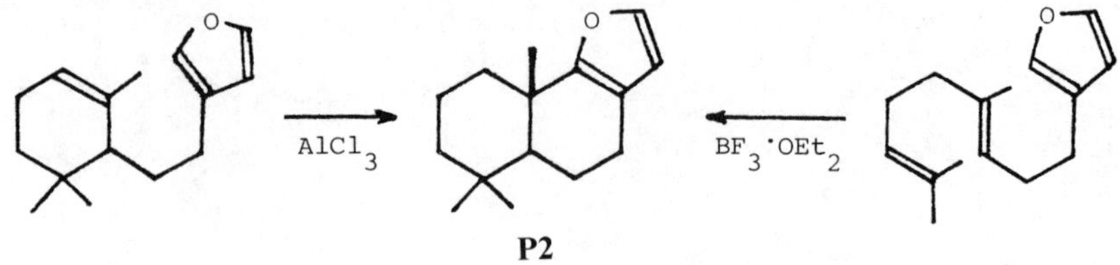

Heating *ent*-copalic acid in 88 % formic acid leads to a tricyclic compound that differs from isoagatholactone (**A18**) only in the oxidation levels at two points. Completion of synthesis of the lactone requires a routine oxidoreduction operation (Imamura, 1981; Nakano, 1982).

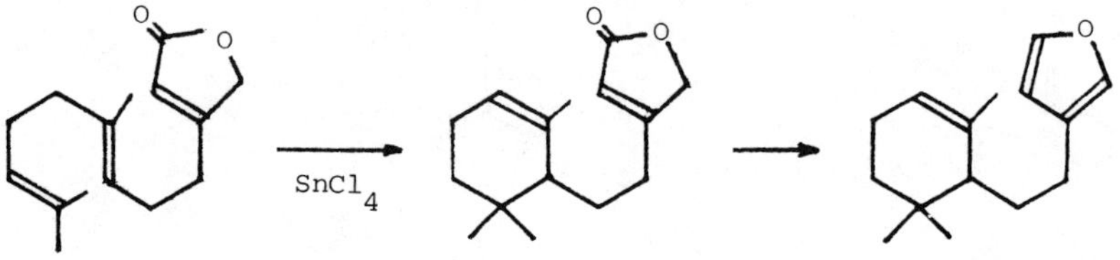

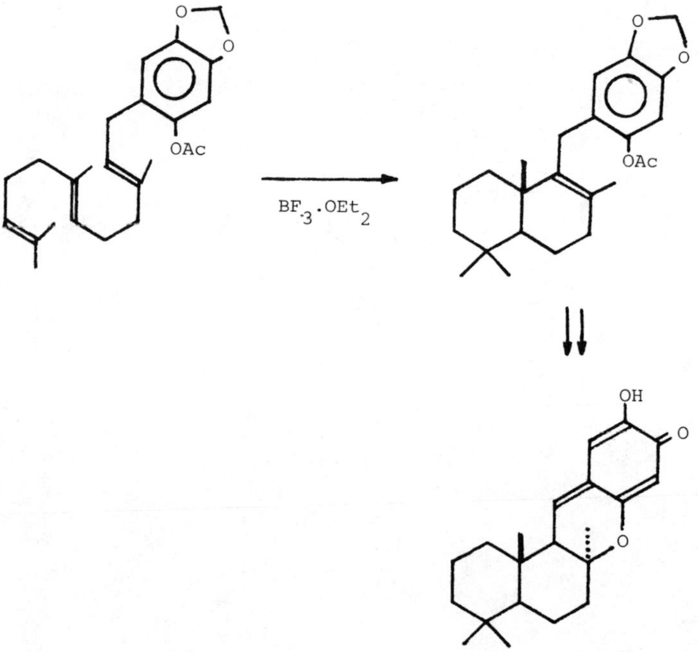

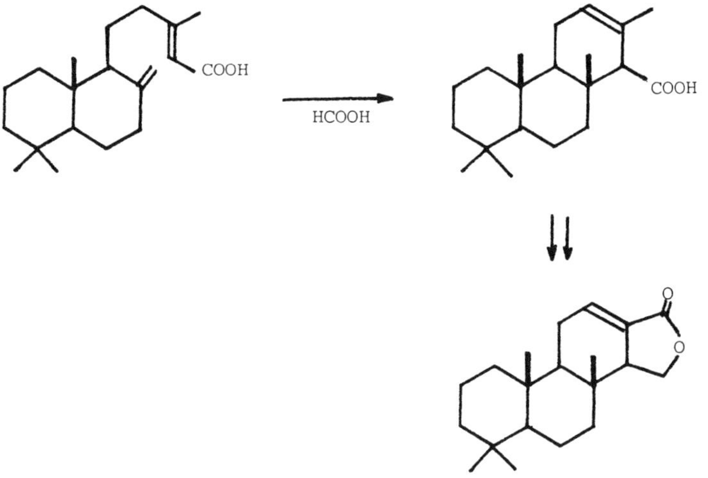

A18

It has been shown that cyclization of farnesyl phenyl sulfone proceeds stereoselectively, although there is evidence for a stepwise pathway (Torii, 1978). The bicyclic product is easily converted into labda-7, 14-dien-13-ol (**L1**) and α- and γ-polypodatetraene (Nishizawa, 1984b).

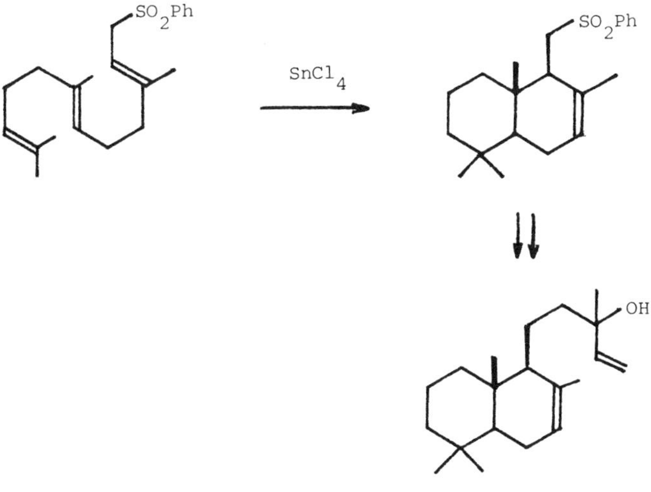

L1

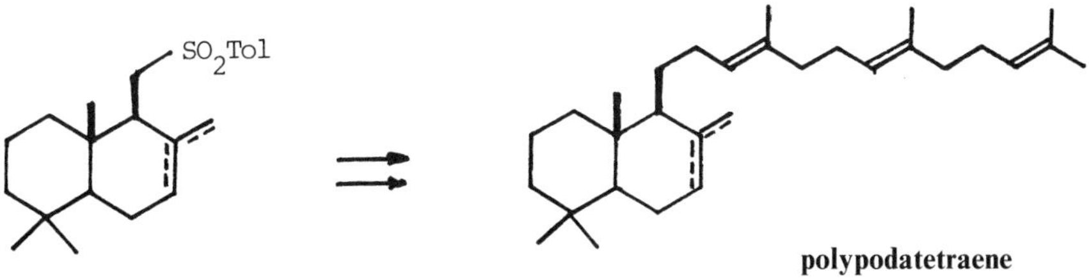

polypodatetraene

As seen in the synthesis of albicanyl acetate, the use of an allylsilane as terminator enables control of the double-bond positioning. Allylsilanes are excellent nucleophiles, which react at the far end of the double bond with concomitant loss of the silyl residue.

An appropriate nucleophile present in the polyene chain may act as the terminator. Thus, the α- and β-levantenolides (**L11** and **L12**) are obtained by treatment of the cyclofarnesyl butenolides with tin(IV) chloride (T. Kato, 1970, 1971).

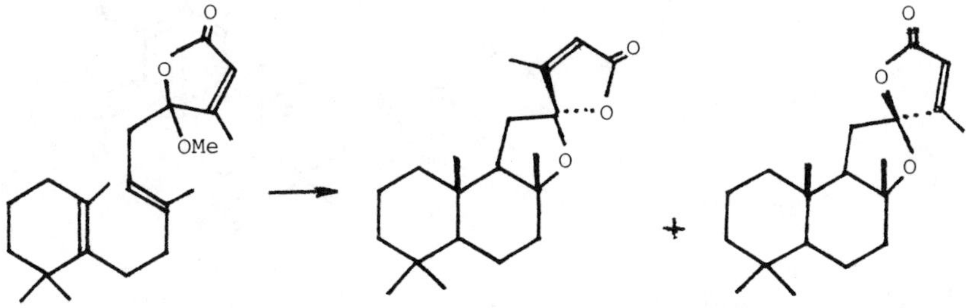

L11   **L12**

While the β-cyclofarnesyl compounds cyclize to *trans*-decalins, the "α" analogs mainly give the cis isomers. Clear evidence is furnished from a study on the ring closure of monocyclohomofarnesic acids (Saito, 1983; cf. Davis, 1978).

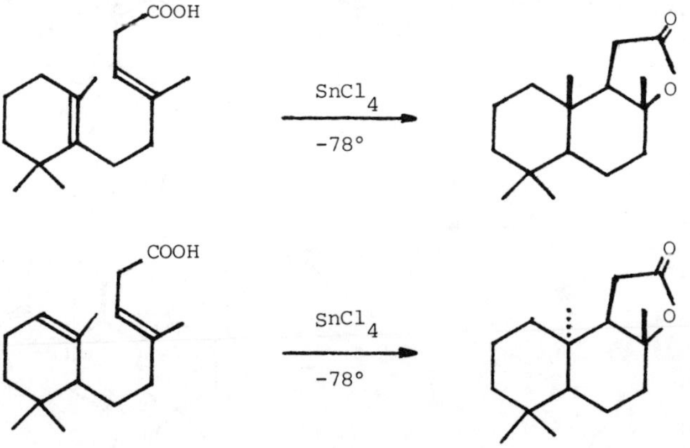

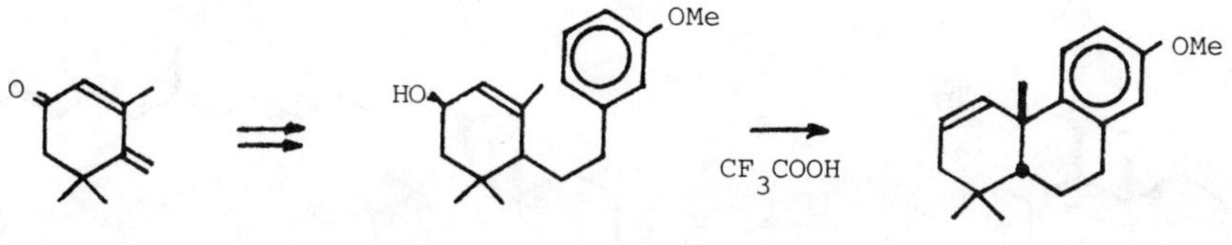

Polyenes containing aromatic end groups cyclize accordingly. There have been modifications in the substrates with the aim of inducing asymmetry (chirality) in the products. Preliminary studies indicate that a pro-C_7 silyl group prefers to maintain an equatorial configuration in the transition state (selectivity, 77:23) (Janssen, 1984).

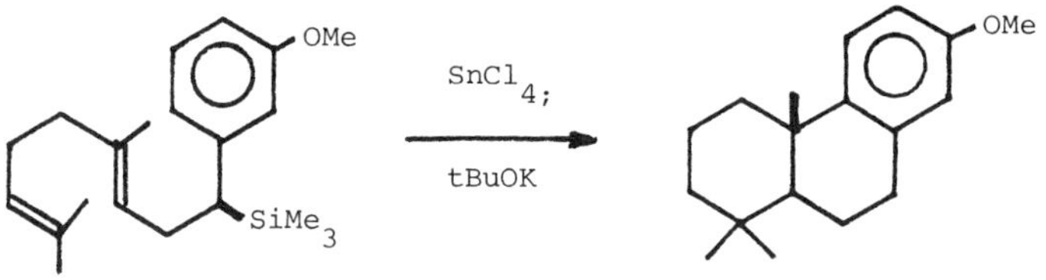

Enolized β-ketoesters are more effective internal nucleophiles for bond formation with a cationic center. The problem associated with regioselective deprotonation of the allosteric cation disappears, and a cyclic ketone results from such a process. Latia luciferin (**L6**) (Sum, 1979) and zonarol (**Z6**) (Skeean, 1976) have owed their synthesis to cyclic intermediates prepared in this manner.

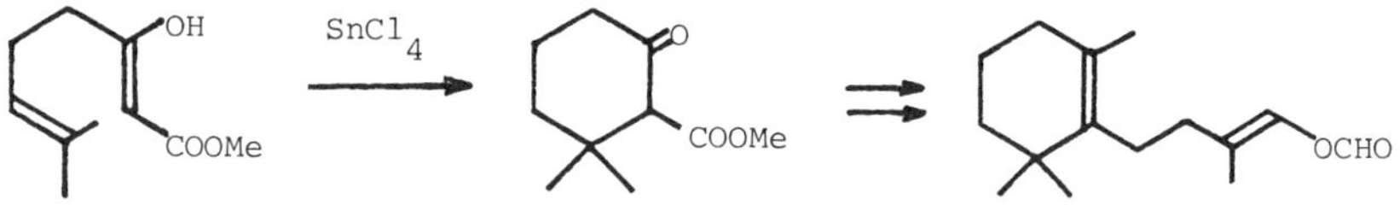

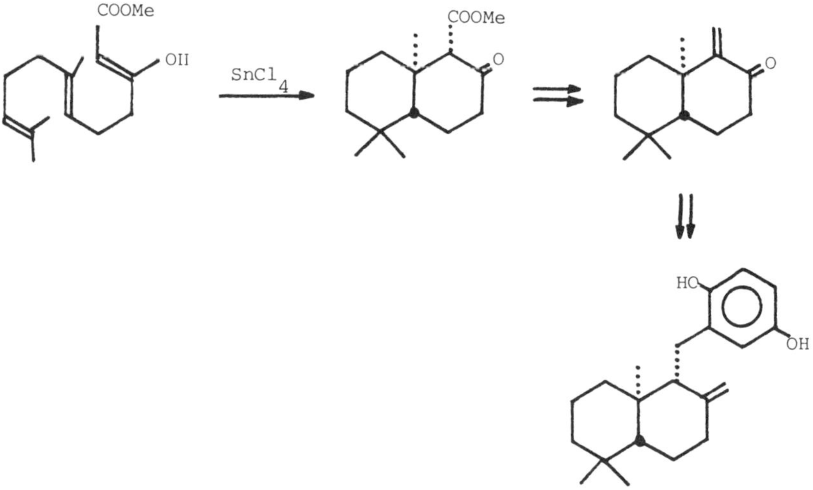

It is interesting to note that in a biogenetic-type synthesis of triptonide (**T29**) (van Tamelen, 1982), the ketone group of a β-ketoester acts as the initiator.

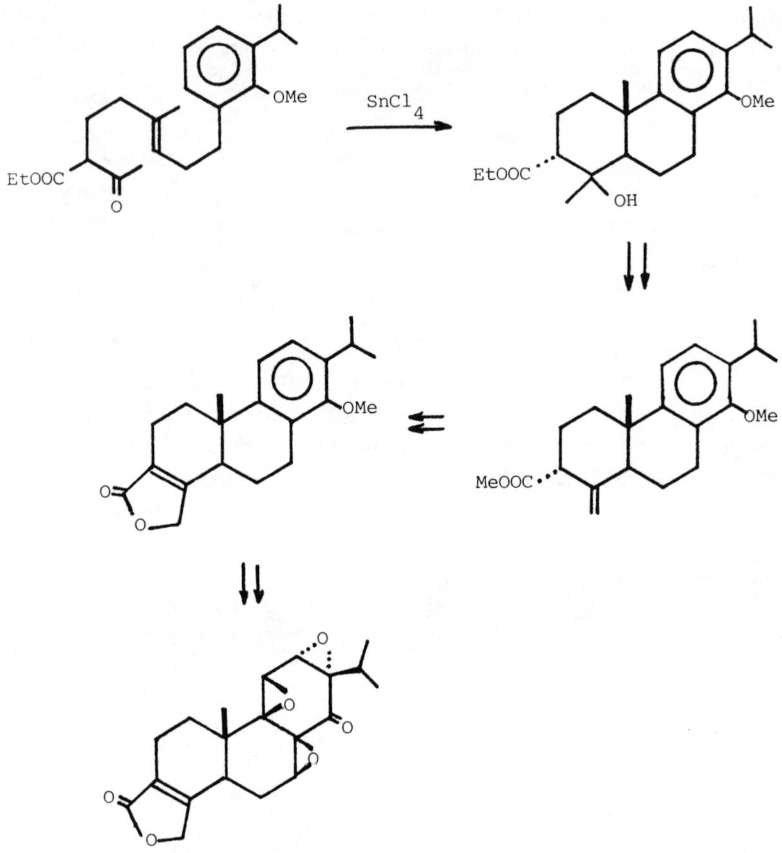

T29

6.2.3 Polyene Cyclizations Initiated by Soft Acids

Polyene cyclization can also be initiated by electrophilic agents containing heteroatoms. These include halogens, pseudohalogens, sulfenyl chlorides, selenyl halides, and mercuric ion. These variants are very beneficial to the synthesis of many halogenated terpenes of marine origin: the synderols (**S39** and **S40**) (Gonzalez, 1976; T. Kato, 1976) 10-bromo-α-chamigrene (**C48**) (Ichinose, 1979; L. Wolinsky, 1976), and 3β-bromocaparrapi oxide (**C20**) (T. Kato, 1980).

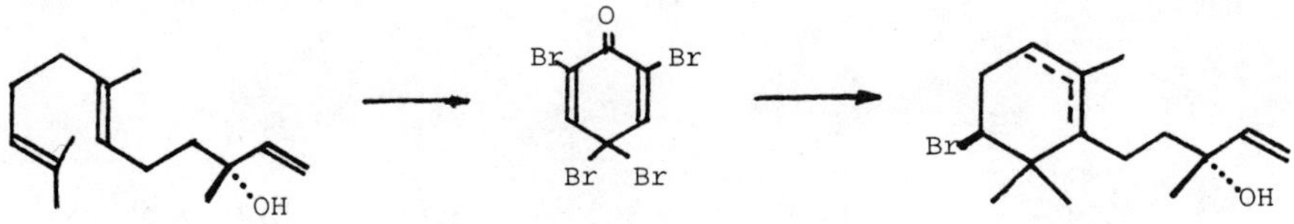

[TBCHD]   **S39**

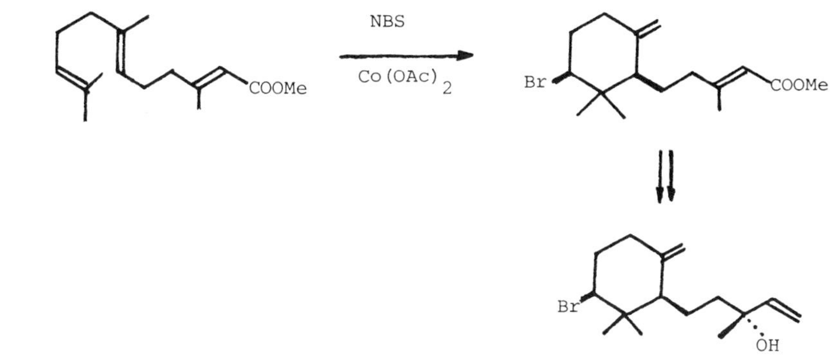

The Br$^+$-induced cyclization of polyenes in acetonitrile provides versatile synthetic intermediates not only for the bromine-containing cyclic terpenes; the carbon-bound bromine can act as a leaving group to further direct ring transformations. By incorporating the cyclization and a subsequent contraction, acoratriene (**A7**) may be secured in a few steps from methyl farnesate (T. Kato, 1984).

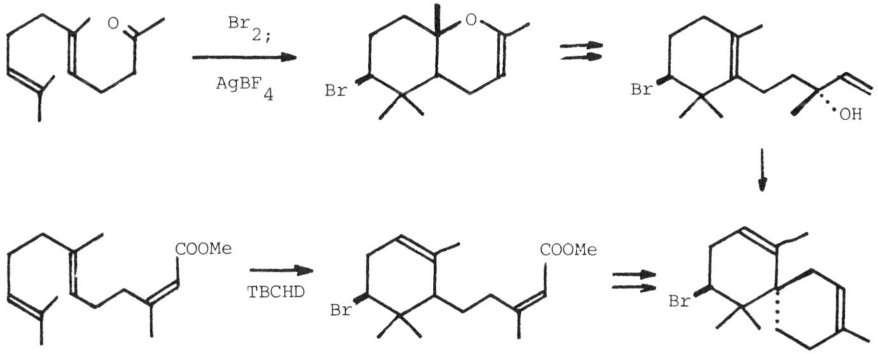

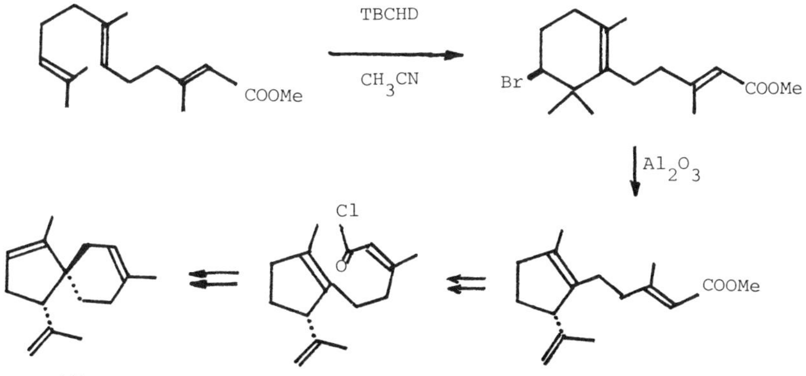

Spirolaurenone (**S47**) differs from bromochamigrene chiefly in the size of the B ring. Like other spirocyclic sesquiterpenes, spirolaurenone lends itself to synthesis by a route embodying intramolecular ketocarbene addition and subsequent cleavage of the extra cyclopropane ring (Murai, 1982).

Bromocyclogeranylacetic acid is the substrate for a cationic cyclization. This process may involve the formation of a bromonium ion intermediate.

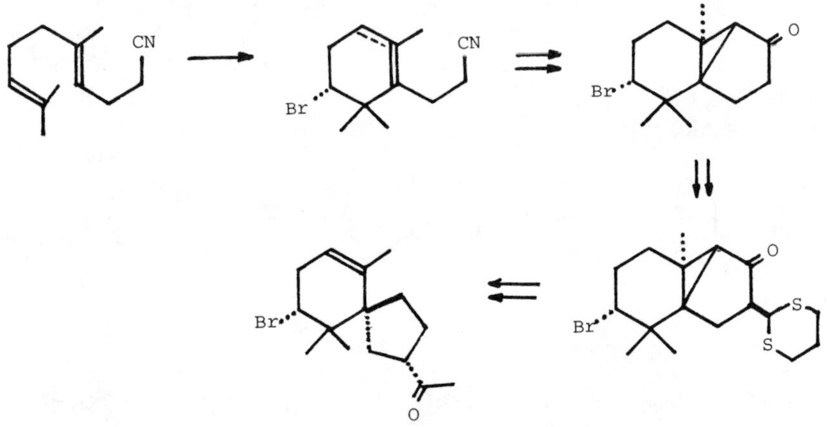

S47

An indirect method for introducing bromine atoms into organic molecules is by bromolysis of mercurioalkanes. Thus, 3β-bromo-8-epicaparrapi oxide (**C20**) has been acquired in this manner (Hoye, 1979).

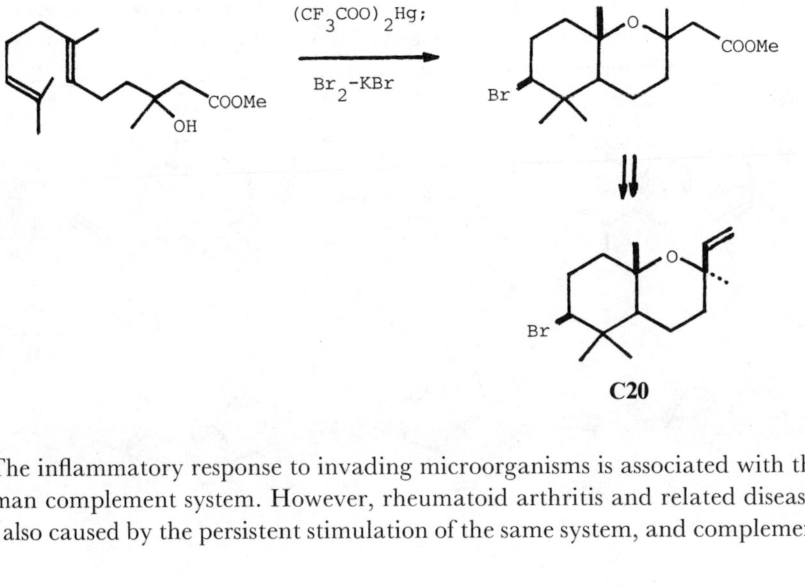

C20

The inflammatory response to invading microorganisms is associated with the human complement system. However, rheumatoid arthritis and related diseases are also caused by the persistent stimulation of the same system, and complement inhibitors such as the powerful fungal metabolite **K76** may be very valuable therapeutic agents.

A synthesis of **K76** (Corey, 1982) starts from a mercury (II)-mediated cyclization of polyene. An alternative synthetic sequence (McMurry, 1985b), which calls for a stereoselective bishydroxylation of the A-ring double bond before coupling of the drimane derivative to the aromatic moiety, actually employs the cyclopropylidene analog in the cyclization. The much-diminished 1,3-diaxial interaction between the angular methyl group and the axial carbon substituent in the form of a cyclopropyl methylene apparently allows the A ring to remain in a chair conformation, and therefore ensures its subsequent hydroxylation from the α face.

A similar cyclization of a 12-siloxy-3-phosphoryloxyfarnesic ester has been employed in the preparation of the AB synthon of aphidicolin (**A36**) (Corey, 1980a). The α-configuration of the siloxymethyl group in the resulting bicyclic product is expected on the grounds that an all-chair conformation is adopted in the transition state.

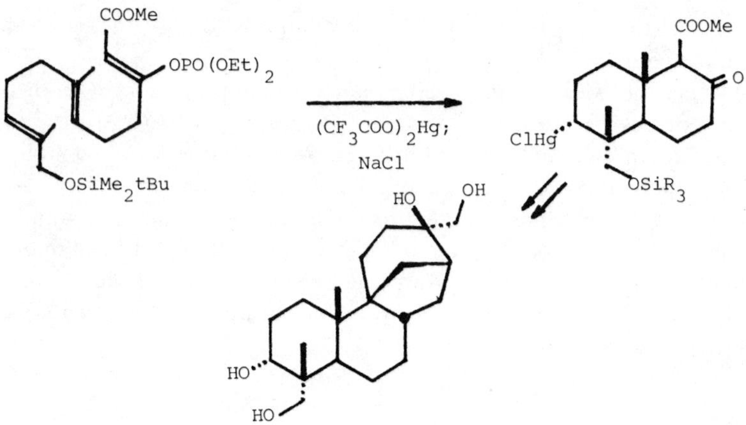

A36

Cyclization of farnesyl derivatives mediated by a mercury(II) triflate–dimethylaniline complex gives excellent yields of 3-mercuriodrimanes. Judicious employment of this reaction on two different occasions permits the synthesis of α,γ-onoceradienedione (**O5**) and thence lansic acid (**L5**) in a concise fashion (Nishizawa, 1984a).

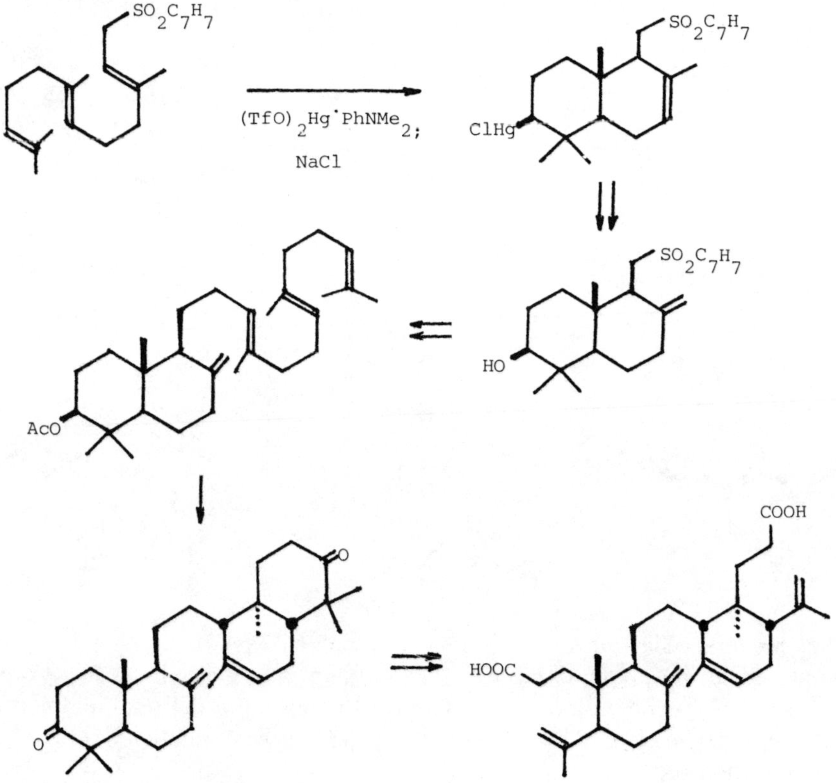

O5

L5

5-Tolylsulfonylambliofuran undergoes cyclization under the influence of the same reagent in the presence of water to afford mono-, bi-, and tricarbocyclic products. These compounds are formed via mercuration at the terminal double bond. The monocyclic product (16 %) furnishes ambliol-A (**A27**) on desulfurization (Nishizawa, 1986b).

Interestingly, cyclization can be initiated by mercuration of the internal double bond of ambliofuran itself when water is excluded from the reaction medium.

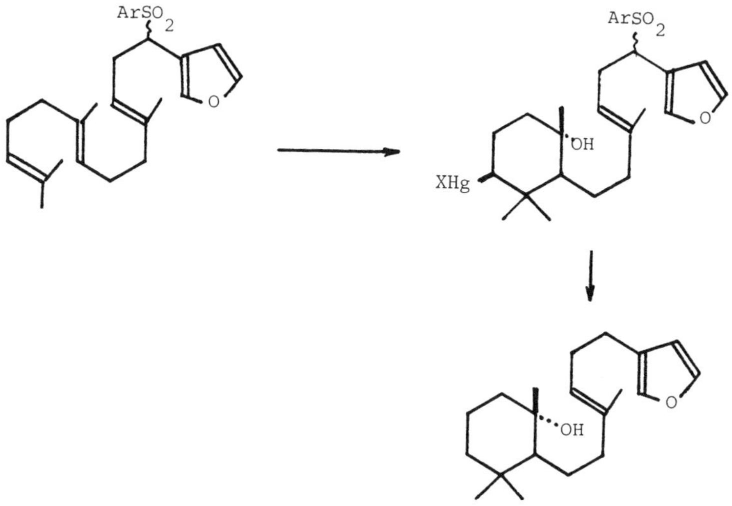

A27

3-Ketodrimanes obtained from cationic cyclizations and oxidation can be selectively cleaved by the Beckmann fragmentation, as shown in the synthesis of lansic acid. The structure of karatavic acid (**K3**) has been settled by means of a synthesis patterned in this manner (Nishizawa, 1984c).

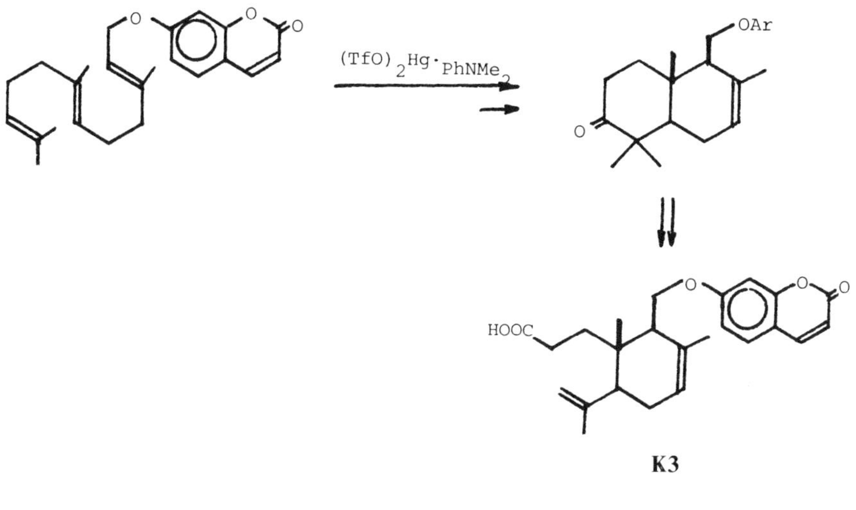

K3

The corresponding reaction with geranylgeranyl esters gives both bicyclic and tricyclic products (Nishizawa, 1983). The tricyclic alcohol is easily converted into isoagatholactone (**A18**) by a redox protocol.

A18

Cyclization of E,E,E-geranylgeranyl acetate under similar conditions has also been studied (Nishizawa, 1986a). After brominolysis and hydrolysis one of the minor products (1.8 %) isoaplysin-20 is produced. Thus the marine diterpene has a perhydrophenanthrene skeleton with a trans-syn-trans fusion. The difficulty of acquiring it by polyene cyclization is understandable, since stereoelectronic factors favor generation of the all-chair trans-anti-trans products.

isoaplysin-20

The absence of nucleophilic species in the mercuration system permits efficient participation of properly situated double bond(s), which results in cycloalkene products. When the anion and/or solvent are of sufficient nucleophilicity, they trap the carbocation(s). The general solvatomercuration–demercuration protocol is now a standard method for alkene hydration in the Markovnikov sense.

There is a useful compromise during mercuration of nerolidol under conventional conditons such that cyclonerodiol (**C93**) can be readily obtained (Matsuki, 1979). Both the proximity of the double bonds in nerolidol and the neopentyl nature of the electrophilic carbon in the mercuronium ion derived therefrom are conducive to cyclization.

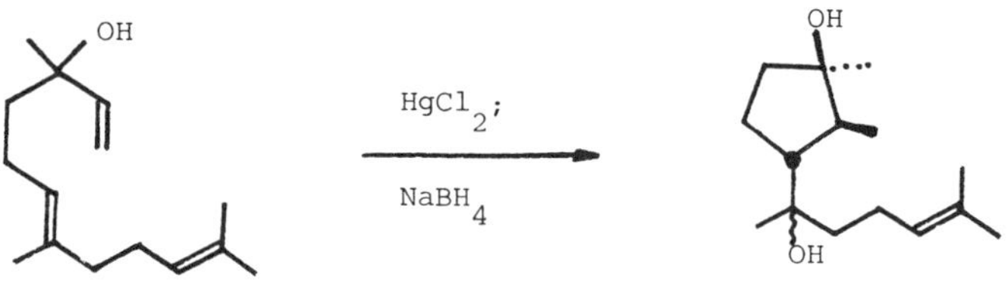

6.2 ALKYLATIONS

C93

Gaining in popularity for synthetic operations is selenylation of alkenes. A synthesis of loliolide (**L21**) (Rouessac, 1983) commences with phenylselenylation of homogeranic acid.

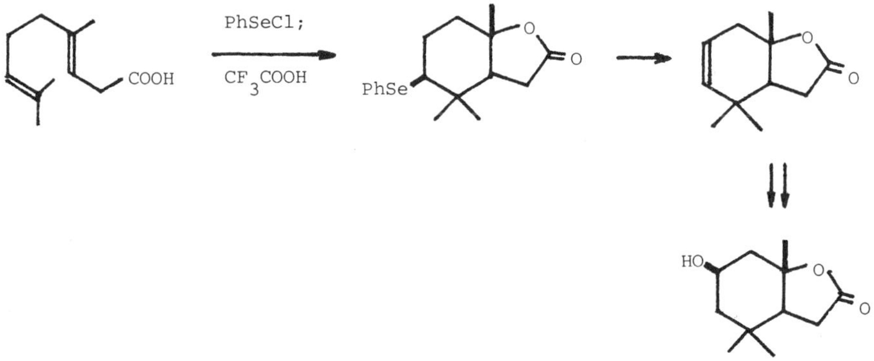

L21

The relatively simple skeleton of caparrapi oxide (**C19**) is surprisingly difficult to make. Ionization of the hydroxy-selenide derived from nerolidol appears to enhance the desirable mode of reaction. The participation of an episelenonium ion intermediate to induce the cyclization process is a possibility (Kametani, 1981).

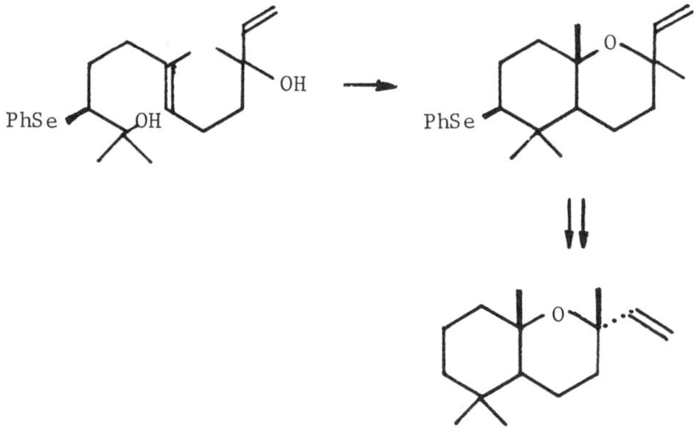

C19

As indicated, the heterosubstituents that are incorporated into the cyclized products can be removed. A synthesis of pleraphysillin 1 (**P38**) (Masaki, 1982) starts from the reaction of myrcene with phenylsulfenyl chloride. The phenylthio group is cleaved off by reduction.

P38

The π-bond is a soft base, and its facile reaction with soft acids such as sulfenyl and selenyl species is readily rationalized. Another soft acid is the thallium(III) ion, and thallation of alkenes has received much scrutiny in recent years. The one-step formation of karahana ether (**K2**) and its endocyclic isomer in a 15:45 ratio by treatment of nerol with thallium(III) perchlorate (Y. Yamada, 1979) demonstrates the unusual utility of such reagents.

K2

A convenient yet nonstereoselective route to the hydrindenone intermediate of valerenal (**V3**) involves a thallium(III)-promoted cyclization of a 1,2,4-triene (Baudouy, 1983). Metal ion coordination of either double bond of the allene may induce an attack by the other terminus of the unsaturated system, which is followed by solvent trapping.

V3

6.2.4 Cyclizations Initiated by Epoxide Opening

In the biosynthesis of 3-oxygenated triterpenes and steroids, squalene oxide is the true intermediate that undergoes enzyme-catalyzed cyclization cascade. *In vitro* cyclization initiated by opening of a terminal epoxide is also more efficient than protonation of the corresponding alkene, as shown by the two routes to tetrahymanol (**T13**) (van Tamelan, 1972). Both routes involve the formation of three rings by virtue of the participation of three similarly substituted and located double bonds.

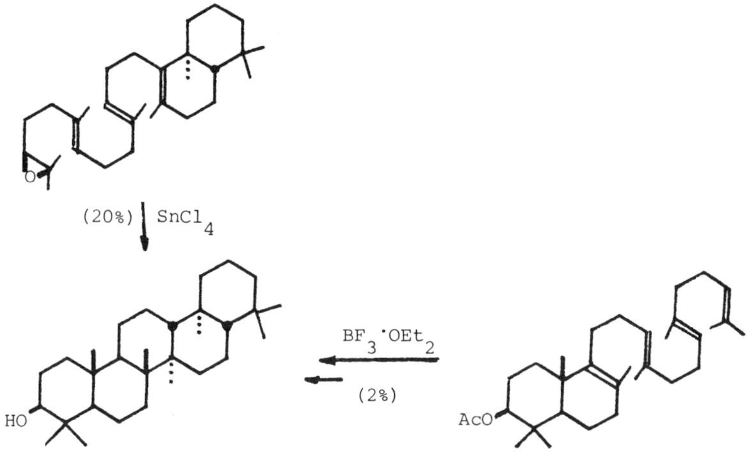

The treatment of a monocyclosqualene oxide fabricated from (+)-limonene with tin(IV) chloride leads to (—)-isotirucallol (**T19**) and parkeol (**P10**) in 18% and 2% yields, respectively (van Tamelen, 1972).

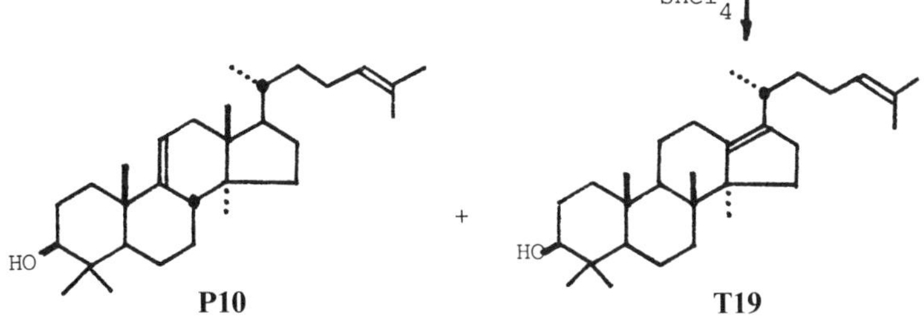

A preparation of δ-amyrin (**A31**) partially follows a biomimetic pathway (van Tamelen, 1972c).

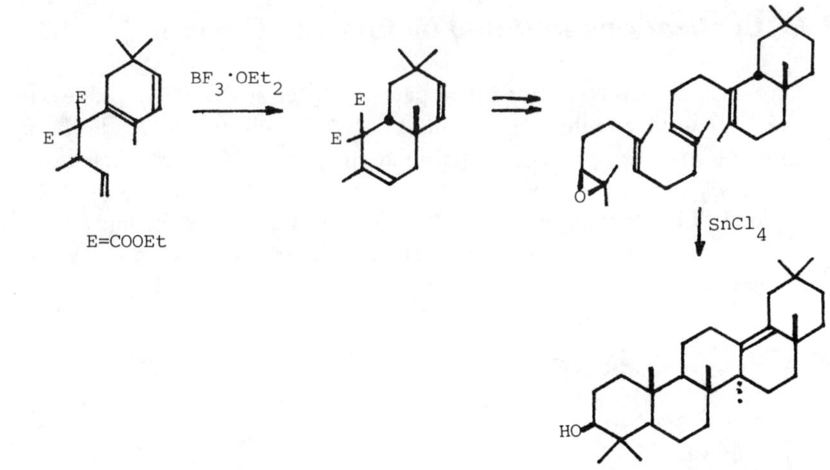

A31

Internally hydroxylated squalene oxides are provided with a new opportunity for cation termination. Malabaricanediol (**M3**) has thus been obtained from cyclization of such an epoxide (Sharpless, 1970), albeit only in 7% yield.

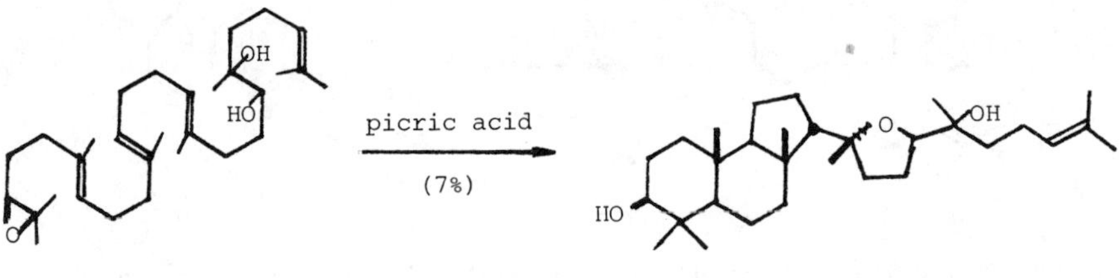

M3

Taondiol (**T4**) co-occurs with δ-tocotrienol oxide. The apparent biosynthetic relationship of the two compounds has drawn circumstantial support from the *in vitro* cyclization of the latter substance with picric acid (Kumanireng, 1973). The smooth formation of taonidiol from the epoxygeranylgeranylphenol derivative (Gonzalez, 1973) is also expected.

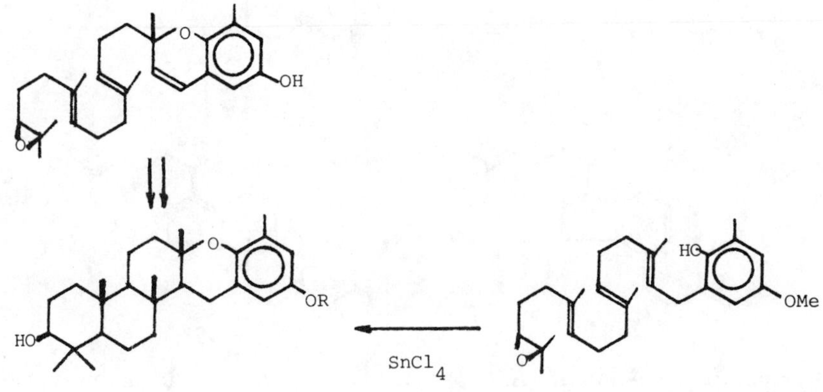

T7

6.2 ALKYLATIONS

This mode of cyclization is followed in the biosynthesis of 3-oxydrimanes and many polycyclic diterpenes. A biomimetic synthesis of farnesiferol A (**F1**) (van Tamelen, 1966) has been achieved.

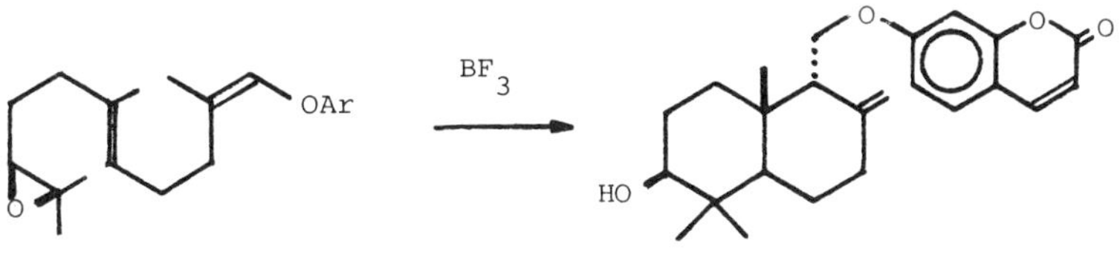

Because of its higher efficiency, the epoxide cyclization route may prove advantageous even in the synthesis of the deoxy analogs. An approach to drimenol (**D22**) (van Tamelen, 1963) is actually based on this plan.

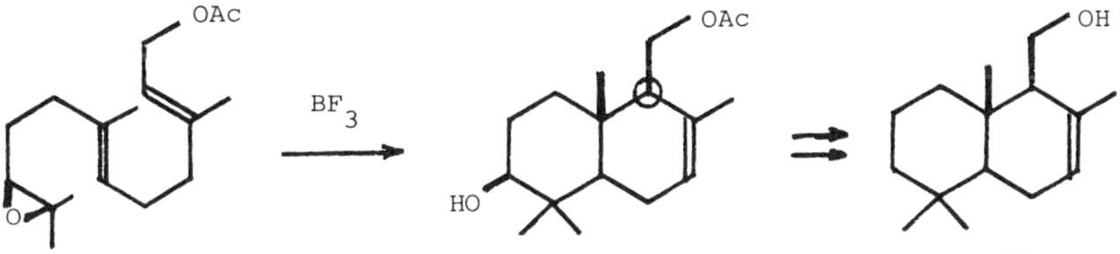

In aphidicolin (**A36**), the A ring possesses four contiguous chiral carbons represents a major problem in synthesis. A cationic cyclization approach in which all these stereocenters are correctly induced from one center (C_3) and an *E*-double bond is definitely attractive. Entry into the tricyclic phase is prepared by the formation of a terminal *erythro*-glycidol and treatment of its benzoate with iron(III) chloride (Van Tamelen, 1983).

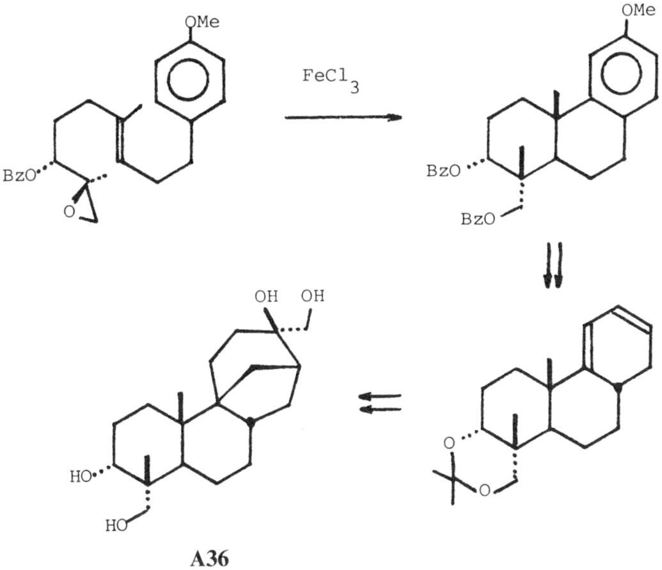

A formal synthesis of aphidicolin (**A36**) (Tanis, 1958a) exploits the cationic bisannulation terminated by a furan nucleus. The furan moiety is a latent 1,4-dicarbonyl function from which the C ring is to be fabricated. Contrary to the preceding route, this approach requires configuration inversion of the secondary hydroxyl group.

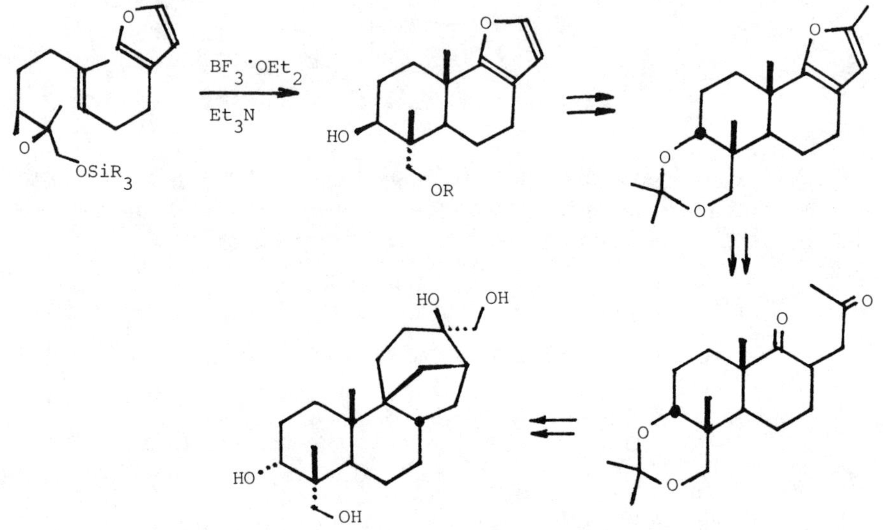

A36

A more conventional variant of cyclization suffices in an approach to maritimol (**M9**) (van Tamelen, 1981) in view of the reduced complication in the A-ring substitution pattern.

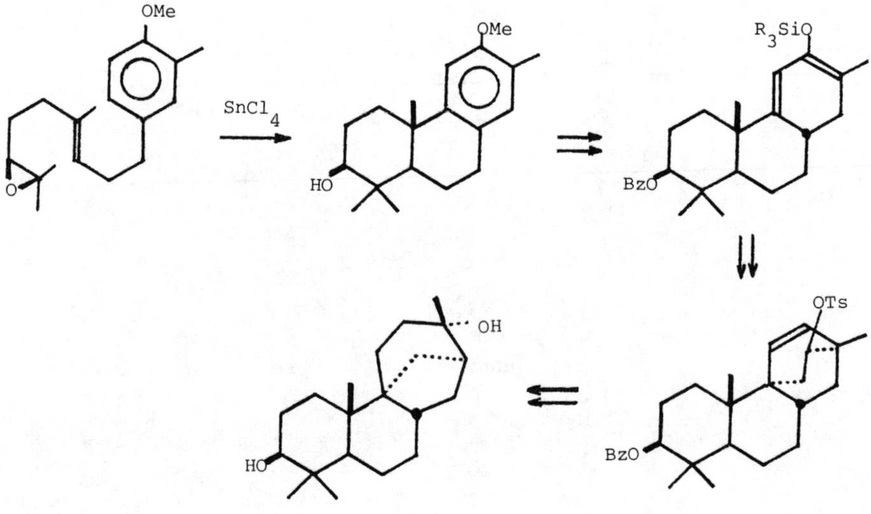

M9

6.2 ALKYLATIONS

Double-bond participation in cationic cyclization is regioselectively dictated by an allylic silane group when present. Exploitation of this positional determinator in a biomimetic synthesis of karahanaenone (**K1**) (Wang, 1984) came logically.

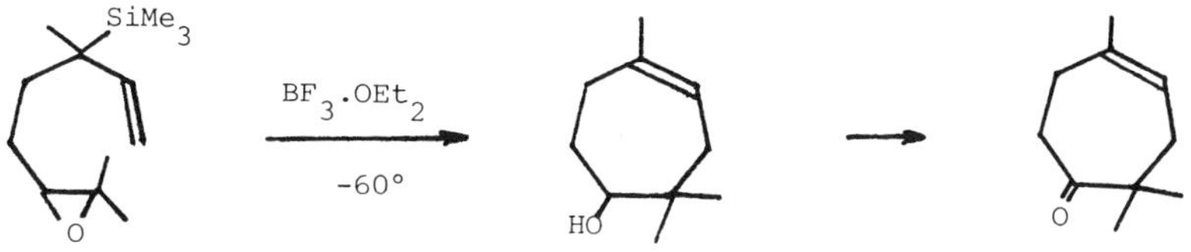

Similarly, the carbocyclic portions of karahana ether (**K2**) and 3-hydroxylabdadienoic acid have been induced by desilylative cyclization (Armstrong, 1986), which simultaneously creates an exocyclic methylene group.

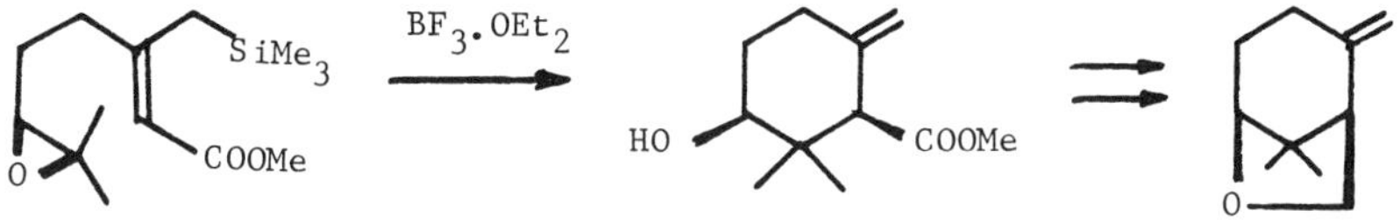

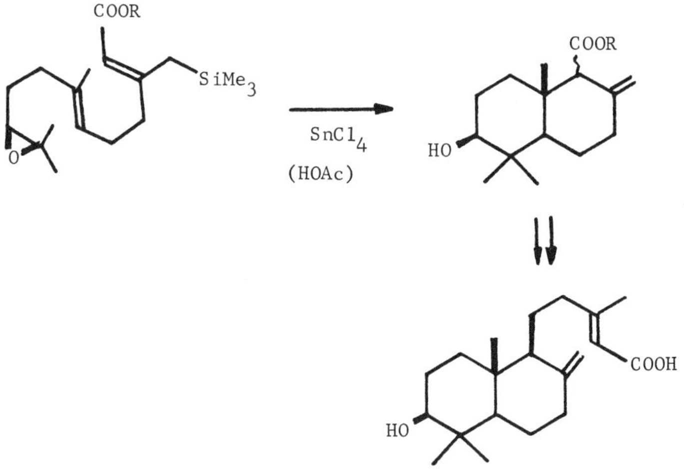

3-hydroxylabdadienoic acid

An entry into the tricyclic skeleton of rhodolauradiol (**R3**) involves epoxidation of elatol acetate and mild solvolysis (SiO$_2$ treatment) (Gonzalez, 1982). Steric constraint seems to dictate the anti-Markovnikov opening of the epoxide ring with participation of the chlorocyclohexene.

An approach to nagilactone F (Burke, 1987) embodies a silicon-directed polyene cyclization which is initiated by ionization of an acetal to form an octalin skeleton. The γ-lactone ring is then fashioned, after functional group modification including shifting the A-ring double bond to the 2,3-position to ameliorate steric requirement for the formation of a tetrahydrofuran nucleus with the axial C_6-oxy radical and the proper methyl group. The Δ^1 isomer and the saturated alcohol favor attack of the radical on the angular methyl substituent.

nagilactone F

6.2.5 Cycloalkylation of Enones

One of the more general schemes for synthesizing the common C-aromatic tricyclic diterpenes consists of coupling of β-cyclocitral with a proper benzyl Grignard reagent, oxidizing the resulting allylic alcohol, and treating the enone with an acid. The presence of a carbonyl group at C_6 permits regulation of the A/B-ring juncture; therefore, it minimizes the waste of valuable intermediates. This route has provided synthetic sempervirol (**S24**) (Matsumoto, 1976), dispermol (**D15**) (Matsumoto, 1977b), ferruginol (**F4**) (Matsumoto, 1977b), and their more highly oxidized analogs.

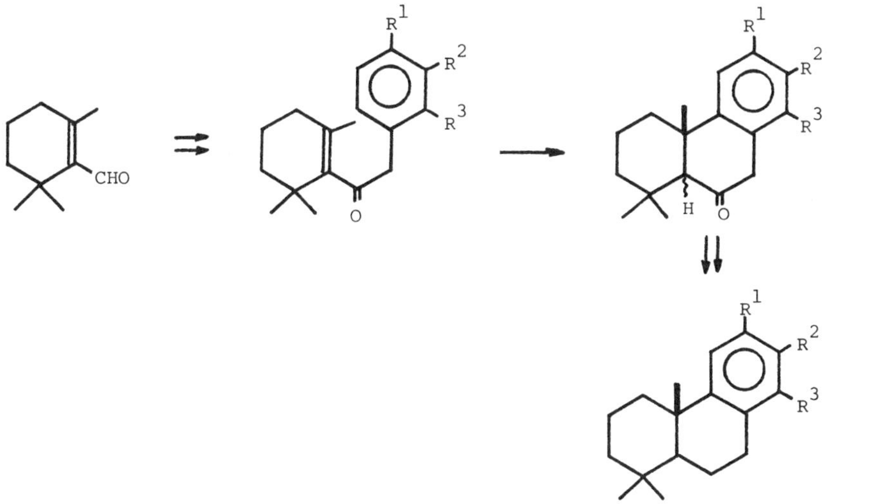

S24 (R^1=iPr, R^2=OMe, R^3=H)
F4 (R^1=OH, R^2=iPr, R^3=H)
D15 (R^1=OMe, R^2=OH, R^3=iPr)

The cyclization step may be critically dependent on structural variations. Thus the precursor of taxodione (**T7**) undergoes closure of the B ring with a mixture of formic and phosphoric acids only (R. Stevens, 1982). Both *trans*- and *cis*-ketones (in a 3:2 ratio) are produced.

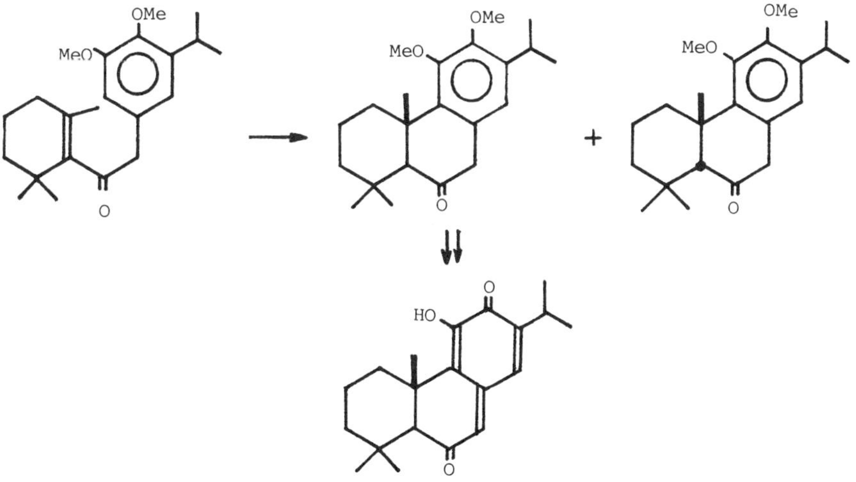

T7

An approach to steviol methyl ester (**S53**) (Ziegler, 1977) involves the formation of a hydrophenanthrenone in which the ketone group becomes quaternary.

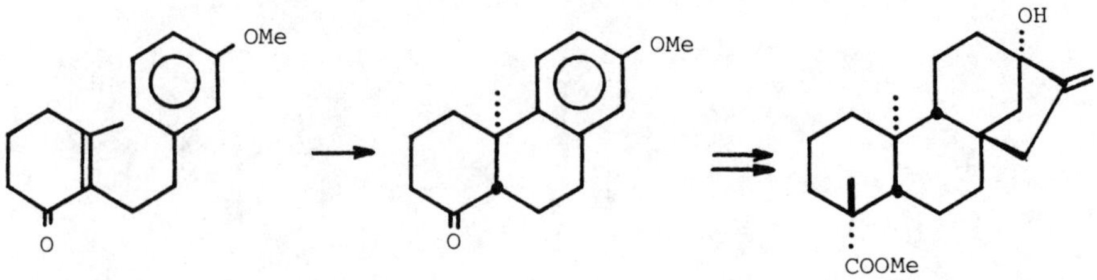

S53

Analogs in which the electrophile is a 2-methylenecyclohexanone give rise to products with different skeletons. Specifically, condensed suberane derivatives related to pisiferin (Matsumoto, 1986) are readily synthesized.

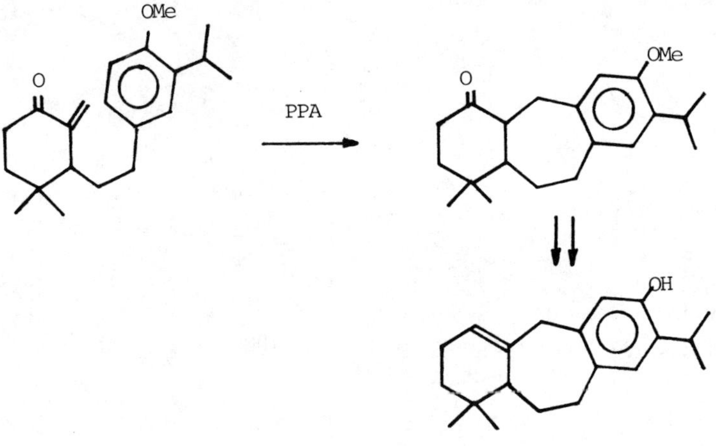

pisiferin

Furans are hypernucleophilic in comparison with homonuclear aromatic compounds. They react very readily with various electrophiles, preferably at an open α position. The intramolecular reaction of an α-methylene cyclohexanone with a furan ring proves to be a practical way to acquire the basic skeleton of pallescensin-G and -F (**P5** and **P4**) (Matsumoto, 1983).

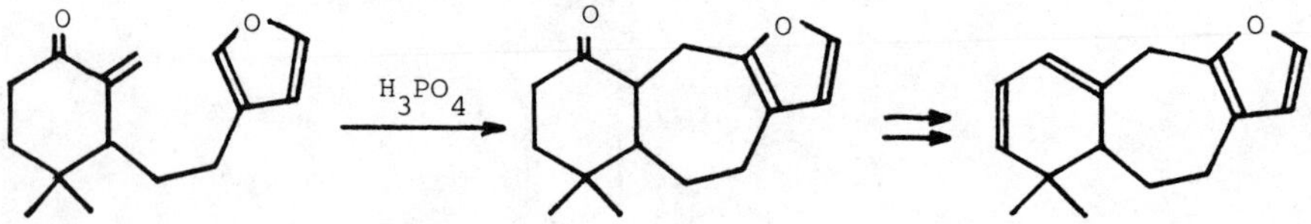

P5

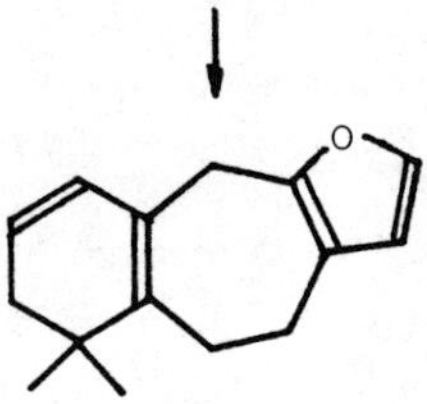

P4

6.2 ALKYLATIONS

A strategic bond in nakafuran-9 (**N1**) is that connecting an α carbon of the furan nucleus to the cyclohexene moiety. This bond can be made via an intramolecular electrophilic substitution reaction, and in a retrosynthetic perspective that takes into consideration the location of the isolated double bond in the target molecule, the best bicyclic precursor would be a cyclohexenone derivative (Tanis, 1985b).

Of course a properly situated multiple bond is capable of interaction with the polarizable enone system. The decalone formation step devised for a synthesis of torreyol (**T20**) (Franke, 1984) is an example.

Interestingly, this cyclization also creates a β-isopropenyl chain, while the analogous cyclization of the lower homolog results in a hydrindanone with the same side chain in the opposite configuration. However, it has been possible to use this product to synthesize oplopanone (**O7**) (Köster, 1981). (However, the synthetic material is most likely 8-epioplopanone, **O8**, as shown: see Piers, 1986.)

Only by synthesis was the constitution of chiloscyphone (Gerling, 1985) settled. The revised structure is most amenable to construction by the cyclization method described.

chiloscyphone

Participation of the isolated double bond follows the Markovnikov rule. Accordingly, 2-(but-3-enyl)cryptone furnishes the enol acetate of an α-decalone upon submission to the same treatment. The bicyclic product is readily converted into ε-cadinene (**C5**), ε-amorphene, and ε-bulgarene (Köster, 1986).

ε-amorphene   **ε-bulgarene**   **C5**

Treatment of ω-methylenedihydro-β-ionone with tin(IV) chloride gives rise to a spirocyclic enone and a tricyclic ketone in a 3:7 ratio. The former compound is generated via a transannular 1,3-hydrogen migration process, whereas a formal [2 + 2]cycloaddition leads to the latter product. The spiroenone has lost an unsaturation in its original ring and is not suitable for synthesis of α- and β-chamigrenes (**C46** and **C47**) as incipiently designed; however, the tricyclic ketone turns out to be a savable intermediate by virtue of the fragmentability of the derived methyl carbinol (Naegeli, 1981).

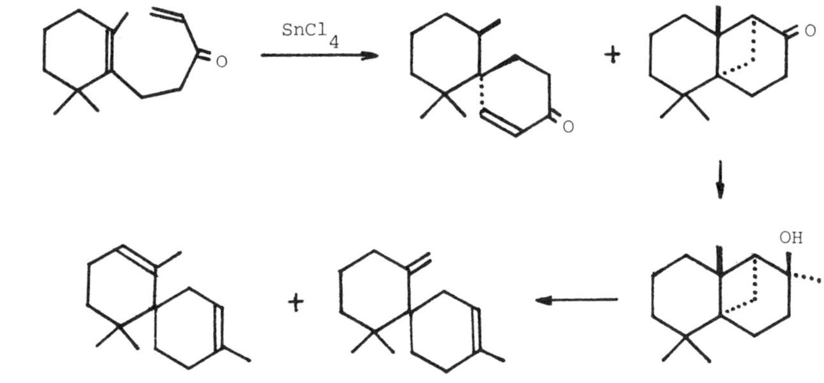

Hydrogen transfer apparently intervenes also in the formation of a conjugated hydrindenone (Snider, 1980), which has subsequently served as an AB synthon for retigeranic acid (**R1**) (Corey, 1985b).

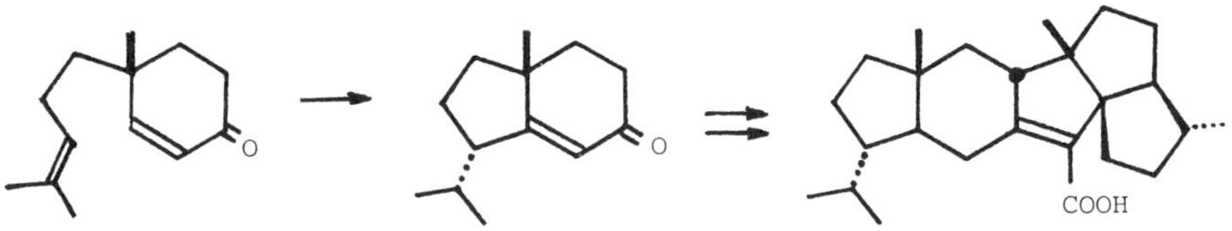

A transannular cyclization initiated by polarization of a cyclopentenone unit fused to a cyclooctene constitutes the key step of a synthesis of pentalenene (**P16**) (Mehta, 1985a). The stereocontrol is exercised at the 1,4-diketone level through equilibration, thereby establishing a trans relationship between the secondary methyl group and the pro-angular hydrogen. The transannulation produces a correct framework by virtue of compulsory formation of two *cis*-diquinane nuclei.

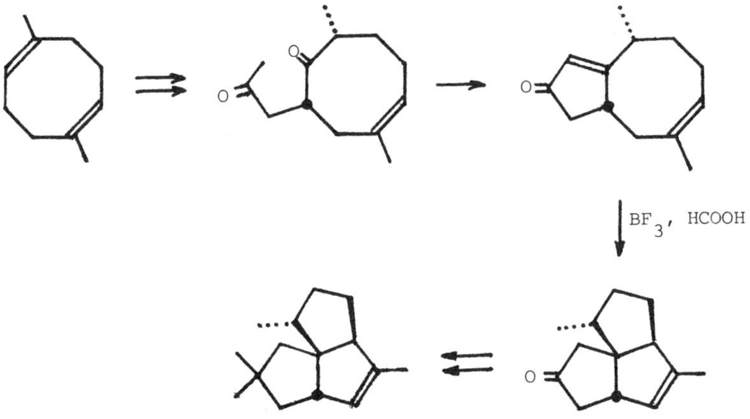

It is notable that the transannulation is quasibiomimetic. In the biosynthesis of pentalenene, the intervening bicyclic cation has a positive charge residing at the subangular, methyl-bearing carbon atom. A 1,2-hydride shift to create the nonangular chiral center precedes the compartmentation of the eight-membered ring.

The β carbon of a cyclopropyl ketone may be rendered electrophilic by reagents capable of coordinating with the carbonyl group. A very delightful route to cedrol (**C38**) (Corey, 1973a) incorporates internal trapping of such a γ-oxocarbenium ion with a double bond and a reflexive cyclization.

A38

Opening of cyclopropyl ketones accompanied by aryl participation is a concerted process (Stork, 1969). As illustrated in the following, the *exo*-(*m*-anisylethyl) derivative gives only the trans products, whereas the endo isomer affords the corresponding cis compounds.

Among several novel features present in the synthesis of cafestol (Corey, 1987f) a stereospecific polyene cyclization induced by ionization of a conjugated cyclopropylcarbinol should be mentioned. In this step the complete skeleton of the terpene molecule is assembled.

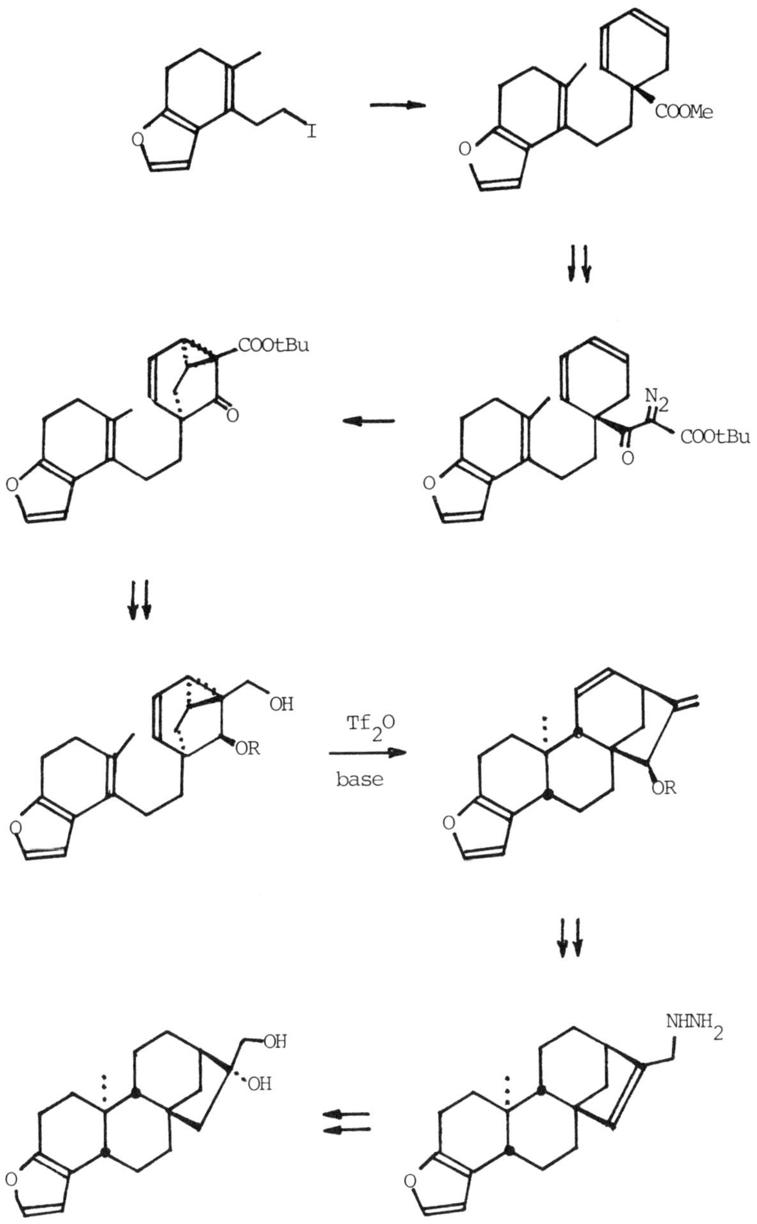

6.2 ALKYLATIONS

cafestol

The same principle has been applied to synthesis of atractyligenin (Singh, 1987). In this case the cyclization results in a cyclohexadienyl A-ring which is suitable for elaboration of the γ-hydroxy acid system via a Δ^4-2α-selenocarbonate by free radical annulation, methanolysis and inversion of the alcohol function.

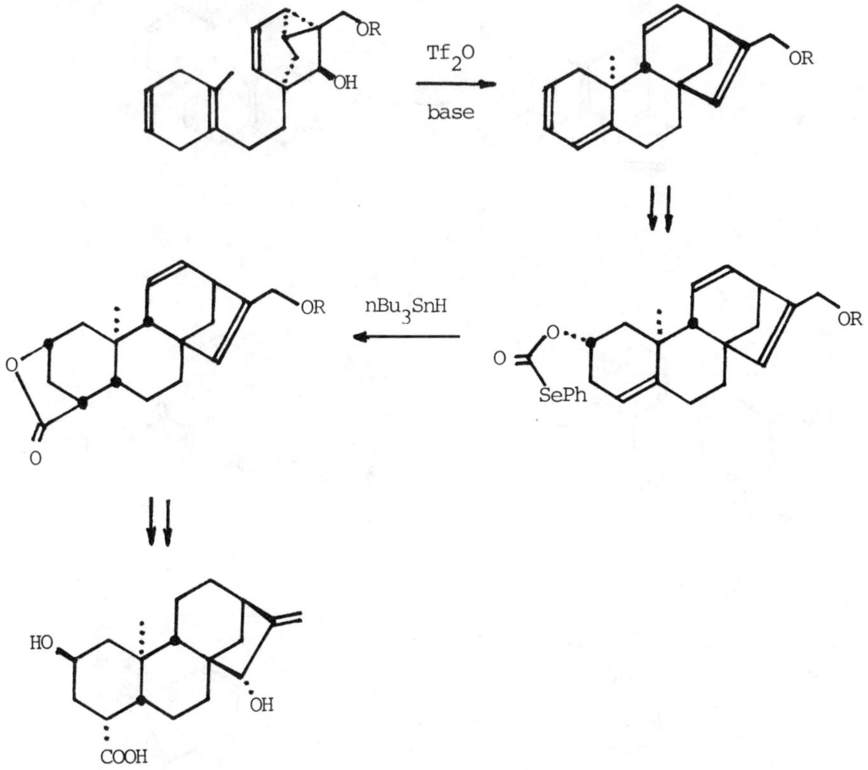

atractyligenin

The transpositional delivery of the allyl residue of allylsilanes to enones in the presence of a Lewis acid (e.g., $TiCl_4$) is known as the Sakurai reaction. An intramolecular version of this process has shown great promise in clerodane synthesis (Tokoroyama, 1984), in which at least three asymmetric centers are established stereoselectively. Since both *cis*- and *trans*-clerodanes occur in nature, eventual fruition of a synthetic effort e.g., toward annonene, (**A34**) from such well-endowed intermediates seems assured.

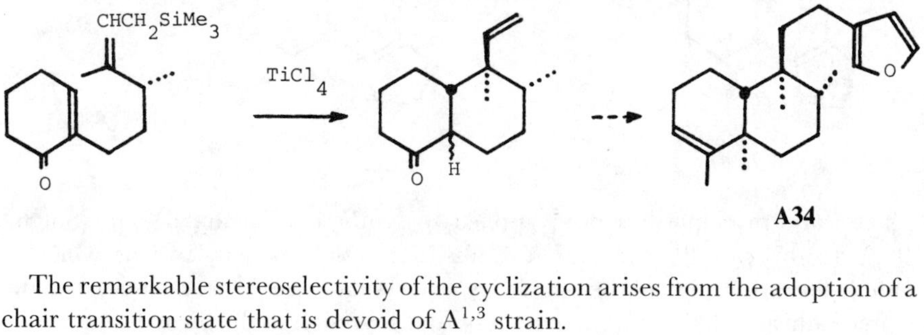

A34

The remarkable stereoselectivity of the cyclization arises from the adoption of a chair transition state that is devoid of $A^{1,3}$ strain.

6.2 ALKYLATIONS

From properly substituted trimethylsilylcyclopentene derivatives the triquinane sesquiterpenes hirsutene (**H17**) and $\Delta^{9(12)}$-capnellene (**C21**) have been assembled (Defauw, 1987). In the hirsutene synthesis an extra acetoxy remnant of the tricyclic intermediate is inherited from an aldehyde. An aldol condensation is much superior to alkylation method for the preparation of the substrate.

Extended conjugate systems accept allyl transfer with equal ease. This variant of C—C bond formation has been exploited in a synthesis of nootkatone (**N11**) (Majetich, 1985b). The origin of the stereoselectivity associated with the cyclization is the same as that of the formolytic route (Dastur, 1974), since the alternative transition state leading to the epimer suffers from mutual repulsion between the nucleophilic chain and the quaternary methyl group.

By the same method, seven-membered rings e.g., that of perforenone A, (**P20**) can also be assembled in good yields (Majetich, 1985a, 1987).

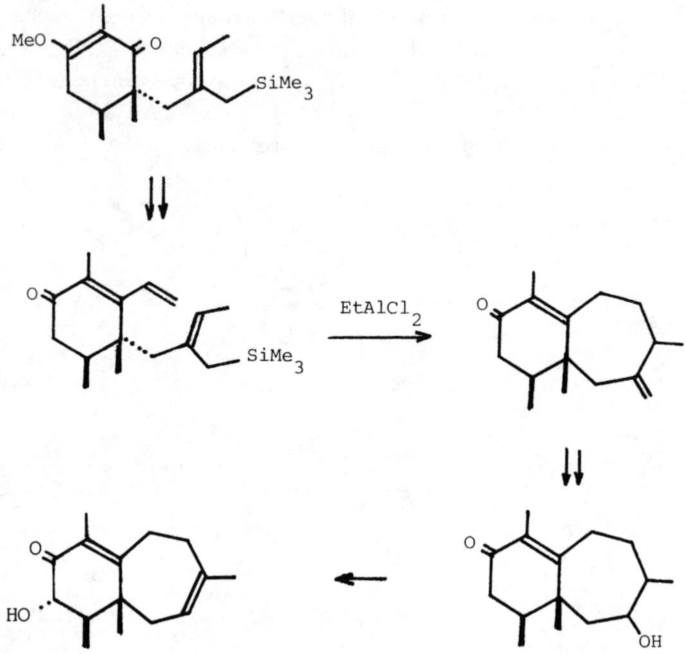

P20

A similar approach to widdrol (**W3**) (Majetich 1985a) is in progress. It is expected that epoxidation of the bicyclic diene (i.e., anhydrowiddrol), proceeds with both regio- and stereoselectivity, and that the hydride reduction of the epoxide will furnish widdrol.

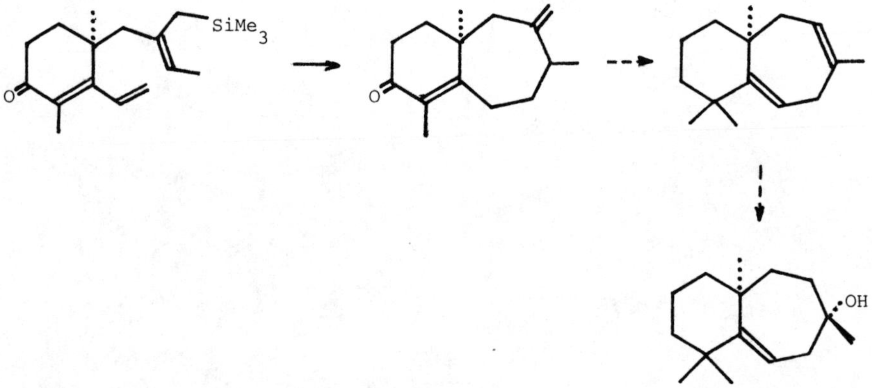

W3

Alternatively, a widdrol precursor has been obtained from cyclization of a homologous silane (Majetich, 1985a). However, it is accompanied by the C_8

6.2 ALKYLATIONS

epimer. Accordingly, the acid intermediate which was obtained previously by Danishefsky is not produced in pure form by the present route. Perhaps the stereochemistry of the quaternary carbon created during the cyclization can be controlled by using a configurationally pure allylsilane, and a stereoselective synthesis of widdrol may be attainable.

W3

A projected synthesis of neolemnane (**N4**) (Majetich, 1985a) is also based on the allylsilane cyclization. Although the key process is no longer a cationic one, its inclusion in this section is deemed appropriate in terms of synthetic significance. It serves to highlight the manifold reactivities of a substrate toward different reagents. Thus, an eight-membered ring is formed via desilylation of an allylsilane by the fluoride ion, which apparently generates a transient allyl anion, and the latter species immediately adds to the accessible acceptor, a conjugated dienone.

N4

6.2.6 Transannular Cyclizations

Transannular participation is frequently observed during solvolysis of *exo*-norborn-5-en-2-yl tosylates. Taking advantage of this phenomenon, a synthesis of cyclosativene (**S13**) has been devised (S. Baldwin, 1975). A tetracyclic structure is generated from the norbornenyl tosylate bearing a 3-*endo*-propargyl chain. Completion of this ingenious synthesis requires only the introduction of an isopropyl group and the removal of the oxygen function.

The longifolene framework that is formed unexpectedly during solvolysis of a cyclopentenol rewarded its discoverers by providing an easy access to the terpene molecule (Volkmann, 1975). The ability to induce two intricate C—C bond formation processes by simple treatment of the substrate with acid is a delight.

Sterpurene (**S52**) probably originates from humulene via cyclization to a protoilludyl cation and subsequent Wagner–Meerwein rearrangements. In a biomimetic synthesis of sterpurene (Murata, 1981), controlled stepwise ring formation is mediated by mercury (II) ion. The demercurated bicyclic ether intermediate furnishes sterpurene on exposure to boron tribromide.

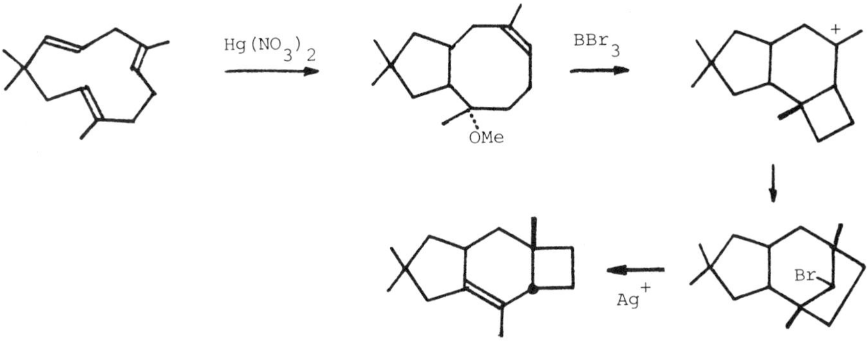

6.2 ALKYLATIONS

S52

Interestingly, the treatment of the 5:8-fused cyclooctadiene with boron trifluoride etherate produces pentalenene (**P16**) (Pattenden, 1984). The transannular reaction is initiated by protonation of the tetrasubstituted double bond.

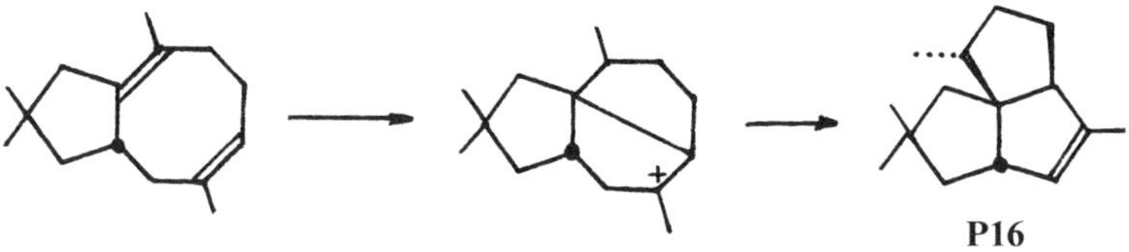

P16

With attention to setting up the proper configuration of two adjacent carbon atoms in a cyclooctene, the solvolytic compartmentation reaction can be used to accomplish an entirely stereoselective synthesis of iridomyrmecin (**I8**) (Matthews, 1975).

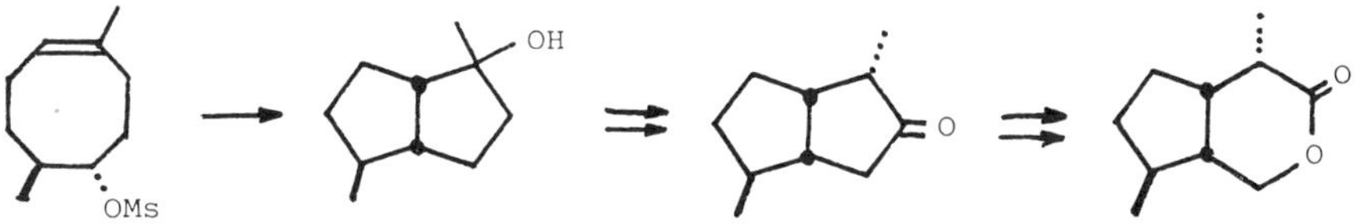

I8

Treatment of humulene with *N*-bromosuccinimide affords a tricyclic substance having an angular 3:8:4-ring system. This transannulated compound undergoes reductive cleavage to yield, among other compounds, caryophyllene (**C32**) (Greenwood, 1965).

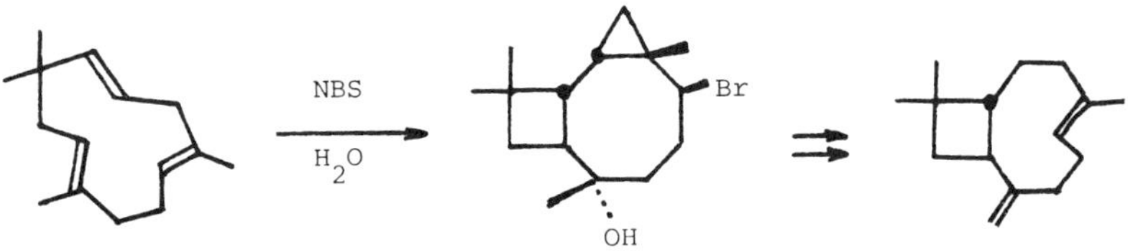

C32

Solvolysis of humulene-4, 5-oxide leads to the ring skeleton of africanolone (**A14**) (I. Bryson, 1979).

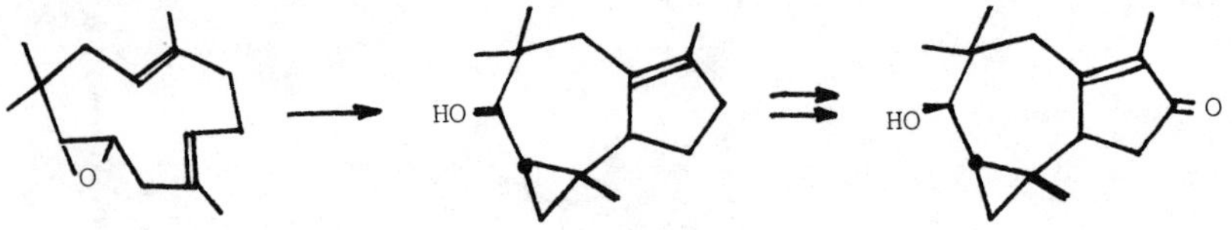

The regioselectivity problem of sensitized, bimolecular [2 + 2]-photocycloaddition of unactivated dienes can be circumvented in well designed substrates. Eight-membered ring compounds can then be generated by Cope rearrangement of the 1,2-divinylcyclobutanes.

A properly constructed 5:8-fused enone derived from a tetraene is ready for transannular closure en route to coriolin (**C70**) (Wender, 1987b). The *cis-anti-cis*-triquinane is the favored product from a less strained ketone conformer, and it is only four steps away from a known precursor of the antitumor antibiotic.

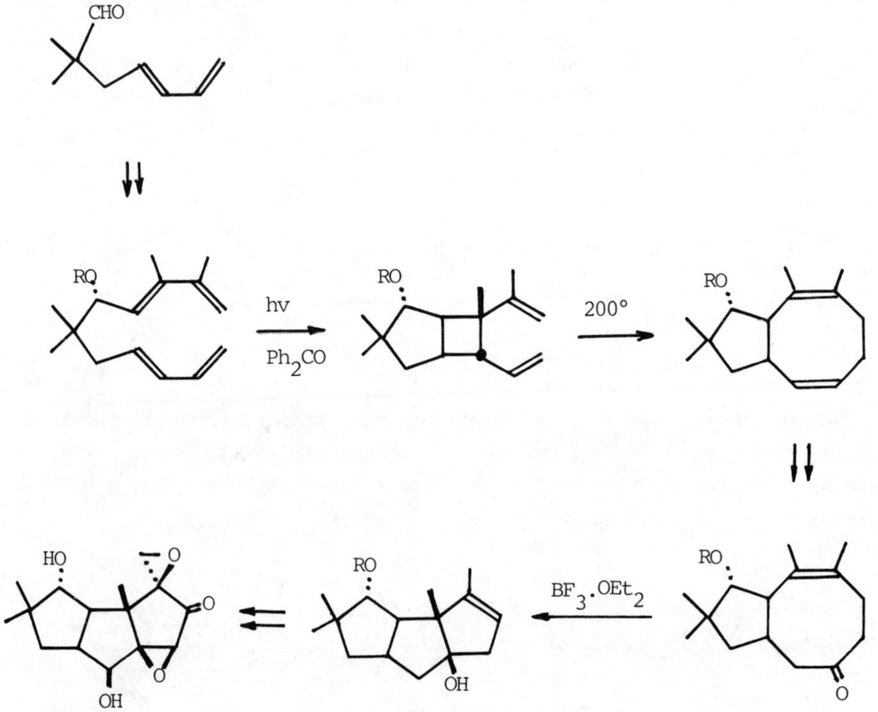

Chapter 7

Diels–Alder Reactions

The Diels–Alder reaction (Wollweber, 1972; Oppolzer, 1984) is a most useful reaction because it forms two bonds in one step with the possibility of establishing four contiguous asymmetric centers in a predictive manner. The absence of reagents that are not incorporated into the adducts (except small amounts of catalysts, which may sometimes be introduced) facilitates product isolation.

The two reactive components of a Diels–Alder reaction are a *diene* and a *dienophile*. New C—C bonds are made at the two termini of the diene with the dienophile to afford a monounsaturated six-membered ring. Normally, the reactivity of the diene is enhanced by electron-donating substituents and that of the dienophile by electron-withdrawing groups. (Reactions with inverse electron demands are also feasible; however, when both diene and dienophile contain substituents of the same electronic nature, the reaction is strongly retarded.)

Since the original observations of Otto Diels and Kurt Alder in 1928, a large assortment of diene-dienophile combinations has been submitted to thermal treatment. Commonly used dienophiles include alkenes and alkynes in which the multiple bond is conjugated to a carbonyl (aldehyde, ketone, carboxylic acid, ester, acyl chloride), nitrile, nitro, or sulfonyl group. Cyclic compounds such as maleic anhydride and quinones are particularly reactive. Benzynes act as dienophiles when they are generated in the presence of dienes. Heterodienophiles give rise to heterocycles. This diversity of the dienophiles indicates the possibility of synthesizing compounds with protean functionalities based on the Diels–Alder reaction.

1,3-Dienes of various substitution patterns are known to undergo the Diels–Alder reaction, as long as they are not fixed in a transoid conformation. Even certain aromatic and heteroaromatic substances have been coaxed to afford cycloadducts (e.g., benzene + benzyne). A styrene may use mixed double bonds (vinyl + aromatic) as the diene unit.

There are occasions when a diene is unstable under conditions prevalent in the accouterment of other parts of the molecule. Protection of the diene in the form of 3-sulfolene is most profitable. The parent heterocycle is a convenient solid precursor of 1,3-butadiene; regeneration being achieved by thermolysis, or under conditions conducive to the Diels–Alder reaction.

The cycloaddition belongs to the pericyclic domain and it proceeds in a cis fashion; that is, the configurations of the substituents are retained in the adduct. The endo transition state is generally favored because secondary π-orbital overlap can develop. For best electron delocalization in the transition state, the diene unit must be cisoid and coplanar or nearly so. Consequently, cyclic dienes are generally much more reactive than their acyclic counterparts.

Regioselectivity is often exhibited in the reaction between unsymmetrical partners. The major product is the isomer that contains the electronically dissimilar groups in an ortho or para relationship. Lewis acids catalysts accentuate such arrangements and enhance the rate of reaction as well.

A spectacular advance in achieving the Diels–Alder reaction (Oppolzer, 1984) uses natural chiral molecules as auxiliaries to transfer chirality nondestructively. Target molecules with up to four chiral centers of predictable relative and absolute configurations are readily assembled.

When the diene and dienophile moieties are placed in the same molecule within reasonable distance, an intramolecular Diels–Alder reaction can often be induced (Brieger, 1980; Ciganek, 1984; Fallis, 1984). In such cases, electronic activation is far less important, since the activation energies can be greatly reduced by virtue of favorable entropy. It is also not uncommon for steric factors (nonbonded interactions) to override the endo transition states. Thus judicious selection of acyclic precursors could shorten synthetic routes appreciably and generate compounds with stereochemistry that is inaccessible otherwise.

The stereochemical outcome of the intramolecular Diels–Alder reaction is best predicted or rationalized by a concerted-but-nonsynchronous-pathway consideration. In this context the bond formation processes at both termini start at the same instant, but they proceed at different rates.

From the following sampling of sesquiterpene syntheses, it can be seen that the cycloadducts do not have to contain all or part of the ring skeleton of the synthetic targets. They are sometimes prepared solely on account of stereocontrol over a small segment of a molecule (cyclic or acyclic!), and the ring size limitation is often not distractive enough to preclude the use of the Diels–Alder reaction.

7.1 ISOPRENE AS DIENE UNITS

The value of the Diels–Alder reaction to the synthesis of bisabolane-type compounds was recognized early. Practically every member of this class has yielded to synthesis by the Diels–Alder approach. For example, the norketone of β-bisabolene (**B10**) is easily derived from the major adducts of isoprene with acrylyl chloride (Manjarrez, 1966), with acrylonitrile (Kuznetsov, 1969), and with 3-buten-2-one (Vig, 1966). The norketone is convertible to β-bisabolene by the Wittig reaction and to bisabolol (**B12**) by the Grignard reaction.

The isoprene–3-buten-2-one adduct is also useful for the synthesis of E-α-atlantone (**A51**) (Babler, 1974; Malanco 1976), α-bisabolene (**B9**) (Givaudi, 1974; Delay, 1979), α-bisabololone (**B13**) (Park, 1977), deodarone (**D12**) (Gopichand, 1974), and lanceol (**L3**) (Manjarrez, 1964).

X= COCl, CN, COMe, CHO

B10 **A51** **B9** **B13**

D12 **L3** **J8**

A convenient route to acorone (**A11**) consists of alkylation of the isoprene–acrolein adduct and formation of the spirocyclic framework (McCrae, 1977; Martin, 1978; Wenkert, 1978b; Altenbach, 1979; Ho, 1982).

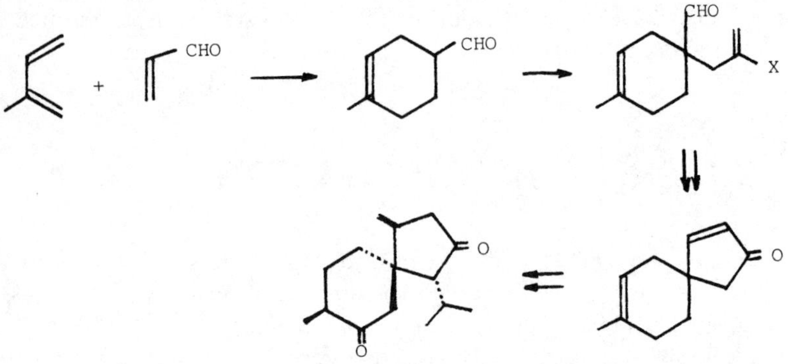

A11

Using the isoprene–methyl acrylate adduct a synthesis of lanceol (**L3**) (Fuji, 1987) has been completed.

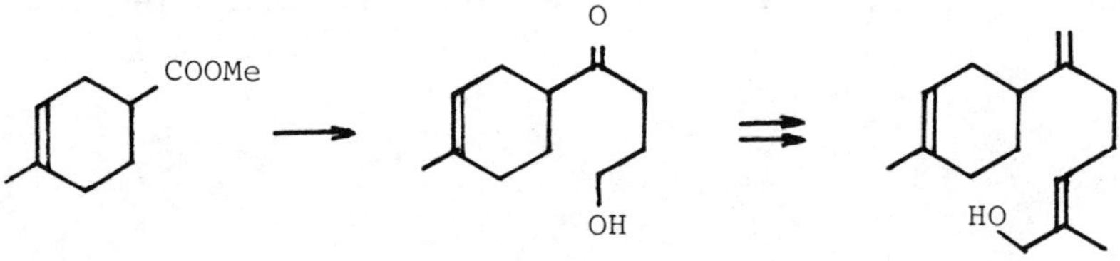

The isoprene-acrylic acid adduct has found use in synthesis of trichodiene (**T26**) (Kraus, 1986; VanMiddlesworth, 1986) and its diastereomer via a modified ester-Claisen rearrangement. This method obviates the difficulty of connecting two tertiary carbon fragments by intermolecular processes.

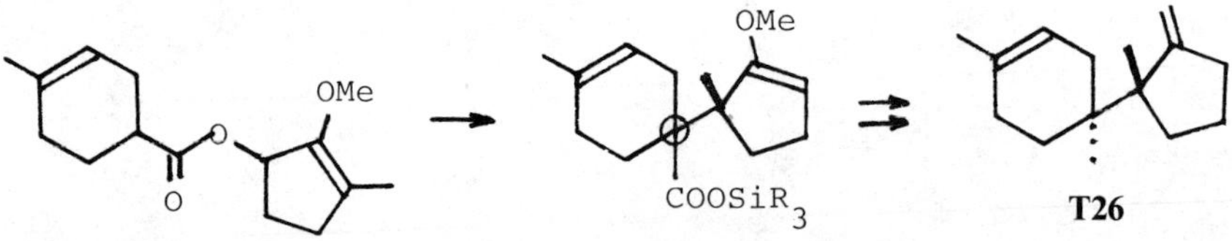

Limonene is an isoprene dimer. The racemate, commonly known as dipentene, is indeed obtained by thermal dimerization of isoprene. Thus, several sesquiterpenes that have been acquired by proper manipulation of limonene are: E-α-atlantone (**A51**) (Crawford, 1972; Alexander, 1973; Plattier, 1974; Adams,

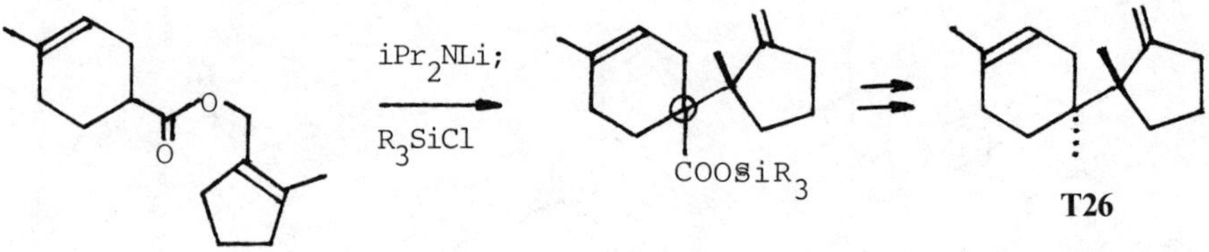

1975; Dauphin, 1979) Z-α-atlantone (**A52**) (Mehta, 1979), β-bisabolene (**B10**) (Vig, 1967; Crawford, 1972; Mehta, 1979; Fukamiya, 1981), α-bisabolol (**B12**) (Knöll, 1975), α-bisabololone (**B13**) (Kergomard, 1977), *erythro*-juvabione (**J8**) (Pawson, 1968, 1970), and *E*-lanceol (**L3**) (Ruegg, 1966; Akutagawa, 1975; Katzenellenbogen, 1976; Cazes, 1978; Tamaru, 1980). The isopropenyl group of limonene can be activated either as an electrophile or as a nucleophile for chain elongation.

A menthone (**M11**) synthesis that is obviously of no economical impact embodies a Diels–Alder reaction of isoprene with a butenolide (Torii, 1973). A lengthy degradation of the adduct is called for.

A one-step preparation of a trimethyloctalone precursor for synthesis of α- and β-himachalene (**H13, H14**) is realizable (Fringuelli, 1986). Using ethylaluminum dichloride as catalyst, ring juncture epimerization is minimized, although regioselectivity is still low.

It is clear that the Diels–Alder reaction is valuable for constructing the α-terpineol moiety of the tetrahydrocannabinols (THCs). Accordingly, condensation of isoprene with the proper cinnamic acid leads to a precursor of Δ^6-THC (**C17**) (Jen, 1967). On the other hand, the aryl-substituted isoprene reacts with 3-buten-2-one to give a methyl ketone adduct with a double bond at a position corresponding to Δ^1-THC (**C16**) (Korte, 1966; Kochi 1967).

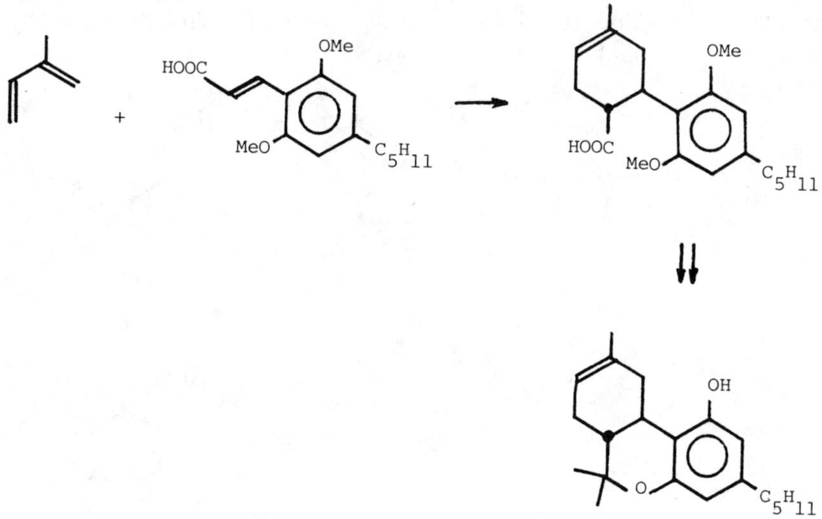

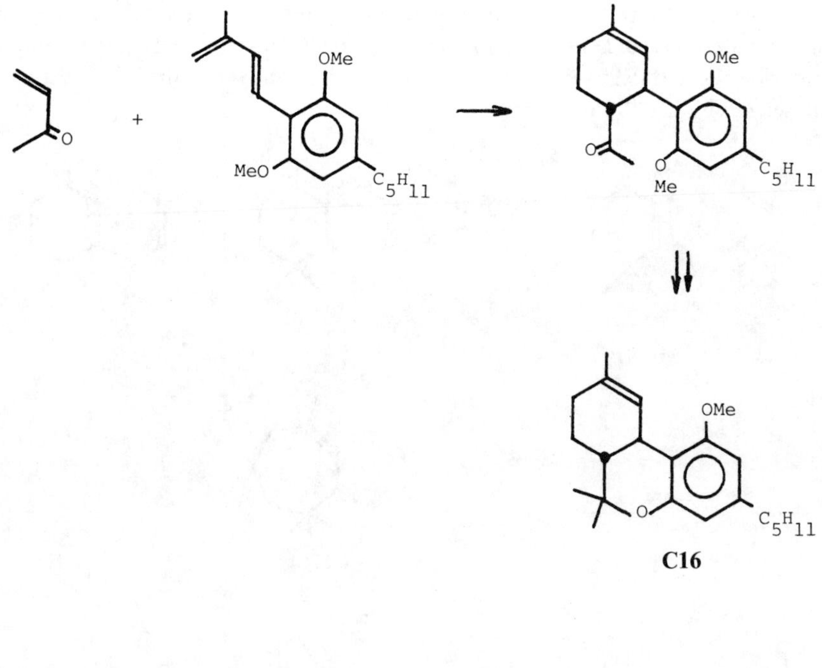

7.1 ISOPRENES AS DIENE UNITS

Isoprene reacts with E- and Z-ocimenes in the presence of $AlCl_3$ to give α- and β-atlantones (**A51** and **A52**) (Ayyar, 1973, 1975), respectively. The terminal alkene unit of the unsaturated ketones is the preferred dienophile.

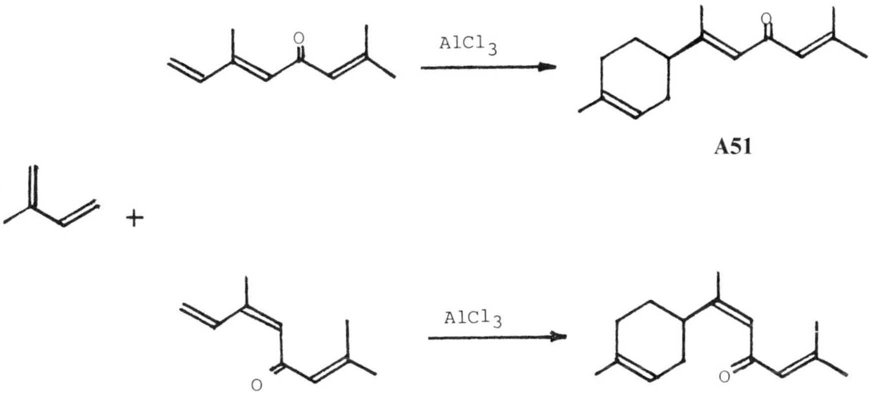

The condensation of isoprene with 7-methyl-1,6-octadien-3-one (available from pyrolysis of nopinone) affords norbisabolenone. A Wittig reaction of the latter compound completes a preparation of β-bisabolene (**B10**) (Ho, 1980).

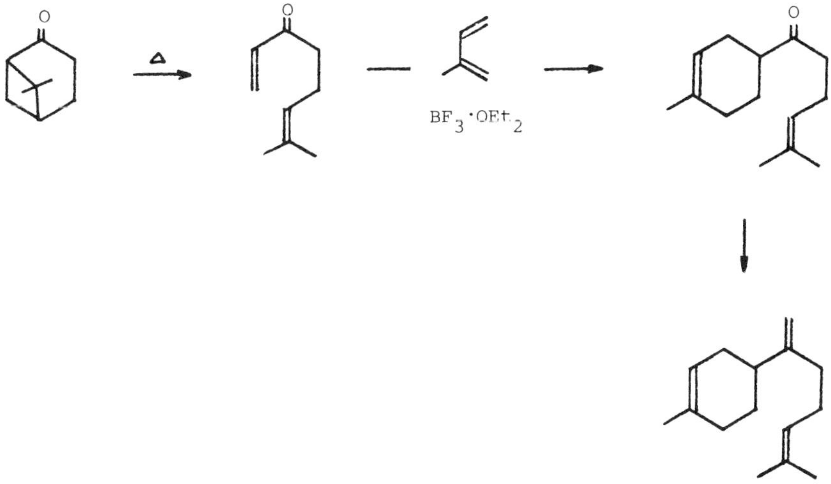

β-Stannyl α,β-unsaturated ketones are adequate dienophiles which condense with isoprene readily. Since γ-stannyl alcohols undergo 1,3-elimination to produce cyclopropanes, the proper Diels–Alder adducts can be employed to synthesize 3-carene (**C26**) and isosesquicarene (**S27**) (C. Johnson, 1987). But because the elimination is stereospecific, an epimer of isosesquicarene is also generated.

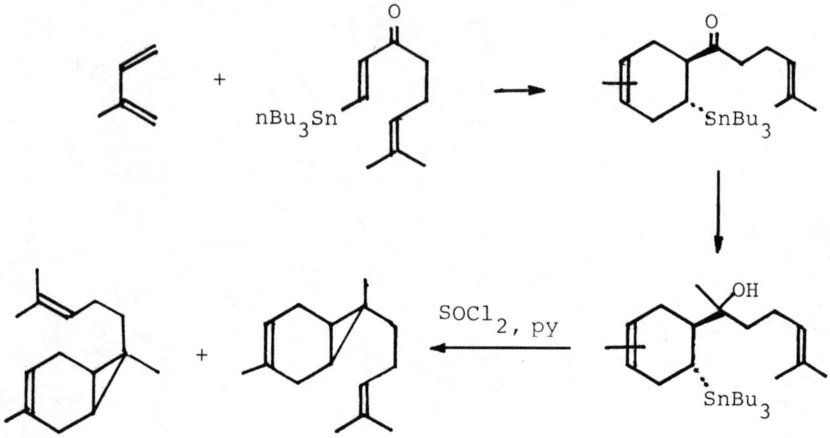

S27

In the first step of a lengthy route to carabrone (**C23**) (Minato, 1967; 1968a), tetrahydrosalicyclic ester is required. The Diels–Alder reaction is the apparent reaction of choice for assembling the intermediate.

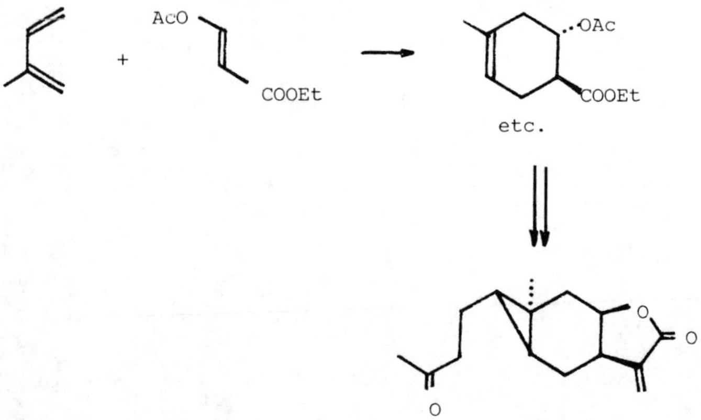

C23

The relatively straightforward transformation of a cyclohexene into a cyclopentanone (e.g., via ozonolysis, aldolization, and Beckmann rearrangement) permits an application of the Diels–Alder methodology to the synthesis of $\Delta^{9(12)}$-capnellene (**C21**) (Liu, 1985). Its use is also connected to the choice of method for the construction of the central ring that delivers a β-ketoester. The dehydro analog of the ketoester is a reactive dienophile.

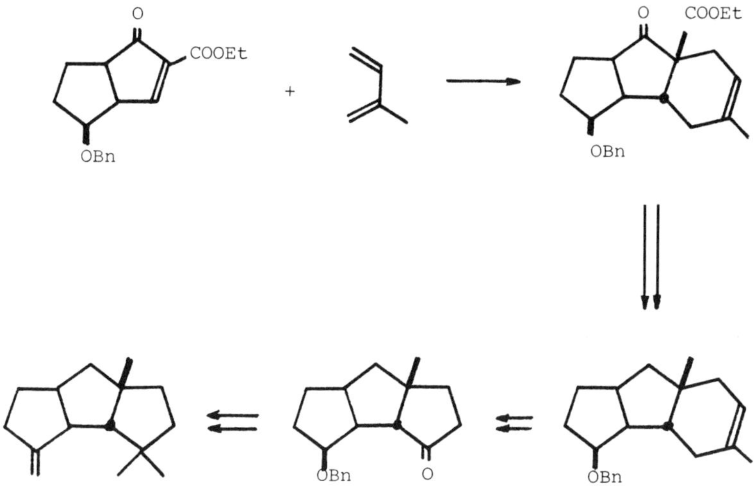

4,4-Dimethyl-2-cyclohexenone is inert to dienes. However, it can be activated by attaching a carboalkoxyl group to C_2. Thus, the isoprene adduct is available and convertible to the himachalenes (**H13**, **H14**) (Liu, 1981) via ring expansion, double-bond shift, and other required reactions.

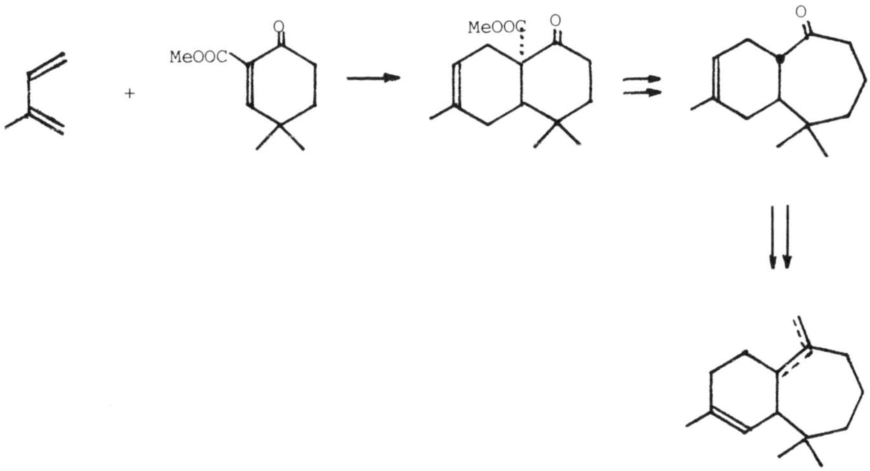

The feasibility of constructing spirocyclic sesquiterpenes such as chamigrene (**C46**) (A. Tanaka, 1967, 1968) and (—)-acorone (**ent-A11**) (Marx, 1973, 1975) by the Diels–Alder route has been tested. For the latter compound, only the enantiomeric isomer has been obtained, owing to the unavailability of (—)-pulegone for preparation of the methylenecyclopentanone.

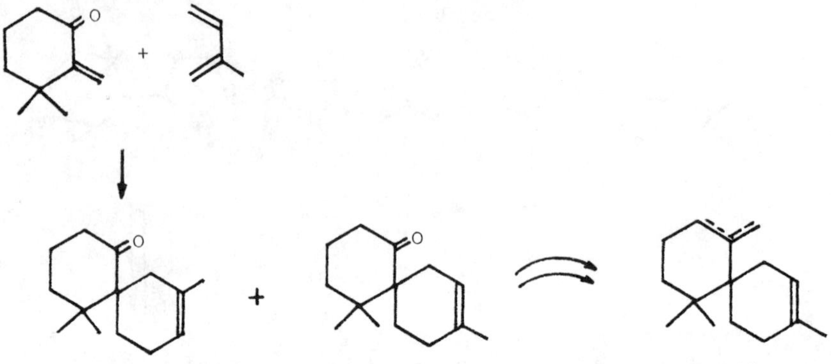

Formation of a spirocyclic intermediate constitutes the second phase of a synthesis of 2-deoxystemodinone (White, 1985, 1987) (See also, ene reaction).

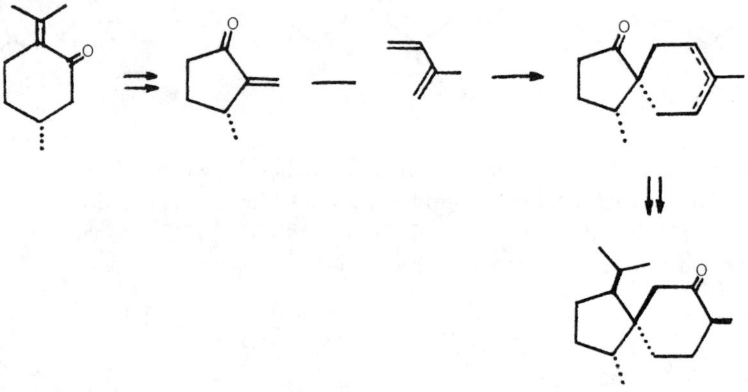

Occasionally, isoprene shows poor regioselectivity during cycloaddition. For example, in a short approach to trichodiene (**T26**) (Schlessinger, 1983), the phenylthio derivative of isoprene proves to be a better addend.

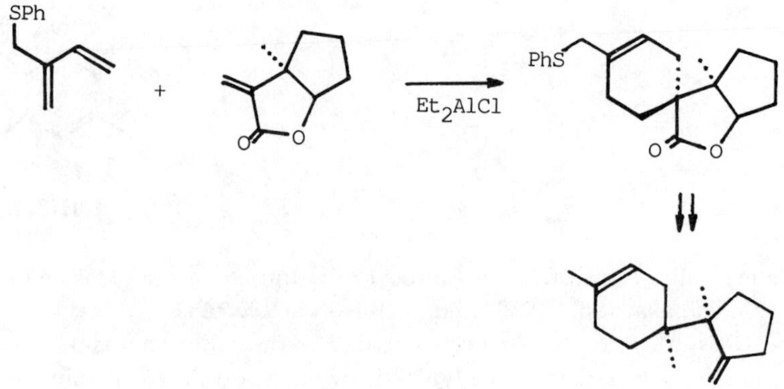

2-(Trimethylsilyl)methyl-1,3-butadiene is a particularly valuable diene because methylenecyclohexanes are readily formed on treating the adducts with an acid. Thus, a simple preparation of δ-terpineol (**T11**) (Wilson, 1979) and the synthesis of ε-muurolene (**M14**) and ε-cadinene (**C5**) (Hosomi, 1982) suffice to show its versatility.

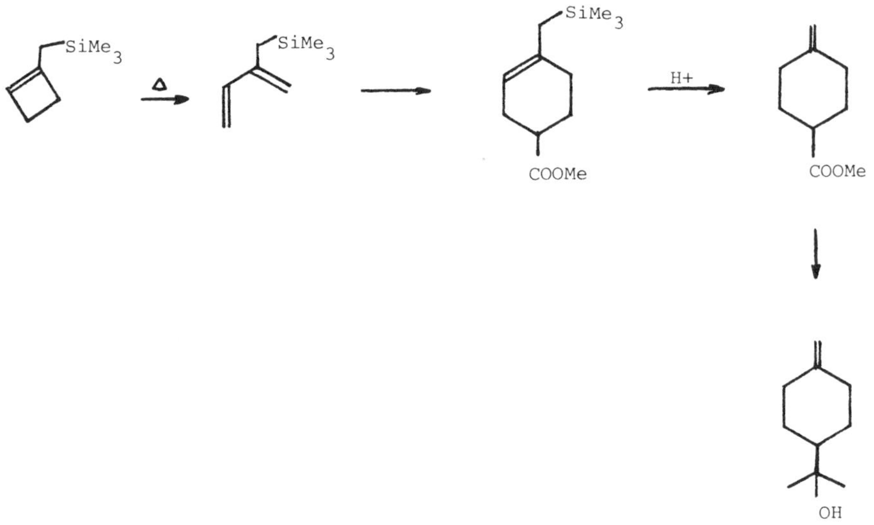

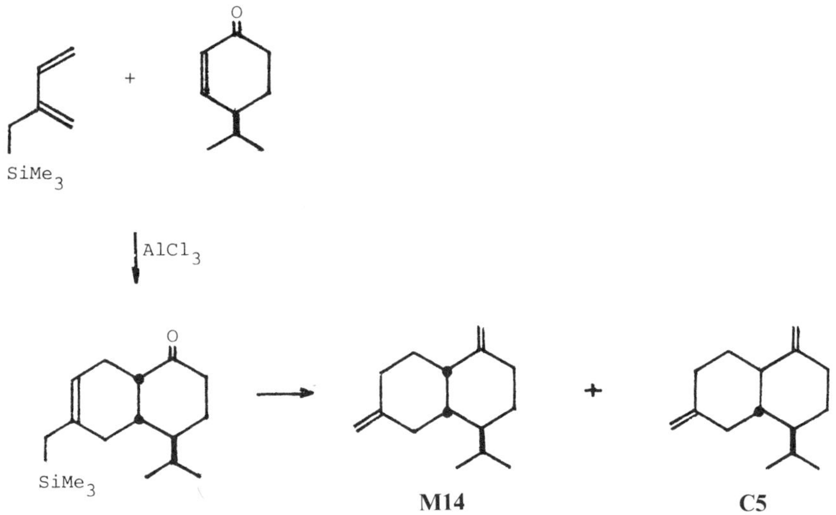

(*Note*: β- and γ_2-Cadinenes (**C2**, **C3**) (Fringuelli, 1985) can be conveniently prepared from isoprene. The primary adduct undergoes epimerization to a large extent (9:1) in the presence of the catalyst $AlCl_3$.

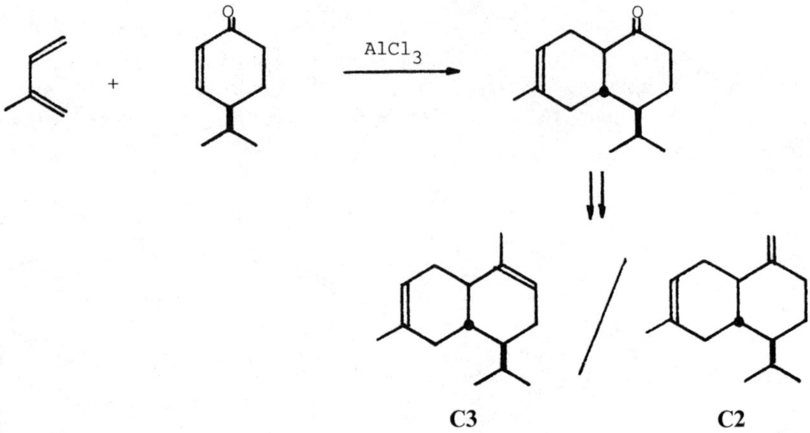

The bicyclic aldehyde nanaimoal, isolated from a dorid nudibranch mollusk, probably originated from the cyclization of γ-cyclofarnesyl pyrophosphate in a manner reminiscent of the C ring of pimarane diterpenes. A very expedient synthesis of the corresponding alcohol (**N2**) (S. Ayer, 1984) together with its regioisomer involves the condensation of myrcene (isopentenylisoprene) with 3-methyl-3-buten-1-ol and treatment with acid.

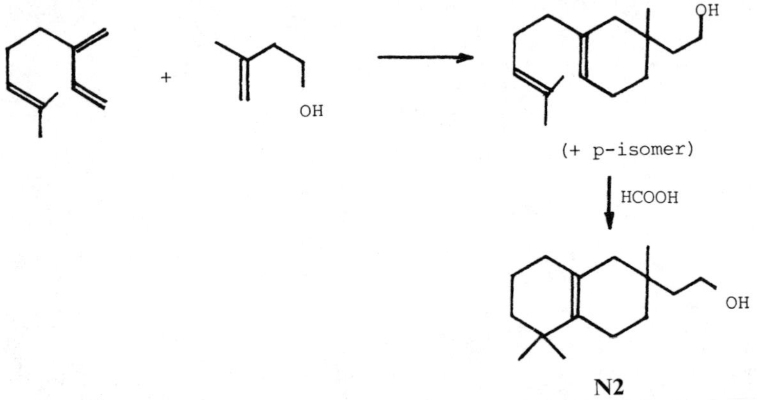

Although at one time in the past chloroprene was a highly regarded building block in synthesis, it is now mainly employed industrially as a monomer in captive form. Mention might be made of its use in a synthesis of perillaldehyde (**P21**) (Tsizin, 1972).

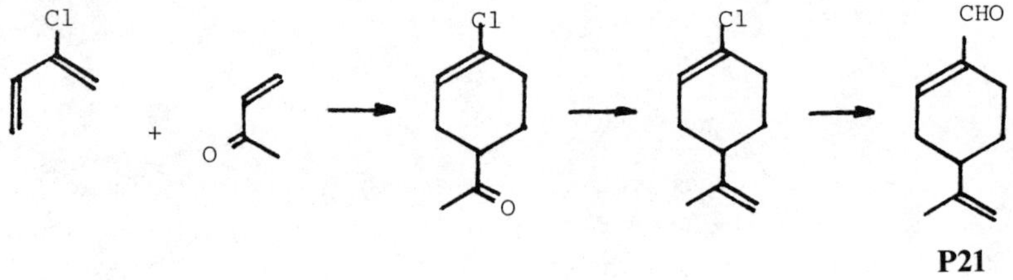

7.2 PIPERYLENE AND ITS HOMOLOGS AS DIENE UNITS

The extensive use of isoprene in terpene synthesis is understandable in view of the structural divisibility of most terpenes into isopentane blocks. On the other hand, piperylene has been employed in only a few synthetic routes. Cinabicol (**C55**) represents one of these rare examples (Bohlmann, 1980).

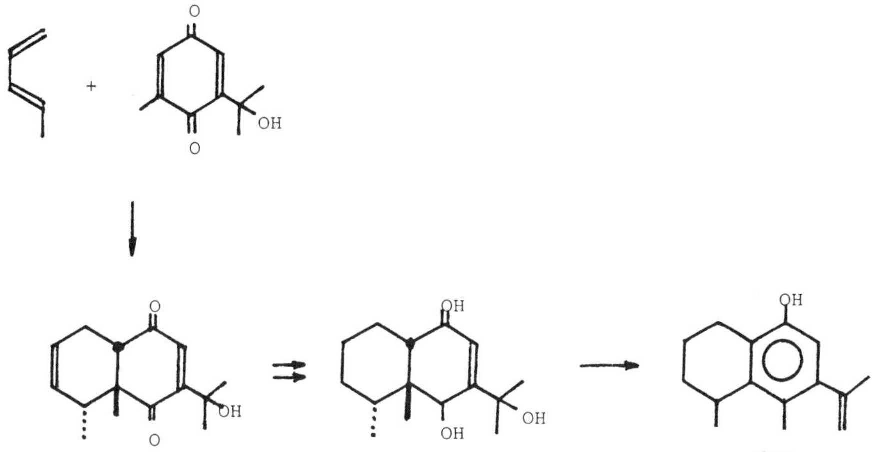

One of the constituents responsible for the exquisite scent of Bulgarian rose oil is β-damascenone (**D2**). A synthesis (Ayyar, 1975) commences with the condensation of piperylene and α-bromomesityl oxide. Another approach to δ-damascone (**D4**) (Dauben, 1975c) consists of a Diels–Alder reaction between piperylene and isopropylidene isopropylidenemalonate.

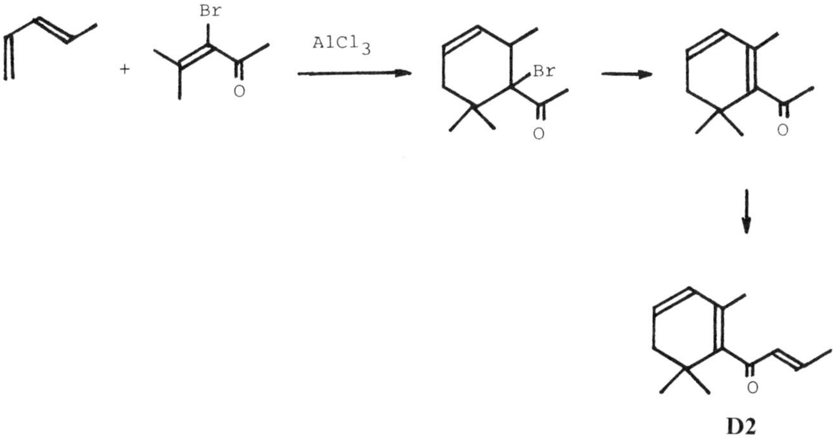

It has been recently discovered that Diels–Alder reactions performed in aqueous media may proceed with enhanced rates. An application to the preparation of a key intermediate of vernolepin (**V10**) (Yoshida, 1984) demonstrates the power of this modification. In comparison, the corresponding ester (of the hexadienoic acid) attains about one-third the reaction in benzene over a much longer period (96 h vs. 17 h).

A similar improvement in the preparation of a quassin precursor has been noted (Grieco, 1983). The rate enhancement may be due to a reduction of entropy by preorientation of the reactants in a micellar environment.

The extensively degraded triterpene bitter principle called quassin (**Q3**) poses a considerable synthetic challenge owing to its seven chiral centers. A perceptive scheme advocates a Diels–Alder method to install the C ring together with an appendage for the δ-lactone. The critical steric consequences of this strategem entail the emergence of three correct stereocenters. The *cis*-B/C juncture and the effective shielding by the acetate chain incorporated in the diene moiety ensure that the subsequent reduction of the C_7 ketone will occur from the β-face. With a *trans*-octalone containing an additional chiral center corresponding to C_4 of quassin as dienophile, a total of six stereocenters are readily organized. The remaining errant configuration at C_9 is rectifiable upon introduction of the 11-ketone (Grieco, 1980).

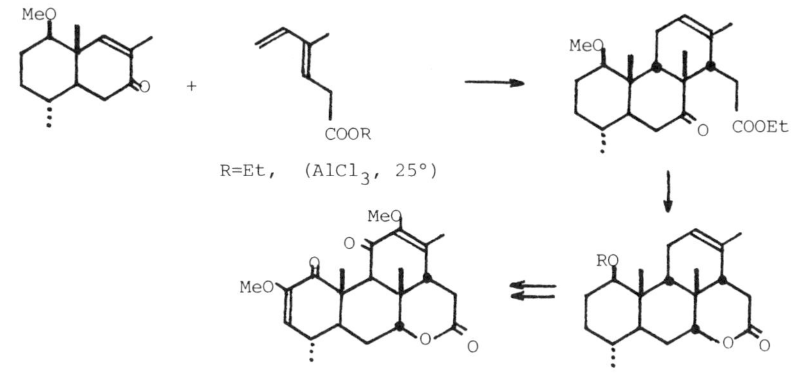

7.2 PIPERYLENE AND ITS HOMOLOGS AS DIENE UNITS

Q3

Nascent benzoquinones undergo Diels–Alder reactions readily. In quest of the BCD region of quassimarin (**Q2**) (Kraus, 1980), the dicarbocyclic portion was made by this method. The quinone component provides oxygen functions at pro-C_7 and the angular substituent between the B and C rings, whereas the diene unit contains a two-carbon chain that eventually is modified into a lactone. After ring juncture epimerization and formation of a γ-lactone to freeze the molecular conformation, allowing a selective reduction of the 7-ketone by a bulky metal hydride reagent (L-selectride), functionalization of the C ring with closure of the tetrahydrofuran moiety can be accomplished with participation of three different alcohol groups.

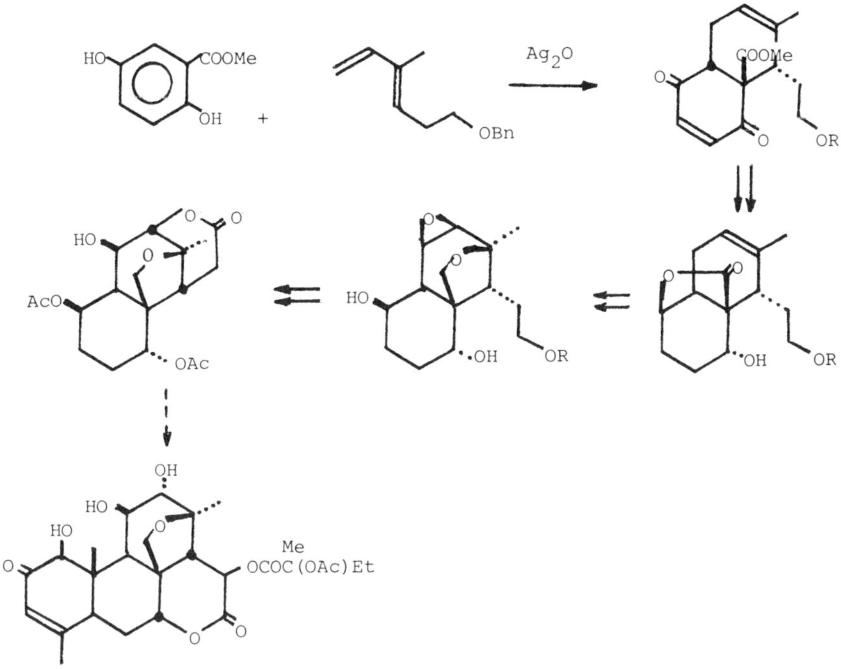

Q2

The heavier portion of lavender oil contains a ketone. Its structure (**L8**) was confirmed by a synthesis (Givaudi, 1980) from *E-β*-ocimene and 3-buten-2-one.

Probably heliocides H_1 (**H4**) and H_4 (**H7**) are produced by the cotton plant from the condensation of *β*-ocimene with hemigossipolone to fight against pests. Heliocides H_2 (**H5**) and H_3 (**H6**) are the corresponding adducts from myrcene (Stipanovic, 1976, 1977, 1978). These insecticides can be synthesized in the laboratory according to the same scheme.

Conjugate trienes generally participate in Diels–Alder reactions with the least substituted terminal diene units. This observation facilitates the synthetic design of fumagillin (**F17**) (Corey, 1972). Using α-bromoacrolein as the dienophile, the spiroepoxide ring can then be constructed stereoselectively via the bromohydrin. Furthermore, the trisubstituted double bond in the side chain is strongly hindered on one side by the ring substituents (Br, CH_2OR) so that only the desired epoxide is produced.

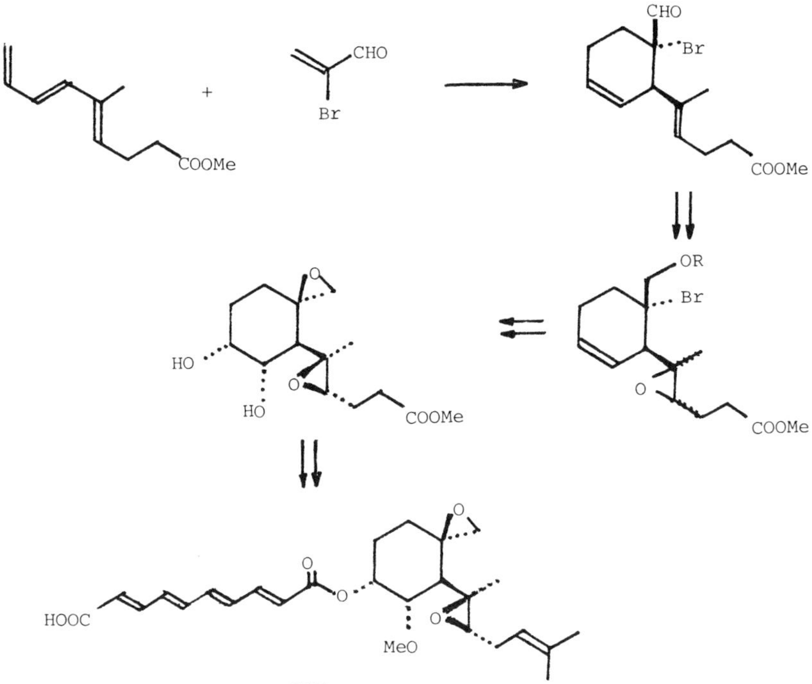

F17

7.3 OTHER ALKYLDIENES AS DIENE UNITS

The particular positioning of the intracyclic double bond in certain cadinanes makes it difficult to apply the Diels–Alder reaction directly on a cyclohexene derivative; a double-bond shift is required from such adducts. The problem can be avoided if the Diels–Alder reaction is designated to form the cyclohexene moiety and other methods are to be employed to complete the second ring. This strategy is embodied in a synthesis of khusitene (**K9**) (Vig, 1975).

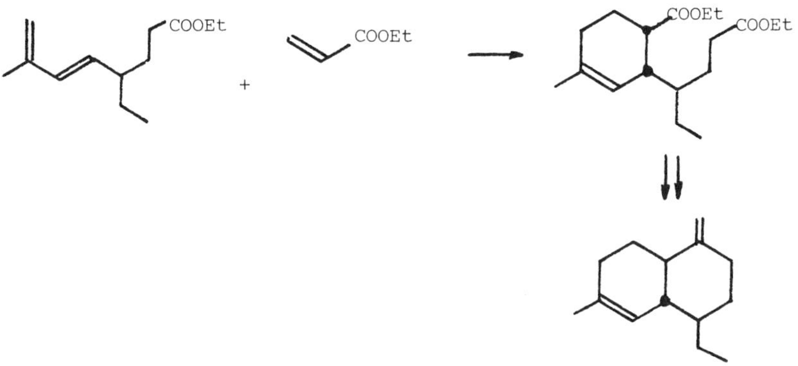

K9

Fraxinellone (**F12**) is probably a degraded limonoid triterpene. Its synthetic design, based on a Diels–Alder strategy (Fukuyama, 1972), requires a slightly oblique analysis. Most crucial to the development of this route is the recognition that it is immaterial whether the double bond of the adduct is in the $\beta\gamma$-position of the carbonyl group.

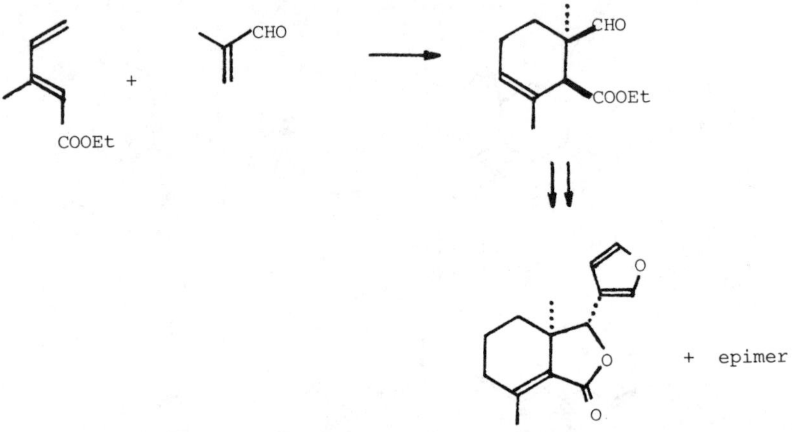

F12

Sometimes polyalkyl dienes are used to afford Diels–Alder adducts, to save alkylation steps which are required. A heavily substituted cyclohexenecarboxylic ester is thus assembled en route to patchoulenone (**P14**) (Erman, 1971).

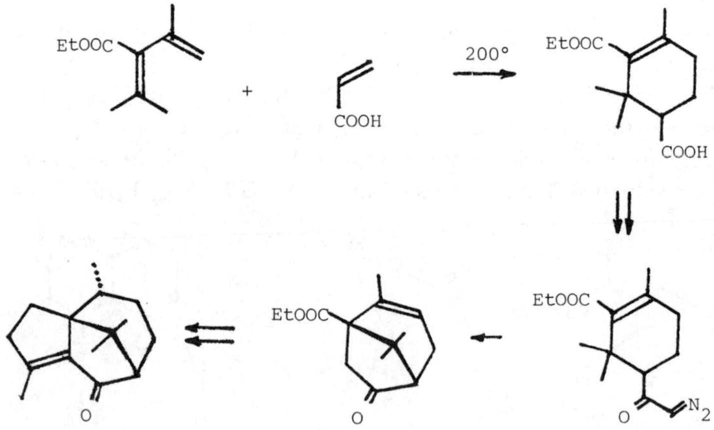

P14

The synthesis of decalins via cycloaddition of vinylcyclohexenes with dienophiles deserves attention. It appears that this route is currently preferred in the approach to various drimane sesquiterpenes, including winterin (**W4**) (Brieger,

1965), warburganal (**W1**) (Tanis, 1979), polygodial (**P44**) (Jallali–Naini, 1981; Howell, 1983), and drimenin (**D20**) (Jallali–Naini, 1981).

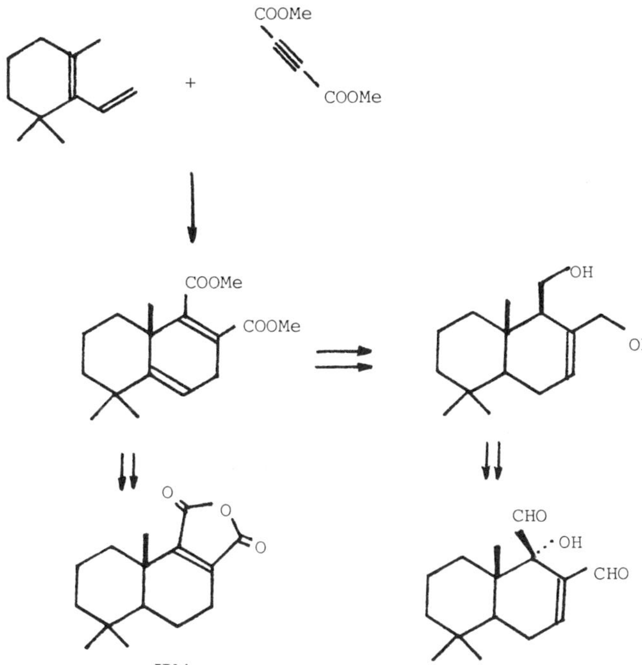

The employment of a chiral vinylcyclohexenol as the diene allows the construction of both natural polygodial (**P44**) and its enantiomer (**ent-P44**) (Mori, 1986b). The Diels–Alder step furnishes a separable mixture of two diastereomers.

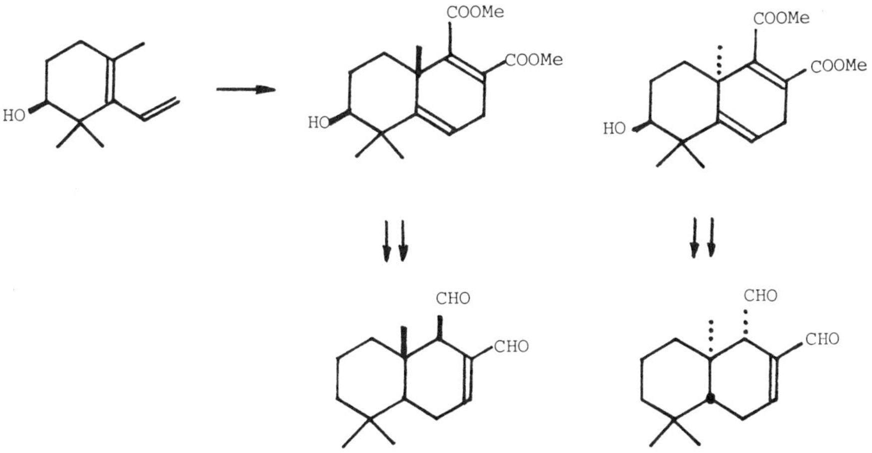

A general method for their synthesis (e.g., tanshinone II (**T2**) and cryptotanshinone (**T3**)) featuring the Diels–Alder reaction (Inouye, 1969) is an intriguing one in view of the high regioselectivity.

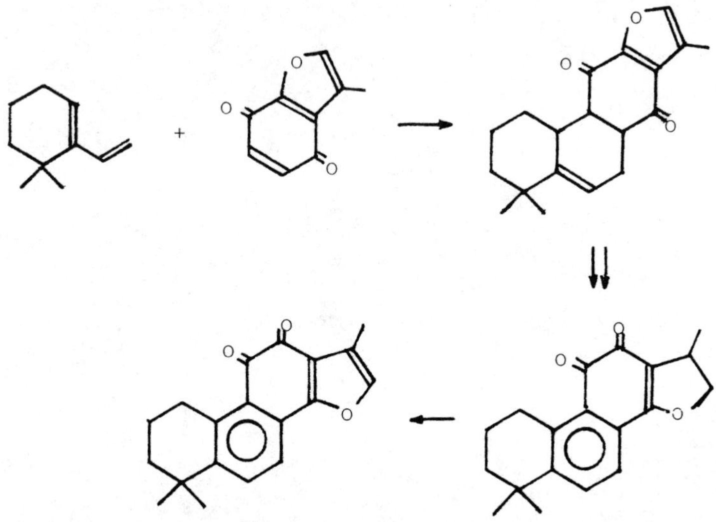

T2   **T3**

Similarly, another norditerpene, rosmariquinone (**R7**), has been assembled in one step (Knapp, 1985). The primary adduct undergoes oxidation *in situ*.

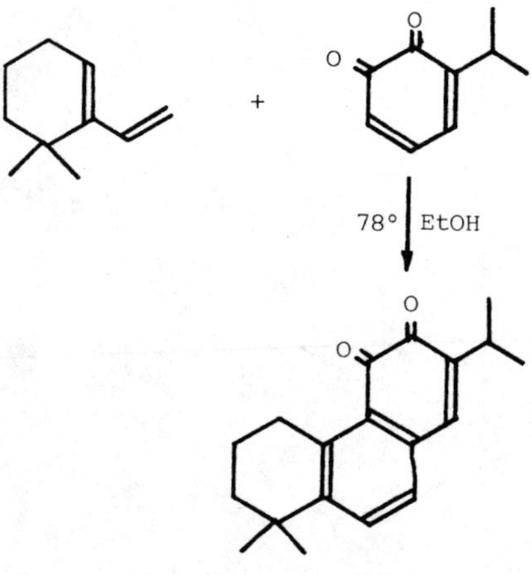

78° | EtOH

R7

A synthesis of the diterpene portulal (**P45**) (Tokoroyama, 1974; Kanazawa, 1975) relies on the generation of an octalin by a Diels–Alder cycloaddition and a ring disproportionation maneuver. The dienophilic reactant, which provides pendants for elaboration of the secondary methyl group, the carbinol residue, and the enediol chain, can be equated to chloromethylmaleic anhydride.

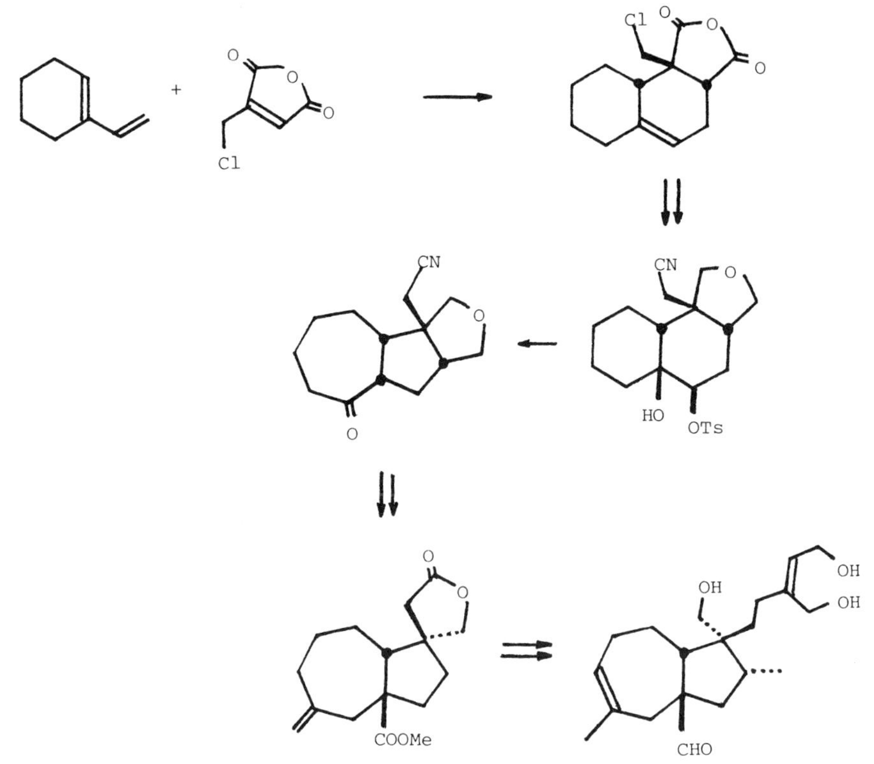

7.3 OTHER ALKYLDIENES AS DIENE UNITS

P45

ent-Aromadendrene (**ent-A41**) (Büchi, 1966, 1969) has been prepared from a cyclopropanated vinylcyclohexene derived from perillyl aldehyde. This approach obviously inspired the synthesis of ε-cadinene (**C5**) (Vig, 1975) using the cycloaddition method.

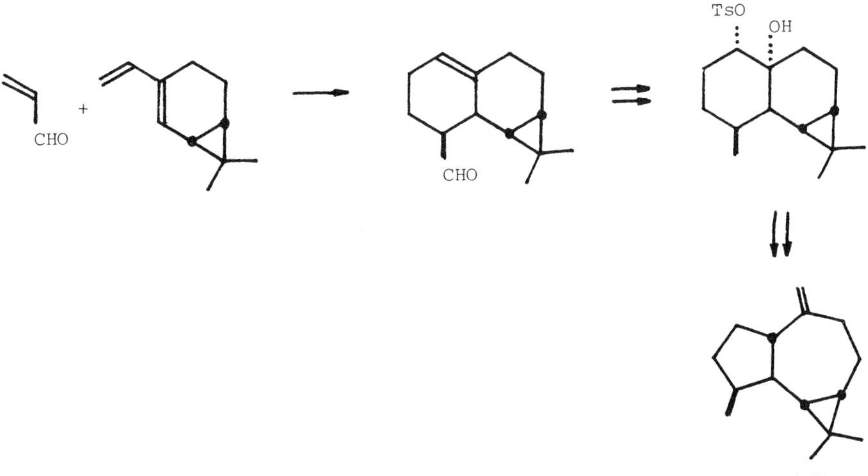

ent-A41

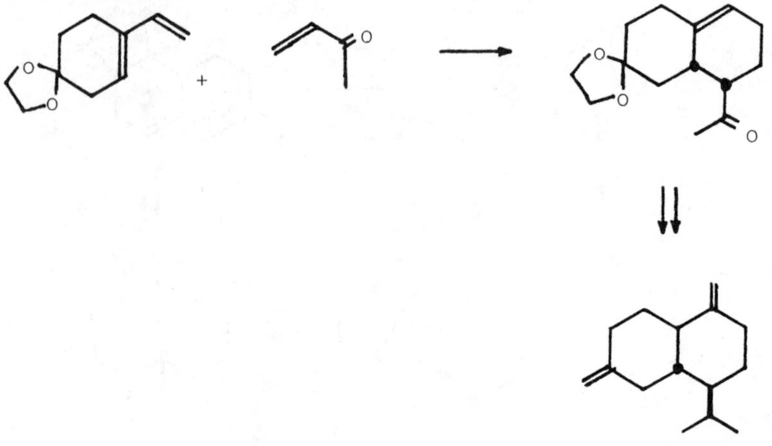

A synthesis of hedycaryol (**H2**) (Wharton, 1972) calls for the preparation of a 1-oxyeudesmane to trigger fragmentation of the intercyclic bond. The required 1-oxyeudesmane derivative may be obtained via the Diels–Alder reaction. It is interesting that angular methylation (at C_{10}) of the post-Diels–Alder octalone proceeds with high stereoselectivity (4:1, cis/trans), thus paving the way to the regio- and stereocontrolled unfolding of the medium ring.

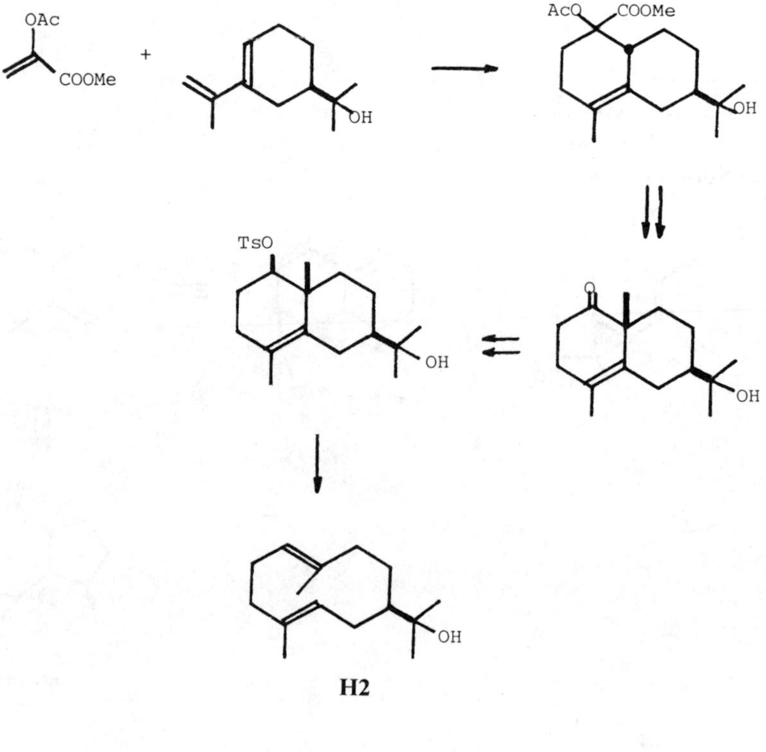

An incisive molecular analysis (Corey, 1985b) of retigerancic acid (**R1**), which forms the basis of its synthesis indicates the gainful employment of annulation by the Robinson, Diels–Alder, and ketene cycloaddition/ring expansion methods to form the B, C, and DE rings, respectively. The C ring is contracted from a cyclohexene to a cyclopentenecarboxylic acid toward the end of the synthesis. The stereochemistry at the junctures of the B, C, and D rings is ushered in during the Diels–Alder process by virtue of the endo rule.

R1

A superb scheme for the synthesis of butyrospermol (**B22**) (Kolaczkowski, 1985; Reusch, 1985) involves an A + CD mode of assemblage. The regioselectivity of the Diels–Alder reaction between the CD synthon and 2-methoxy-5-methylbenzoquinone can be controlled by appropriate Lewis acids. Polarization or activation of a particular carbonyl group is effected by catalysts capable of complexation with monodentate or polydentate ligands. Since boron trifluoride coordinates preferentially with the carbonyl at C_4, rendering C_6 more electron-deficient, the cycloaddition proceeds in the desired manner.

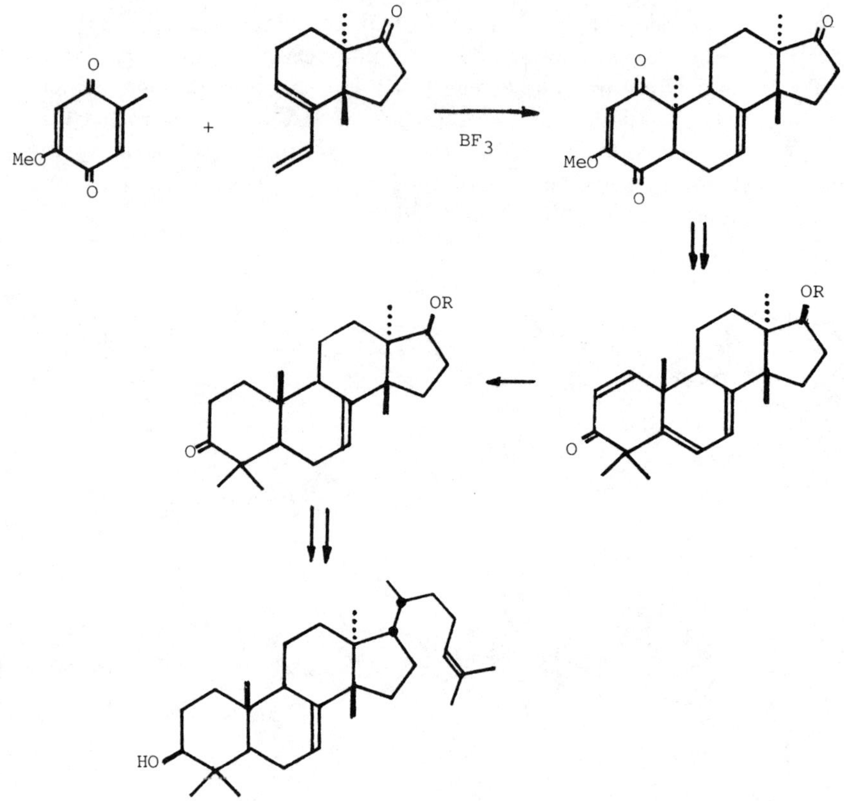

B22

Sterepolide (**S51**) is a tetrahydrophthaladic acid linearly condensed with five- and three-membered rings. With regard to its synthesis, the subtending relation between the cyclopropane with the double bond is translated into a Diels–Alder approach involving halomethylmaleic anhydride (Trost, 1985a) (cf. marasmic acid, **M6**, synthesis, Section 7.11.1).

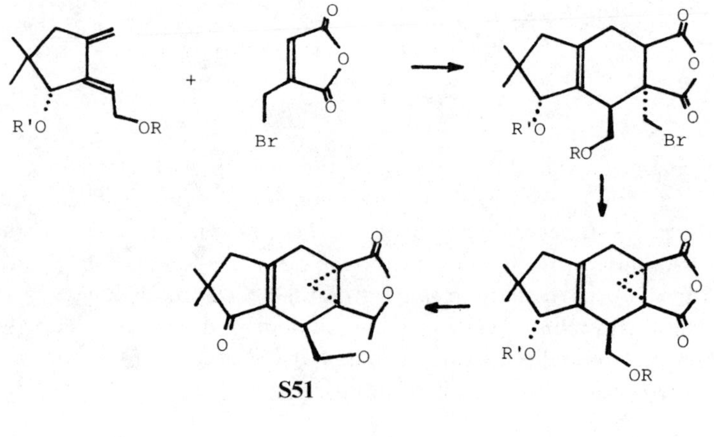

S51

7.3 OTHER ALKYLDIENES AS DIENE UNITS

o-Quinodimethanes may be generated and intercepted *in situ* by various dienophiles. A short route to occidol (**O3**) (Ho, 1972) consists of treating 2,3-bisbromomethyl-1,4-dimethylbenzene and methyl acrylate with zinc dust to furnish the tetralincarboxylic ester, and reacting the latter compound with methyllithium.

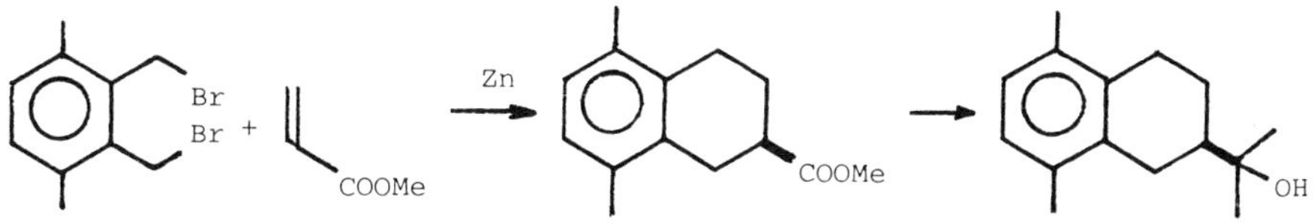

An intriguing route to the diterpene alkaloid chasmanine (**C50**) (Tsai, 1977) is initiated by a Diels–Alder reaction between an indenecarboxylic ester and maleic anhydride. The active diene is the *o*-quinodimethane isomer, which is attainable (in small concentrations) via a 1,5-hydrogen shift. The benzonorbornene adduct forms a sulfonylaziridine, which opens spontaneously with attendant skeletal rearrangement to reveal a properly functionalized molecule.

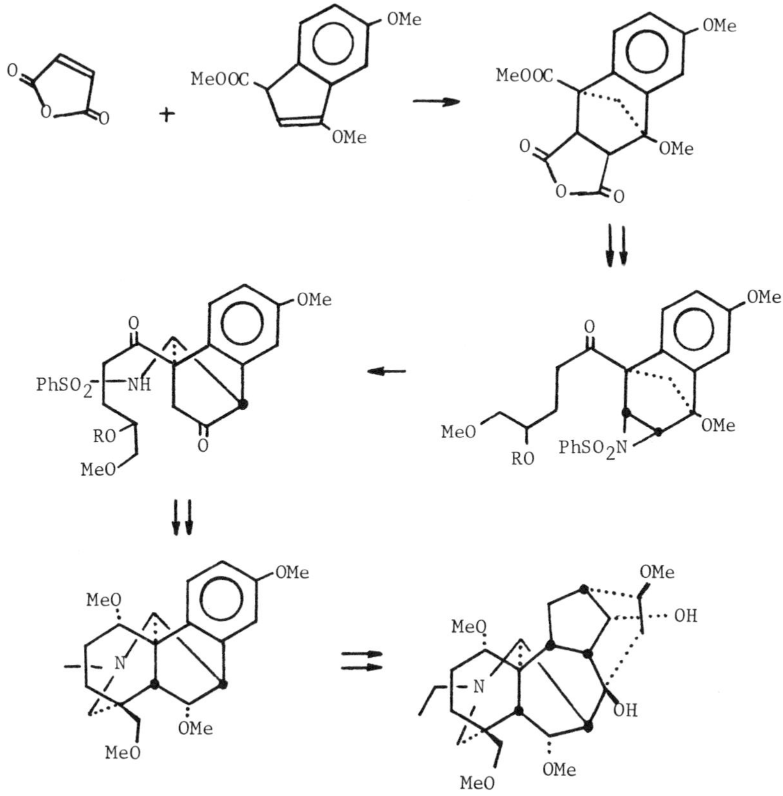

7.4 1,3-BUTADIENE AS DIENE UNIT

An astonishingly large number of sesquiterpene syntheses use butadiene. Occidol (**O3**) (Ho, 1973) starts with the well-known 1:1 adduct of butadiene and benzoquinone. The enedione moiety is rearomatized after methylation, while the isolated double bond is the anchoring point of the isopropanol chain.

Tetrahydrophthalic anhydride is a convenient building block for the trisnor-sesquiterpene geijerone (**G2**) (Wakamatsu, 1980), which belongs to the elemane family. Thus lactonization serves to introduce the oxygen atom into the cyclohexane ring and allows for selective methylation at the α position of the other carboxyl group.

An advanced synthetic intermediate for the complex antitumor agent vernolepin (**V10**) (Wakamatsu, 1985) can also be built from tetrahydrophthalic anhydride. One of the carboxyl groups is used, after its epimerization, to direct the oxygenation of the cyclohexene in a stereoselective fashion, and also to assist the introduction of the angular vinyl substituent.

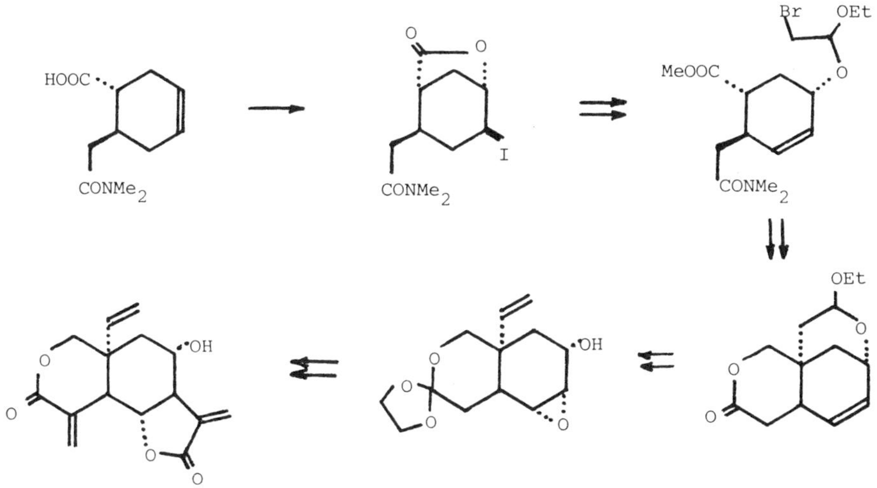

V10

β-Costol (**C74**), β-costal (**C71**), and arctiol (**A38**) (Torii, 1980) have been synthesized from methyl 3-oxo-1-cyclohexenecarboxylate. The ester group becomes the angular methyl, and the ketone is converted into the exocyclic methylene unit by a Wittig reagent.

For the synthesis of arctiol, the angular carboxylic acid directs the oxygenation at C_8 via iodolactonization, which in turn activates C_7 to accept a carbon chain by aldolization of the derived ketone. Both substituents in the natural product have stable equatorial configurations.

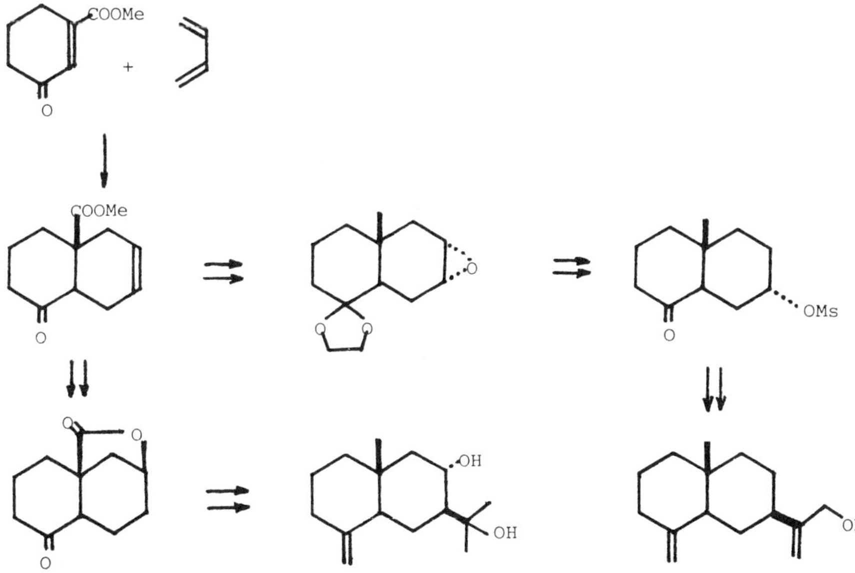

A38   **C74**

The same ketolactone also finds application in an approach to isopetasol (**P25**) (Torii, 1979a). The vicinal *cis*-dimethyl segment is elaborated from the lactone via reductive methylation of the derived α-cyclopropyl ketone. In the *trans*-fused system, a peri substituent is more stable when it is cis to the angular group because it is equatorially oriented.

The butadiene adduct with methyl 6-methyl-3-oxo-cyclohexenecarboxylate has two one-carbon pendants cis to each other because the approach of butadiene to the dienophile is sterically favored from the opposite side of the *C*-methyl group. The ester group again must be reduced totally; its initial presence is important to the reactivity enhancement of the enone. The adduct is convertible to dehydrofukinone (**F16**) (Torii, 1979b). [*Note*: The same dimethyloctalone has also been prepared via internal alkylation during a synthesis of ishwarone (**I12**: Piers, 1977a).

Both functional groups can be fully exploited, as is indicated in a synthesis of valencene/nootkatone(**V1/N11**) (Torii, 1982). The A-ring carbonyl is protected and reserved for later introduction of the unsaturated system, while the alkene is oxygenated to perform its assigned duty in the chain-grafting process.

An intriguing aspect of eremophilanes is that while all natural substances possess a vicinal *cis*-dimethyl array, certain members exhibit a trans ring juncture and others a cis juncture. From the synthetic viewpoint, both situations represent a challenge.

In an eremophilenolide (**E12**) synthesis (Nagakura, 1975), a Diels–Alder adduct obtained from butadiene and 2-methylcyclohexenone serves as the starting material. The cyclohexanone unit of the adduct is transposed with attendant methylation. Since the cis ring juncture has already been fixed (during the cycloaddition step), equilibration of the methylated ketone leads to the desired stereochemistry at the three contiguous asymmetric carbon centers. These intermediates prefer a steroid conformation, hence the secondary methyl group in them is equatorially oriented. At this stage the stereochemical problems pertaining to the synthesis are essentially solved; the remaining chiral center of eremophilenolide at C_8 carries an equatorial oxy substituent.

The reductive alkylation of the hydrindenone derived from 2-phenylthio-2-cyclopentenone and butadiene provides an assortment of angular substituents. The acetic ester thus generated proves to be an excellent precursor of vernolepin/vernomenin (**V10/V11**) (Lee, 1985a). The properly functionalized condensed δ-lactone is assembled without wasteful steps for functional group protection and demasking.

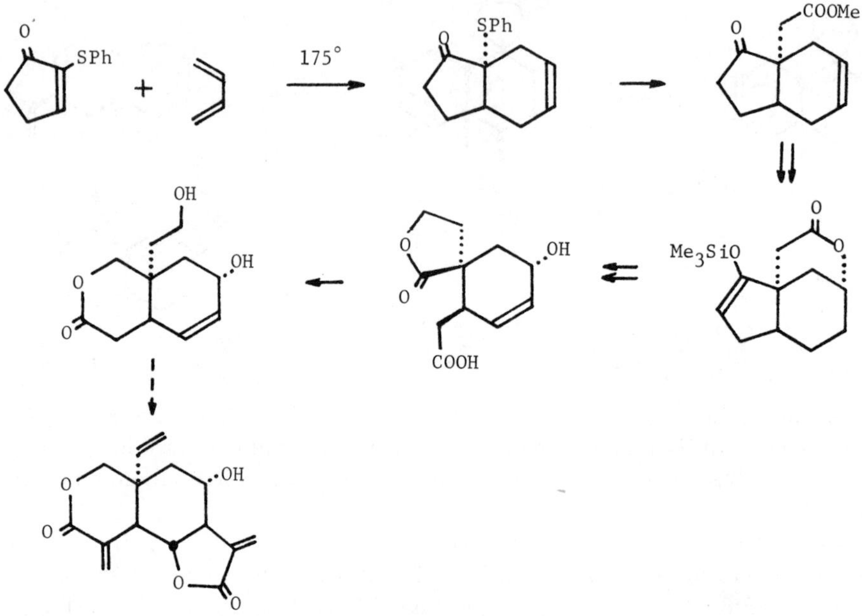

V10

Many approaches to the alkaloid dendrobine (**D11**) are based on Diels–Alder cycloaddition. The one that starts with a benzoquinone and butadiene and involves stepwise ring contraction (Kende, 1974) is relatively short, although it suffers from lack of regioselectivity in the contraction step(s) (via cleavage and aldol condensation).

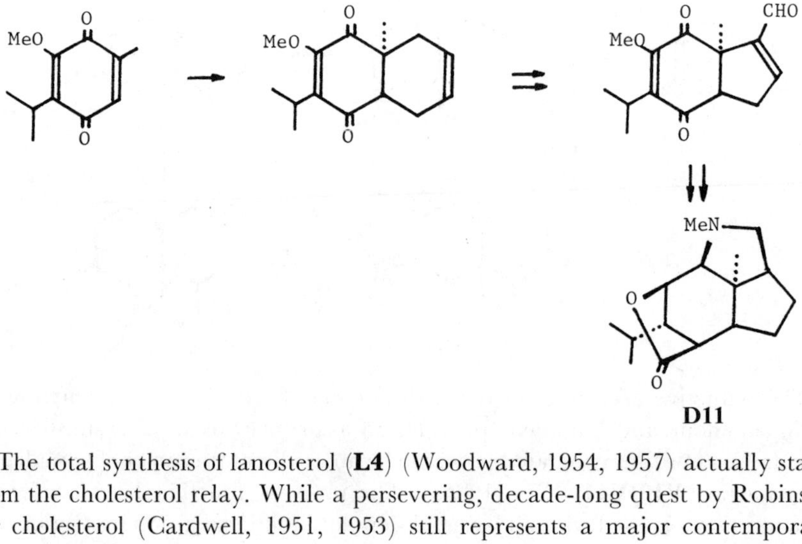

D11

The total synthesis of lanosterol (**L4**) (Woodward, 1954, 1957) actually starts from the cholesterol relay. While a persevering, decade-long quest by Robinson for cholesterol (Cardwell, 1951, 1953) still represents a major contemporary achievement, a more brilliant effort (Woodward, 1951, 1952) embodies a CD + B + A approach, in which the CD segment arises from a Diels–Alder

7.4 1,3-BUTADIENE AS DIENE UNIT

process. The D-ring is retained as a six-membered unit until all annulation operations are complete (by Robinson annulation, and an alkylation–aldolization sequence).

Advantages of this plan include an easy access of the CD synthon and ring juncture equilibration permitted by the 6:6 system.

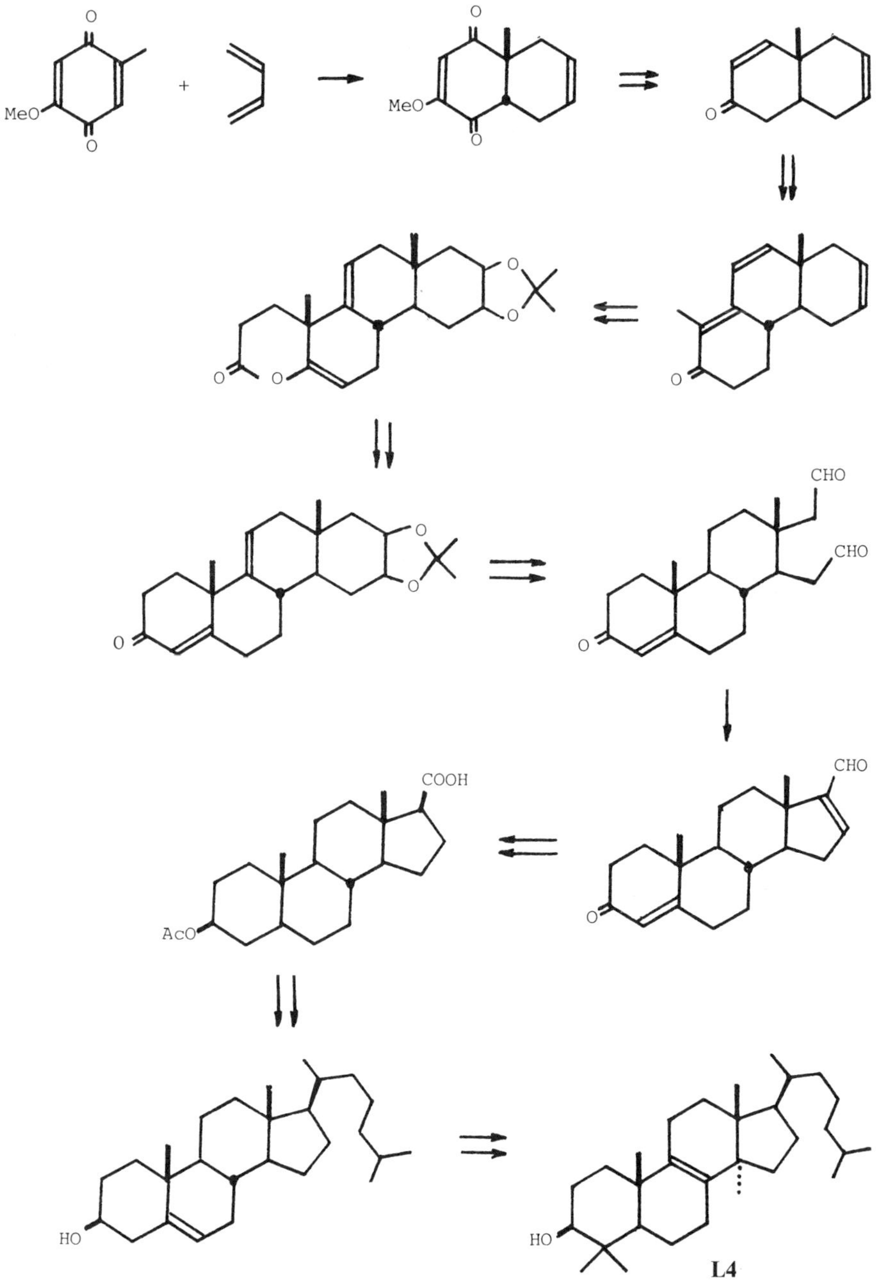

L4

A synthesis of cantharidin (**C18**) (Schenck, 1953) is unique in that furan is not used as a diene, although the scheme embodies a Diels–Alder reaction. The butadiene–dimethylmaleic anhydride adduct is dehydrogenated via bromination–dehydrobromination, and then reacted with singlet oxygen. The endoperoxide is systematically transformed into a tetrahydrofuran ring.

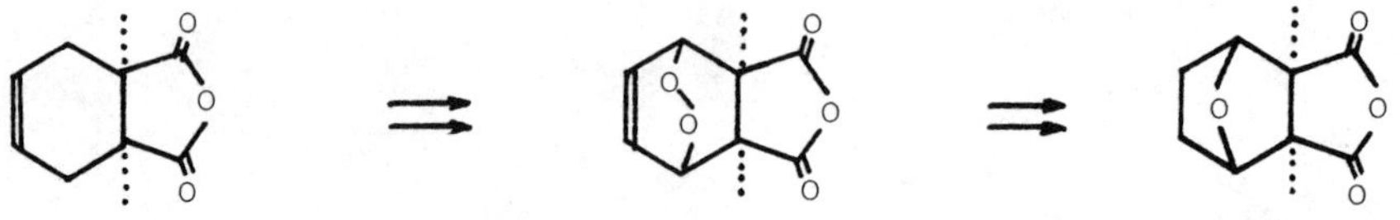

C18

7.5 OXYGENATED DIENES

Alkoxylated butadienes (and homologs), and siloxylated analogs in particular, are generally more reactive than the alkyl analogs, and they are most suitable for preparing cyclohexanone derivatives. An early synthetic route to the cadinanes (**C3**, **C5**) (Soffer, 1963, 1965, 1970; Burk, 1976) consists of condensation of 2-ethoxy-1,3-butadiene and cryptone, and further elaboration of the adduct at the carbonyl and the enol ether groups. Epimerization at the ring juncture also occurs.

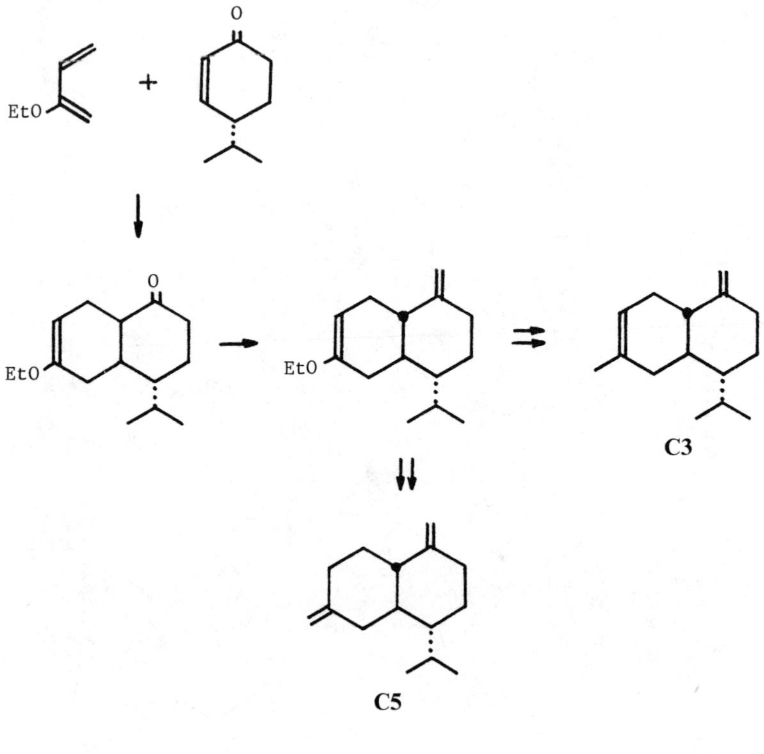

C3

C5

7.5 OXYGENATED DIENES

The simple 4-isopropenylcyclohexanone, which is used in nootkatone (**N11**) (Pesaro, 1968), appears to be most conveniently prepared by the cycloaddition method.

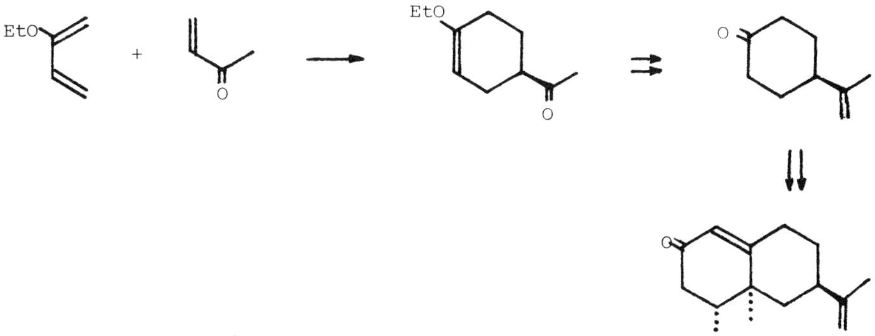

2-Trimethylsiloxy-1,3-butadiene is transformed into a dienophile unit for synthesis of norbisabolide (Zschiesche, 1987) based on the Diels–Alder reaction.

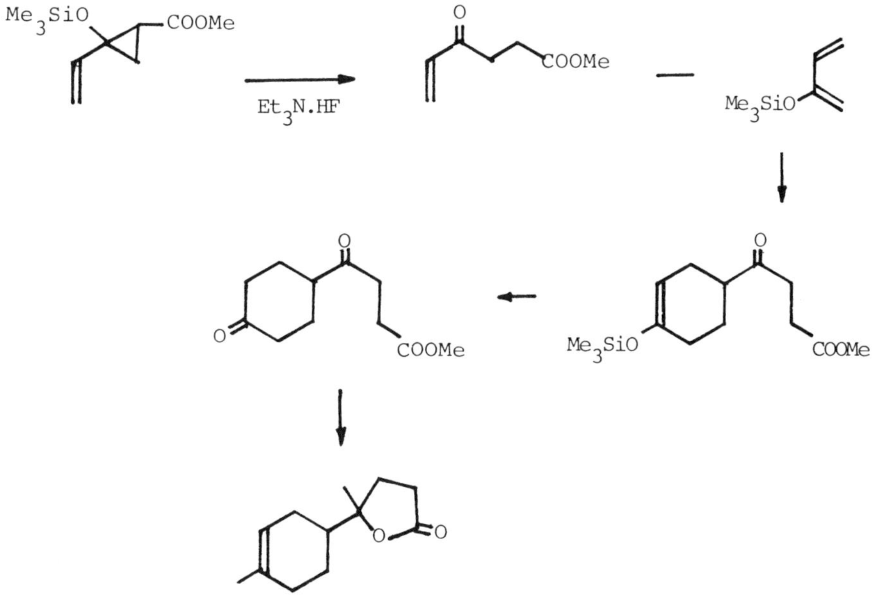

norbisabolide

The favored ortho orientation of alkyl groups in the Diels–Alder adducts derived from 1-alkyl-1,3-dienes and unsymmetrical dienophiles enables an assemblage of ligularone (**L13**) (Bohlmann, 1976; Yamakawa, 1977a, b, 1978) in a few steps. 3-Ethoxy- or 3-acetoxy-1,3-pentadiene may be used to form the complete carbon framework of the terpene. Because of the cis ring juncture, the undesirable configuration of the secondary methyl group is less stable and corrigible via equilibration of the 3-ketone.

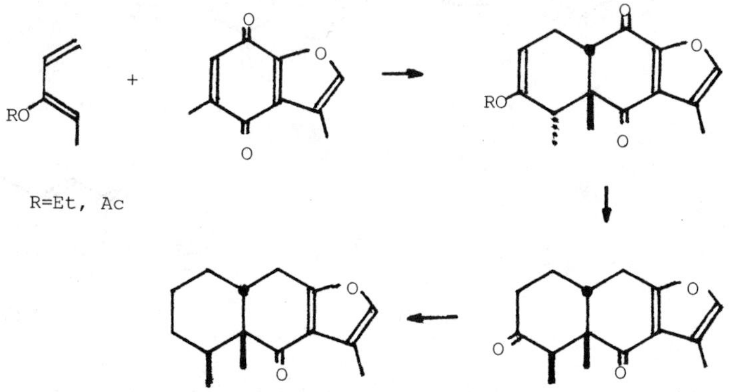

A very similar strategy is adopted in the synthesis of petasitolone (**P24**) (Liu, 1984).

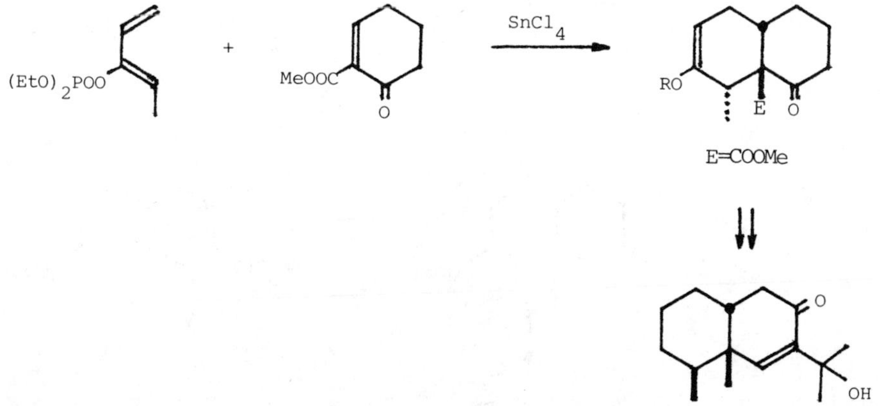

The presence of a discrete cyclohexene moiety in various trichothecanes makes them particularly suitable for the Diels–Alder maneuver. With proper addends, the major cycloadducts can be directed to form the heterocycle and the cyclopentane ring. The synthesis of calonectrin (**C12**) (Kraus, 1982) reflects the value of the approach, which also demonstrates stereocontrol by means of an intramolecular alkylation procedure.

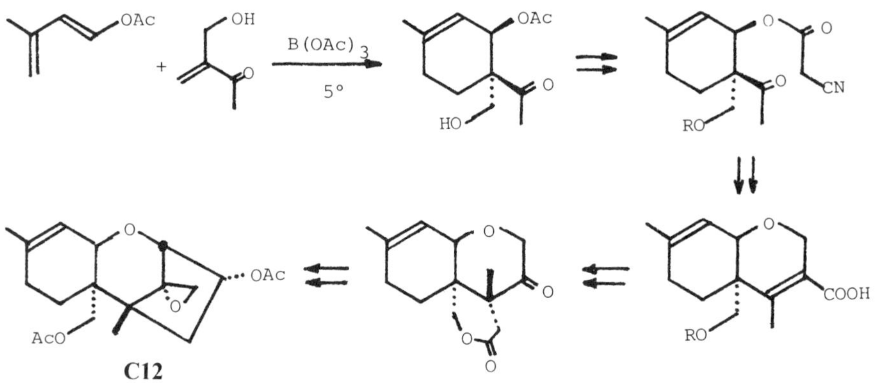

7.5 OXYGENATED DIENES

C12

(+)-Phyllanthocin (**P26**) is the methyl ester of the aglycone of an antileukemic substance. In one synthetic route (Burke, 1985), a cyclohexene is constructed and carefully transformed into a trihydroxyketone. The spiroketalization is controlled in such a way that it results in the natural conformation.

In an earlier stage an asymmetric epoxidation is effected by the Sharpless method in the presence of (+)-diethyl tartrate. Stereoselective aldolization of the epoxyketone with an optically active aldehyde furnishes the trihydroxyketone. After the complete cyclic system has been assembled, the cyclohexene undergoes hydroformylation from the exo side (regioselectivity, 4:1) to acquire the last skeletal carbon atom of phyllanthocin.

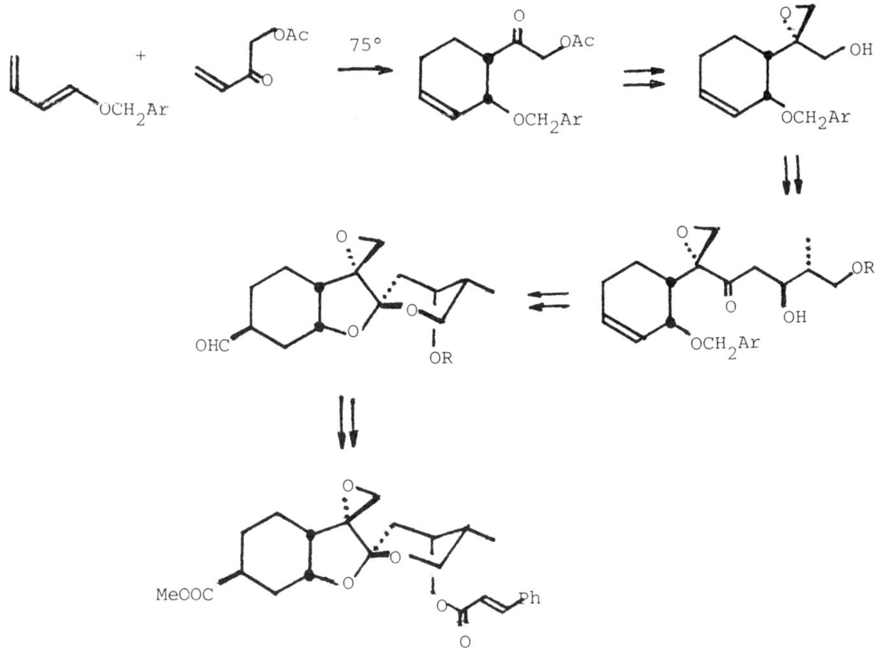

P26

By using a *m*-phosphinylbenzoyl ester of the secondary alcohol to direct the hydroformylation an increased regioselectivity can be effected. The resulting α-aldehyde is epimerizable and therefore convertible to (+)-phyllanthocin (**P26**) (Burke, 1986).

The key intermediate for periplanone B (**P22**) (Still, 1979) can be prepared starting from a Diels–Alder reaction (Hauptmann, 1986).

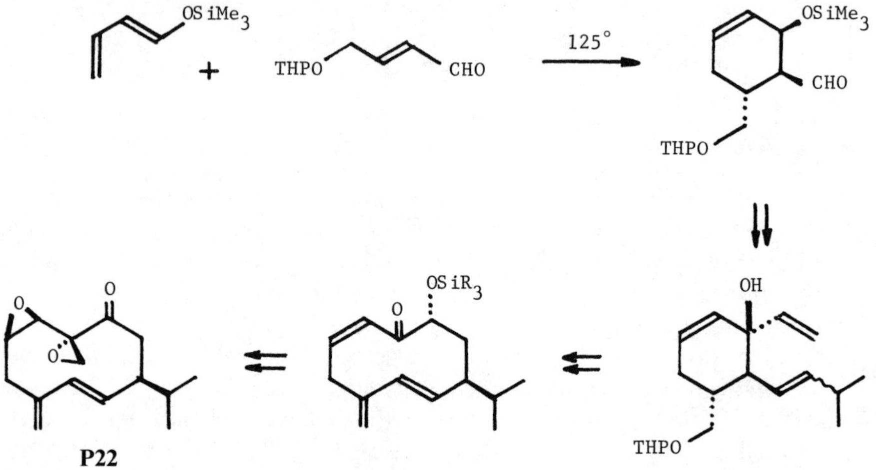

Simple but important monocyclic precursors for nootkatone (**N11**) (Hiyama, 1979) and hinesol (**H15**) (Ibuka, 1979a, b) have been assembled from alkoxy-1,3-dienes.

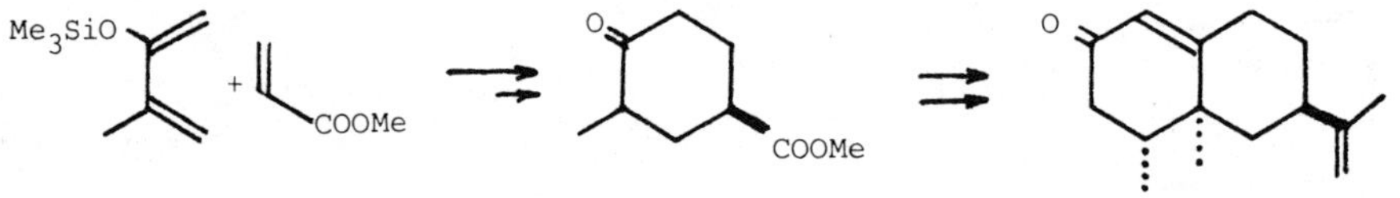

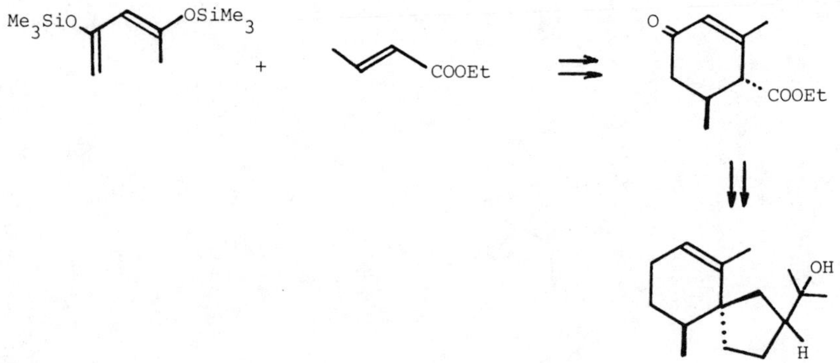

7.5 OXYGENATED DIENES

The Diels–Alder reaction is featured in the synthesis of a forskolin (**F9**) precursor (Bold, 1987). The desired regiomer is obtainable as the major (10:1) product which contains an α-oxygen function at C_1. The two ketone groups are easily differentiated to allow establishment of the *gem*-dimethyl group. Furthermore, the intermediate has been resolved using porcine pancreatic lipase.

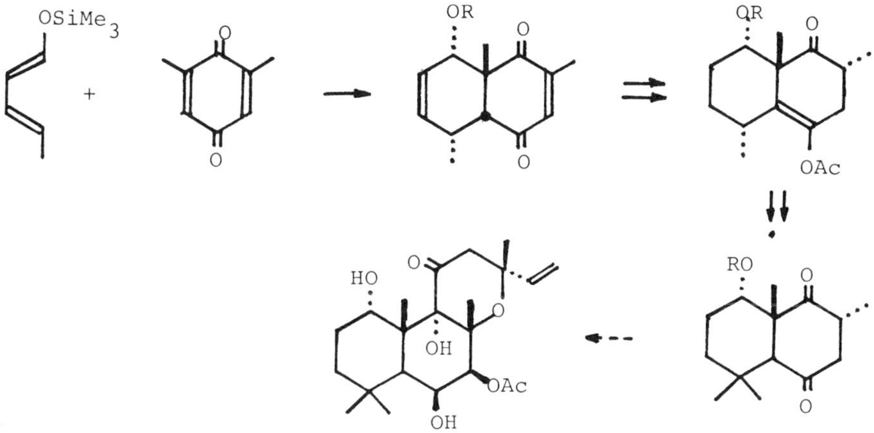

F9

A promising intermediate for synthesis of aphidicolin (**A36**) (V. Bell, 1986) is available from a Diels–Alder reaction. Model studies have shown that the bridged ring system can be constructed without difficulty.

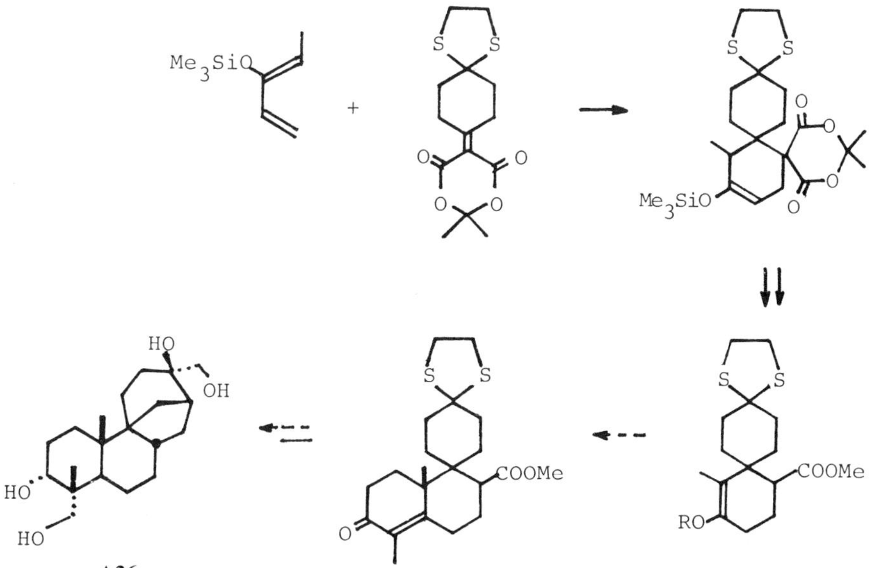

A36

1-Methoxy-3-trimethylsiloxy-1,3-butadiene (Danishefsky diene) is an extremely versatile Diels–Alder addend owing to its high reactivity and its functional group distribution, which facilitates many chemical manipulations. In one approach to ligularone (**L13**) and isoligularone (**L14**) (Miyashita, 1979), the adduct of Danishefsky diene and 2-methylcyclohexenone is formed and hydrolyzed to the enone, which is ready to accept methyl transfer from the cuprate reagent on the less hindered face.

In an approach to deoxynivalenol (vomitoxin) (Colvin, 1986), a siloxydiene is condensed with methyl coumalate to establish the *cis*-AB ring system. The undesired configuration of the siloxy group can be corrected via equilibration of the corresponding alcohol. After conjugate methylation of the unsaturated lactone, 1,2-transposition of the carbonyl group takes advantage of the enolization tendency of the α-ketolactone intermediate. An allyl ether of the latter compound not only keeps the ketone in a protected form, its stereoselective (exo) Claisen rearrangement provides a latent acetaldehyde chain for eventual cyclization of the cyclopentane nucleus.

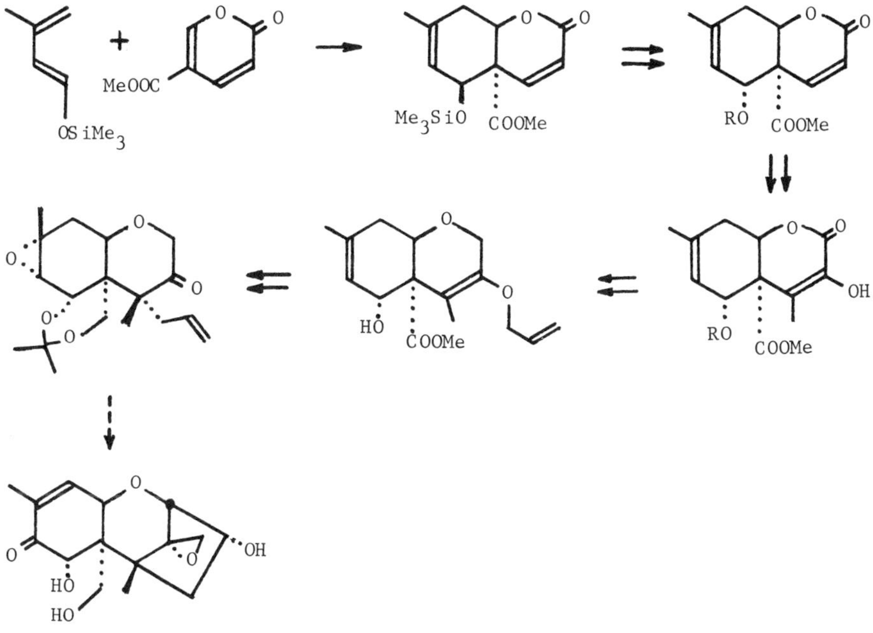

7.5 OXYGENATED DIENES

deoxynivalenol

In an elegant synthesis of vernolepin (**V10**) (Danishefsky, 1976, 1977) two consecutive Diels–Alder reactions are employed to build a *cis*-hexalone in which the angular carboxyl group is assigned the additional duty of managing the functionalization of B ring before its transformation into a vinyl substituent. The A ring is easily degraded by one carbon atom by virtue of the hydrolytic sensitivity of the substituents in the adduct. In other words, a cyclohexeneone emerges when the adduct is treated with mild acid, and its α carbon is ready for removal. Thus when B-ring modification reaches an adjournment stage, the δ-lactone is installed.

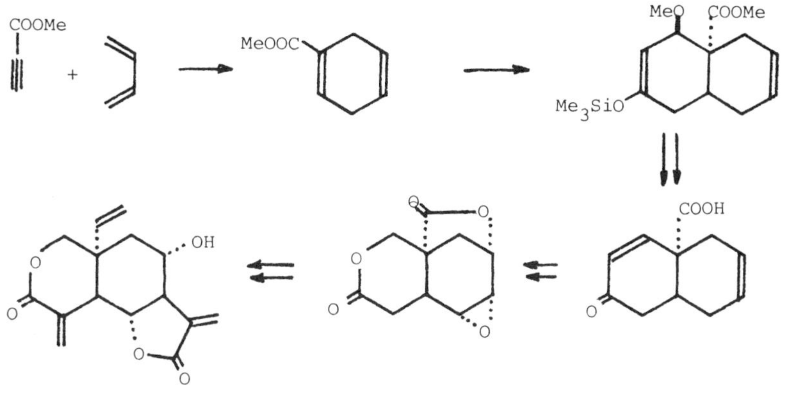

V10

The strategy recurs in the work on pentalenolactone (**P18**) (Danishefsky, 1978, 1979). Based on a convoluted analysis, the synthesis starts with a Diels–Alder reaction. From the bicyclo [2.2.1]heptadiene, a δ-lactone is fastened to one bridge by the same technique used in vernolepin, although decarboxylation is allowed to proceed at the enone stage. The other bridge is then cleaved, and each carboxyl thus released is fashioned accordingly.

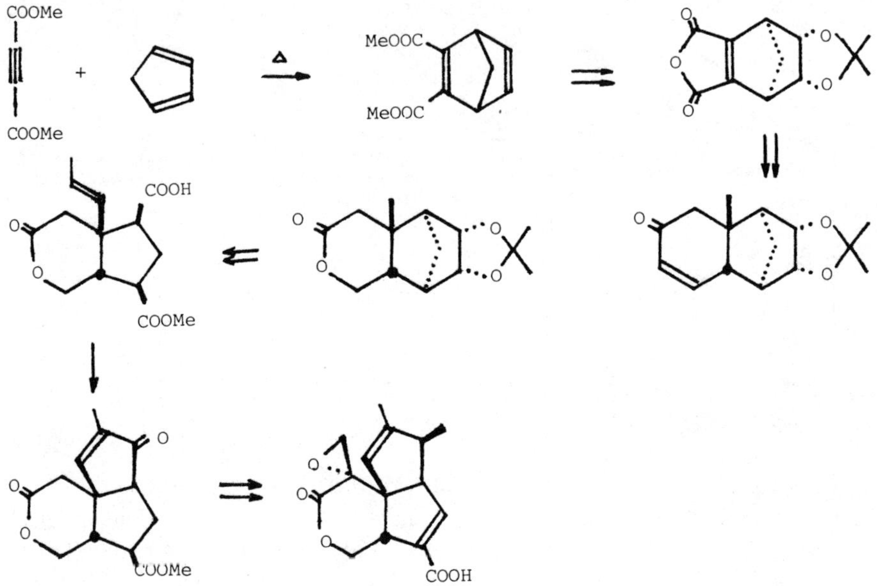

P18

A newer design for the synthesis of the trichothecane verrucarol (**V12**) (Schlessinger, 1982) involves a Diels–Alder process to form the cyclohexene unit. Stereocontrol is attained upon incorporating the dienophile (an α-methylene-δ-valerolactone) into a bridged system.

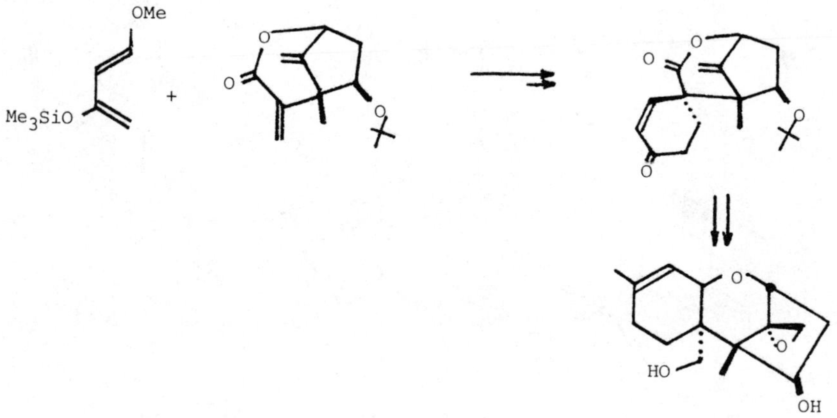

V12

7.5 OXYGENATED DIENES

While the pursuit of a Diels–Alder strategy for the construction of verrucarol (**V12**) (Trost, 1982a) based on an α-cyclopentanylacrylic ester is logical and apparent, the ene cyclization encountered immediately after the cycloaddition is a rather unexpected bonus. It provides temporary masking of one of the carbonyl groups without employment of external reagents, and renders the ring system rigid and sterically biased such that the free ketone can be reduced to a homogeneous product. The stereoselectivity is most advantageous, despite the need to invert the hydroxyl group at a later stage.

V12

The ready access of cyclohexenones via Diels–Alder reactions with Danishefsky's diene (and analogs) and the cis mode characteristic of the cycloadditions are features that can be exploited to great advantage when chamaecynone (**C45**) is contemplated as a synthetic target (Harayama, 1977). Although it turns out that other bicyclic isomers are also produced, this route still represents the shortest one to the norsesquiterpene.

C45

An ABCD-ring precursor of bruceantin (**B18**) has been fabricated through a Robinson annulation (A + B), a Diels–Alder reaction (AB + C), and a Michael-type cyclization initiated by iodotrimethylsilane (Voyle, 1983a). The intramolecular nature of the last reaction ensures C—C bond formation from the α-face.

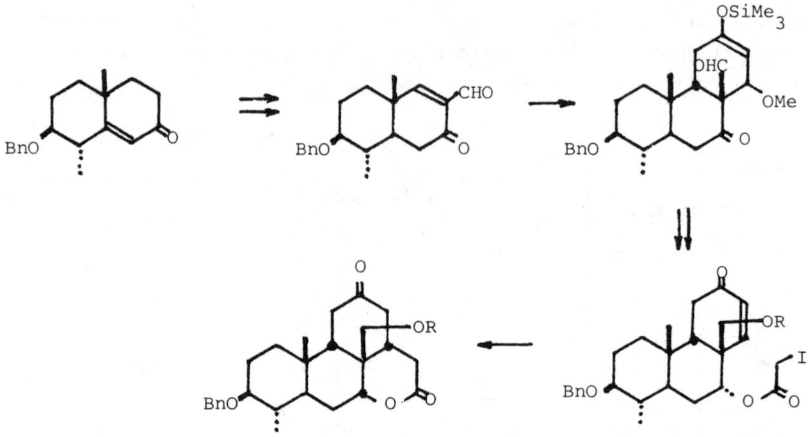

The cis-dihydrohomo Wieland–Miescher ketone may be obtained via a Diels–Alder cycloaddition. The compound serves as a starting material for synthesis of aphidicolin (**A36**) (Ireland, 1979).

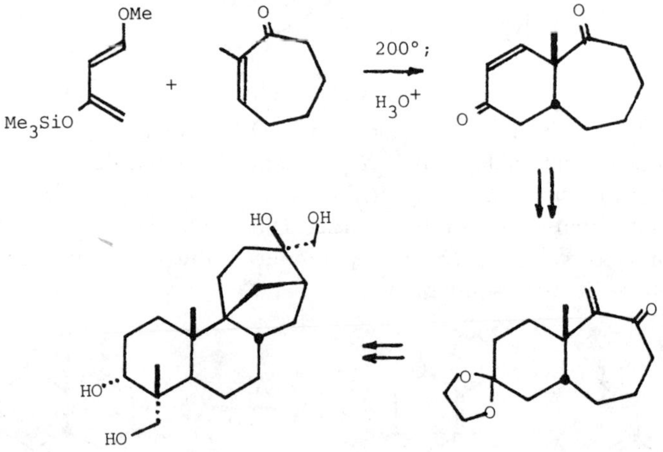

A36

An obviously useful intermediate for the cyathins is a tricyclic diketone that needs only expansion of the C ring with inversion of its hydroxyl group, isopropylation at the A ring, and removal of two ketone functions to reach cyathin A_3 (**C89**) (W. Ayer, 1981).

Since the B-ring of the cyathins is six-membered with two para-substituted methyl groups, a Diels–Alder methodology involving 2,5-dimethylbenzoquinone is a logical proposition for the preparation of the key intermediate. Furthermore, the enedione that emerges provides a platform for attaching the A-ring elements by means of a photocycloaddition with allene.

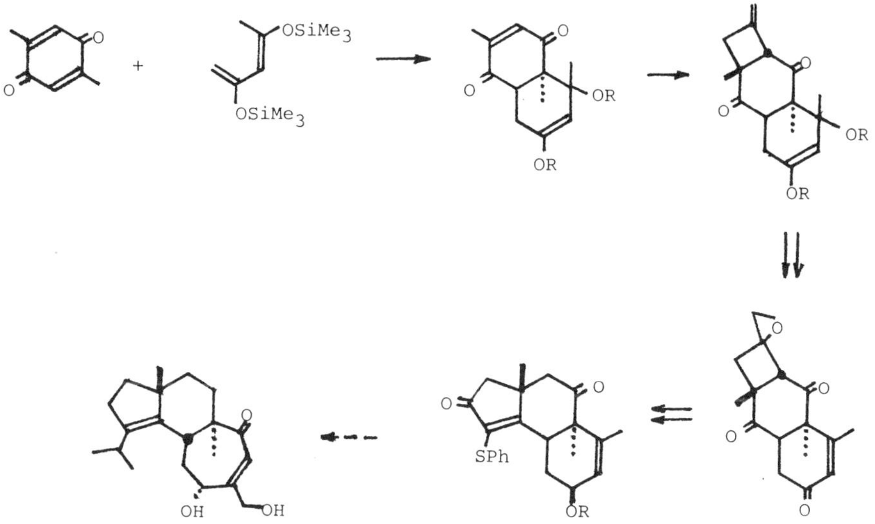

A synthesis of talatisamine (**T1**) (Wiesner, 1974c) consists of assembling a tricyclic intermediate by a Diels–Alder reaction whose nitrile function is a latent amino group for subsequent closure of the heterocycle.

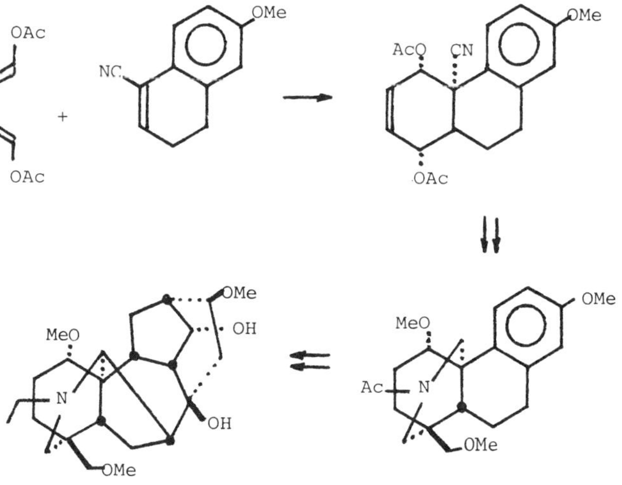

Rather unexpectedly, sulfur substituents have a more pronounced influence on conjugate dienes than their oxygen counterparts when it comes to directing the Diels–Alder reaction. This observation finds application in a synthesis of carvone (**C31**) (Trost, 1976).

C31

The synthesis of illudol (**I3**) (Semmelhack, 1980a) is admittedly a challenging task. There are five contiguous asymmetric centers that require close attention. The recognition of the feasibility of a Diels–Alder method with its stereochemical implications (i.e., endo addition) is the most crucial aspect in determining the successful termination of the venture.

The anti relationship established for the two newly created angular substituents in the tricyclic adduct from a cyclobutene and a vinylcyclopentene has consequences for the steric outcome of the remaining chiral centers. Thus the hydrogen (β) directs protonation of the enol to afford the more stable cis ring juncture. In turn this juncture governs the steric course of reduction of the two ketone groups that are to be unmasked. The angular methyl substituent is replaced by an ester in the dienophile because of its activating ability, its orientation effect, and the ease of preparation of the dienophile.

I3

F8

The tricyclic intermediate is also convertible to fomannosin (**F8**) (Semmelhack, 1981a).

A retrosynthetic analysis reduces the complex carbocyclic framework of secodaphniphylline (**D7**) to a cyclohexeneone, a vinylcyclopentene, and an allyl group as construction units (Orban, 1983). The isopropyl chain can be temporarily ignored because its introduction by alkylation is feasible at a later stage of the synthesis; it should also be possible to close the amino bridge by a reductive linkage of the diketone intermediate with ammonia.

A key intermediate has been assembled in four steps: Diels–Alder reaction, allylation, ozonolysis, and aldolization.

D7

An excellent strategy for synthesizing a potential intermediate for the kaurane-type diterpenes including corymbol is germane to the Diels–Alder reaction of bridgehead enones (Kraus, 1986). In the demonstration of exo addition, it is of interest to find out whether the methylated diene will similarly undergo cycloaddition. A successful reaction would result in a product with a correctly oriented methyl group at the ring juncture, and the oxidative desilylation would place a double bond between C_4 and C_5 facilitating introduction of the remaining methyl substituents.

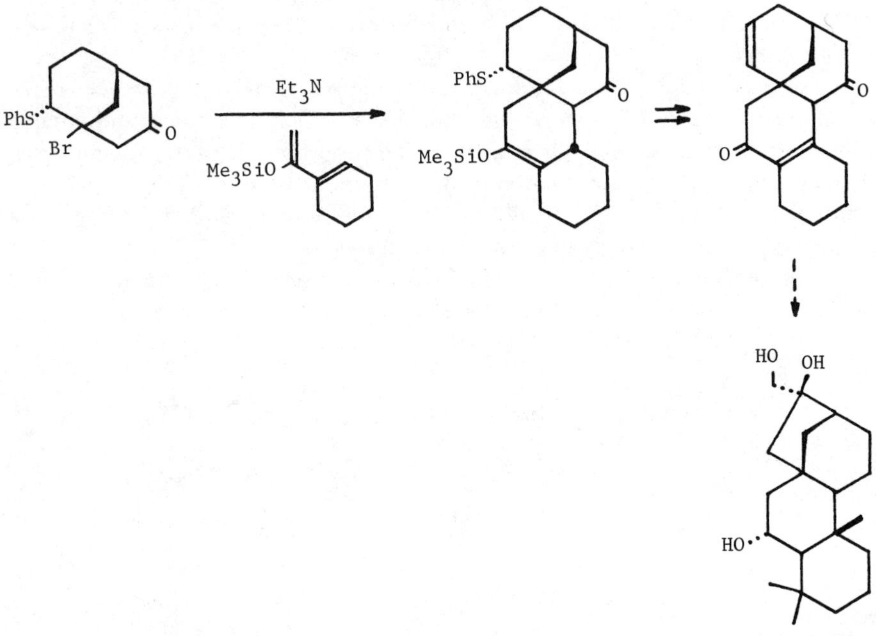

7.6 CYCLOPENTADIENES

Bridged skeletons are formed from Diels–Alder reactions of homonuclear dienes. The cyclopentadiene dimer (*endo*-dicyclopentadiene) has been converted into sesquifenchene (**S28**) (Grieco, 1975b) by semihydrogenation and elaboration of the nonbridged cyclopentene moiety.

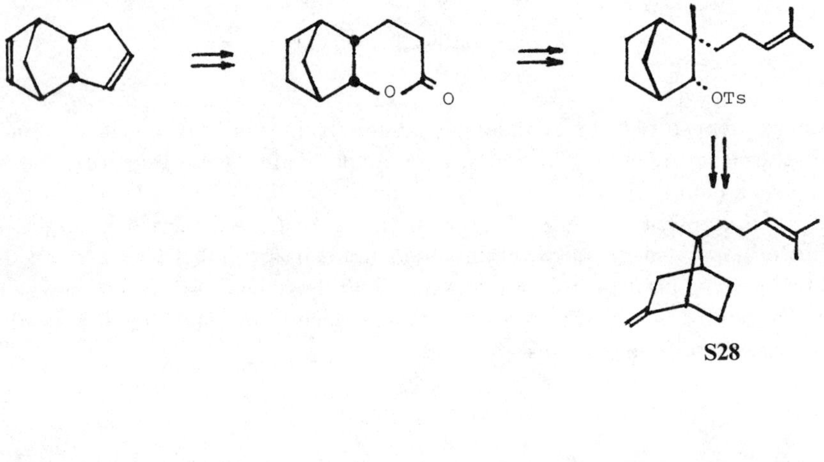

7.6 CYCLOPENTADIENES

It has been stated that any molecule containing a six-membered ring can be made with the aid of a Diels–Alder reaction. It can also be said that bicyclo-[m.n.p]alkanes, where $0 < m, n, p < 3$, are most expediently constructed by the same method. Since there is a particularly high degree of stereoselectivity (endo rule) in reactions associated with cyclic dienes, molecules with stereochemically defined functional groups can be made almost at will. By well-conceived transformations, even more complex systems are within reach.

One of the earliest applications of the Diels–Alder reaction to natural product synthesis concerned the preparation of camphor (**C15**) and epicamphor (**C15e**) (Alder, 1939).

Camphene (**C13**) is obtainable from the adduct of cyclopentadiene and methylmaleic anhydride. The two carbonyl groups are converted to a methyl and an exocyclic methylene group, respectively (Vaughan, 1963).

The route to cyclosativene (**S13**) (Heissler, 1980) is illustrative of functional group deployment from a Diels–Alder adduct. The two carboxyl groups originated from methylmaleic anhydride are respectively to become member of a new cyclohexane ring and a tertiary methyl group. The reductive process involving the latter is coupled to a cyclopropanation maneuver via homoconjugate addition which also lays an activating anchor at the ethano bridge to enable the closure of the new cyclohexane at a later stage.

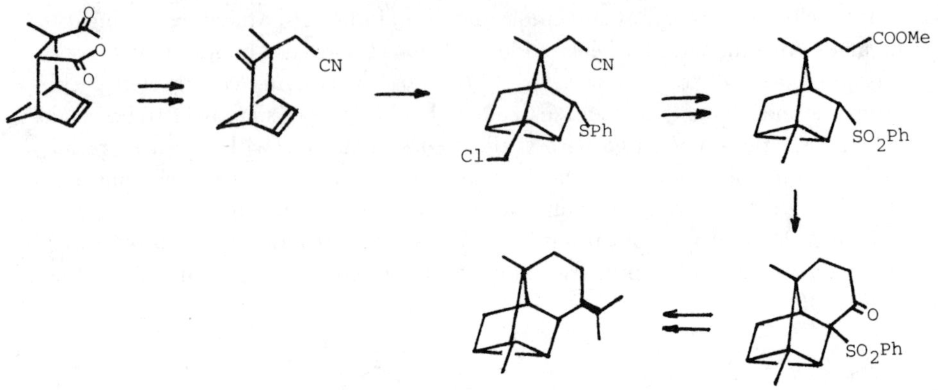

S13

From the practical point of view, the reaction of cyclopentadiene with dimethylmaleic anhydride to provide starting material for albene (**A22**) (J. Baldwin, 1981) is not reliable, as only the minor exo adduct conforms to the required steric disposition. On the other hand, the preferred exo addition with methacrylic acids can be exploited with advantage in the syntheses of both albene (**A22**) (Manzardo, 1983) and sinularene (**S37**) (Collins, 1979). In the latter work the exo adduct undergoes acid-catalyzed rearrangement whereby the quaternary carbon atom is shifted to the bridge apex while functionalizing one of the ethano bridges. The chain and ring building process can now begin.

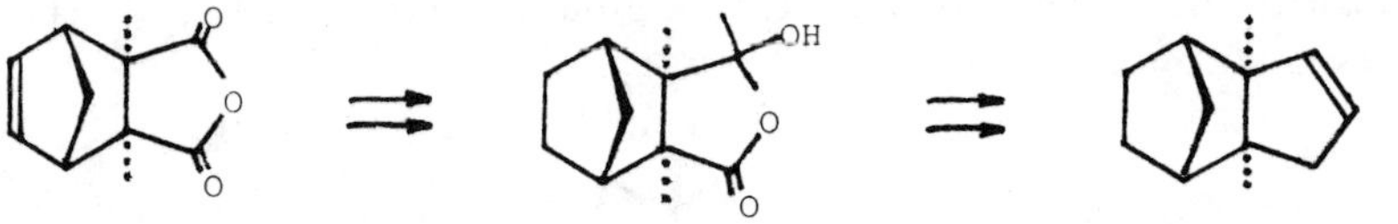

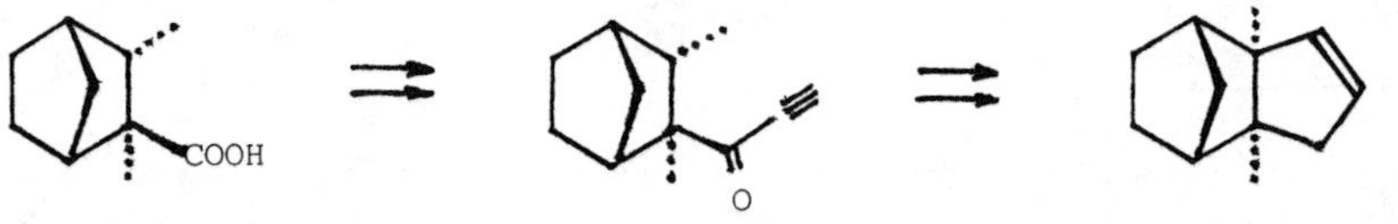

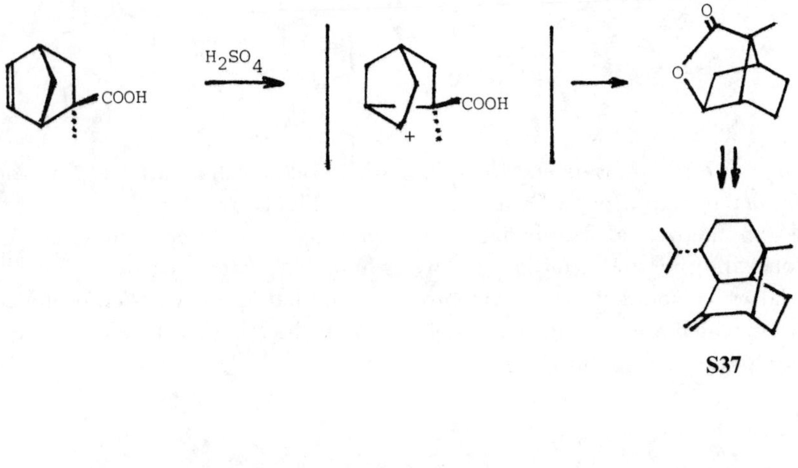

S37

7.6 CYCLOPENTADIENES

The 1:1 adduct of cyclopentadiene and cyclopentadienone, which is more simply prepared by oxidation of dicyclopentadiene, serves as a convenient building block for coriolin (**C70**) (T. Ito, 1982). After introduction of the gem-dimethyl group into the A-ring the etheno bridge is broken. The carbon atom closer to the oxygen function is retained and eventually converted to a methyl group. Conventional annulation (alkylation-aldol) follows.

An essentially identical strategy was mapped out in a synthesis (Schuda, 1984) that starts from the monoketal of the cyclopentadienone dimer. Differences of the two syntheses are only in the details of operation as mandated by the presence or absence of the extra functional group.

R= OCH$_2$CH$_2$O

C70

A similar use of the cyclopentadiene-cyclopentenone adduct appears in a synthesis of silphinene (**S34**) (Tsunoda, 1983). However, all the carbon atoms in the original adduct are retained. The broken bridge becomes the focal point at which the tertiary methyl groups and the elements of the third ring are to be attached.

S34

The utility of the adducts of cyclopentadiene and substituted benzoquinones is amplified by their facile intramolecular photocycloaddition and cyclo-reversion at the alternate C—C bonds. The reaction sequence has emerged as a most efficient route to linear triquinanes such as hirsutene (**H17**) (Mehta, 1981), and $\Delta^{9(12)}$-capnellene (**C21**) (Mehta, 1983).

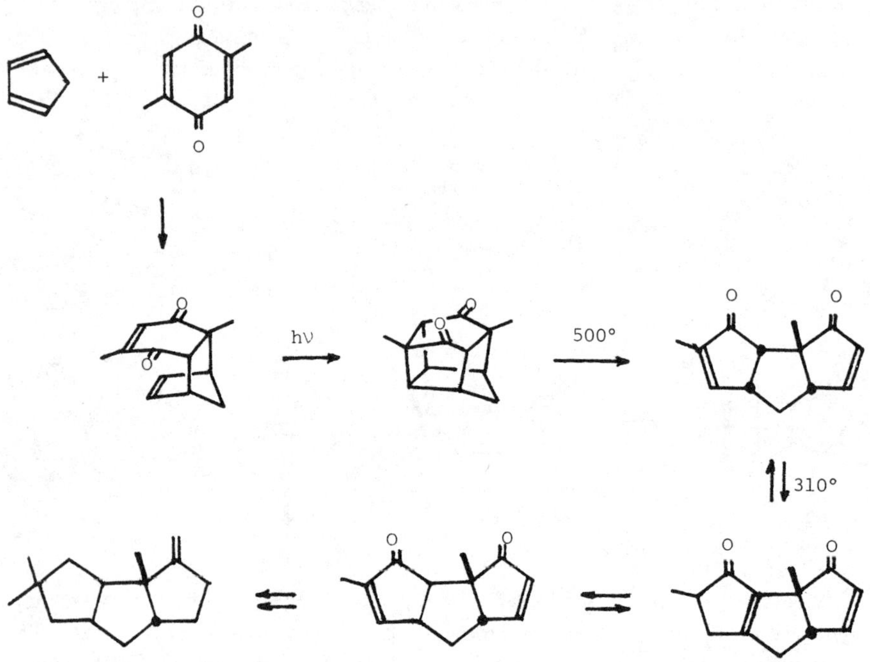

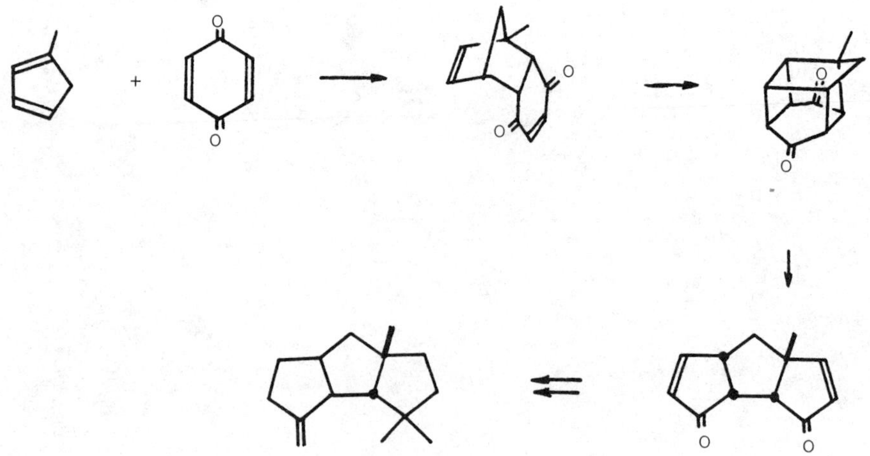

It does not require a quantum leap of imagination to perceive the suitability of Diels–Alder adducts from cyclopentadiene for constructing the santalanes: β-santalene (**S4**) (Bertrand, 1979) and β-santalol (**S6**) (Baumann, 1979; Sato, 1981; H. Monti, 1982).

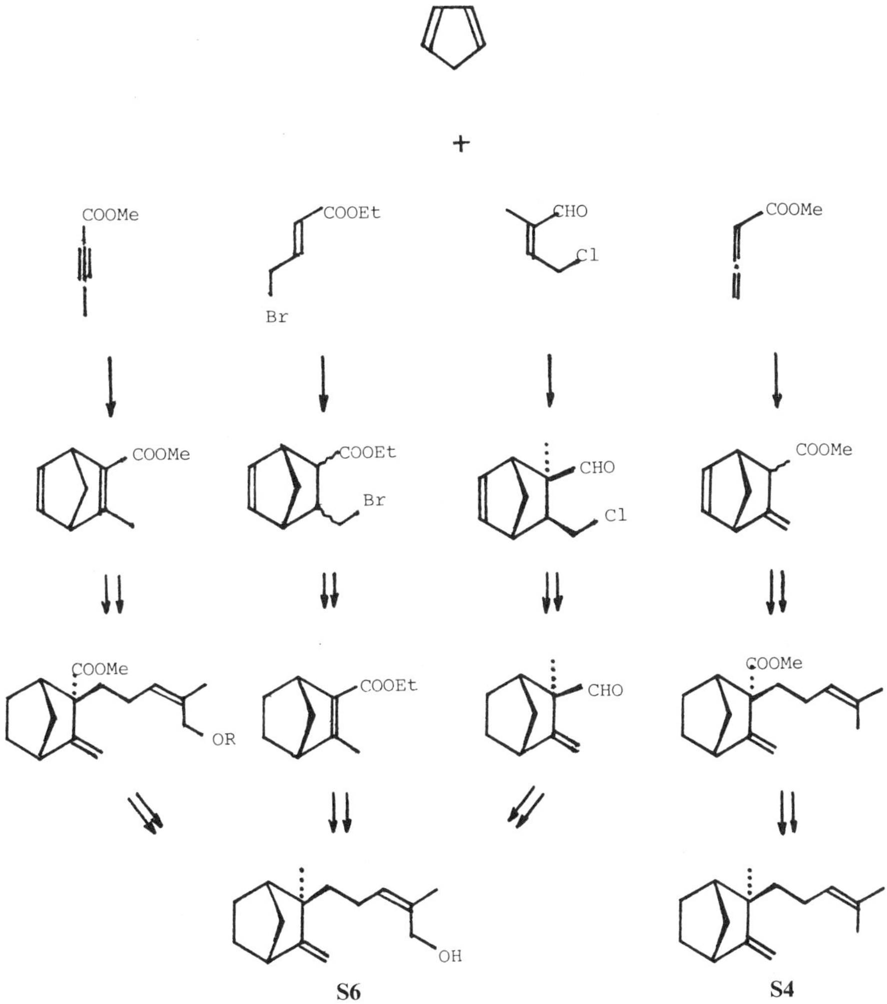

The salient features associated with this work include the preferred exo alkylation (Corey, 1962) and the unusual adoption of an exo transition state in the cycloaddition involving methacrylic esters and methacrolein derivatives.

(—)-β Santalene (**S4**) has been obtained from the cyclopentadiene adduct with a 2,3-butadienoate ester of 3-*exo*-hydroxy-(—)-isobornyl neopentyl ether (Oppolzer, 1985).

A short but stereorandom synthesis of β-santalene (**S4**) (Brieger, 1963a) consists of heating geraniol with cyclopentadiene and dehydrating the adducts.

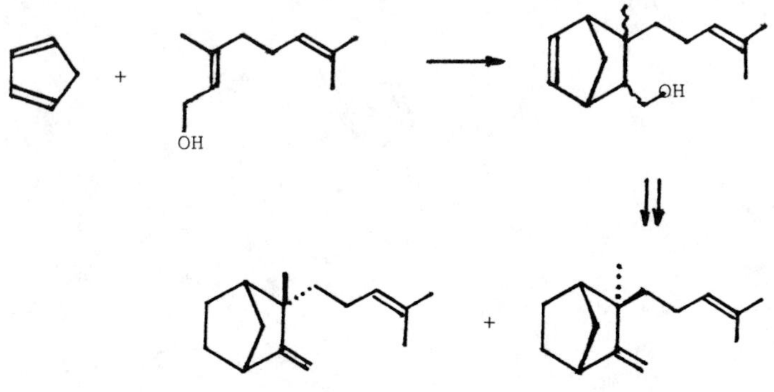

Two different Diels–Alder processes are put into service in an efficient synthesis of verrucarol (**V12**) (Roush, 1983). Although regioisomers are produced from the first cycloaddition, both are subsequently transformed into a homogeneous hydroxyester via desilylative rearrangement. The second Diels–Alder reaction is employed to complete the cyclohexene moiety (cf. Schlessinger, 1982).

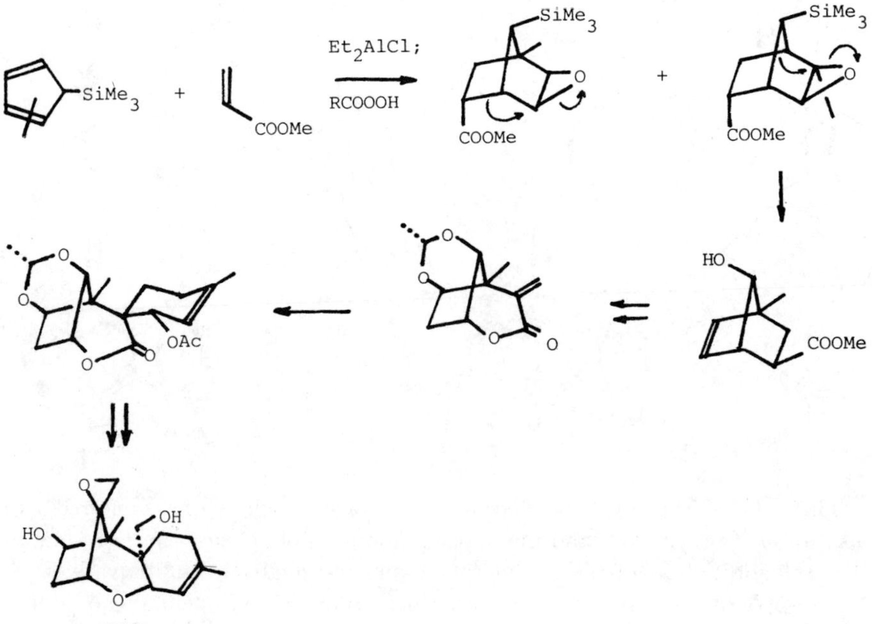

During the treatment of the isomeric epoxides with acid, the direction of opening is influenced by the substitution pattern (Markovnikov rule). In the absence of local dissymmetry, the 2,6-interaction (H:COOMe) becomes an important factor, such that uptilting of the *endo*-hydrogen as the C—O bond breaks relieves a certain amount of strain.

A precursor of β-cuparenone (**C86**) (Jung, 1980b) has been prepared from the dimethyl ketal of tetrachlorocyclopentadienone. It requires sacrifice a large portion of the carbon sources of adducts that are formed in low yields. Furthermore, vast amounts of reducing agents are needed for the removal of the chlorine atoms.

The Diels–Alder reaction employed in the synthesis belongs to the inverse electron demand class, and such a reaction is generally much less efficient.

7.7 CYCLOHEXADIENES

Despite their lower reactivity, cyclohexadienes are adequate partners in Diels–Alder reactions. For this reason, they have been frequently summoned to serve as building blocks for sesquiterpenes containing a bicyclo[2.2.2]octane nucleus, including seychellene (**S29**) (Schmazl, 1970; Mirrington, 1972b; Jung, 1980a), patchouli alcohol (**P15**) (Danishefsky, 1968; Mirrington, 1972a; Bertrand, 1980), and norpatchoulenol (**P13**) (Teisseire, 1974).

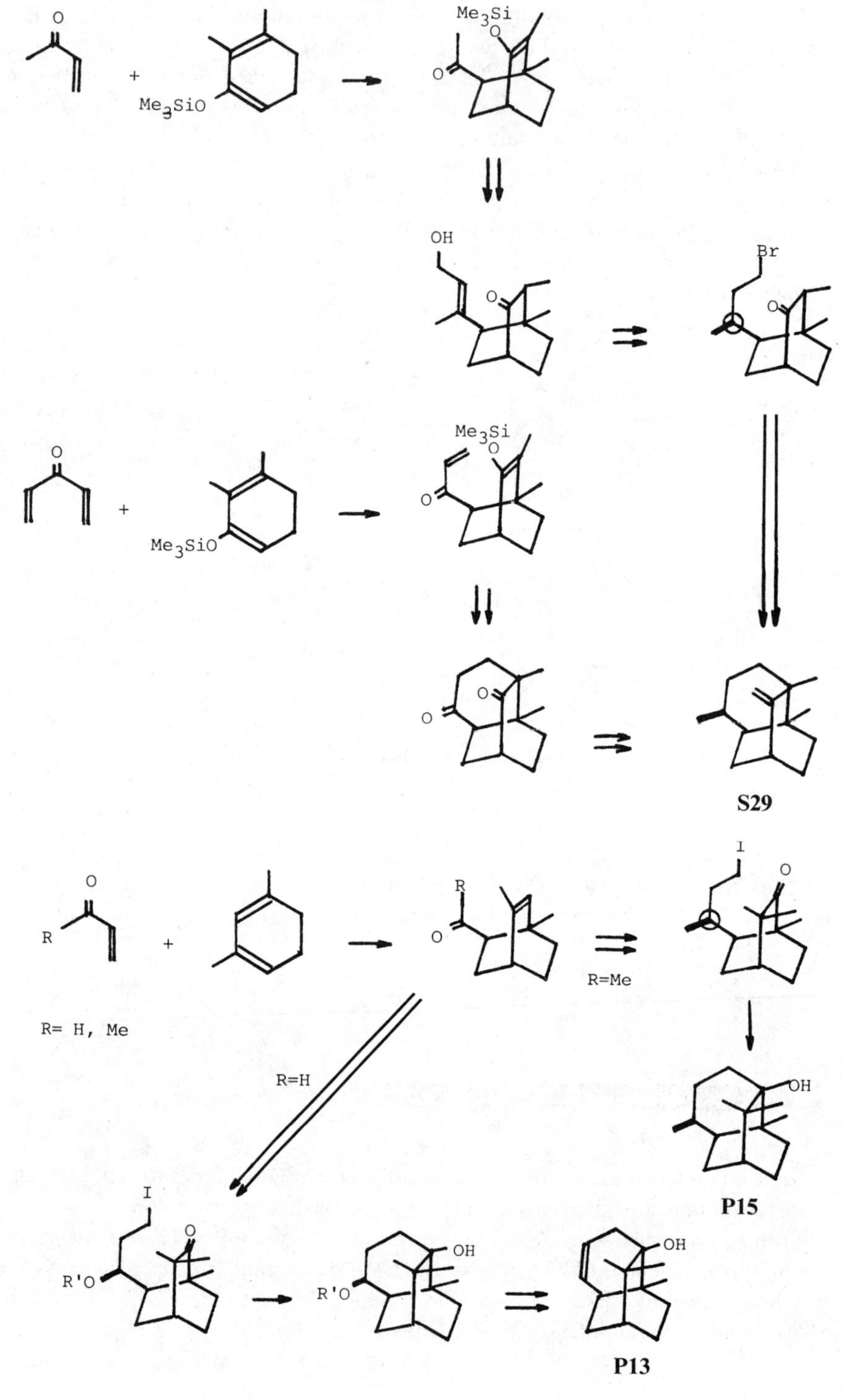

7 DIELS—ALDER REACTIONS

S29

P15

P13

R= H, Me

R=H

R=Me

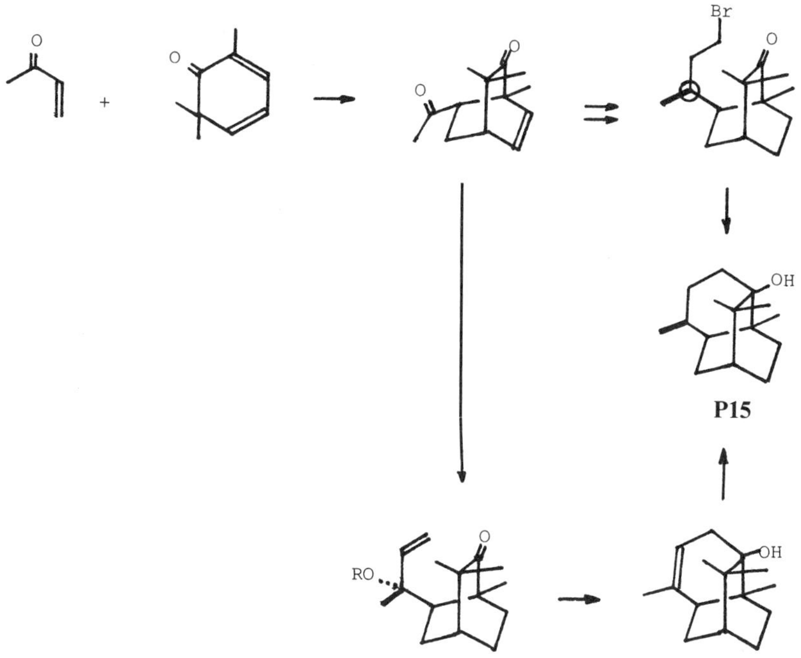

7.7 CYCLOHEXADIENES

In all the routes, the carbonyl function from the dienophiles is the locus at which an appropriate carbon chain is added as a prelude to completing the annulation process. Even the reaction sequence that highlights a radical cyclization to gain entry to the seychellene skeleton relies on the Diels–Alder method for acquiring the bicyclo[2.2.2]octane intermediate (Stork, 1985) for seychellene (**S29**).

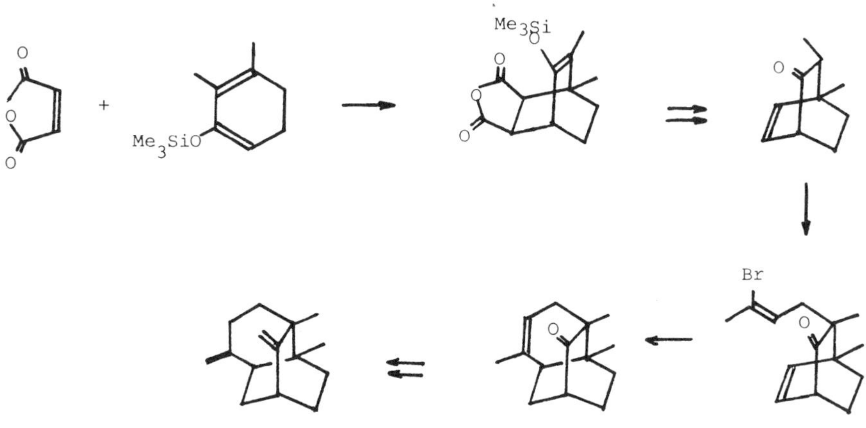

It is even more apparent that eremolactone (**E7**) is accessible via a Diels–Alder adduct. In practice, the regioselectivity of the cycloaddition can be greatly improved by placing an alkoxy group on the diene moiety. However, production of a diasteroisomeric mixture about the secondary methyl group must be accepted as an unavoidable shortcoming.

The desired diastereomer is formed in larger amounts (2:1 ratio). With this major isomer, the elaboration of the target follows a relatively uncomplicated path (Asaoka, 1983).

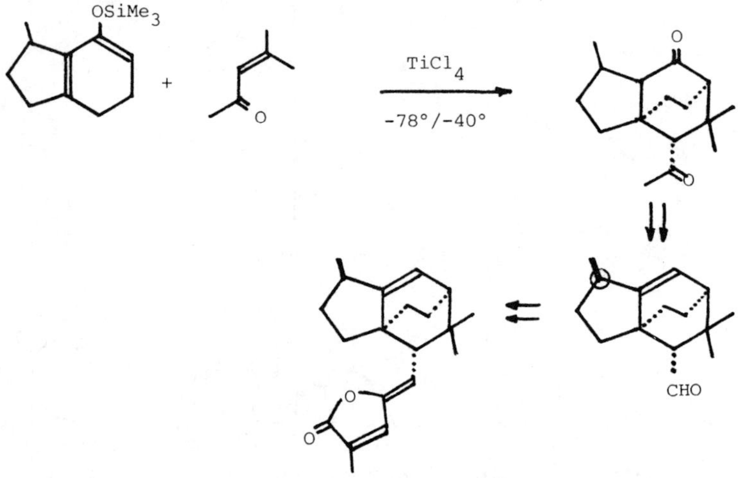

E7

The bridged ring system of antheridiogen-An (**A35**) is assembled in the last phase of its synthesis (Corey, 1985c) by a Diels–Alder reaction of a cyclohexadiene with nitroethene. The all-*cis* arrangement of the existing cyclic array ensures an exo approach of the dienophile.

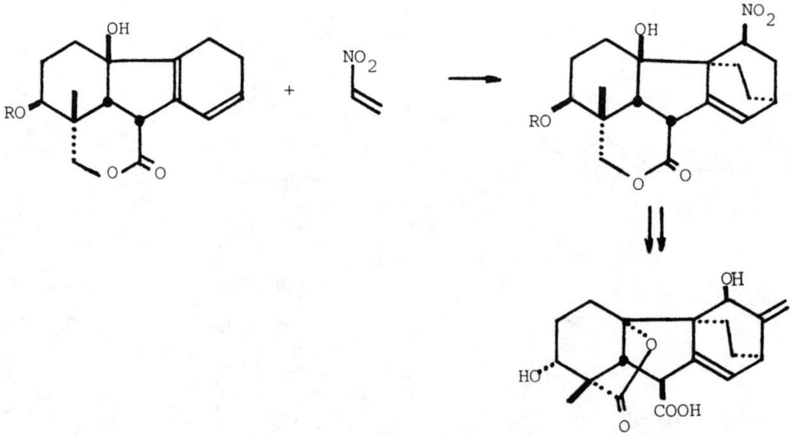

A35

Functionalized bicyclo[2.2.2]octanes are capable of undergoing specific rearrangements. Consequently, these substances offer unique opportunities for entry into uncommon carbon frameworks. An unobvious but charming synthesis of α-cedrene (**C37**) (K. Stevens, 1980) makes use of a photoinduced oxadi-π-methane rearrangement of a partially hydrogenated cycloadduct of 6,6-dimethylcyclohexa-2,4-dienone and dimethyl acetylenedicarboxylate. The semibullvalane derivative contains all the essential substituents with which the isolated methyl group and elements of the remaining six-membered ring may be introduced.

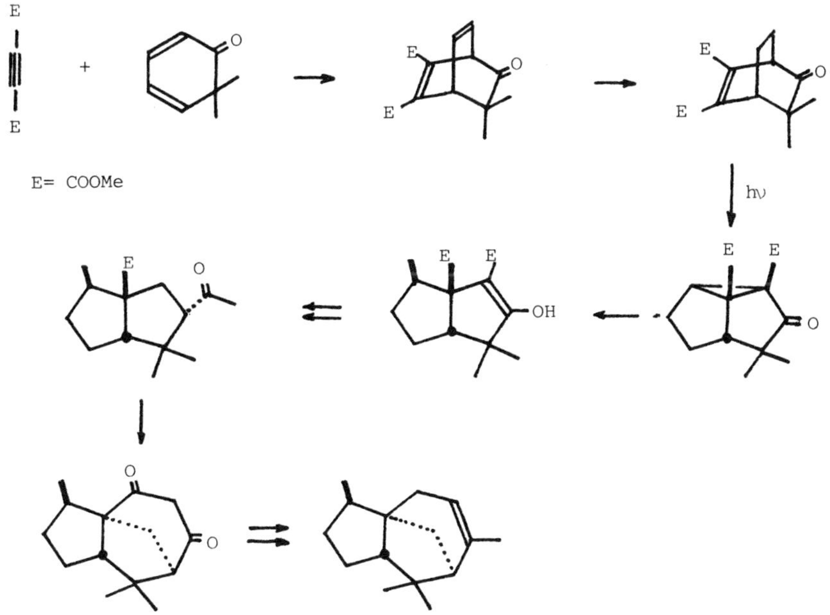

E = COOMe

C37

The oxadi-π-methane rearrangement is also induced during synthesis of modhephene (**M12**) (Mehta, 1985b), pinguisone (**P32**) (Uyehara, 1985a), and the potential precursors of hirsutene (**H17**) and oxosilphiperfolene (**S36**) (Demuth, 1985).

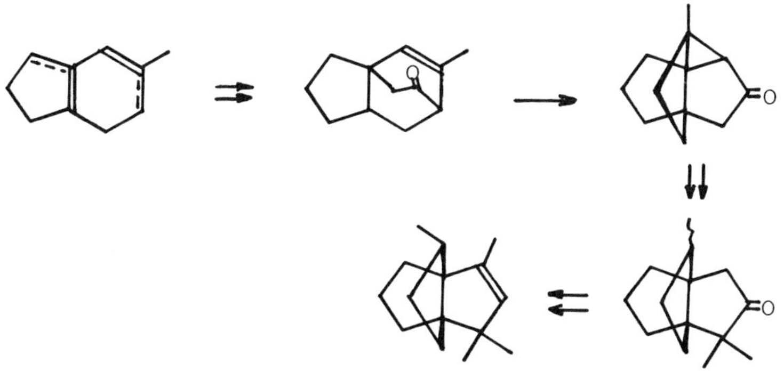

M12

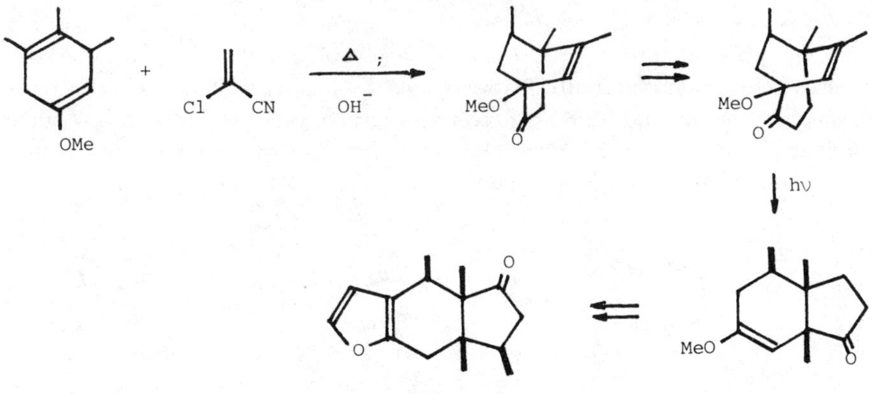

P32

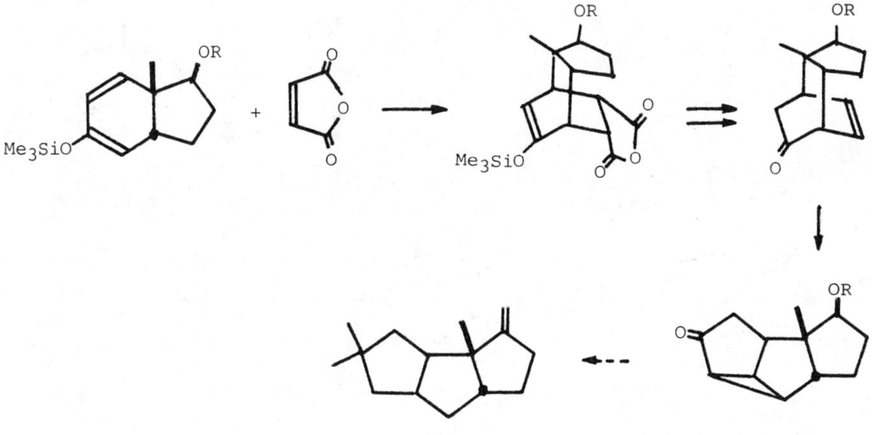

H17

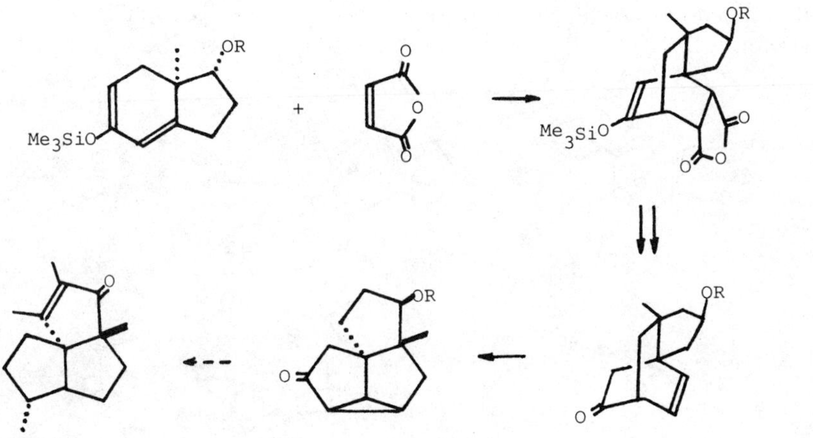

S36

7.7 CYCLOHEXADIENES

On the basis of the bicyclic rearrangement from a [2.2.2] to a [3.2.1] system, many natural products embodying the less symmetrical carbon framework may be secured indirectly from Diels–Alder adducts. The placement of an oxygen substituent at the bridgehead has a dramatic effect on the regioselectivity of the adduct formation as well as rate enhancement of the rearrangement step.

The carbocyclic nucleus of quadrone has been acquired (S. Monti, 1982) from a bicyclo[2.2.2]octenone via a pinacolone rearrangement and an intramolecular aldolization. Interestingly, the presence of a siloxy group at the γ position of the aldol is conducive to another pinacolone rearrangement that ultimately gives rise to the tricyclic diketone.

The synthesis of gymnomitrol (**G21**) and the barbatenes (**B3** and **B4**) (Kodama, 1979) also exploits the [2.2.2]-to-[3.2.1] skeletal transformation.

B3 (X=H, endo db)
B4 (X=H, exo db)
G21 (X=OH, exo db)

Alkoxydihydroaromatic substances obtained from the Birch reduction may be induced to react with dienophiles in the presence of a conjugation catalyst such as dichloromaleic anhydride. A short route to zizaene (**Z1**) (Pramod, 1982) has been developed based on the cycloaddition of a hydroindane and rearrangement of the properly modified adduct.

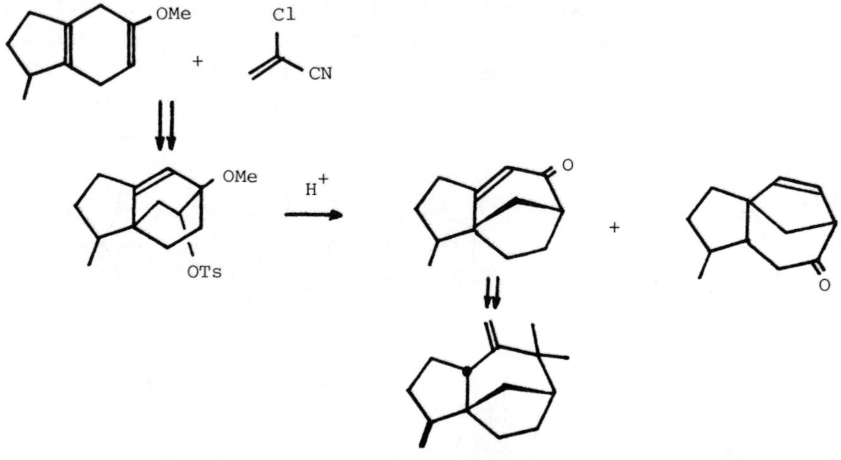

More complex molecules that have been assembled according to the cycloaddition–rearrangement principle include aphidicolin (**A36**) (van Tamelen, 1983) and maritimol (**M9**) (van Tamelen, 1981). Because of the inverse steric relationship of the CD segment of the two compounds, a nucleofugal group for triggering movement of a specific bond must be placed in two different bridges.

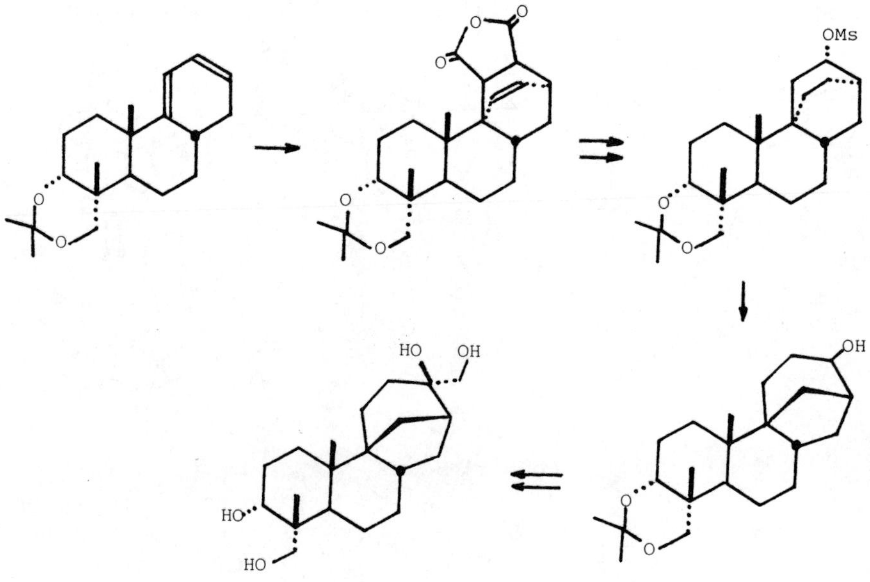

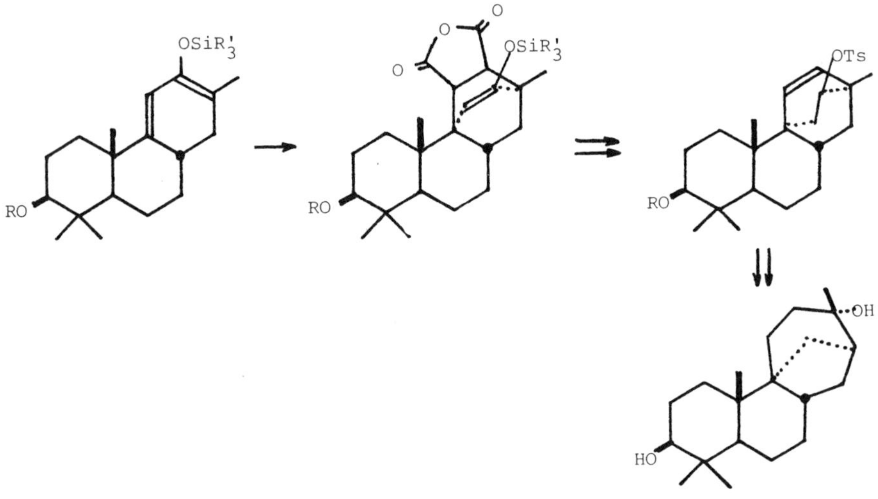

M9

In an improved synthesis of chasmanine (**C50**) (Wiesner, 1977) the construction of the BCD ring system also is predicated on the skeletal rearrangement step. A Diels–Alder reaction of a monoprotected *o*-quinone with a vinyl ether gives an adduct that is easily modified into a rearrangement-prone species.

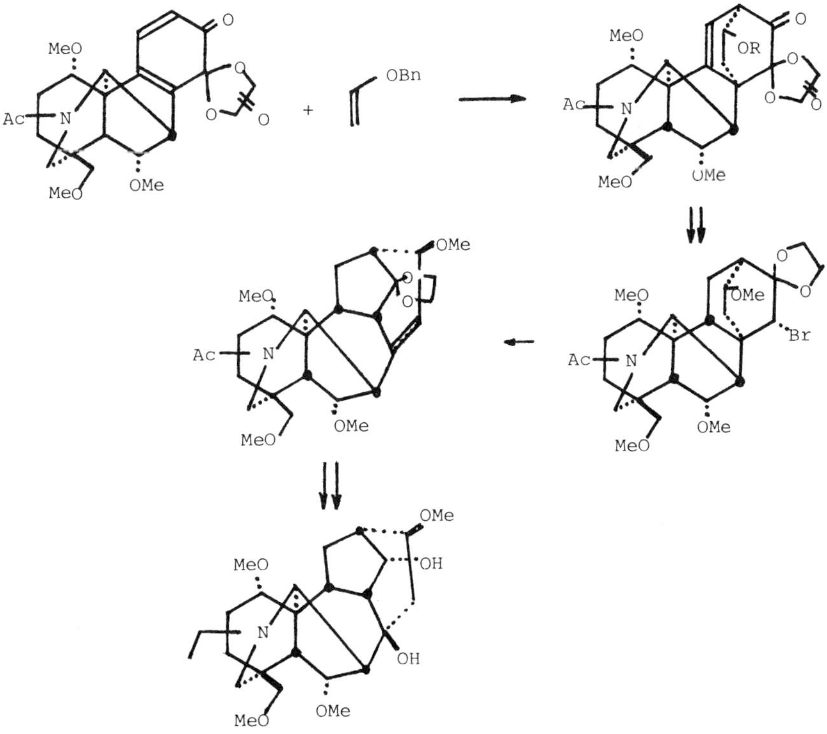

C50

The same type of condensation starts a synthesis of ryanodol (**R11**) (Belanger, 1979). The newly created cyclohexene ring must be contracted, whereas the preexistent cyclopentene is expanded (via a cleavage–aldolization operation) at the expense of the other six-membered ring. Together they provide the BC portion of the diterpene molecule. Two intramolecular aldol condensation processes furnish the elements of the lactol (initially) and then the A ring. Extensive oxidative adjustment of functional groups ensues.

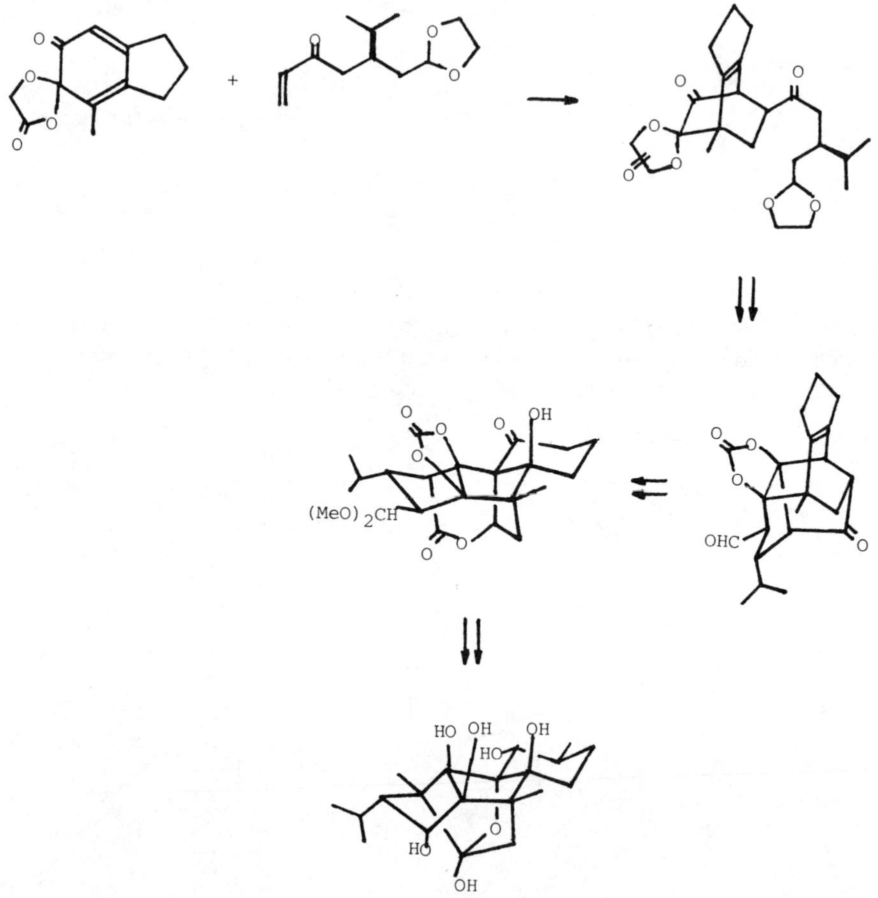

R11

An essentially complete skeleton of quassin (**Q3**), that is, an A-seco derivative, has been elaborated (Stojanac, 1979). By a Lewis acid-catalyzed Diels–Alder reaction, the dimethylbenzoquinone is destined to become the C ring, whereas a cyclohexadiene provides the B-ring elements, the two-carbon pendant for the lactone unit, and two anchoring chains for a yet-to-be-performed closure of the A ring.

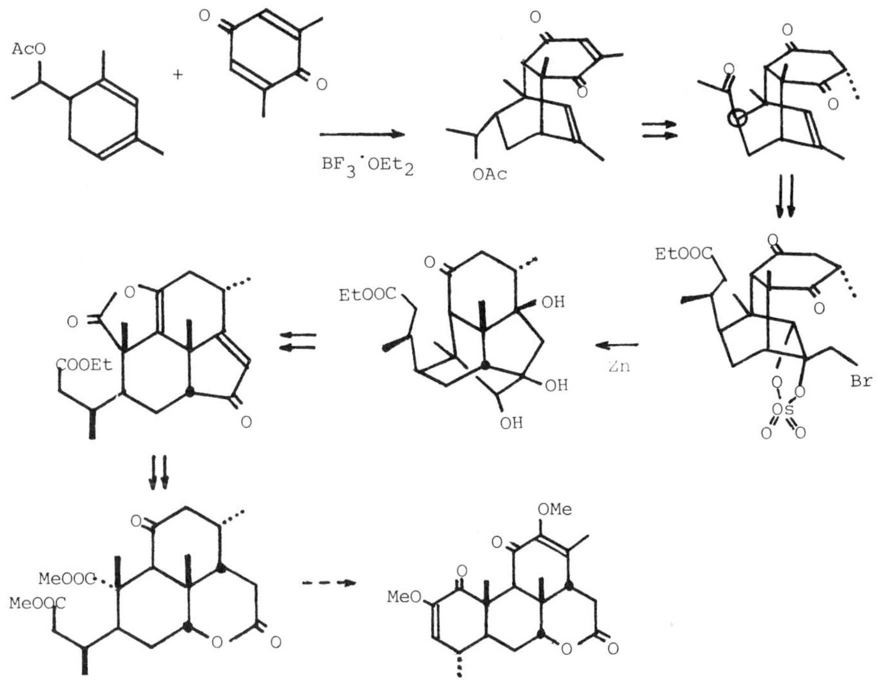

7.7 CYCLOHEXADIENES

Q3

The enantiomer of methyl trachylobanate (**ent-T23**) may be derived from methyl levopimarate (Herz, 1968) via a Diels–Alder reaction with a crotonic ester. The isopropyl substituent present in the original terpene is removed oxidatively after it has undergone double-bond isomerization, thereby introducing an oxygen function at the position to be involved in the cyclopropanation.

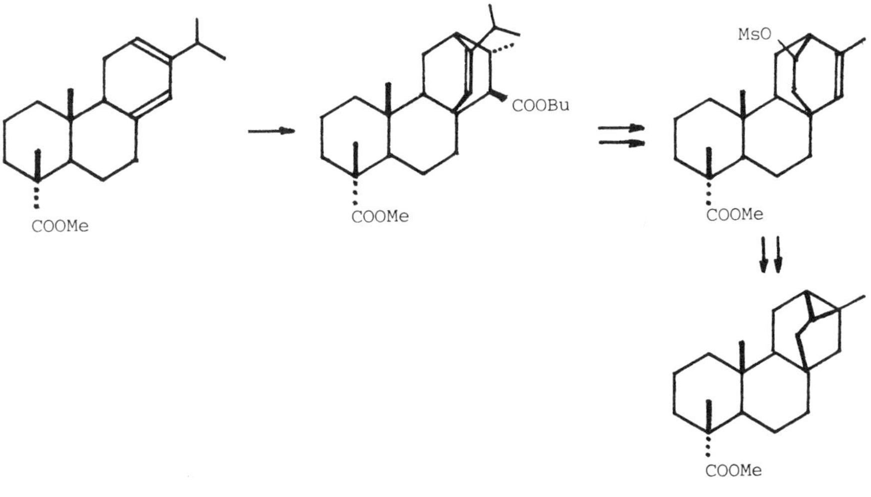

ent-T23

α-Amorphene (**A29**) is characterized by a *cis*-fused hexalin. A very attractive method for its acquisition (Gregson, 1973, 1976) involves an oxy-Cope rearrangement. The building block is formed from the reaction of 1-methyl-1,3-cyclohexadiene with α-chloroacrylonitrile.

(+ isomer)

A29

7.8 CYCLOHEXENONE SYNTHESIS

A powerful method for the elaboration of 4-substituted cyclohexenones is by way of the Diels–Alder adducts from dihydroanisoles, as first demonstrated in a synthesis of *erythro*-juvabione (**J8**) (A. J. Birch, 1969, 1970).

The dihydroanisoles obtained by Birch reduction are usually conjugated *in situ* by dichloromaleic anhydride and then condensed with the dienophiles. After adjustment of functional groups, cleavage of the C—C bond of the bicyclooctenes between the methoxylated bridgehead and the adjacent carbon is completed on acid treatment, and the cyclohexenones are recovered. The nonbridgehead carbon atom involved in the bond cleavage process is usually activated by an electron-withdrawing group originally present in the dienophile or a leaving group on an adjacent atom, making the deannulation very smooth.

The deannulation step may be accompanied by secondary reactions if the intermediates are specially modified. For example, in the synthesis of nootkatone (**N11**) (Dastur, 1973, 1974) and glutinosone (**G15**) (Murai, 1980) the monocyclic intermediates are trapped by a nearby double bond to furnish products with an octalone nucleus. On the other hand, in an approach to lubimin (**L27**) (Murai, 1982) the released carbon chains are destined to combine into the spirocycle.

A significant pressure-dependence of the syn/anti product ratio has been witnessed during the condensation of methyl acrylate with (2,6-dimethyl-4-methoxy-1,3-cyclohexadien-1-yl)acetaldehyde ethyleneacetal. At atmospheric

7.8 CYCLOHEXENONE SYNTHESIS

pressure (190°C), almost equal amounts of syn and anti adducts are formed. At 15,000 atm (20°C), the syn isomer is favored by a factor of 3.2. The high-pressure reaction affords only the endo compounds, whereas the atmospheric reaction is totally nonselective (Murai, 1981).

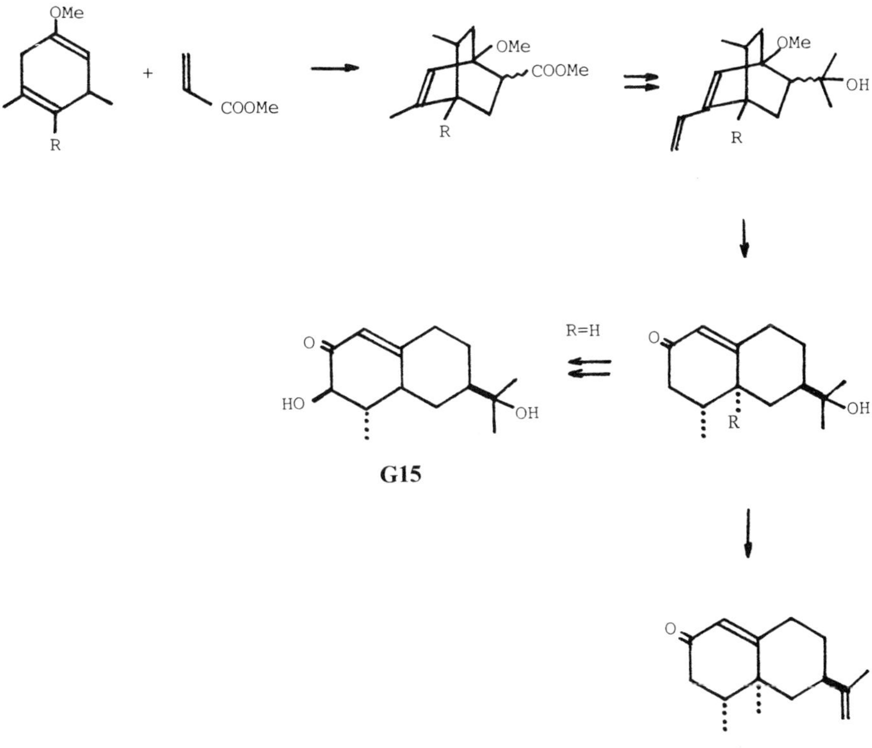

G15

N11

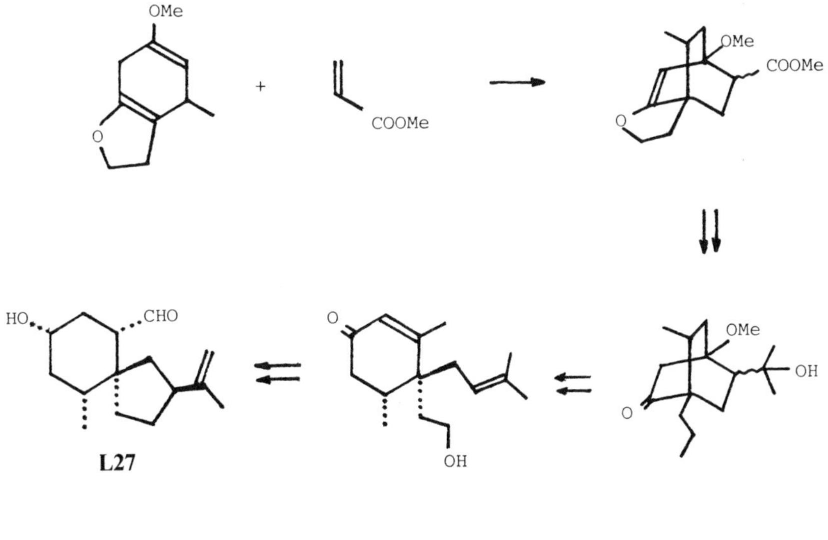

L27

For the synthesis of solavetivone (**S41**), the use of the corresponding acetonitrile and 3-buten-2-one fares better because the anti adduct is now predominant (Murai, 1981).

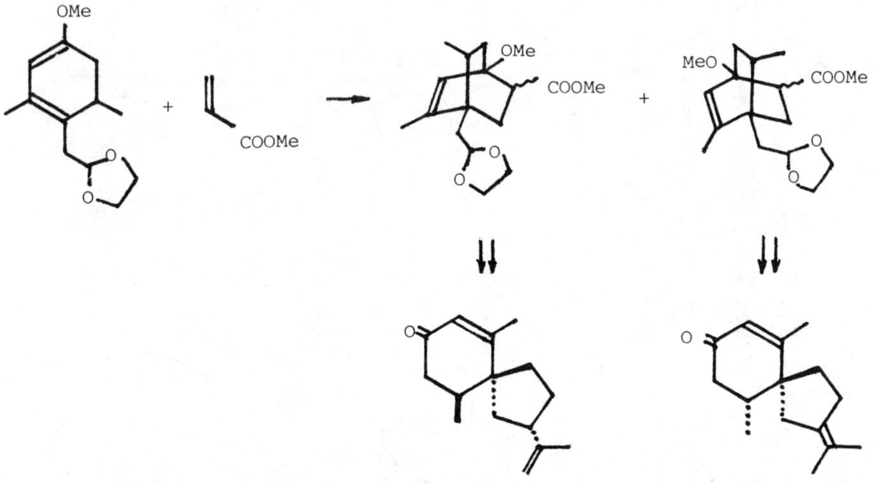

During a synthesis of norpatchoulenol (**P13**) (Niwa, 1984), the bicyclo[2.2.2]octane skeleton is made, degraded, and then re-formed in a subsequent step by a Michael–aldol tandem. The initial Diels–Alder process serves to link all the necessary skeletal carbon atoms and functional groups for eventual transformation into the terpene molecule.

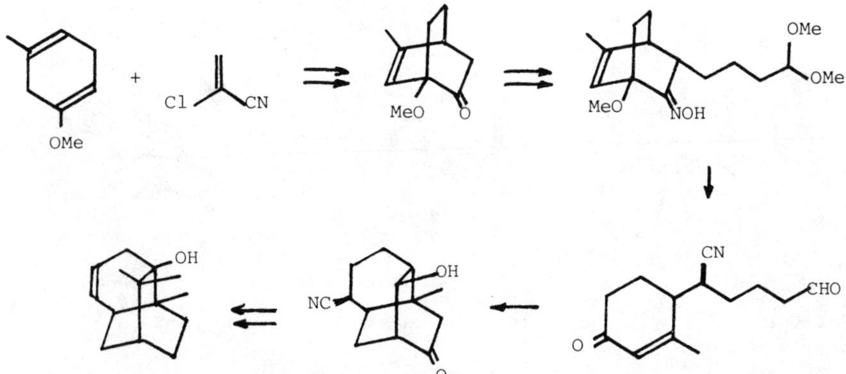

The fragmentative route to cyclohexenones is the preferred method for the preparation of large quantities of a fused γ-butyrolactone unit from which trichodermin (**T24**) is built (Colvin, 1971, 1973).

T24

A later synthesis of the same molecule (Still, 1980), which does not rely on subsequent closure of the five-membered ring, also embodies the cycloaddition and recovery of a cyclohexenone derivative. The use of benzoquinone as the dienophile is a good choice because it can be regioselectively contracted to a cyclopentane ring while retaining (the capacity for introduction of) two oxygen functional groups in exo configurations. These functions are chemically differentiated: one acts as a fragmentation inducer, the other guides the stereochemical course of epoxidation after the unfolding of the cyclopentenol intermediate. Herz ring contraction also supplies the tertiary methyl group in the form of an ester.

T24

7.9 FURANS AS DIENES

Generally, furans participate in the Diels–Alder reaction only with the extremely active dienophiles. Their utility is further diminished by the facile decomposition of the adducts into the original components.

Cantharidin (**C18**), the notorious vesicant from "Spanish fly" (a cantharides beetle), prompts a direct synthetic scheme centered on the Diels–Alder reaction. However, furan fails to condense with dimethylmaleic anhydride. Only recently, this direct approach succeeded in a modified version wherein the two methyl groups of the dienophile are connected to a common sulfur atom, and the reaction carried out under high pressure (Dauben, 1980).

Two earlier cantharidin syntheses also employ the Diels–Alder strategy. These routes are necessarily more protracted. Thus, in one of these (Stork, 1951, 1953) the furan/dimethyl acetylenedicarboxylate adduct is partially hydrogenated and condensed with butadiene. The two ester groups are then converted into the methyl substituents and the cyclohexene moiety is degraded to give the succinic anhydride unit.

In one of the triptonide (**T29**) syntheses (Garver, 1982), the designated C ring is annulated by means of a Diels–Alder reaction of a furan derivative. This step produces a salicylic ester.

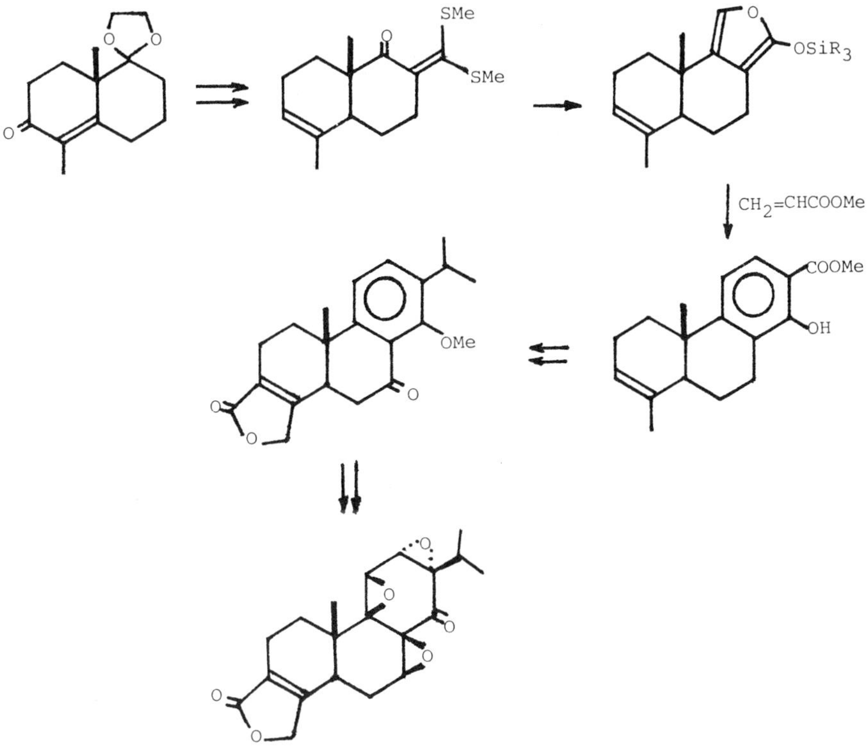

7.9 FURANS AS DIENES

T29

Another application of the high-pressure technique to Diels–Alder condensation involving furans is illustrated in a synthesis of (+)-jatropholone A and B (**J1** and **J2**) (Smith, 1985). The adduct is transformed into a phenol on treatment with acid.

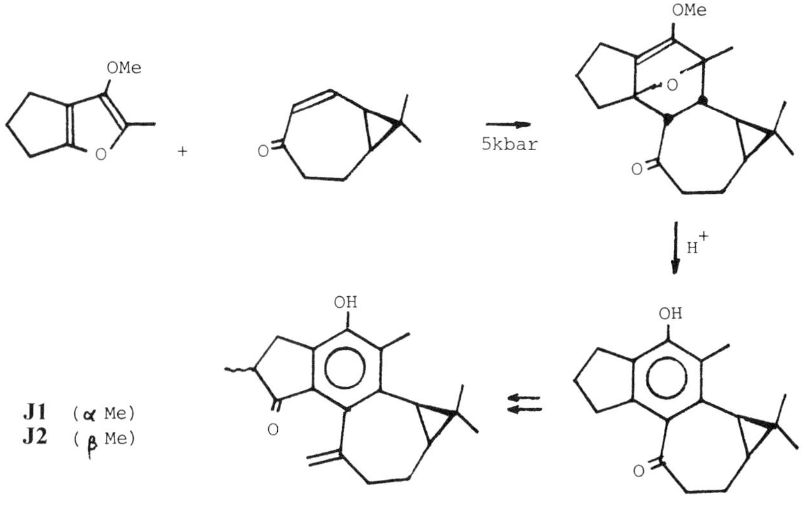

J1 (α Me)
J2 (β Me)

7.10 OTHER APPLICATIONS OF DIELS–ALDER ADDUCTS

The stereocontrol endowed by bridged ring systems has been exploited in establishing the two contiguous quaternary carbon atoms in trichodiene (**T26**). The synthesis (Snowden, 1984) also depends on a thermal fragmentation of homoallylic alkoxides.

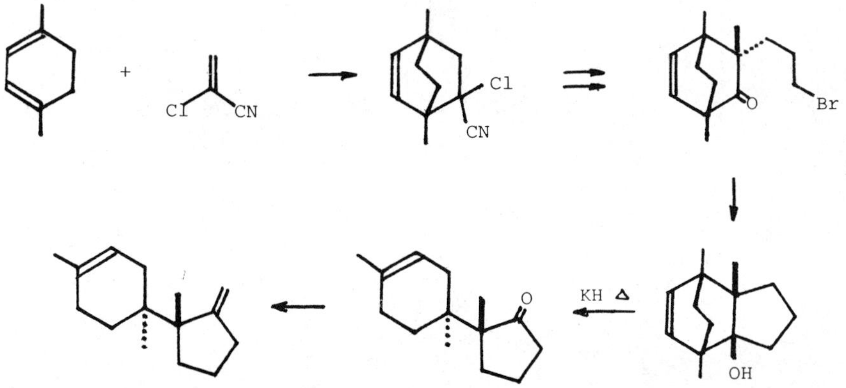

Benzoquinone virtually always acts as a dienophile. However, its latent potential as a diene is manifested in the reaction of hydroquinone with maleic anhydride. Bisdecarboxylation of the diacid leads to bicyclo[2.2.2]oct-7-ene-2,5-dione, which serves as a starting material for coriolin (**C70**) (Demuth, 1984). An oxa-di-π-methane rearrangement is a crucial step for converting the bicyclic system into the linear skeleton of the terpene.

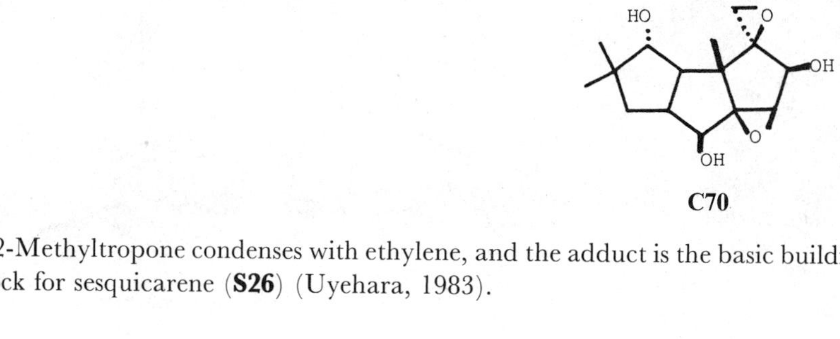

2-Methyltropone condenses with ethylene, and the adduct is the basic building block for sesquicarene (**S26**) (Uyehara, 1983).

7.11 INTRAMOLECULAR DIELS–ALDER REACTIONS

7.11.1 Condensed Rings

The most exciting recent development in synthesis is the extensive exploitation of the intramolecular Diels–Alder reaction. The benefits of the process are enormous: it effects closure of two rings in one step from an acyclic triene, and it offers heightened reactivity of the substrates (in comparison with the intermolecular reaction components) as a consequence of entropic activation. Often the geometric constraints imposed on such systems cause the adoption of unusual transition states, and normally inaccessible stereoisomers may be produced.

The intramolecular Diels–Alder reaction represents the most convenient method by which α- and β-himachalenes (**H13** and **H14**) are assembled (Wenkert, 1973b; Oppolzer, 1981d). The cis ring juncture required for α-himachalene and the double bond in the six-membered ring are correctly established in this operation.

In the synthesis of torreyol (**T20**) (Taber, 1979) (see also γ-muurolene, (**M13**), and sclerosporal, **S16**: Katayama, 1983) a boat transition state is selected by the molecule to avoid severe nonbonded interactions. The predominant product (9 : 1) is a *cis*-fused octalone in which the isopropyl chain is cis to the angular hydrogen atoms.

T20 **M13** **S16**

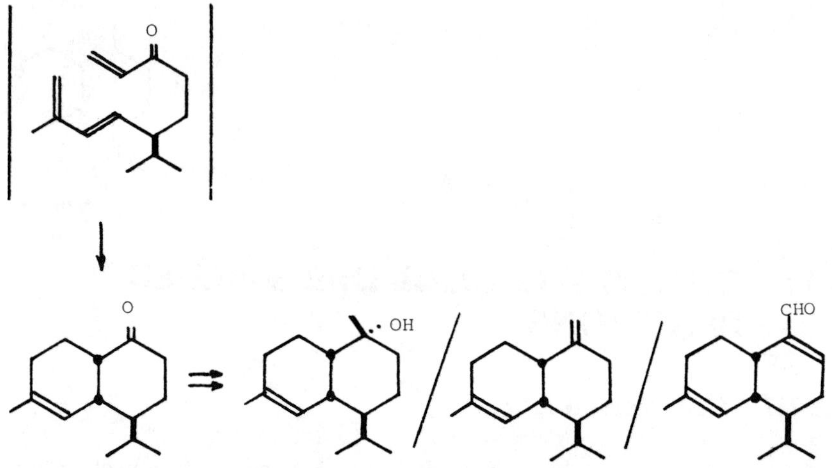

Clarification of the absolute configuration of sclerosporin awaited its synthesis (T. Kitahara, 1984) from (−)-carvone. The (+)-isomer shows strong sporogenic activity and therefore corresponds with the natural product.

sclerosporin

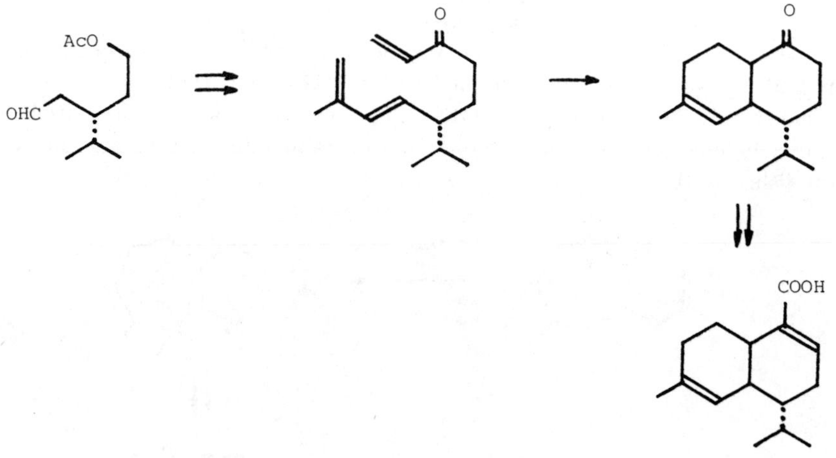

(+)-7,20-Diisocyanoadociane has a very unusual perhydropyrene skeleton for diterpenes. Two intramolecular Diels–Alder reactions are involved in a synthesis (Corey, 1987c) which also determines its absolute configuration.

7.11 INTRAMOLECULAR DIELS–ALDER REACTIONS

The *threo*-selective Michael reaction of menthyl ester enolates to methyl crotonate establishes two chiral centers which induce two more to be created during the first cycloaddition step. Presently, the second Diels–Alder reaction is nonstereoselective, giving the desired product in 54% yield along with a 36% yield of the diastereomer.

One of the angular carbon is incorrectly configured. However, when the functionalized pendent of the tetracyclic adduct is degraded to a ketone, correction is readily achieved. The all trans stereochemistry is the most stable one.

7,20-diisocyanoadociane

The sensitivity of the intramolecular Diels–Alder reaction to nonbonded interactions is further revealed in a synthesis of selina-3,7(11)-diene (**S18**) (Wilson, 1978). The cisoid (endo) transition state is strongly destabilized by the presence of the methyl group on the diene moiety, which is in close proximity to the substituent (OR or H) at C_7. The ratio of *trans*- to *cis*-octalols produced by this reaction is 94:6.

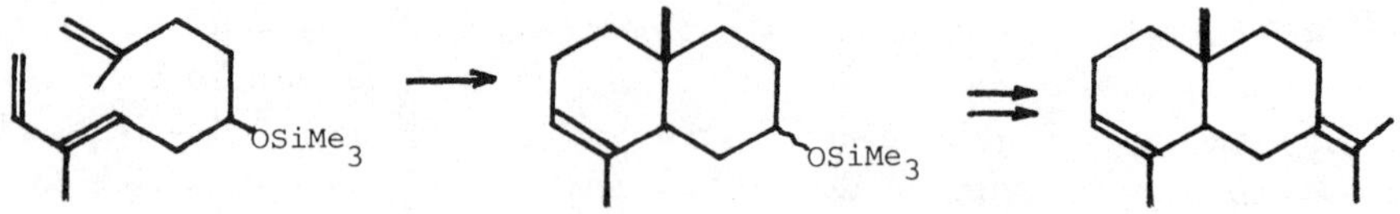

Similarly, α-eudesmol (**E17**) is obtained as the major product (82% of four isomers) directly from an intramolecular Diels–Alder reaction (Taber, 1982). Thus one asymmetric center induces the relative stereochemistry of two others at the ring juncture as they emerge.

The acyclic precursor of epizonarene (**Z5**) (Wilson, 1980) also undergoes cycloaddition via an exo transition state. The target molecule does not need the full complement of the three asymmetric centers generated from the process, but the two that remain till the end are in the correct configurations. It should also be noted that only the 3*E*-isomer undergoes cycloaddition; the *Z*-isomer prefers a sigmatropic 1,5-hydrogen shift.

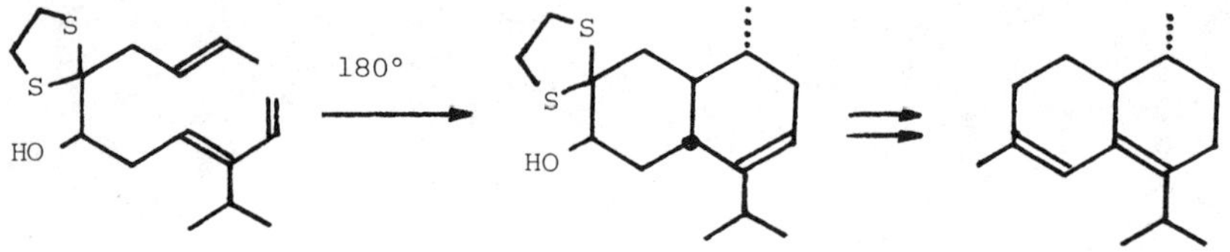

Ionic Diels–Alder reactions provide excellent stereocontrol and an apparent precursor of α-cadinene and β-cadinene (**C2**) (Gassman, 1986).

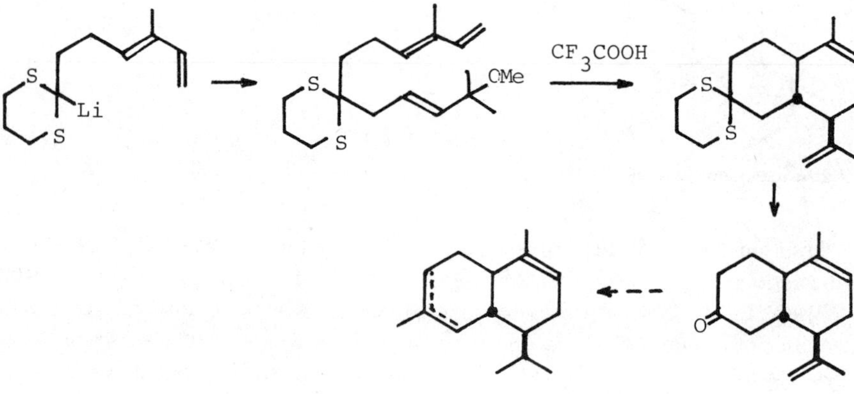

7.11 INTRAMOLECULAR DIELS–ALDER REACTIONS

In a synthesis of biflora-4,10(19),15-triene (**B8**) (Grieco, 1986) an intramolecular Diels–Alder process is central to the construction of the cyclohexane ring to which the longer alkyl chain is attached. The major adduct contains five contingous asymmetric centers, three of which to be retained while a fourth one is to be epimerized. The two carbonyl pendants can be differentiated and fashioned accordingly into a 1,7-dialdehyde which undergoes aldolization to complete the skeletal assembly.

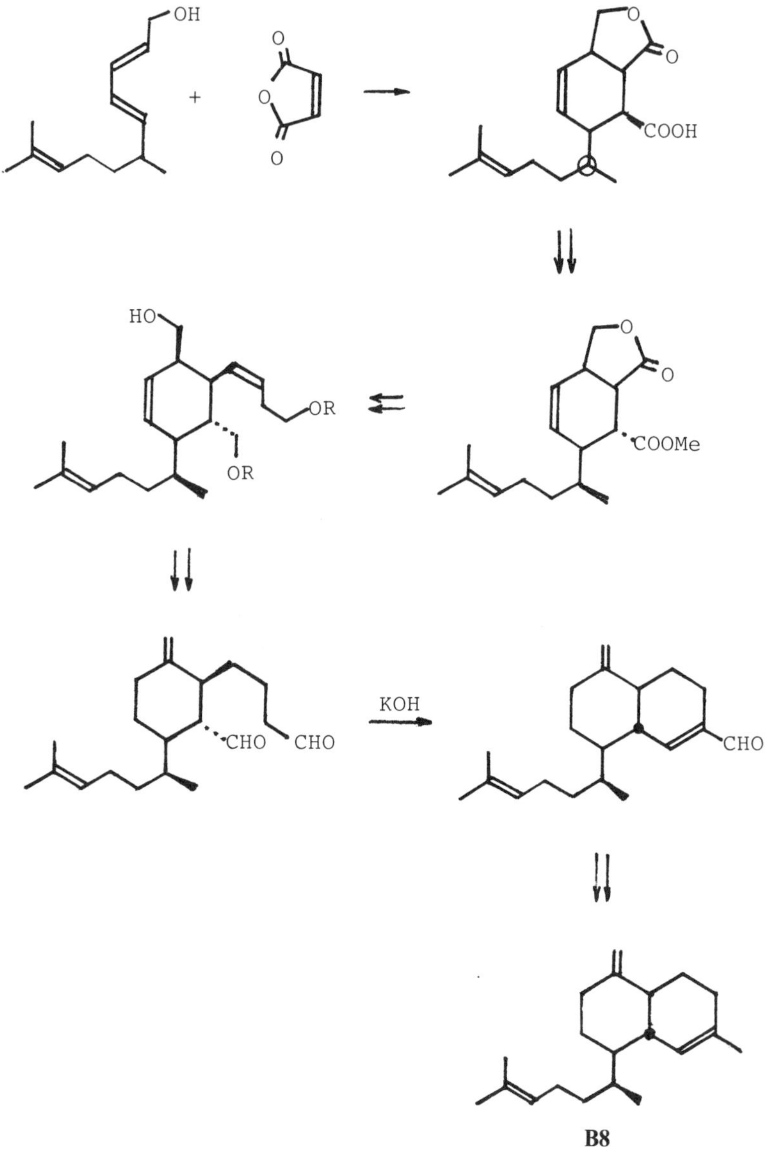

An alternative and more conventional Diels–Alder approach identifies the norketone as precursor (Parker, 1986). Although previous observations indicate a *cis*-fused octalone to be the predominant product, the opportunity for epimerization trivializes the stereochemical problem.

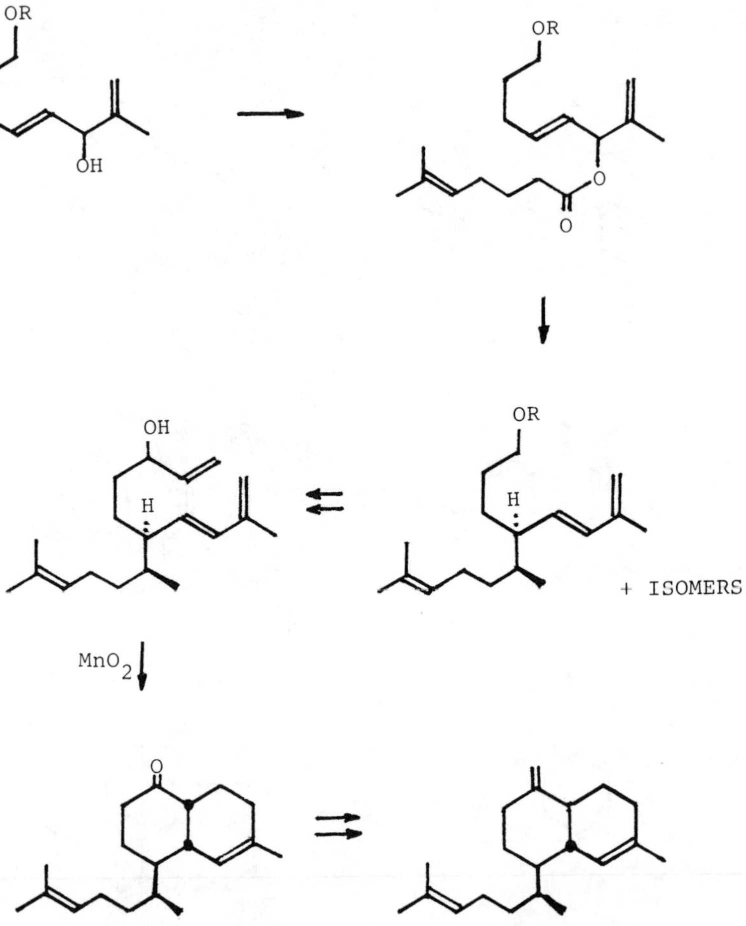

Intramolecular [4 + 2]cycloaddition with full stereocontrol is perhaps never more impressively demonstrated in the sesquiterpene arena than in the cycloaddition leading to eremophilone (**E13**) (Näf, 1979). Although the primary *trans*-fused product is unstable due to a 1,3-diaxial interaction between the CH$_3$ and COOEt groups, its partial epimerization causes no harm because this particular chiral center eventually is destroyed by trigonalization.

7.11 INTRAMOLECULAR DIELS–ALDER REACTIONS

The use of abundant and optically active natural substances as synthons is a fashionable pursuit in current synthesis. The conversion of (+)-glucose to (+)-tuberiferine (**T34**) (Georges, 1985b) is but one of the more recent successes. This synthesis proceeds through the diacetone glucose, and from the latter compound two carbon chains, each containing a diene and an olefin linkage, respectively, are constructed, with the goal of inducing their participation in an intramolecular Diels–Alder reaction. The two chiral centers directly inherited from glucose govern the three additional ones that emerge during the cycloaddition and methylation steps.

An organization of the norterpene fichtelite (**F5**) by an intramolecular Diels–Alder reaction (Taber, 1980) has a stereochemical outcome governed by an endo transition state. The existing asymmetric centers favor the formation of a trans-anti-trans skeleton by selecting the conformation of the incipient rings.

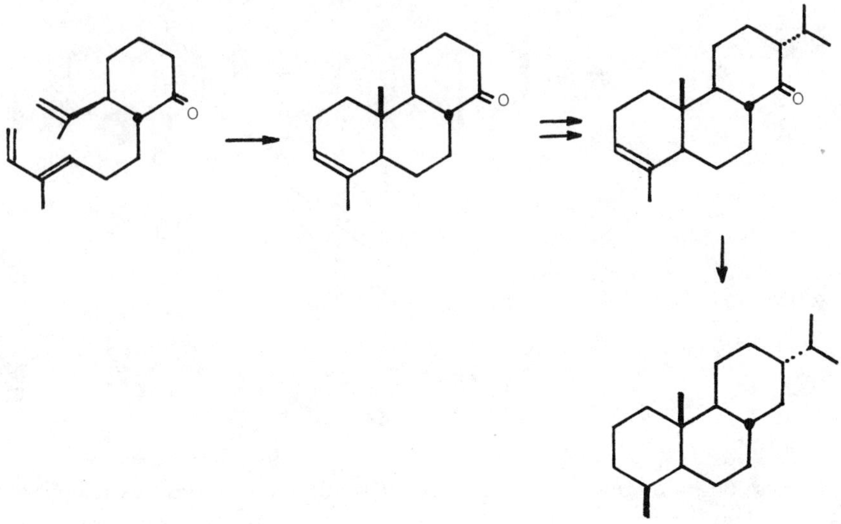

Trans-fused products still prevail from reaction of trienes separated by more or less than four atoms. Thus, a bicyclic synthon for retigeranic acid (**R1**) has been assembled from the appropriate triene (Attah-Poku, 1985).

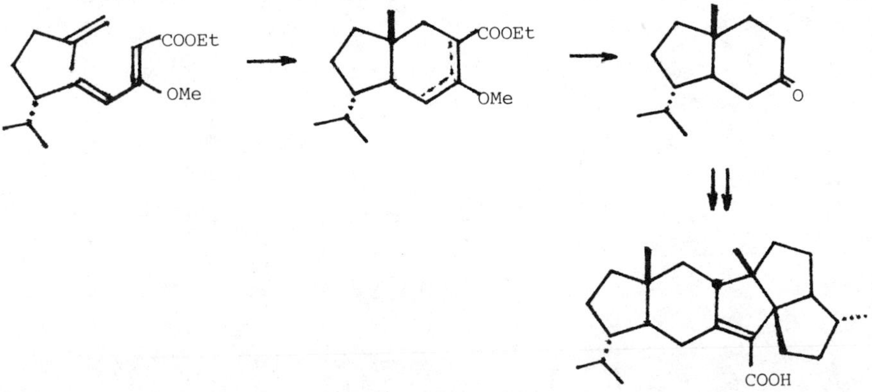

The basic strategy for synthesis of phorbol (Wender, 1987c) as demonstrated in the acquisition of an analog embodies an internal Diels–Alder reaction of a 2-ethylidenetetrahydropyran-3-one. The adduct contains a double bond for annulation of the three-membered ring, an oxa bridge to rigidify the B-ring to enable stereoselective transformations and to serve as latent hydroxyl at C_9.

In the model study the A-ring is assembled by an aldol condensation and the attachment of the D-ring calls for reaction with a dibromcarbenoid species (In the synthesis of phorbol, different reaction sequences are necessary).

7.11 INTRAMOLECULAR DIELS–ALDER REACTIONS

The versatile design of this approach is aimed at maximizing the utility of the BC-synthon so that diterpenes of the related tiglane, ingenane and daphnane skeletons may be elaborated.

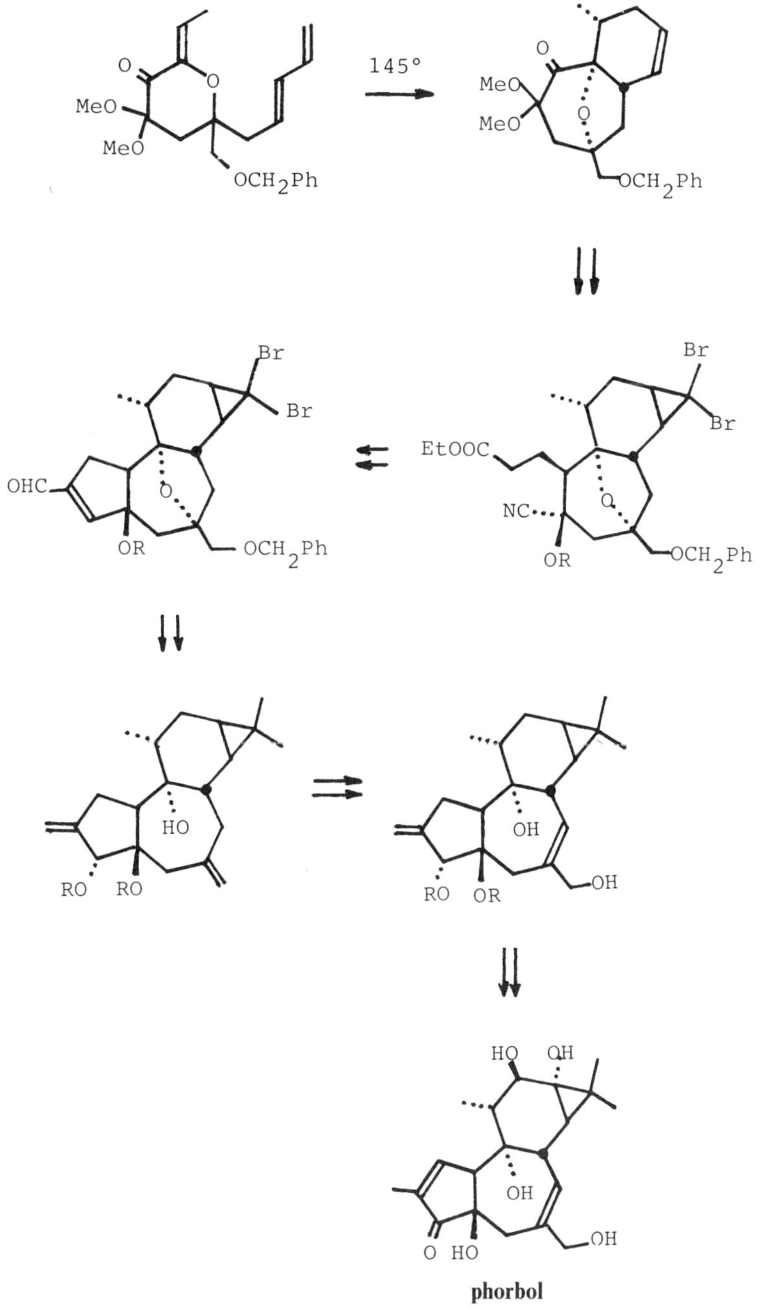

phorbol

Ircinianin is apparently derived in nature from a furanosesterterpene by an intramolecular Diels–Alder reaction. Duplication of this biogenetic step has been realized in high yield in refluxing benzene (Takeda, 1986). Four chiral centers are created in the process.

ircinianin

A model study of taxane synthesis is based on an intramolecular Diels–Alder process to unite the **BC**-ring system in one step (K. Sakan, 1983). Interestingly, in the presence of dimethylaluminum chloride, the predominant product has a cis juncture, whereas uncatalyzed reaction gives the desired *trans* adduct. It should be possible to find a suitable precursor from which the extra bridged carbocycle is liable to undergo cleavage so that naturally occurring taxanes such as taxusin (**T6**) may be elaborated following this scheme.

T6

A valuable bicyclic synthon for resin acids of the abietic type is available from a route highlighting an intramolecular Diels–Alder cycloaddition and a stereoselective axial methylation at C_4 (Burke, 1983). The corresponding olefinic ester (methyl group at the α carbon) undergoes cycloaddition to give the methylated decalonecarboxylate together with the *cis*-fused and C_4-epimeric decalone in ratio of about 2 : 1.

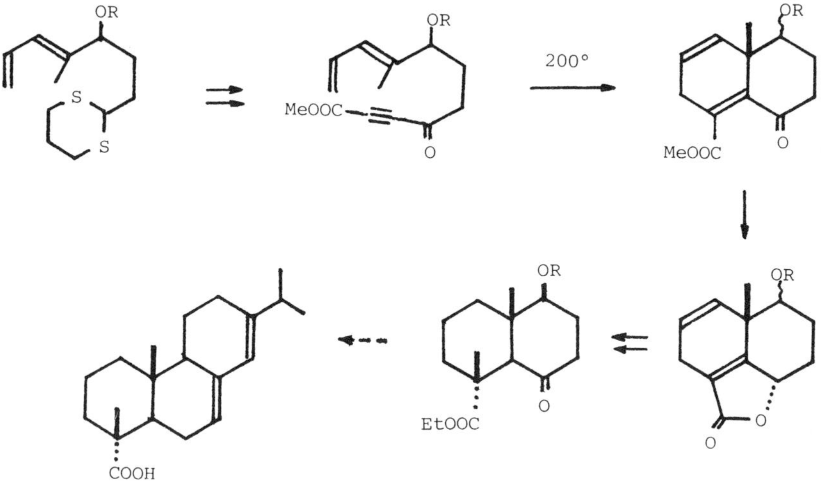

The mild conditions for effecting the formation of a hydrindene ester en route to valerenal (**V3**) (Bohlmann, 1980a) are noteworthy. The configuration of the ester group in the heteronuclear diene determines the conformation of the six-membered ring and hence the direction of the partial hydrogenation that establishes the secondary methyl substituent.

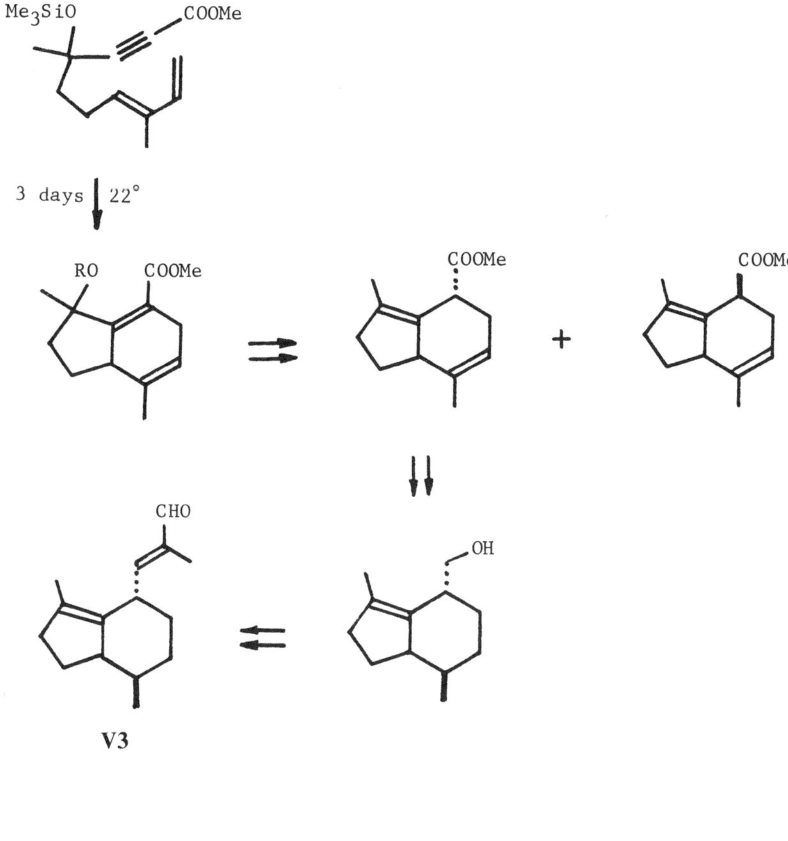

An apparently useful intermediate for forskolin (**F9**) has been obtained via an intramolecular Diels–Alder reaction (Ziegler, 1985a; 1987; cf. Nicolaou, 1985). This process establishes two critical stereocenters at C_1 and C_{10} in addition to providing the complete carbocyclic framework of the terpene.

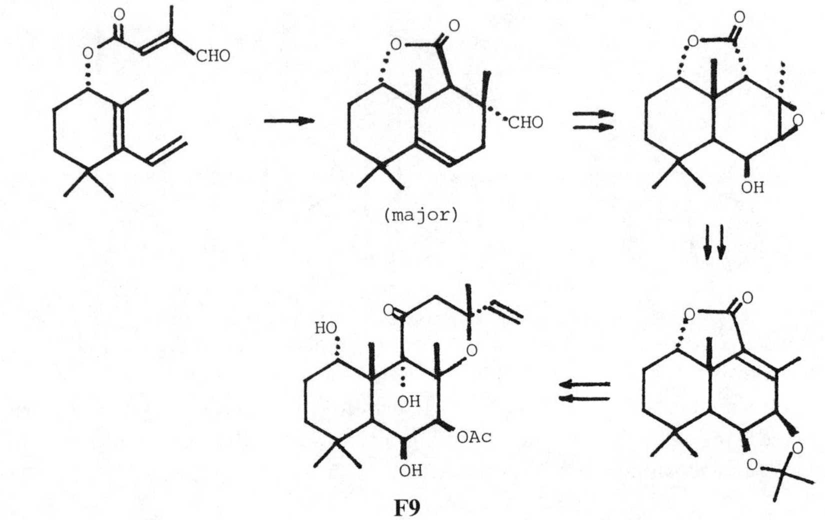

Propargyl ethers undergo equilibration with allenyl ethers on exposure to a base (e.g., tBuOK) at slightly elevated temperatures. These allenyl ethers are more reactive than propargyl ethers in cycloaddition, and in the presence of properly juxtaposed conjugate diene units, intramolecular Diels–Alder reaction of the two systems could occur. This protocol has been put to good use in a synthesis of platyphyllide (K. Hayakawa, 1986).

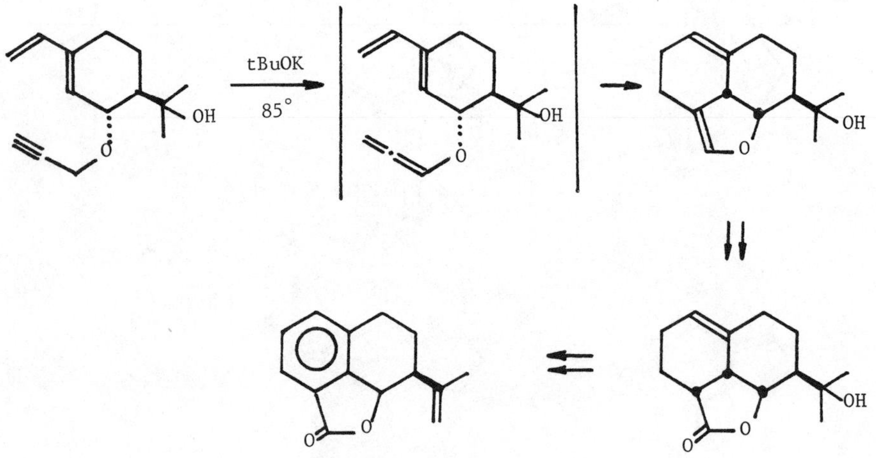

7.11 INTRAMOLECULAR DIELS–ALDER REACTIONS

Conjugated allenes may act as dienes. Naturally, their Diels–Alder adducts contain a conjugated diene moiety. The ready availability of 3-siloxy-1,2,4-alkatrienes from a reaction of α-chloroalkenyl silyl ketones with alkenyllithium reagents opens a new path to α-alkylidenecyclohexanones via their cycloadditions. Eremophilane and eudesmane sesquiterpenes bearing such a structural segment are easily obtained by this route.

The Diels–Alder reaction of such substrates has a drawback with respect to poor stereoselectivity. However, for the synthesis of dehydrofukinone (**F16**) and selina-4(14),7(11)-dien-8-one (**S19**) (Reich, 1984), the problem disappears upon trigonalization and equilibration of the ring juncture, respectively.

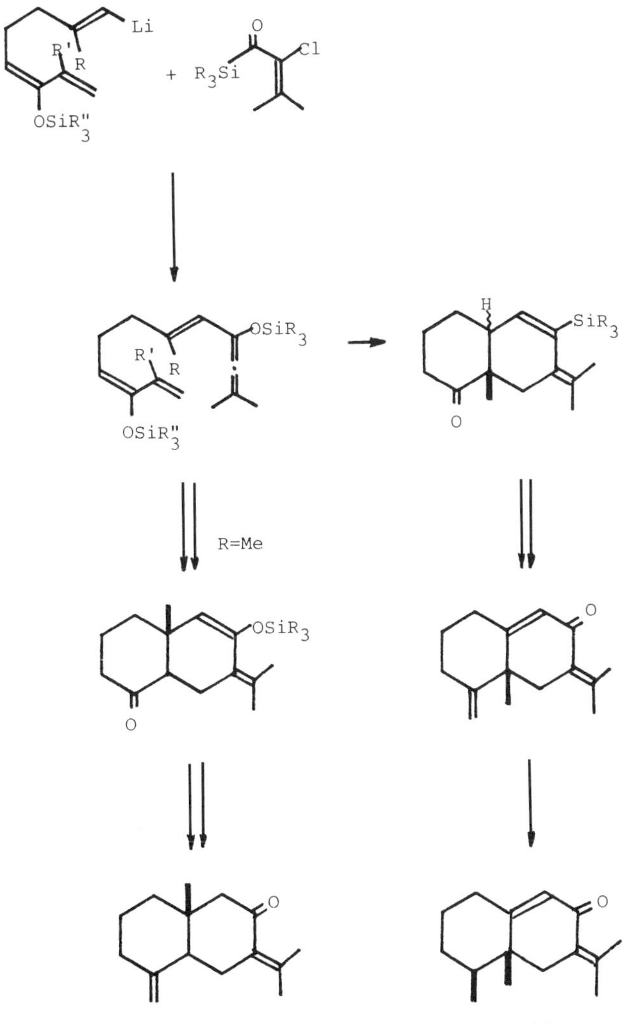

Among the reported syntheses of dendrobine (**D11**), two of which use the Diels–Alder process. Both inter- and intramolecular versions are represented. The intramolecular cycloaddition approach (Roush, 1978) involves the preparation of a well-defined triene ester precursor and depends on the adoption of an endo transition state. Indeed the endo products (hydroxy epimers) predominate (39% + 36%), and their separation from the minor isomers allows attainment of the goal. Accordingly, oxidation of the hydroxyl group, angular methylation, and introduction of an aminomethyl pendant are called for, and the double bond generated during the cycloaddition is functionalized (epoxidized) to create the opportunity for heterocyclization on two occasions.

D11

A very rewarding exploitation of the intramolecular Diels–Alder reaction is witnessed in two different routes to marasmic acid (**M6**) (Greenlee, 1976; Boeckman, 1980). The one starting from alcoholysis of bromomethylmaleic anhydride with the dienol already containing a five-membered ring is extremely facile, and the endo adducts are generated at room temperature (Greenlee, 1976). On the other hand, the process in which two carbocycles are formed simultaneously requires more vigorous conditions (200°C, toluene), and both exo and endo transition states are adopted (Boeckman, 1980). The vast difference in reactivity of the two trienes may be due to the normal polarization of addends in the former compound, whereas both the diene and the dienophile contain electron-withdrawing groups in the latter.

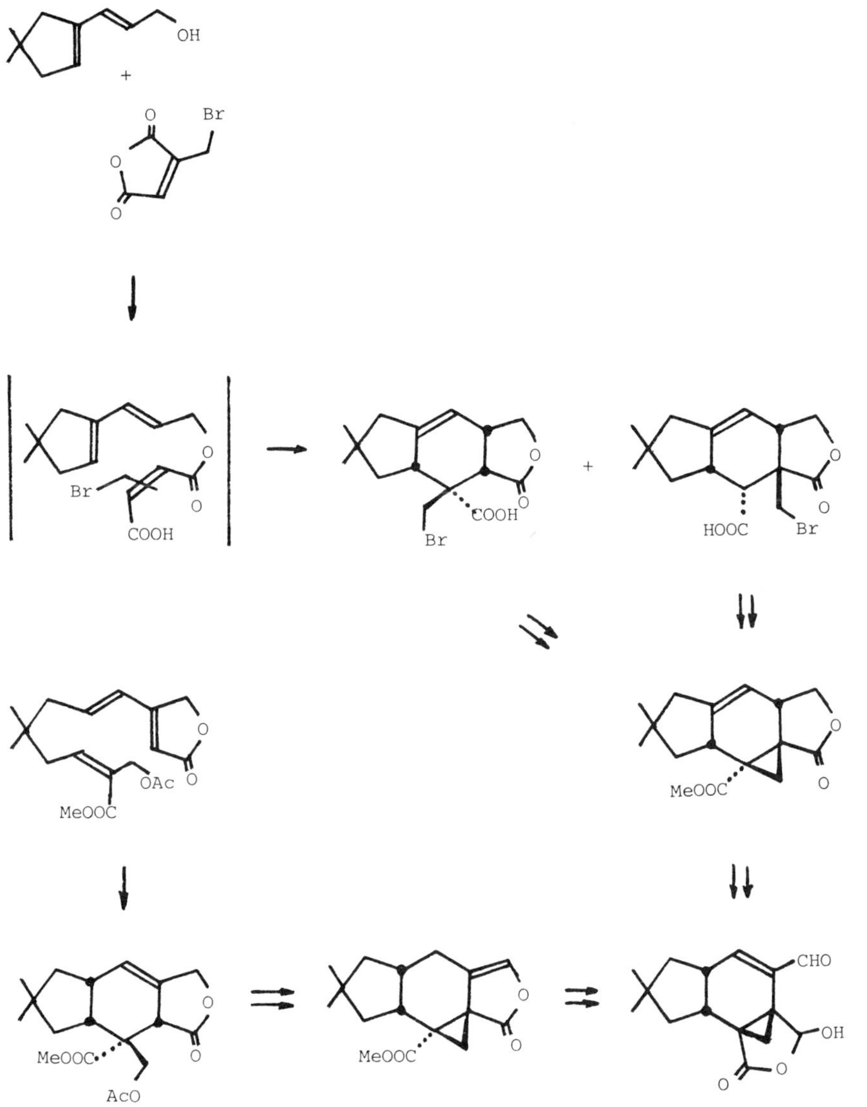

7.11 INTRAMOLECULAR DIELS–ALDER REACTIONS

A penetrating cerebration on the synthesis of gibberellic acid (**G12**) (Corey, 1978a,b) has focused the A-ring construction by an intramolecular Diels–Alder reaction. The lactone carbonyl present in a *cis*-fused 6:6 ring portion of the adduct permits stereoselective introduction of the last methyl group at C_4 by means of alkylation. With the tertiary carboxyl function in place, the A ring is ready to be fashioned into the allylic alcohol unit and to anchor the γ-lactone. The carboxylic acid at C_6 already existed in latent form in the epimeric configuration, which is inherent in its genealogy from the hydroxymethyl group of 2,4-pentadien-1-ol. Since the β-configuration of this carboxyl substituent is known to be the more stable one, the incorrectly oriented α-carboxyl group in the precursor has no lasting consequences.

The components for the B, C, and D rings are assembled by a Diels–Alder reaction involving a substituted benzoquinone. The quinone oxygen atoms are removed, leaving a cyclohexanone, which can then be coupled to the acetaldehyde chain extending from the ring juncture with a low-valent titanium species.

G12

Intramolecular Diels–Alder reaction is suitable for synthesis of 3-oxosilphinene (Ihara, 1987b) because contraction of the cyclohexene furnishes a handle for installation of the *gem*-dimethyl group. The carbonyl function of the terpene molecule can be incorporated in the educt as an activator for the cycloaddition. Activation of the diene unit such as by a thio substituent is also beneficial.

The Diels–Alder reaction proceeds with one of the exo transition states which minimizes nonbonding interactions. This transition state leads to a tricyclic ketone containing all the chiral centers in the correct relative configurations.

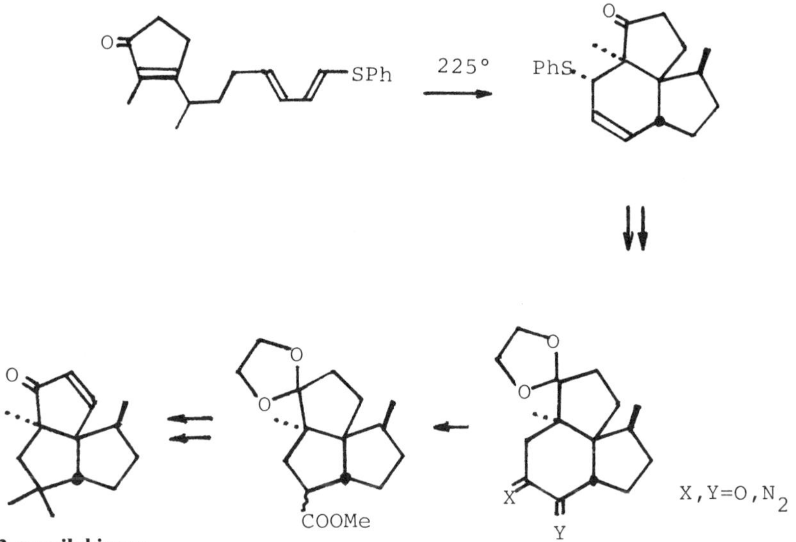

7.11 INTRAMOLECULAR DIELS–ALDER REACTIONS

3-oxosilphinene

X, Y = O, N$_2$

An intramolecular Diels–Alder approach to quassinoids (Voyle, 1983b) is thwarted, probably due to a high kinetic barrier in the transition state in which the π-electron overlap of the ester must be destroyed.

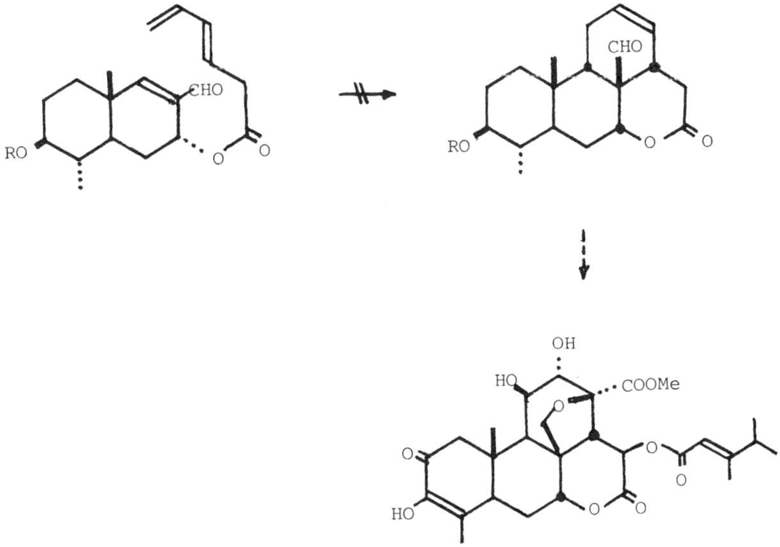

On the other hand, the exo-selective cycloaddition has found an application to the construction of the BCE fragment of quassimarin (**Q2**) (Schlessinger, 1985).

Q2

The *s*-indacene skeleton of the sesquiterpene cinncassiol D_1 is perhaps amenable to construction by intramolecular Diels–Alder reaction of a macrocyclic triene lactone (Ensley, 1987) because five of the chiral centers can be established in one step. Furthermore, the tricyclic product is shaped like an umbrella in which the double bond is exposed to epoxidizing agents on one side. On alcoholysis the released lactonic oxygen can then participate in epoxide opening in the desired fashion.

cinncassiol D_1

The elegant synthetic method involving intramolecular cycloaddition with *o*-quinodimethanes generated *in situ* from benzocyclobutenes has been extended to synthesis of an aromatic precursor of diterpene alkaloids (Kametani, 1976). The salient feature of this cycloaddition is the stereoselectivity with which the ester and the cyano groups become cis in the adduct. The *cis*-A/B juncture is inverted via the enone.

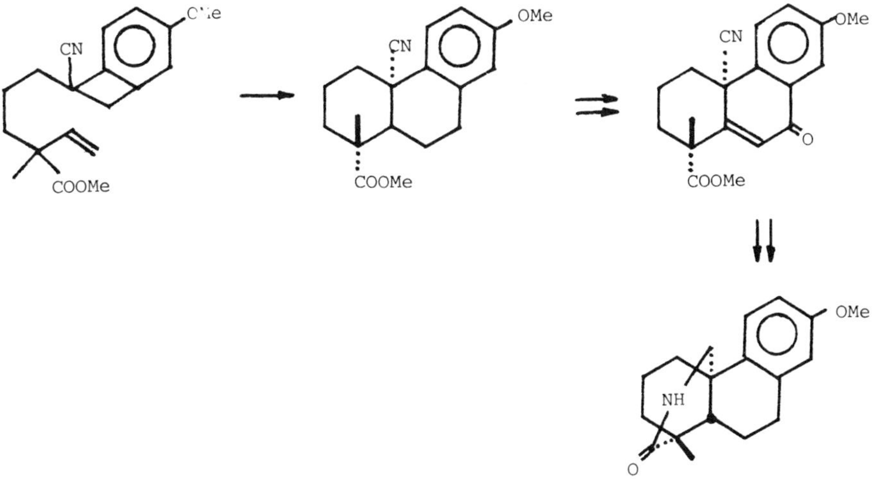

A new route to the intermediate for synthesis of atisine (**A48**) has been developed (Shishido, 1987). It involves electrocyclic isomerization of a benzocyclobutene to en enol carbonate of β-tetralone and modification of the two carbon chains after cleavage of the enol ester double bond. An internal Diels–Alder reaction completes a tricyclic skeleton with an angular cyano group which serves as latent amine. Heterocyclization is most unusual because the iminium ion attacks the double bond in the anti-Markovnikov sense.

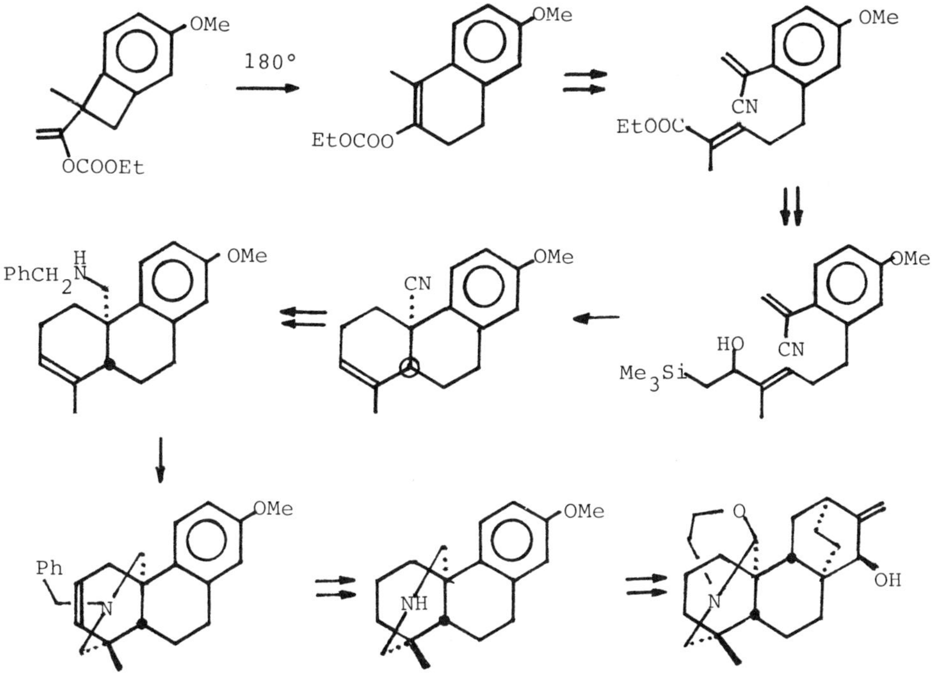

A48

An analogous process is pivotal to a most impressive access to the pentacyclic intermediates for alnusenone (**A26**) and friedelin (**F13**) (Kametani, 1977, 1978). An exo chair transition state with minimal steric compression is favored, and it leads to a *trans-anti-trans* skeleton.

Another model study for taxane synthesis (Bonnert, 1987) based on the Diels–Alder approach has culminated in the tricyclic system which lacks only several oxygen functions and an exocyclic carbon atom on the C-ring.

7.11.2 Bridged Rings

The coupling of a cyclic diene with an alkene tethered from one of the ring atoms results in a tricyclic stucture. A number of complex sesquiterpenes have been obtained via this strategy. Frequently, the most arduous part of the task is the preparation of the proper precursors.

Seychellene (**S29**) (Fukamiya, 1971, 1973; Frater, 1974a), patchouli alcohol (**P15**) (Näf, 1974, 1981), and norpatchoulenol (**P13**) (Oppolzer, 1978c) are excellent targets for testing the strategy.

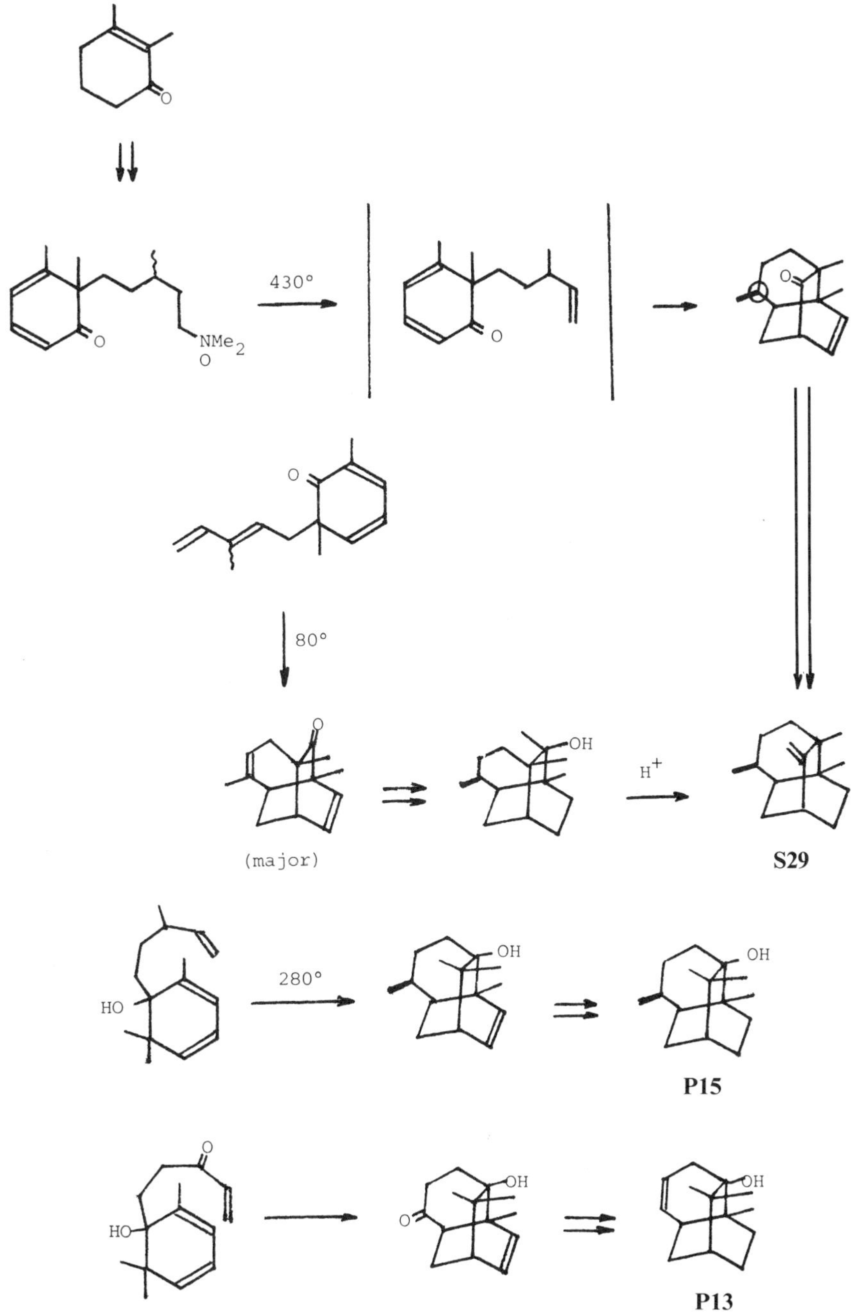

7.11 INTRAMOLECULAR DIELS—ALDER REACTIONS

Despite the moderate yield (30 %) from the cyclization of the methylpentenyl-cyclohexadienol (Näf, 1974), the high degree of stereoselectivity is gratifying. The exclusive formation of dehydropatchouli alcohol may be due to the strong destabilization of the transition state leading to its epimer by 1,3-diaxial-type interactions.

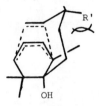

Interestingly, when the side chain of the cyclohexadiene also contains a conjugate diene unit, the condensation gives rise to two regioisomers in a 3:1 ratio. (Frater, 1974a). The major isomer is a twistane, which undergoes a Wagner–Meerwein rearrangement to afford seychellene after hydrogenation and methylation.

Several groups of chemists independently pursued the synthesis of 9-isocyano-pupukeanane (**P54**) (Yamamoto, 1979; Schiehser, 1980; Piers, 1982b) by the same strategy. Only the details of precursor preparation vary.

P54

The advanced ketonic intermediate for 2-isocyanopupukcanane (**P53**) (Frater, 1984), is similarly assembled. The preincorporation of the three-carbon chain (as an isopropylidene) facilitates the access of the precursor by alkylation of a phenolate while saving steps and avoiding the chemoselective manipulations that otherwise would be required.

P53

As indicated previously, bicyclo[2.2.2]octenes are susceptible to rearrangements. The fused system derived from an intramolecular cycloaddition of zingiberene norketone has been converted into a carbon skeleton characteristic of khusimone (**K8**) (Büchi, 1977).

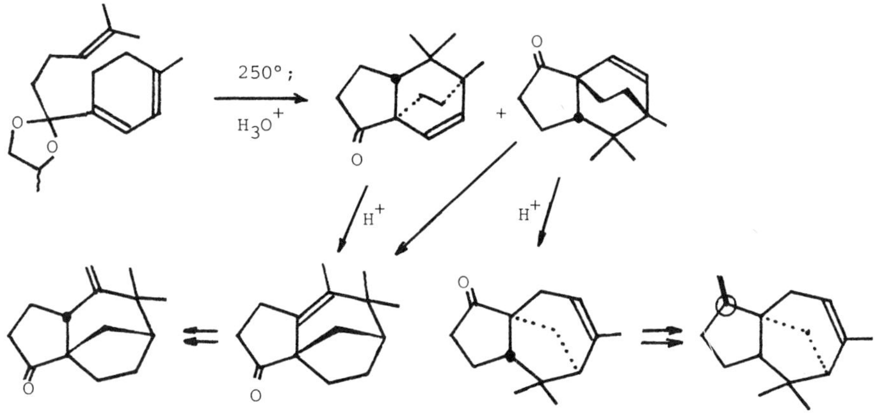

The minor component of the cycloadducts is also convertible to dehydrokhusimone and a ketone having the framework of α-cedrene (**C37**) (Büchi, 1977).

An outstanding plan for the synthesis of quadrone (**Q1**) (Wender, 1985d) develops the potential of the assisted ring expansion/contraction theme and identifies an intramolecular Diels–Alder approach. Stereocontrol and efficiency of the cycloaddition step are maximized by the presence of a Lewis acid, ethylaluminum dichloride.

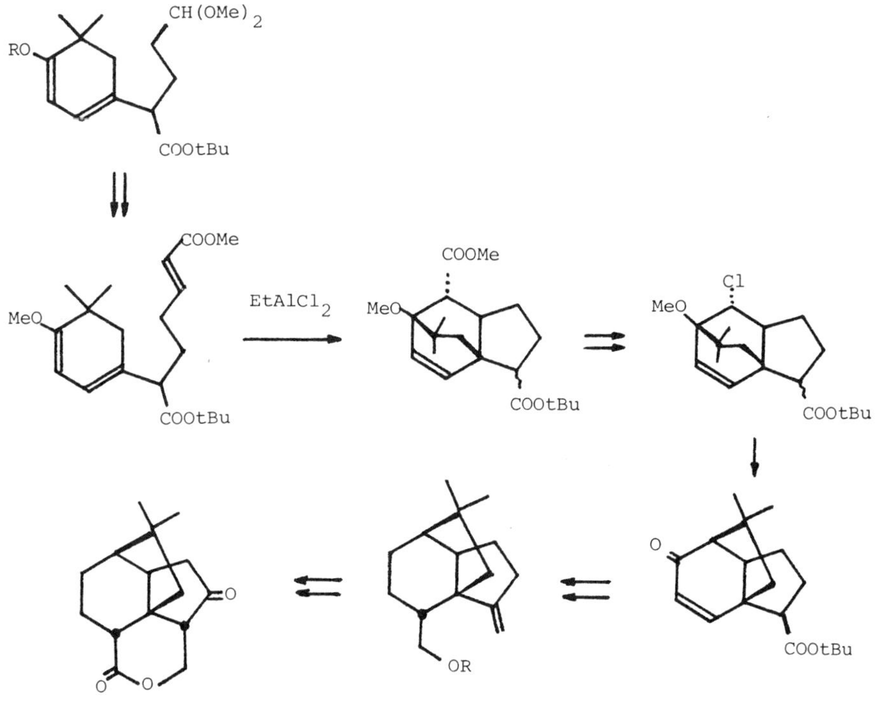

Bicyclic systems spanned by two odd-numbered bridges cannot be prepared directly from a Diels–Alder reaction involving neutral addends. However, by means of post-cycloaddition modification, such systems should become available. α-Cedrene (**C37**) (Breitholle, 1976, 1978) and 8*S*,14-cedranediol (**C36**) (Landry, 1983) have been synthesized from intramolecular Diels–Alder adducts via ring expansion.

With respect to the latter synthesis, stereocontrol at the quaternary carbon atom, which bears the hydroxymethyl substituent, is provided by an ester group in the dienophilic moiety (endo addition). This oxygen function also participates in the ring expansion step as an internal nucleophile. Consequently, only the desired 8-keto derivative is produced.

The formation of this type of gross structure is due to the favorable existence of alkenylcyclopentadienes in the 1-substituted state, the slightly less favored 2-alkenyl isomers are not capable of forming the cycloadducts (Bredt's rule). In

7.11 INTRAMOLECULAR DIELS–ALDER REACTIONS

this context one should be reminded of an early attempt to synthesize longifolene (**L24**) (Brieger, 1963b), which resulted in the skeletal isomer.

Since more and more sesquiterpenes are being discovered with 2,7-bridged norbornane skeletons, it would be most helpful to find ways to preclude the premature isomerization of the 5-alkenyl isomers. Blocking the 5-position of cyclopentadiene with removable groups is impractical, since even a chlorine atom is prone to sigmatropic migration. However, when two alkyl groups are bound together in the form of a cyclopropane, the alkyl shift is suppressed.

The strategy is most suited for sinularene (**S37**) (Antczak, 1985), since severance of the cyclopropane ring in the appropriate cycloadduct will generate a tertiary methyl group in the desired position. By virtue of the endo transition state with the least nonbonded interactions, the R,R,S-isomer of the pentenoic ester substituted with a spiro[2.4]heptadiene undergoes cyclization to give a product corresponding to the configuration of sinularene only. The conditions for the reaction (THF, 67°C) are very mild.

Blocking with immovable groups is not applicable in the synthesis of other 2,7-spanned norbornanes in which C_7 must remain tertiary. Maintaining the integrity a 5-alkenyl-1,3-cyclopentadiene with special substituents in the cyclopentadiene portion while at the same time lowering the reaction temperatures should favor formation of the desired product. The siloxy group appears to be a useful stabilizer and activator, and its effectiveness has been demonstrated in an approach to sativene (**S11**) (Snowden, 1981).

Intramolecular Diels–Alder reaction of α-allenyl ketone to an α-cyclopentadienyl moiety gives rise to a brexanone. The process is well suited for elaboration of sativene (**S11**) (Sigrist, 1986) by properly placing alkyl substituents in the precursor, and addressing the need for a subsequent ring expansion. Such "addends" are conveniently generated from a brominated biocyclo[3.2.0]heptanone by reaction with an allenyltitanium reagent. For unclear reasons the isopropyl group of the brexanone acquires the desired configuration.

It is apparent that the cleavage of the 1,6-bridged norbornenes represented by the cedrene intermediate give rise to condensed systems. Specifically, diquinanes may be thereby obtained.

Pursuing this strategy has culminated in a formal synthesis of $\Delta^{9(12)}$-capnellene (**C21**) (Sternbach, 1984b, 1987).

If active pendant groups are present in the adducts, the released chain(s) may be coaxed to form new rings with them. This concept has brought to fruition in a quest of silphinene (**S34**) (Sternbach, 1985).

If this scheme is to be adapted to a synthesis of spirocyclic systems, the norbornene double bond must be reduced and a new double bond introduced in the other bridge. The strategy of employing substrates with an acetylenic side chain and selective hydrogenation of the less higly substituted double bond of the norbornadiene product provides a simple solution to this problem. An access to hinesol (**H15**) (Helquist, 1985; Nystrom, 1985) serves to demonstrate the validity of this concept, and the complete utilization of all the carbon resources from the substituted cyclopentadiene is particularly satisfying.

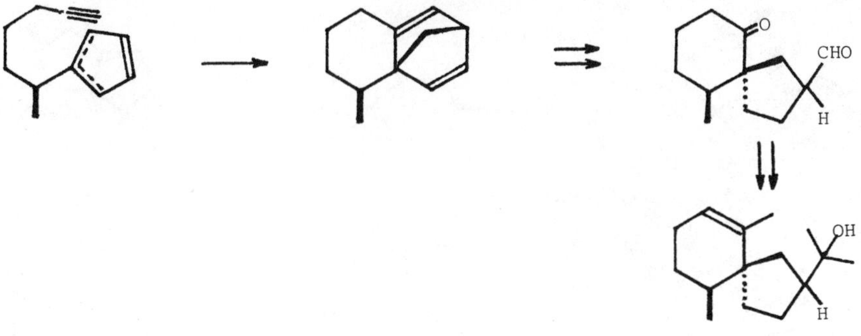

A nonclassical norbornene cleavage with concomitant nucleophilization of one of the trigonal carbon atoms for interlinking with an ester pendant in the other bridge is the crucial step in an approach to $\Delta^{9(12)}$-capnellene (**C21**) (Stille, 1986). The synthetic potential of Tebbe reagent ($Cp_2TiCH_2AlMe_2Cl$) as a methylene group donor and for methathesis with strained alkenes is fully exploited. The regioselectivity of the titanacyclobutanation is most rewarding.

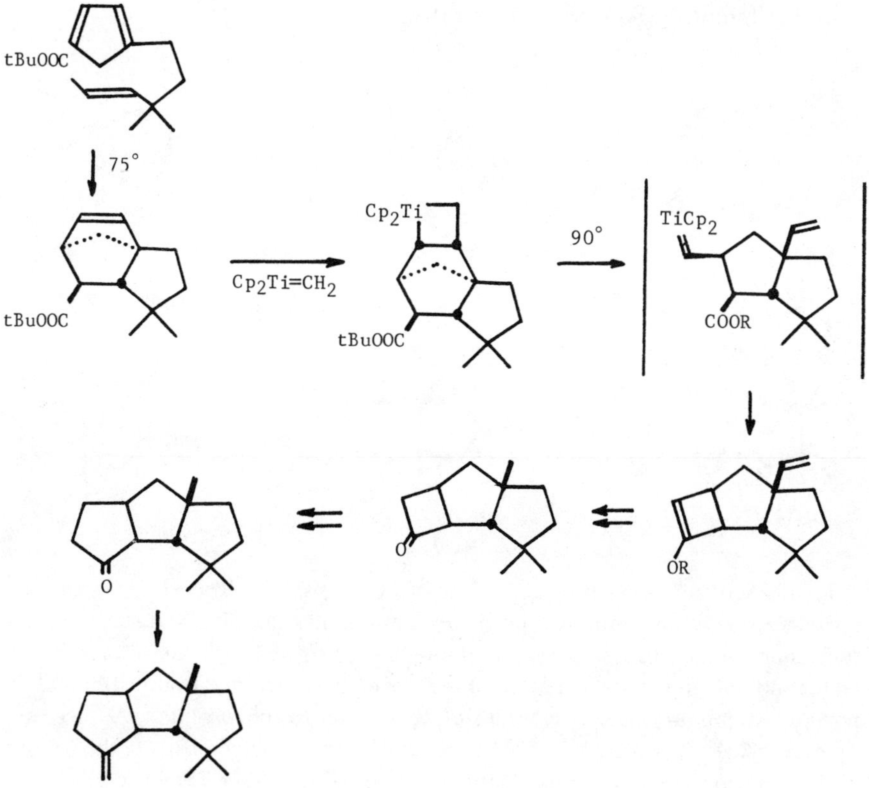

7.11 INTRAMOLECULAR DIELS—ALDER REACTIONS

An exciting approach to $\Delta^{9(12)}$-capnellene (**C21**) (Wu, 1987) consists of tandem oxidation–Diels–Alder reaction of a furylethylcyclopentadiene. Thus, a cyclopentanonorbornene derivative has been obtained.

An interesting means of constructing the bridged rings of quadrone (**Q1**) (Dewanckele, 1983; Schlessinger, 1983b), involves a sidelong Diels–Alder reaction. An extraneous cyclohexene ring is salvaged for parts of the lactonic carboxyl and those to be used in the formation of the cyclopentanone.

The intramolecular Diels–Alder reaction of properly designed 2-(alkenyl)tropone can lead to tricyclic substances which resemble the cedrene skeleton. Actually, selective removal of a carbon atom from the etheno bridge is the only major requirement for "naturalizing" the carbon framework. Such adducts are particularly valuable for synthesis of α- and β-pipitzols (**P35** and **P36**) (Funk, 1987b) because of the functional pattern. Thus, cleavage of the nonconjugated double bond followed by aldolization accomplish the crucial transformation.

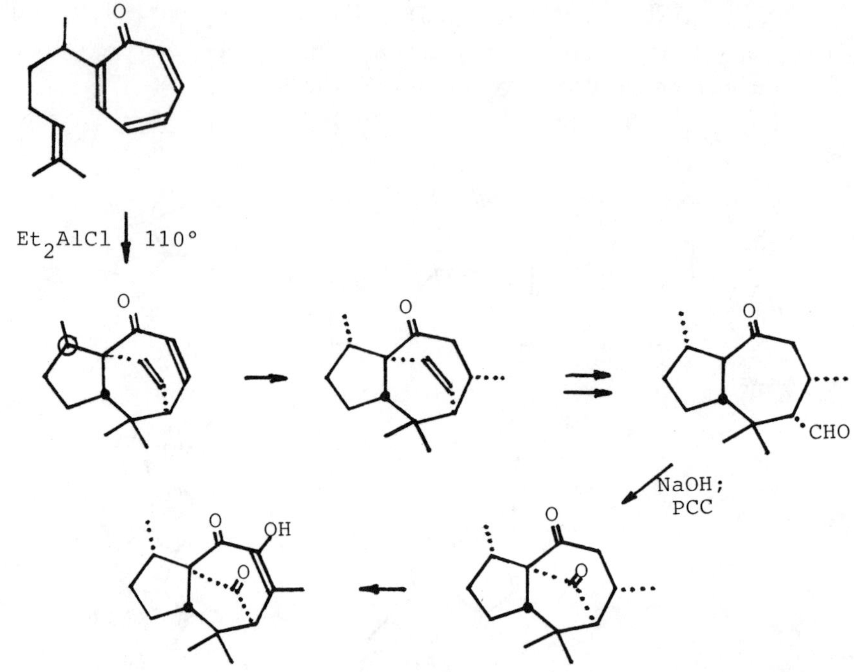

P35

7.11.3 Reactions Involving Heterocyclic Dienes

The applicability of the intramolecular Diels–Alder process is not at all apparent with respect to mansonone E (**M5**) (Best, 1981). However, the reaction sequence involving interception of a benzyne with a furan and rearomatization of the adduct to furnish the naphthalene precursor has proven very successful. The regiochemistry is enforced by the short connecting chain.

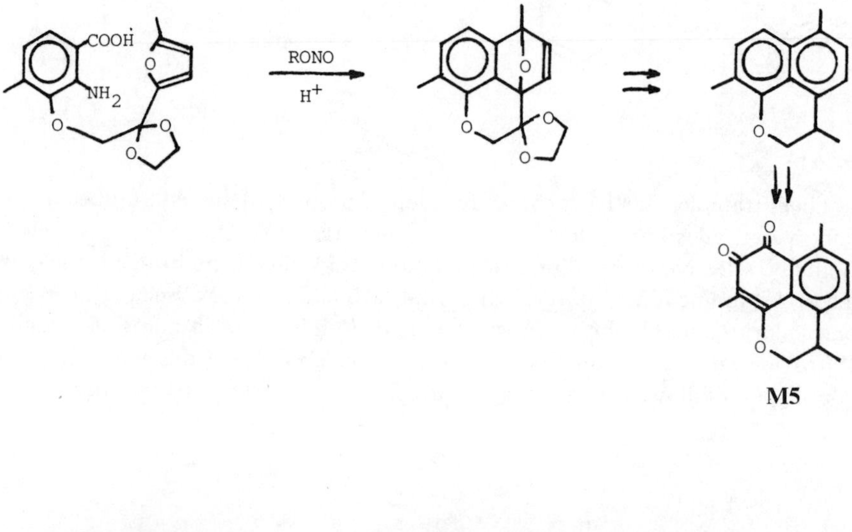

M5

7.11 INTRAMOLECULAR DIELS–ALDER REACTIONS

Furans in general react with very reactive dienophiles only, and such adducts easily revert to the addends. However, the entropically favored intramolecular reactions may be considered to be viable synthetic processes. (−)-Farnesiferol C (**F2**) (Mukaiyama, 1981) probably cannot be constructed any better than by employing this method. Linking the amidic nitrogen atom to an optically active substituent renders the cycloaddition diastereofacially discriminative.

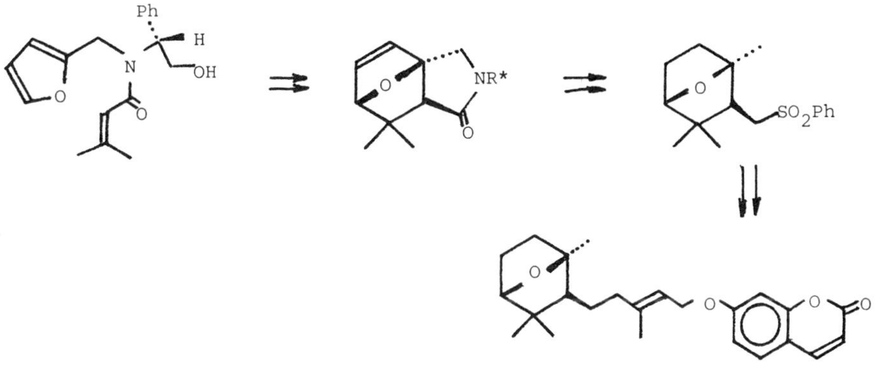

F2

Another entry into the gibberellin skeleton by intramolecular Diels–Alder reaction features the condensation of a furan with a propargylic ester (Grootaert, 1982). The AB-ring system evolves in one step.

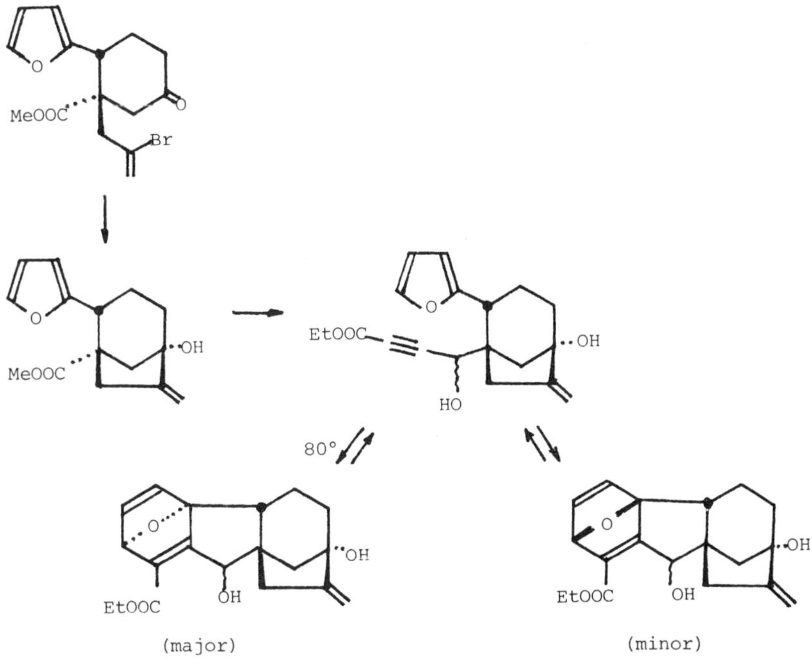

(major)   (minor)

The major (75 %) tetracarbocyclic diol has been converted into gibberellin A_5 (**GA5**) by a stepwise reduction, homologation at C_6, and manipulation of the heterocyclic moiety (Grootaert, 1986).

GA5

Methods capable of forming condensed heterocyclic–carbocyclic ring systems in one operation are very desirable, especially when the heterocyclic unit can be further modified with ease. It has been shown that oxazoles undergo thermal cycloaddition with alkynes and the adducts decompose to give furans. The intramolecular version of this reaction leads to [*b*]cycloalkanofurans, and therefore all the furanosesquiterpenes may be synthesized. The ready conversion of furans to γ-lactones extends the synthetic utility of the method further.

Several rather complex sesquiterpenes have been approached on the basis of this key reaction. They include paniculide A (**P9**) (Jacobi, 1984a), gnididione (**G16**) (Jacobi, 1984c), and ligularone (**L13**) and petasalbine (**P23**) (Jacobi, 1984b).

P9

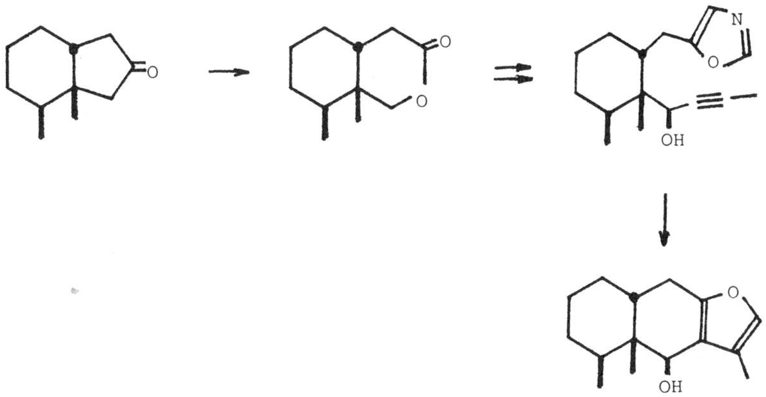

7.11 INTRAMOLECULAR DIELS–ALDER REACTIONS

P23

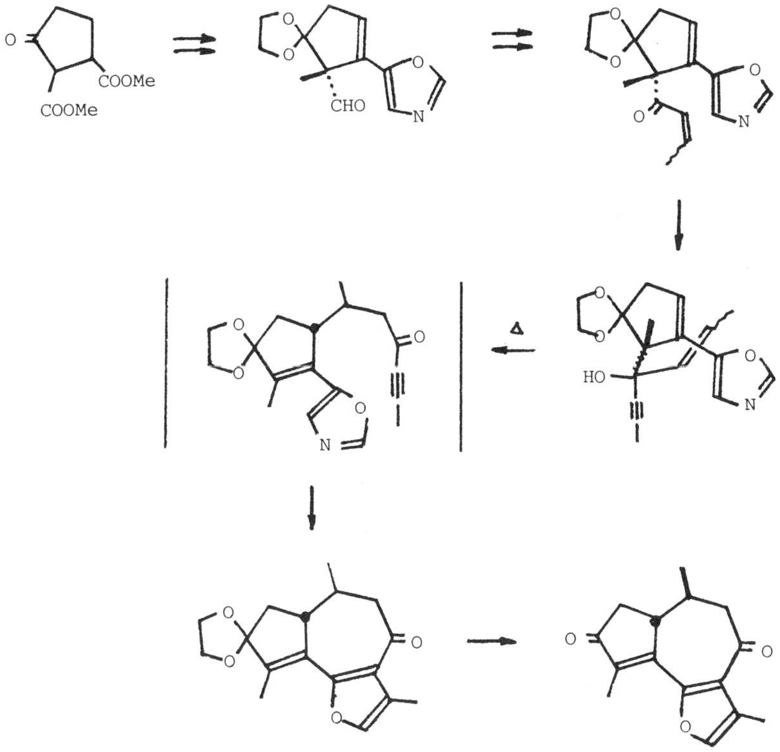

G16

The extrusion of carbon dioxide from the Diels–Alder adducts of α-pyrones is energetically favorable because the liberation of a small molecule is accompanied by the relief of strain. Furthermore, a conjugated diene system is restored.

The reaction of methyl α-pyrone-3-carboxylate with 4-methyl-3-cyclohexenone at 150°C gives a bicyclic homonuclear diene that is suitable for elaboration into occidentalol (**O2**) (Watt, 1972) and copaene/ylangene (**C66/Y1**) (Corey, 1973). The regioselectivity of the condensation is sponsored by the ketone group through enolization, such that C_4 becomes more nucleophilic (anti-Markovnikov).

7.12 HETERODIENES

Hetero-Diels–Alder reactions are seldom encountered in terpene synthesis. In connection with a model spirocyclization study, the condensation of α-methylene cycloalkanones with acrylic esters has been effected. The spirocyclic intermediate is so closely related to β-chamigrene (**C47**) that it was natural to use a ring contraction maneuver to complete the synthesis of the latter compound (Ireland, 1984).

The adduct from the inexpensive compounds norbornene and acrolein may be transformed into epi-β-santalene (**S5**) and epi-β-santalol (**S6′**) (Snowden, 1981b).

Chapter 8

Other Thermal Processes

In addition to the Diels–Alder reaction, other thermal processes have found increasing and frequently ingenious applications in synthesis. The mechanistic insights and hence regio- and stereochemistry pertaining to these concerted reactions afforded by the Woodward–Hoffmann rules have contributed to the enormous success of their employment. In this chapter we shall touch on cycloadditions besides the Diels–Alder reaction, and the ene reaction, electrocyclization, as well as sigmatropic rearrangements including those of Cope, Claisen, and their variants.

8.1 [3 + 2]CYCLOADDITIONS

8.1.1 Nitrones

The [3 + 2]cycloadditions are dominated by the concerted 1,3-dipolar additions, which are extremely important in the preparation of five-membered heterocycles. For terpene synthesis the usefulness of these reactions is quite limited, since elimination of the heteroatom(s) must be accomplished afterward, and for carbocyclic synthesis an intramolecular version is usually required.

The "trivial" use of the 1,3-dipolar addition includes the formation of cyclopropane moieties via the pyrazolines (see Section 10.1.4). The more serviceable reaction is the addition of nitrones to double bonds to yield isoxazolidines. This is an indirect method for C—C bond formation, and in the intramolecular case, the removal of the heteroatoms unveils a carbocyclic nucleus.

The (heterocyclic) ring formation process introduces oxygen functions into the molecules indirectly, as shown by a synthesis of α-bisabolol (**B12**) from 6Z-farnesal (Iwashita, 1979; Schwartz, 1979).

α- and β-Eudesmols (**E17** and **E18**) are produced from a substituted cyclohexenecarboxaldehyde via a similar reaction sequence (Schwartz, 1985). An interesting observation is that during extrication of the nitrogen atom from the heterocyclic adduct(s) (in fact the tetrahydrooxazines), quaternization in anhydrous sulfolane followed by reduction led to mainly α-eudesmol (**E17**), whereas the same reaction sequence gave rise to β-eudesmol (**E18**) if the quaternary ammonium salt were formed in technical sulfolane.

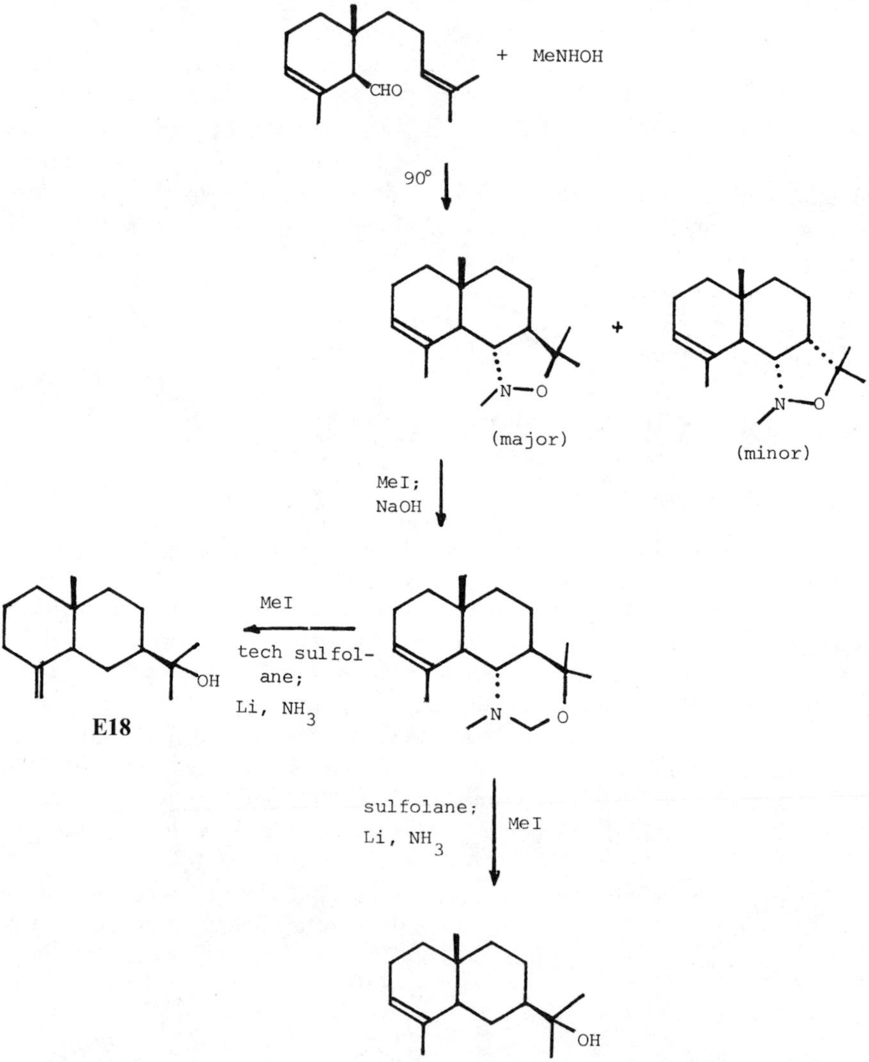

When 3- (or 4-) alkenylcycloalkanones are heated with an *N*-alkylhydroxylamine with continuous removal of water, isoxazolidines fused to bridged bicyclic nuclei are generated. A concise synthesis of the antifertility agent 7,12-seco-ishwaran-12-ol (**I13**) (Funk, 1983) is based on this approach.

I13

A successful application of the intramolecular nitrone–alkene cycloaddition in the triquinane area indicates how a concerted process benefits stereoselection. Under the circumstances, the closure of the central ring of hirsutene (**H17**) (Funk, 1984) by this method gives the desired adduct only. The formation of the unwanted cis-syn-cis isomer is sterically impossible and therefore excluded, and the uncyclized nitrone precursor is depleted only by equilibration. Consequently, all diastereomers are eventually consumed for the good cause.

H17

Starting from optically active pyranose templates (sugars), fused isoxazolines suitable for elaboration into iridoids may be acquired. Thus, a pyranocyclopentanone that differs from O-ethylverbenalol (**V9**, R = Et) by having a hydroxyl group instead of the unsaturated ester has been reported (Curran, 1985a).

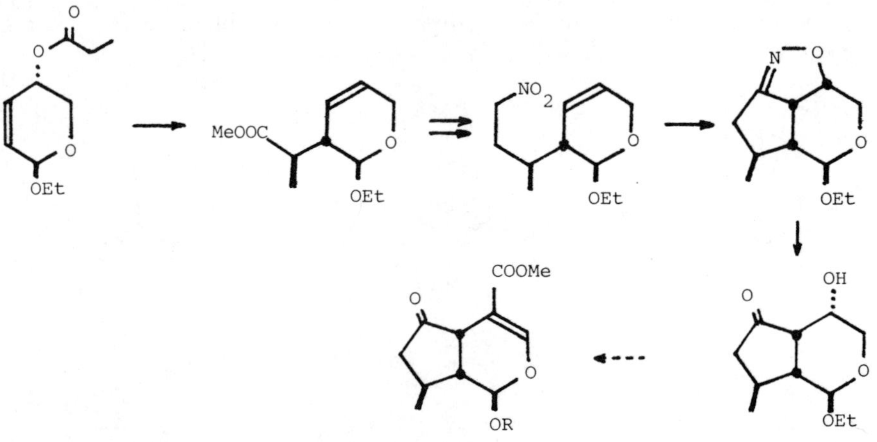

V9

The guaiane skeleton may be assembled by an intramolecular nitrile oxide addition to an alkene linkage (Kozikowski, 1983). It should be possible, probably after some structural modifications, to convert such intermediates to the natural sesquiterpenes.

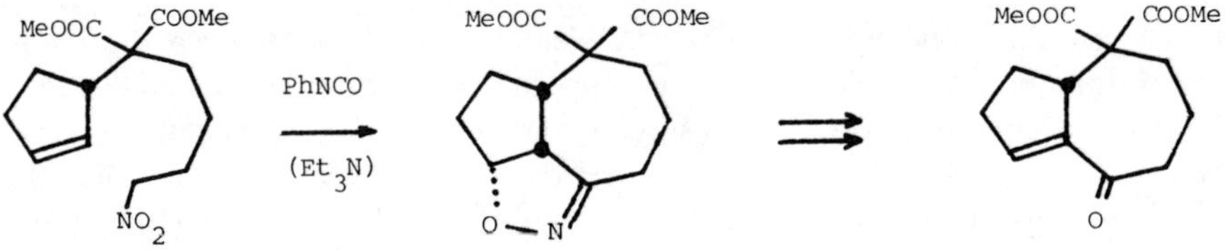

Cycloaddition is an excellent method for constructing a forskolin (**F9**) synthon from α-damascone (Baraldi, 1986). Conversion of the original ketone group into an acetate increases the efficiency of the cyclization step 4.5-fold.

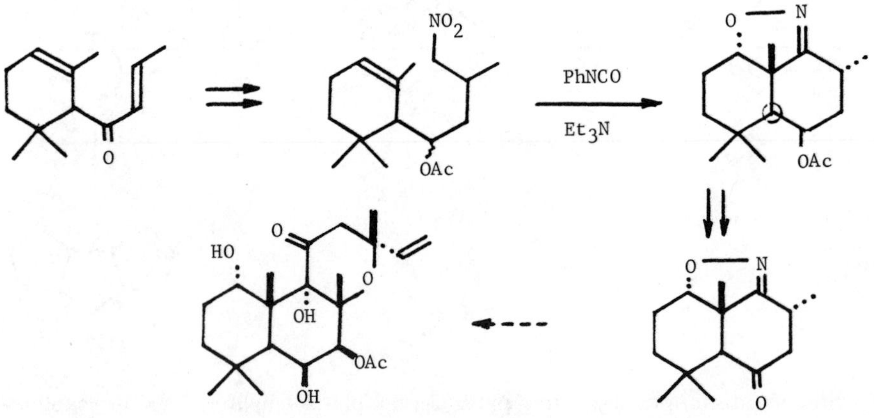

F9

Two examples of intermolecular alkene-nitrile oxide cycloaddition as applied to terpene synthesis are mentioned now. The optically active lactone derived

8.1 [3 + 2]CYCLOADDITIONS

from cyclohex-3-enecarboxylic acid has been converted into (+)-phyllanthocin (**P26**) (S. Martin, 1987) via the isoxazoline which acts as a latent 3-furanone.

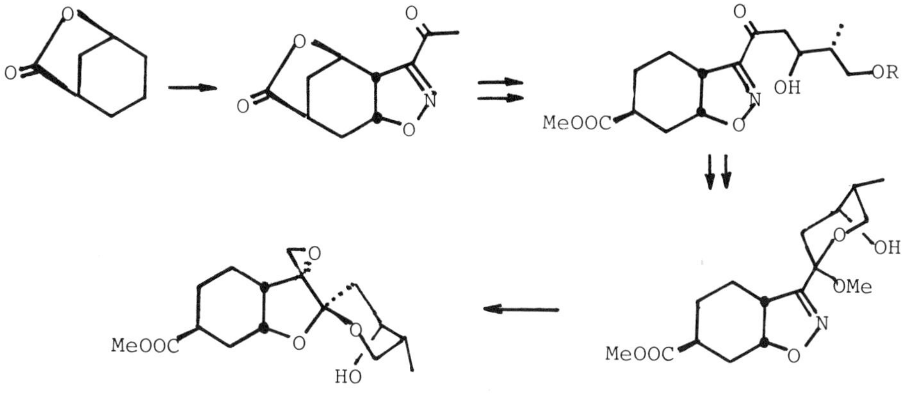

Similarly, in a synthesis of bisabolangelone (Riss, 1986) the cis ring juncture of the hydrobenzofuran framework is established by the method.

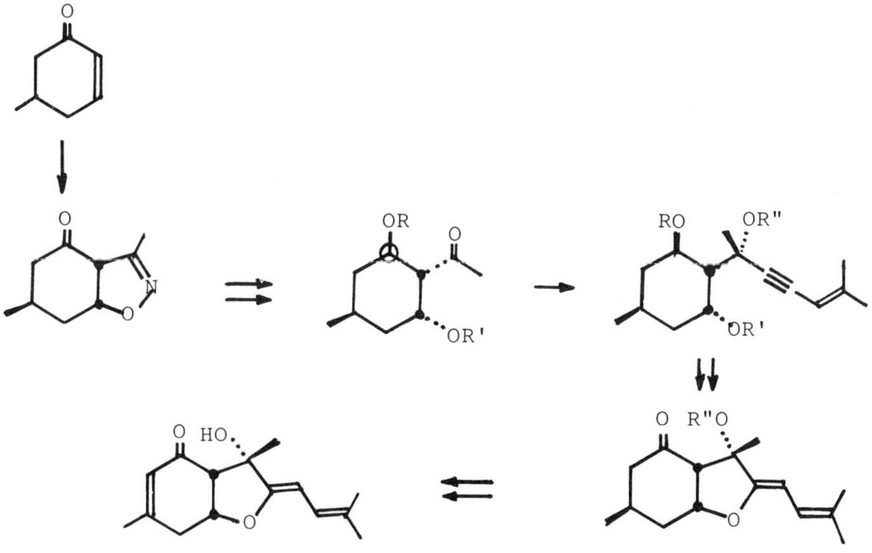

8.1.2 Oxyallyl Cations

Cyclopentanation by means of cycloaddition methods is underdeveloped. The restriction inherent in such reactions is the necessity of using ionic addends.

A process that is isoelectronic with the Diels–Alder reaction must involve allyl anions, a simple example of which has yet to be discovered. On the other hand,

[3 + 2]cycloadditions between allyl cations and alkenes are not concerted because the four-electron transition state is symmetry-forbidden. Yet a number of examples are known.

The most useful [3 + 2]cycloaddition process involves trapping of oxyallyl–iron(II) cations with alkenes. The allyl species are conveniently generated by treatment of α,α'-dibromoketones with an iron carbonyl reagent. Alternatively, the ionization of 3-bromo-2-siloxyalkenes in the presence of a Lewis acid such as zinc chloride also leads to the allyl cations. Since these species are very reactive, interceptors (alkenes) must be present *in situ*. When intramolecular trapping is induced, a polycyclic adduct ensues.

The synthesis of α-cuparenone (**C85**) (Hayakawa, 1978) by a one-step reaction demonstrates an unexpected regiochemistry. Perhaps the precyclized intermediates are formed reversibly, and only the more stable one collapses in the forward direction.

It is instructive to compare the debrominative method to the ionization route (Sakurai, 1979). In the latter case, the C—C bond formation at the initial cationic site is competitive to some extent with charge delocalization. As a result, slightly more than one-third of the total cyclopentanone product has the 2,2,4,4-tetraalkyl substitution pattern. [However, this rationale is only tentative, given the low yield of α-cuparenone from the oxyallyl–iron(II) reaction and the unknown nature of its coproducts.]

While camphor (**C15**) has been synthesized from an α,α'-dibromoketone by the action of diiron enneacarbonyl, campherenone (**C14**) and its epimer are obtained in a 2:1 ratio by treatment of the C_{15} analog derived from farnesol with iron pentacarbonyl at 100°C (Noyori, 1979).

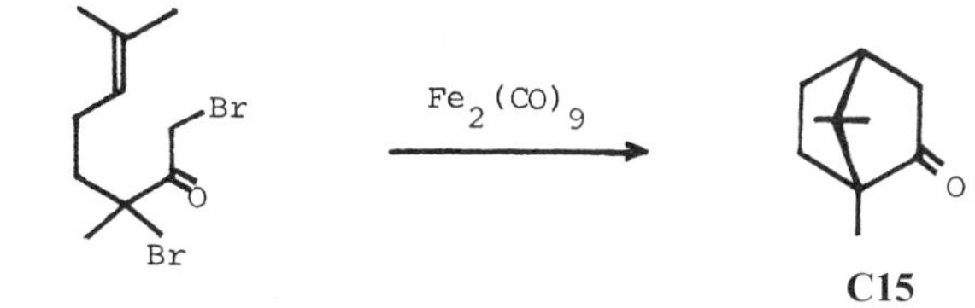

8.1.3 Trimethylenemethane Complexes

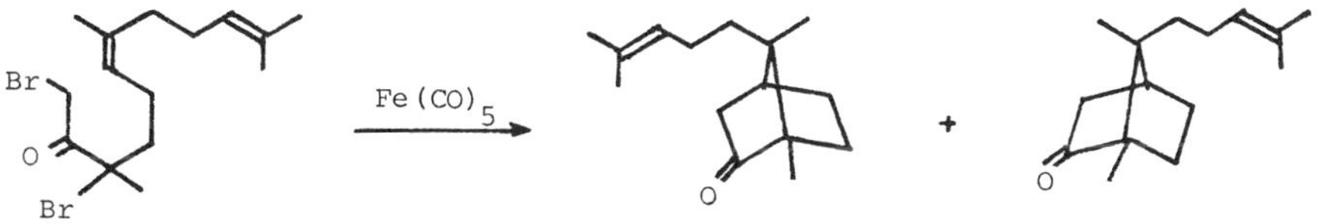

Conjunctive "1,3-dipolar" reagents based on allylsilanes also effect [3 + 2]cycloadditions with activated alkenes. Thus the palladium(II)-catalyzed reaction of 2-(trimethylsilylmethyl)-2-propen-1-yl acetate with dimethyl bicyclo[2.2.1]oct-2-ene-2,3-dicarboxylate to give a methylenecyclopentane derivative is a most direct method for synthesis of albene (**A22**) (Trost, 1982b).

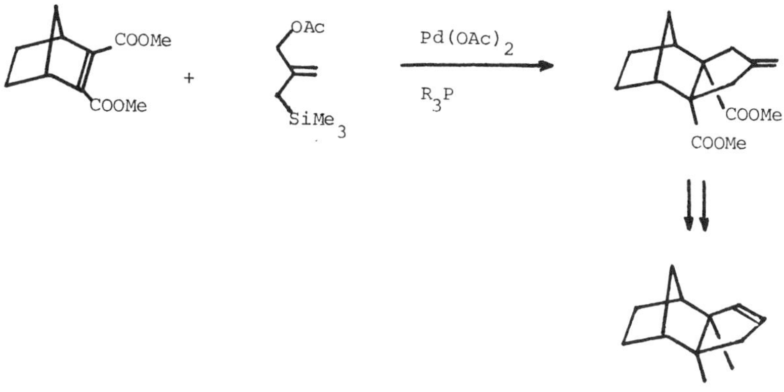

Loganin aglucone (**L20'**) is another good candidate for studying the scope of the [3 + 2]cycloaddition. Cyclopentenone and the proper conjunctive reagent are induced to form a 5:5-fused ketone. The exocyclic methylene group is an excellent latent carbonyl function whose unraveling is needed at a later stage of the synthesis for equilibrating the secondary methyl substituent and as a precursor of the hydroxyl group. The existing ketone offers a handle for chain attachment and modification of the carbocycle into the characteristic dihydropyranyl unit (Trost, 1985b).

The same [3 + 2]cycloadduct is also a convenient intermediate for chrysomelidial (**C54**) (Trost, 1981).

Methylenecyclopropane undergoes ring opening on exposure to certain nickel complexes, and the resulting species is a source of trimethylenemethane. The adduct of the latter entity with 2-cyclopentenone is easily transformed into hirsutene (**H17**) (Binger, 1983).

8.2 [4 + 3]CYCLOADDITIONS

These [4 + 3]cycloadditions enable rapid assembly of seven-membered rings. As a method for the synthesis of karahanaenone (**K1**) the cycloaddition is direct, although a regioisomer is also formed (Chidgey, 1977; Sakurai, 1979; Shimizu,

1979). The three-carbon addends are the same as those encountered in Section 8.1.2, only the other components are changed into conjugated dienes.

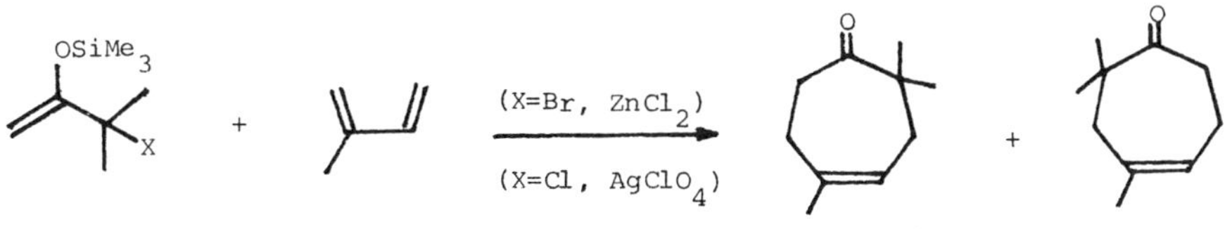

K1

When furans are used as the dipolarophile, more highly unsaturated cycloheptenones are generated. Nezukone (**N8**) is a typical example of compounds that have yielded to this convenient construction (Hayakawa, 1975; Takaya, 1978).

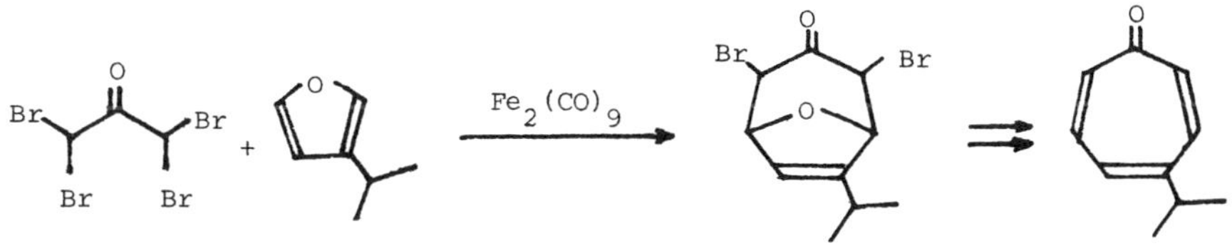

N8

8.3 [5 + 2]-, [4 + 4]- AND [6 + 4]CYCLOADDITIONS

The [5 + 2]cycloaddition is isoelectronic with the Diels–Alder reaction and hence blessed by an aromatic transition state. An interesting application of this cationic cycloaddition in the terpene area concerns the synthesis of gymnomitrol (**G21**) (Büchi, 1979a).

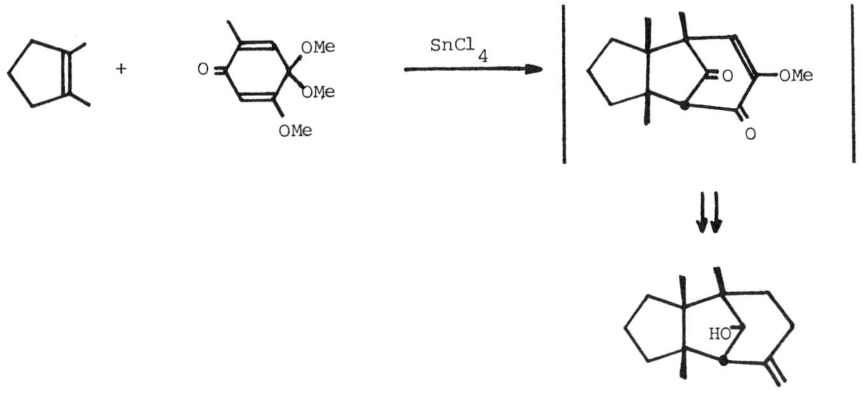

G21

Cyclic pentadienyl cations may also be generated by the electrochemical oxidation of appropriate phenols. Interception of the active species with styrenes gives rise to bicyclo[3.2.1]octenones. Helminthosporal (**H9**), which contains a such skeleton, is therefore amenable to construction by the method (Shizuri, 1986).

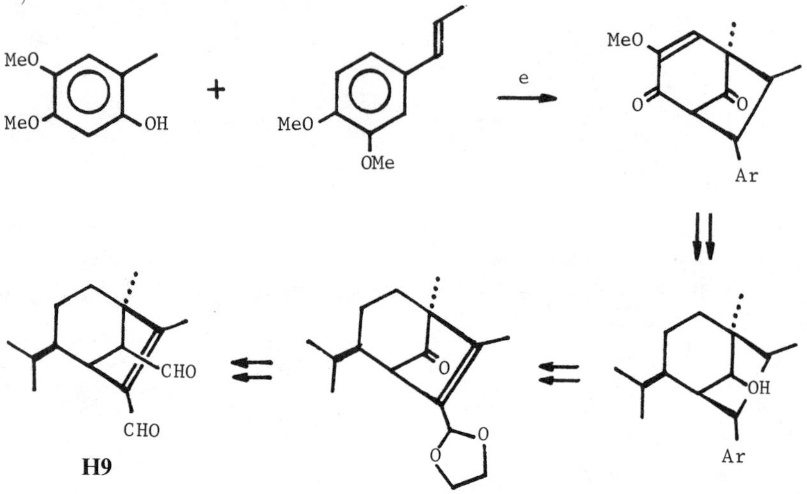

Ionization of a methoxy group from the benzoquinone monoketal affords the pentadienyl cation, which is intercepted by the cyclopentene, mainly in the endo fashion. The distribution of the substituents in the quinone derivative is such that only a few more steps are needed to transform the tricyclic adduct into the sesquiterpene.

The biosynthesis of the pipitzols (**P35** and **P36**) (Walls, 1965) probably is not mediated by enzymes (Woodward, 1971). Thus the hydroxybenzoquinone perezone is a legitimate precursor whose tautomerization into a cyclopentadienyl cation and its reaction with the side-chain double bond may account for the generation of two isomeric triketones.

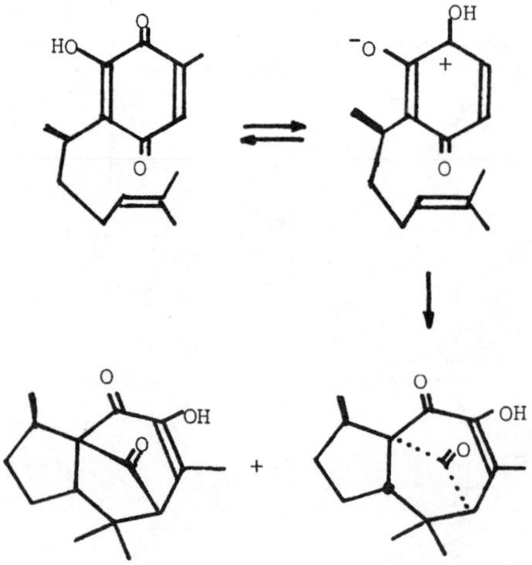

Enolization and ionization of pyranulose acetates can be induced by heat. In the presence of an unsaturated carbon chain, the resulting pyrylium oxides can be intercepted to form fused cycloheptenones, which are also bridged by an oxygen atom. This process offers a valuable entry into the guaiane sesquiterpenes. For example, β-bulnesene (**B20**) has been acquired rapidly by adopting this method (Sammes, 1983; Bromidge, 1985).

The use of a dienic side chain is critical for stereoselective production of the natural terpene. The monoolefin analog gives rise largely to the 4-epimer as a result of the assumption of an exo transition state with the least nonbonding interactions.

3-Hydroxypyridinium ylides undergo cycloadditions with electron-deficient alkenes. The adducts are susceptible to bridge destruction via quaternization and β-elimination, leading to α-aminotropones. The inevitable invitation to synthesis of natural tropolones such as β-thujaplicin (**T16**) has been accepted (Y. Tamura, 1977).

Intramolecular cyclodimerization of 1,3-dienes with nickel(O) species provides an unusually short route to 8-membered ring compounds. A synthesis of asteriscanolide (Wender, 1987a) is the harbinger of a systematic study aiming at acquisition of functionalized natural products that contain a cyclooctane nucleus.

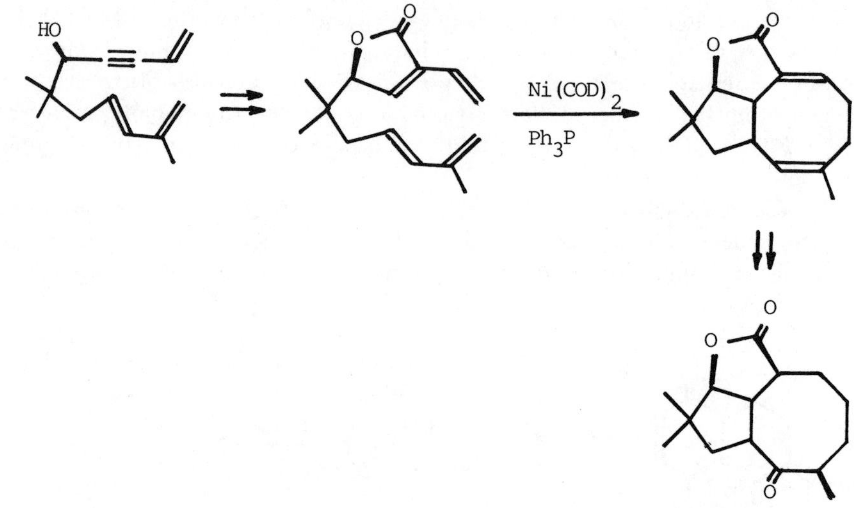

asteriscanolide

The [6 + 4]cycloaddition is thermally allowed as predicted by Woodward and Hoffmann. This very interesting pericyclic reaction found no application in natural product synthesis until the problem of the ingenanes was brought to light (Rigby, 1985, 1986). Since these diterpenes contain a bicyclo[4.4.1]undecanone segment (BC-rings), they are well suited for essaying a strategy that embodies the cycloaddition of tropone with a butadiene.

A model study involving the reaction of tropone and 1-acetoxybutadiene has reached a tricyclic intermediate with reasonably equipped functionalities for eventual elaboration into ingenol (**I4**). The use of more highly substituted addends may facilitate operations after the ABC-ring system is assembled.

The formation of the A ring proves feasible via an alkylation at the bridgehead and aldolization of the appropriately constructed 1,4-diketone.

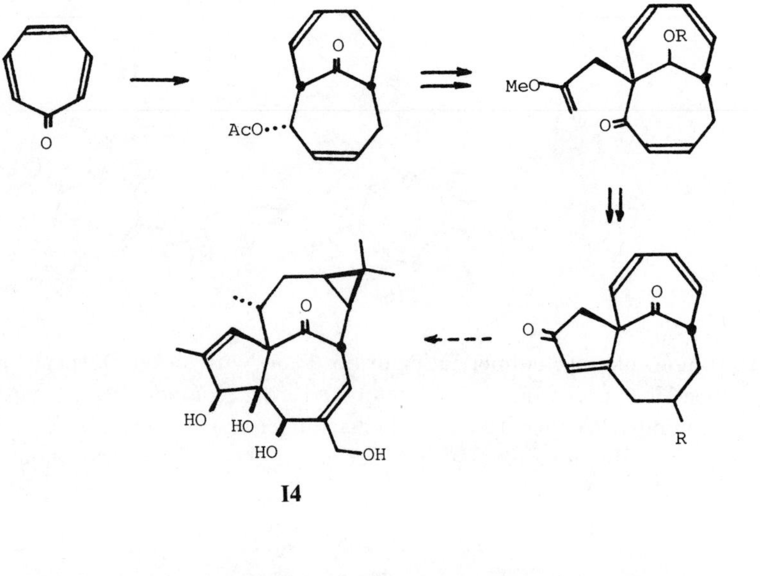

I4

8.4 CYCLOACYLATION

Carbon monoxide undergoes reductive addition to a diene or enyne to give a cyclopentanone or cyclopentenone adduct in the presence of a metal carbonyl. These reactions are mediated by organometallic species.

α-Cuparenone (**C85**) is obtained from 3-methyl-3-p-tolyl-1,4-pentadiene under the hydroformylation conditions [CO, $H_2/Co_2(CO)_8$] and methylation of the product (Eilbracht, 1984). Formally, the C—C bond formation processes involve one terminal and one internal carbon atom each of the two alkene linkages.

Strained alkenes cocyclize with alkynes in the presence of dicobalt oxtacarbonyl (Khand, 1973). A potential precursor of furanether B (**F18**) has been assembled (LaBelle, 1985).

The formation of a dicobaltatetrahedrane from the catalyst and the alkyne and its cycloaddition to the alkene molecule is implicated. Insertion of carbon monoxide, bond reorganization, and extrusion of the catalyst follow.

Annulation of bicyclo[3.2.0]hept-6-enes gives rise to tricyclic enones. With proper functionalization of the substrate to allow eventual fragmentation a precursor of damsin (**D5**) has been prepared (Sampath, 1987).

The Pauson–Khand reaction of 1,6-enynes to give condensed cyclopentenones shows a high degree of stereoselectivity as governed by allylic and propargylic substituents. The cobalt metallocycle intermediate formed is the one with an exo configuration for the internal allylic and/or propargylic substituent. This is necessary to avoid a strong steric interaction with the endo pendant originated from the terminal proparagylic group. Ensuing carbonyl insertion and oxidative demetallation complete the annulation.

In pursuit of quadrone (**Q1**) by this strategy (Magnus, 1987b) the hydropentalenone thus obtained is submitted to angular alkylation via photocycloaddition with allene and fragmentation of the four-membered ring.

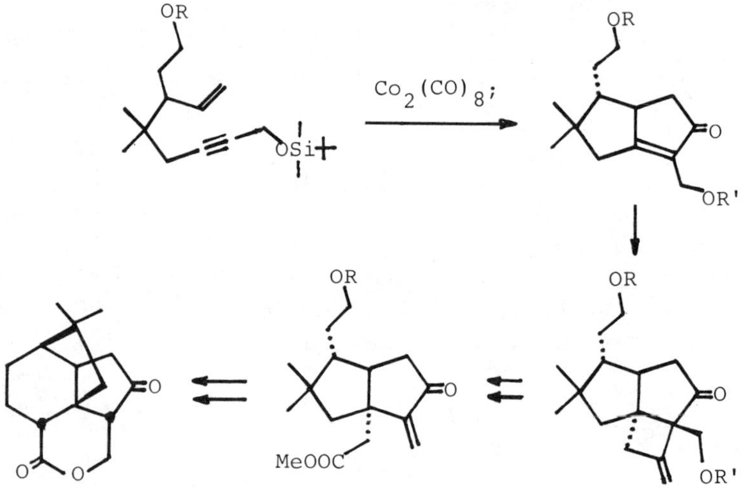

Q1

For synthesis of pentalenolactone H (Magnus, 1987a) adjustment of the substrate and carbomethoxylation of the diquinane intermediate are required.

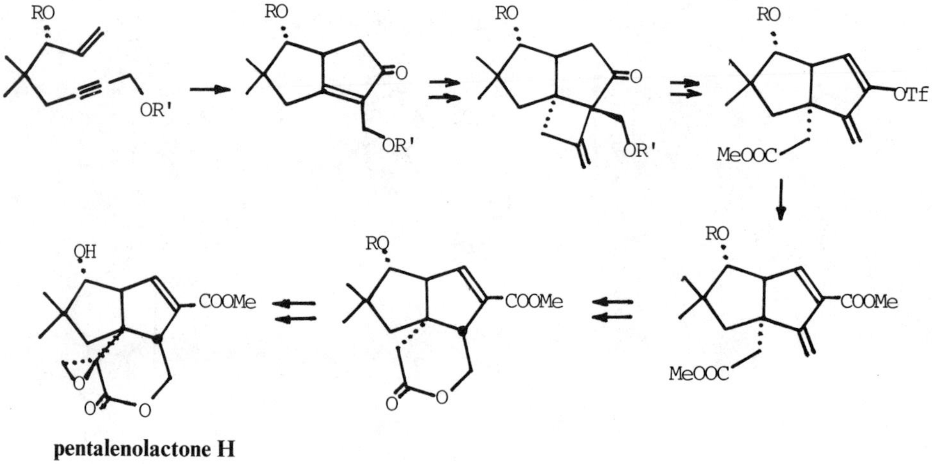

pentalenolactone H

The stereoselective construction of a crucial precursor for coriolin (**C70**) (Exon, 1983; Magnus, 1985), and the key intermediate of hirsutic acid C (**H19**) (Magnus, 1985) has been reported. However, in the latter work the remote asymmetric center with its substituents of insufficiently different bulkiness exerts no control over the reaction course. Two diastereomers are produced.

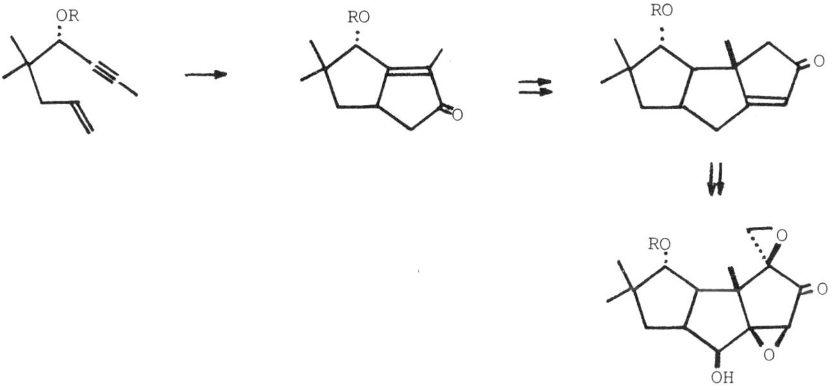

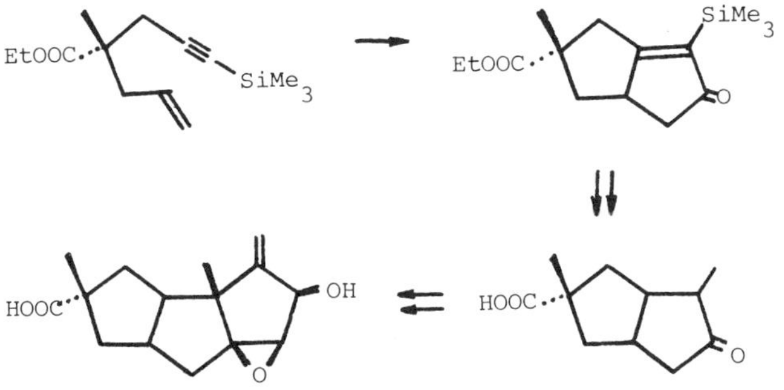

Synthesis of isocomene (**I14**) by the same method was precluded by the degree of substitution on the unsaturated carbon atoms. However, the bisnor analog of isocomene can be made (Knudsen, 1984).

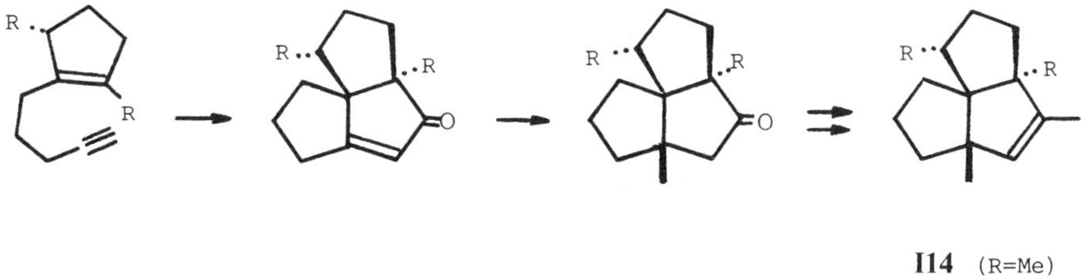

I14 (R=Me)

Reductive acylation of 1-alkenes with aldehydes by a free-radical reaction is not efficient in most cases, and the intramolecular addition is practically useless. For preparation of cyclopentanones from certain 4-alkenals, tris(triphenylphosphine)rhodium chloride (Wilkinson's catalyst) proves to be a good mediator. Thus, the optical isomers of 3-methyl-3-(*p*-tolyl)-pent-4-enal, which are obtainable from a common intermediate, have been successfully transformed into (+)-α-cuparenone (**C85**) and (—)-α-cuparenone (**ent-C85**) (Kametani, 1985b).

8.5 THE ENE REACTION

The ene reaction (Hoffmann, 1969; Oppolzer, 1978b) is that of an alkene having an allylic hydrogen with an unsaturated species (enophile). A new bond is created between the ene and the enophile while the allylic hydrogen is transferred from the former to the latter. The intramolecular ene reaction is subject to considerable geometric constraint and therefore is highly stereoselective. The synthetic utility of the intramolecular ene reaction has been extensively exploited in recent years, and it is now recognized as a special but important method for ring formation.

Three different modes of cyclization are possible, depending on the relative disposition of the participating molecular segments. Type I reactions are most common; type III reactions are rarely encountered.

8.5 THE ENE REACTION

TYPE I, **TYPE II**, **TYPE III** [schematic reaction diagrams]

The cyclization of 1,6-dienes is very facile and fairly stereoselective. A nonactivated double bond is often reactive enough to partake the ene process, thanks to the entropic gain in the activation energy.

One of the oldest known intramolecular ene reactions is the formation of plinols from linalool (Ikeda, 1936; Strickler, 1967). The two diastereomers in which the methyl and the isopropenyl groups on the adjacent carbon atoms are *cis*-related are favored. At higher temperatures the other two isomers are also produced, presumably via a cycloreversion pathway. The major plinol is useful in synthesizing the fungal sesquiterpene cyclonerodiol (**C93**) (Nozoe, 1971).

[structures showing linalool → plinol (+ isomers) → ketone intermediate → C93]

C93

The alkyl pattern of the plinols corresponds to the cyclopentane moiety of the acoranes. Provided the central ethylidene fragment can be changed into a functionalized unit, plinols should be an excellent building block for the spirosesquiterpenes.

Dehydroplinols deriving from dehydrolinalool contain the key elements for completing the spirocycle. For example, a Claisen rearrangement may be employed to extend the central vinylidene group by an acetone unit which is accompanied by double bond migration into the ring; the tertiary allylic alcohol obtained from the ketone can now act as an electrophile to induce an internal attack by the cyclopentene. Consequently, acoratriene (**A7**) emerges (Naegeli, 1972).

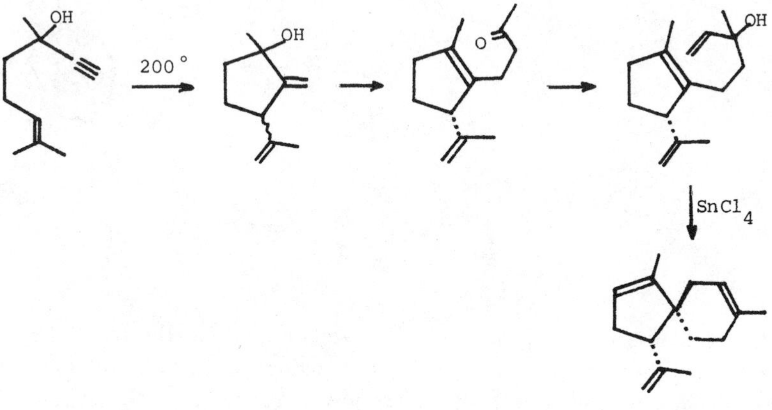

A strategy for the synthesis of phyllanthocin (**P26**) (Smith, 1984a, 1987), entails coupling the cyclohexane moiety with a dihydropyranone and subsequently closing the tetrahydrofuran ring. The functionalized cyclohexanecarboxaldehyde is prepared stereoselectively via an ene reaction.

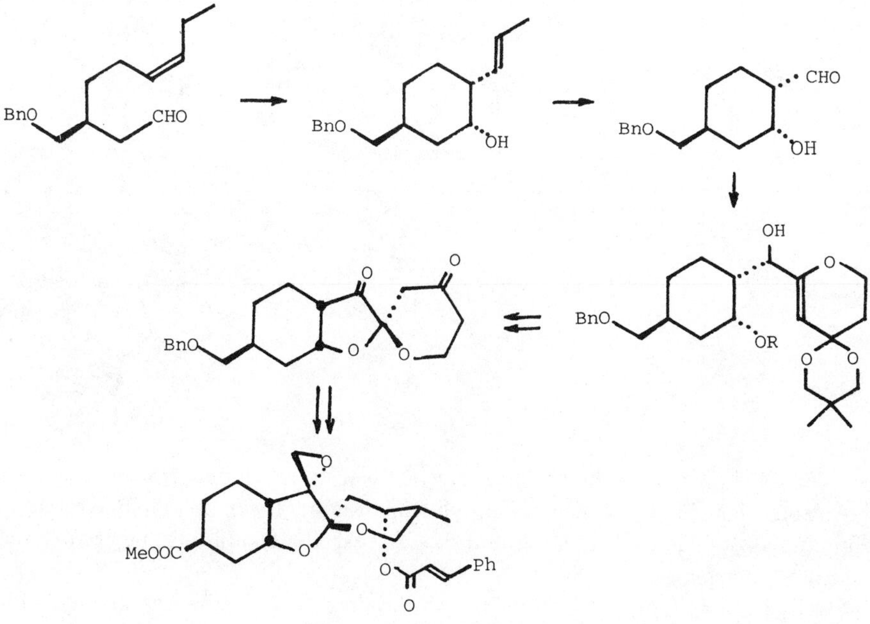

Spirocycles are also formed from pentenylcycloalkenes on thermolysis. The newly created secondary methyl group bears a syn relationship with the (shifted) cycloalkene unsaturation, as dictated by a concerted delivery of the allylic hydrogen atom to the chain terminus. Three different acoranes have been fashioned from the spiro[4.5]decenecarboxylic ester intermediate, which is the ene reaction product of the 2-(1-cyclohexenyl)-5-hexenoic ester: β-acoradiene (**A5**) (Oppolzer, 1973), acorenone B (**A10b**) (Oppolzer, 1975), and acorenone (**A10**) (Oppolzer, 1977).

This powerful method can be extended to the syntheses of isocomene (**I14**) (Oppolzer, 1979) and modhephene (**M12**) (Oppolzer, 1981b, c). The cyclization leading eventually to the former hydrocarbon must be carried out in a homologous system to overcome the excessive strain. However, once the tricyclic skeleton has been established, the cyclohexanone can be contracted by conventional procedures.

There is no such complication in the synthesis of modhephene. It should be emphasized that the application of the ene reaction must be well conceived in stereochemical terms. Once the ring has been made, there is little opportunity for altering the configuration of the methyl substituent.

The cyclization of the ene adduct of β-pinene and acryloyl chloride by base treatment is probably mediated by an ene-type reaction involving a ketene. The product, which has undergone conjugation, is a useful precursor of (+)-β-selinene (**S21**) (Moore, 1983).

A synthesis of 2-deoxystemodinone (White, 1987) consists of the following key steps: (a) polyene cyclization terminated by a β-ketoester unit to construct the AB-synthon, (b) Diels–Alder reaction to give a spirocyclic ketone, and (c) a hydroxyl-assisted intramolecular ene reaction. Regarding the ene reaction, the hydroxyl group is essential; it forms a hydrogen bond with the aldehyde that results in a conformation in which π-overlap between the unsaturations is favored. The corresponding deoxy compound fails to undergo cyclization at 250° or in the presence of a Lewis acid.

An outstanding contribution to the quassinoid area concerns an approach to bruceantin (**B18**) (Ziegler, 1985b). The reaction sequence is inherent of acyclic

8.5 THE ENE REACTION

diastereoselection by a Claisen rearrangement to establish the stereochemistry at C_8 and C_{14}, taking advantage of the preference of such rearrangement for a chair transition state, with bond making from the less hindered side. Consequently, five contiguous asymmetric centers are set within the bicyclic framework. The reaction also supplies an isopropenyl group for eventual formation of the C ring by an ene reaction and the possibility of oxygenating C_{13} and its pendant. Although the model investigation dispenses with the necessary functionalities in the A ring, there are no foreseeable difficulties in adapting the reaction sequence to an actual access to bruceantin.

B18

The observation of cyclic transfer of a halomagnesium residue from an allylic position to the enophile has broadened the outlook of functionalized carbocycle synthesis. Furthermore, the transposed Grignard reagent may then be urged to combine with another carbon chain or ring structure. Some captivating research based on this particular ene reaction has culminated in the synthesis of sinularene (**S37**) (Oppolzer, 1982d), khusimone (**K8**) (Oppolzer, 1982c), $\Delta^{9(12)}$-capnellene (**C21**) (Oppolzer, 1982b), and (+)-iridomyrmecin (**I8**) (Oppolzer, 1986b),

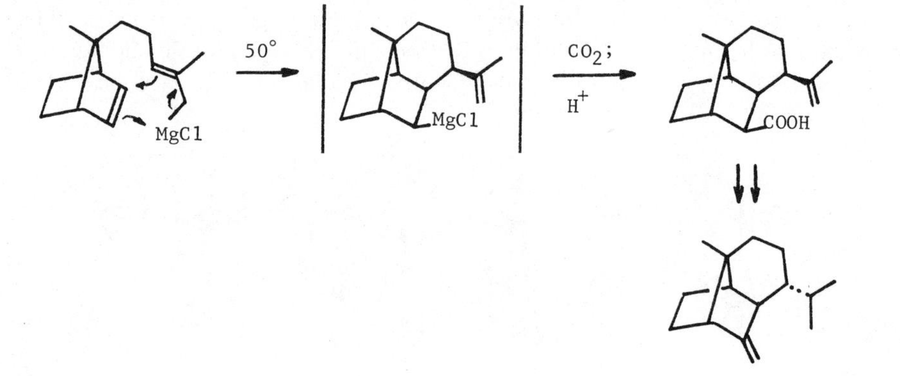

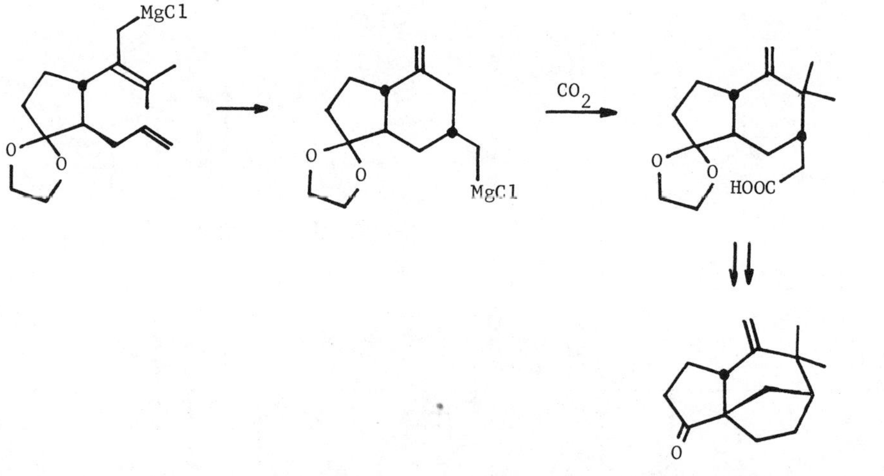

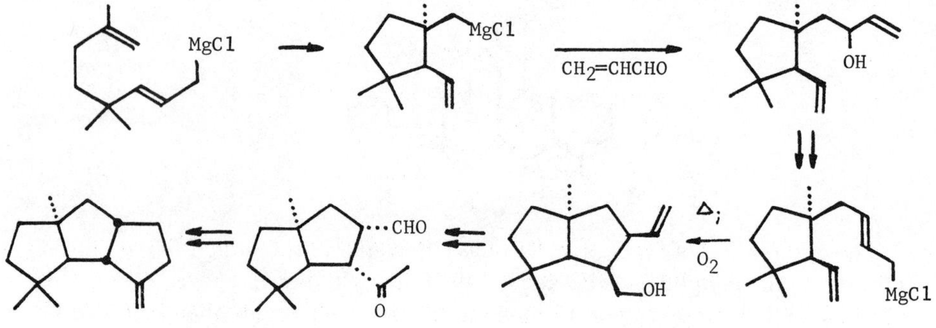

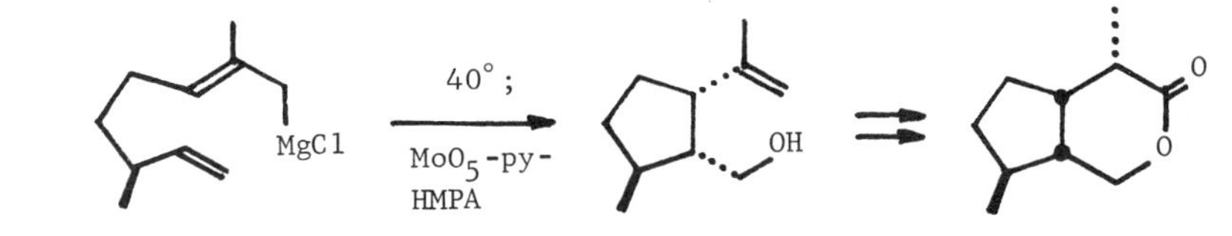

8.5 THE ENE REACTION

and chokol A (Oppolzer, 1986a).

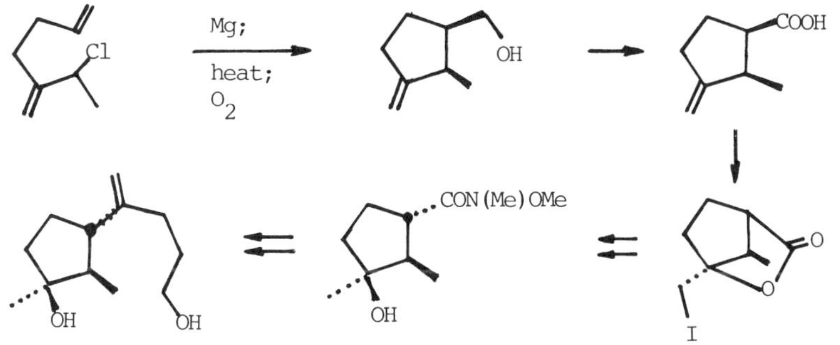

chokol A

and β-necrodol and 1-epi-β-necrodol (Oppolzer, 1986d).

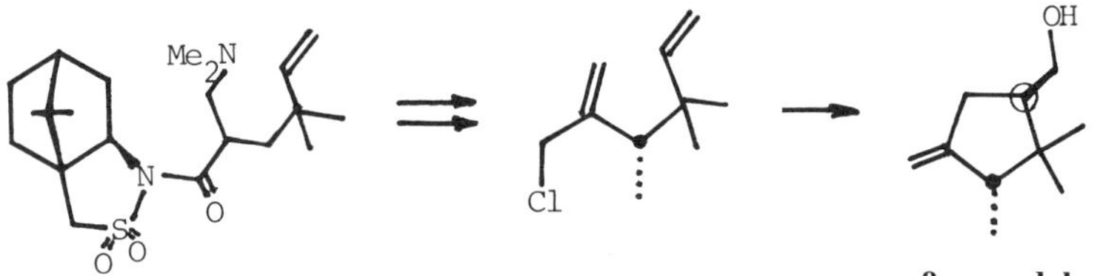

β-necrodol

Further application of the Mg-ene reaction includes closure of the cyclopentane moiety during synthesis of 6-protoilludene (Oppolzer, 1986c). An intramolecular ene/ketene [2 + 2]cycloaddition is incorporated in the synthetic pathway for assembly of the 4:6-ring segment. Unfortunately, the major cyclobutanone has the syn stereochemistry owing to spatial interaction between two methyl groups in the transition state.

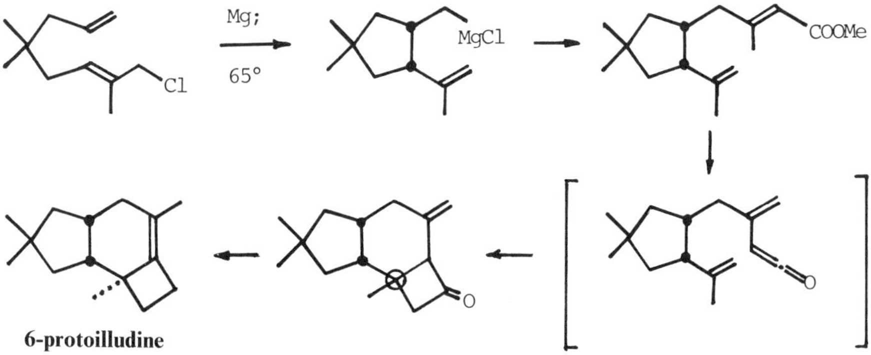

6-protoilludine

The assembly of the picrotoxane skeleton demonstrates an effective palladium(II)-catalyzed ene type reactions. The compound obtained from (—)-carvone via this process lacks only a carboxyl group to be useful for ultimate conversion into the natural terpenes of this series (Trost, 1987).

P29

Ketones containing an unsaturation several atoms apart may undergo cyclization via their enols. A C—C bond is formed between the β carbon of the enol (or the α carbon of the ketone) and an unsaturated carbon, if these atoms are within bonding distance (Conia, 1975). This reaction is essentially an ene reaction in which the ordinary roles of the reactants are switched. Normally a carbonyl group acts as the hydrogen acceptor and C—C bond formation takes place at the ipso carbon (cf. citronellal → isopulegol).

A simple example of this process is a preparation of the menthones (**M11**) (Brocard, 1973). The formation of (+)-camphor (**C15**) from (+)-dihydrocarvone by thermal action at 400°C (Conia, 1978) is more intriguing.

R=H, COOEt

M11

C15

8.5 THE ENE REACTION

Thermolysis of (+)-2-(2-methyl-1,6-heptadien-3-yl)-5-methylcyclohexanone leads to a mixture of acorane-type compounds (Conia, 1971). Presumably because only a minor portion of the product (20% + 7.5%) is potentially convertible to acorone (**A11**)/cryptoacorone, and because of the difficulty of isolating these substances, the actual transformation into the terpene molecule(s) (e.g., via carbonyl transposition, double-bond hydration–oxidation, and equilibration) has not yet been carried out.

A nice scheme for synthesis of quadrone (**Q1**) (Hackett, 1986) involving the Conia reaction is as follows. Successful execution of this concise plan is awaited.

The most ambitious application of the Conia reaction to date is in a synthesis of modhephene (**M12**) (Schostarez, 1981a, b). Although the original form (i.e., that involving the ketoolefin) is doomed to failure because it leads only to the methyl epimer, a reaction of the corresponding bicyclic alkynone leaves a stereoregulatory outlet. The primary cyclized product would be a methylenecyclopentane, and its reduction is controllable.

Actually the cyclized product undergoes a double-bond shift *in situ* to minimize steric compression. However, the correct stereochemistry of the methyl group can be installed by epoxidation and rearrangement of the epoxide involving an internal hydride transfer.

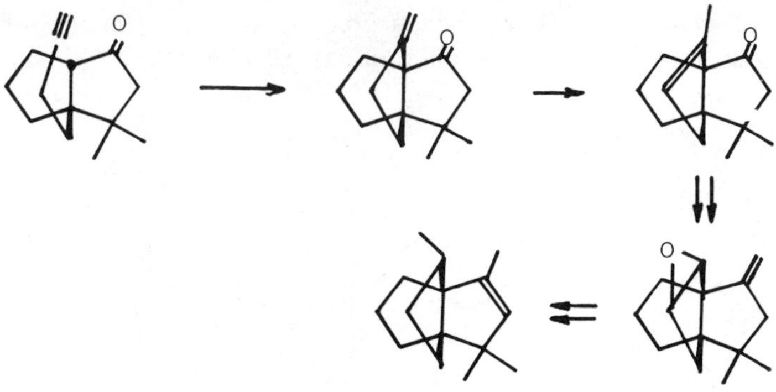

M12

Although the ene reaction does not contribute to ring-forming action in the synthesis of verrucarol (**V12**) described by Trost (1982a), its occurrence serves to protect one of the ketone groups and permits a stereoselective reduction of the other by virtue of the facially discriminative structure produced by the ene process. The original carbocyclic skeleton must be restored by a retrograde ene reaction.

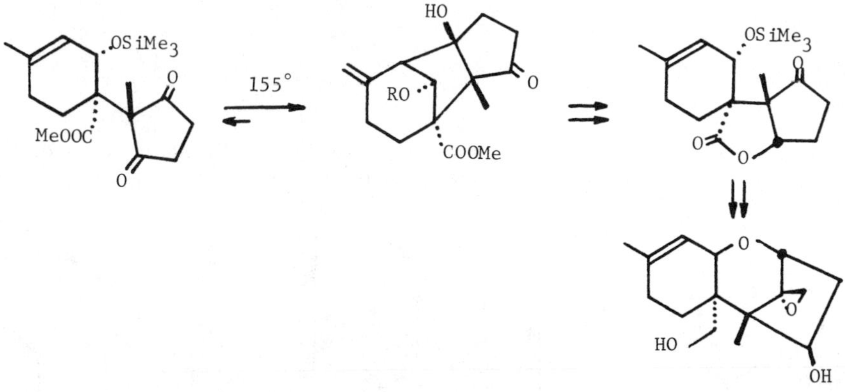

V12

8.6 COPE REARRANGEMENT AND RELATED PROCESSES

The thermal reorganization of 1,5-dienes is known as Cope rearrangement. The reaction is reversible, and it frequently results in an equilibrium mixture. Since there is no net gain in C—C bonding in sigmatropic shifts, the inclusion of Cope-type processes must be in the context of cyclic interconversion.

8.6 COPE REARRANGEMENT AND RELATED PROCESSES

Cope rearrangement often intervenes during the synthesis of mesocyclic 1,5-dienes. In an apparent attempt at synthesizing hedycaryol by coupling a bisallylic bromide with nickel carbonyl, elemane-type products were isolated. Consequently, a synthesis of elemol (**E4**) (Corey, 1969d) materialized. The only isolable 10-membered ring diene is an E,Z-isomer. The Cope rearrangement of this latter isomer is less favorable because of geometrical constraints on its transition state.

E4

A synthesis of isobicyclogermacrenal (**G8a**) (Magari, 1987) from piperitenone via isocaran-2-one hinges on a Cope rearrangement at the last stage. The bicycloelemane precursor is available from a sequence of alkylation and degradation reactions. The Cope rearrangement proceeds smoothly at room temperature in the presence of silica gel, but it cannot be induced thermally at temperatures up to 200°.

G8a

1,2-Dialkenylcyclopropanes are thermolyzed to give 1,3-cycloheptadienes. Sesquiterpenes that contain a seven-membered ring are suitably poised to emerge from such homo-Cope rearrangements, with necessary attention to other molecular details. The construction of β-himachalene (**H14**) (Piers, 1979b) and damsinic acid (**D6**) and confertin (**C62**) (Wender, 1979) suffice to show the usefulness and flexibility of these rearrangements.

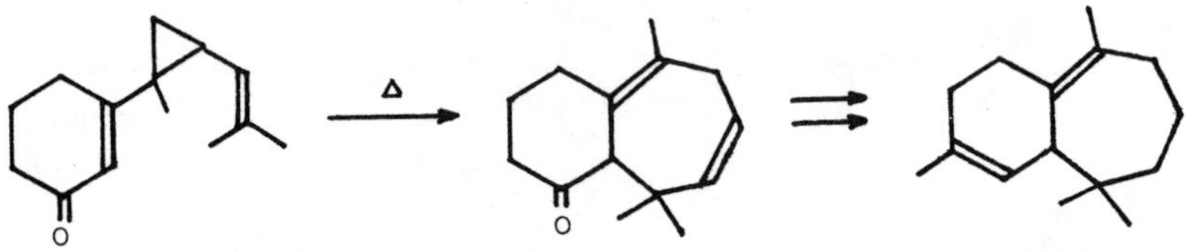

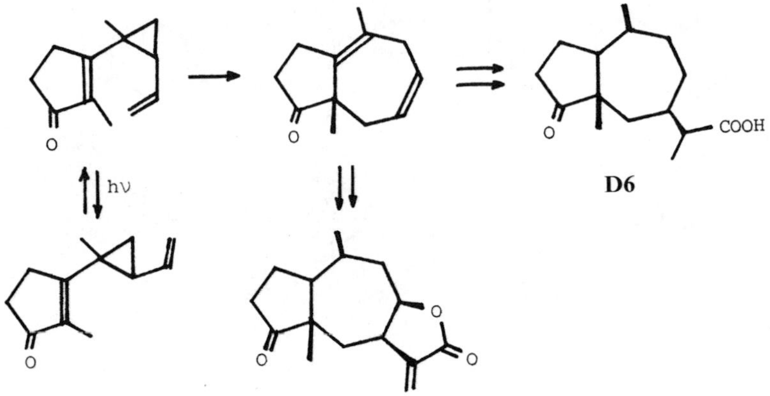

A simple version of this homo-Cope rearrangement provides a short pathway to karahanaenone (**K1**) (Wender, 1976; Cairns, 1982). A heterosubstituent (X = Me$_3$SiO, PhS) facilitates the rearrangement as well as the development of a ketone group in the correct position.

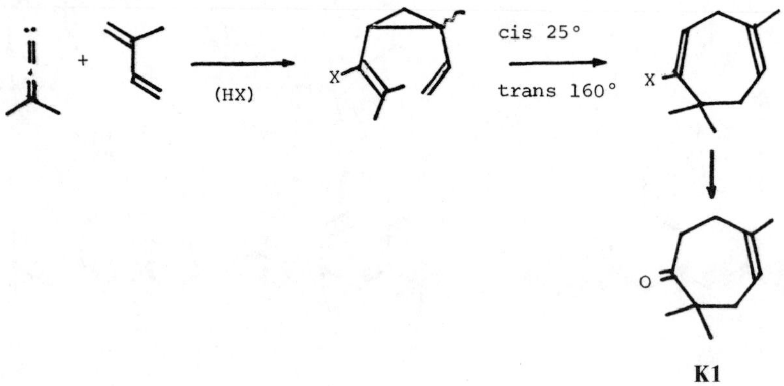

8.6 COPE REARRANGEMENT AND RELATED PROCESSES

When the alkenylcyclopropane unit is fused with its far edge to a cycloalkene ring the rearrangement leads to a bicyclo[n.3.1]alkadiene.

Such ring system becomes even more readily accessible if the intracyclic unsaturation of the precursor is an enol ether. The outstanding applications of the rearrangement process are syntheses of sinularene (**S37**) (Piers, 1985b) and quadrone (**Q1**) (Piers, 1985c).

Although the concerted reorganization requires a cis relationship of the two unsaturated substituents, experience indicates that *trans*-1,2-divinylcyclopropanes also afford 1,3-cycloheptadienes at higher temperatures. It is fortunate for the synthesis that a homo-[1.5]-sigmatropic hydrogen migration is totally absent. The crucial requirement for sinularene is actually the *Z*-configuration of the alkenic side chain rather than its endo orientation, although the endo isomer must be formed *in situ* before the sigmatropic rearrangement can occur.

The same bicyclo[3.2.1]octane unit is hidden in the carbon framework of quadrone; therefore, this architectural principle can be applied to a compound derived from a diquinane.

Similarly, prezizaene (**Z2**) is accessible from a bicyclo[3.1.0]hexene derivative (Piers, 1987a). Underlying this route is the development of a method for synthesis of trisubstituted alkenes by palladium(O)-catalyzed coupling of organozinc reagents with alkenyl iodides. It is also interesting that the intramolecular alkylation gives predominantly the ester with the desired configuration which is probably less congested.

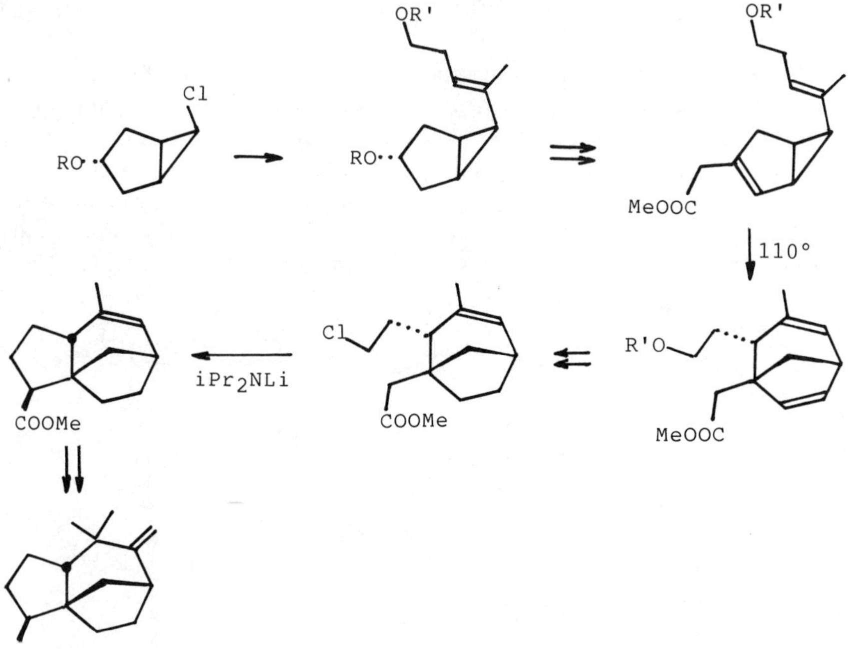

Z2

The value of substituting an allylic hydrogen in a 1,5-diene with a hydroxyl (or siloxy) group in arresting the reversal of a Cope rearrangement has been recognized. The oxy-Cope rearrangement has found use in a synthesis of α-amorphene (**A29**) (Gregson, 1973, 1976). This transformation of a bridged bicyclic system into a *cis*-octalone is highly satisfying.

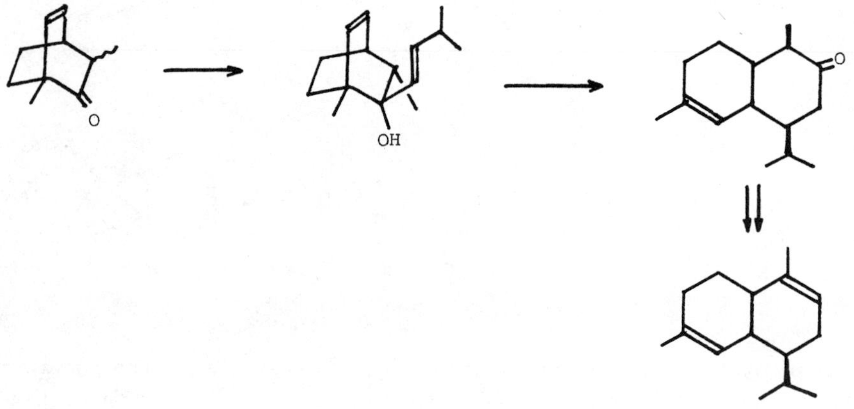

A29

8.6 COPE REARRANGEMENT AND RELATED PROCESSES

There are four contiguous chiral centers in dihydronepetalactone (**N6**). It has been observed that all four stereocenters may be created during the formation of a hydrindanone by an oxy-Cope rearrangement (Fleming, 1984). The surplus carbon atom in the six-membered ring is removed by an oxidative manipulation. The norbornenol (the oxy-Cope precursor) is stereoselectively prepared by desilylative oxygenation of the silylpropylidene norbornene upon treatment with a peracid.

A simple route to the tricyclic ketone intermediate of gibberellic acid (**G12**) is based on the devolution of a Diels–Alder adduct by an oxy-Cope rearrangement into a *cis*-hydrindanone (Corey, 1982b). The plan also calls for a coupling between the angular 2-bromopropenyl chain with the cyclohexanone (effected by treatment with lithium dibutylcuprate).

An elegant approach to pleuromutilin (**P39**) (Boeckman, 1987) embodies the preparation of a bridged octalone via 1,6-addition with lithium dimethylcuprate and trapping the resulting enolate with an angularly substituted ethyl tosylate. The key step leading to the tricyclic skeleton of the terpene molecule is an anionic oxy-Cope rearrangement whereby ring expansion to give a cyclooctanone subunit is accomplished. The double bond shift allows functionalization of the five-membered ring.

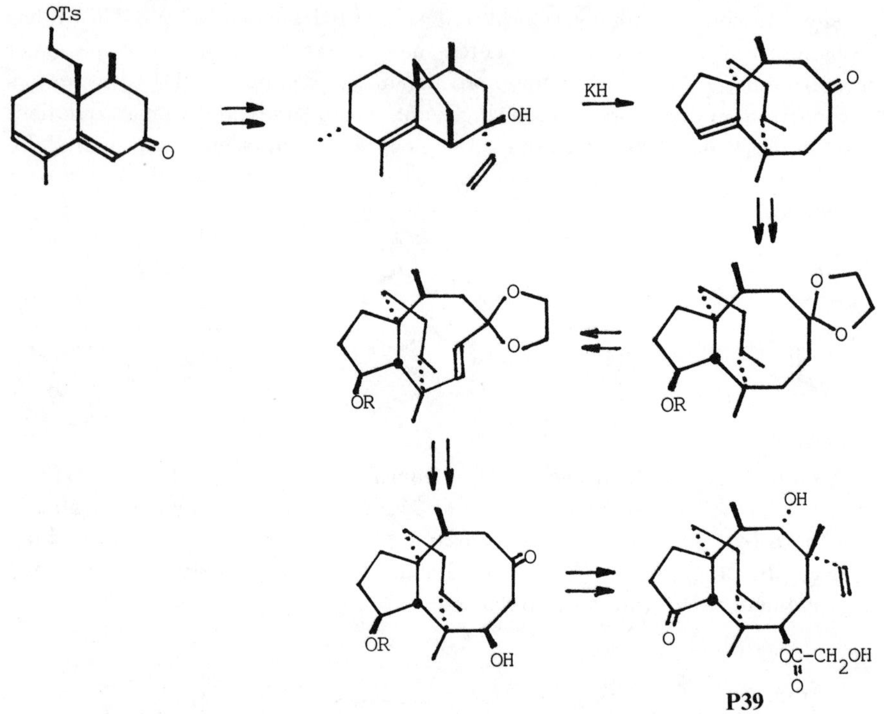

An efficient construction method for the tricyclic framework of the ophiobolins involves an *in situ* anionic oxy-Cope rearrangement (Paquette, 1983). A model study demonstrates the accessibility of functionalized intermediates which hold great promise to the elaboration of the natural terpenes.

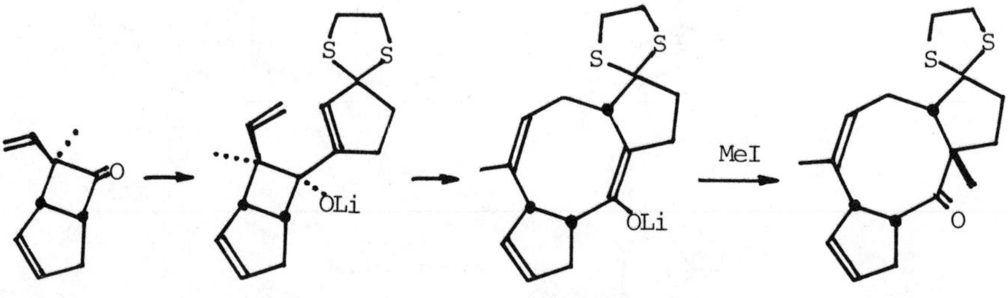

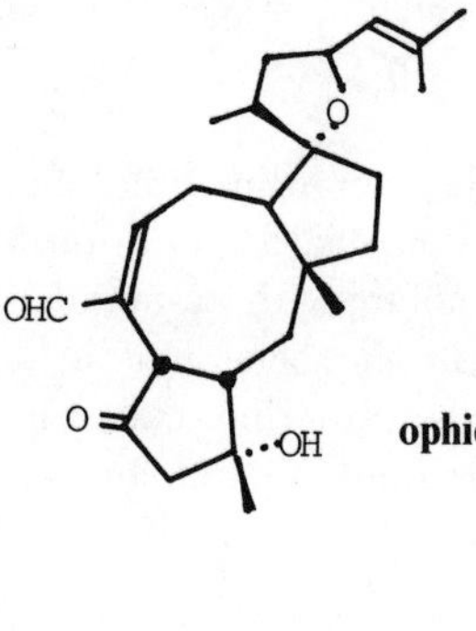

8.6 COPE REARRANGEMENT AND RELATED PROCESSES

Anionic oxy-Cope rearrangement also proves to be a very attractive method for transforming a bicyclo[2.2.2]octenone system into the forskolin (**F9**) skeleton (Oplinger, 1987). Although a model study indicates the generation of a diastereomer requiring configurational correction at two angular centers, it is expected that the problem will be resolved when the C_{11} carbonyl is set in place.

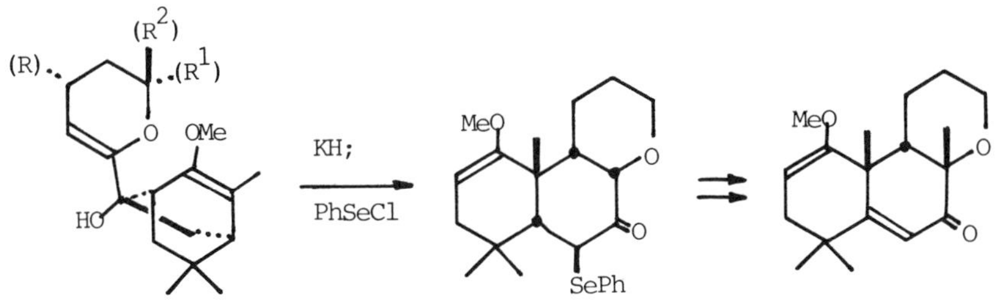

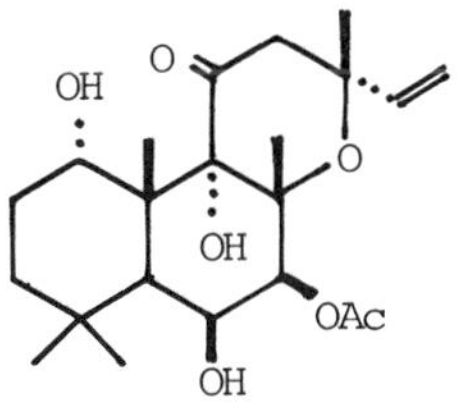

Anionic oxy-Cope rearrangement transforms a substituted carveol into a cyclodecadiene and thence heliangolide (Nakamura, 1986).

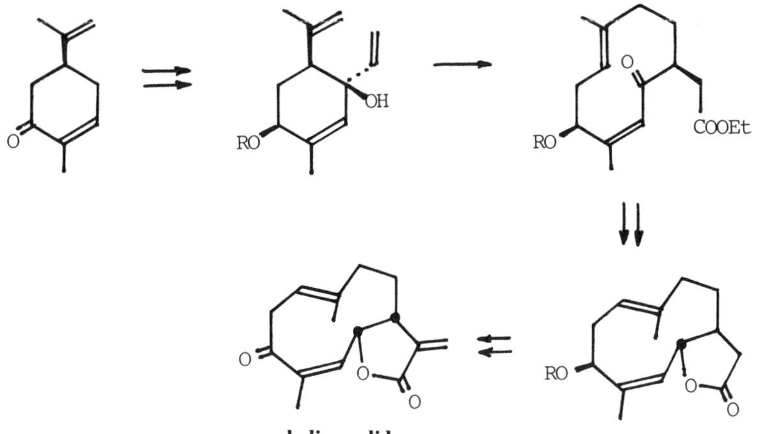

heliangolide

Pyrolysis of 2-*endo*-vinyl-3-*endo*-propenyl-2,3-bornanediol gives a 1,6-diketone that cyclizes *in situ*. One of the enone products may be reduced to β-patchoulene (**P12**) (Leriverend, 1970).

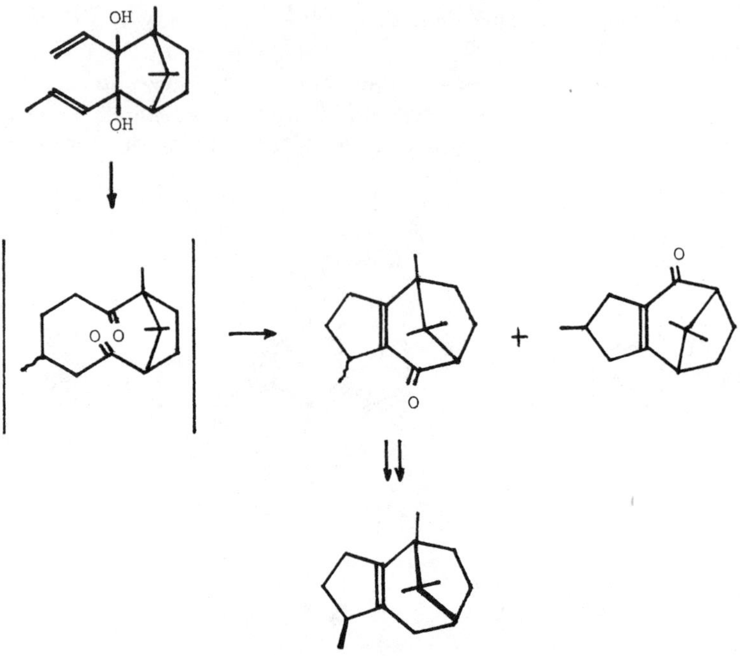

P12

The possibility of achieving consecutive thermal reactions is always aesthetically and economically appealing. An example is the combination of the Cope and Claisen rearrangements that provides elements for the seven-membered ring of aromatin (**A43**) (Ziegler, 1981).

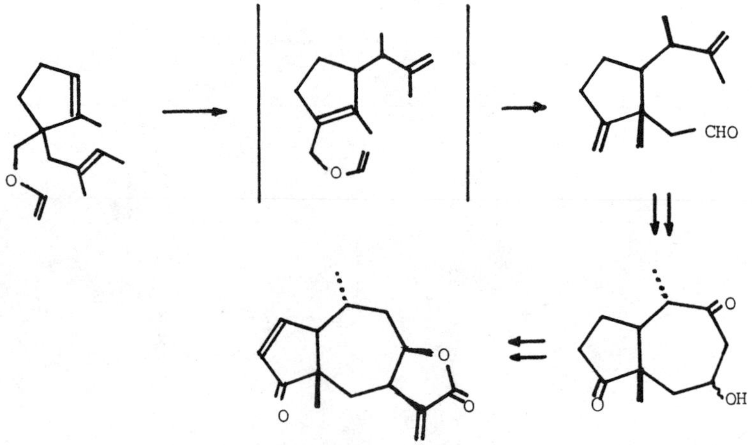

A43

Tandem Cope–Claisen rearrangements are the crux of a synthesis of (+)-dihydrocostunolide (**C76**) (Raucher, 1986) from (4S)-dihydrocarvone. The route is slightly lengthened by the necessity of protecting the more reactive trisubstituted double bond of the cyclodecadiene intermediate before the allylic lactone unit can be established by reaction with osmium tetroxide, inversion of the hydroxyl group and a dehydration sequence.

C76

8.7 CLAISEN AND WITTIG REARRANGEMENTS

The aliphatic Claisen rearrangement of allyl vinyl ethers (Rhoads, 1975; Ziegler, 1977) renders excellent service in the synthesis of carbocycles of different sizes. Besides providing a simple way to form karahanaenone (**K1**) from linalool (Demole, 1969, 1971; Uneyama, 1985), the reaction has helped set up precursors of perforenone A (**P20**) (Gonzalez, 1978), widdrol (**W3**) (Danishefsky, 1980a), and quadrone (**Q1**) (Funk, 1985a, 1986). A potential intermediate of dictyol-C (**D13**) has also been obtained (Begley, 1984). The Ireland version of this rearrangement involving ketene silyl ethers is particularly efficient and mild. The possibility of acquiring either the E- or the Z-isomer enables the chemist to achieve stereocontrol of the final product.

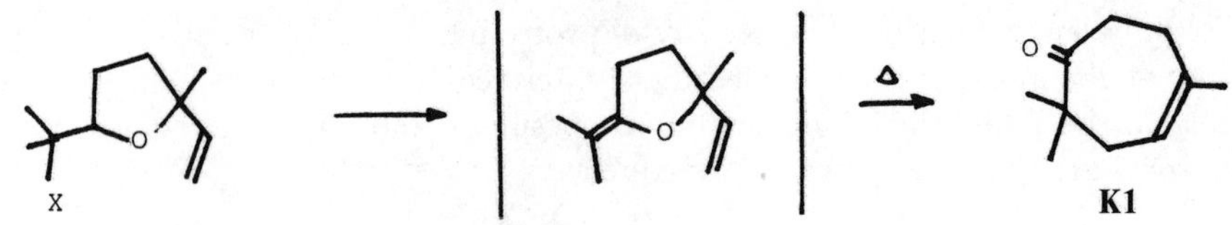
X
O
O
Δ
O

K1

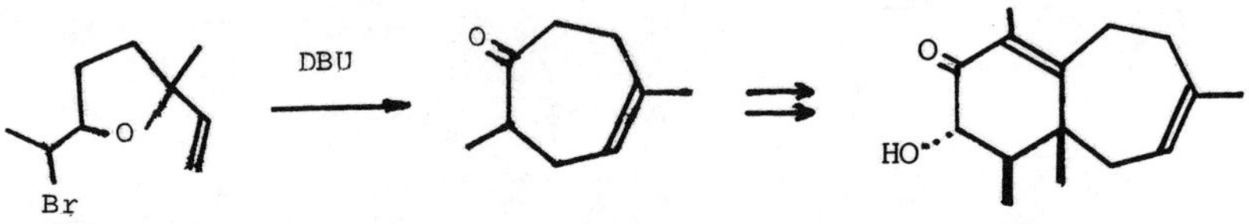
Br
O
DBU
O
O
HO

P20

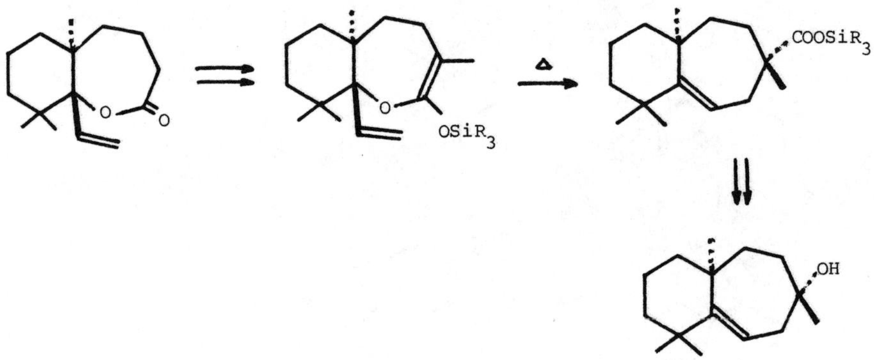
O
O
O
OSiR3
Δ
COOSiR3
OH

W3

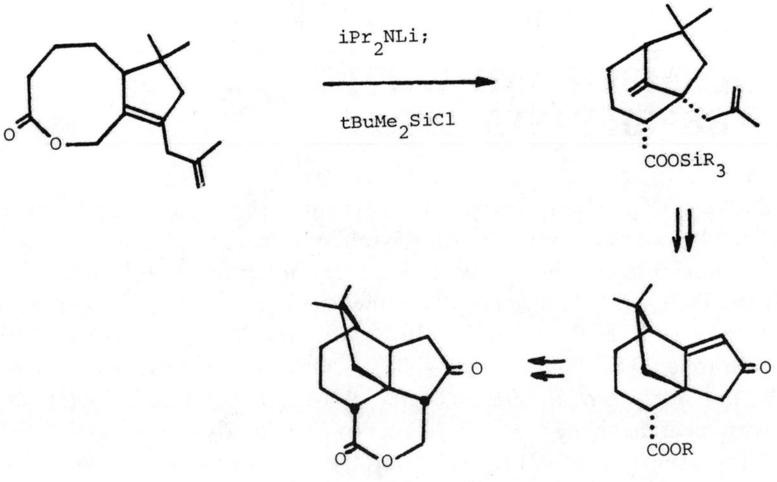
O
O
iPr2NLi;
tBuMe2SiCl
COOSiR3
O
COOR
O
O
O

Q1

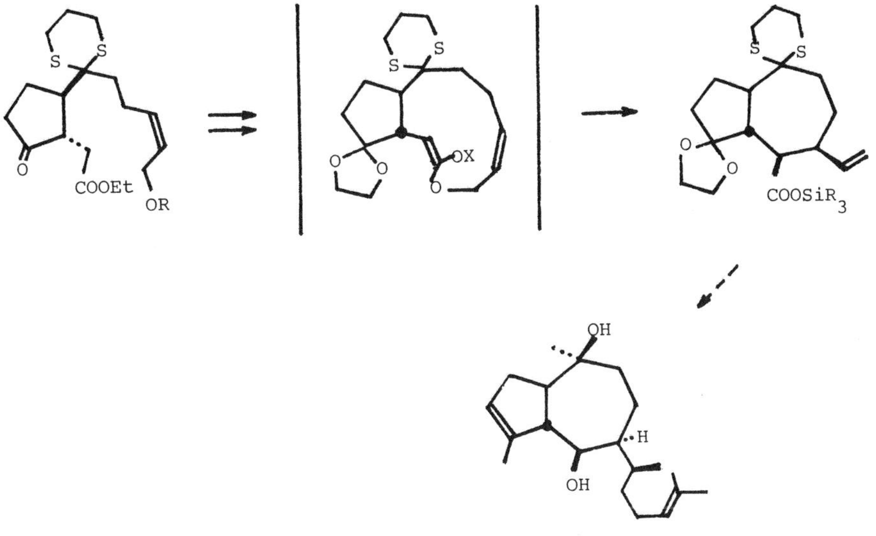

8.7 CLAISEN AND WITTIG REARRANGEMENTS

D13

It is also possible to form cyclooctenones in a manner analogous to the preparation of the perforenone precursor. The expedient conversion of a lactone carbonyl into the vinylidene group makes the Claisen rearrangement route much more attractive. Indeed the soft coral metabolite precapnelladiene (**P46**), which contains a fused cyclooctadiene skeleton, is amenable to this approach (Kinney, 1984).

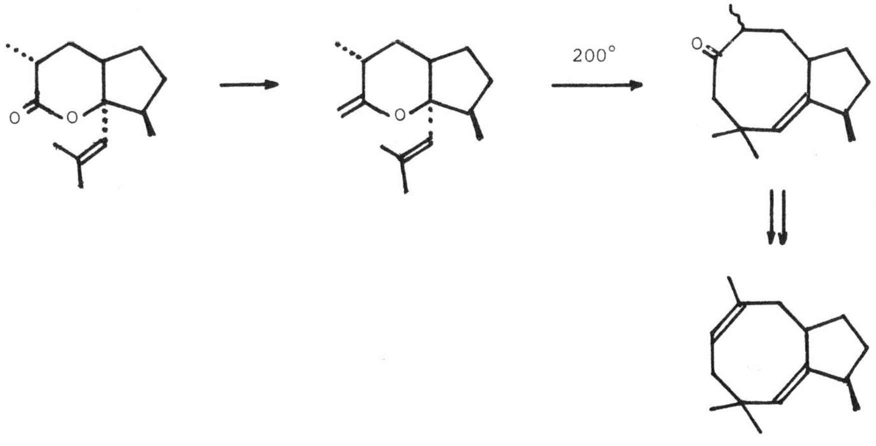

P46

A spirocyclization via the Claisen rearrangement has been demonstrated in a synthesis of β-chamigrene (**C47**) (Ireland, 1984) and the preparation of a pivotal intermediate for aphidicolin (**A36**) (Ireland, 1984).

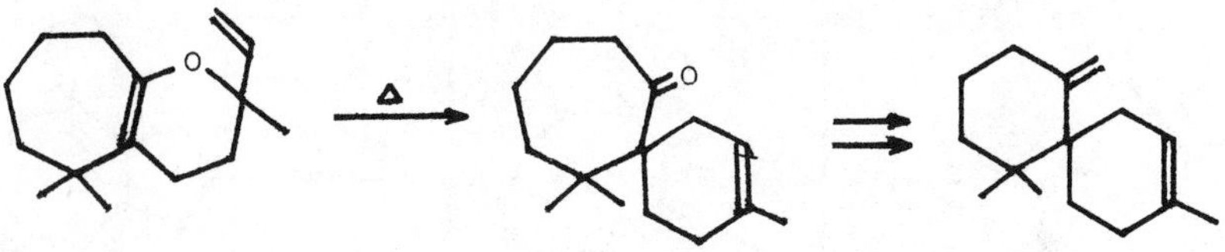

C47

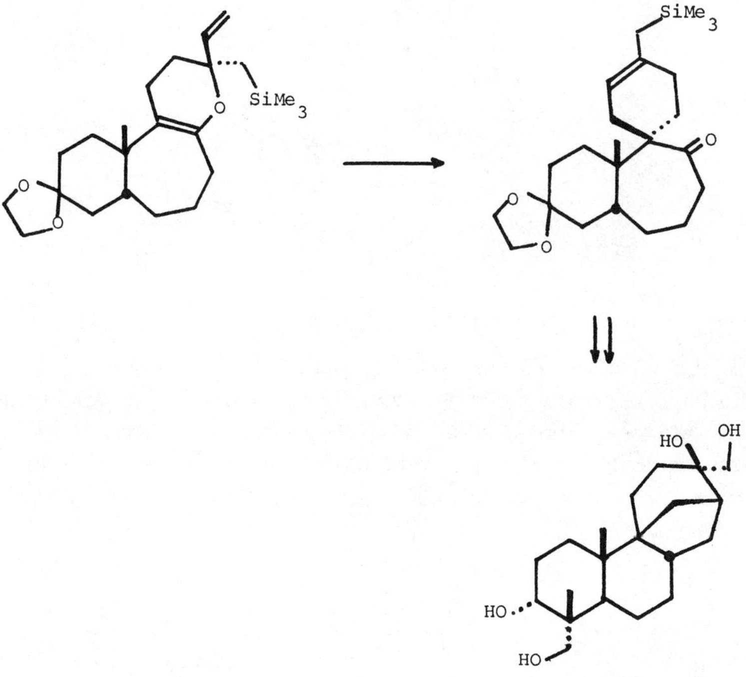

A36

The Claisen rearrangement of the ketene silyl ethers derived from $\Delta^{(\omega-2)}$-lactones opens a new contractive pathway for carbocyclization. Furthermore, the 2-vinylcycloalkanecarboxylic acid products possess a defined stereochemistry that correlates with the transition state of the rearrangement. The facile access of *cis*-chrysanthemic acid (**C51c**) (Funk, 1985b) and dihydronepetalactone (**N6**) (Abelman, 1982) and isodihydronepetalactone (**N6'**) attests to the usefulness of the process.

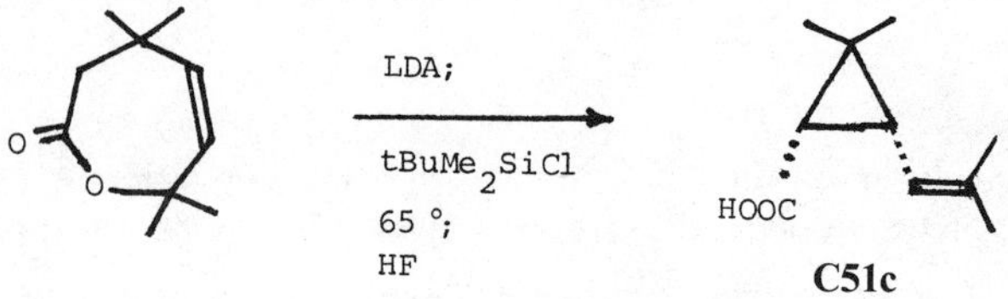

C51c

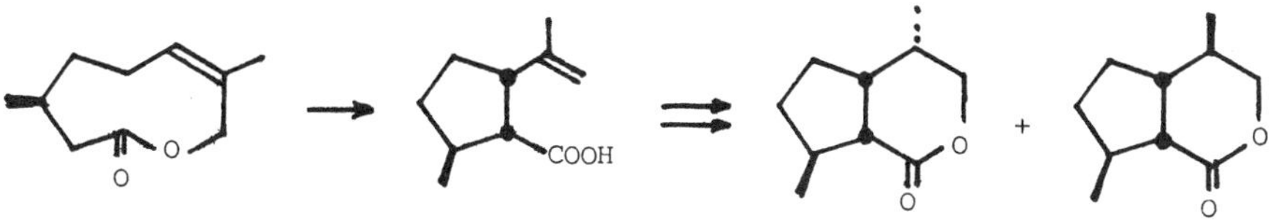

8.7 CLAISEN AND WITTIG REARRANGEMENTS

This efficient and versatile approach to carbocyclic construction has now been extended to a model of taxinine (Funk, 1987a). The desired rearrangement reflects the adoption of a transition state in which severe transannular interaction between the angular methyl group and the cyclohexadiene ring is avoided.

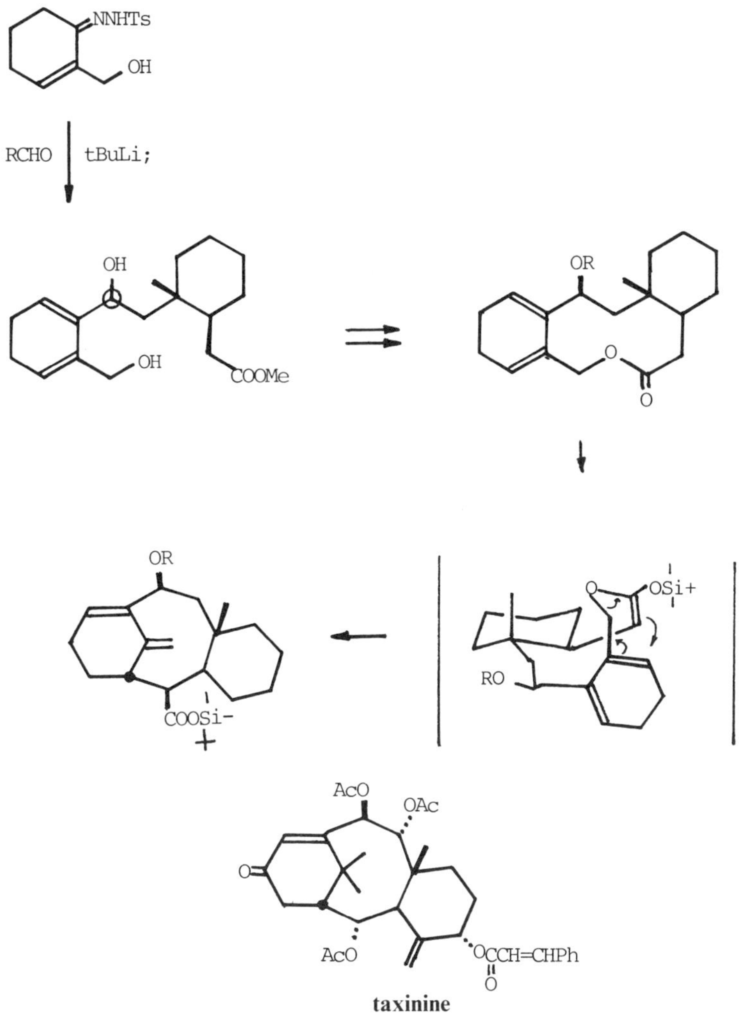

taxinine

Another intriguing possibility concerns the synthesis of ingenol (**14**) (Funk, 1987a). Thus, the B ring has been closed in the following exquisite manner from a suitably functionalized intermediate. The crux of this elegant approach starting from 3-carene is the facile bridging of a relatively unstrained inside-outside macrobicyclic lactone to avoid direct carboannulation of a smaller and strained ring from two dangling carbon chains. Stereocontrol of the β-ketoester dianion alkylation is given by the cyclopropane. The alkoxybutenyl pendant just introduced in turn directs the Michael reaction at the β-face of the 7-membered ring.

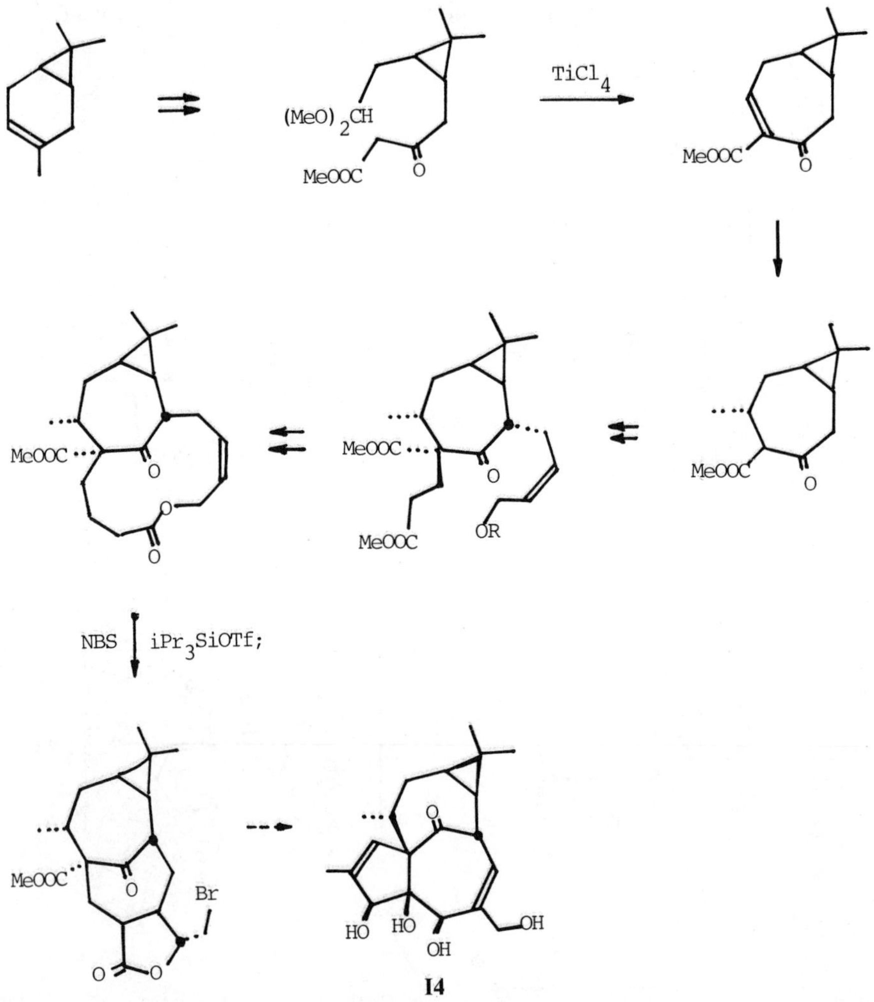

An aromatic version of the Claisen rearrangement of lactonic silyl enolates has been developed to annulate a tetrahydrobenzofuran moiety onto a six-membered ring, thus providing a novel pathway to eudesmane sesquiterpenes (Nemoto, 1987).

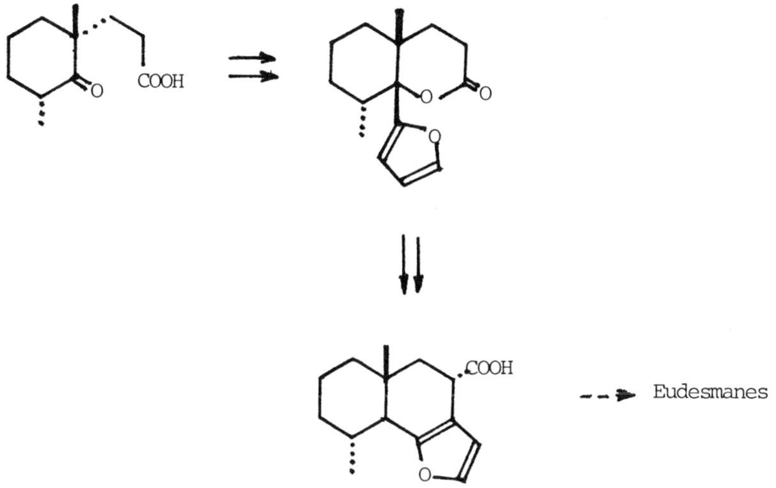

The [2.3]Wittig rearrangement has not been explored synthetically until recently. It seems that terpenes containing medium-sized rings are amenable to synthesis by ring contraction based on the rearrangement. Furthermore, stereoselectivity of the process turns out to be much higher than initially anticipated.

Syntheses of costunolide (**C75**) (Takahashi, 1986b), haagenolide (Takahashi, 1987), epimukulol (Marshall, 1986d), and aristolactone (Marshall, 1987c, d) serve to highlight the efficiency of the reaction and the suitability in terpene synthesis since an isopropenyl chain can be generated from a trisubstituted allyl ether.

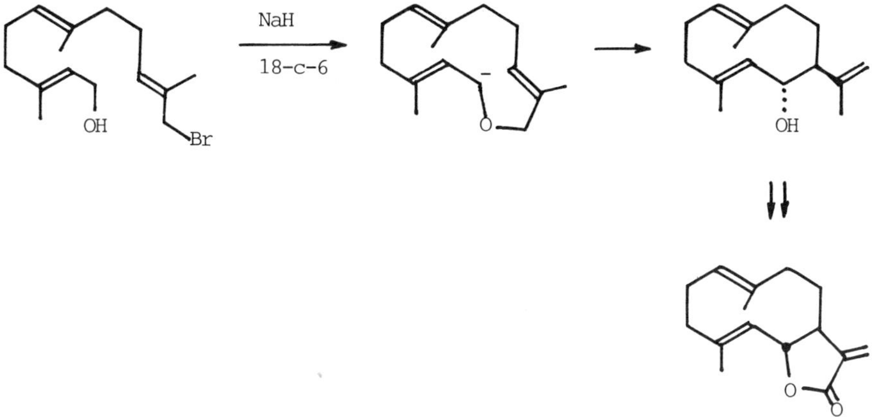

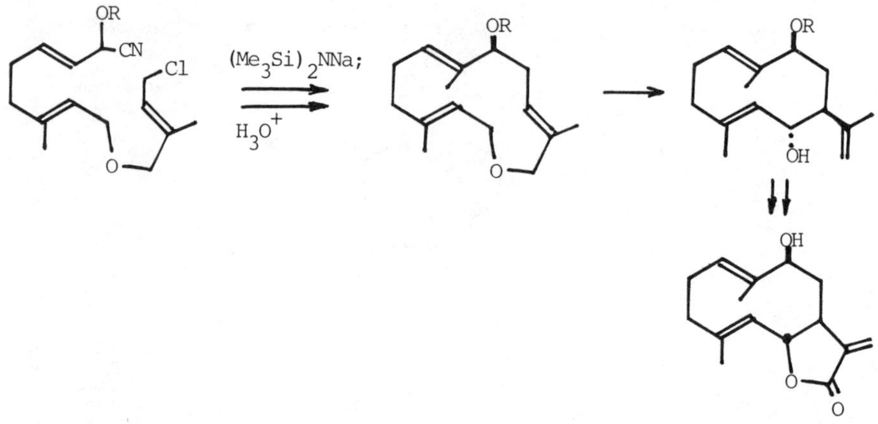

haagenolide

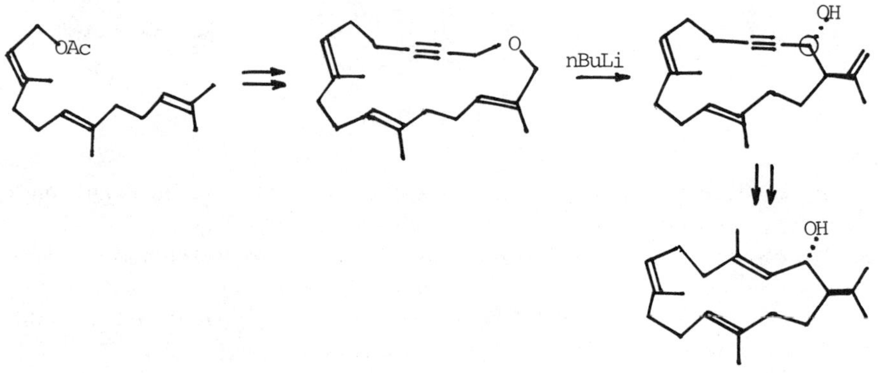

eqimukulol

Enantioselective Wittig ring contraction of propargyl allyl ethers has been effected using optically active amide bases. Thus, (+)-aristolactone (Marshall, 1987c) is available through a reaction sequence embodying a lithium (S,S)-bis-(1-phenethylamide) induced rearrangement.

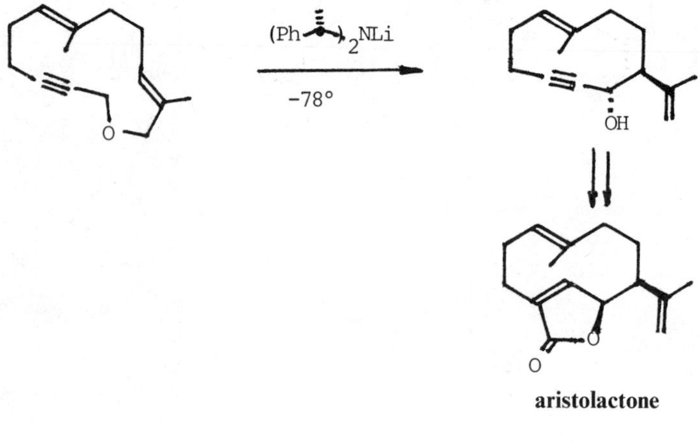

aristolactone

8.8 ELECTROCYCLIZATION

Electrocyclic reactions (Woodward, 1971) are also useful for ring transformations. The conjugated retro-tetraene derived from vinyl-β-ionol cyclizes into a spirocyclic molecule on heating at 100–110°C. Selective hydrogenation of the cyclohexadiene unit leads to α-chamigrene (**C46**) (Frater, 1977).

On the other hand, drimane-type compounds are generated by thermolysis of β-ionylideneacetic esters (Frater, 1974b).

A synthesis of pallescensin A (**P2**) (Liotta, 1987) calls for electrocyclization of a β-furyldiene. *In situ* [1,5]sigmatropic hydrogen migration gives rise to a product with cis ring juncture. However, equilibration of this product to the more stable trans isomer occurs in the presence of certain catalysts such as silica gel. The synthesis requires preparation of a silylated furan in order to prevent saturation of the heterocycle during reduction of the triple bond.

The structure of genipin (**G3**) suggests for its synthesis an aldol condensation for closure of the five-membered ring and hence a cyclohexene precursor (with the dialdehyde as intermediate). The choice of starting material ultimately rests on the availability of suitably substituted carbocyclic substances (Büchi, 1967). It is noted that a dihydroindenecarboxylic ester is electrocyclically related to the monoadduct of cyclooctatetraene and carbethoxycarbene, the use of which in synthetic studies is logical.

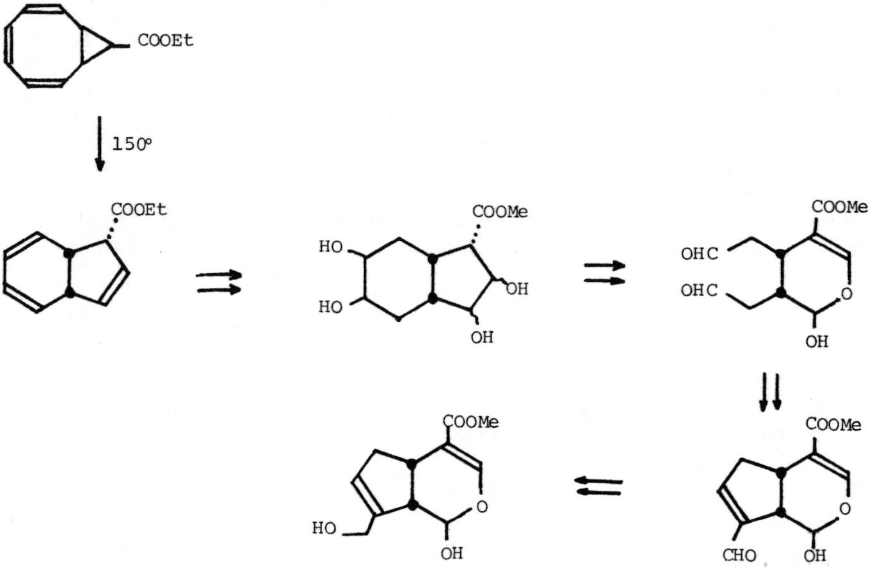

Photoinduced electrocyclization of eucarvone forms a critical step in the synthesis of grandisol (**G18**) reported by W. Ayer (1974). With the problem of establishing a properly substituted cyclobutane resolved, the two side chains are elaborated by a systematic degradation of the five-membered ring.

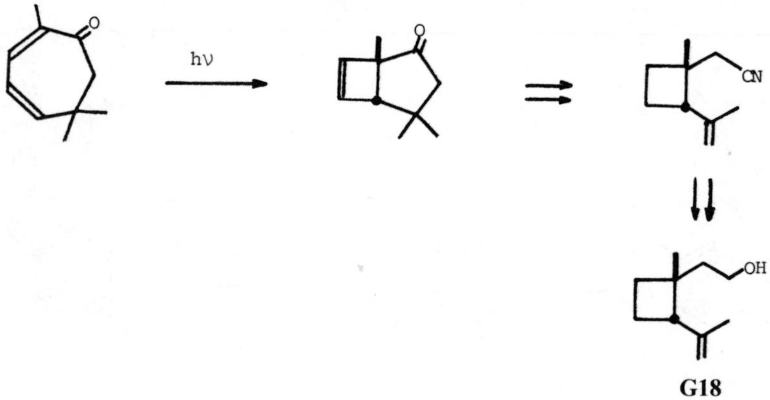

8.8 ELECTROCYCLIZATION

The light-induced conrotatory electrocycloreversion of cyclohexadienes has been used to assist in the acquisition of dihydrocostunolide (**C76**) (Corey, 1965a). The *trans*-fused hexalin derivative opens to an E,Z,E-cyclodecatriene; therefore hydrogenation of the least substituted central double bond accomplishes the synthetic goal.

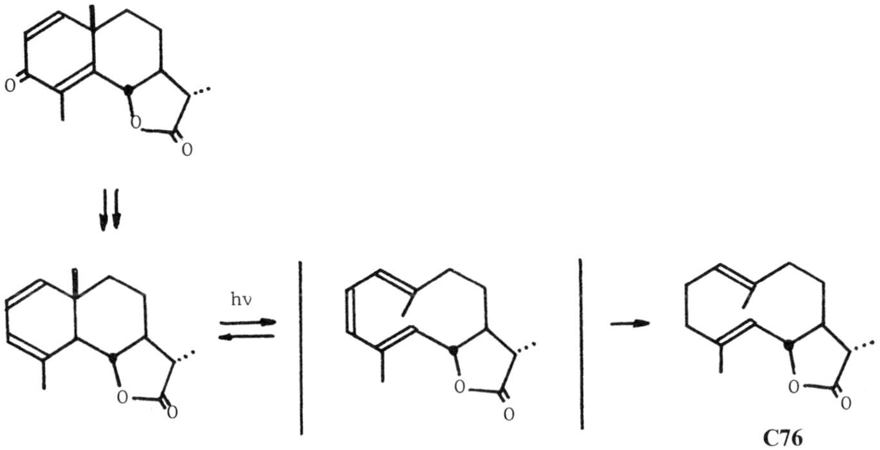

An alternative synthesis strategy involves formation of a trienol intermediate that ketonizes immediately to arrest unwanted thermal reactions (Fujimoto, 1976). However, the E-configuration of the double bond that needs to be reintroduced could not be guaranteed.

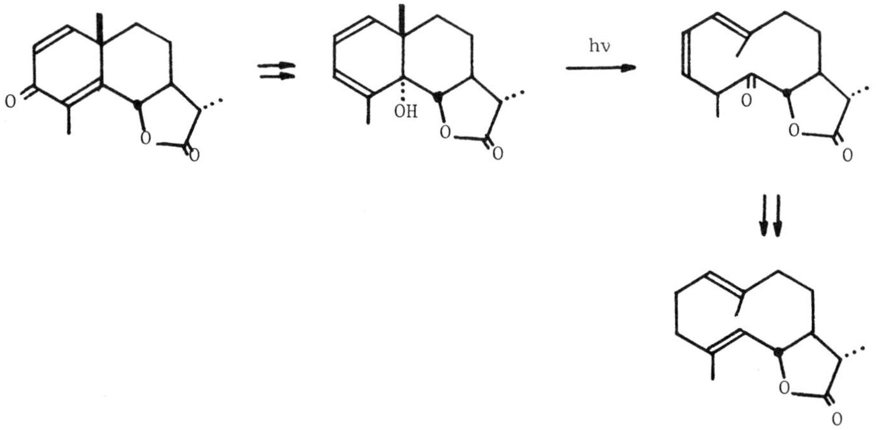

On deliberately allowing the E,Z,E-cyclodecatriene intermediate to recyclize on warming to $-20°C$ (which proceeds by a disrotatory pathway), a *cis*-fused homonuclear hexalin is produced. The reaction sequence is ideal for synthesizing occidentalol (**O2**) (Hortmann, 1973).

Both the natural (+)-occidentalol and its enantiomer are available from (+)-carvone. The E,Z,E-cyclodecatrienecarboxylic ester can undergo two distinct disrotatory cyclization processes leading to bicyclic compounds in which the ester appendage is either trans or cis to the substituents at the ring juncture. The former product is a direct precursor of (+)-occidentalol, whereas the latter compound can undergo epimerization of the ester group and be converted into (—)-occidentalol.

Alkenyl(methoxy)carbene-chromium(O) complexes react with alkynes to form 2,4-cyclohexadienones including spiroannulated derivatives. Application of the reaction to a synthesis of (—)-acorenone (**A10**) (Wulff, 1987) shows a high diastereoselectivity with which electrocyclization of the complexed dieneketene is directed to the face of the cyclopentane ring away from the existing alkyl substituents.

8.8 ELECTROCYCLIZATION

A10

The Nazarov reaction (Nazarov, 1949; Santelli-Rouvier, 1983) is an electrocyclization involving four electrons spreading among five carbon atoms. Accordingly, the stereochemistry of the β (and β′) substituents in the resulting cyclopentenones can be predicted because it correlates with conrotation of the two alkenes.

The Friedel–Crafts reaction between tigloyl chloride and acetylene embodies a Nazarov cyclization. The reaction intermediate adds hydrogen chloride to stabilize the unstable cyclopentadienone. The adduct has been used in a synthesis of aplysin (**A37**) (Ronald, 1976).

A37

The frequently encountered intermediate for cuparene/β-cuparenone (**C84/C86**) is easily prepared by the Nazarov cyclization. The cross-conjugated dienone in turn is an acylation product of β-silyl-p-methylstyrene (Paquette, 1980).

C86

The Nazarov reaction is also the most convenient method for constructing the diquinane synthons of $\Delta^{9(12)}$-capnellene (**C21**) (K. Stevens, 1981) and modhephene (**M12**) (Oppolzer, 1981b, c; Schostarez, 1981a, b).

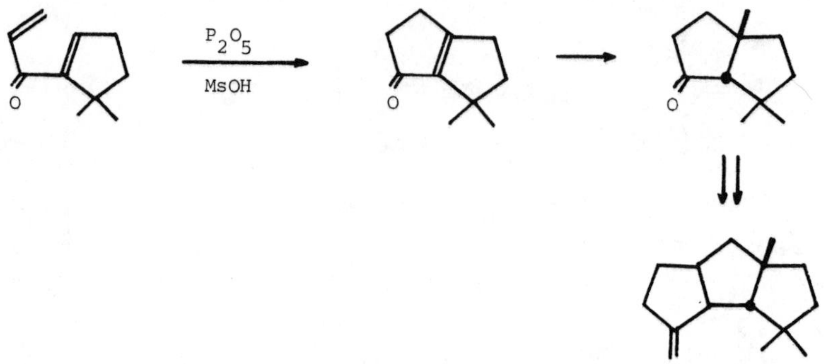

Iterative Nazarov cyclizations are the key steps in another approach to $\Delta^{9(12)}$-capnellene (**C21**) (Crisp, 1984). The cross-conjugated dienones are formed by carbonylative coupling of vinyl triflates with a vinylstannane.

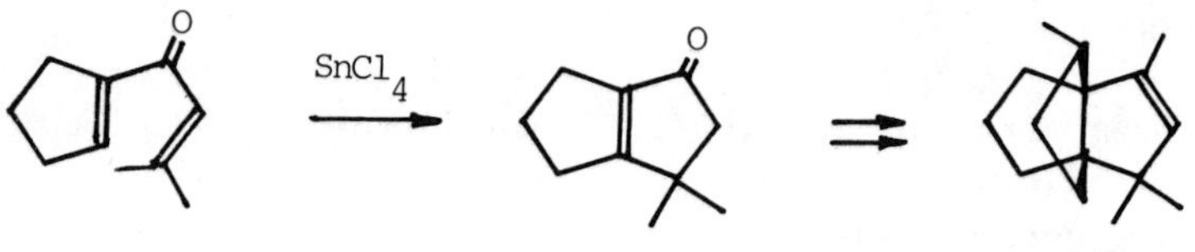

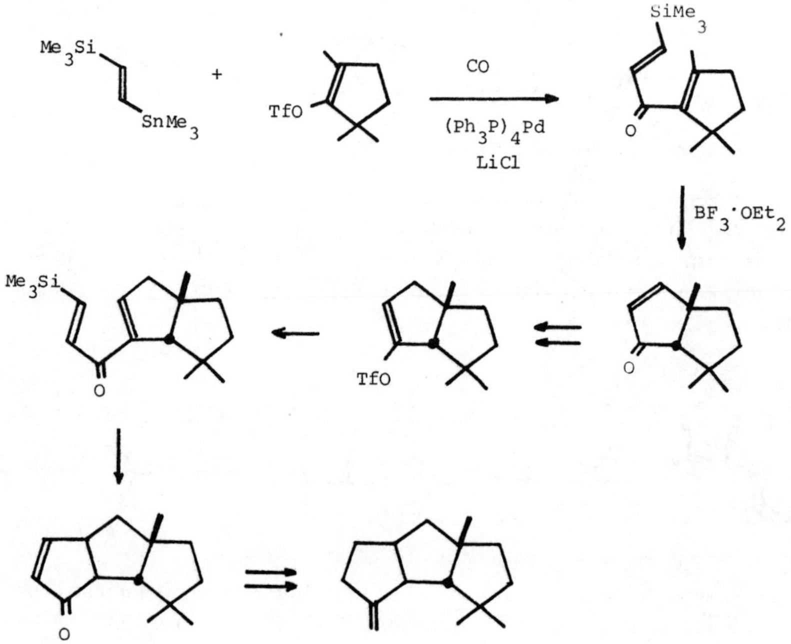

8.8 ELECTROCYCLIZATION

The bicarbocyclic portion of strigol (**S55**), a potent germination factor of witchweed, can be prepared from 2,2-dimethylcyclohexanone via the alkynediol and a 1,4-pentadien-3-one (MacAlpine, 1974).

S55

A similar route has been designed for building the A ring of nootkatone (**N11**) (Hiyama, 1979). By placing methyl groups at both the β and β' carbon atoms of the dienone intermediate, the vicinal dimethyl group having a cis relationship evolves during the cyclization. The limitation of forming only five-membered rings by this method of course necessitates a ring expansion protocol to reach nootkatone.

N11

A Nazarov cyclization is employed in a synthesis of norsterepolide (n-**S51**) (Arai, 1985). It is conceivable that kinetic α-methylation will lead to the natural terpene.

S51

In pleuromutilin (**P39**) there is a 1,5-ketol unit that can be related through a Michael reaction (via the diketone). In other words, it should be feasible to close the eight-membered ring from a bicyclic diketone intermediate. The synthesis of this diketone (Paquette, 1985a, c) has been effected commencing with the Nazarov cyclization.

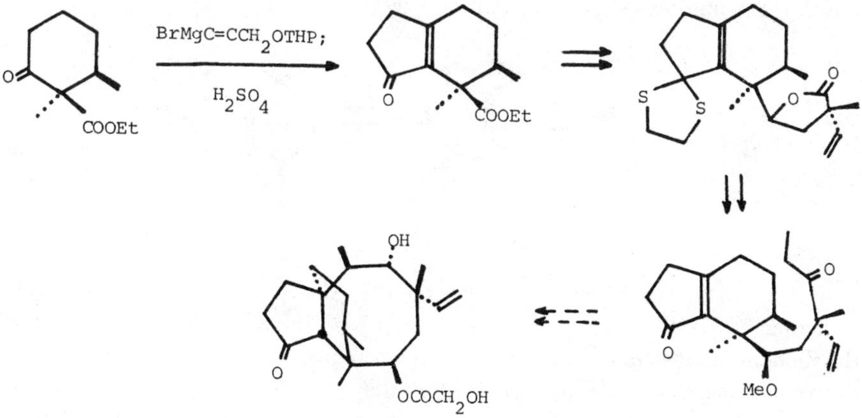

P39

The difficult task of stereocontrol at the two quaternary carbon atoms during synthesis of trichodiene (**T26**) is resolvable by the formation of a tricyclic ketone via the Nazarov reaction. Because both conjugate double bonds are in the Z-configuration, the conrotation automatically generates a product in which the methyl substituents are anti to each other. Only the regioselective cleavage of the tricyclic compound awaits accomplishment. Fortunately, the relative strain of cyclohexene versus cyclopentene favors the location of the double bond at the former portion of the molecule, and the scission of the C—C bond linking to the cyclohexene is now biased electronically (Harding, 1984).

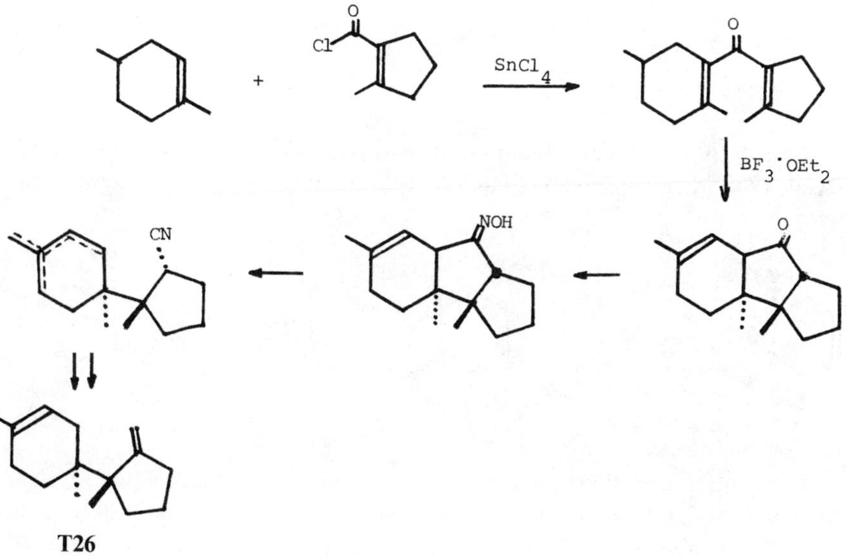

T26

8.8 ELECTROCYCLIZATION

The linear tricyclic ketone precursor of illudinine (**I2**) (Woodward, 1977), which is obtained together with its isomer by acid treatment of the β-chloroketone, may be considered to be a Nazarov product.

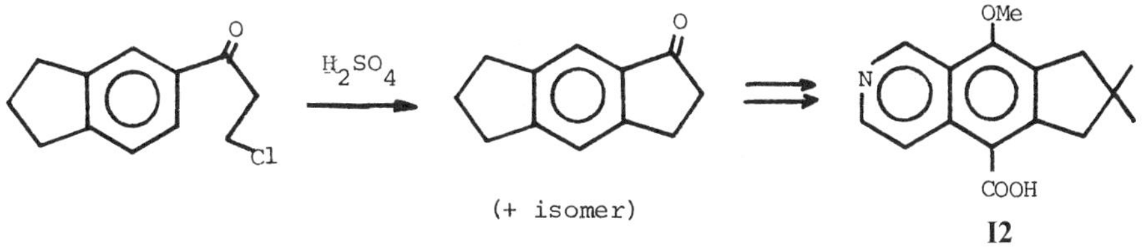

The phenol synthesized by condensation of alkynes with cyclobutenones has been exploited in an approach to barbatusol (Kwasigroch, 1987). Thus, the A ring, constructed with a methoxypropargyl pendant which reacts with 3-isopropylcyclobutenone to give a resorcinol monomethyl ether, has been completed. Plan calls for C—C coupling at the phenol carbon with another side chain of the cyclohexene nucleus to form the 7-membered ring.

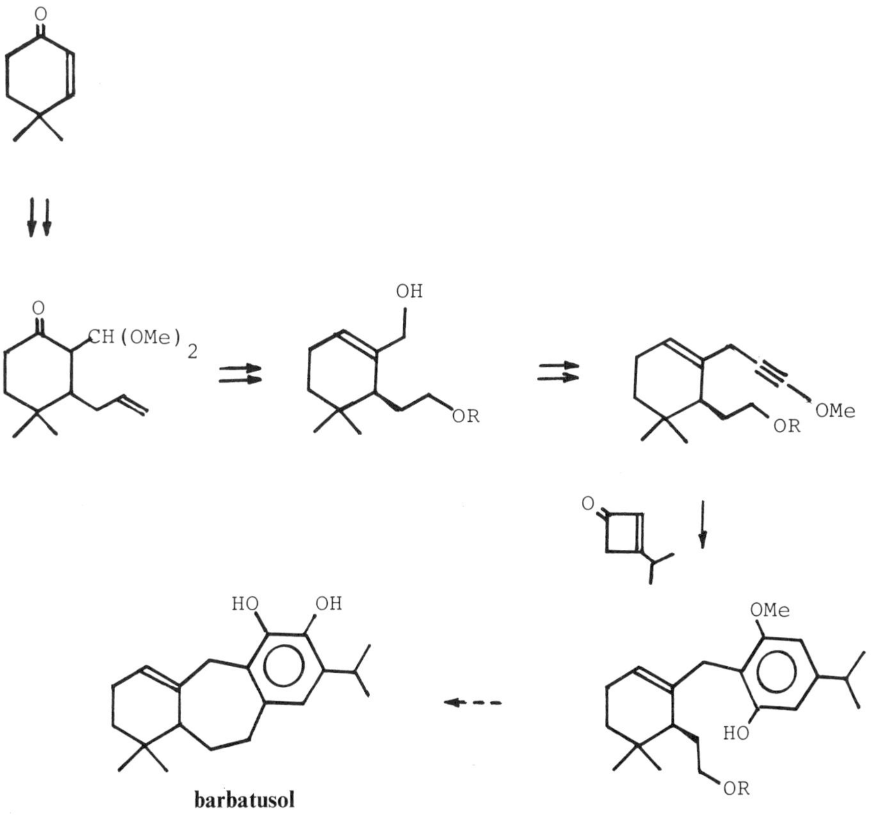

barbatusol

This phenol annulation involves electrocyclic cleavage of the cyclobutenone, cycloaddition of the resulting ketene to the alkyne molecule, another electrocyclic opening followed by a π6e cyclization.

8.9 THERMAL REORGANIZATION OF VINYLCYCLOPROPANES

Vinylcyclopropanes are converted to cyclopentenes on heating. Despite the nonconcertedness of this reorganization, its utility is still substantial. For example, the carbon frameworks of *cis*-sativenediol (**S12c**) (McMurry, 1976) and longifolene (**L24**) (Schultz, 1985) are established by this method.

Synthetic approaches to vetispiranes embodying this ring expansion technique in forming the spirocycle are very expedient. A 1,2-asymmetric induction by the secondary methyl group present in the six-membered ring dictates the C—C bond formation from the opposite face, leaving the cyclopentene double bond in a most desirable syn relationship with the methyl group. This unsaturation becomes a handle for annexing the three-carbon chain. Syntheses of β-vetivone (**V16**) (Piers, 1977b) and α-vetispirene (**V14**) (Yan, 1982) have been reported.

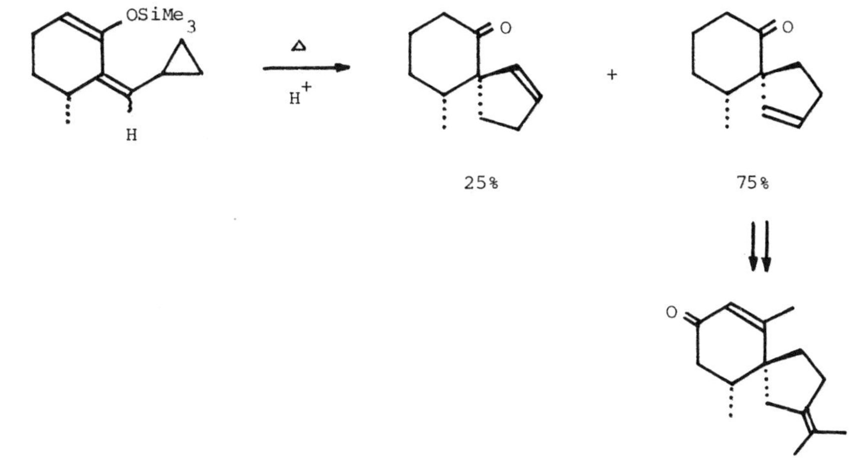

As far as synthesis of β-vetivone (**V16**) is concerned, the presence of an oxygen substituent at the cyclopropyl carbon closest to the cyclohexene ring is very beneficial. Its thermal product is hydrolyzable to the ketone intermediate, which is directly correlated with the terpene (Barnier, 1984).

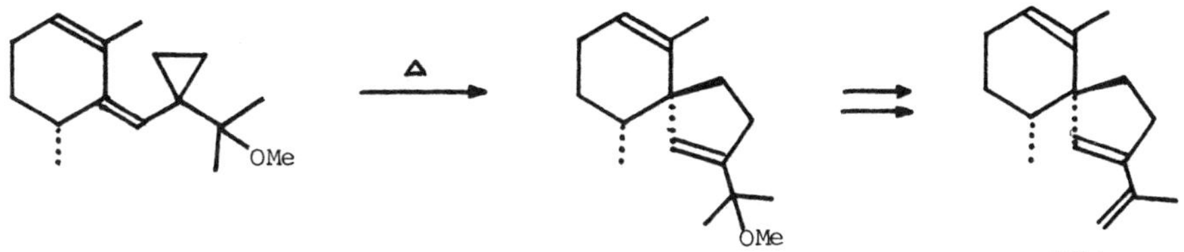

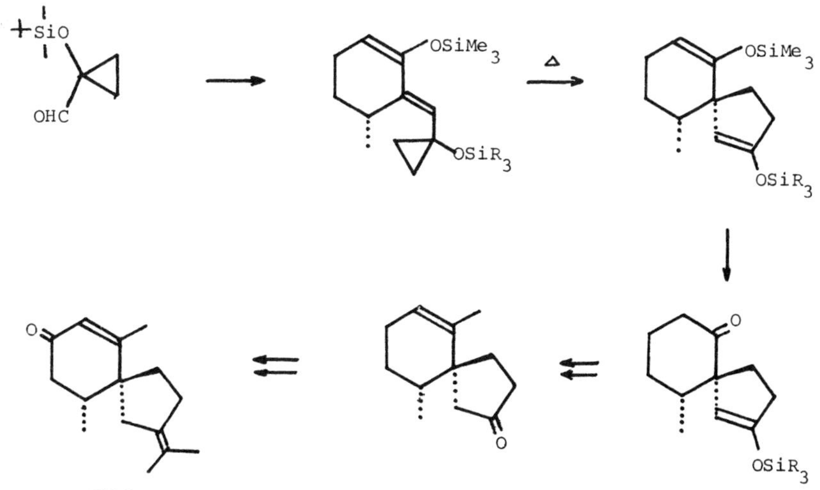

A condensed version of the ring expansion is a crucial step in a synthesis of aphidicolin (**A36**) reported by Trost (1979a). The required siloxycyclopropyloctalin is available from the oxaspiropentane, which is turn is made by the reaction of the decalone with diphenylsulfonium cyclopropylide. The thermolysis generates a stereoisomeric mixture, but it can be rendered homogeneous via an oxidation–reduction sequence (enone intermediate).

Vinylcyclopropanes are accessible by way of the insertion of carbenoids into conjugated dienes. It is well known that the adducts are very susceptible to thermal reorganization to give cyclopentene derivatives. The intramolecular version is readily adaptable to a synthesis of hirsutene (**H17**) (Hudlicky, 1980).

The use of $(C_2H_4)_2Rh(acac)$ catalyst enables the decrease in reaction temperature from 590 to 180°C while increasing the stereoselectivity to the exclusive formation of the anti triquinane. Apparently, participation of the rhodium atom changes the rearrangement into an orbital symmetry-allowed process.

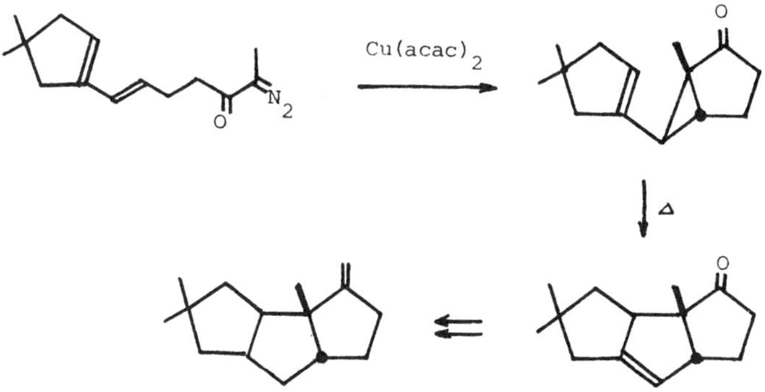

8.9 THERMAL REORGANIZATION OF VINYLCYCLOPROPANES

H17

Isocomene (**I14**) has also been assembled based on the same theme (Ranu, 1984). The entry to the secofenestrane skeleton requires a different arrangement of the diene and the diazoketone units. And since the secondary methyl group must be established by hydrogenation from the more hindered face of the molecule, any hope of completing the synthesis depends on the installation of an epimerizable C_1 unit until the stereochemical issue is settled.

A potential entry into the guaiane sesquiterpene skeleton (Hudlicky, 1985) has been delineated.

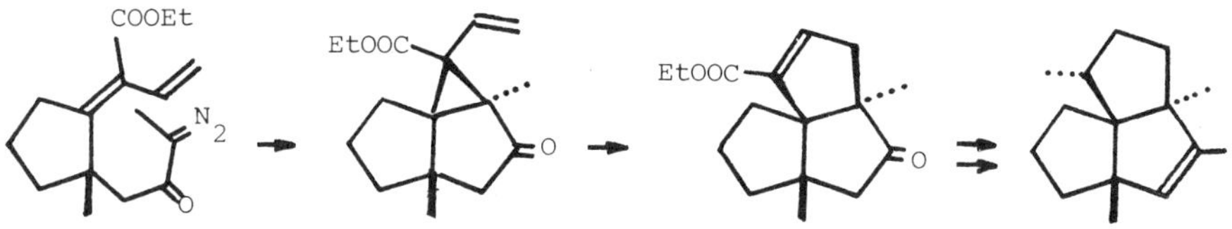

I14

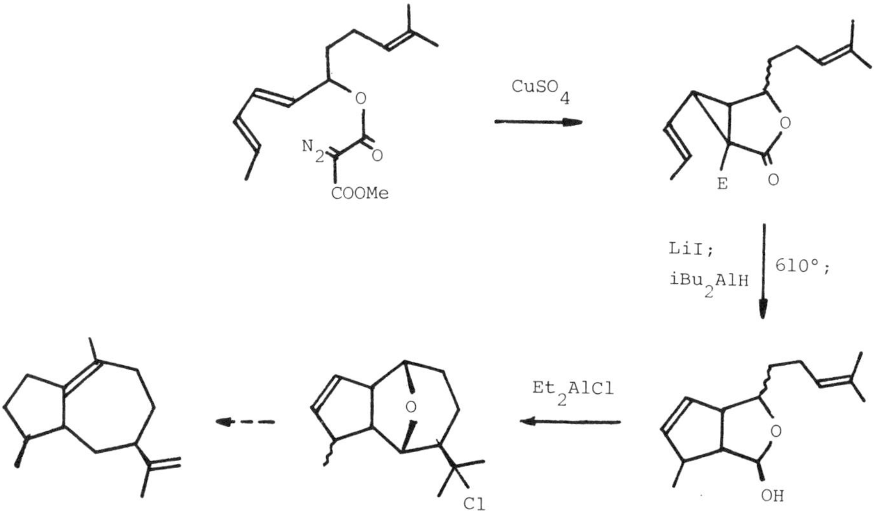

B19

The application to synthesis of zizanoic acid (**Z3**) (Hudlicky, 1983) is interesting because the intermediate has a bridged/spirocyclic skeleton.

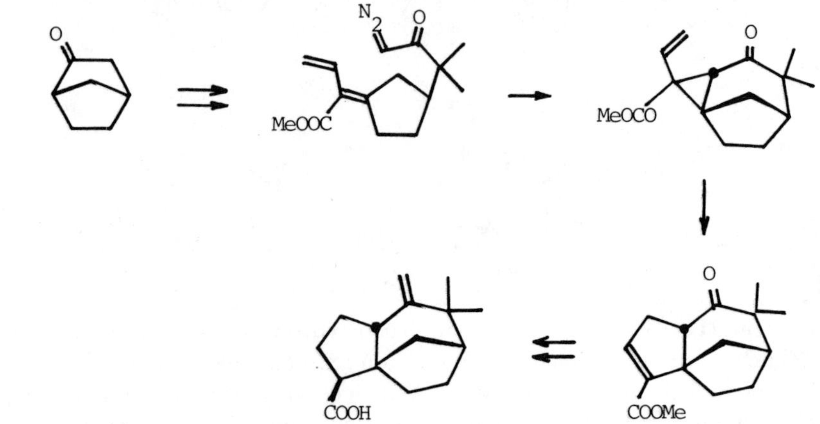

In a zizaene (**Z1**) synthesis (Piers, 1979a) that follows the strategy of elaborating and closing the ethano bridge onto an existing hydrindenone, the required intermediate is prepared by pyrolysis of a β-cyclopropylcyclohexenone.

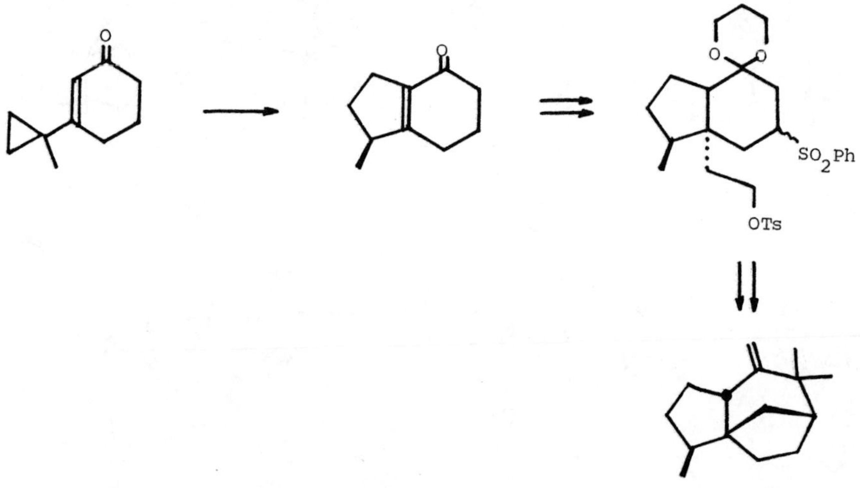

Chapter 9

Radical Cyclization

9.1 SINGLE RADICALS

The radical process has been generally neglected in considering controlled C—C bond formation, especially that constituting ring closure, except perhaps the photochemical reactions. Organic chemists have finally overcome the fear and misconceptions that uncontrollable chain reactions that result in polymers always predominate, and they are beginning to appreciate the potential offered by the radical processes.

A convenient method for generating radicals is by reacting alkyl halides with triorganotin hydrides. The resulting radicals can add to multiple bonds. Intramolecular addition is particularly favorable when it leads to five- and six-membered rings.

The radical can be primary, secondary, or tertiary. It also can be located at a bridgehead. Since radicals are much less prone to rearrangement, even on the periphery of norbornyl systems, the regioselectivity associated with such reactions is very high. Examples involving radical addition as the key annulation step include syntheses of β-agarofuran (**A16**) (Büchi, 1979b), sativene/copacamphene (**S11/C63**) (Bakuzis, 1976), seychellene (**S29**) (Stork, 1985), $\Delta^{9(12)}$-capnellene (**C21**) (Curran, 1985a), and hirsutene (**H17**) (Curran, 1985c).

It is important to note that while regioselectivity is not lost during the radical reactions, the configurations of alkenyl radicals can undergo inversion (cf. seychellene). This property is advantageous from the synthetic viewpoint.

Bisannulation (see **C21** and **H17**) shows that a nascent radical originating from the first addition can attack another multiple bond that is separated by a chain of proper length so that another common ring may be created.

The great potential of this polyannulation technique has been further demonstrated by its involvement in assembling the angular triquinane skeleton of silphiperfol-6-ene (**S35**) (Curran, 1986). It is notable that the tricyclic product with an *exo*-methyl group predominates when the secondary radical center generated during the initial addition step is flanked by a quarternary carbon atom (e.g., as a ketal function). On the other hand, a ketone group at that site favors formation of the *endo*-methyl isomer (ratio, $2.5x : 1n$ to $1x : 3n$).

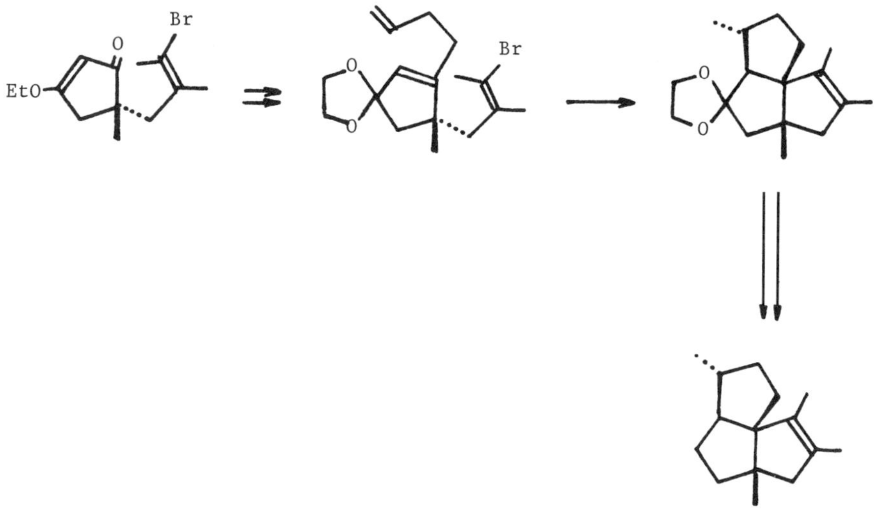

The hydroxy-α-*trans*-bergamotene, which has been synthesized by an intramolecular ketene–alkene cycloaddition and hydride reduction, is convertible to a cyclobutyl radical by treating its imidazolylthiocarbonate with tributyltin hydride. The radical is trapped by the side chain, thereby producing β-copaene (**C67**) and β-ylangene (**Y2**) (Snider, 1985), albeit in low yields. Allylic oxidation of β-ylangene gives lemnalol (**L9**).

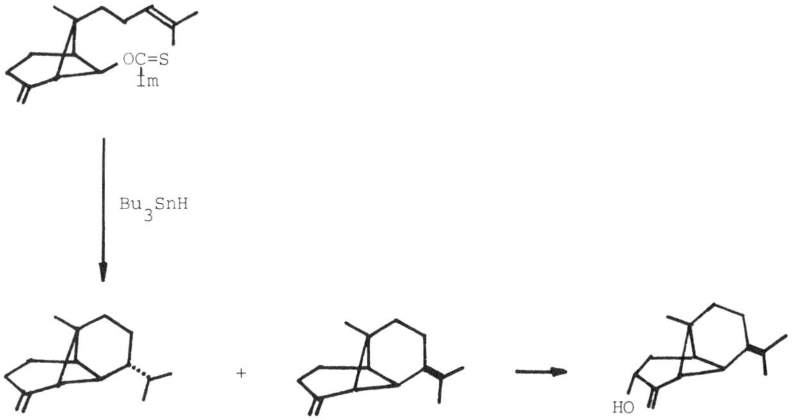

A carbohydrate-based synthesis of 1-α-*O*-methylloganin aglucone is mediated by intramolecular radical addition to a terminal vinyl group in a stereoselective manner (Hashimoto, 1987). The diastereomer corresponding to the terpene configuration is preferentially formed.

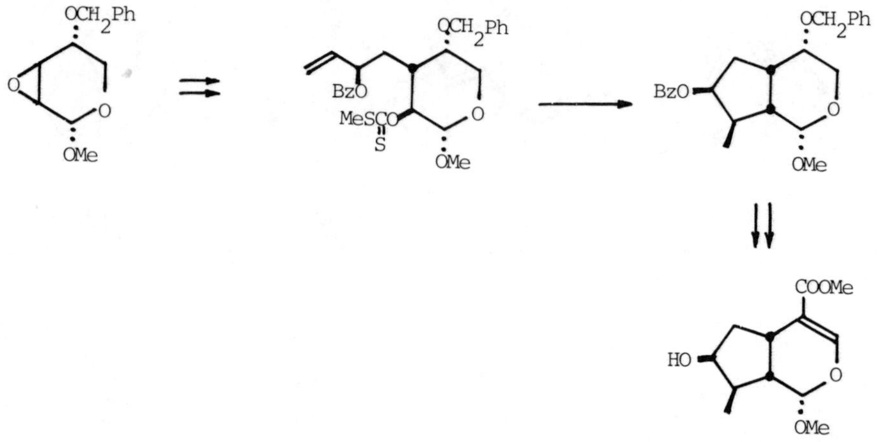

Intramolecular conjugate addition of a radical species to a β-hydroxybutenolide is the pivotal step in an approach to alliacolide (**A25**) (Ladlow, 1985). A cyclopentenone formed from an acyclic ketoaldehyde is converted into a spirobutenolide on reaction with the carbethoxyacetylide anion (and necessary functional group adjustments), which is then ready for cyclization. The final epoxidation is directed by the homoallylic hydroxyl group.

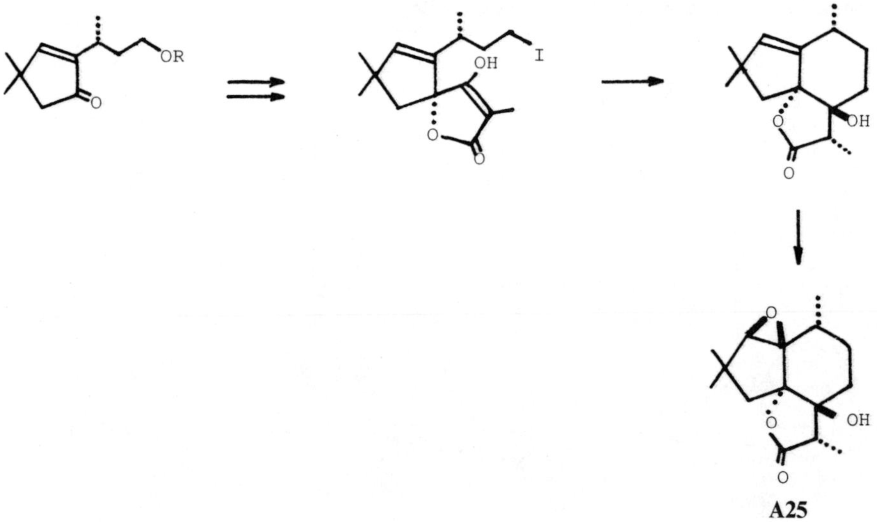

Methylenecyclopentannulation initiated by radical addition of 3-butyn-1-yl to conjugated systems is a valuable method in organic synthesis. Its complementarity to the [2 + 3]cycloaddition using conjunctive polar reagents (cf. Trost, 1982b) is best illustrated in a concise synthesis of albene (**A22**) (Curran, 1987).

A mixture of dehydroiridodiol (**I7**) and isodehydroiridodiol (**I7**′) has been synthesized (K. Nozaki, 1987) by employing a hydrostannylation reaction to link a 1,6-enyne system. This process is different from the previous examples because the tin radical participates in an addition reaction instead of atom abstraction.

Current cyclization schemes based on radical reactions generally suffer from termination by hydrogen atom transfer, thereby reducing the possibility of renewed manipulation at the former radical sites. However, β-ketoesters lend themselves to ready oxidation to give stabilized radicals that may attack a nearby double bond may be the exception. They lead to cyclic molecules while retaining the versatile functionalities.

In a new route to podocarpic acid (**P43**) (Snider, 1985b), not only are two six-membered rings made according to this concept, but the stereohomogeneity of the tricyclic product (with an axial ester group) is truly remarkable.

Another example of radical generation via addition process is shown in a synthesis of karahana ether (**K2**) (Coates, 1970b), which embodies an attack of a benzoyloxy radical on the more accessible double bond to start the ring closure.

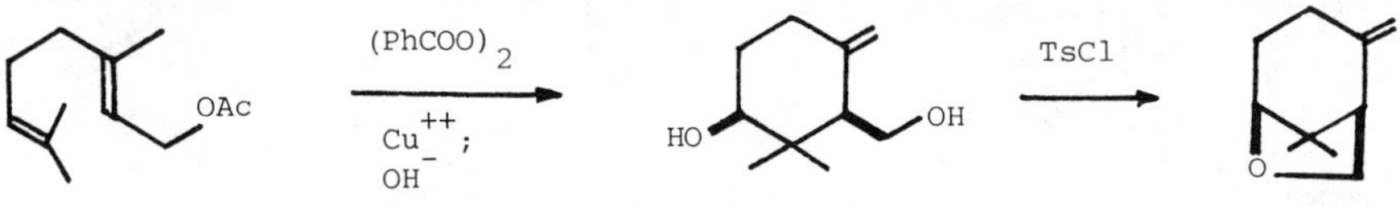

The intramolecular reductive cyclization of acetylene ketones by electrolysis or using sodium naphthalenide leads to allylic alcohols. The reaction probably involves addition of a vinyl radical to the carbonyl group. This process has received a good test from capnellenediol (**C22**), the additional hydroxyl functional is properly introduced after the annulation (Pattenden, 1982, 1985).

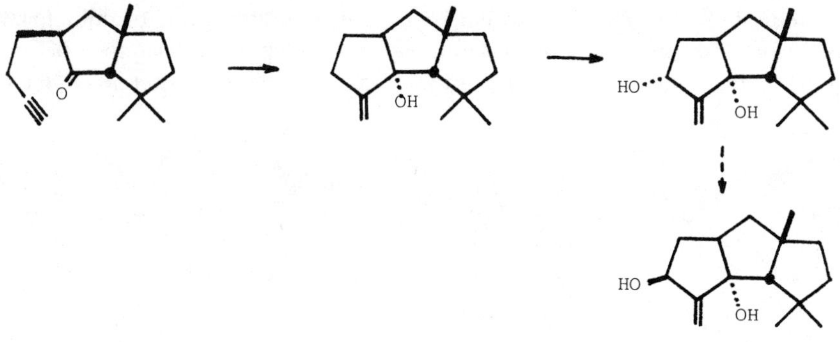

The cyclization is apparently under thermodynamic control, and an exploitation of this tendency embodies the annulation of the six-membered ring of isoamijiol (**A28**) (Pattenden, 1986).

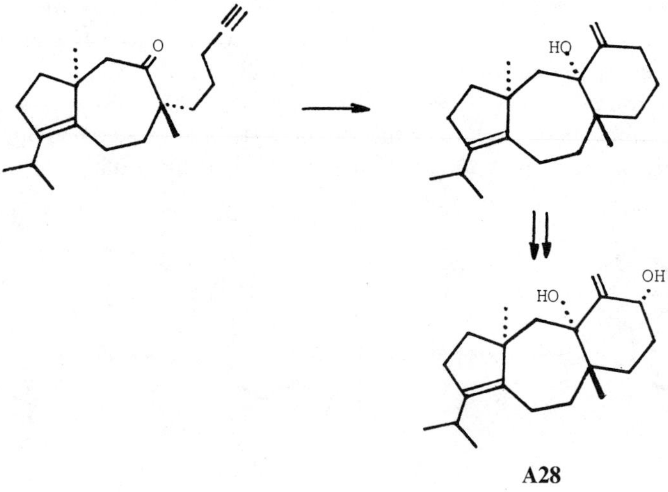

An example of reductive cyclization initiated by an α-alkoxy radical is the generation of loganin tetracaetate (**L20'**) from secologanin tetraacetate by treatment with Mg/Me₃SiCl (Ikeda, 1985).

L20'

A new synthesis of hirsutene (**H17**) (Cossy, 1987) is based on photoreductive cyclization of alkynyl ketone to form an exocyclic methylenecyclopentane. The allylic angular hydroxyl group undergoes replacement by a nickel-doped Grignard reagent. The cyclization step is subject to steric control since simple reduction of the carbonyl group of the epimeric ketone with an endo butynyl chain is observed.

H17

The 14-epimer of upial has been secured by an intramolecular oxidative γ-lactonization of a cyclohexene using manganese(III) acetate (Paquette, 1987c). Because upial (**U3**) has an exo methyl group which must assume an axial orientation in the transition state of the annulation, the step by which the bicyclo[3.3.1]nonane skeleton is formed proceeds in very low yield.

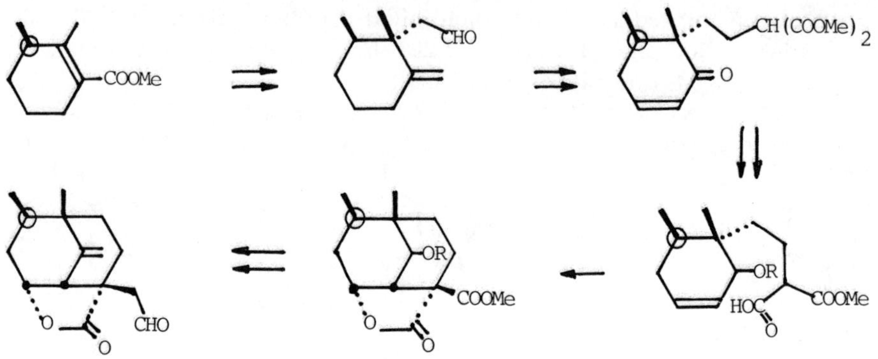

U3

The fungal antibiotic punctatin A is biogenetically closely related to caryophyllene. However, the presence of three hydroxyl groups, particularly a tertiary one at the ring juncture, demands special synthetic planning (Paquette, 1986c). In that respect the Norrish Type II photocyclization of a hydrindenone proves to be the ideal process for closing the four-membered ring. The stereoselective formation of the *trans*-fused 4 : 6 ring system is due to strong steric compressions between the *gem*-dimethyl group and the angular substituent in the alternative mode of cyclization.

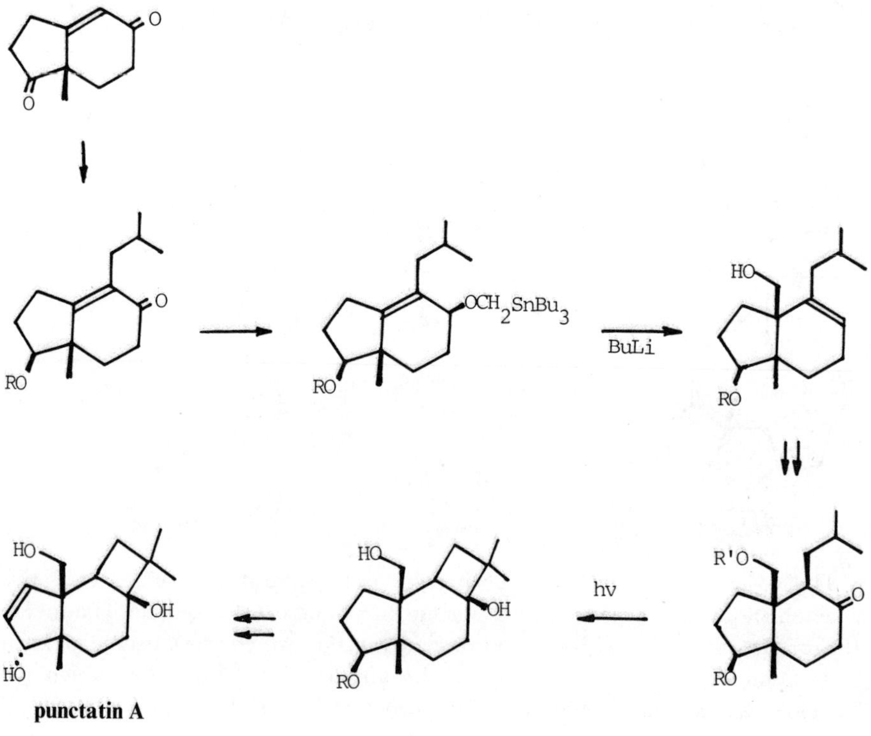

punctatin A

9.2 1,1-DIRADICALS

The 1,1-diradicals are better known as carbenes (both singlet and triplet species). Although carbenes are employed most often in synthesis for their facile addition to double bonds to form cyclopropanes (see Chapter 10), their ability to insert into a C—H linkage to form cyclic products has been realized in the syntheses of cuparene (**C84**) (Mane, 1973), (+)-α-cuparenone (**C85**) (Taber, 1985a), the methyl esters of pentalenolactones E (**P19**) (Cane, 1984), and a few others.

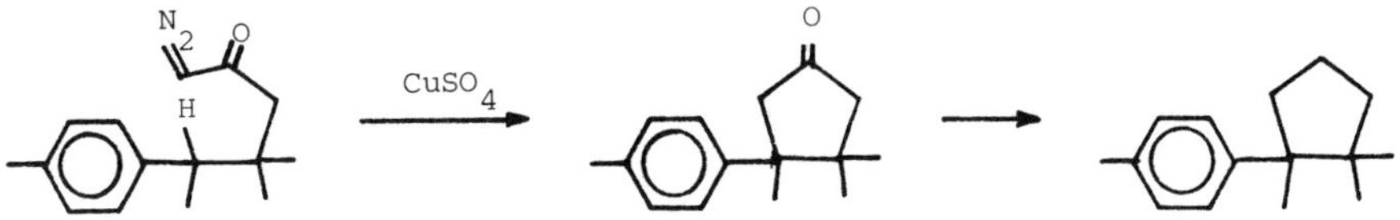

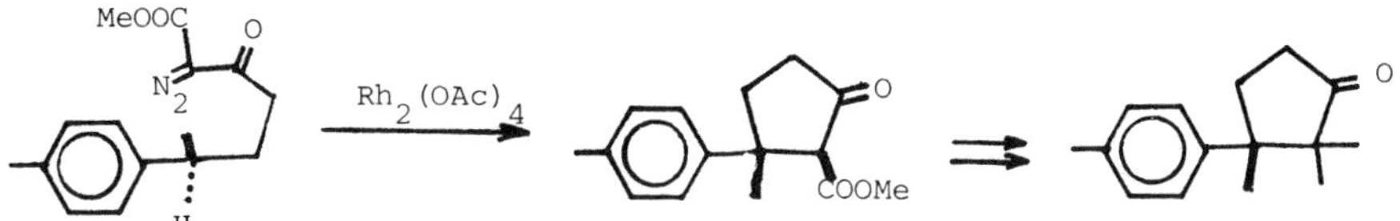

It is evident that the insertion proceeds with retention of configuration (Taber, 1985a). Furthermore, carbene insertion into secondary C—H bonds is also very efficient, as demonstrated in a facile construction of the framework of pentalenolactone E methyl ester (**P19**) from an oxaspiro precursor (Taber, 1985b).

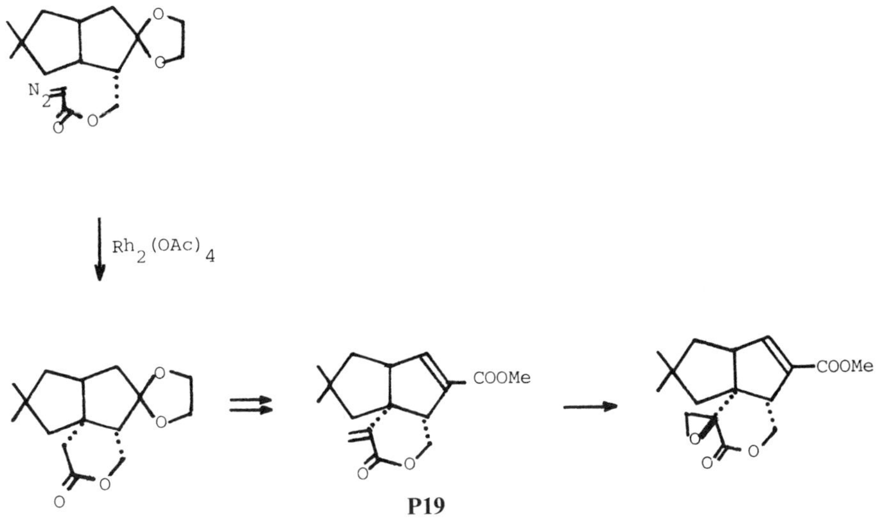

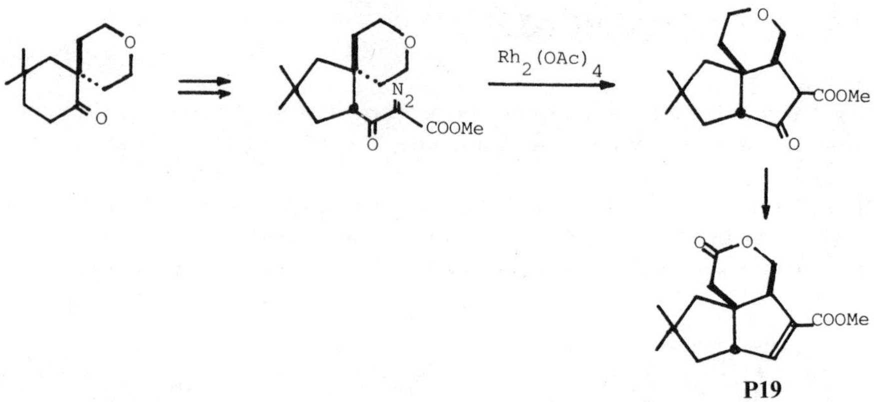

P19

A hibaene (**H11**) synthesis based on a C—H insertion to form the bridged system also hinges on an electrophilic attack of a η^3-allylnickel species by methylmagnesium bromide. The attack is highly regioselective at the more highly substituted carbon atom, and it proceeds from the axial direction. The vinyl group is then transformed into the required diazoketone function (Buckwalter, 1978).

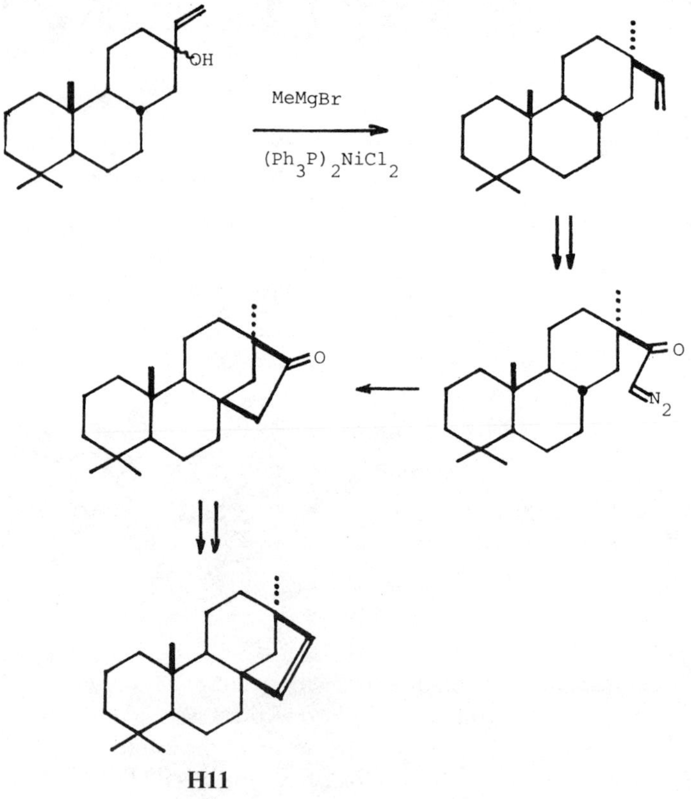

H11

A method for C—C bond formation via a formal group displacement at a quaternary center by a carbenoid has been applied to the synthesis of cuparene (**C84**) (Kametani, 1985). Intervention of a sulfonium ylide in this reaction is likely.

C84

In the syntheses of ishwarane (**I11**) (Cory, 1977) and Ishwarone (**I12**) (Cory, 1979) three C—C bonds (two rings) are established in essentially one operation. This is a truly outstanding annulation scheme.

I11/I12

X= H,H; O

The new cyclopentenone synthesis via pyrolysis of ethynyl ketones is a versatile method that has been extensively developed. The reaction proceeds via migration of the acetylenic hydrogen to the α carbon to form a vinylidene carbene intermediate. The abstraction of a β' hydrogen by the carbene and coupling of the resulting 1,5-diradical concludes the reaction. As expected, the relative stability of the β'-carbon radical (i.e., 3° > 2° > 1°) determines the preferred site of bond formation, if choices are available. Cyclopentanoid sesquiterpenes that have been synthesized by this method include modhephene (**M12**) (Karpf, 1980), $\Delta^{9(12)}$-capnellene (**C21**) (Huguet, 1982), albene (**A22**) (Manzardo, 1983), and acorone (**A11**) (Ackroyd, 1985).

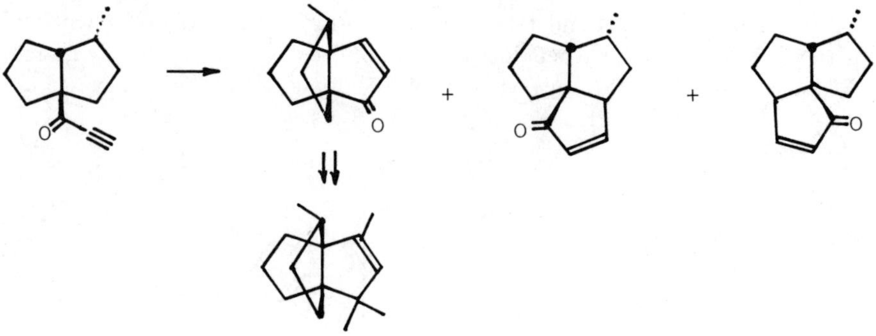

M12

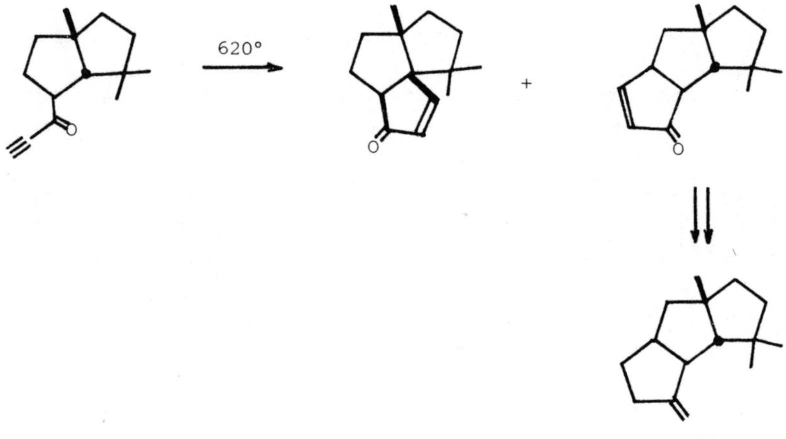

C21

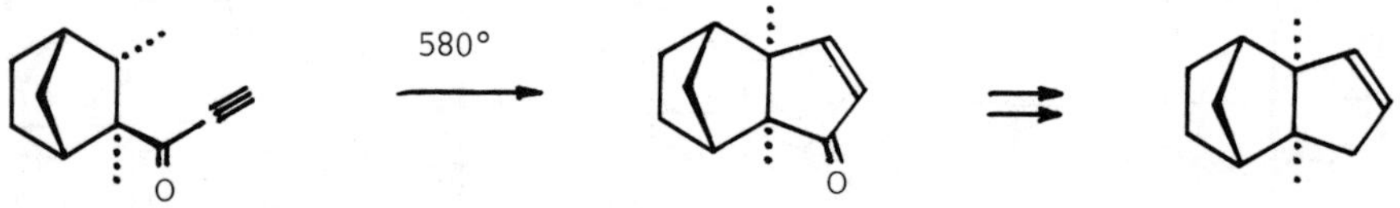

A22

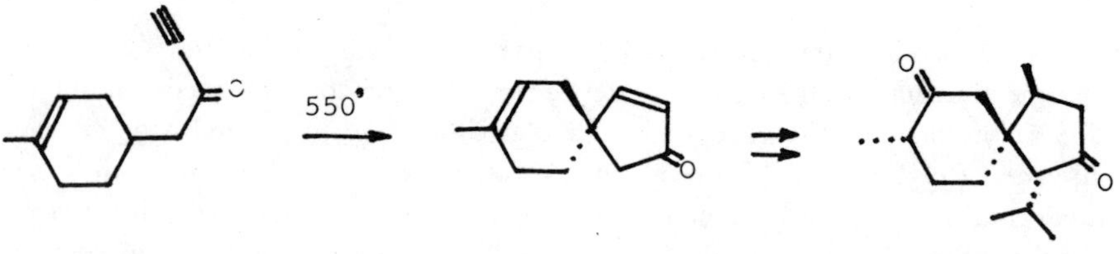

A11

The lessened regioselectivity (a welcome one) observed during the synthesis of the capnellene is due to steric hindrance by the *gem*-dimethyl group at the γ′ carbon; therefore the formation of the angular triquinane is discouraged.

The spirocyclic intermediate for acorone is unstable at higher temperatures because a retro-Diels–Alder reaction occurs. There is also a competing ene reaction, which gives rise to the bicyclo[3.3.1]nonane derivative and its dimer.

An attempt has been made to synthesize ptychanolide (**P50**) (Huguet, 1983). The all-*cis* tetramethyl diquinane system has been assembled; however, only a stereoisomer of the natural product was obtained. The point at which the stereoisomer differs from ptychanolide is not clear.

In the elaboration of isocomene (**I14**) (Manzardo, 1986) the crucial intermediate is a bicyclic acid in which the β-carbon atom is *cis*-substituted to prevent subsequent reaction at that site. While the quaternary chiral center of the acid is easily established by a reaction sequence of conjugate additions, erection of the methyl blocking is only partially successful (3 : 1 after equilibration).

9.3 1,3-DIRADICALS

1,3-Diradicals or 1,3-diyls can add to a double bond, generating cyclopentanes. These 1,3-diyls are stabilized when the central atom is trigonal, and they are also easy to acquire from bridged pyrazolines derived from fulvenes via Diels–Alder cycloaddition with azodicarboxylic esters. Intramolecular trapping of the diyls with a properly substituted and positioned alkene linkage can lead to carbon frameworks of triquinane sesquiterpenes such as hirsutene (**H17**) (Little, 1981a), $\Delta^{9(12)}$-capnellene (**C21**) (Little, 1981b), and coriolin (**C70**) (Van Hijfte, 1985).

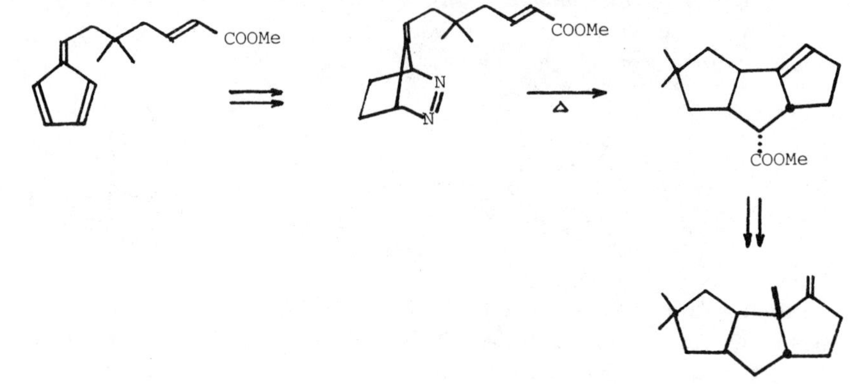

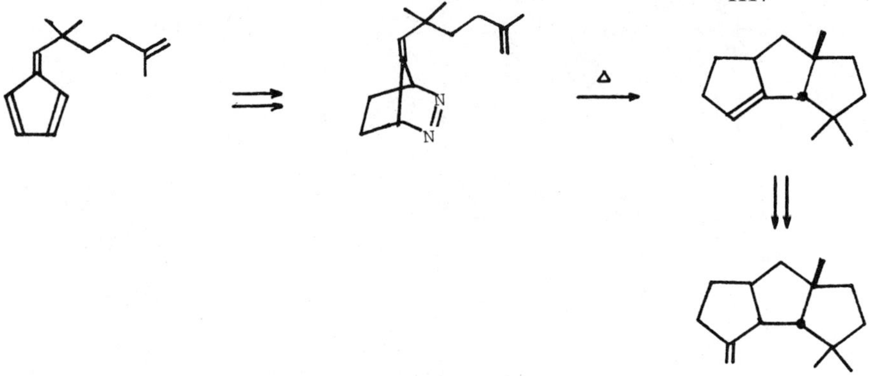

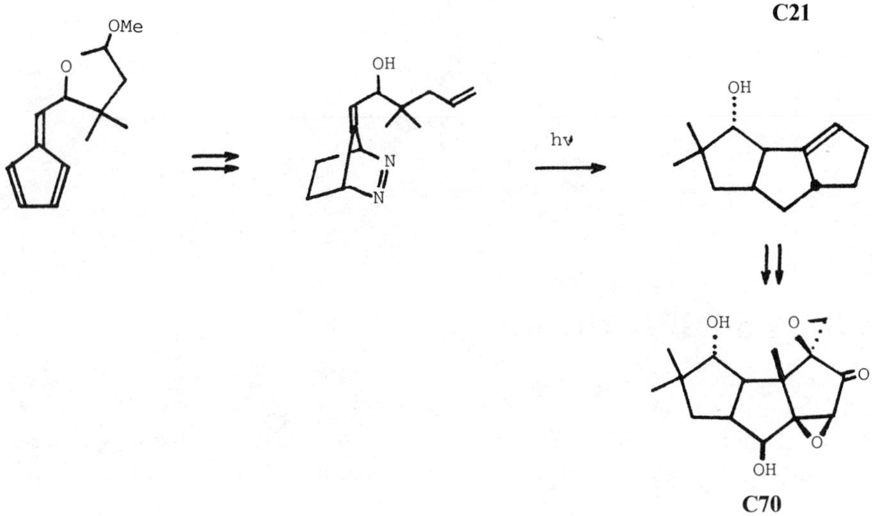

Generally, conformational effects rather than electronic factors play a dominant role in determining the stereoselectivity of these reactions. In the entry to capnellene, the conformation leading to the desired cis-anti triquinane is somewhat destabilized by an $A^{1,3}$-interaction such that it is favored by a factor of only 1.6.

Photoexcited benzene and derivatives undergo *meta* bonding. The diradical intermediates can be intercepted by alkenes to give dihydrosemibullvalenes (Bryce-Smith, 1976, 1977).

Recognition of the dynamics of this cycloaddition has spurred intense activities toward its application to polycyclic synthesis (Wender, 1984), and a brilliant orchestration of the various intra and intermolecular arene–alkene cycloadducts into terpenes of diverse structural families continues.

The meta mode of addition is favored on a molecular orbital correlation basis, especially when the frontier orbital energies of the addends are similar. Moreover, regiocontrol often can be exerted—for example, by a methoxy group in the aromatic ring, which directs a particular pair of meta carbon atoms to engage in the C—C bond formation process within allowable geometrical constraints.

In the intramolecular version of the cycloaddition, steric factors dictate the orientation of the carbon chain containing the alkene moiety. In a cedrene (**C37**) synthesis (Wender, 1981b), total regio- and stereocontrol are attained during photolysis of a methoxycurcumene. The exo exciplex has a better orbital overlap in the transition state, and the chain folding is such that crowding between the secondary methyl group and the methoxy substituent is avoided.

C37

Further fragmentation of the tricyclic products may be implemented in a designated fashion by incorporating a potential leaving group at a proper locale. This added flexibility of the photocycloaddition methodology is amply rewarded in an approach to the acoranes, for example, β-acoradiene (**A5**) (Wender, 1984).

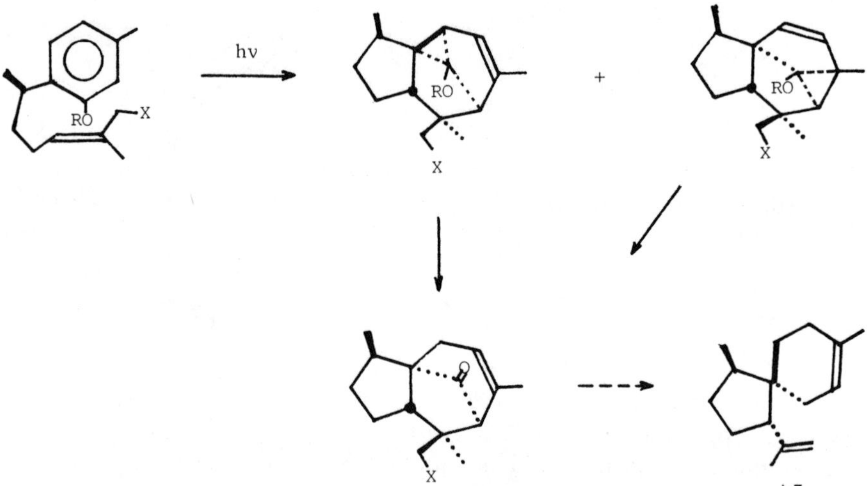

In a synthesis of rudmollin (**R10**) (Wender, 1986). A stepwise scission of two C—C bonds unfolds a cycloheptane nucleus while enabling regioselective functionalization of the latter. But before its final disengagement, the intermediate containing a bicyclo[3.2.1] octane moiety is structurally attuned to alkylation at the β-side (exo) in preparation for eventual lactonization.

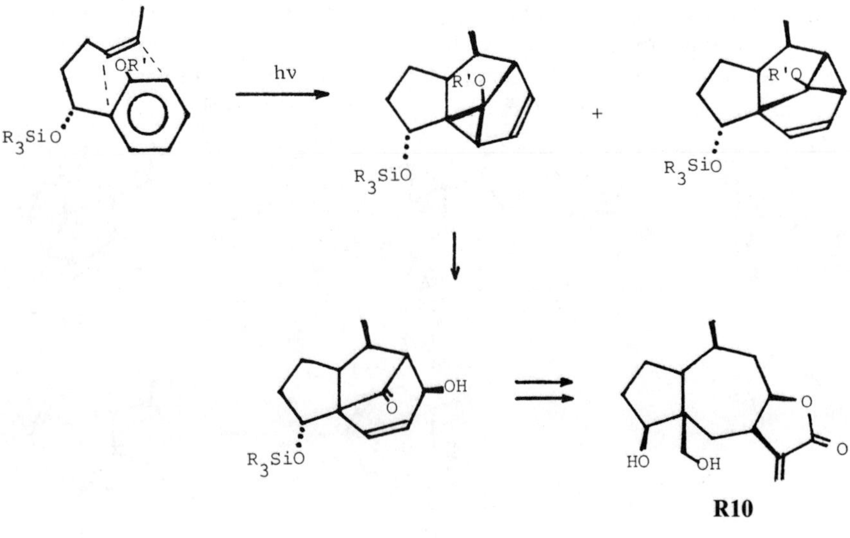

Isocomene (**I14**) has also been assembled in a similar manner (Wender, 1981a). The major difference of this synthesis from the others lie in the ring-opening step after the cycloaddition. The direction of the cyclopropane cleavage is no longer imposed by a methoxy group (cf. cedrene synthesis), and thus it becomes possible to break the more exposed peripheral bond selectively. Although a thermally induced sigmatropic hydrogen shift converts only one of the two photoadducts into dehydroisocomene, the unchanged isomer retains its value by virtue of its photoconvertibility into the useful compound.

I14

Noting that cleavage of the peripheral cyclopropane bond of one of the tetracyclic adducts would unveil an angular triquinane skeleton characteristic of silphinene (**S34**), Wender (1985c) assayed a synthesis of the terpene, starting from *o*-curcumene. The desired cleavage has been achieved, using lithium in methylamine.

S34

A logical extension of the general scheme to a synthesis of silphiperfol-6-ene (**S35**) (Wender, 1985b) further confirms the notion that the preferred exciplex geometry for cyclization is the one in which the more bulky allylic substituent is aligned away from the arene subunit. Other salient features of this work include a regio- and stereoselective attack of the vinylcyclopropane moiety by an acetyl radical, thus permitting the introduction of the missing methyl group.

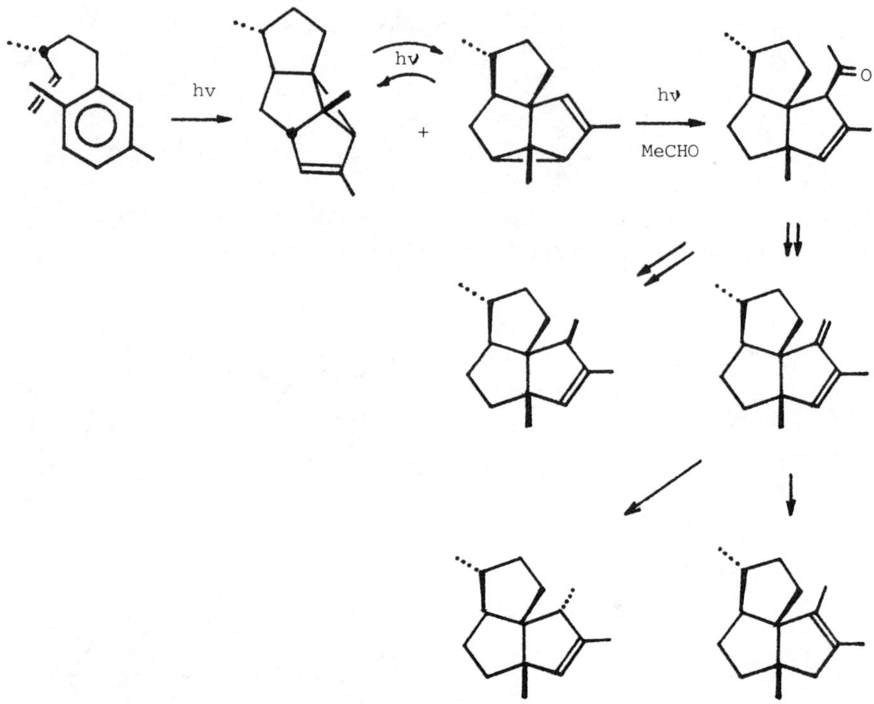

S35

The angular triquinane portion of retigeranic acid (**R1**) is in principle derivable from the same process. A suitable intermediate for the sesterterpene has been prepared in three steps (Wender, 1984). The task yet to be accomplished consists of grafting a 10-carbon chain onto the tricyclic nucleus and inducing a Diels–Alder reaction of the resulting triene. This approach is as elegant as it is brief.

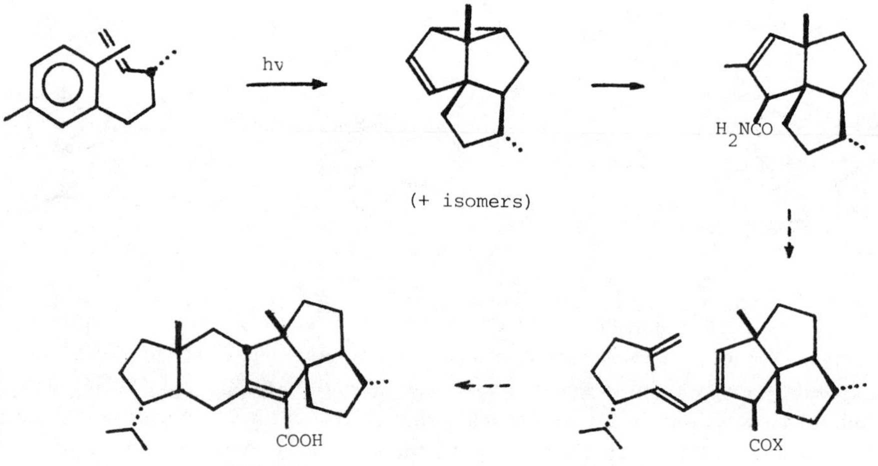

(+ isomers)

R1

The framework of quadrone (**Q1**) is also accessible following a relatively short reaction sequence (Wender, 1984). While bisnorquadrone has been synthesized, the tetracyclic lactone intermediate should be useful for elaboration into quadrone itself via oxygenation of the cyclopentene double bond. Both cyclopentanone isomers derived from this operation are theoretically convertible to quadrone.

Q1

Linear triquinanes such as hirstutene (**H17**) (Wender, 1982c) and coriolin (**C70**) (Wender, 1983b) may be unraveled from the same type of tetracyclic photo-adducts by selective scission of the cyclopropane bond that spans the two terminal cyclopentane nuclei. Scission can be achieved by exposing the adducts to acid, or in the process of acquiring coriolin, to thiophenol. The allylic attack by the latter reagent results in a functionalized intermediate amenable to further elaboration.

C70 **H17**

Intermolecular photoreaction of vinyl acetate with indane produces a propellane as the major adduct. With the demonstration that the derived ketone (from the acetate) is capable of permethylation via an equilibrating semibullvalene oxide ion, modhephene (**M12**) (Wender, 1982a) has been created.

M12

The establishment of the secondary methyl group by anionic opening of the cyclopropyl ketone is aided by the steric environment, since the 1,7-addition (at the double bond) is inhibited by the *gem*-dimethyl group.

In the same vein isoiridomyrmecin (**I9**) has been prepared from the photoadduct of benzene and vinyl acetate (Wender, 1983a).

I9

9.4   1, *n*-COUPLING

Whereas many methods are available for achieving the fission of a double bond to two carbonyl groups, the reverse transformation (i.e., the direct reductive coupling of two carbonyl groups to form a double bond) remained unknown until the use of the titanium(0) species was discovered (McMurry, 1983a). More recently, the intramolecular version of this coupling has been further explored in the synthesis of terpenes, among others. The great advantage of this process is the seeming insensitivity to the size of the ring that can be formed. This phenomenon is now understood on the ground that ketyl intermediates are generated on metal surfaces, where coupling can take place between two radical termini of the same molecule. The reaction is therefore akin mechanistically to the acyloin condensation.

A reasonable extension of the intramolecular dicarbonyl coupling is the synthesis of cyclic ketones from ketoesters. The primary products (enol ethers) may find other synthetic applications.

While a neat synthesis of α-chamigrene (**C46**) (Koft, 1987b) from α- or β-ionone and the preparation of the carbocyclic system of strigol (**S55**) (Berlage, 1987) amply show the synthetic usefulness of the coupling method, the true test is in the application to more complex molecules. The 15-membered ring diterpene flexibilene (**F7**) was the first compound with inherent difficulty in synthesis to be conquered by this coupling (McMurry, 1982b). More recently, (+)-casbene (**C34**) has yielded to synthesis by this method (McMurry, 1987).

However, the 11-membered ring compound humulene (**H22**) is perhaps a more challenging target for authenticating the efficacy of this outstanding protocol. A three-step synthesis of the hop constituent (McMurry, 1982a) represents one of the milestones in medium-sized ring closure.

The formation of cyclodecadienes has been equally successful using the 1,n-coupling method. Thus, bicyclogermacrene (**G7**), isobicyclogermacrene (**G8**), lepidozene (**L10**), and the previously unknown isolepidozene (**L10′**) have been obtained (McMurry, 1985a). Contrary to observations during the synthesis of humulene, both *E*- and *Z*-isomers are produced at closure of the 10-membered rings.

E,E-Germacrenes are notoriously difficult to prepare, owing to their tendency to isomerize via Cope rearrangement. Their synthesis has been attempted from the ketoaldehyde obtained from geranylacetone. However, the two hydrocarbon products prove to be helminthogermacrene (**H8**) and β-elemene (**E1**) (McMurry, 1985a). Apparently the *E,E*-product (germacrene A) has undergone thermal reorganization.

The obvious approach to caryophyllene/isocaryophyllene (**C32/C33**) by the reductive coupling technique is logistically problematic, because the ketoaldehyde precursor containing a cyclobutane moiety requires much effort to prepare. However, a C_{14}-cyclobutanone ester is readily available from a homogeranic ester via a ketene addition reaction. This ketoester cyclizes, and upon hydrolysis of the resulting enol ether affords the norketone of isocaryophyllene (McMurry, 1983b). The configuration of the *E* double bond has changed during the coupling operation, apparently because of ring strain. Consequently, caryophyllene is not available directly, although the terpene itself is stable to the titanium reagent.

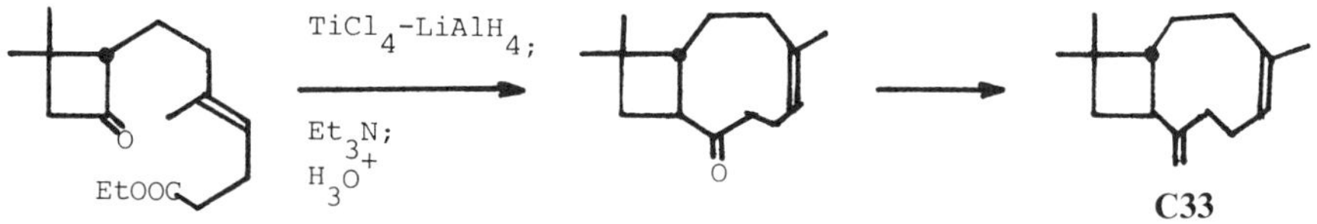

The ketone synthesis has also been used to form the norketone of $\Delta^{9(12)}$-capnellene (**C21**) (Iyoda, 1987b).

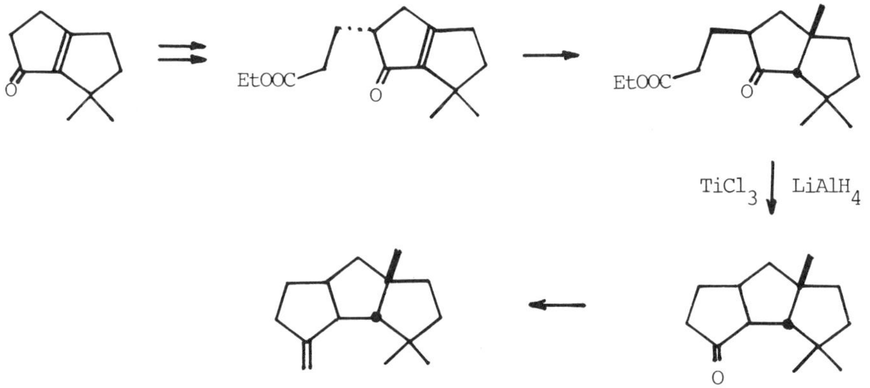

A synthesis of hirsutene (**H17**) based on the de Mayo reaction also embodies a reductive transannular cyclization of a diketone (Disanayaka, 1985).

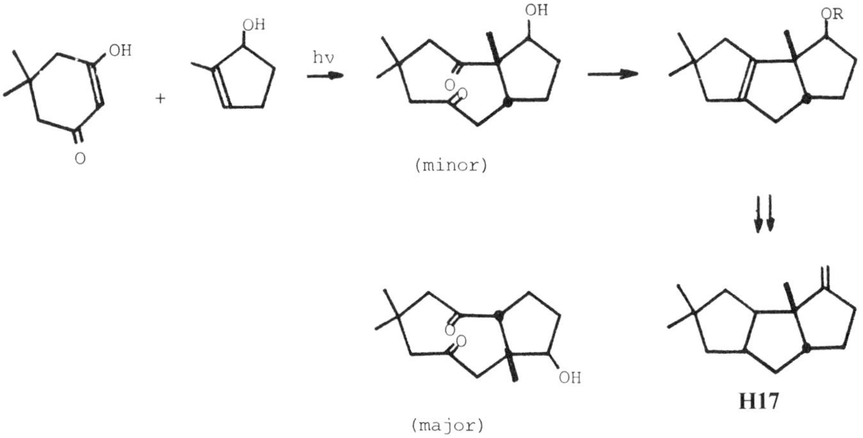

In a synthesis of gibberellic acid (**G12**) (Corey, 1978a, b) the D ring is bridged by means of the titanium (O)-mediated condensation of a ketoaldehyde. Deoxygenation of the resulting diol is sterically inhibited (Bredt's rule).

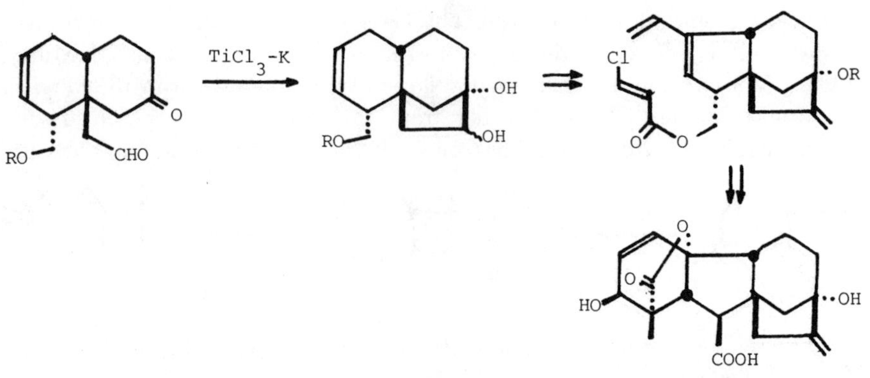

On other occasions when the diols are desired, milder reagents may be employed. Substituted 2,6-dioxoalkanoic esters, which are obtainable from the DeMayo reaction, may be cyclized (e.g., with $TiCl_4/Zn$) to give 1,2-cyclopentanediols. Accordingly, such intermediates have been prepared and elaborated into terpenes such as chrysomelidial (**C54**) and dehydroiridodial (**I6**) (Takeshita, 1982), and cuparene (**C84**) (Takeshita, 1984b).

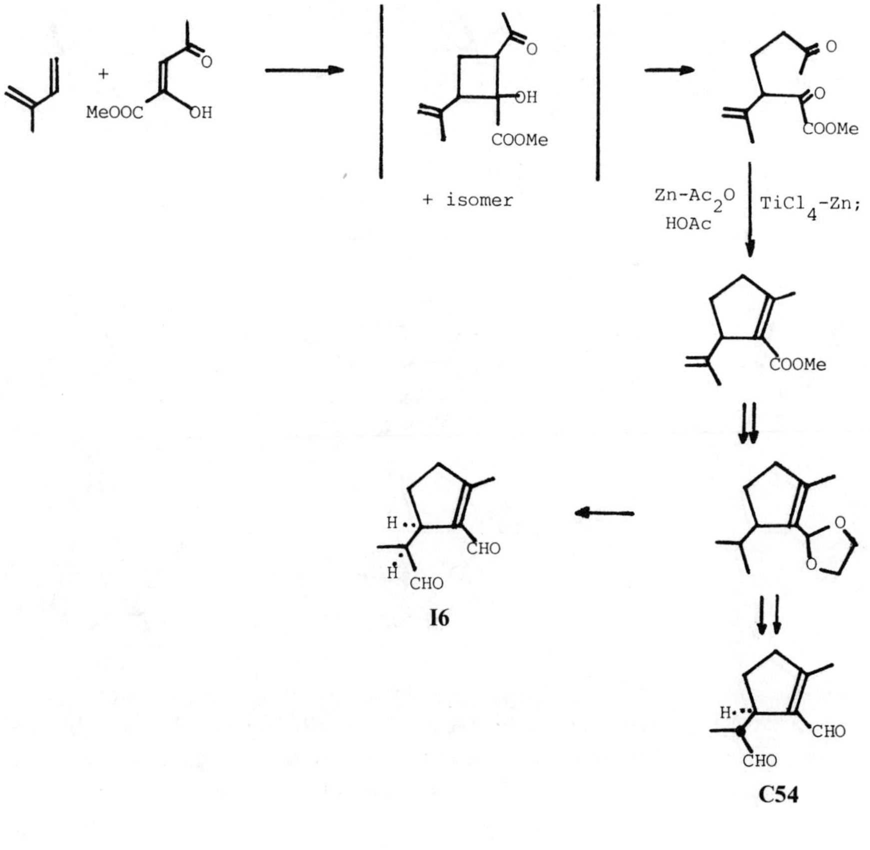

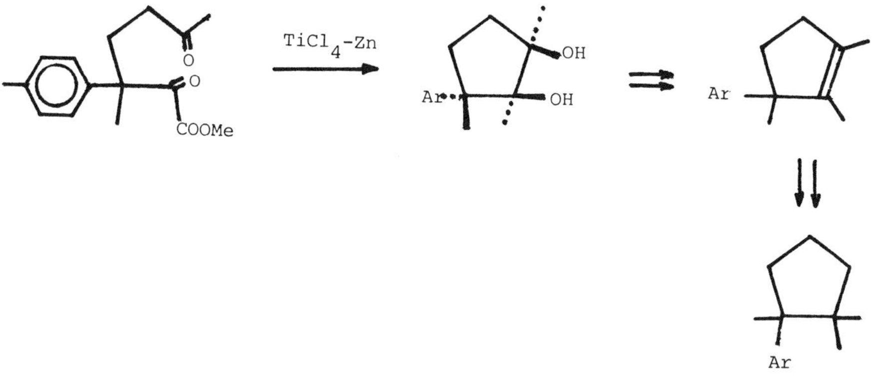

Similarly, closure of the eight-membered ring of the fusicoccane skeleton (cf. fusicoccin H, **F22**) has been effected (Takeshita, 1984a). The secodialdehyde precursor is derived from a 1,5-diene by a Cope rearrangement, which in turn is obtained from the coupling of two iridoid synthons.

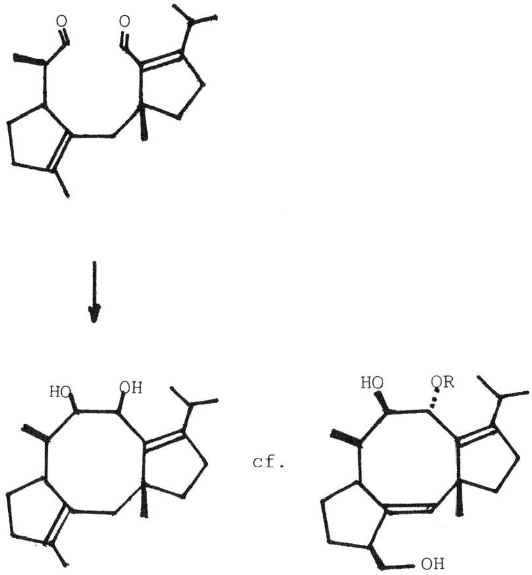

The McMurry coupling is the crucial step for synthesis of a taxane triene, easily recognized as a useful precursor for taxusin (**T6**) (Kende, 1986). The reaction forms an 8-membered ring which contains all the methyl substituents and unsaturation required for oxidative maneuvers. The only drawbacks of this approach are the relatively low yield (20 %) of the cyclization step and the formation of a useless diastereomer during the inital linkage of the A, C ring components.

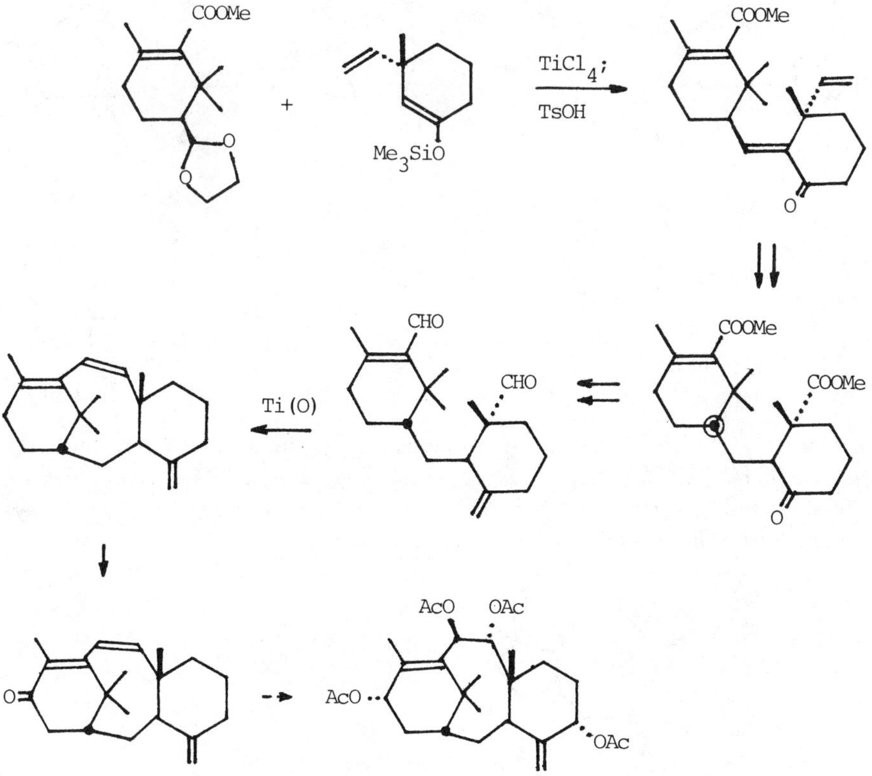

Electroreductive cyclization has been used twice in an approach to quadrone (**Q1**) (Little, 1987). First, a cyclopentanol is formed from a ζ-oxo-α,β-unsaturated ester, and at a later stage, a bridged cyclohexanol.

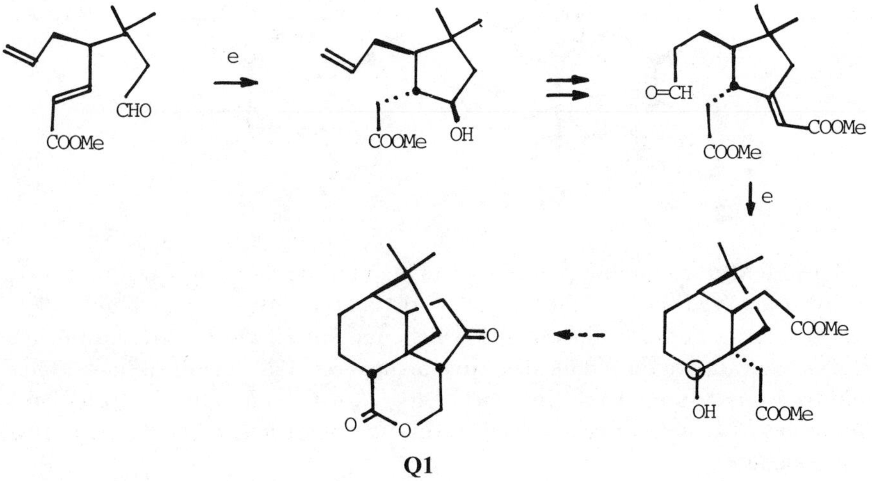

9.5 ACYLOIN CONDENSATION

Acyloin condensation (McElvain, 1948; Bloomfield, 1976) is one of the most powerful annulation techniques. The serious limitations that plague the Dieckmann, Thorpe–Ziegler, and related cyclizations with respect to ring size do not seem to apply. Cyclic structures ranging from four-membered rings upward can be prepared by the acyloin method.

The effectiveness of the reaction is the result of the electron transfer process on the surface of sodium metal, whereby two carbon radicals are held in close proximity, thus facilitating intramolecular coupling.

A recent development that calls for the addition of a chlorosilane in the reaction medium improves significantly the efficiency of acyloin condensation. The added silane serves as a trap for the enediolate ions and the alkoxide ions that are coproduced. The removal of these bases suppresses side reactions such as the Dieckmann and Claisen reactions.

So far, very few examples of acyloin cyclization are known in the area of sesquiterpene synthesis. This is probably because common rings are easily made by most other methods. A rare example is found in a synthesis of the vetispiranes including agarospirol (**A17**), hinesol (**H15**), and α-vetispirene (**V14**) (Ibuka, 1979a, b).

[See also, synthesis of sterpurene by Moens, et al. 1986].

Chapter 10

Synthesis of Small-Ring Compounds

10.1 CYCLOPROPANES

The classical 1,3-elimination procedure for cyclopropanation (see Section 5.3) is limited by the availability of suitable precursors. Consequently, the use of carbene and carbenoid reagents to effect both intra- and intermolecular cyclopropanation has gained widespread popularity.

The techniques for these reactions are so well developed that they are frequently employed to acquire structures other than intact cyclopropanes.

10.1.1 Simmons–Smith Reaction

The Simmons–Smith reaction (Simmons, 1973) involves the treatment of an alkene with methylene iodine in the presence of a zinc–copper couple. A zinc–silver couple (Denis, 1972) and dialkylzincs (Furukawa, 1966) may also be used in place of Zn–Cu. The methylene transfer reagent is not carbene but an organometallic complex. The existence of a concrete C–Zn bond in the reagent explains the directing effect of a proximal hydroxyl group through its coordination with the zinc atom. This complexation ensures delivery of the "CH_2" species from the same side as the hydroxyl function, even if that side is (moderately) more hindered.

The directing effect or syn stereoselectivity of an allylic hydroxyl group was first investigated in connection with a synthesis of thujopsene (**T18**) (Dauben, 1963). With the establishment of the hydroxyl group cis to the angular methyl substituent, the major obstacle to synthesis is resolved.

T18

10.1 CYCLOPROPANES

While the preparation of sabina ketone and hence sabinene (**S1**) (Fanta, 1968) is easily achieved based on the Simmons–Smith reaction, the synthesis of debromolaurinterol acetate (**L7**) (Feutrill, 1973) using the same reaction as a key step is plagued by the production of inseparable mixtures at both the olefin and the bicyclo[3.1.0]hexane levels. Furthermore, the bulkier aryl substituent disfavors the formation of the desired diastereomer.

Recently a number of modified procedures for the Simmons–Smith reaction have appeared. It is particularly interesting that the very powerful reagent CH_2I_2-Bu_3Al shows a reverse chemoselectivity for double bonds other than those situated at the β position of a hydroxyl group (Maruoka, 1985).

10.1.2 Dihalocarbene Addition to Alkenes

The chemistry of dihalocarbenes has been thoroughly studied (Kirmse, 1971). The halocyclopropanes are valuable intermediates capable of many useful transformations. The stepwise, regioselective replacement of the two chlorine atoms in the dichlorocarbene adduct of 2-cyclohexeneone ethyleneketal enables a stereoselective synthesis of sesquicarene (**S26**) (Kitatani, 1976, 1977).

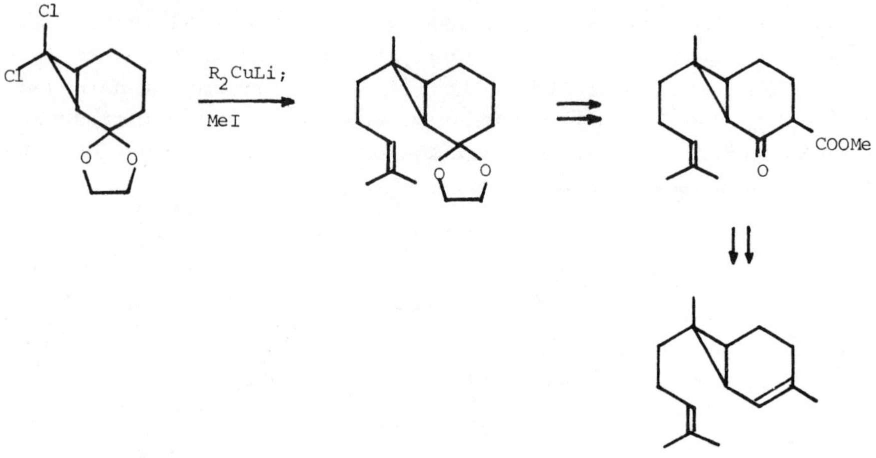

A conventional synthesis of a mixture of 2-carene (**C25**), 3-carene (**C26**), and β-carene (Cocker, 1978) depends on the dibromocarbene addition and a bromine–methyl exchange reaction mediated by lithium dimethylcuprate.

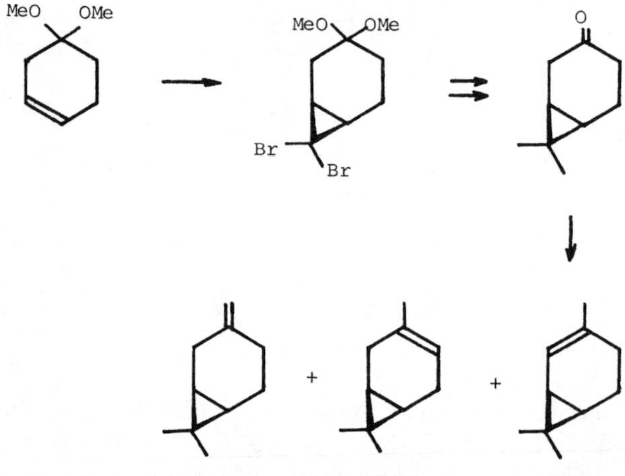

The final operations of a globulol (**G14**) synthesis (Marshall, 1974) are delegated to the same reaction sequence.

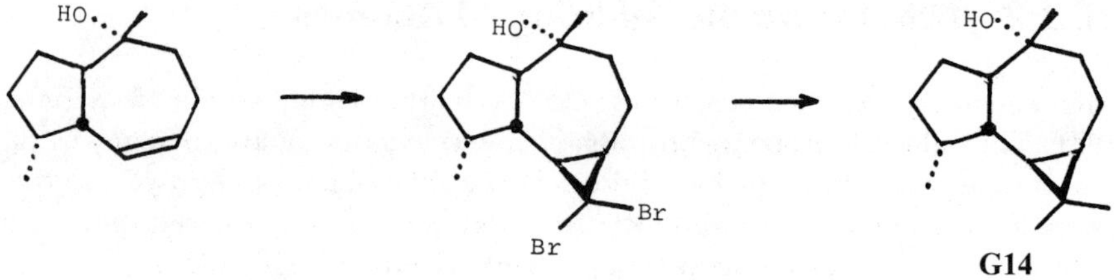

For the generation of dihalocarbenes, the reaction of haloforms with strong bases is the most convenient method. Nowadays, the effectiveness of dihalocarbene formation under phase transfer conditions (in the presence of a quaternary ammonium salt) with aqueous sodium hydroxide has been recognized, and this procedure is routinely used. The approach has practically superseded all other methods except when nonhydroxylic conditions are critically required. In that case, thermal decomposition of halocarbene precursors such as sodium trichloroacetate is preferred.

A remarkable reaction sequence leading to ishwarane (**I11**) (Cory, 1977, 1979) involves the trapping of dibromocarbene by a tetramethyloctalin, and the *in situ* debrominative insertion of the dibromocyclopropane product into a nearby methyl group in one operation.

10.1.3 C-Substituted Carbene and Carbenoid Addition to Alkenes

Carbenoids, especially those stabilized by an α substituent are employed frequently in cyclopropane formation by their reaction with proper alkenes. The closure of the third and fourth rings of ishwarone (**I12**) (Piers, 1977a, 1980) calls for the decomposition of dimethyl diazomalonate in the presence of an octalone and modification of the ester groups of the adduct into chloromethyl pendants.

The applicability of the carbenoid addition reaction to syntheses of carabrone (**C23**) (Minato, 1967, 1968a) and precarabrone (**C24**) (Bohlmann, 1982) is apparent to all synthetic chemists.

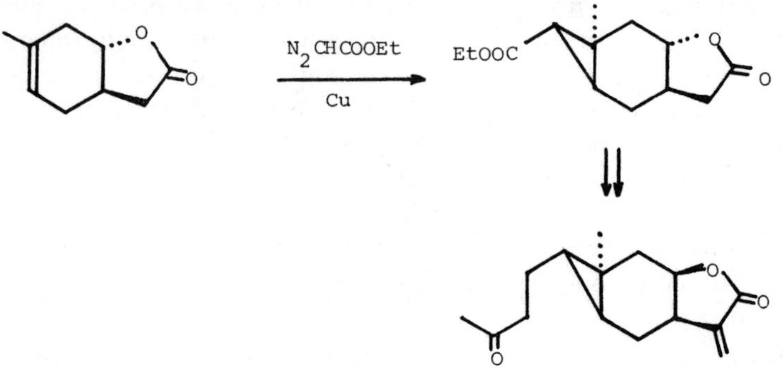

C23

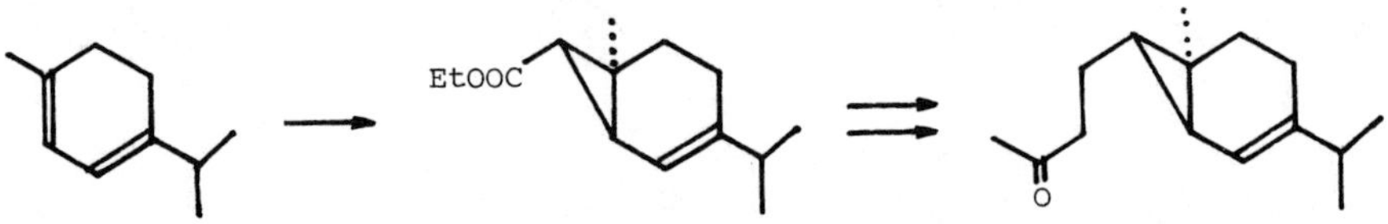

C24

The reaction of ethyl diazoacetate with 2,5-dimethyl-2,4-hexadiene, originally studied by Staudinger in 1924, has been greatly improved (Campbell, 1945) and has become a practical method of chrysanthemic acid (**C51**) synthesis. Using chiral copper complexes as catalyst and menthyl diazoacetate as reagent, a reaction attaining an enantiomeric excess of 90% may be achieved (Aratani, 1977).

α-Sulfenylcarbenes are more nucleophilic and they may add to electron-deficient double bonds with facility. This behavior enables a short synthesis of *cis*-chrysanthemic acid (**C51c**) (Franck–Neumann, 1979).

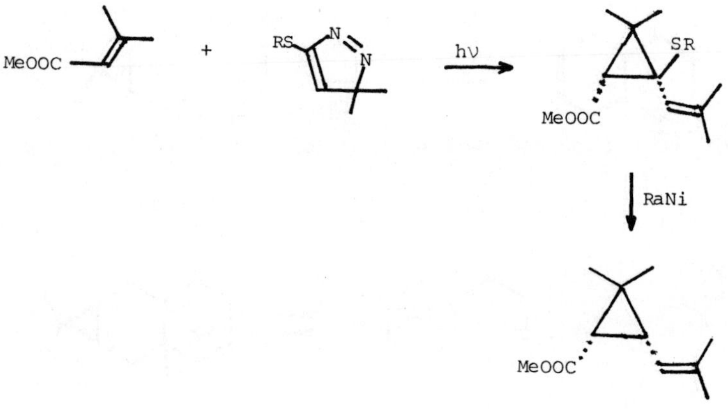

C51cE

10.1 CYCLOPROPANES

A very nice approach to *cis*-chrysanthemic acid (**C51c**) is through an intramolecular acylcarbene addition (review: Burke, 1979). Beckmann fragmentation of the bicyclic product unfolds the two carbon chains.

C51

Optically active 2-carene (**C25**) is accessible from the 2-cyclohexenyl ester of alanine via the corresponding α-diazopropionate (S. Yamada, 1975).

C25

Sabina ketone, which is synthetically related to sabinene (**S1**) and sabina hydrates (**S2**) by the Wittig and Grignard reactions, respectively, is conveniently prepared from the unsaturated α-diazoketone (Vig, 1969; Mori, 1970d).

S1

The tricyclic structure of aristolone (**A40**) may be constructed from a cyclohexene with appendices containing a diazoketone and an alkene linkage, respectively (Piers, 1969a). The ring-forming process unfortunately suffers from a high degree of stereorandomness because of the free rotating chains. In fact the undesirable diastereomer is produced in larger quantities.

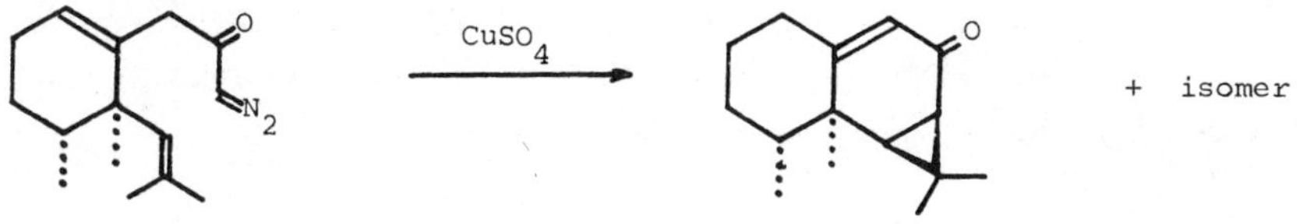

A40

When the alkene moiety is embedded in a ring or protruded therefrom, thereby eliminating one level of molecular freedom, the intramolecular carbene addition is stereoselective. The facile assemblage of the skeletons of β-cubebene (**C81**) (Piers, 1969c; A. Tanaka, 1969, 1972), cubebol (**C81a**) (Torii, 1976), cycloeudesmol (**E20**) (Chen, 1982), and thujopsene (**T18**) (Mori, 1970b; McMurry, 1974; Branca, 1977) attests to the unique usefulness of this approach in dealing with fused cyclopropanes.

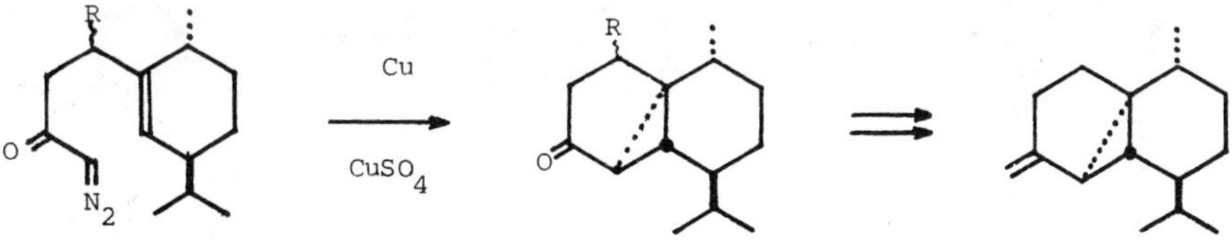

C81

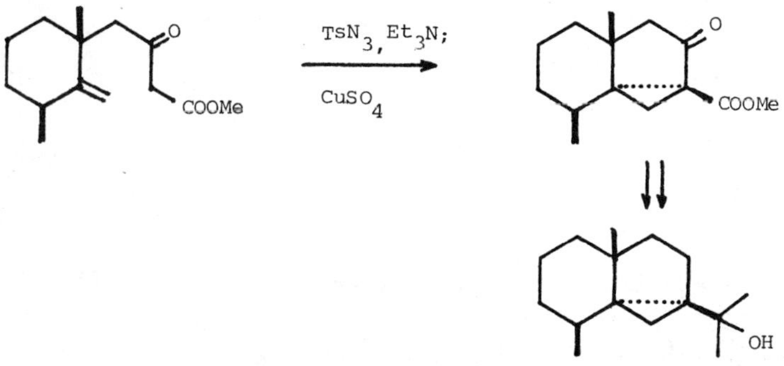

E20

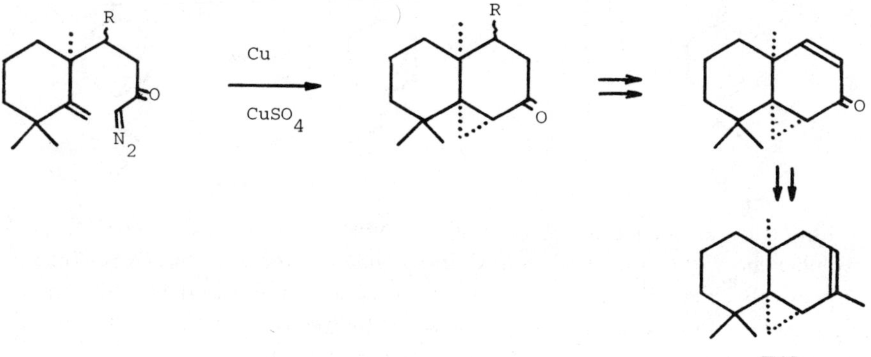

T18

10.1 CYCLOPROPANES

Extensive exploitation of the chemistry of α-diazoketone is associated with one of thujopsene synthesis (Branca, 1977). A vinylogous Wolff rearrangement, a photo-Wolff rearrangement, and cyclopropanation are involved.

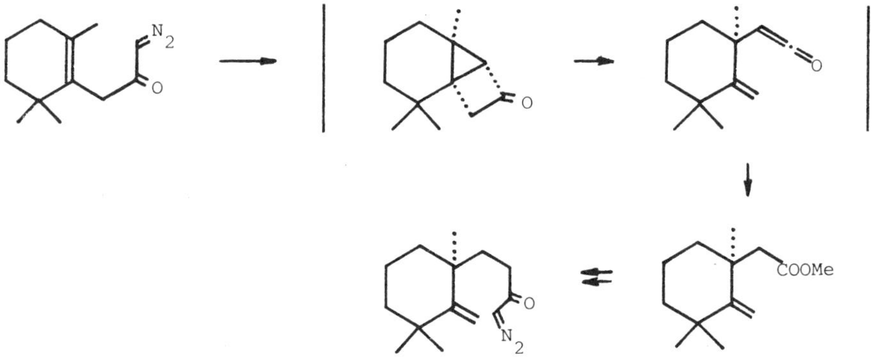

(An earlier synthesis of thujopsene (Büchi, 1964b) using the direct method is seriously complicated by side reactions.)

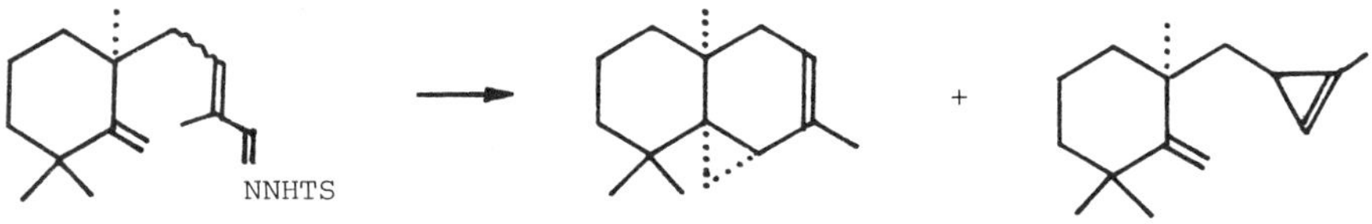

T18

The decomposition of farnesyl diazoacetate in the presence of copper gives rise to a cyclopropanated γ-lactone that can be hydrolyzed, oxidized, epimerized, coupled with a Wittig reagent, and finally reduced to afford presqualene alcohol (**P48**) (Coates, 1971).

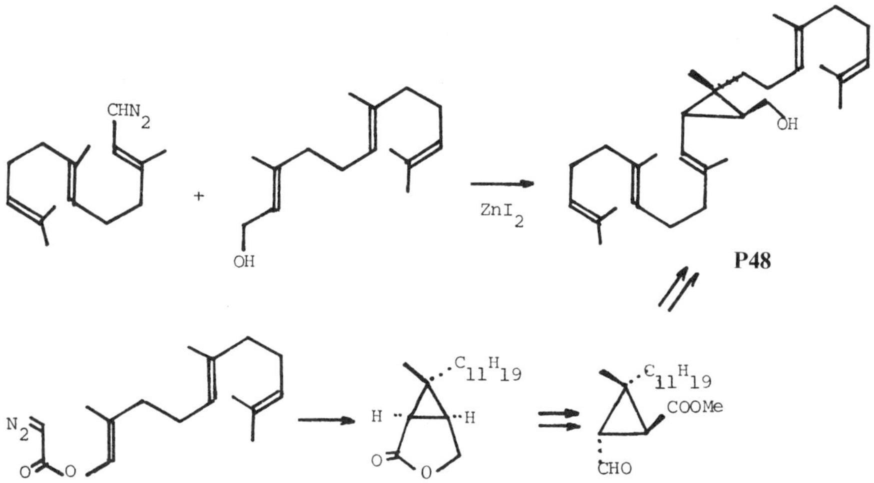

This elusive intermediate just preceding squalene in the biosynthesis of triterpene has also been prepared by a method involving an intermolecular carbene addition (Altman, 1971).

The structures of sesquicarene (**S26**) and sirenin (**S38**) would plainly convince any organic chemist that the most direct and fruitful synthetic pathways toward them must involve an intramolecular carbene addition. Starting from farnesal tosylhydrazone or hydrazone, sesquicarene has been generated in moderate to low yields (Coates, 1969; Corey, 1969b, 1970; Nakatani, 1969). The efficiency of the process is affected by the inability of the 2E-isomer to undergo cyclization. However, mercury(II) iodide has been found to catalyze double-bond inversion, and in its presence the yield of sesquicarene increases (Corey, 1970).

The more rewarding but lengthy alternative routes to sesquicarene consist of using the diazoketone precursors (Coates 1969, 1970a; Corey, 1969a, Mori, 1970a).

The fungal pheromone sirenin is accessible analogously, although different methods must be employed to prepare suitable progenitors (Corey, 1969c, 1970; Grieco, 1969; Plattner, 1969, 1971; Mori 1970a; Garbers, 1975). A slightly different and most recent route is the following (Mandai, 1983; see also; Ono, 1984).

10.1 CYCLOPROPANES

The generation of casbene (**C34**) by the decomposition of the diazo compound derived from geranylgeranial (Toma, 1982) in the presence of copper(I) iodide indicates that a folded conformation may be quite favorable.

C34

The addition of an allenecarbene to prenol and subsequent redox transformation constitute another nice approach to *trans*-chrysanthemic acid (**C51**) (Mills, 1971).

C51

10.1.4 Cyclopropanes from 1,3-Dipolar Adducts

Because of their instability, hence a manifold reaction pathway, the utility of unstabilized carbenes is quite limited. Usually it is advisable to pursue a stepwise cyclopropanation sequence via 1,3-dipolar addition of the diazoalkane to an unsaturated compound, extruding nitrogen from the pyrazoline. A synthesis of aristolone (**A40**) (Berger, 1968) illustrates this tactic.

A40

[In connection with this work it should be noted that cisoid enones form pyrazolines with hydrazine, thus making available a reductive cyclopropanation, as used, e.g., in the syntheses of 2-carene (**C25**) (Naves, 1942) and aristolene (**A39**) (Coates, 1968, 1970b).]

Application of the two-step method to the synthesis of *cis*- and *trans*- hemicaronic aldehydes, the precursors of chrysanthemic acids (**C51** and **C51c**) requires a properly chosen acrylic ester substituted at the β position with a masked aldehyde group. A useful latent function for the aldehyde appears to be a metal-coordinated 1,3-diene (Franck–Neumann, 1982).

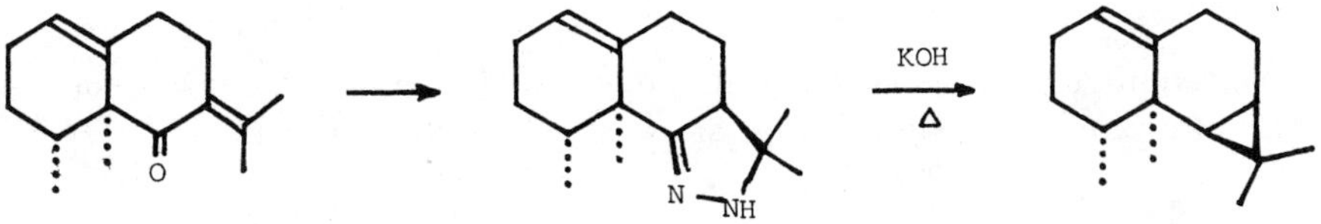

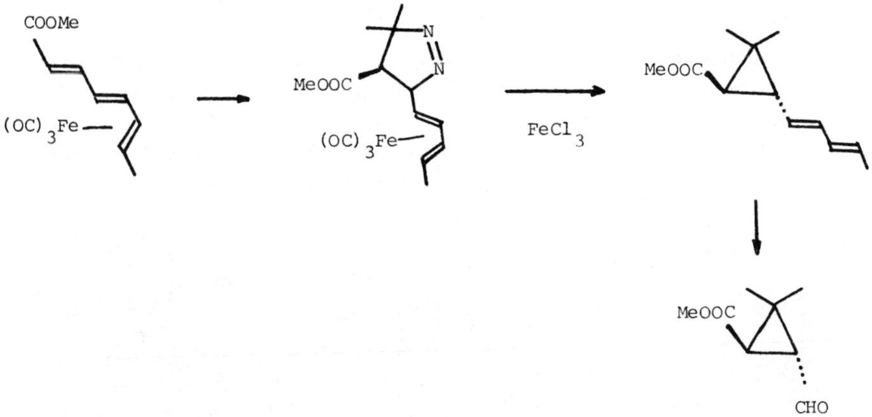

When reacted with 2-diazopropane, an optically active γ-isopropylbutenolide forms adducts that on sensitized photolysis yield a precursor of *cis*-chrysanthemic acid (**C51c**) (Franck–Neumann, 1985).

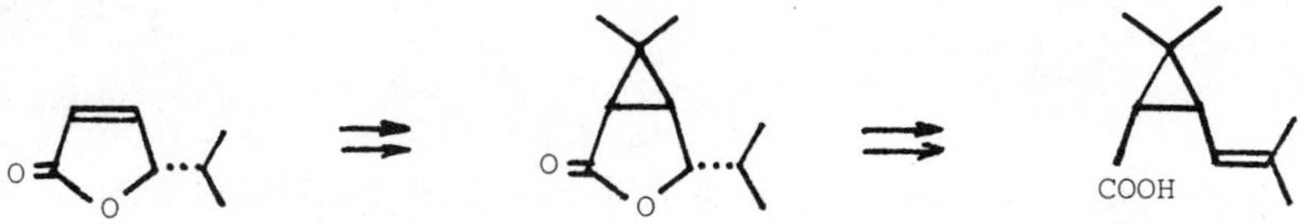

Pyrazoles undergo photodecomposition to give cyclopropanes. A synthesis of *cis*-chrysanthemic acid ester (Franck–Neumann, 1980) is realizable when selective hydrogenation of the cyclopropene ring is achieved.

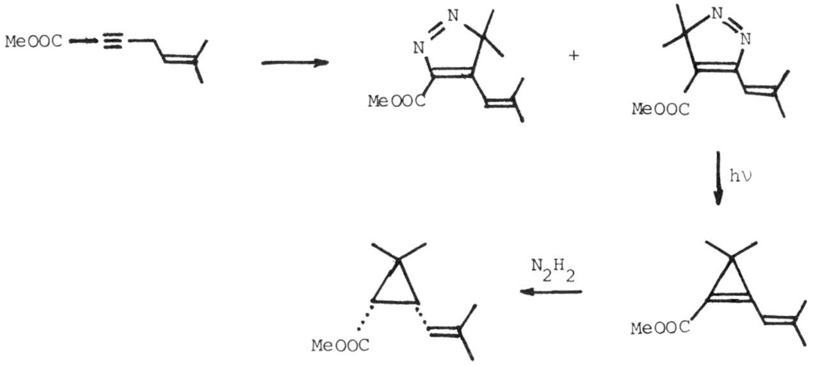

C51cE

Thermolysis of aziridinyl imines also causes a fragmentation to give diazoalkanes. An intramolecular reaction of the diazoalkane with an alkene upon the liberation of the former is the first of two key transformations during the pursuit of longifolene (**L24**) (Schultz, 1985).

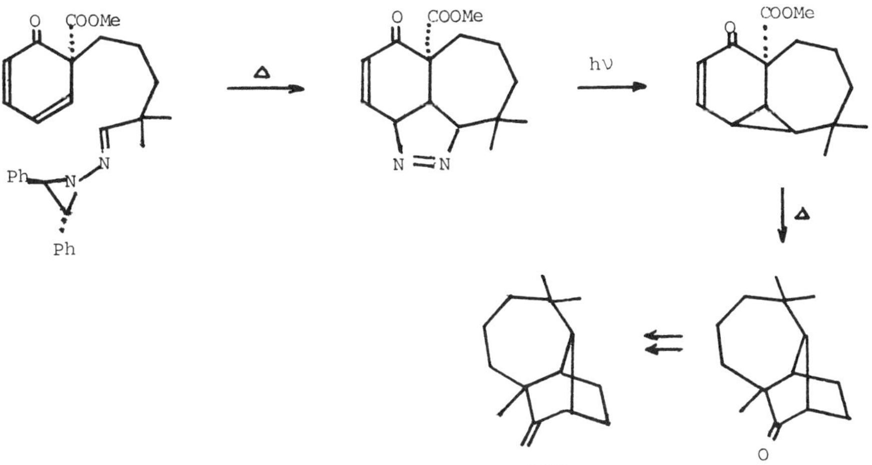

L24

Decomposition of tosylhydrazone by a strong base in the presence of an alkene leads to pyrazolines, and sometimes cyclopropanes. This process constitutes the final stage of the quests for (—)-cyclocopacamphene (**C64**) (Piers, 1971, 1975b), (+)-cyclosativene (**S13**) (Piers, 1973a, 1975c), and cycloseychellene (**S30**) (Niwa, 1983).

It must be emphasized that nitrogen extrusion, either by thermal or photochemical means, is not a concerted process.

Oxidation of a hydrazone with mercury(II) oxide affords the corresponding diazoalkane, which may lose nitrogen to give the carbene. Insertion of the carbene into a β C—H bond results in a cyclopropane. π-Bromotricyclene is available from π-bromocamphor hydrazone, and its transformation into α-santalene (**S3**) (Corey, 1957) has been documented.

10.1.5 Solvolytic Ring Closure

Trachylobane (**T22**) is a 12,16-cyclokaurane. Synthetically it is expedient to approach it by first constructing an atisirene derivative; a cyclopropanation maneuver accomplishes the task.

A viable route (R. Kelly, 1973a) involves photocycloaddition of allene to a tricyclic enone, degradative opening of the cyclobutane ring of the adduct, and aldol condensation. Since experience indicates that the cycloaddition would lead to compounds with B/C-ring juncture, the cyclic ketone group is the site to receive the final methyl substituent, and then connect to the hydroxyl-bearing carbon during cyclopropanation.

An interesting cyclopropanation involves solvolysis of the tetracyclic mesylate in dimethyl sulfoxide. Participation of the double bond and solvent trapping of the cyclopropylcarbenium ion lead to trachylobanone.

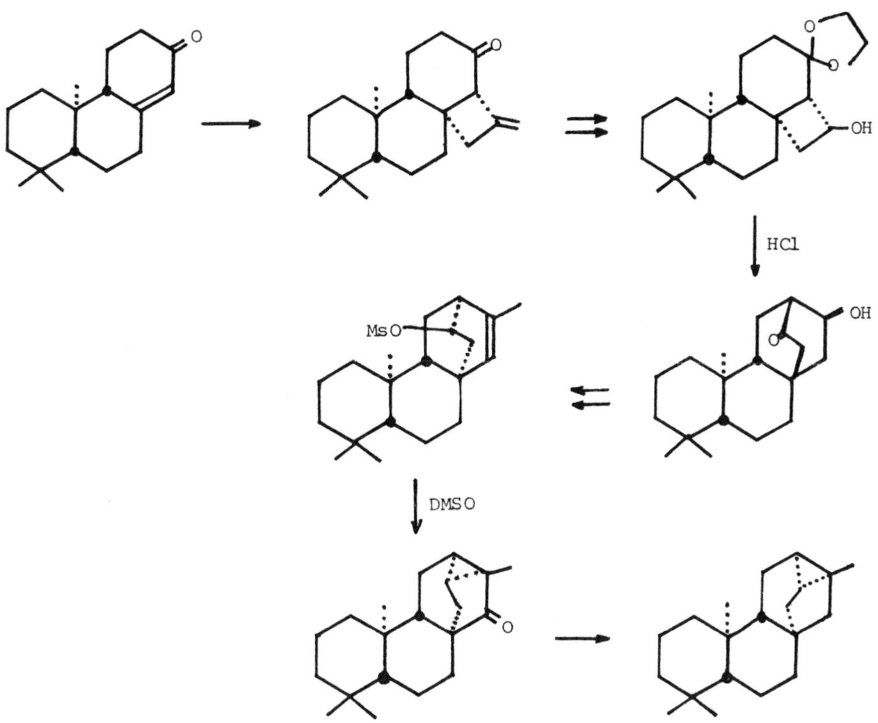

T22

Two epimeric carboxylic acids based on the trachylobane sketeton are known. The enantiomer (**ent-T23**) of methyl trachylobanate has been elaborated from methyl levopimarate via a Diels–Alder reaction and solvolytic cyclopropanation (Herz, 1968).

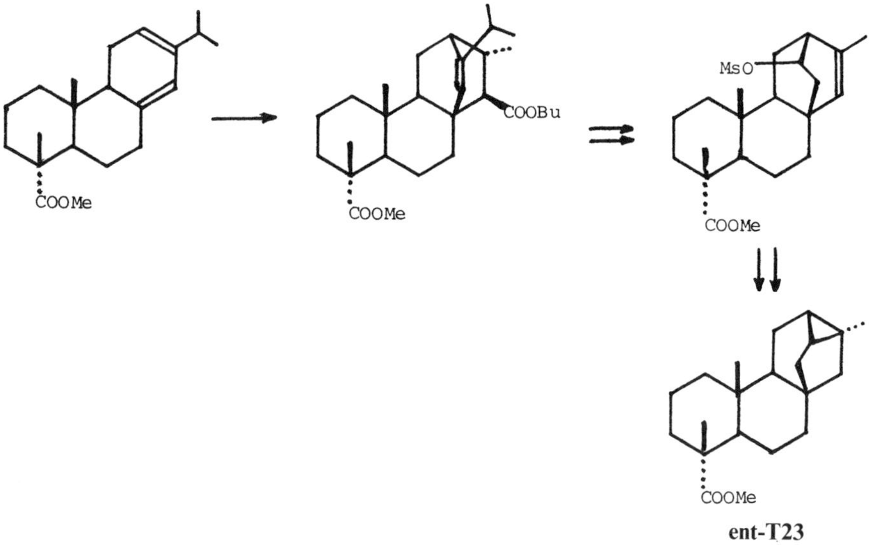

ent-T23

10.1.6 Cyclopropanation Using Other 1,1-Dipolar Reagents

Although cyclopropanation by the Michael–alkylation tandem was discussed in Chapter 4, one more example is mentioned here, just to remind us of the importance of this technique for making three-membered rings.

As dimethylsulfoxonium methylide converts α,β-unsaturated ketones into the corresponding cyclopropyl ketones, and saturated ketones into epoxides, amalgamation of these observations indicates the possibility of directly synthesizing molecules containing both a cyclopropane and an epoxide ring that are interconnected. As reduction of the epoxide gives a cyclopropyl methyl carbinol preparation of sabinene hydrate (**S2**) (Higo, 1978) is a logical application of this method.

Optically active (+)-chrysanthemum dicarboxylic acid (**C53**) is available from a pathway starting from an anhydro sugar synthon (Fitzsimmons, 1984). Cyclopropanation is achieved by reacting an epoxide with an α-diethoxyphosphonopropionic ester. Interestingly, the annulation must proceed either by equatorial attack of the nucleophile (to allow a C → O transfer of the phosphoryl group before cyclization can occur) or, alternatively, diaxial opening of the epoxide ring must be followed by intermolecular transphosphorylation.

10.1.7 Miscellaneous Cyclopropanations

A di-π-methane rearrangement of an α,β;δ,ε-diunsaturated ester induced by light leads to methyl chrysanthemate (**C51E**) as a geometrical isomer mixture, albeit in only 12% yield (Bullivant, 1976).

The choice of eucarvone for the synthesis of *trans*-chrysanthemic acid (**C51**) (Welch, 1977) was apparently based on a previous observation that methylation of the dienone gives rise to a bicyclo[4.1.0]heptenone in which the two carbon chains of chrysanthemic acid lie latent in the six-membered ring. The functional group distribution is highly favorable for the required elaboration.

The fungal antibiotic mycorrhizin A (**M15**) contains a modified isopentene unit that is spirocyclically united to a blocked phenol nucleus. In its synthesis (Koft, 1982a) an internal alkylation is performed on a hydrobenzofuran derivative that differs from its biosynthesis in the order of the two C—C bond attachments to the six-membered ring.

10.2 CYCLOBUTANES

The chemistry of the four-membered ring compounds is less developed than that of the cyclopropanes because methods for making the former are scarcer. This phenomenon is paradoxical, because cyclobutane is less strained than cyclopropane. Presently there are only two major techniques by which cyclobutane derivatives may be confidently acquired. These are the [2 + 2]cycloadditions of alkenes, induced photochemically or thermally, with the latter processes being limited to ketene partners.

10.2.1 Photoinduced Cycloaddition

The potential of photocycloaddition (Sammes, 1970; Dilling, 1977) in the synthesis of complex molecules was recognized by Corey, who demonstrated it in a classic synthesis of caryophyllene (**C32**) and isocaryophyllene (**C33**) (Corey, 1963a, 1964b). The orientation of the reaction partners has also been delineated.

Much later, the same reaction was applied to a cyclononadienone to produce the 4:9-fused ring system directly, thereby permitting a shorter synthesis of isocaryophyllene (**C33**) to be achieved (Kumar, 1976).

The formation of a fused 4:n-ring system ($n = 5, 6$) and subsequent cleavage of the larger ring represents a popular scheme for the synthesis of grandisol (**G18**). This method ensures a stereochemical pattern corresponding to that in the target.

Several photochemical routes have been recorded. The use of a cyclopentenone (Cargill, 1975; Rosini, 1979) has the advantage of preserving all the skeletal carbon atoms. [In one synthesis ethyl 3-oxo-1-cyclopentenecarboxylate is employed (Mori, 1978). Optical resolution of the cycloadduct is achieved.]

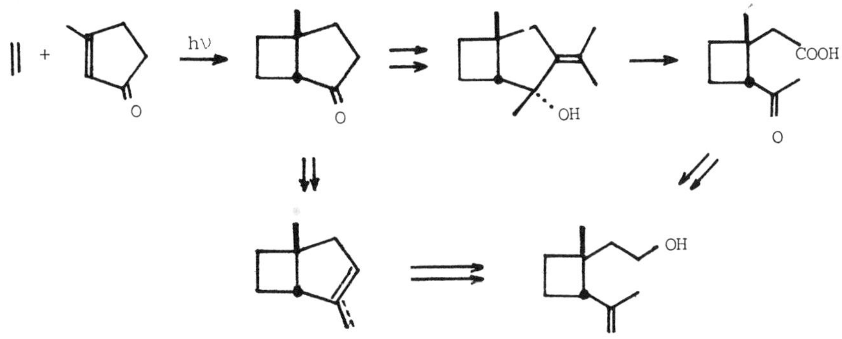

The homologous photoadduct from 3-methyl-2-cyclohexenone requires a different degradation protocol accordingly (Zurflüh, 1970). In terms of chemical source utilization, anhydromevalonolactone is a superior photoaddend for grandisol synthesis (Gueldner, 1972).

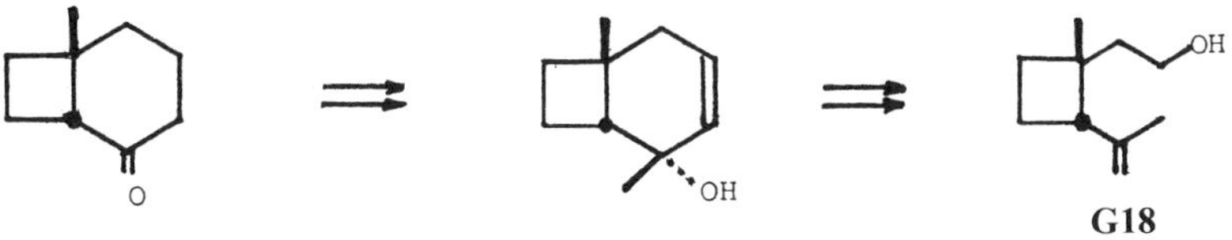

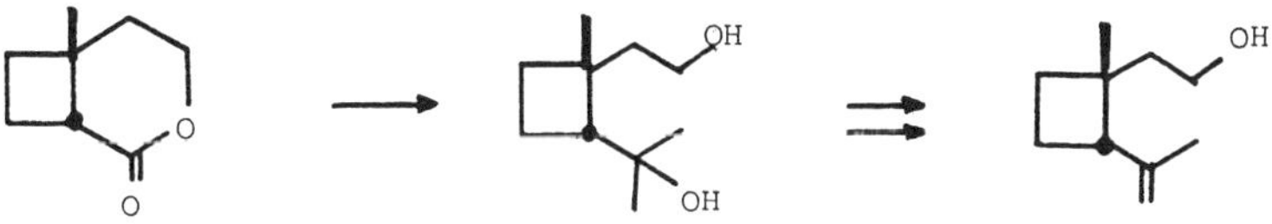

ent-Grandisol (**ent-G18**) may be obtained from the photocycloadduct of ethylene and a bicyclic lactam derived from (+)-valinol (Meyers, 1986a).

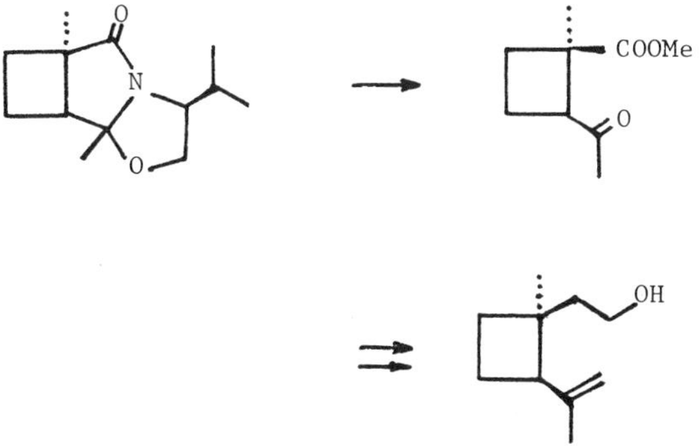

Anhydromevalonolactone also forms with acetylene a photoadduct that is useful for the synthesis of lineatin (**L19**) (White, 1982). The deficiency of this approach is the production of regioisomers during oxygenation of the cyclobutene via hydroboration. The use of allene as addend also gives two photoadducts in a 2:3 ratio favoring the unwanted isomer. Processing of this mixture leads to lineatin and its isomer (McKay, 1982).

The photocycloaddition of 2,4,4-trimethyl-2-cyclopentenone with vinyl acetate is not regioselective either. In this case a carbonyl transposition is needed as well as a Baeyer–Villiger reaction in the pursuit of lineatin (**L19**) (Mori, 1979).

A significant aspect of the synthesis of sterpurene (**S52**) (Moens, 1986) is asymmetric induction during photocycloaddition which results from the approach of the ethylene molecule to the *trans*-fused hydrindenone on the same side of the β'-hydrogen at the angle. The corresponding *cis*-fused enone is unreactive toward the cycloaddition due to steric hindrance by the methylene group of the five-membered ring.

Synthesis of sterpuric acid from a five-membered ring synthon is almost invariably plagued by the lack of stereocontrol during annulation unless it is

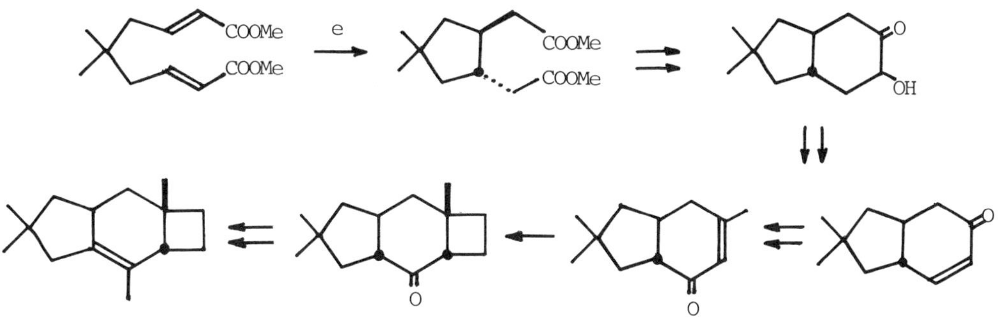

sterpurene

made intramolecular. In an accomplished work (Paquette, 1987a) the formation of the six-membered ring is done by a Diels–Alder reaction which leaves a double bond for a subsequent [2 + 2]cycloaddition and an easily removable sulfonyl group at the angular position.

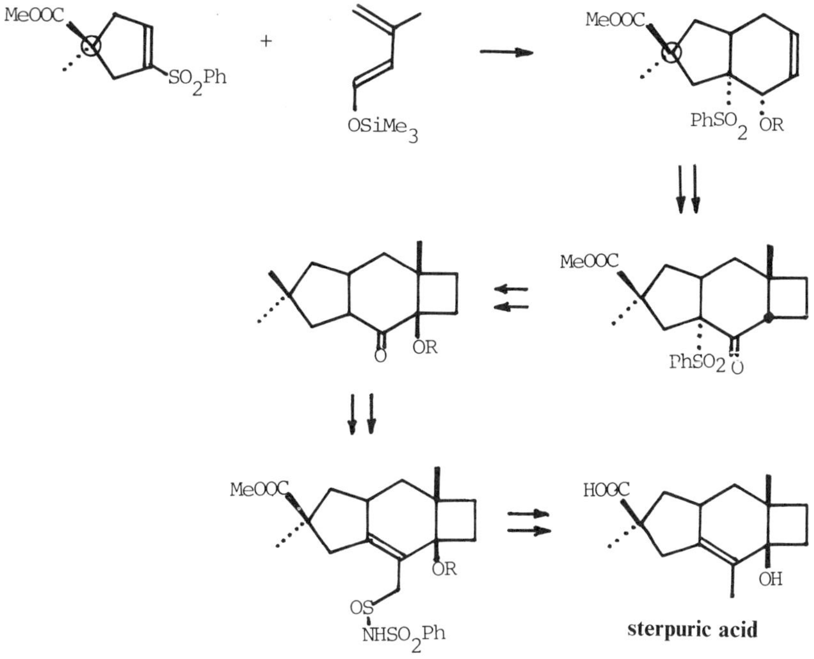

sterpuric acid

A synthesis of illudol (**I3**) (Kagawa, 1969; Matsumoto, 1971) requires an oxygenated ethylene derivative as the addend. For reactivity reasons and to avoid formation of excessive amounts of unwanted isomers, 1,1-diethoxyethene is used in the photocycloaddition step. The choice of a cyclopentenone derivative as the other building block is based on the consideration that cis ring fusion can be guaranteed, and perhaps on the relative ease with which such compounds can be acquired. Necessarily, the photoadduct must undergo ring expansion at some stage of the synthesis.

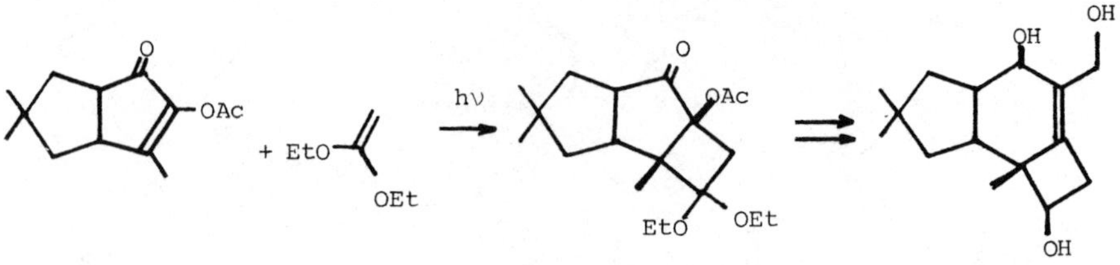

Two reported syntheses of the bourbonenes (**B16** and **B17**) also are characterized by intermolecular photocycloaddition. Thus, cyclopentenone is added to 1-methyl-3-isopropylcyclopentene to give two adducts in 1:1 ratio. One of these adducts is converted into β-bourbonene (White, 1966, 1968). The addition is stereoselective (cis-anti-cis) but nonregioselective.

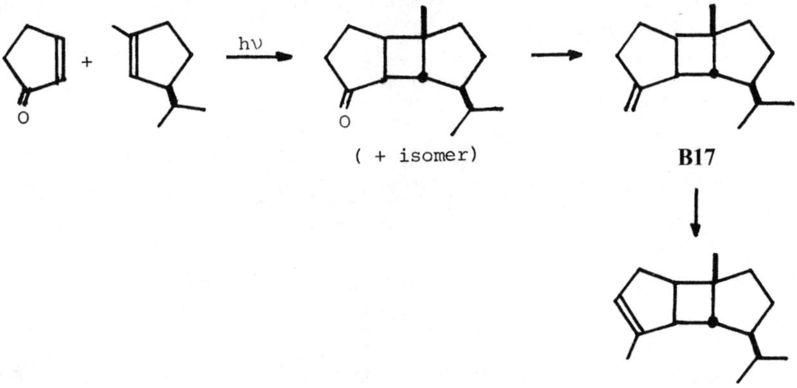

The other synthesis leading to optically active (—)-β-bourbonene (**B17**) (Tomioka, 1982) employs a chiral butenolide as the addition partner for cyclopentenone ethyleneketal.

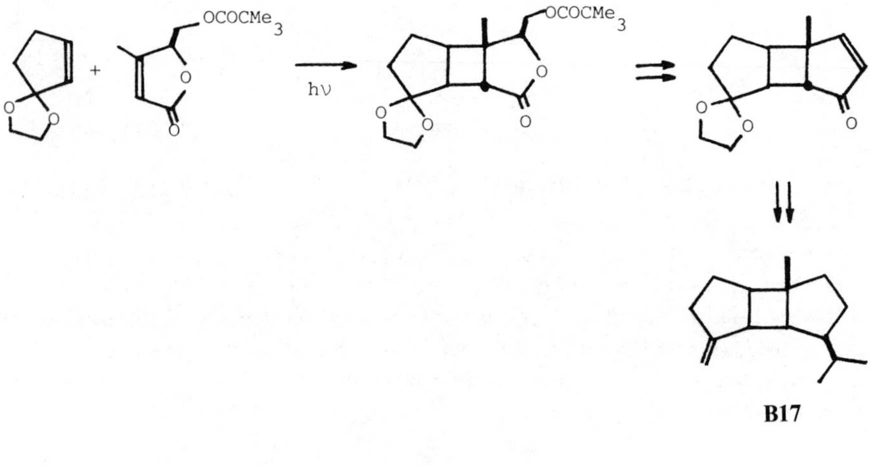

The butenolide is not suitable for synthesis of the spatane diterpenes in strictly the same manner because they are in the opposite (to the bourbonenes) optical series. However, by delegating to the chiral substrate the role of A-ring synthon, the problem is resolved. Greater control in subsequent functionalization of the C ring can be exerted by performing the cycloaddition intramolecularly. The synthesis of stoechospermol (**S54**) (M. Tanaka, 1985a) and spatol (**S42**) (M. Tanaka, 1985b) successfully exploits this principle.

For synthesis of racemic spatanes via an intermolecular [2 + 2]cycloaddition, the crucial features include attention to the stereochemistry of the nonangular chiral carbon atoms. The secondary methyl group, which has an endo configuration, should be easily established from a ketone. Thus the A-ring synthon is identified as 2-cyclopentenone. The hydroxyl function may be tied back onto a carbon chain, most conveniently as a lactone unit; consequently, a proper synthetic equivalent for the C ring is a norbornenone. Necessarily the configuration of the carbon chain must be inverted during elabortation of the 8-carbon side chain.

A synthesis of stoechospermol (**S54**) (Sálomon, 1984) has been realized as planned. The conversion of the norbornanone moiety into a 3-hydroxycyclopentylidenemalonate is instrumental to the said configuration inversion, that is, by oxygen-directed hydride reduction of the conjugated double bond.

A third synthesis of α-bourbonene (**B16**) (Brown, 1968) features an intramolecular photocycloaddition. The research was apparently directed toward copaene, hoping that a head-to-tail mode of cycloaddition would prevail. However, the overwhelming production of head-to-head cycloadducts forced a change of target.

The otherwise disastrous result has of course been turned into a successful venture by extending the two ester chains and cyclizing the diketone. Due to the epimerizability of the pendants that are to be united, the thermodynamically favorable cis-anti-cis tricyclic skeleton of the bourbonene is ensured. The regioselectivity of the aldol condensation step has been adequately discussed (Chapter 3).

B16

Copper(I) triflate is an excellent catalyst for inducing unactivated alkenes to unite photochemically. An efficient synthesis of grandisol (**G18**) (Rosini, 1985) attests to the usefulness of the method.

G18

The structure of the panasinsenes (**P7** and **P8**) pleads eloquently for the employment of a photocycloaddition method in their synthesis. While isobutene fails to add to the proper hydrindenone, an intramolecular version proves adequate (McMurry, 1980; Johnson, 1981).

P7
P8

Italicene (**I17**) and isoitalicene (**I18**) are apparent candidates for synthesis based on intramolecular [2 + 2]cycloaddition (Leimner, 1984).

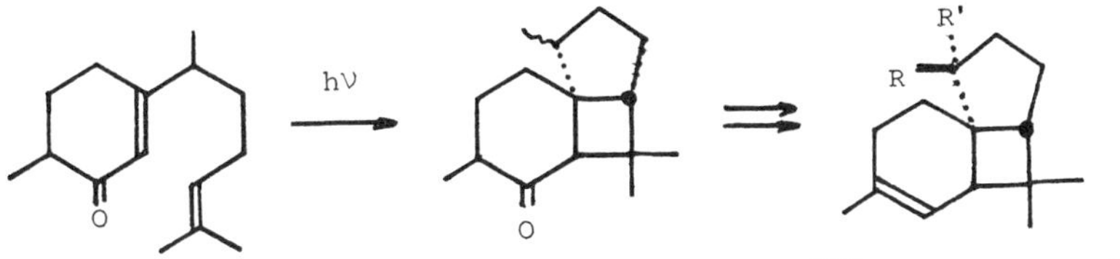

I17 (R= Me, R'= H)
I18 (R= H, R'= Me)

Intramolecular photocycloaddition of an allene linkage to cyclopentenone features prominently in an attractive approach to pentalenolactone G (Pirrung, 1986, 1987). By this method the ring juncture stereochemistry is established unambiguously. Furthermore, the regioselective ring expansion can be effected via reduction and Sharpless epoxidation. Homologation of the less hindered ketone group follows.

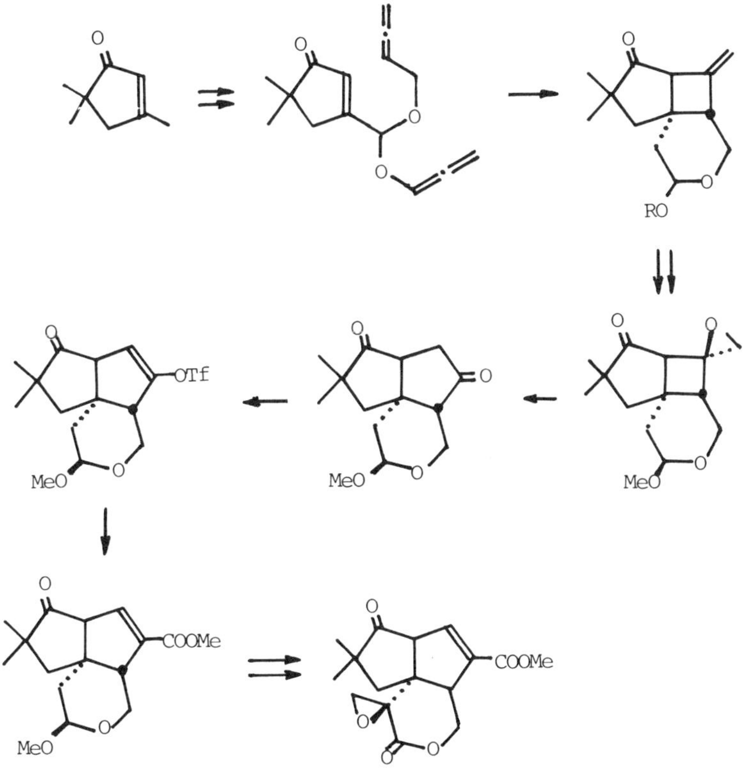

pentalenolactone G

A formal synthesis of trihydroxydecipiadiene (**D10**) consists of an intramolecular photocycloaddition of a cyclohexenone with an allene side chain (Dauben, 1984). To avoid the formation of *trans*-decalin products, a hydroxyl group is placed in the proangular position. The natural configuration of that carbon atom can be established at the tricyclic level by means of a dehydration–hydrogenation sequence.

D10

Intramolecular photocycloaddition in the crosswise fashion can be induced by conjugating one of the alkenes with another unsaturated unit. The syntheses of α-*trans*-bergamotene (**B5**) (Corey, 1971) and the longipinenes (**L25** and **L26**) (Miyashita, 1971, 1972, 1974) are further examples. In both cases, a ring expansion is required, and the exocyclic methylene group provides a handle for that operation.

B5

L25, L26

The major reasons for using these substrates, in which the two unsaturated moieties are separated by a methylene group, and for performing the reaction in the presence of a photosensitizer, are based on experience with myrcene. Direct photolysis of myrcene gives mainly a cyclobutene as a result of electrocyclization, whereas a bridged cyclobutane ring is the sole product arising from relaxation of the triplet species.

10.2.2 Thermal Cycloaddition

The thermal addition of ketenes to alkenes to form cyclobutanones proceeds by the $[\pi^2 s + \pi^2 a]$ mechanism, which has yet to realize its full potential in the synthesis of complex natural products, despite its noteworthy application in the alkaloid and prostanoid areas. With regard to sesquiterpene synthesis, the following examples are significant: (+)-isocaryophyllene (**C33**) (Bertrand, 1974), hirsutene (**H17**) (Greene, 1980), ivangulin (**I20**) (Grieco, 1977d), eriolanin (**E14**) (Grieco, 1978), bicyclogermacrene (**G7**) and lepidozene (**L10**) (McMurray, 1985a), pentalenene (**P16**) (Annis, 1982), the methyl ester of pentalenolactone E (**P19**) (Cane, 1984), and β-cuparenone (**C86**) (Leriverand, 1973; Greene, 1984). Only in the case of isocaryophyllene, however, is the cyclobutane ring retained throughout the synthesis.

C33

An aggregation pheromone for the ambrosia beetle is a monoterpene called lineatin (**L19**). The presence of the four-membered ring in this molecule tends to invoke strategies for its synthesis based on ketene addition to alkenes. Indeed, two methods have been developed involving isoprene (Mori, 1983) and 2,2,4-trimethyl-5,6-dihydro-2H-pyran (Johnson, 1985) as addends.

(+ isomer)

L19

Diastereofacial differentation for [2 + 2]cycloaddition involving ketenes has been demonstrated by using an enol ether with effective shielding on one face of the alkene linkage. The combination of this cycloaddition with ring expansion provides a crucial intermediate for both (—)-α-cuparenone (**C85**) and (+)-β-cuparenone (**C86**) (Greene, 1987a).

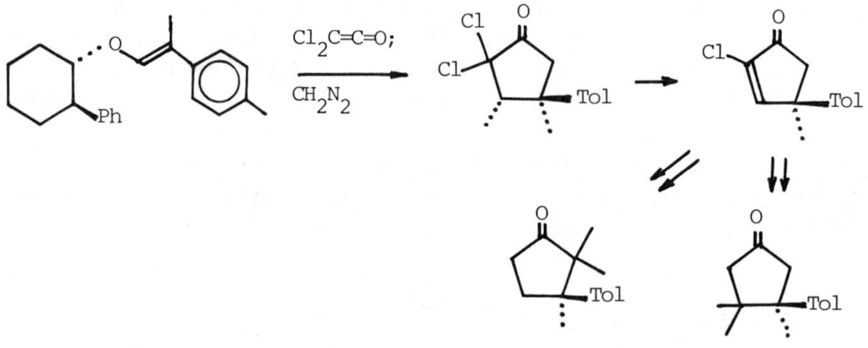

An authentic [2 + 2] approach to filifolone (**F6**) has been belatedly realized (Stadler, 1984). 6,6-Dithiofulvene derivatives appear to be uniquely qualified for participating in the desired reaction.

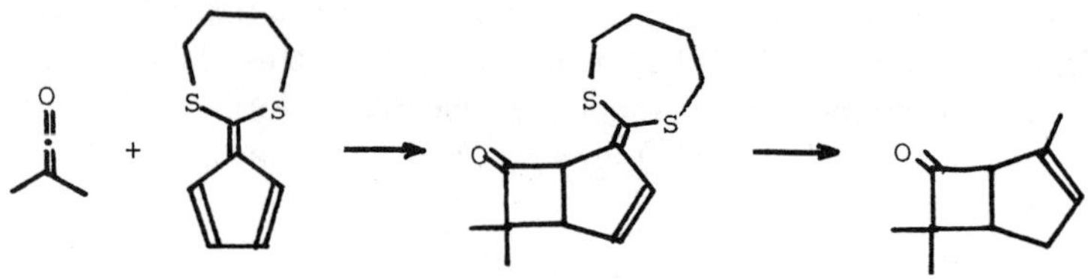

There are only sporadic reports on the intramolecular [2 + 2] cycloaddition reactions involving ketenes. Among these are an approach to β-cis-bergamotene (**B7**) (Kulkarni, 1985) and β-trans-bergamotene (**B6**) (Corey 1985a, Kulkarni, 1985), from 6Z- and 6E-geranic acids, respectively.

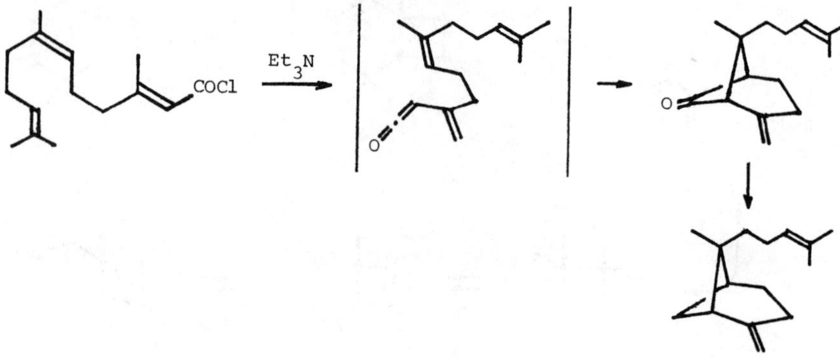

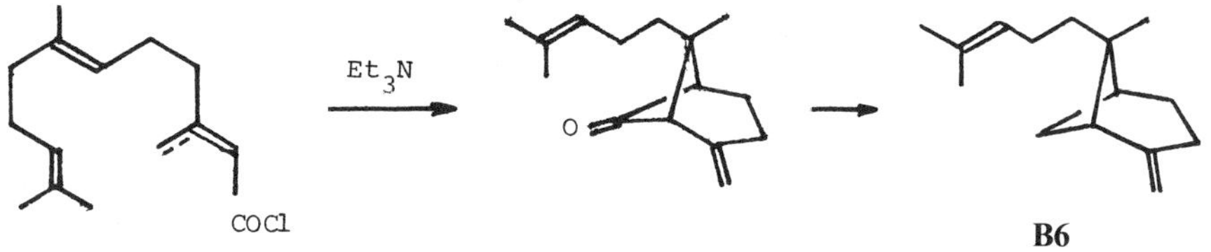

Heating ocimenone(s) with aluminum chloride gives filifolone (**F6**) and other products (Weyerstahl, 1986). Formation of the strained ketone likely proceeds via a ketene intermediate.

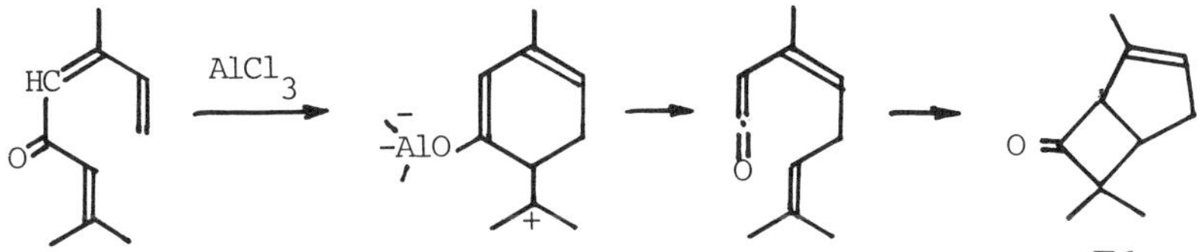

Trihydroxydecipiadiene (**D10**) provides an excellent opportunity for testing the synthetic potential of [2 + 2]cycloaddition with ketenes. While in most other applications the cyclobutanone is left untouched or is prodded to undergo simple cleavage reactions, here it serves as the donor in an aldol process to create a tricyclic framework (Greenlee, 1981).

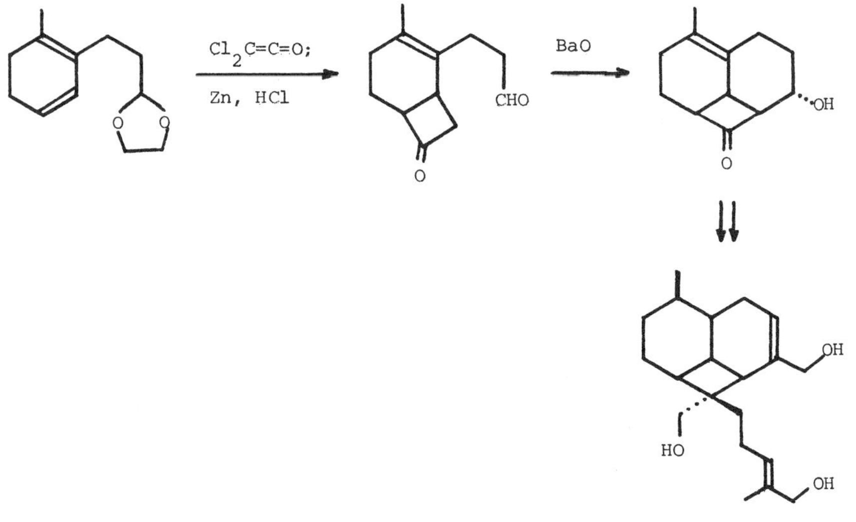

Intramolecular thermal [2 + 2]cycloaddition of an isolated double bond with an allene affords ring-fused alkylidenecyclobutanes. Apparently the reaction is ideal for the elaboration of lineatin (**L19**) (Skattebol, 1983).

Perhaps the most practical synthesis of grandisol (**G18**) is that consisting of nickel(0)-catalyzed dimerization of isoprene and selective hydroboration of the cyclobutane product, which constitutes about 12–15 % of the dimer mixture (Billups, 1973).

It should be noted that photochemical dimerization produces a 2.4 % yield of an inseparable cyclobutane derivative mixture. This mixture has been converted into grandisol (**G18**) and fragranol (**F11**) (Tumlinson, 1971). A selective formation of the trans dimer in 20 % yield has been observed (Corey, unpublished) when benzophenone is used as photosensitizer. Thus fragranol is as easily accessible in two steps as grandisol.

10.2.3 Ring Expansion and Cycloalkylation

It is logical to apply the cyclobutanone formation method starting from carbonyl substances via the sulfenylcyclopropylcarbinols to a synthesis of grandisol (**G18**) (Trost, 1975b). In fact, two such reaction sequences leading to a spiro[3.3]heptanone and the cleavage of the cyclobutanone ring to release a one-carbon and a two-carbon chain are central to this approach.

10.2 CYCLOBUTANES

The zirconium-catalyzed carboalumination of alkynes can proceed further to afford cycloalkenes when the alkyne is substituted with a silyl group at the *sp*-hybridized carbon and a halogen at the other end of the alkyl chain. The utility of this reaction is demonstrated in a synthesis of grandisol (**G18**) (Negishi, 1983).

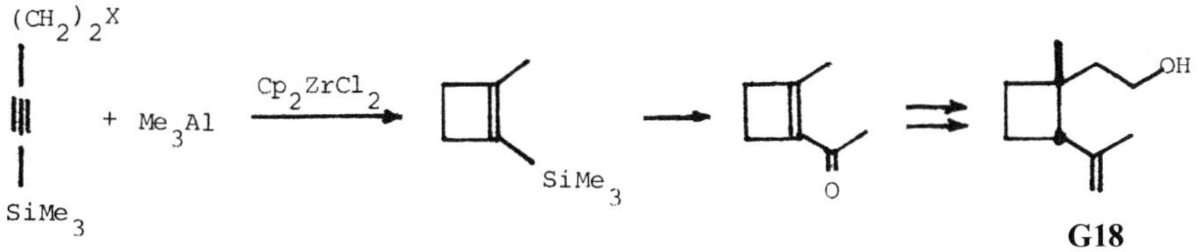

G18

Chapter 11

Ring Expansion and Contraction

Ring expansion and contraction maneuvers are employed to adjust ring size in accordance with final structural requirements. The need arises chiefly because precursors with homologous cyclic arrays are sometimes more readily available. For example, [2 + 2]cycloaddition generates cyclobutane derivatives that may undergo enlargement. On the other hand, strain factors may militate against the closure of certain smaller rings, and a ring contraction strategy that can take advantage of intramolecular reactions must be implemented. Finally, a very effective method for synthesizing hydrazulenes by rearrangement of appropriate decalins involves simultaneous ring expansion and contraction.

11.1 RING EXPANSION

11.1.1 Expansion of Three- and Four-Membered Rings

Common methods for ring expansion (Gutsche, 1968) include treatment of cycloalkanones with diazoalkanes, the Demjanov–Tiffeneau rearrangement involving diazotization of aminomethylcycloalkanols, solvolysis of halomethyl-cycloalkanols (or direct treatment of an epoxide with lithium iodide; the iodohydrin often undergoes ring enlargement), and stepwise ring cleavage and re-formation via the dicarbonyl compounds.

Before listing some examples of these stepwise homologation reactions, mention is made of the more exciting $3 \to 5$ and $3 \to 6$ expansion processes. Thus, in a synthesis of antheridiogen-An (**A35**) (Corey, 1985c), two six-membered rings are linked by a Lewis acid-catalyzed vinylcyclopropane-to-cyclopentene rearrangement.

11.1 RING EXPANSION

A35

Two imaginative routes to quadrone (**Q1**) (Piers, 1985b) and sinularene (**S37**) (Piers, 1985b) are highlighted by a three-carbon bridging of siloxycyclopentenes through a thermal reorganization.

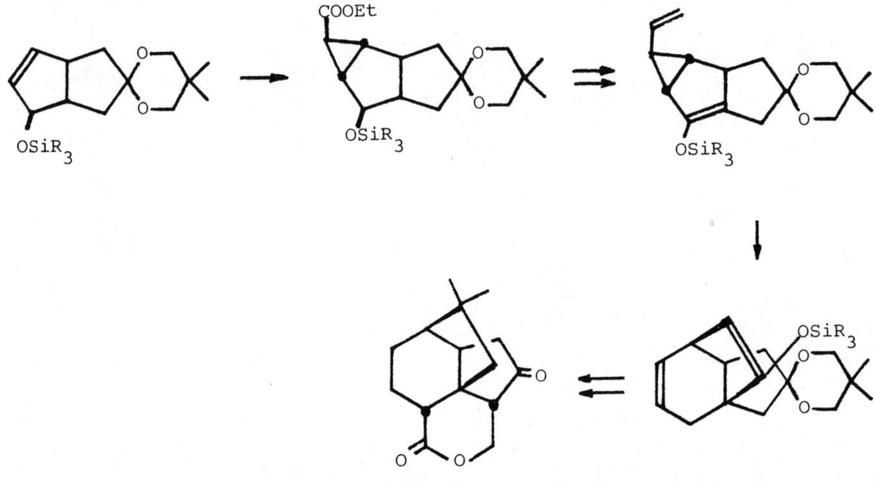

Q1

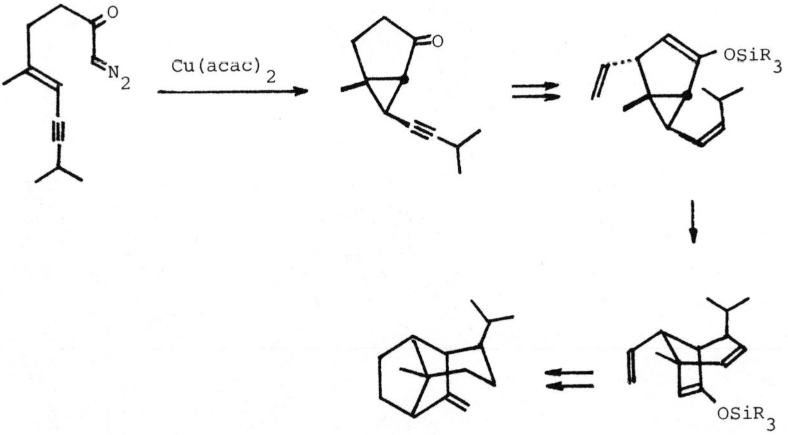

S37

The vinylcyclopropane reorganization reaction is employed in a synthesis of pentalenene (**P16**) (Hudlicky, 1987b). The required substrate can be secured by two routes: carbenoid/alkene insertion and Michael-alkylation tandem. The key step is improved by converting the cyclopentanone function into an excocyclic methylene group. Apparently, formation of a diallyl radical is favored. The final stage of the synthesis involves epimerization of the saturated ester before its deoxygenation.

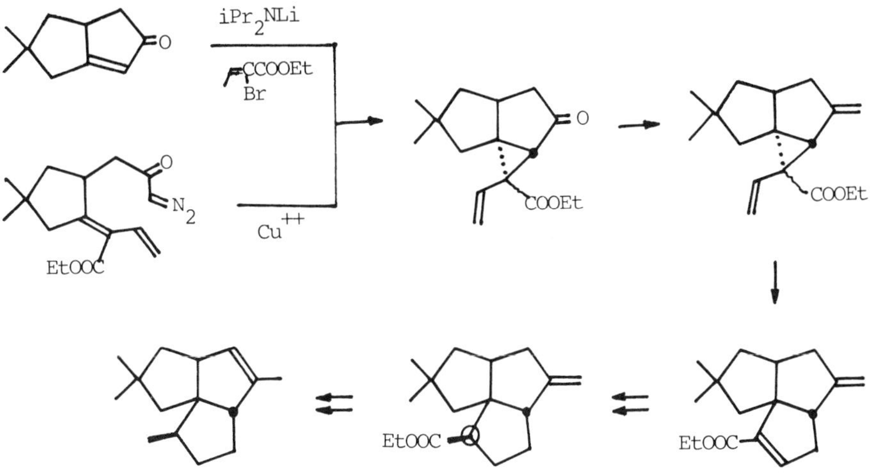

11.1 RING EXPANSION

P16

Further applications of the methodology include very concise approaches to modhephene (**M12**) and retigeranic acid (**R1**) (Hudlicky, 1987a).

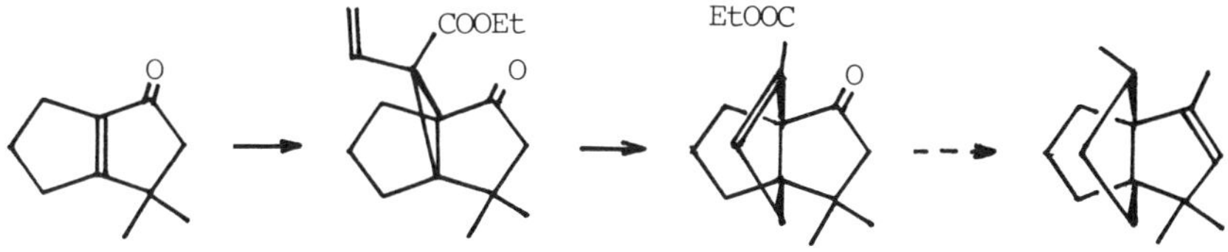

M12

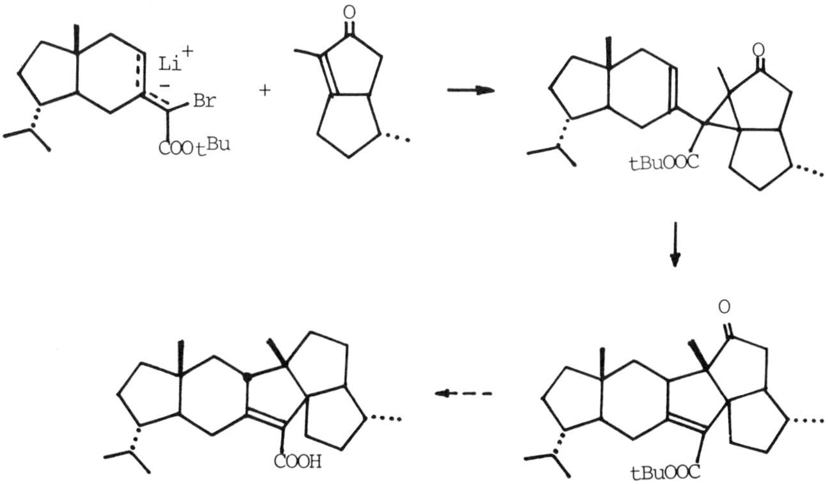

R1

A versatile method for cyclobutanone synthesis (Trost, 1974a) involves the reaction of 1-phenylthiocyclopropyllithium or cyclopropyl diphenylsulfonium ylide with carbonyl compounds and subsequent rearrangement catalyzed by a Lewis acid. The expansion process is stereoselective, and its application to acoreneone B (**A10b**) (Trost, 1975a) and the spirovetivanes such as hinesol (**H15**) (Trost, 1975b) amply demonstrates the power and ramifications of the method in organic synthesis.

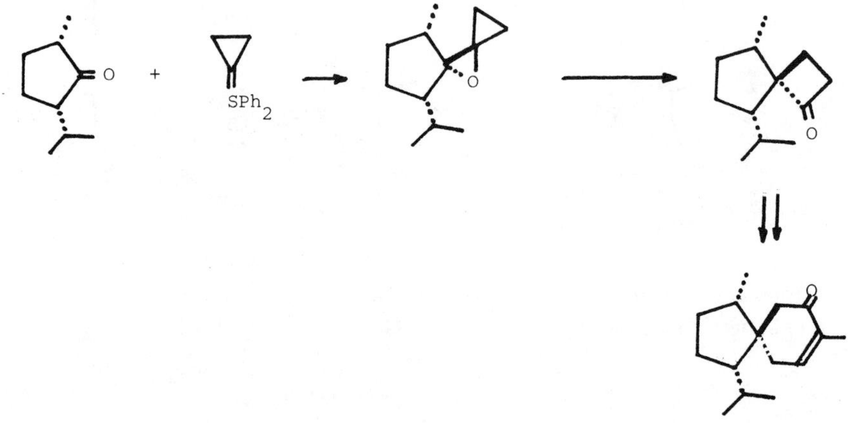

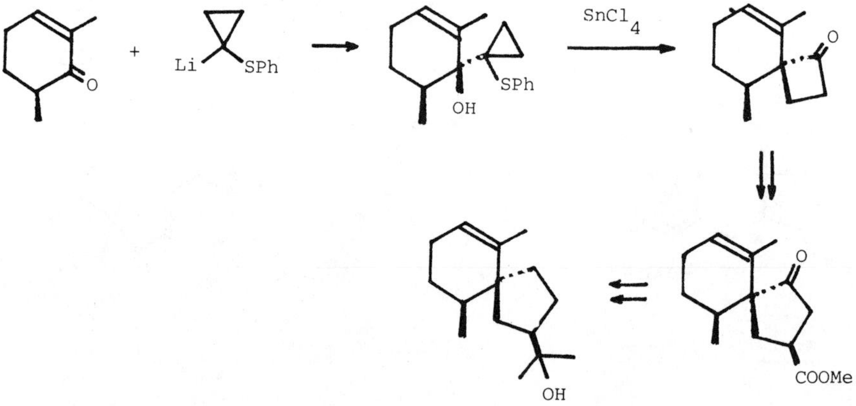

A strategy for synthesis of lineatin (**L19**) (Slessor, 1980) involving cyclopropanation and ring expansion has been formulated. However, this approach suffers somewhat from the production of both the required *exo* and the unwanted endo bicyclic intermediates (in a 2.1:1 ratio), as well as two regioisomeric ketones (4:1) from the ring expansion step.

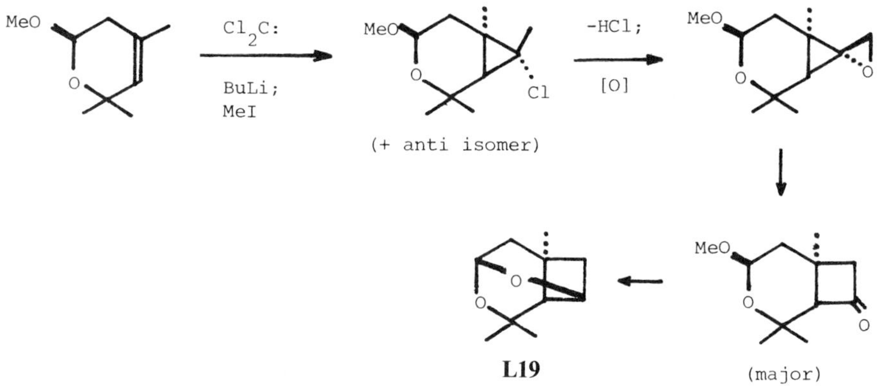

11.1 RING EXPANSION

It is possible to prepare α- and β-cuparenones (**C85** and **C86**) by two consecutive ring expansion operations (Halazy, 1982; Gadwood, 1983) starting from cyclopropane derivatives.

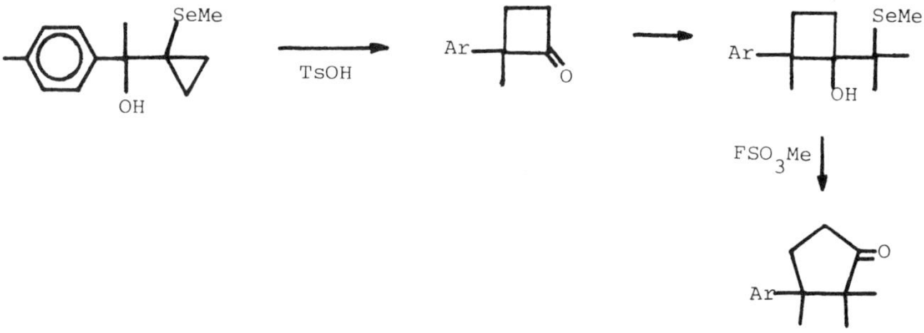

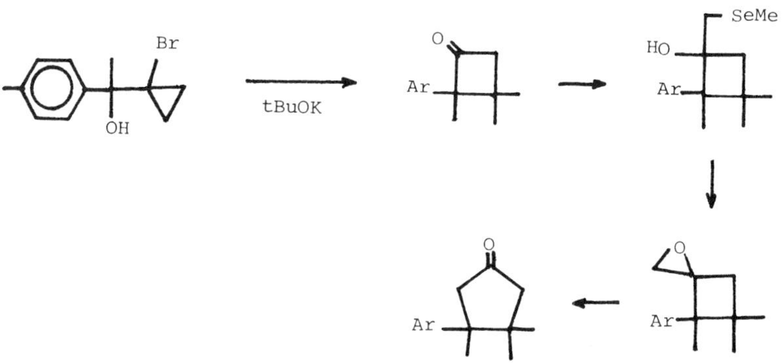

The Demjanov–Tifeneau rearrangement on the tetrasubstituted cyclobutanone furnishes both α- and β-cuparenones in a 2 : 1 ratio (Leriverend, 1973).

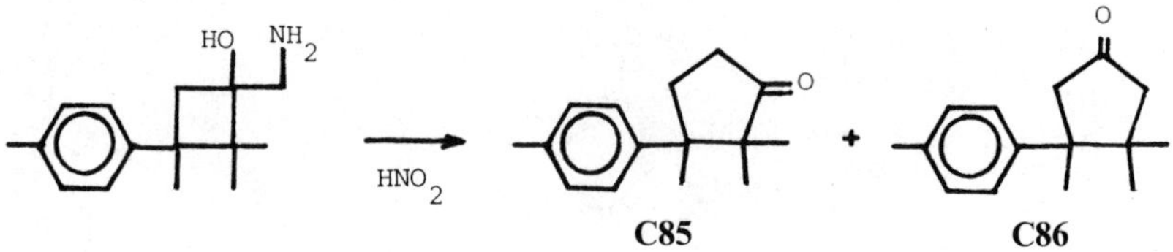

Expansion via a double shift is witnessed in the preparation of a precursor of cuparene (**C84**) (Fadel, 1985) by treatment of a tertiary cyclobutanol with FeCl$_3$–SiO$_2$.

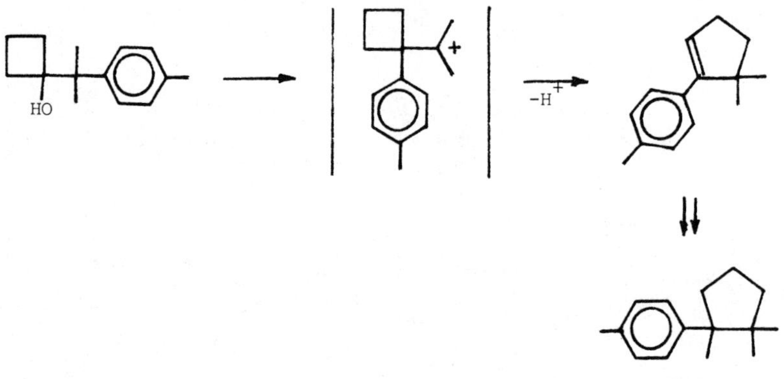

The cyclobutanone obtained from aldol condensation of 1,2-bis(trimethylsiloxyl)cyclobutene with acetals undergoes facile ring expansion to give 2,3-dialkylcyclopentenones, including a precursor of cuparene (**C84**) after the Wittig reaction (Shimada, 1984).

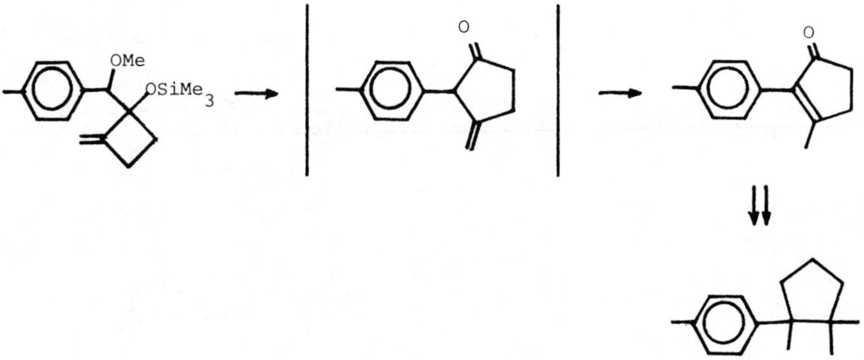

In a deMayo route to β-bulnesene (**B20**) (Oppolzer, 1980), the required substrate is also conveniently accessible by this method.

11.1 RING EXPANSION

B20

Through a series of manipulations on small rings about the periphery of a hydrindane and, later, a diquinane framework, isocomene (**I14**) is obtained (Wenkert, 1983). In the later stage, expansion of a methylenecyclobutane oxide is called for.

I14

In a different route to isocomene (**I14**) (Tobe, 1985), the presence of a *syn*-hydroxyl group, which helps chelate the lithium ion and renders the conformation of the bromohydrin intermediate unfavorable for migration by the quaternary carbon atom, ensures configuration of the secondary methyl group by means of conjugate addition of the derived enone. It should be noted that the more highly substituted C—C bond usually takes precedence in the migratory process, as shown in the example immediately above.

I14

In essentially the same manner, the [3.3.2]propellane derivative obtained from photocycloaddition of a dimethyl diquinane enone with allene has been transformed into modhephene (**M12**) (Tobe, 1984). The regioselective migration of the methylene group during the expansion of the four-membered ring is now mediated indirectly through the ketal oxygen.

M12

Nucleophilic species that also possess a potential leaving group are very useful for effecting cycloalkanone homologation. The reaction of tris(methylthio)methyllithium with these ketones and treatment of the resulting alkoxides with copper(I) salts lead to the homologous cycloalkanedione monothioacetals. The blocking of one side of the ketone facilitates regioselective transformations, such as alkylation, which a synthetic route may require. Furthermore, there is the option of retaining a carbonyl group at one of two positions. The BC-ring synthon for coriolin (**C70**) is available via this method (Knapp, 1984).

C70

One of the classical and best known methods for ring expansion is by treatment of a cyclic ketone with diazoalkanes. Both inter- and intramolecular reactions are feasible. For unsymmetrical ketones, usually the more highly substituted carbon atom would joint the new member. However, exceptions are known.

Since α-chlorocyclobutanones have become readily available via cycloaddition of alkenes with α-chloroketenes, their synthetic utility has been examined. They undergo regioselective ring expansion upon treatment with diazomethane; the electron-deficient α carbon to which the chlorine is attached remains stationary. This behavior is profitable for the synthesis of triquinanes because the reaction sequence may be repeated (with or without modification of the alkene and/or the chloroketene). Hirsutene (**H17**) (Greene, 1980) and pentalenene (**P16**) (Annis, 1982) have been successfully assembled on the basis of this strategy.

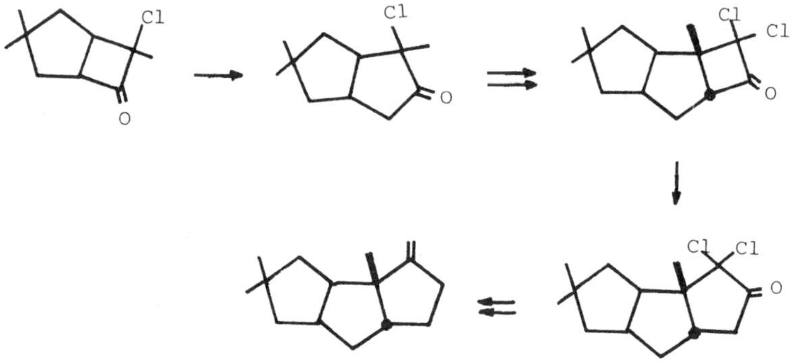

11.1 RING EXPANSION

H17

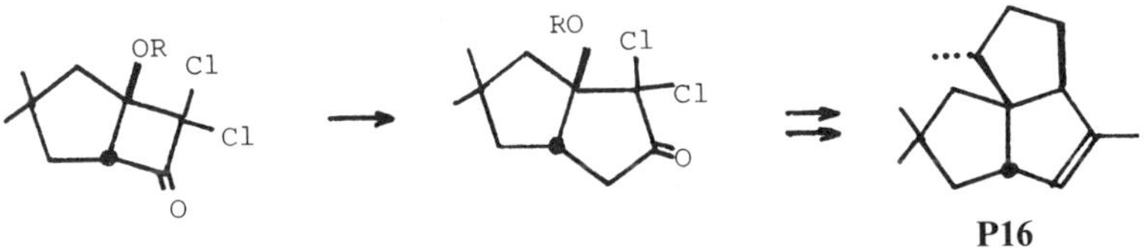

P16

Not only does the α-chlorine atom influence the regiochemistry of ring expansion, it also contributes to a more favorable reduction of the cyclopentanone during a synthesis of boonein (**B14**) (Lee, 1985b) through a dipole–dipole interaction, although the desired (major) alcohol must be formed by an approach of the hydride reagent from the more hindered, concave side of the molecule.

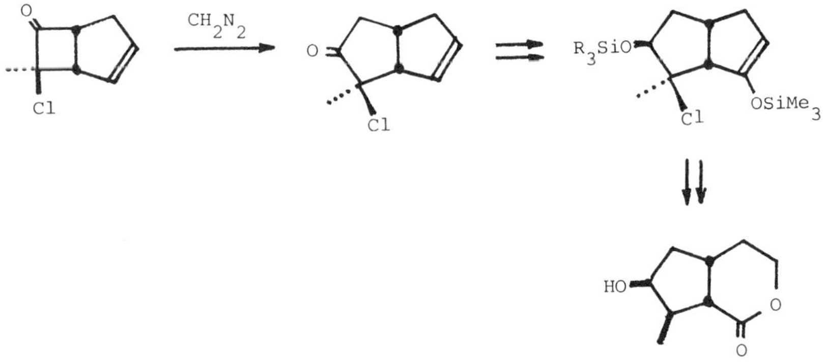

B14

A synthesis of (+)-hirsutic acid C (**H19**) (Greene, 1985b) via the ring expansion route proceeds from the symmetrical cyclobutene derivative, which undergoes an asymmetric hydroboration with (+)-diisopinocampheylborane and oxidation. The expansion is best carried out with ethyl diazoacetate in the presence of antimony(V) chloride to give the optically active, unsymmetrical ketone.

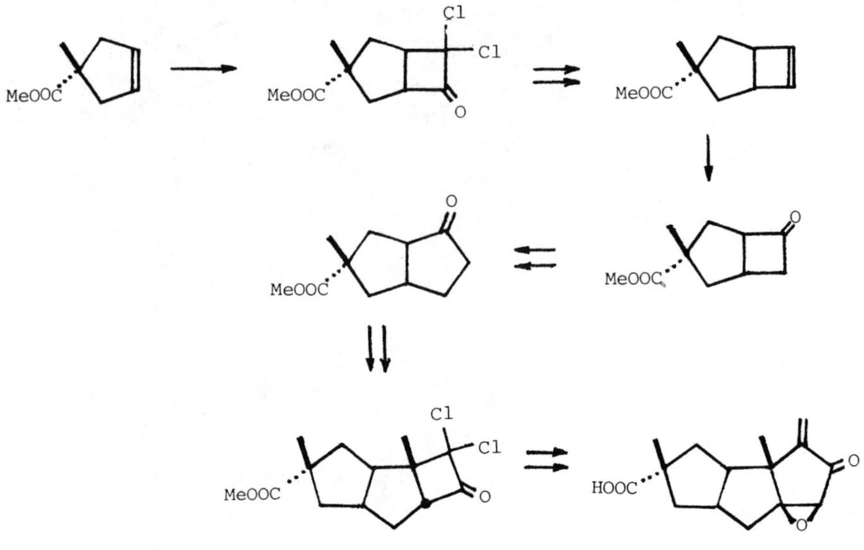

11 RING EXPANSION AND CONTRACTION

H19

In one of the more recent syntheses of loganin aglucone acetate (**L20′**) (Au-Yeung, 1977), the basic protocol also embodies the cycloaddition/ring expansion scheme. The allylsilane remaining in the five-membered ring offers an opportunity to achieve a transpositional carboxylation (via reaction with chlorosulfonyl isocyanate), leaving a double bond in the correct location for a subsequent oxidative cleavage.

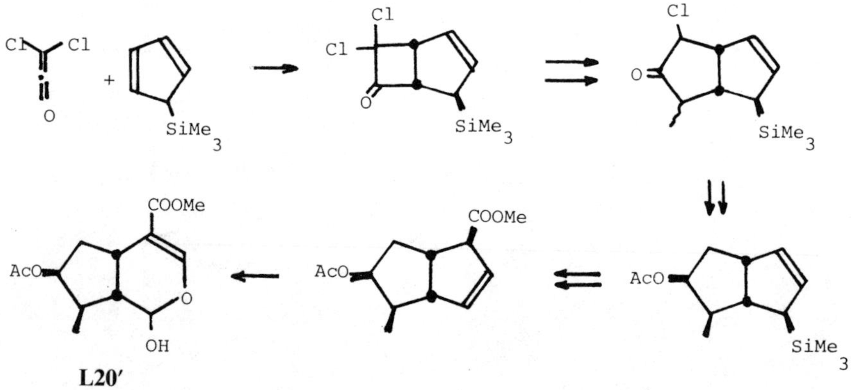

L20′

Condensed cyclic ketones in which the carbonyl group is adjacent to the angle undergo ring expansion when treated with diazoacetic esters in the presence of boron trifluoride etherate. Remarkably, the nonangular α carbon migrates preferentially. This phenomenon is suitable for synthesis of such compounds as khusimone (**K8**) (Liu, 1979), in which the ketone group must remain at the subangular position.

11.1 RING EXPANSION

In a quest for $\Delta^{9(12)}$-capnellene (**C21**) (Liu, 1985) starting from cyclopentenone and involving consecutive annulations, the ketone function was identified as the seat of the exocyclic methylene group. Thus, annexation of a cyclobutanone moiety by means of a photocycloaddition and expansion of the four-membered ring conclude the first phase of the synthesis.

The ring expansion results in a β-ketoester whose dehydro analog is capable of undergoing Diels–Alder reaction. Modification of the newly created cyclohexene subunit in the adduct via cleavage and aldolization permits entry into the triquinane system. Both the photo- and thermal cycloaddition steps lead to *cis*-fused adducts, and the preferential exo approach of isoprene to the diquinane dienophile ensures the stereochemistry of the four chiral centers conforming to capnellene.

The cyclohexene ring is easily fashioned into a cyclopentanone via ozonolysis, aldolization, and Beckmann rearrangement.

On another occasion, the ring expansion procedure is employed in fashioning the terminal cyclopentane ring of $\Delta^{9(12)}$-capnellene (**C21**), which contains an exocyclic methylene group (Stille, 1986). The cyclobutanone precursor is assembled by a novel method involving titana methylenative fragmentation of a norbornene double bond and subsequent intramolecular condensation of the organometallic residue with the ester pendant at the other bridge.

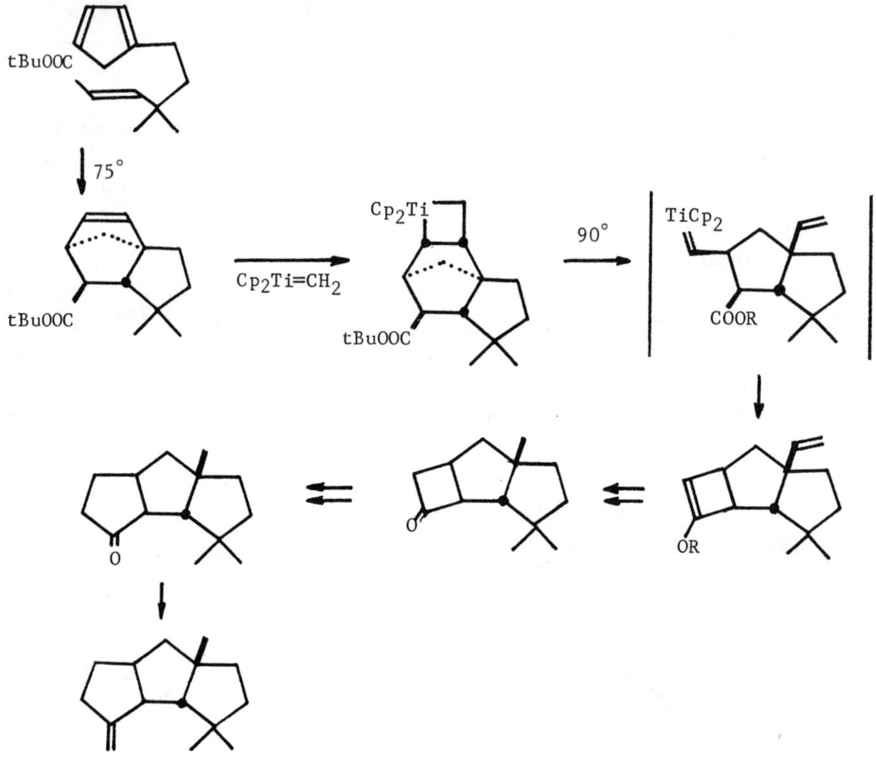

C21

For 3 → 5 ring expansion involving the thermal reorganization of vinylcyclopropanes, see Section 8.9.

Small ring compounds are prone to undergo ring expansion by the Wagner–Meerwein rearrangement if a cationic site is created at an extracyclic atom. This process mediates a synthesis of (+)-quadrone (**ent-Q1**) (Smith, 1984a) from a [4.3.2]propellanelactone. At the conclusion of the synthesis, it was found that (+)-quadrone is the enantiomer of the natural substance.

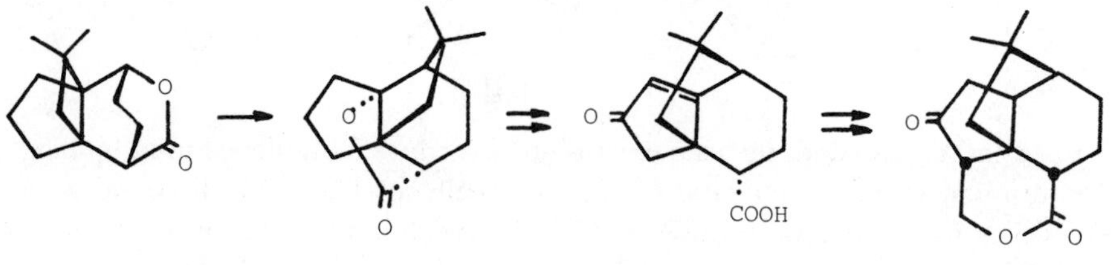

ent-Q1

11.1 RING EXPANSION

The bicyclo[3 2.1]octane moiety of quadrone, with the essential one-carbon appendage on the cyclohexane ring, can be erected from Corey's ketone via a skeletal rearrangement. This last step unfortunately produces a mixture (K. Takeda, 1983).

A well-conceived route to (+)-phyllocladene (**P27**) consists of treating the photoadduct of a tricyclic enone and allene with an acid (TsOH), carbonyl group removal, and double-bond shift (Duc, 1975). The critical aspects of this synthesis are the high stereoselectivity of the photocycloaddition and the susceptibility of the highly strained adduct to rearrangement.

An analogous approach to (−)-hibaene (**H11**) (Duc, 1978) requires that the two-carbon addend be convertible into an olefin and that a methyl group be present at the bridgehead. The use of vinyl acetate fulfills these needs, albeit imperfectly. A mixture of products is obtained (only the major components are shown in the following equation). The solvolytic rearrangement, which is preceded by dehydration of the tertiary alcohol, is smooth but slow.

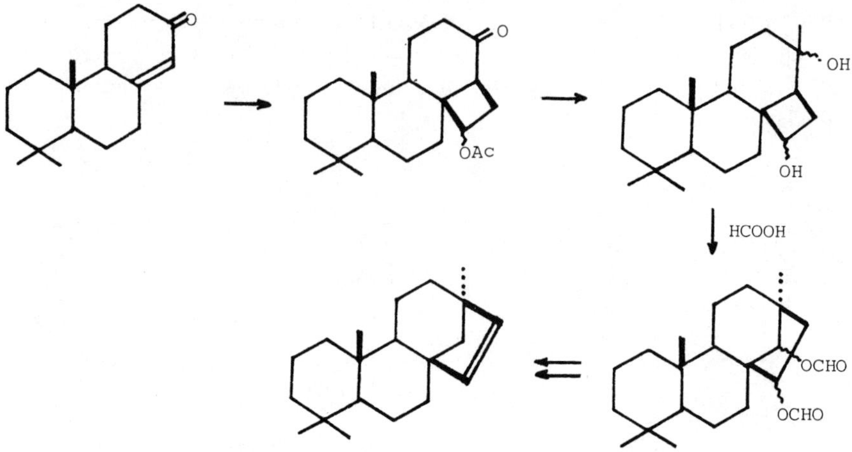

H11

Steviol methyl ester (**S53**) has been assembled by a slightly different method (Ziegler, 1977b) in which a ring expansion is initiated by ionization of an angular primary mesylate in the bicyclo[3.2.0]heptane segment.

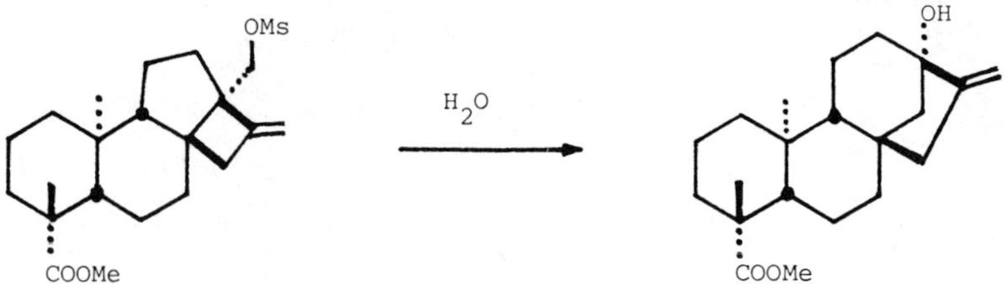

S53

A rapid synthesis of the terpene skeleton of verrucarol (**V12**) embodies a Diels–Alder reaction of 2-ethoxybutadiene with methyl coumalate, introduction of a methyl group at the β position of the dihydropyrone, photocycloaddition with acetylene, and acid-induced rearrangement of the derived lactol acetate (White, 1981). The most crucial stereochemical problem pertains to the photocycloaddition. The roof-shaped substrate enforces an exo approach of acetylene as desired.

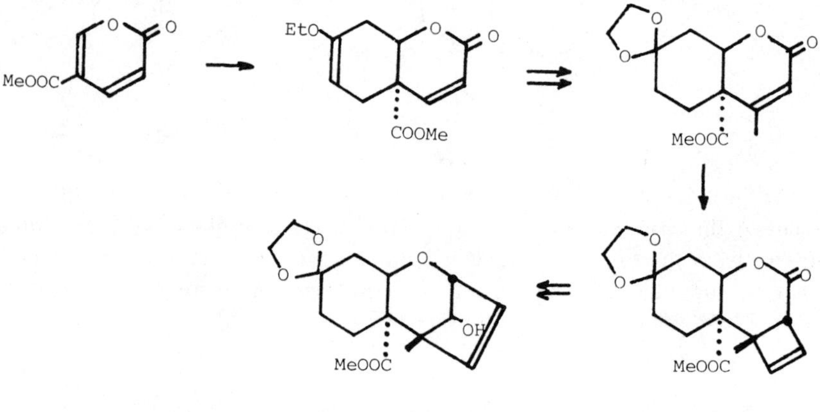

Photoadducts from piperitone have been elaborated into sativene (**S11**) and copacamphene (**C63**) (Yanagiya, 1979). The rearrangement of the condensed system into a bridged bicyclic readies the intermediate to accept an ethano span to complete the final architecture of the molecules.

S11 (αMe)
C63 (βMe)

Noting that the oxy-Cope rearrangement enables a ring enlargement by four carbon atoms, the 5:8-fused skeleton of pitoediol (**P37**) (Gadwood, 1984) was generated from a fused cyclobutanone. The ability of a propargyl alcohol unit to participate in the rearrangement amplifies the potential of such a method. Cyclic dienones with one conjugated double bond and one at the δ,ε-position are formed.

P37

A fascinating ring expansion (with rearrangement) process is the type II ene cyclization that follows the thermolysis of the photoadducts of 3-methylcyclohexenones with cyclobutenes. In the favorable conformation of the cyclodecadienone intermediates, the carbonyl oxygen and the vinylic methyl group are in such proximity that the ene reaction is inevitable. Calameon (**C9**) (Wender, 1980a; Williams, 1980), an angularly hydroxylated cadinane, is most expediently prepared from piperitone and methyl cyclobutenecarboxylate.

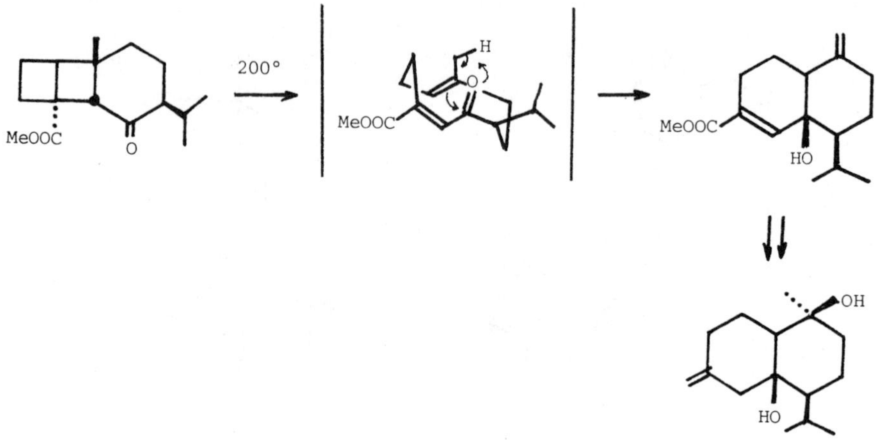

C9

It is discernible that if the tricyclic substrate contains a methylene group instead of a ketone, pyrolysis of the compound affords a β-eudesmol precursor (Wender, 1980c).

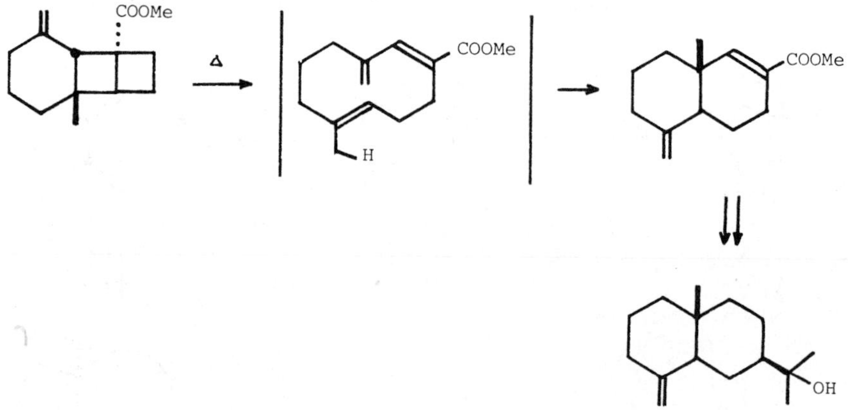

E18

In a synthesis of the germacranolide isabelin (**I10**) (Wender, 1980b), linear tricarbocyclic intermediates are constructed and retained until the final step, when heating at 200°C generates the cyclodecadiene segment. Some isoisabelin is also formed as a result of the free-radical mechanism.

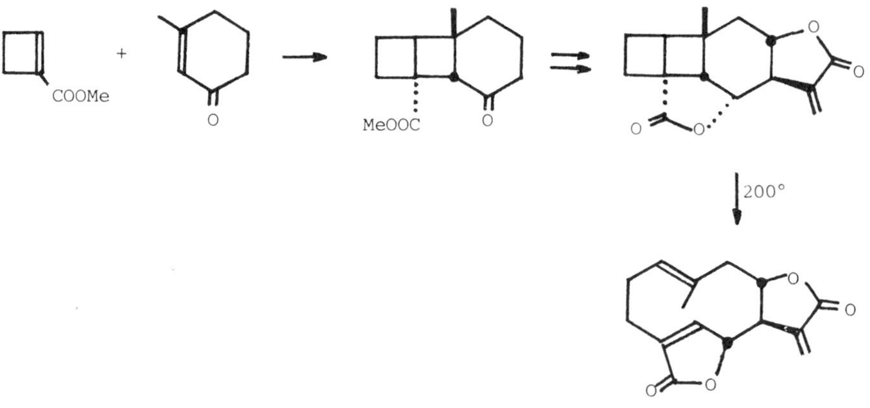

11.1 RING EXPANSION

[Scheme leading to **I10** at 200°]

A very unusual skeletal rearrangement of a fused cyclobutanone on the hydrophenanthrene nucleus leads to a bridged cyclopentanone that is suitable for elaboration into diterpene alkaloids such as atisine (**A48**) (Ghatak, 1976).

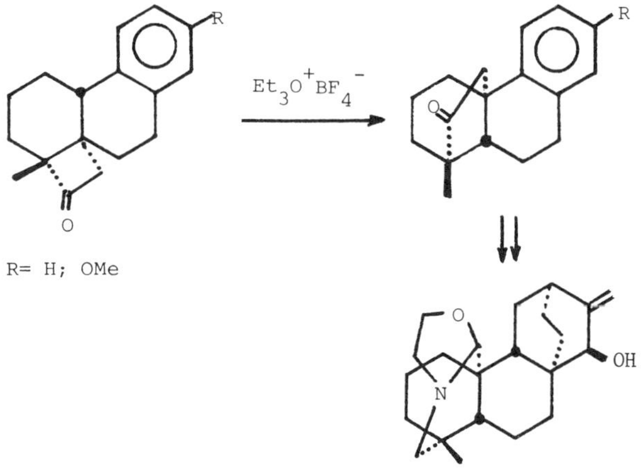

[Scheme with Et$_3$O$^+$BF$_4^-$, R = H; OMe, leading to **A48**]

11.1.2 Expansion of Five- and Six-Membered Rings

The general methodology for the enlargement of five- and six-membered rings does not differ from that employed for the smaller cycles. The following equations show the use of various expansion procedures during the syntheses of the thujaplicins (**T15**, **T16**, and **T17**) (Cook, 1951), α-cedrene (**C37**) (Breitholle, 1976, 1978), widdrol (**W3**) (Enzell, 1962), α- and β-himachalenes (**H13** and **H14**) (Liu, 1981), α-*trans*-bergamotene (**B5**) (Corey, 1971), and longifolene (**L24**) (Corey 1961, 1964c).

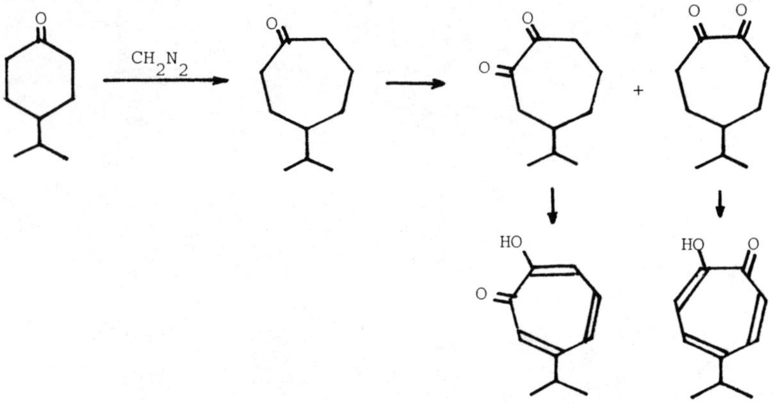

11 RING EXPANSION AND CONTRACTION

T16   **T17**

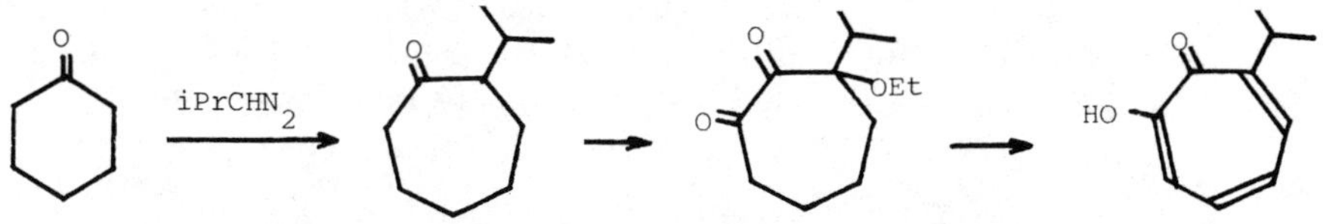

T15

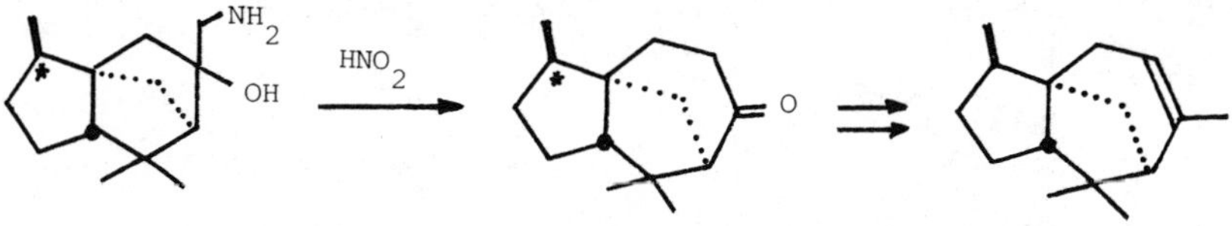

(major)   **C37**

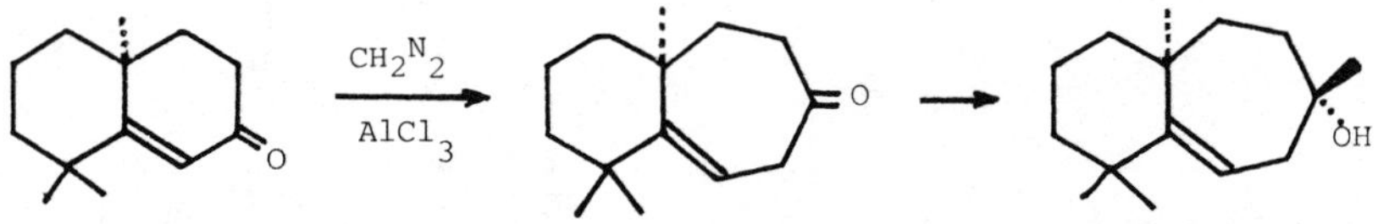

W3

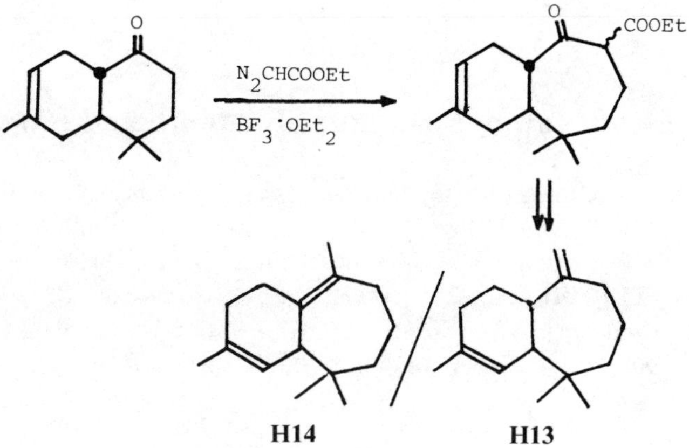

H14   **H13**

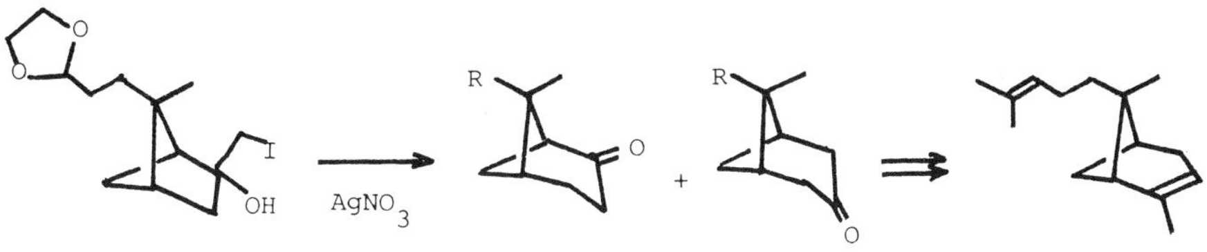

11.1 RING EXPANSION

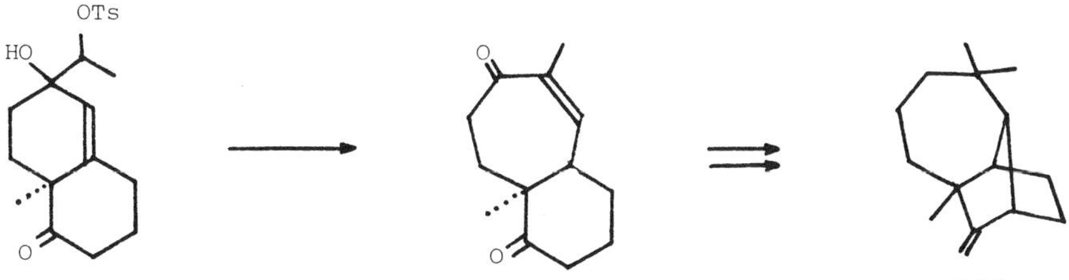

Ring expansion of oxaspiranes has been mentioned in an earlier section in connection with a cyclobutanone synthesis. Terpinolene-4,8-oxide undergoes a similar reaction on treatment with boron trifluoride etherate to give karahanaenone (**K1**) and its double-bond isomer, in addition to other compounds (Klein, 1971).

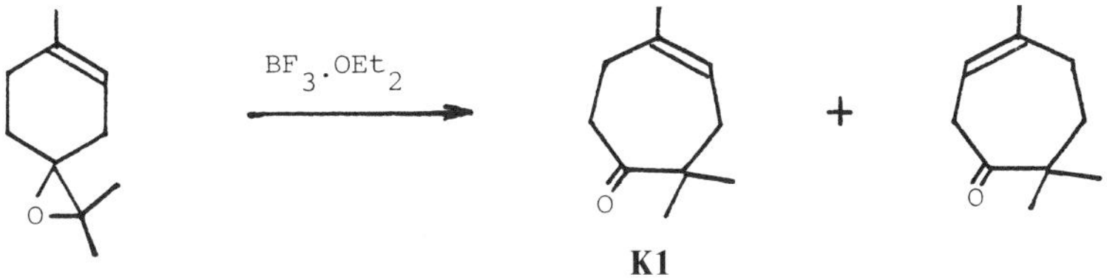

In a synthesis of veatchine (**V5**) (Wiesner, 1968), a ring expansion is in fact induced intramolecularly through reaction of a diazo group with a ketone. (*Note*: The diazoethyl chain contributes a methylene group to the cyclic ketone while retaining one CH_2, which becomes the single bridge of the C/D-ring system.]

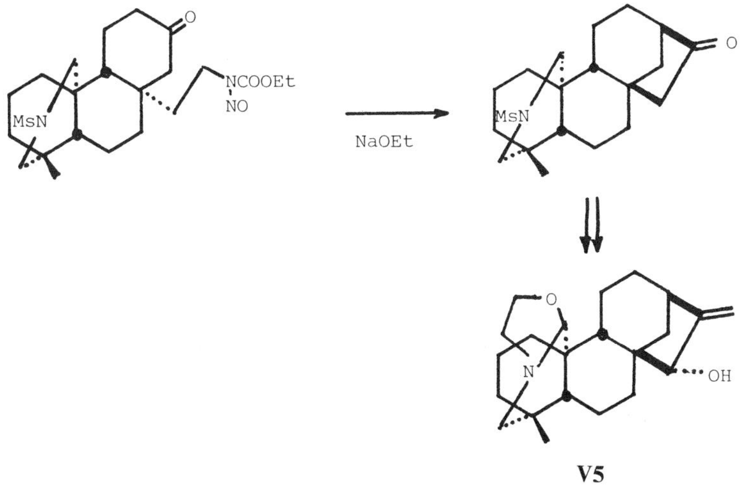

As a complement of the classical method involving diazoalkanes for cyclic ketone expansion, the use of carbenoids derived from halomethanes is adequate, provided no other functional groups that are sensitive to organometallic reagents are present in the molecule. A novel approach to nootkatone (**N11**) (Hiyama, 1979) is based on two key steps: Nazarov cyclization and carbenoid-mediated ring expansion.

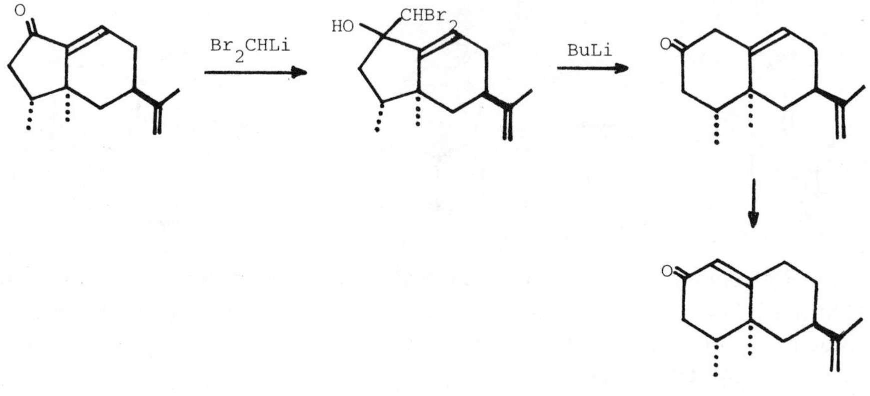

N11

Expansion of a hydrindanone to a hydrazulenone is the key step in the early stage of a synthesis of carpesiolin (**C30**) (Nagao, 1981).

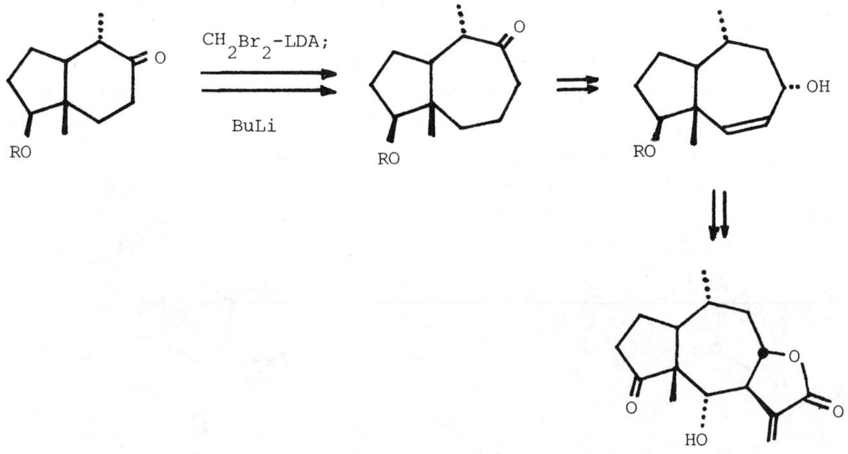

C30

The solvolytic expansion of the six-membered ring of a hydrindane with homoallylic participation holds the key to one successful synthesis of guaiol (**G20**) (Marshall, 1971b, 1972). The seven-membered ring unfolds as solvent attacks the most stable cyclopropylcarbenium ion. Simultaneously, the double bond is fixed between the two rings.

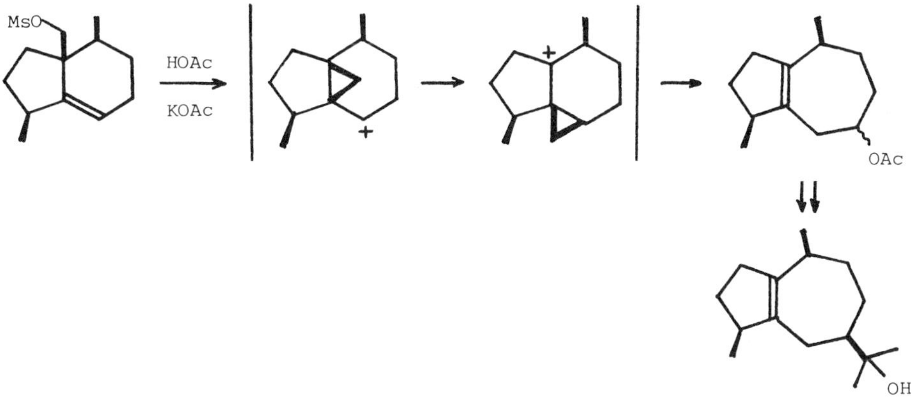

11.1 RING EXPANSION

G20

Intramolecular interception of the cyclopropylcarbenium species by a proximal carboxylic acid constitutes a crucial transformation during a confertin (**C62**) synthesis (Marshall, 1976). Geometrical constraints dictate the direction and topology of the attack, which in effect retrieves the steric information laid down in a simple hydrindenol many steps ago.

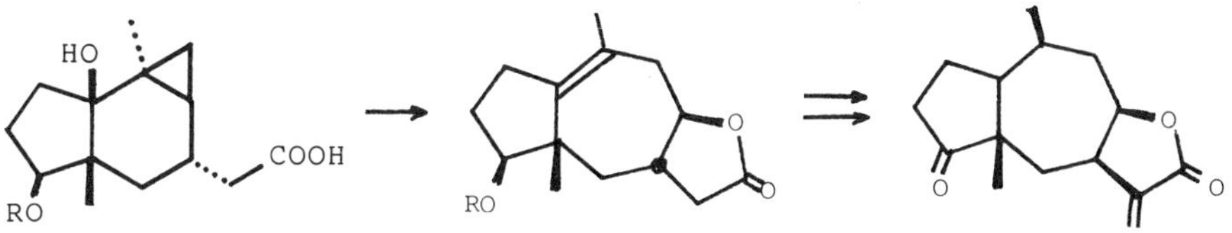

C62

The ring homologation protocol via cyclopropanation and regioselective cleavage of the intercyclic bond has yet to be fully exploited as a general method for acquiring cycloheptanones. However, a synthesis of parthenin (**P11**) (Heathcock, 1982b) has taken advantage of this procedure to build the seven-membered ring.

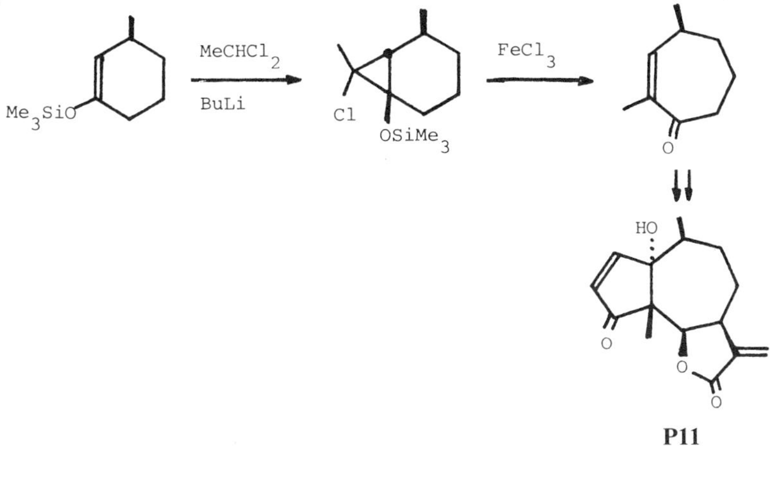

P11

Carbonyl insertion into a C—C bond between a ketone group and the α-carbon atom is possible via dichlorocarbene addition to the enol derivative and solvolytic opening of the cyclopropane ring. Conceivably the reaction sequence is most useful in converting appropriate cyclohexanones into naturally occurring tropolones such as nezukone (**N8**) (A. J. Birch, 1968) and γ-thujaplicin (**T17**) (Macdonald, 1978).

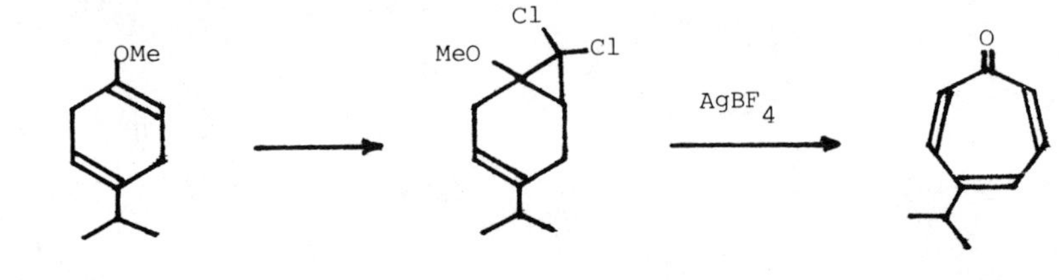

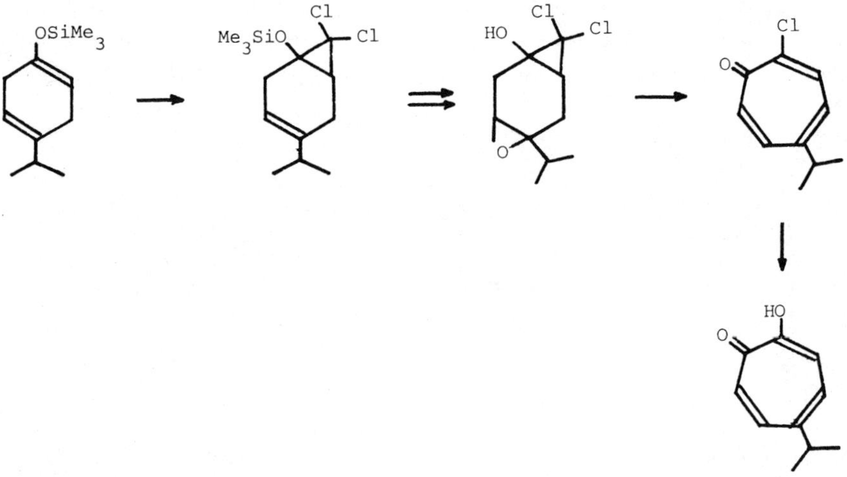

The enol silyl ether of a brexanone has been submitted to the reaction sequence of Simmons–Smith reaction and oxidative fragmentation to arrive at the correct carbon skeleton of sativene (**S11**) (Snowden, 1981a).

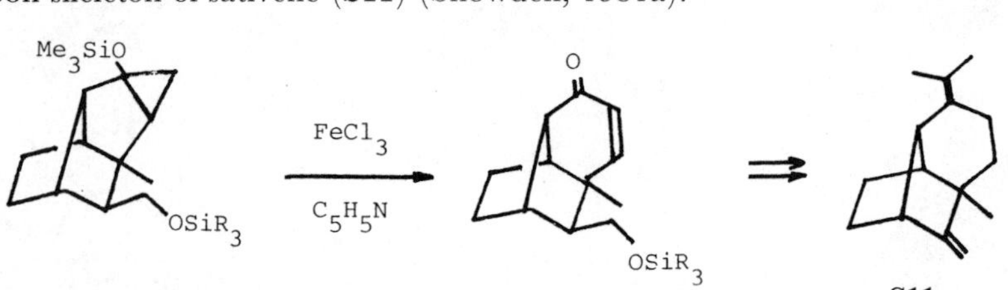

A similar situation concerning longifolene (**L24**) (McMurry, 1972) demanding a 6 → 7 ring expansion is somewhat different, however, because the formation and solvolysis of a dibromocyclopropane are involved.

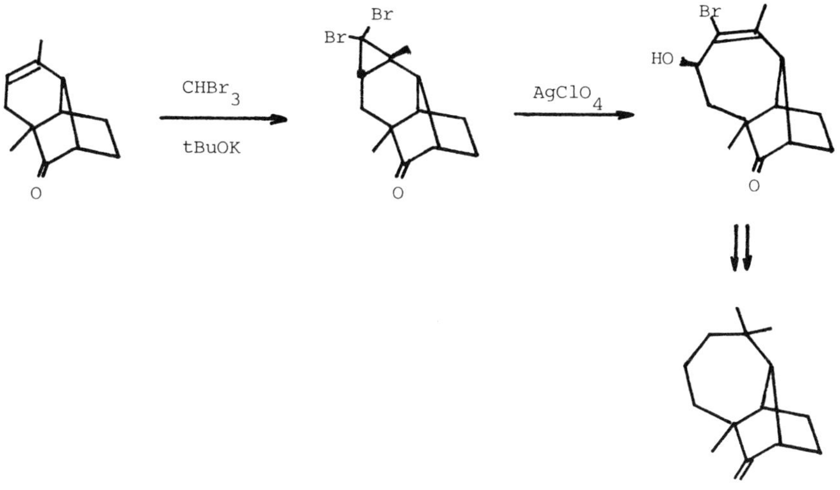

11.1 RING EXPANSION

L24

During a synthesis of 8S,14-cedranediol (**C36**) (Landry, 1983), the ring expansion via solvolysis of a dichlorocarbene adduct is assisted by a carboxyl group, thus permitting oxyfunctionalization at the desired position.

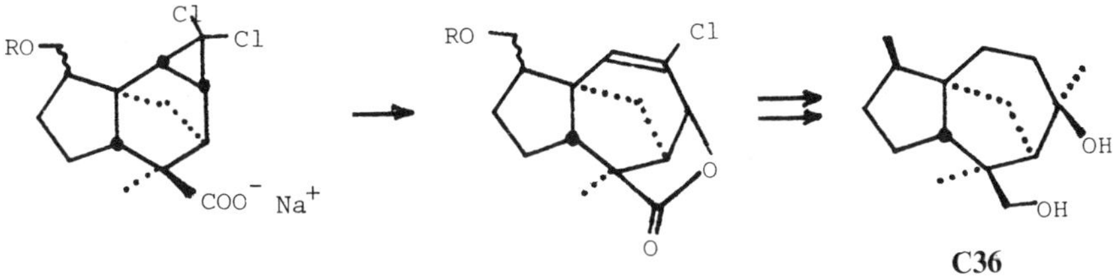

C36

Another ring expansion method is illustrated in a synthesis of β-dolabrin (Evans, 1978). Oxosulfonium methylide delivers a methylene group to the p-benzoquinone monoketal, and the isopropenyl group can then be appended to the cyclopropyl ketone product before the intercyclic bond is broken.

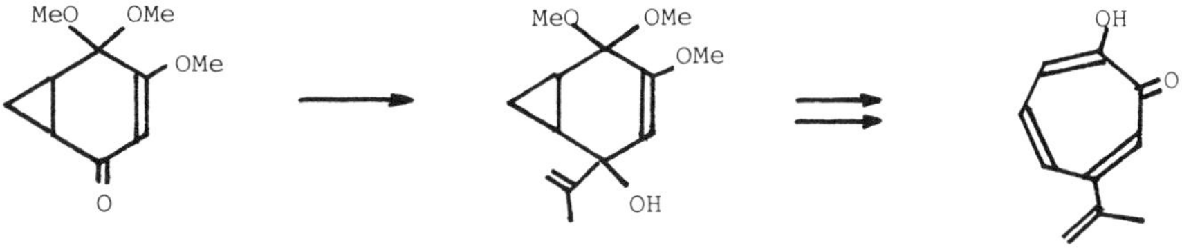

β-dolabrin

The adducts of enamines with acetylene carboxylic esters often undergo ring opening, which in effect results in a two-carbon ring expansion. Application of this method to a synthesis of velleral (**V6**) and vellerolactone (**V7**) (Froborg, 1978) is shown below.

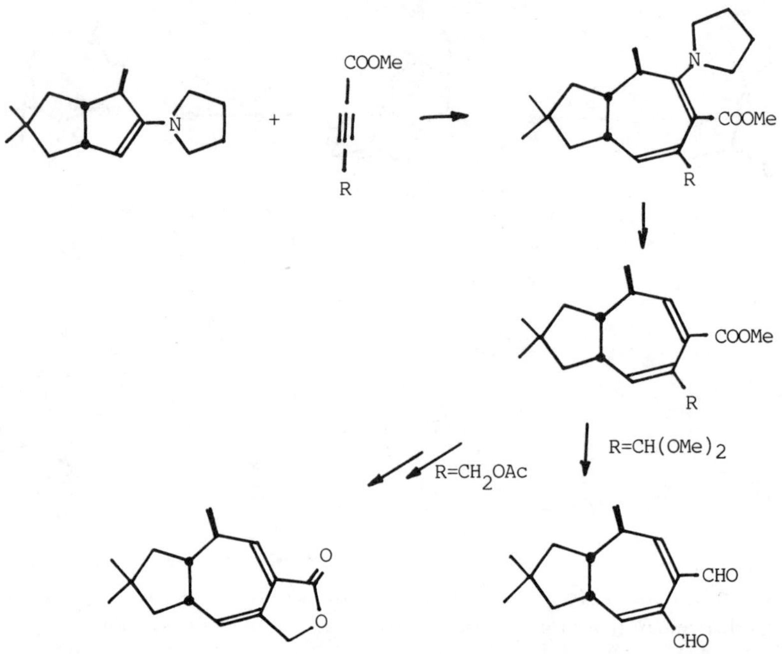

A classical method for ring expansion is the severance of an intercyclic double bond using reagents such as potassium permanganate, ruthenium tetroxide, or ozone. A simple example found in sesquiterpene synthesis is the conversion of a triquinene to the 5:8-fused system of precapnelladiene (**P46**) (Mehta, 1984).

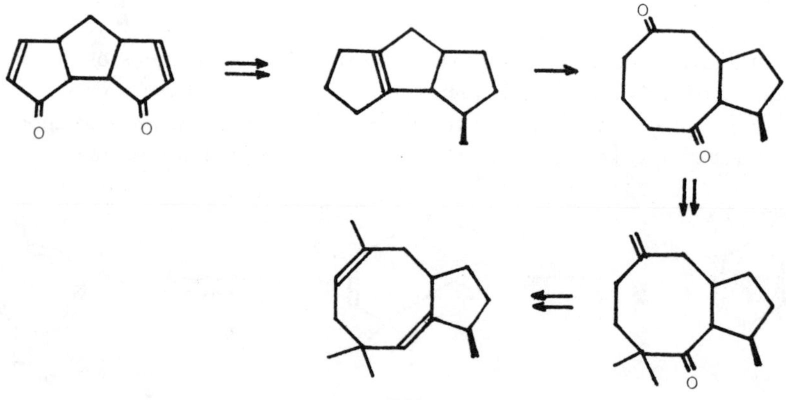

Ring expansion by two carbon atoms via [2 + 2]cycloaddition and cleavage of the intercyclic bond of the adduct may be illustrated in the convenient route to the β- and γ-thujaplicins (**T16** and **T17**) (K. Tanaka, 1971). Dichloroketene and isopropylcyclopentadiene are the thermal addends.

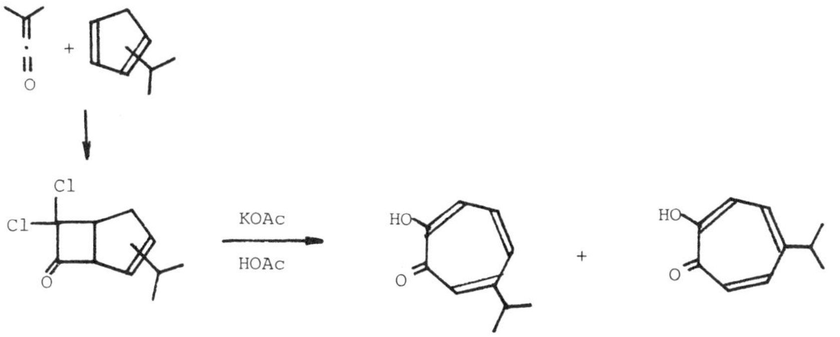

11.1 RING EXPANSION

T16   **T17**

Presently, none of the naturally occurring ophiobolins or fusicoccins has been successfully synthesized. However, many model studies have reached the characteristic 5:8:5-tricyclic nuclei. The one based on hydrindene annulation by an intramolecular photocycloaddition with subsequent reductive fission of an intercyclic bond is an interesting essay (Coates, 1985). To demonstrate this approach, a synoptic reaction sequence is shown below (cf. ceroplastol II, **C44**).

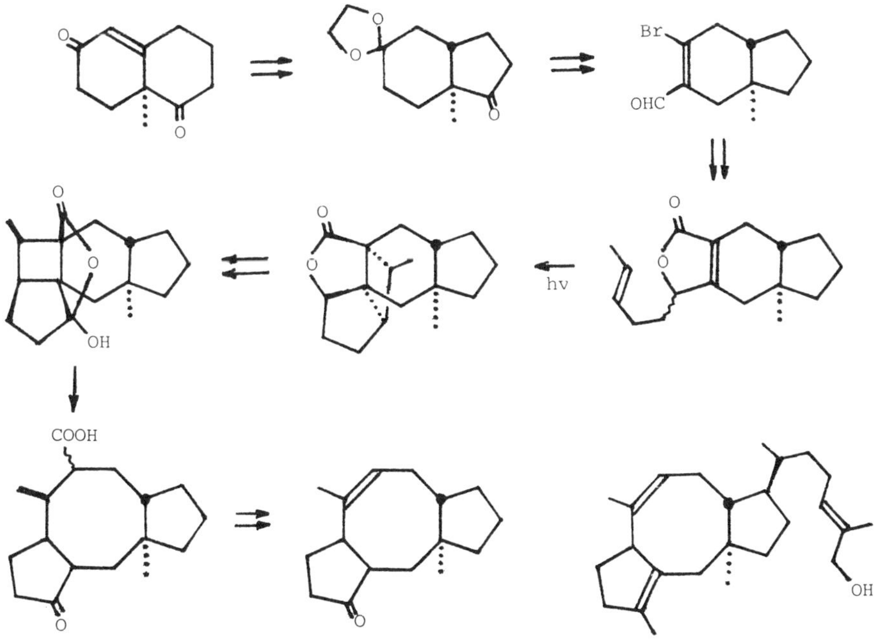

C44

A different stategy embodies an internal aldol condensation to form a bicyclo[3.3.1]nonanolone derivative and cleavage of the carbonyl bridge to reveal an eight-membered ring and an ester group that participates in a Dieckmann cyclization for closure of the third ring (Boeckman, 1977).

ophiobolin F

11.1 RING EXPANSION

The Cope rearrangement is a [3.3]sigmatropic change. Although not a ring-forming process, it can be employed to expand an existing ring structure by four atoms. Consequently, the Cope rearrangement, particularly the oxy-Cope version, occupies a significant place among methods for generating germacrane-type compounds.

Linderalactone (**L16**) and its congeners have been synthesized (Gopalan, 1980) without relying on any stereocontrol elements. Isolinderalactone (**L17**) can be equilibrated with linderalactone at 160°C, and its diastereoisomer is transformed into neolinderalactone at the same temperature. The possibility of converting neolinderalactone to isolinderalactone at 260°C in effect completes the recycling of the unwanted diastereomer.

(+)-Costunolide (**C75**) can be isolated in 20 % yield from the pyrolysate of dehydrosaussurea lactone (**S15**) (Grieco, 1977a). The latter compound is obtainable by a systematic degradation of α-santonin at the A ring.

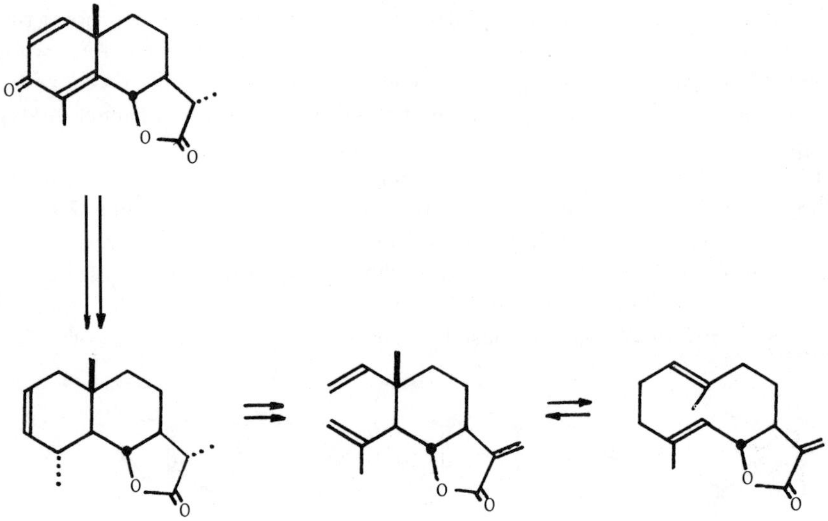

The Cope rearrangement often produces an equilibrium mixture of 1,5-dienes. An excellent way of inhibiting the backward reaction is to destroy the diene unit of the product upon its formation by combining it with a hydroxyl group so that it ketonizes rapidly. The oxy-Cope rearrangement plays a crucial role in the assemblage of acoragermacrone/preisocalamendiol (**G10/P47**) (Still, 1977a), periplanone B (**P22**) (Still, 1979; Schreiber, 1984), eucannabinolide (**E16**) (Still, 1983), and germacrene D (**G6**) (Schreiber, 1985).

The oxy-Cope rearrangement is greatly facilitated by preionization of the alcohol group (as potassium alkoxide). The application of this reaction in two versions of imaginative synthesis of periplanone B attests to its unique potential.

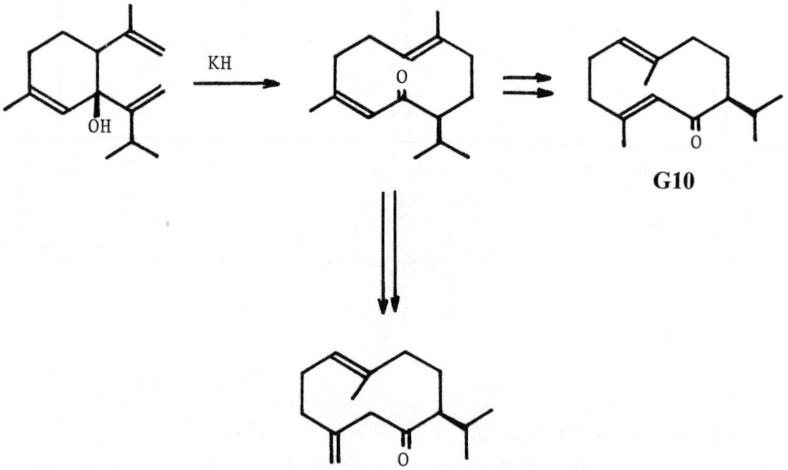

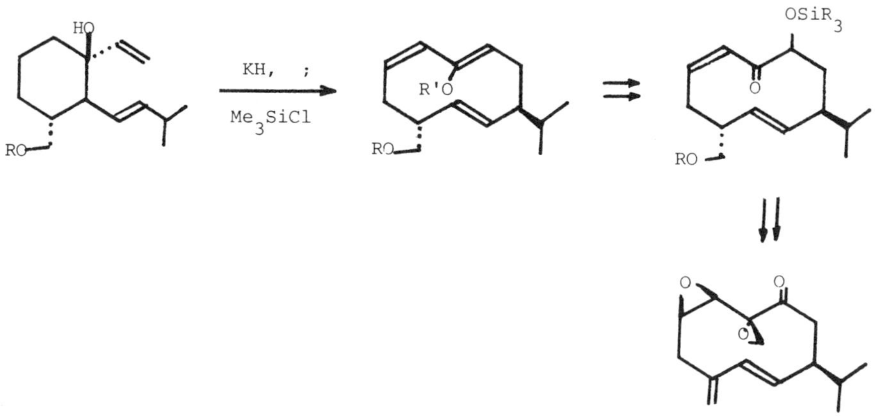

11.1 RING EXPANSION

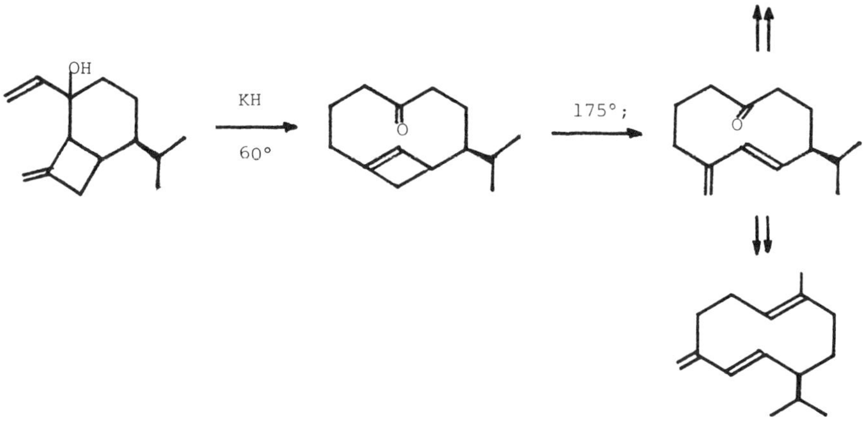

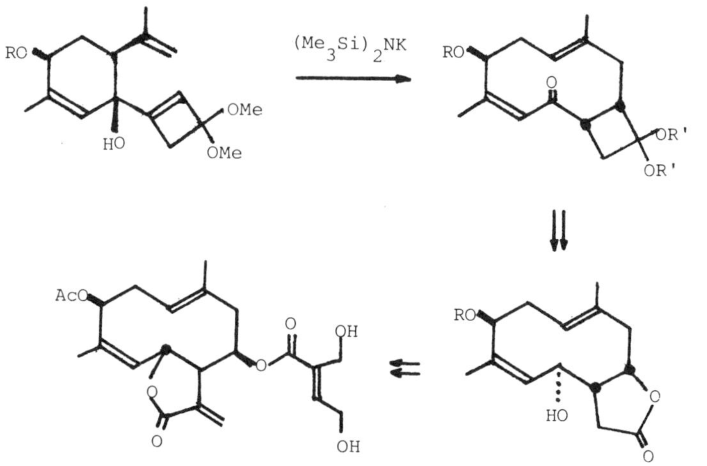

P22

G6

E16

The bisvinylogs of 1,5-hexadien-3-ols undergo similar ring expansion, by eight carbon atoms instead of four. This variant is the basis of an impressive synthesis of (−)-3Z-cembrene A (**C41**) (Wender, 1985a) from (+)-carvone.

11.2 RING CONTRACTION

The most frequently encountered ring contraction operations (Redmore, 1971) in the terpene area involve (6 → 5)-membered rings. α-Diazocycloalkanones, α-alkoxycycloalkanones, epoxides, and α-glycol monosulfonates are susceptible to the contracting action.

In a synthesis of isocomene (**I14**) (Oppolzer, 1979), it proved impossible to prepare the tricyclic system directly by an ene reaction. The homologous 5:6:5 skeleton is accessible, however. Additional steps are therefore needed to address the problem, and a photolytic decomposition of the corresponding α-diazocyclohexanone unit constitutes the key maneuver.

2-Methyl-bicyclo[3.2.1]oct-2-en-7-one undergoes an oxadi-π-methane rearrangement on exposure to ultraviolet light to give a symmetrical tricyclic ketone, which is then contracted via the α-diazoketone. The resulting carboxylic acid is easily transformed into α-santalene (**S3**) (S. Monti, 1978).

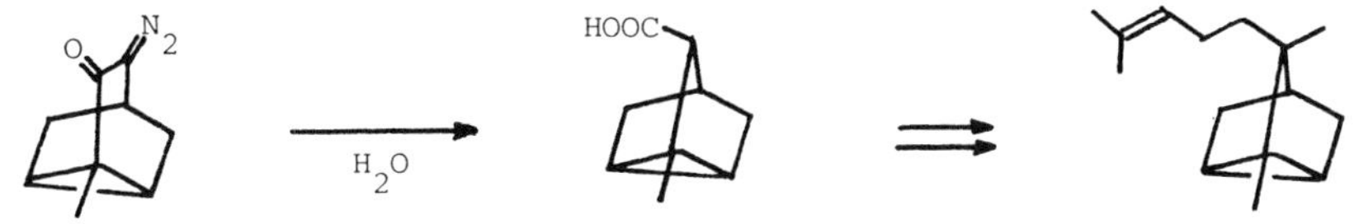

11.2 RING CONTRACTION

The use of *R*-(+)-3-methylcyclohexanone [originated from (+)-pulegone] in a synthesis of α-acorenol (**ent-A8**) and β-acorenol (**ent-A9**) leads only to antipodes of the natural terpenes (Guest, 1973). After spiroannulation, the original cyclohexanone ring is contracted via the α-diazoketone by photolysis (Wolff rearrangement).

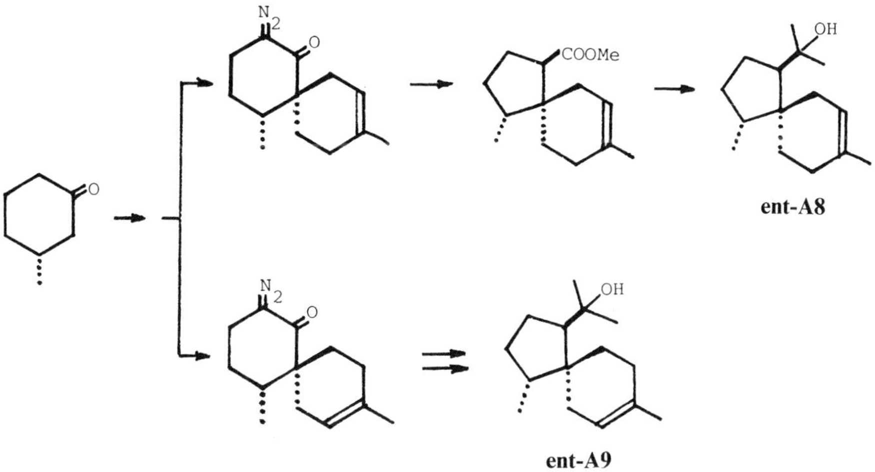

For synthetic expedience, an oxaspirannulated cyclopentanecarboxylic acid precursor of pentalenolactone E methyl ester (**P19'**) is preferentially formed via the corresponding cyclohexanone (Table, 1985b).

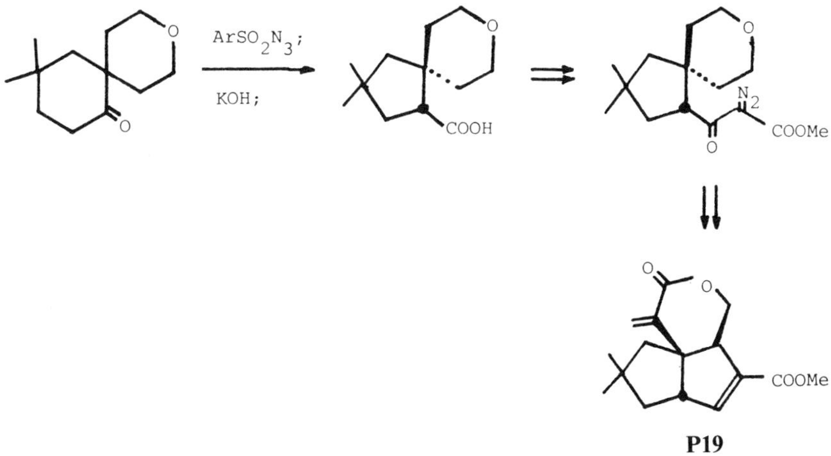

A synthesis of gibberellic acid (**G12**) following the BC + D + A annulation strategy (Lombardo, 1980) is interposed by a B-ring contraction operation by photodecomposition of an α-diazoketone. This processs is preordained by the use of anisole as a donor in bridging with a diazoketone.

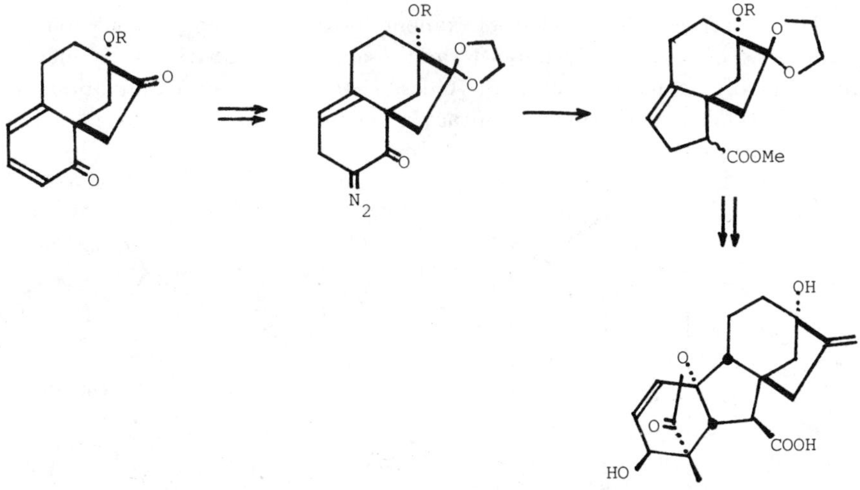

The potential precursors of grandisol (**G18**) and fragranol (**F11**) have been obtained from a substituted cyclopentanone (U. K. Banerjee, 1983).

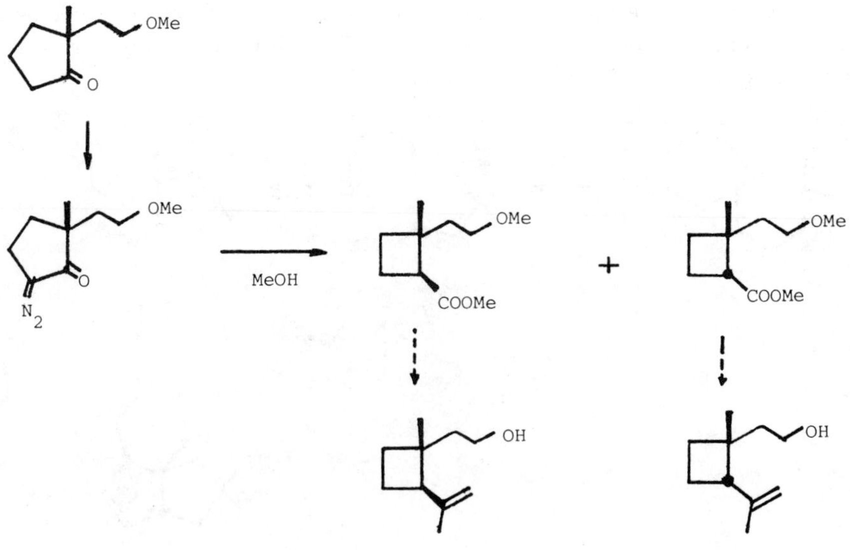

11.2 RING CONTRACTION

Cuparene (**C84**) is available from a trisubstituted cyclohexene oxide by treatment with boron trifluoride etherate and reduction of the aldehyde group (Bird, 1974). However, there is a competing hydride shift process leading to a cyclohexanone in a slightly larger amount than the aldehyde.

Exposure of α-cyperone 4β,5β-oxide to boron trifluoride etherate gives rise to a contracted ketol. Several oxidation–reduction steps complete the conversion to cyperolone (**C94**) (Hikino, 1966).

A "biomimetic" approach to bakkenolide A (**B1**) (K. Hayashi, 1973) starts from the epoxidation of fukinone. Treatment of the resulting epoxyketone with base effects a stereoselective ring contraction to a carboxylic acid, which is readily transformed into the target molecule.

The roof-shaped fukinone avails itself of epoxidation at the convex side, and the subsequent S_N2-like opening of the epoxide by the migrating methylene group places the three-carbon chain in the concave side of the molecule. The same stereochemical consequence arises if a cyclopropanone intermediate intervenes.

In an elegant approach to trichodermol (**T25**) (Still, 1980), a Favorskii-type contraction of an 2,3-epoxy-1,4-cyclohexanedione provides the complete set of carbon atoms except the spirocyclic epoxy ring of the final product. The extra C—C bond present in the bridged ring is severed on further adjustment of the functional units.

T25

The Favorskii rearrangement (Kende, 1960) proper—that is, the transformation of α-haloketones to carboxylic acid derivatives—effectively contracts cycloalkanones. With the popularization of chloroketene cycloaddition with alkenes, cyclopropylcarboxylic acids (hence the aldehydes, ketones, etc.) become readily available in a few steps. The synthesis of bicyclogermacrene (**G7**) and lepidozene (**L10**) (McMurray, 1985a) depends on this reaction sequence to build up the penultimate precursors.

G7   **L10**

The Favorskii rearrangement of a monochlorocyclobutanone derived from dichloroketene and 2,5-dimethyl-2,4-hexadiene (with subsequent monodechlorination) leads to *trans*-chryanthemic acid (**C51**) (Brady, 1983).

Cleavage of the double bond of (+)-α-pinene and elaboration of the monocyclic compound gives rise to an α-bromocyclobutanone. Methyl (+)-*trans*-chrysanthemate (**C51E**) is produced upon treatment of the latter compound with sodium methoxide (Mitra, 1977).

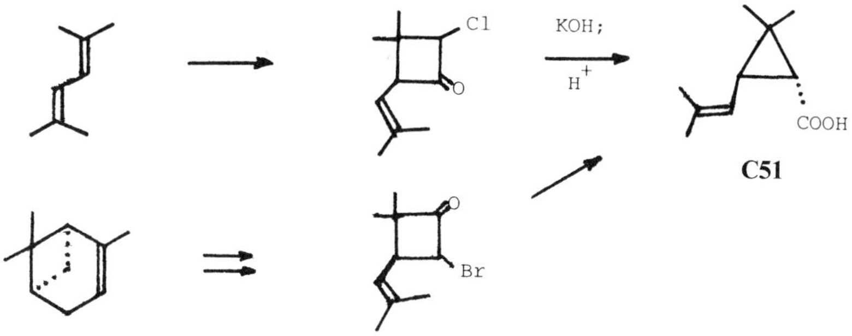

11.2 RING CONTRACTION

A nice synthesis of the fungal pheromone sirenin (**S38**) (Harding, 1987; Strickland, 1987) involves cycloaddition of methylchloroketene to a substituted 1,3-cyclohexadiene and stereoselective ring contraction (Favorskii type). Stepwise chain lengthening of the exo carboxylic ester with appropriate redox manipulation completes the synthesis.

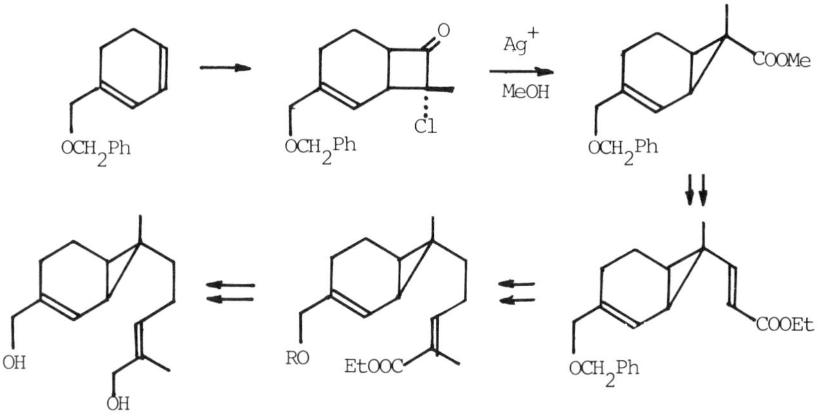

A rather readily available tricyclic ketone derived from *exo*-dicyclopentadiene is convertible to a linear triquinene ketone via homoenolization and reketonization at an alternative carbon. This intriguing approach to hirsutene (**H17**) (Dawson, 1983) has been realized.

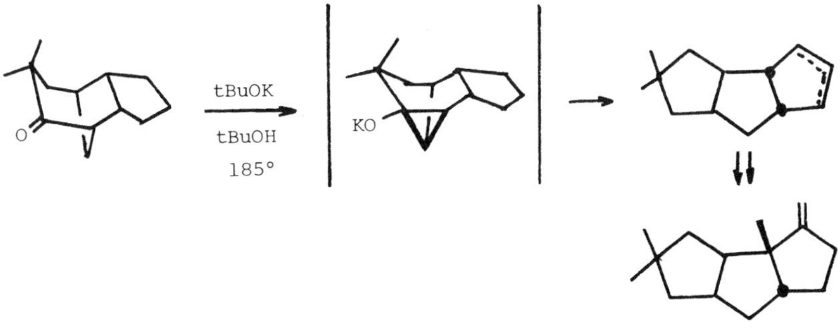

A serendipitous synthesis of herbertene (**H10**) (Frater, 1982) arises from the solvolysis of dehydroepidrimenol. A cyclopropyl carbenium ion is formed, and it undergoes ring contractive and circumambulatory motions before aromatization.

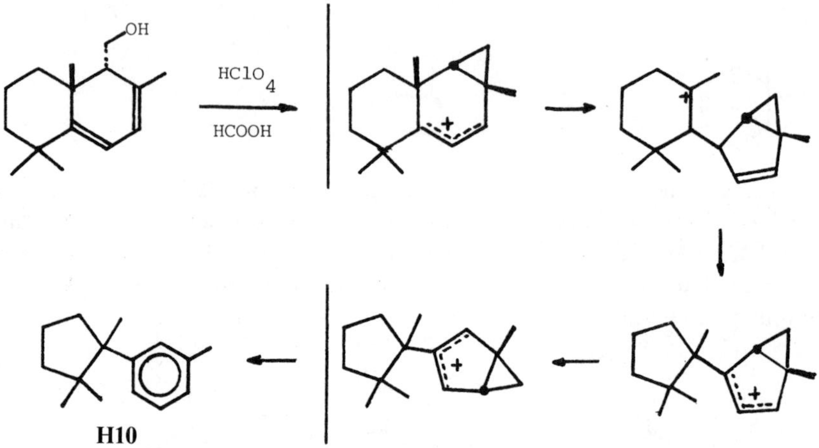

Ionization of homoallylic cyclohexenyl bromides often is anchimerically assisted by the double bond. Consequently, ring contraction may result if the substitution pattern is favorable. It has been observed that the bromocyclogeranyl derivatives are convertible to the iridienes and thence to acoratriene (**A7**) (T. Kato, 1984).

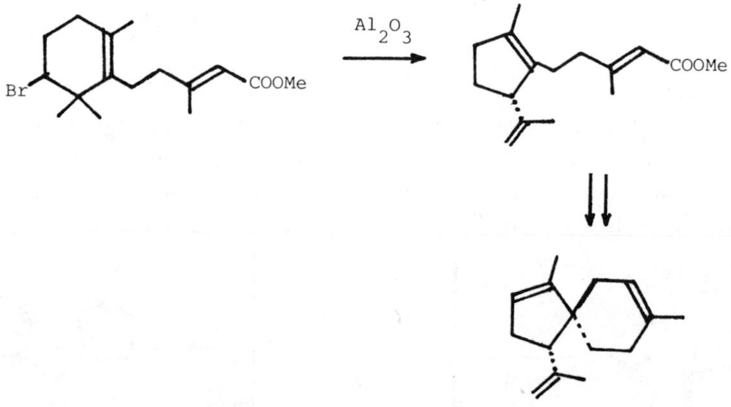

A tricyclic intermediate employed in the synthesis of (+)-longifolene (**L24**) is also applicable to (+)-sativene (**S11+**) (Oppolzer, 1981a). Ring contraction of the cycloheptene moiety by the action of thallium(III) nitrate brings the molecule very close to the target structure.

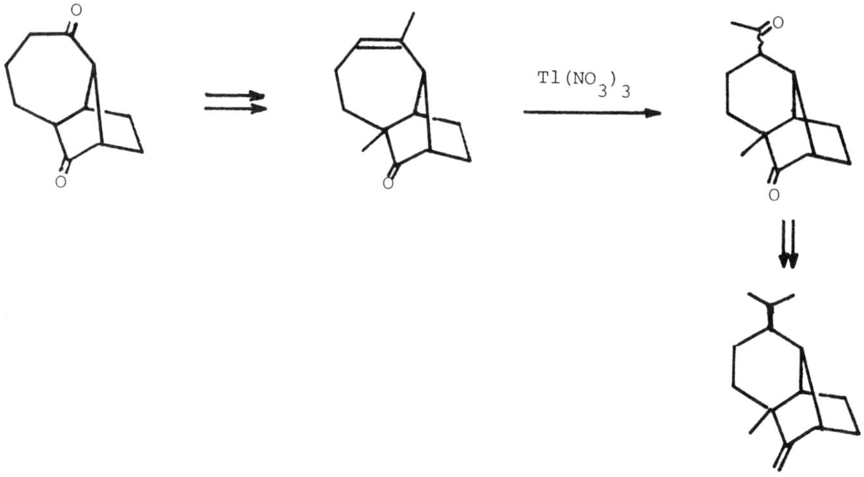

11.2 RING CONTRACTION

S11 +

A humulenyl mesylate suffers both reduction and ring contraction when treated with diisobutylaluminum 2,6-di-*t*-butyl-4-methylphenoxide. Germacrene A (**G5**) is formed along with humulene (**H22**) (Itoh, 1978).

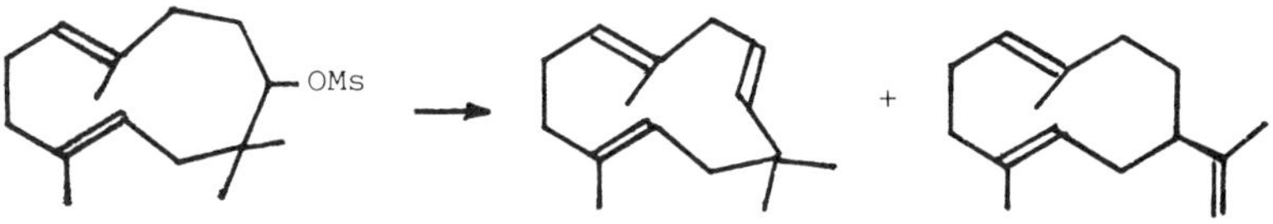

H22   **G5**

The conversion of bridged to condensed-ring systems is characterized by syntheses of bulnesol (**B21**) (Marshall, 1968d, 1969b) and pyrovellerolactone (**V8**) (Froborg, 1975).

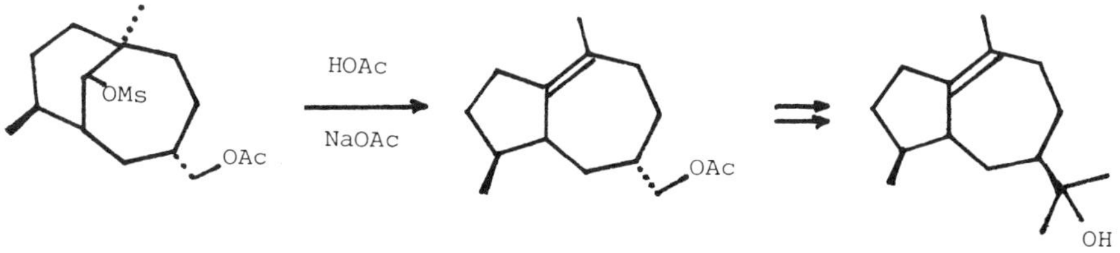

B21

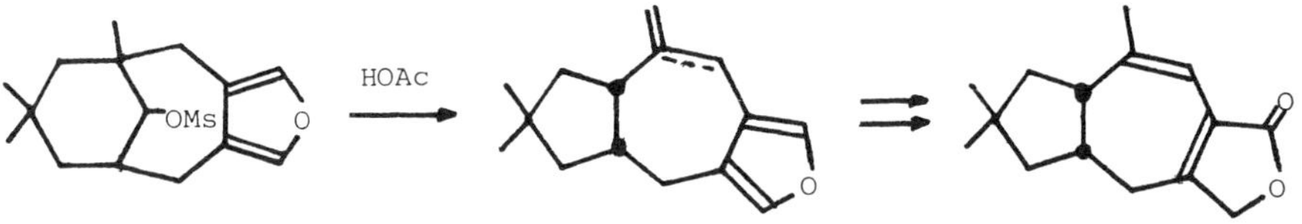

V8

The bond migration is stereoselective; therefore the leaving group must be antiperiplanar to the migrating bond. Such a steric relationship is ensured in the bulnesol precursor because it arises from a rational, systematic manipulation.

The intermediate for pyrovellerolactone must also have the correct stereochemistry. One side of the bicyclic ketone must be strongly shielded by the furan ring (i.e., a boat-shaped, seven-membered ring) so that its reduction occurs axially from the side opposite the furan.

An acid-catalyzed rearrangement of the bicyclic tertiary alcohol is planned for unraveling the C/D-ring system of stachenone (**S48**) (S. Monti, 1979), while simultaneously setting up the 1,5-diketone unit for B-ring construction.

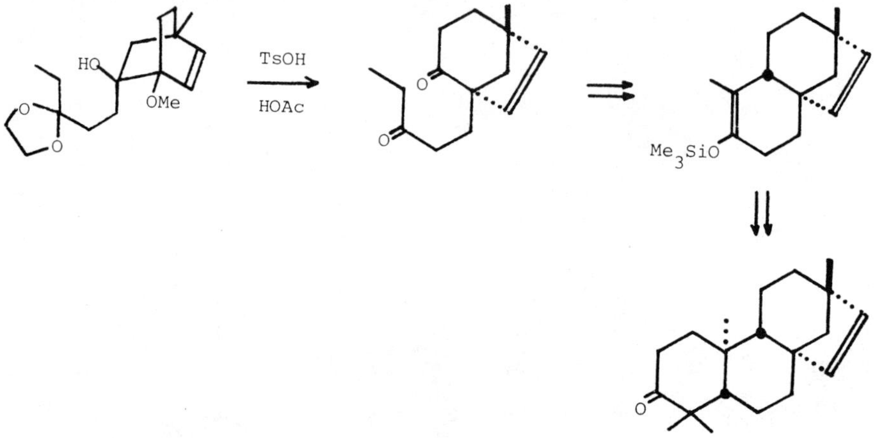

S48

Stemarin (**S49**), having a bicyclo[3.2.1]octane system, is expected to be mutated from the more symmetrical [2.2.2]analog, provided the carbon adjacent to a bridgehead bears a proper leaving group. The [2.2.2] precursor is most easily constructed by an aldol process (R. Kelly, 1980).

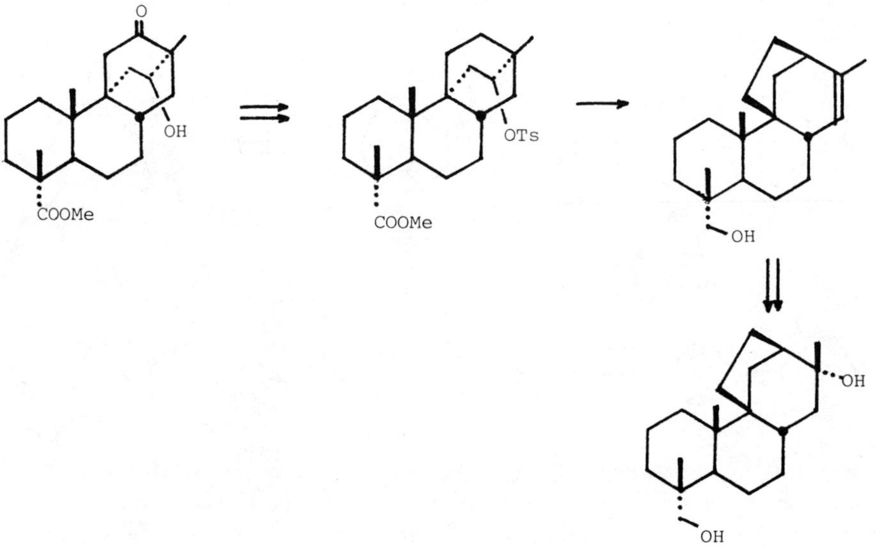

S49

11.2 RING CONTRACTION

The orientation of the leaving group dictates the selective migration of an antiperiplanar bridge in the conversion of the [2.2.2] to the [3.2.1] skeleton. Thus, solvolysis of the epimeric mesylates gives rise to the stemodinone (Marini–Bettolo, 1983a; Lupi, 1984) and aphidicolin (**A36**) (Marini–Bettolo, 1983b) frameworks, respectively.

stemodinone

A36

Aspterric acid (**47**) is conveniently approached by building a 6:7-fused ring system by the Robinson annulation and contraction of the A ring at a later stage. This is accomplished by reacting a neopentyl alcohol with phosphorus pentachloride (Harayama, 1983).

A47

A synthesis of hop ether (**H20**) (Imagawa, 1979) from the Diels–Alder adduct of furan and maleic anhydride is fascinating in that the monoester of the dihydro compound undergoes a Wagner–Meerwein rearrangement during anodic oxidation. A cyclopentane ring appears as a result.

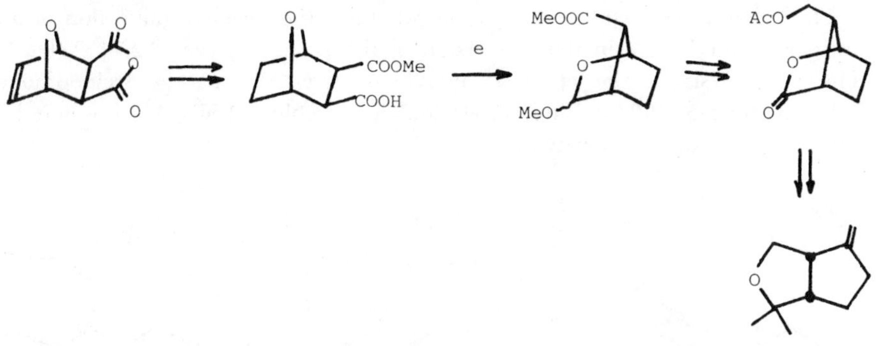

11 RING EXPANSION AND CONTRACTION

H20

The Cope rearrangement can lead to either ring expansion or ring contraction, depending the molecular environment. 1,5-Cyclodecadienes such as germacrene A undergo facile thermal reorganization to afford divinylcyclohexanes. Synthesis of the mesocyclic dienes is often thwarted by the *in situ* rearrangement: elemol (**E4**) (Corey, 1969d) and β-elemene (**E1**) (McMurry, 1985c) are obtained instead of the desired primary products.

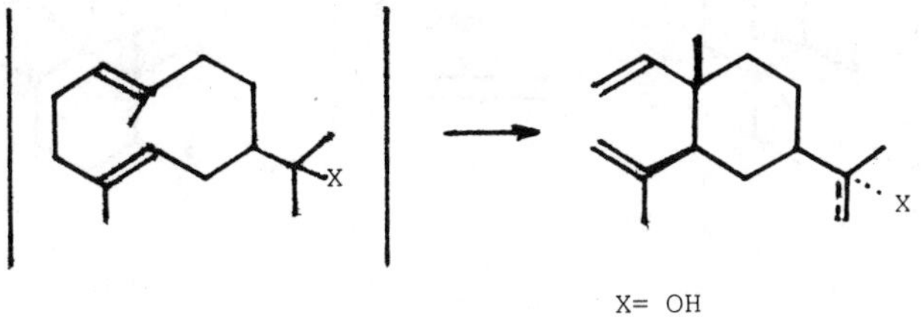

X= OH

Certain chromophores embedded in special carbon frameworks absorb light energy and incur structural changes. For example, β,γ-unsaturated ketones are subject to di-π-methane rearrangement to give novel ring systems (cf. cedrene synthesis: K. Stevens, 1980).

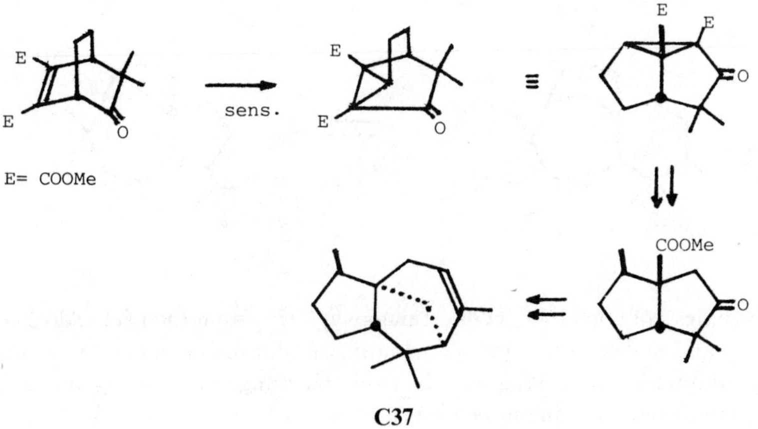

E= COOMe

C37

11.2 RING CONTRACTION

A potential intermediate for pentalenolactone G (Demuth, 1984a) has been acquired from a 1-methoxybicyclo[2.2.2]oct-5-en-2-one. The photoisomer is readily fragmented to provide a diketone. According to molecular modeling, the two carbonyl functional groups should be differentiable.

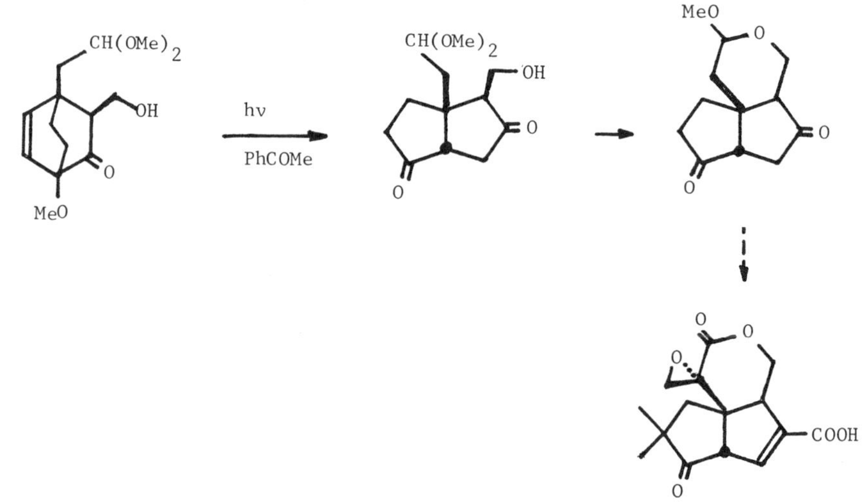

pentalenelactone G

The oxa-di-π-methane rearrangement is also the key operation for reaching the skeletons of certain triquinane sesquiterpenes (Demuth, 1985). Thus, two differently fused cyclopentanobicyclo[2.2.2]octenones have been assembled and their photochemical transformations lead to precursors of hirsutene (**H17**) and oxosilphiperfolene (**S36**), respectively. The lateral opening of the cyclopropane ring can be achieved by Li–NH$_3$ reduction.

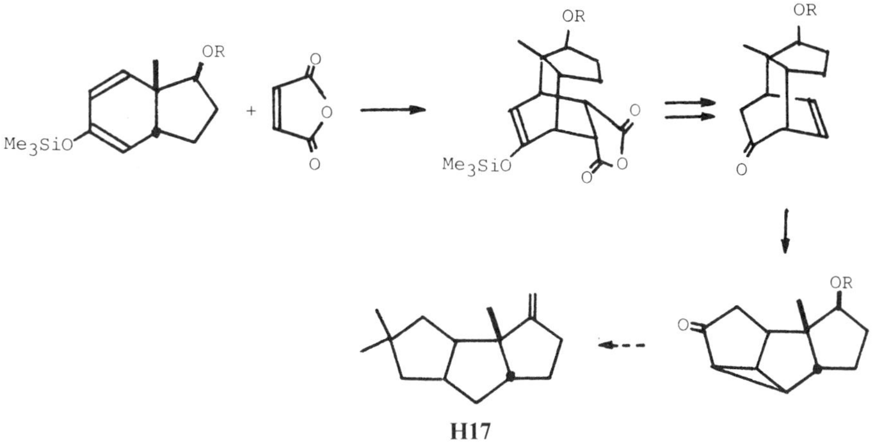

H17

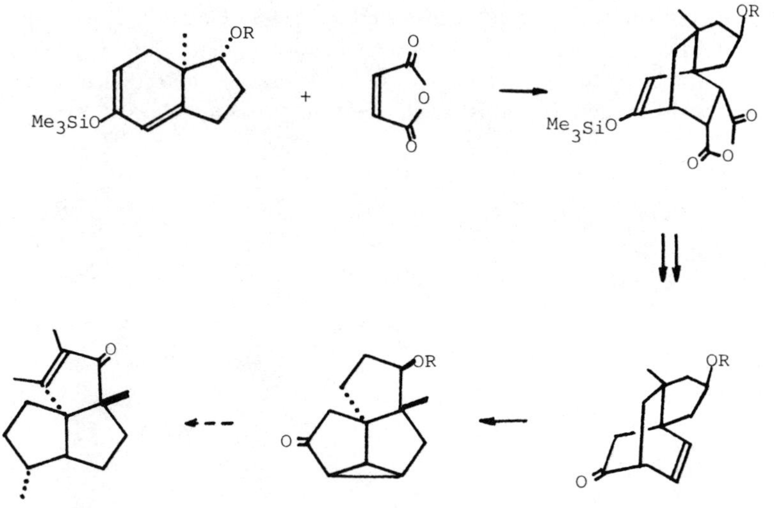

S36

Cross-conjugated cyclohexadienones are photochemically transformed into cyclopropanated cyclopentenones. These products are easily solvolyzed, sometimes with opening of the three-membered ring. The net result is a migratory ring contraction.

Extensive application of this reaction in terpene synthesis has emerged. For β-vetivone (**V16**) (Marshall, 1968c), the attachment of a methoxy group to the α position facilitates the onward processing of the spirocyclic intermediate (Caine, 1976a). Instead of requiring regioselective 1,2-transposition of the carbonyl group, a more straightforward removal of the same is performed. Further improvement of the synthesis involves photolysis of the *cis*-dimethyl dienone (Caine, 1976b).

V16

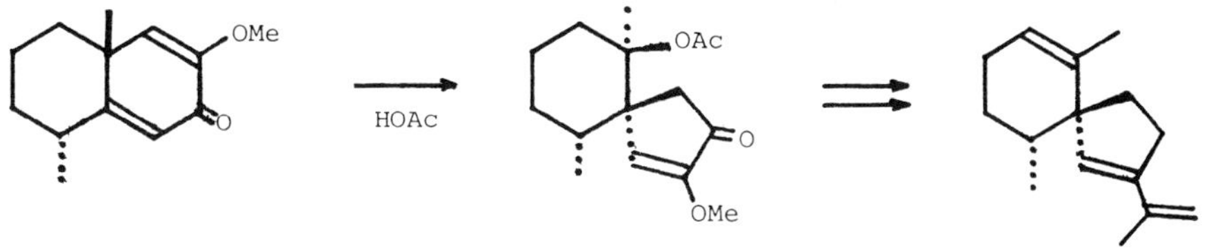

V14

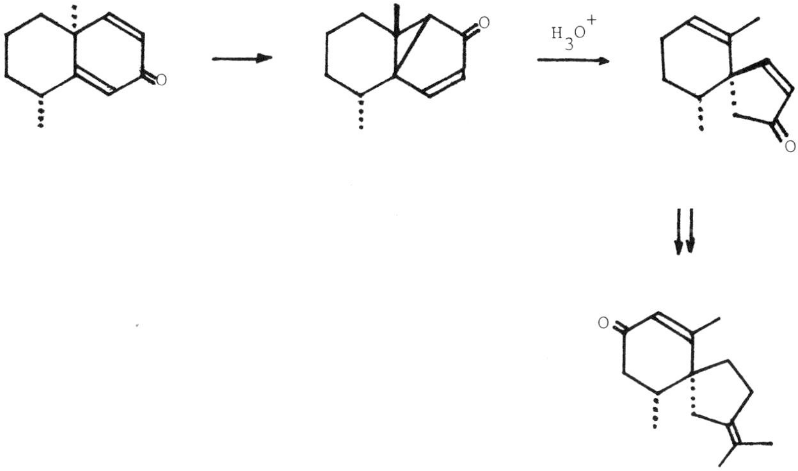

V16

The direction of ring opening depends on solvents and on photochemical conditions. The hydrazulene skeleton can be obtained when the reaction is carried out in aqueous acetic acid. Numerous partial syntheses of guaianolides have been accomplished based on this method. This popular aproach is due to the easy access of various derivatives of α-santonin, which are very suitable starting compounds. A synthesis of (−)-cyclocolorenone (**C92**) (Caine, 1972) is outline below.

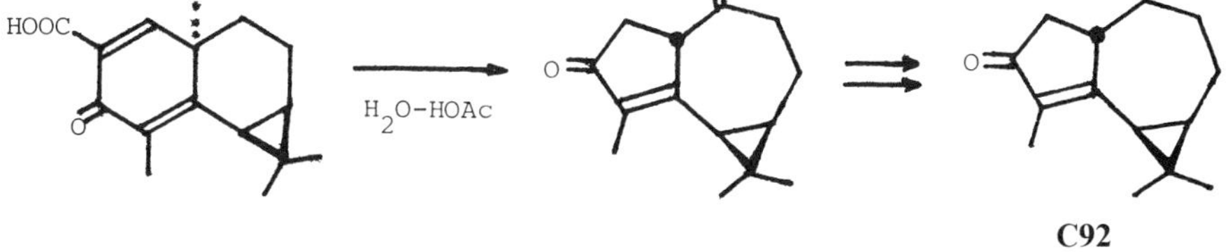

C92

A totally synthetic tricyclic dienone has been transformed into the linear 5:7:6 ring system for further elaboration into grayanotoxin II (**G19**) (Gasa, 1976; Hamanaka, 1972).

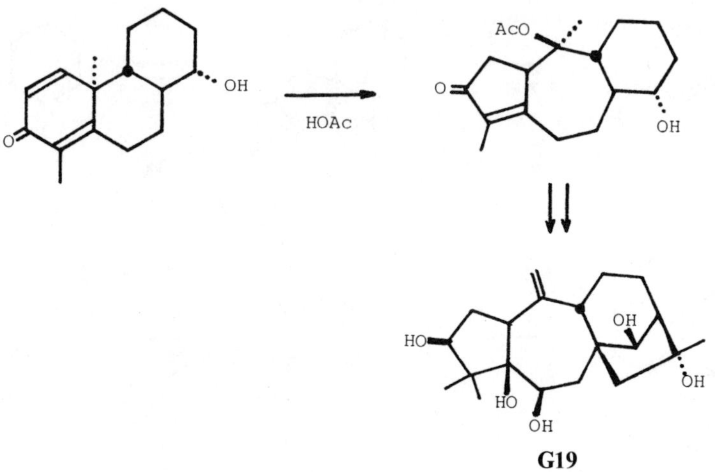

G19

The photoinduced reorganization of bicyclo[3.2.x]alkadienones to bicyclo [x + 4.1.0]alkenecarboxylic esters in an alcohol medium is formally a [3.3]sigmatropic rearrangement that generates ketene intermediates. The stereoselectivity associated with such a reaction is valuable in synthesis, and neat applications of the process to the construction of sesquicarene (**S26**) (Uyehara, 1983) and isosesquicarene (**S27**) (Uyehara, 1985b) have appeared.

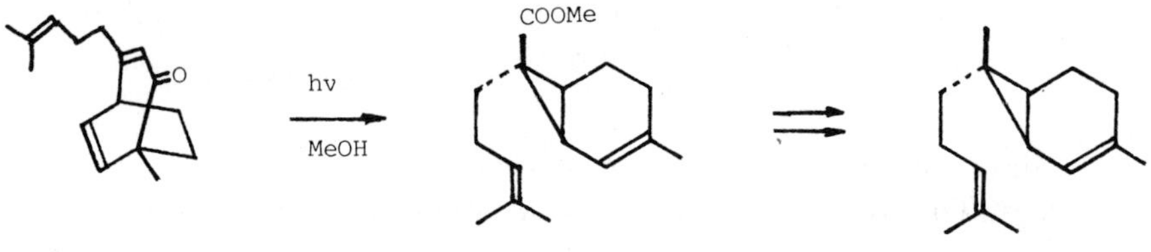

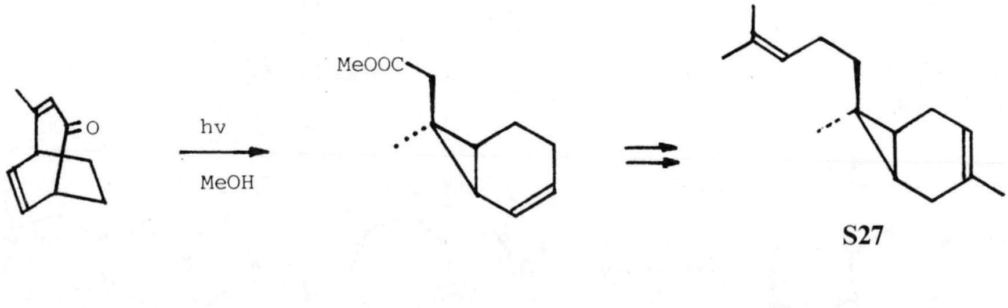

11.3 BICYCLIC DISPROPORTIONATION

11.3.1 Concerted Processes

The transformation typified by that of santonin to guaianolide embodies a simultaneous ring contraction and expansion. This process can be extended to the acquisition of 5:6-fused systems. Moreover, it constitutes the key step in the syntheses of oplopanone (**O7**) (Caine, 1973) and α-cadinol (**C6**) (Caine, 1977). In the latter endeavor, the cyclopentanone ring is remodeled to install a cyclohexene unit. The merit of the photochemical transformation consists of stereocontrol at several asymmetric centers.

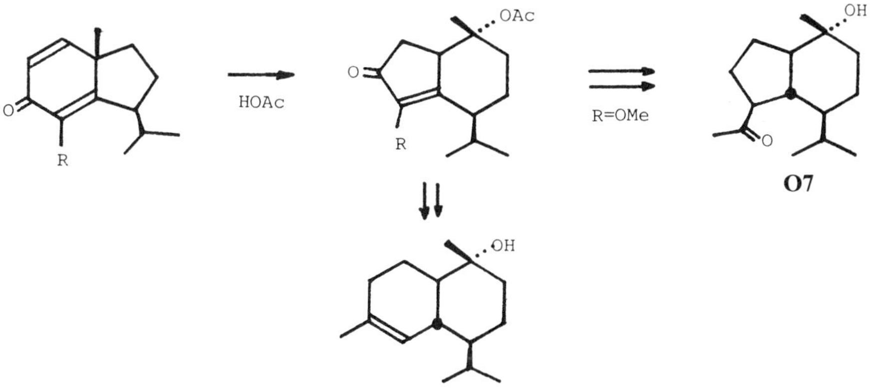

A tricyclic skeleton of isoingenol has been synthesized starting from 6-methoxy-α-tetralone (Paquette, 1984a, Ross, 1987) via an intramolecular alkylation step and a photoinduced 6:6 to 5:7 ring transposition. The product contains all the oxygen functions of ingenol while it lacks a secondary methyl group, a cyclopropane ring and it is epimeric at the ring junction of the bridged system.

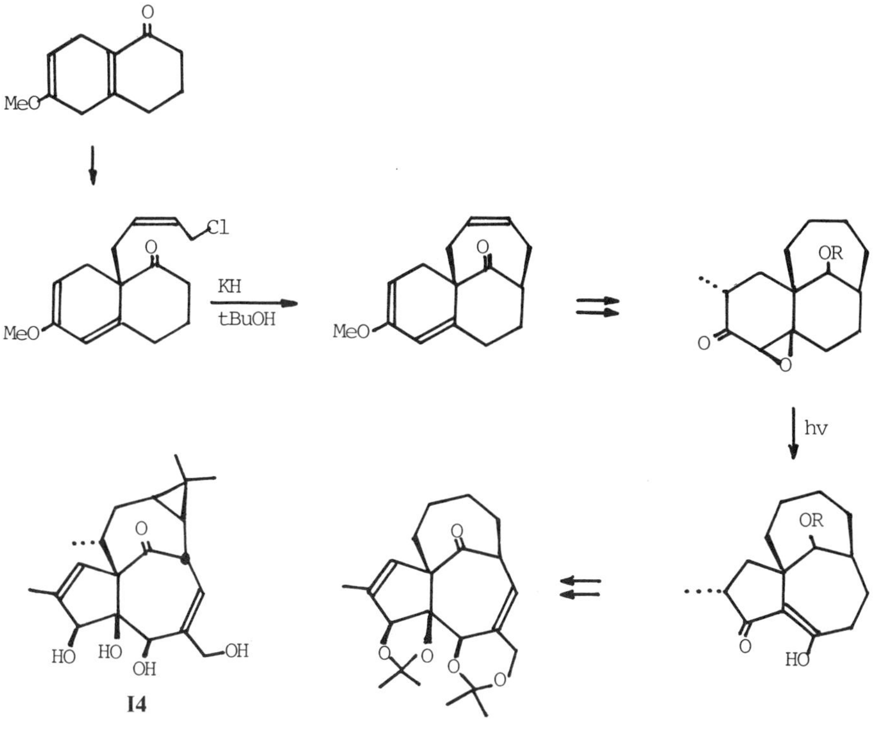

It is rare that a photocycloaddition reaction involving 3-acetoxy-2-cycloalkenones is designed without a subsequent fragmentation (DeMayo reaction) in mind. A hirsutene (**H17**) synthesis (Tatsuta, 1979) affords such an exception, however. The oxygen function, which usually acts as a trigger for fragmentation, instead serves to neutralize the positive charge on an adjacent carbon when solvolytic rearrangement is called for to unveil a triquinane nucleus.

By far the most popular scheme for entry into the hydrazulene sesquiterpenes is via a solvolytic rearrangement of 1-hydroxydecalin derivatives. The great advantage inherent in this approach is the stereocontrol that can be obtained in six-membered rings. After a successful demonstration of the ring disproportionation in a (−)-aromadendrene (***ent*-A41**) synthesis (Büchi, 1966, 1969), many others followed: α-bulnesene (**B19**) (Heathcock, 1971; Mehta, 1975), bulnesol (**B21**) (M. Kato, 1970a; Heathcock, 1971), kessane (**K5**) (M. Kato, 1970b), confertin (**C62**) (Heathcock, 1982a), zizanoic acid (**Z3**) (MacSweeney, 1971), and epizizanoic acid (**Z4**) (Kido, 1969).

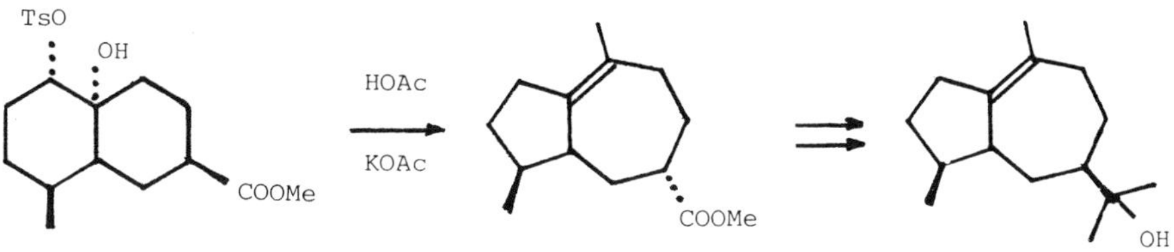

In the confertin synthesis, it has been observed that the successful anionic pinacol rearrangement is critically dependent on the presence of a free hydroxyl group at C_4 (generated *in situ* from the acetate), which forms a hydrogen bond with the alkoxide trigger. The bonding stabilizes a nonsteroid conformation in which the C_5—C_{10} bond is antiperiplanar to the leaving tosyloxy group. Without this stabilization, the steroid conformation prevails and the rearrangement leads to a bridged ketone.

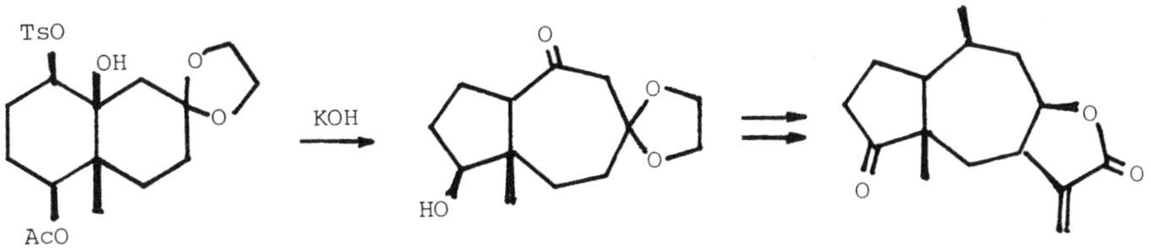

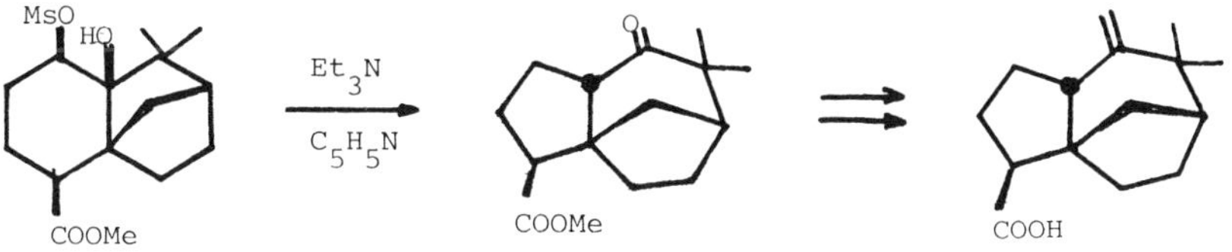

A synthesis of portulal (**P45**) (Tokoroyama, 1974; Kanazawa, 1975) consists of the transformation of a decalindiol derivative into a hydrazulenone, which can then be processed further by introduction of two different one-carbon chains at the α- and α'-carbon atoms of the ketone function.

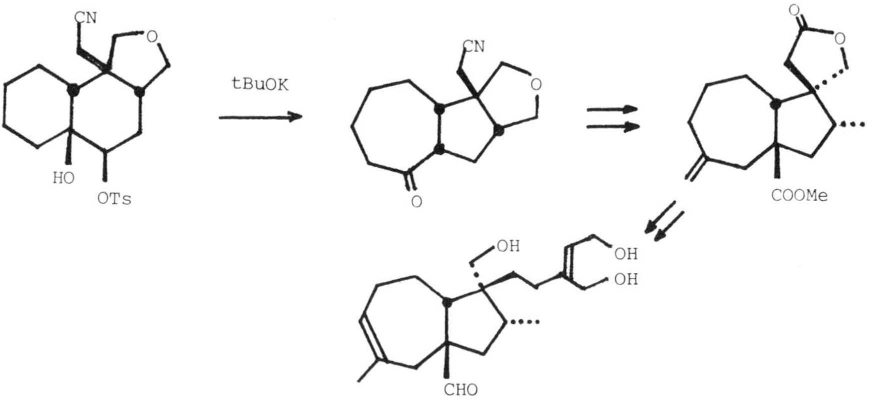

A key operation in the synthesis of talatisamine (**T1**) (Wiesner, 1974c) is the rearrangement step that converts a bridged decalin system into the corresponding hydrazulene. As seen locally, the transformation also belongs to the [2.2.2] → [3.2.1] category.

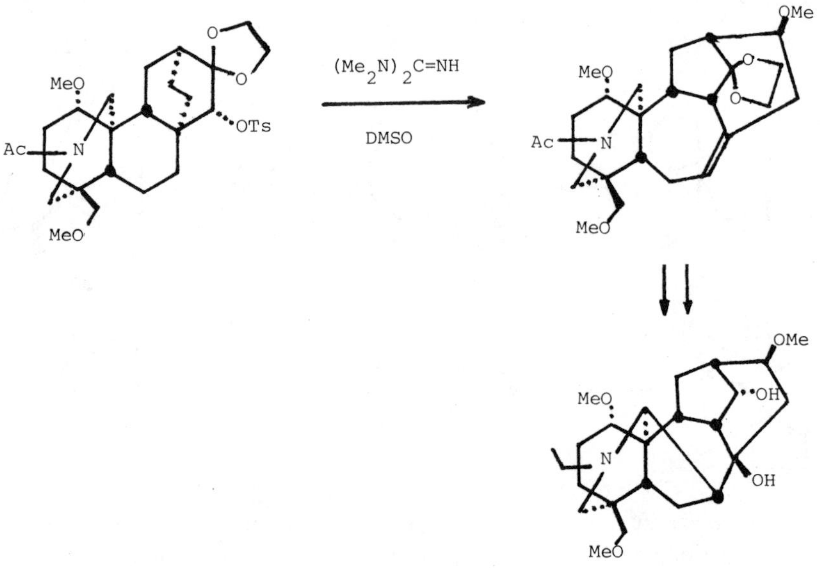

A singular case in which 5:7-to-6:6 skeletal rearrangement is effected helps achieve a synthesis of cryptofauronol (**F3**) and valeranone (**V2**) (Sammes, 1983). The hydrazulene compound is obtained from an intramolecular [5 + 2] cycloaddition of a pyranulose acetate.

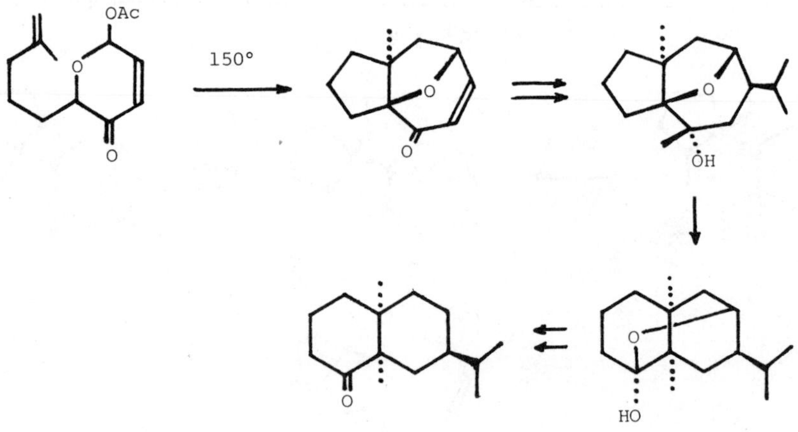

11.3 BICYCLIC DISPROPORTIONATION

Skeletal rearrangements with formal ring expansion and contraction are essential in the establishment of the final carbon frameworks of (+)-sesquifenchene (**S28**) (Bessière-Chrétien, 1972), (−)-β-santalene (**S4**) (Hodgson, 1973), isocomene (**I14**) (Pirrung, 1979, 1981; Wenkert, 1983), khusimone (**K8**) (Büchi, 1977), α-cedrene (**C37**) (Büchi, 1977), and hirsutene (**H17**) (Tatsuta, 1979) under solvolytic conditions, and zizaene (**Z1**) (Coates, 1972) by intramolecular diazoalkane insertion.

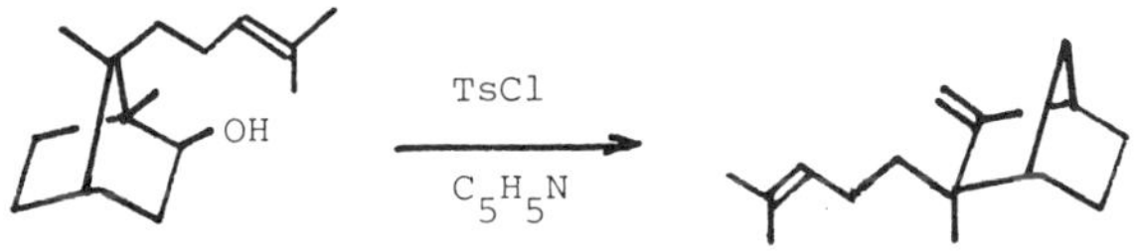

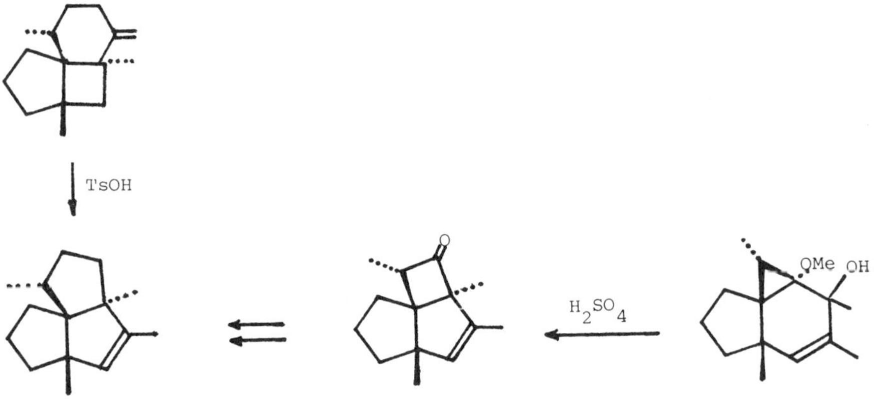

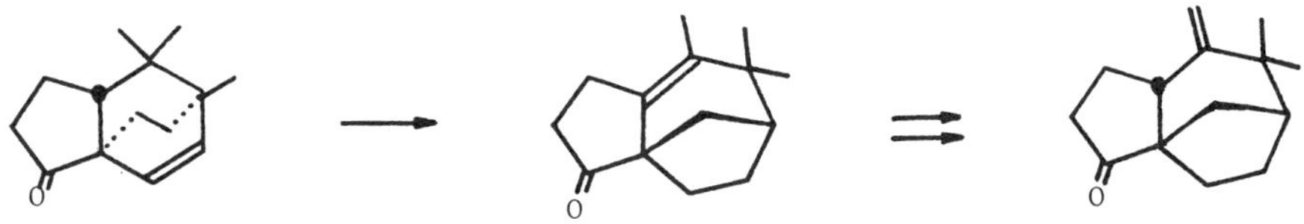

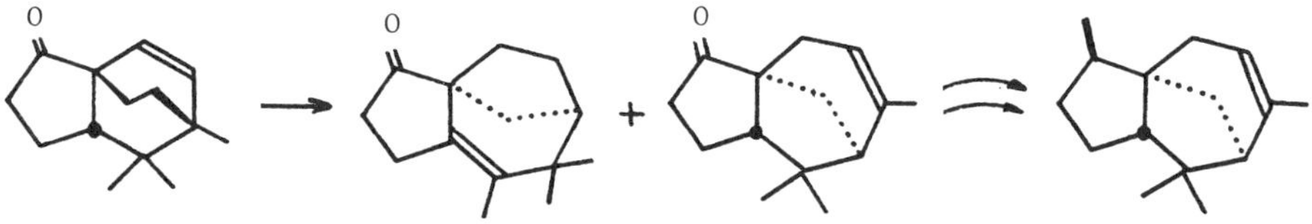

In a synthetic precursor of widdrol (**W3**) connecting the tertiary hydroxyl group to the angular methyl substituent ensures the cis relationship. Thus a bridged ketone is a proper candidate from which the two functionalities may be generated by means of, among others, a Baeyer–Villiger reaction. However, a one-carbon oxo bridging element is not as ideal as expected in view of its flanking on each side by quaternary carbon atoms and consequently insignificant regioselectivity. A functionalized two-carbon bridge is a compromise that could lead to the desired lactone, but the latter must be further degraded.

Accordingly, a bicyclo[3.2.2]non-2-ene derivative represents a subgoal for the synthesis of widdrol (Uyehara, 1986), and an expedient intermediate has been obtained by a rearrangement route.

The photoadduct of 4,4-dimethylcyclopentene and 2-cyclohexene-1,4-dione is susceptible to rearrangement on exposure to iodotrimethylsilane. The resulting triquinane enone is most readily converted into hirsutene (**H17**) (Iyoda, 1986).

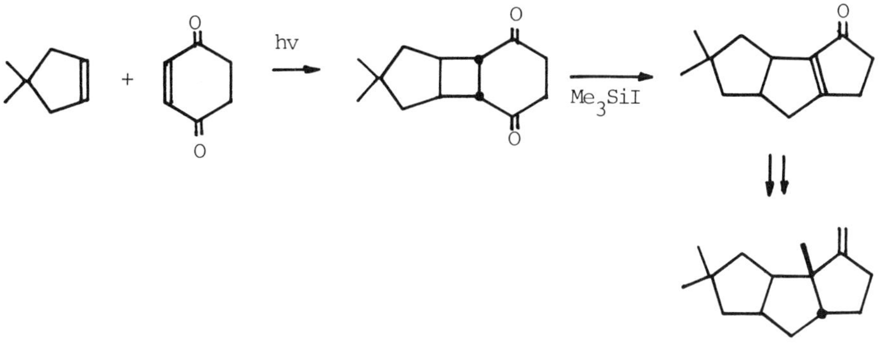

11.3 BICYCLIC DISPROPORTIONATION

H17

A Diels–Alder route to quadrone (**Q1**) (Wender, 1985d) is also critically dependent on a selective ring expansion/contraction step. This process is achieved by degrading the *endo*-ester pendant into a chlorine substituent and the ionization of the latter upon exposure to silver ion.

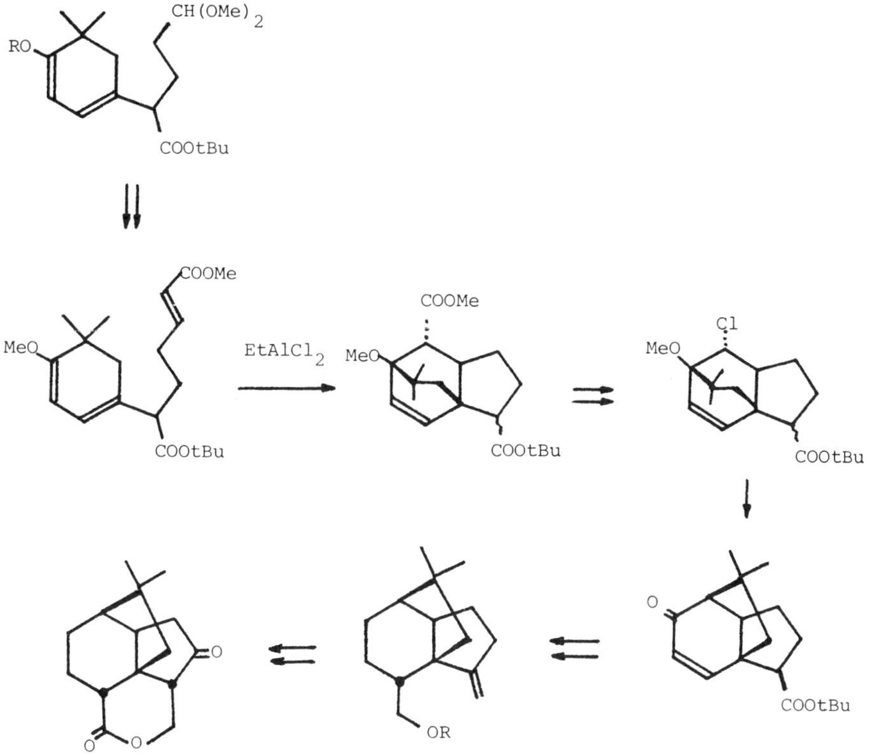

Q1

The contrathermodynamic skeletal rearrangement of bornane to a pinane framework is possible when a donor atom is placed in the bridgehead of the bicyclo[2.2.1]heptane nucleus. Nopinone, a precursor of α- and β-pinenes (**P30** and **P31**), is obtained in high yield from 1-aminoapobornyldiazonium ion (Kirmse, 1972).

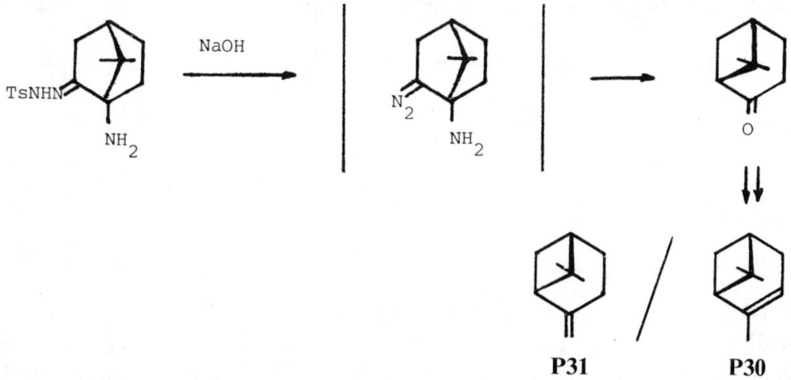

An excellent scheme for the synthesis of grandisol (**G18**) centers on cyclopropanation of a methoxycyclohexadienone and acid-catalyzed rearrangement of the bicyclo[4.1.0]heptenone. It remains to reduce the cyclobutanone carbonyl and convert the cyclopentanone unit to an oxime to reach a known synthetic intermediate at that stage (Wenkert, 1978a).

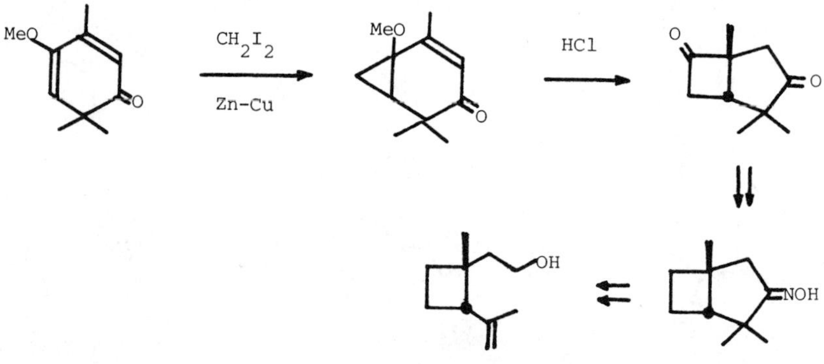

This ring expansion/contraction protocol has been included in a synthesis of pitoediol (**P37**) (Gadwood, 1984). As the fate of only one ring is concerned, the steady expansion from three to four to eight members is witnessed.

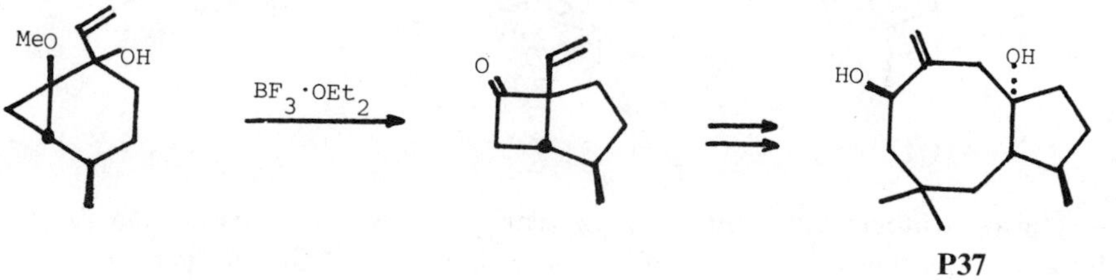

11.3 BICYCLIC DISPROPORTIONATION

Interestingly, photolysis of (+)-2-carene gives a racemic bicyclo[3.2.0]heptene whose structure is transparently correlatable to grandisol (**G18**). The actual transformation is effected by functionalization of the double bond and cleavage of the resulting cyclopentanone (Sonawane, 1984).

A synthesis of filifolone (**F6**) from geranic acid (Beereboom, 1965) proceeds via a ketene. However, the latter compound undergoes a [2 + 2]cycloaddition to afford chrysanthemone but not filifolone directly. It has been shown that chrysanthemone is subject to acid-catalyzed isomerization to filifolone at room temperature (Erman, 1971).

The formation of filifolone from ocimenone on treatment with aluminum chloride (Adams, 1975) plausibly involves similar intermediates.

Research concerned with the biogenetic implications of illudane-type sesquiterpenes has culminated in the conversion of an angular 5:6:4 system (illudane) to the linear 5:5:5 triquinane skeleton characteristic of hirsutene (**H17**) (Hayano, 1978).

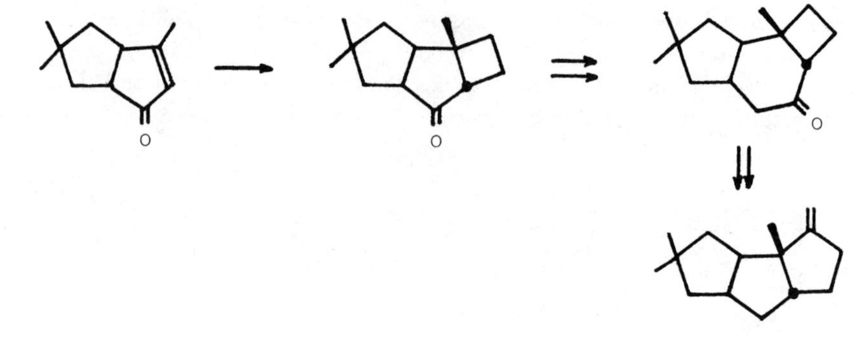

H17

The approach to dendrobine (**D11**) (Connolly, 1985) based on a cyclodisproportionation scheme leading from a cyclobutanopiperideinium skeleton to the cyclopentenopyrrolidine intermediate has yielded valuable information concerning the particular C—C bond that would undergo 1,2-migration. The following route should have a good chance to succeed.

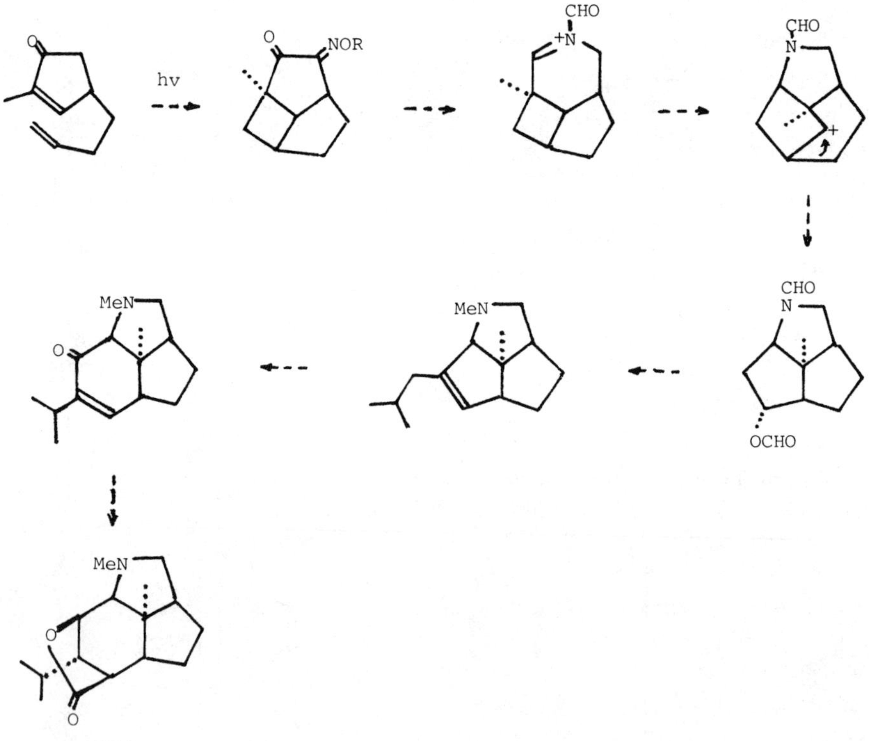

D11

[m.n.2]Propellenones are accessible via photocycloaddition of bicyclo-[m.n.0]alk-$(m+n)$-en-1-ones with an acetylene equivalent. By means of a double rearrangement, the photoadducts may be transformed into the conjugated

11.3 BICYCLIC DISPROPORTIONATION

[*m* − 1.*n*.3]propellenones based on these key reactions, the synthesis of modhephene (**M12**) was pursued (Smith, 1981c). The only drawback of this approach is the stereorandomness of the photocycloaddition step.

Strain relief is the major factor of the rearrangement. The specific migration of the sp^2-hybridized carbon is favored by thermodynamic considerations in that a conjugate enone system is produced, and the migration of the alternative cyclobutene bond would generate a bridgehead carbenium ion that could not profit by delocalization of the charge (Bredt's rule).

Thermal elimination of methanol from the photoadduct(s) of piperitone and 1,1-dimethyoxyethene is also accompanied by a double rearrangement. In this case a bridged bicyclic system ensues. The utility of the rearranged compound in the syntheses of helminthosporal (**H9**) and sativene (**S11**) (Yanagiya, 1979) is evident.

11.3.2 Stepwise Processes

In a synthesis of carotol (**C29**) (deBroissia, 1972a,b) based on decalin derivatives, systematic B-ring expansion is induced, and followed by A-ring contraction. However, the simultaneous disproportionation scheme is not appropriate here because of different substitution patterns.

A compound containing the ophiobolin nucleus (cf. ceroplastol II, **C44**) has been obtained from a route involving ring contraction and expansion of an decalindione monoketal (Dauben, 1977a). The contraction of the silyl enol ether with an arylsulfonyl azide consists of a 1,3-dipolar addition and nitrogen extrusion from the resulting triazoline, which is assisted by the siloxy group. The expansion step involves reaction of the cyclohexanone enamine with dimethyl acetylenedicarboxylate.

11.3 BICYCLIC DISPROPORTIONATION

A 6:6-to-5:7 ring dismutation in a damsin (**D5**) synthesis (Kretchmer, 1976) involves a double-bond cleavage and recyclization of the 10-membered triketone internally. The strategic placement of another ketone (which is essential) in the molecule ensures regioselectivity in the formation of the hydrazulene.

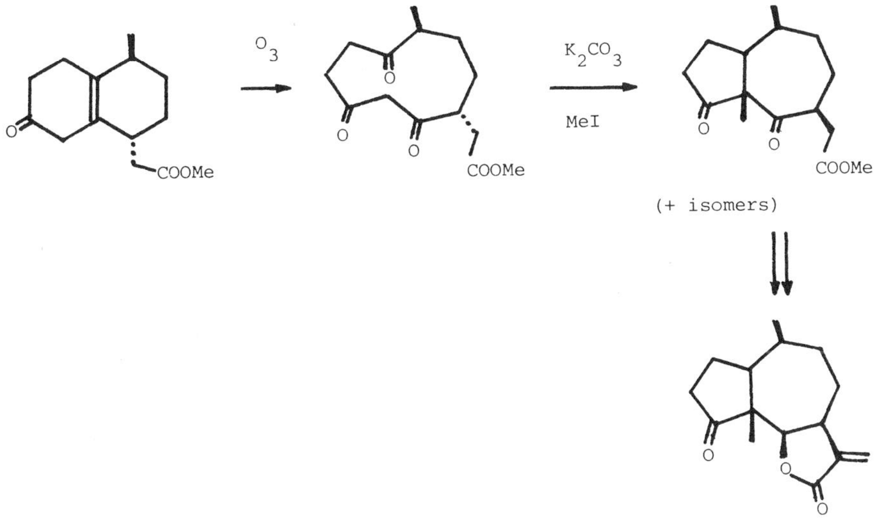

$$\text{[structure]} \xrightarrow{O_3} \text{[structure]} \xrightarrow[\text{MeI}]{K_2CO_3} \text{[structure]}$$

(+ isomers)

⇓⇓

D5

Chapter 12

Transitory Annulation

Transitory annulation is defined as a ring-forming process that does not contribute to the overall cyclic array of the synthetic goal. While such a structure is of no permanent significance, the incorporation of the transitory annulation operation in a synthesis often solves the problem of carbon chain introduction, when nonannulative methods are cumbersome or unavailable. More important is the stereocontrol bestowed by ring structures, particularly the six-membered ring, which permits the placement of substituents in proper configurations. On other occasions, extra rings are made inadvertently and they must be removed.

12.1 CYCLOPROPANES

12.1.1 Cyclopropanes as Latent Methyl and Functionalized Methylene Groups

Introduction of the *gem*-dimethyl group to a carbon skeleton is a perennial problem that has only recently evoked a promising solution (Reetz, 1981). Before this development, a reaction sequence involving the formation of a methyl-substituted alkene or vinylidene, a Simmons–Smith reaction, and hydrogenation was the standard method for fulfilling the need. Fortunately, this reaction sequence is quite efficient: it has gained universal acceptance, and many terpene syntheses have relied on it to make necessary structural amends. Examples include β-himachalene (**H14**) (Challand, 1967), β-cuparenone (**C86**) (Casares, 1976), herbertene (**H10**) (Chandrasekaran, 1982), longifolene (**L24**) (Oppolzer, 1978), warburganal (**W1**) (Wender, 1982b), $\Delta^{9(12)}$-capnellene (**C21**) (Mehta, 1983; Piers, 1984b), and cedrene (**C37**) (Horton, 1983, 1984).

H14

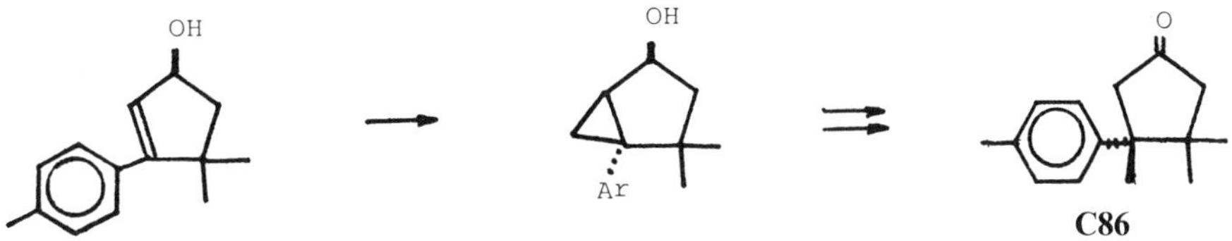

12.1 CYCLOPROPANES

C86

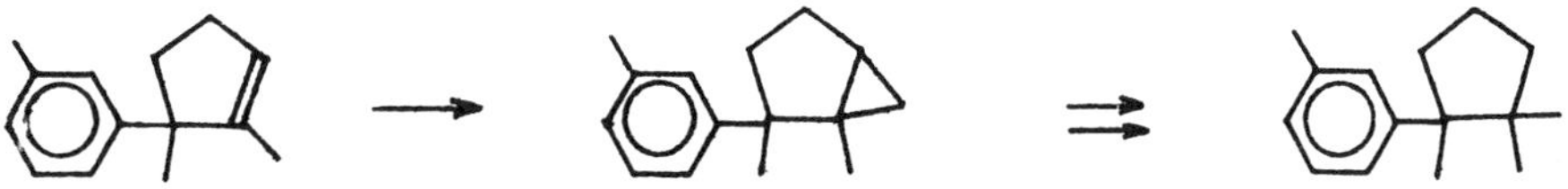

H10

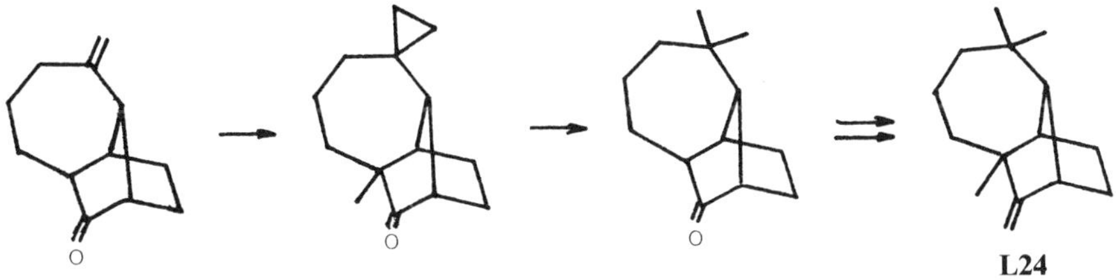

L24

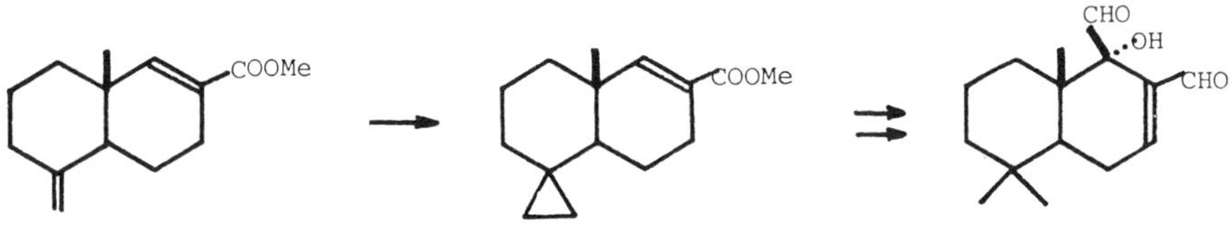

W1

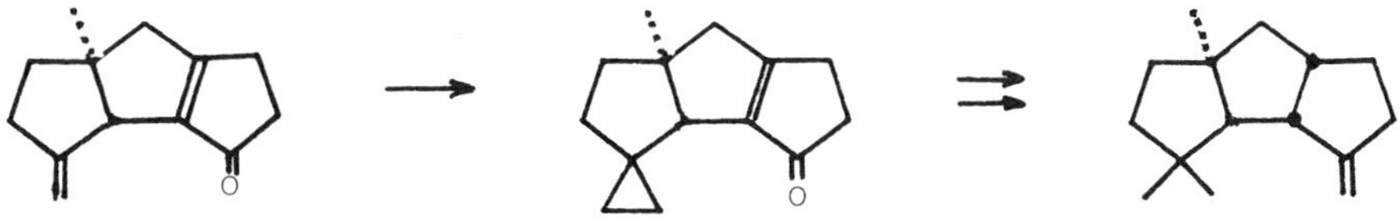

C21

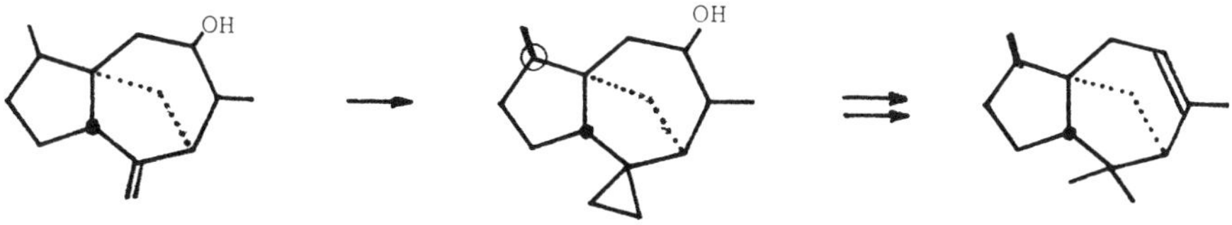

C37

In a route to pentalenene (**P16**) (Piers, 1984), the hydrogenation step accomplishes two tasks (cf. sinularene). Unfortunately the saturation of the exocyclic methylene group is nonstereoselective (42:58 in favor of the undesired epimer) as a result of the relatively exposed position of this group.

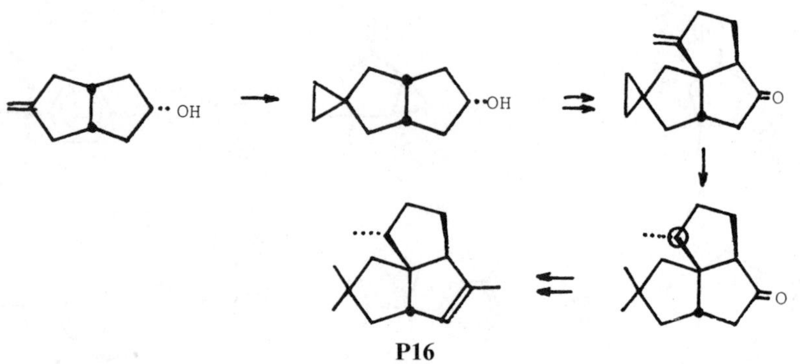

P16

In a synthetic context the least accessible point of hirsutic acid C (**H19**) is the quaternary carbon atom bearing the carboxyl function. Two approaches involve cyclopropanation and subsequent opening of the small ring. By virtue of steric bias, the addition of a methoxycarbene to an exocyclic methylene group of the *cis*-fused bicyclo[3.3.0]octane nucleus is favored at the exo face by a factor of 2.6. The major adduct(s) have the desired configuration (F. Sakan 1971; H. Hashimoto 1974; Yamazaki, 1981).

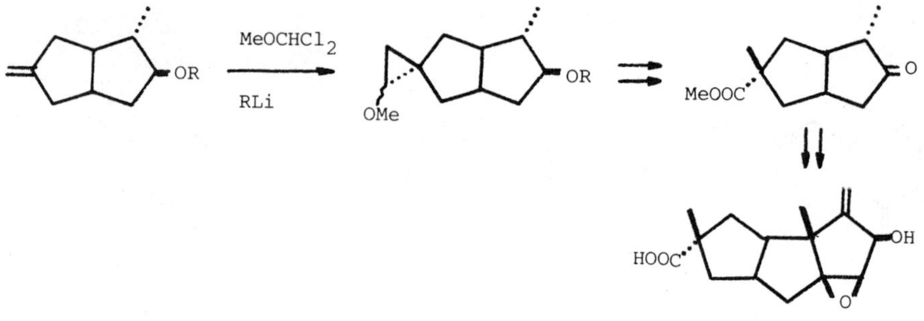

H19

An alternative route in which higher stereoselectivity is attained involves the Simmons–Smith addition to an existing enol ether. Stereoguidance for the reagent is provided by a remote hydroxyl function (Shibasaki, 1982); methylene delivery is from the more hindered endo side.

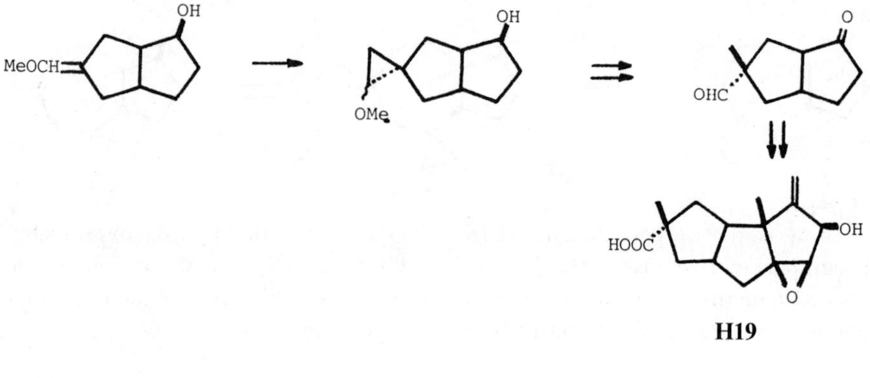

H19

12.1 CYCLOPROPANES

The indirect methylation protocol is not restricted to the erection of *gem*-dimethyl groups. Several variations show other uses in solving nontrivial architectural problems. The introduction of an angular methyl group in the synthesis of (—)-valeranone (**V2**) (Wenkert, 1967, 1978a), the vinylic methyl in isocaryophyllene (**C33**) (Kumar, 1976), and a secondary methyl group at the γ position of an octalone for the synthesis of ivangulin (**I20**) (Grieco, 1977d) are some of the better known examples.

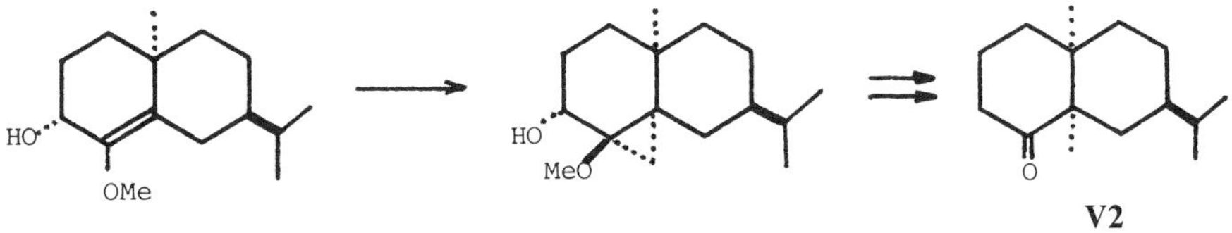

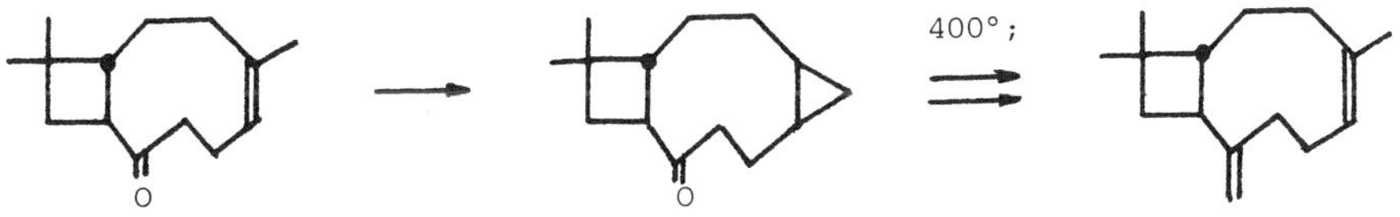

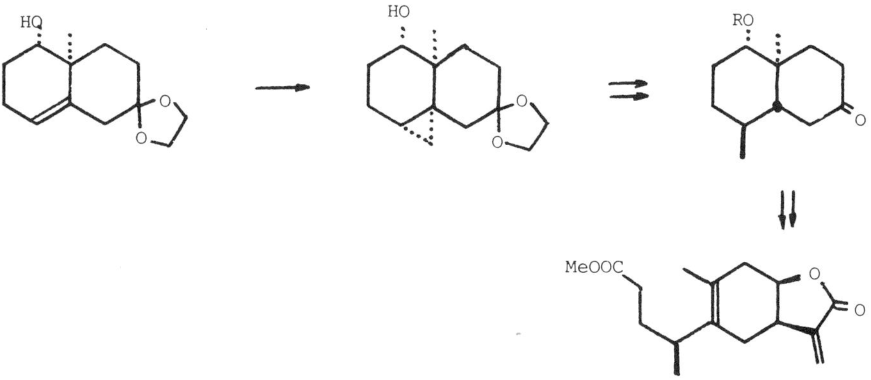

The intramolecular Diels–Alder route leading to sinularene (**S37**) (Antczak, 1985) calls for the preparation of a spirannulated cyclopentadiene, to avoid untoward sigmatropic rearrangements. Hydrogenation of the cyclopropane moiety of the adduct disentangles the tetracyclic framework and reveals the tertiary methyl group.

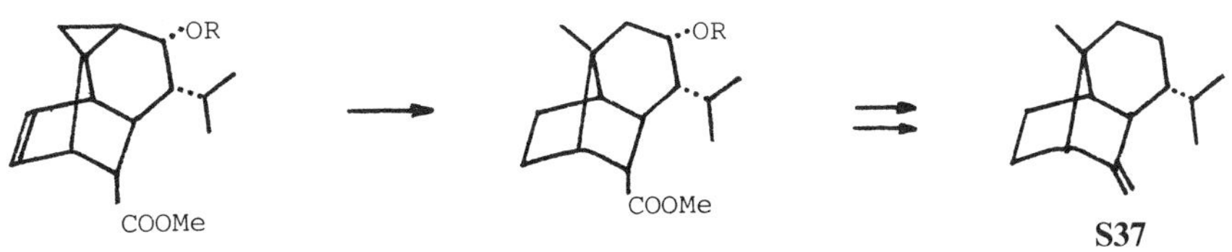

The cationic cyclization of a cyclopropylidene analog of methyl farnesate is crucial to a stereoselective bishydroxylation during a synthesis of the human complement inhibitor **K76** (McMurry, 1985b). The retention of the A ring in a chair conformation permits the hydroxylation to occur from the α face. The cyclopropane ring devolves into the *gem*-dimethyl group in a subsequent operation.

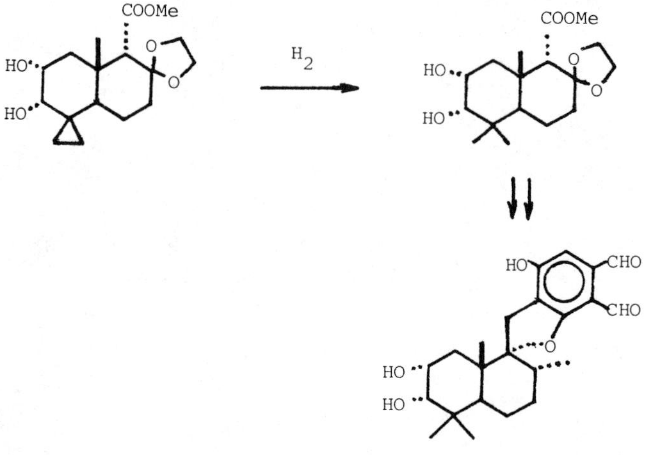

In the first successful synthesis of alnusenone (**A26**) (Ireland, 1970) via a dibenzopentacyclic precursor, the E ring is reduced selectively and converted into a cyclopropyl ketone unit as a latent β-methyl cyclohexanone. This tactic also ensures an uneventful *gem*-dimethylation at C_{20}.

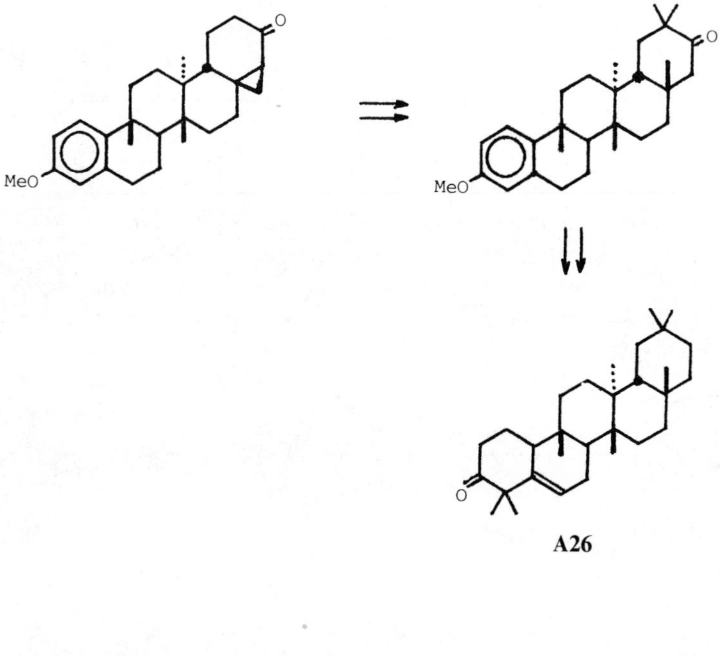

12.1 CYCLOPROPANES

In an EDC + B + A approach to lupeol (**L28**) (Stork, 1971), the stereoselective methylation at C_8 and C_{14} is one of the focal points. The C_{14} methyl group arises from a conjugate hydrocyanation and transformation of the primary mesylate derived from the cyano group. Instead of direct abrogation of the mesyloxy function by reduction, an internal displacement is executed so that a subsequent cleavage of the cyclopropyl ketone (with Li–NH$_3$) generates an enolate and enables methylation at C_8 also.

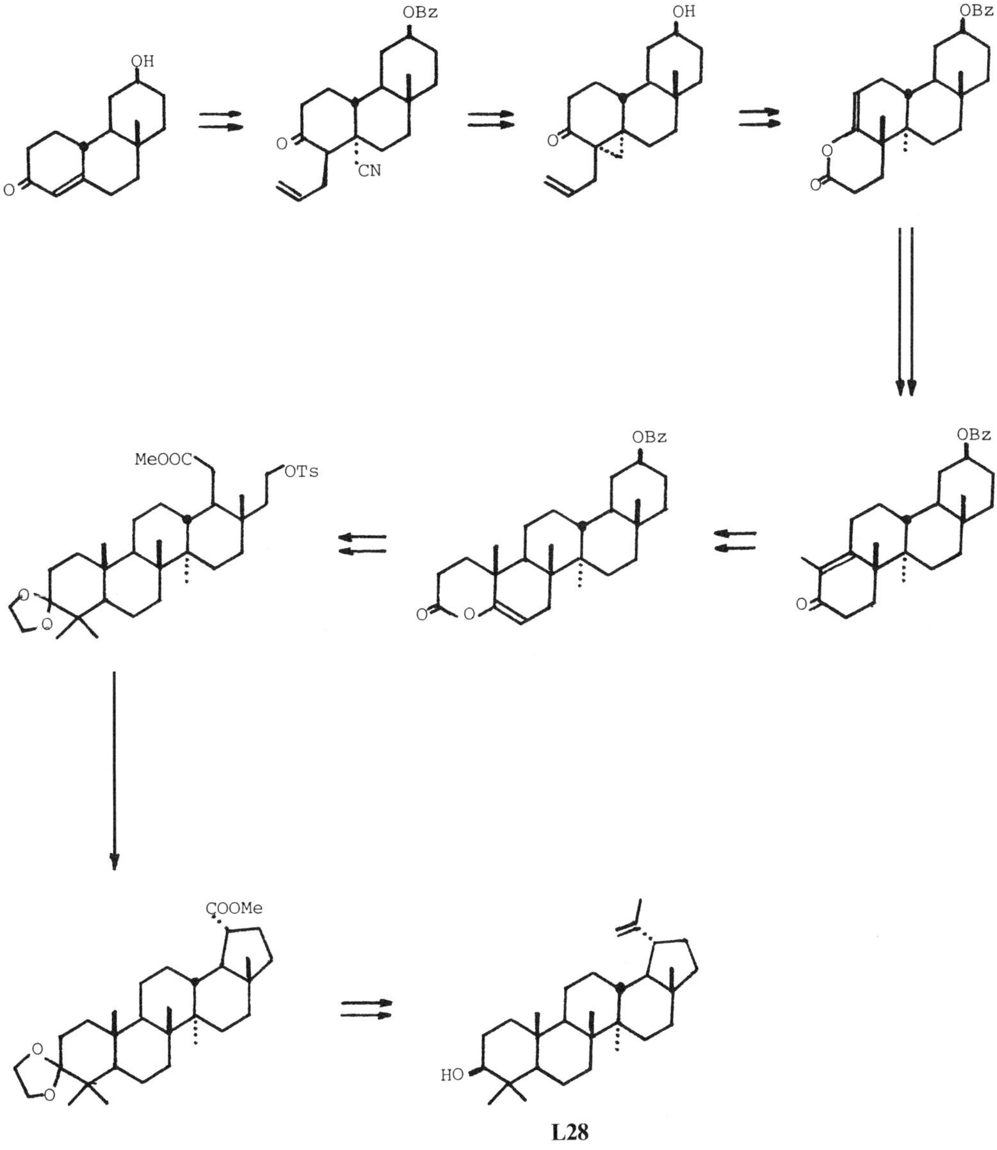

L28

Cyclopropanation by 1,3-elimination as an intermediary stage of deoxygenation appears in the synthesis of dehydrofukinone (**F16**) (Torii, 1979a), isopetasol (**P25**) (Torii, 1979b), and α-onocerin (**O6**) (Tsuda, 1981).

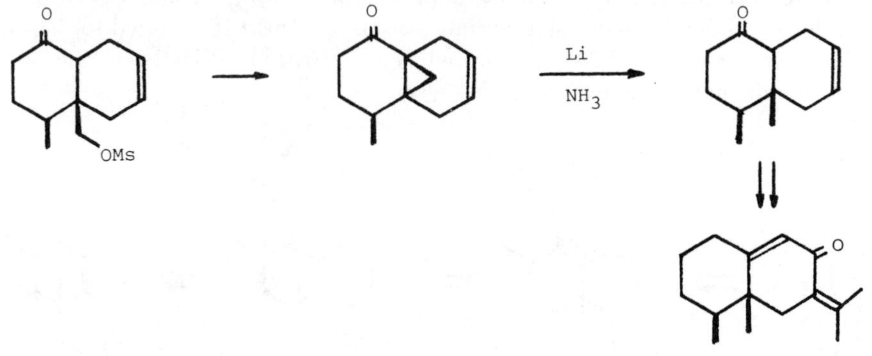

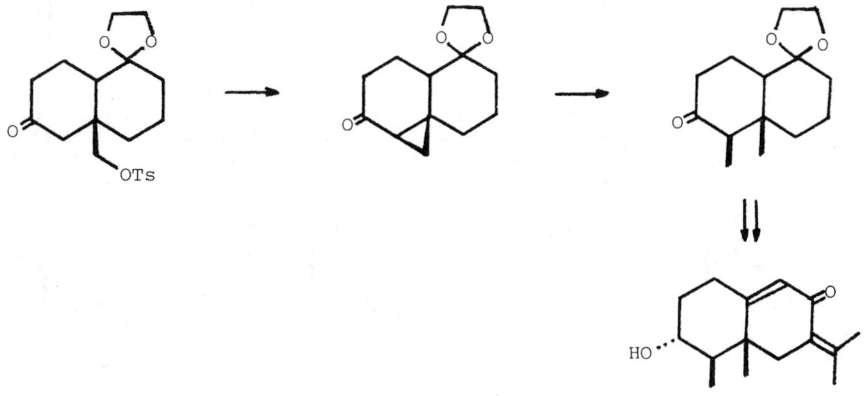

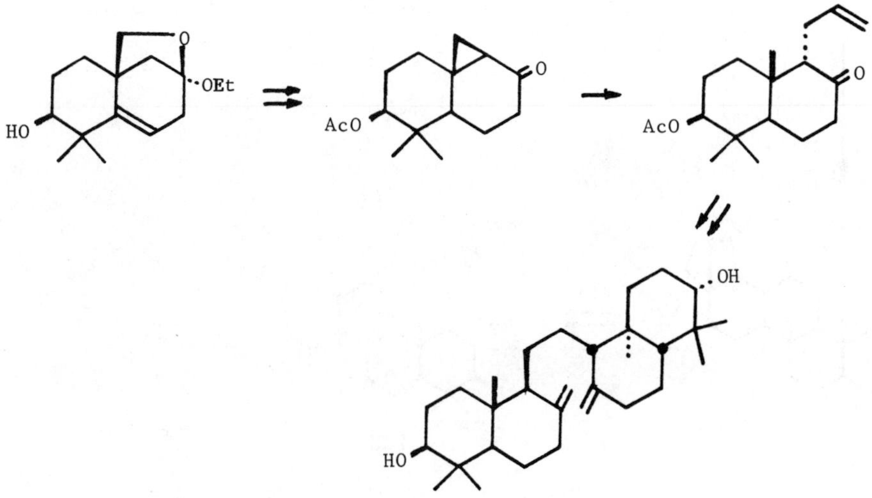

Asymmetric induction by a chiral ketal group in the allylic position allows synthesis of optically active cyclopropyl ketones. A synthesis of (+)-β-eudesmol (**E18**) (Mash, 1987a) has been performed from a such ketone by subsequent reductive cleavage and Wittig methylenation.

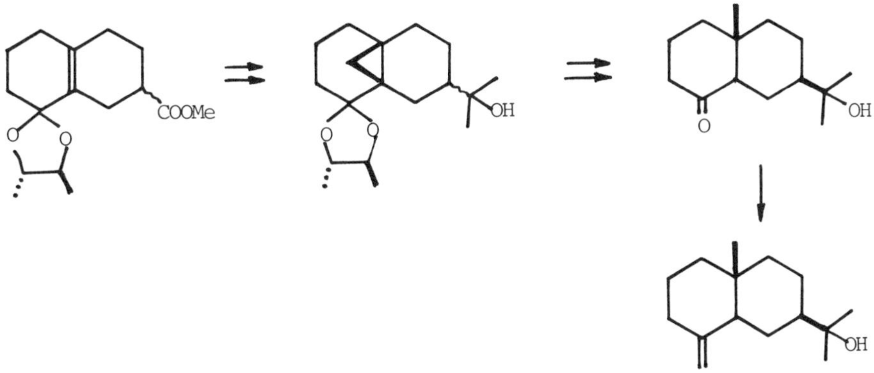

E18

A similarly indirect reductive methylation is involved in an approach to bonandiol (Mash, 1987c).

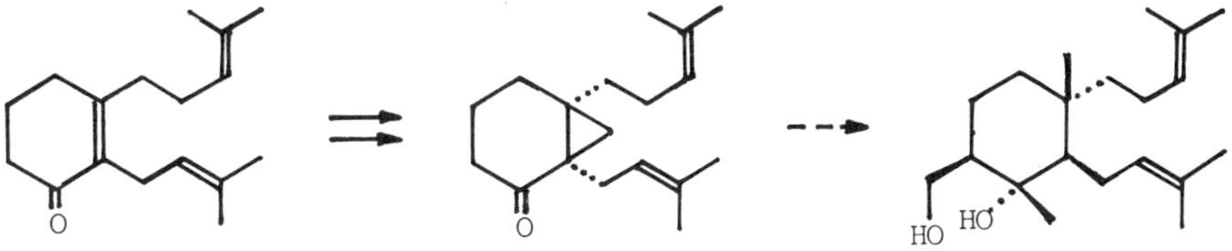

bonandiol

The demesylation of an angular mesyloxymethyl substituent in an α-decalone intermediate during synthesis of petasitolone (**P24**) (Liu, 1984) with Zn–NaI gives a rather stable cyclopropanol. Regeneration of the ketone group is effected by heating the tricyclic compound with alkali.

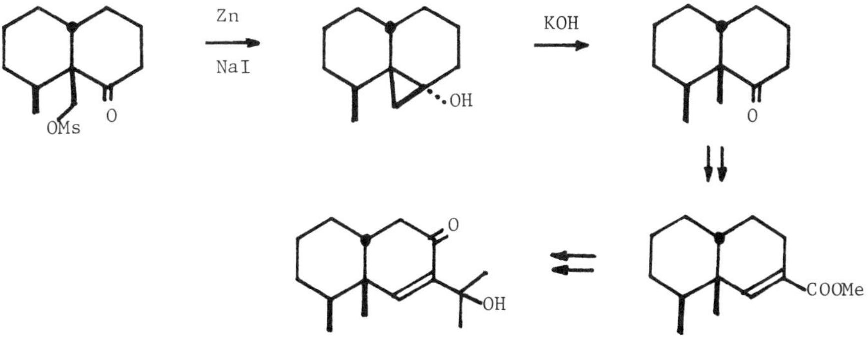

P24

Formation of cyclopropanol from an enol derivative followed by its opening represents an indirect α-methylation of carbonyl compounds. This method is extremely useful when the methyl group to be introduced is nonepimerizable and its stereochemistry can be controlled in the cyclopropanation step by a remote substituent (e.g., a hydroxyl group). In the synthesis of shionone (**S31**) (Ireland, 1975), the configuration of the methyl group at C_{13} is ultimately determined by the equatorial attack of dimethyloxosulfonium methylide on a methoxyoctalone, as shown in the following equation.

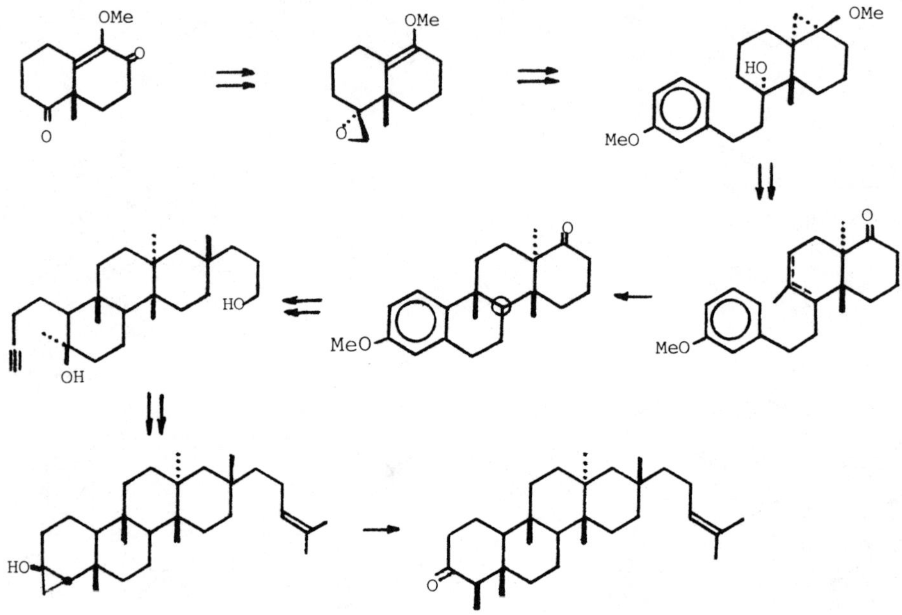

S31

Ring opening is particularly facile when the cyclopropane is β to a carbonyl group. Further assistance is furnished by the presence of an additional electron-withdrawing substituent on the small ring. Thus, the ketoester intermediate for trichodermin (**T24**) (Colvin 1971, 1973) is readily produced from the carbenoid adduct of a cyclohexene.

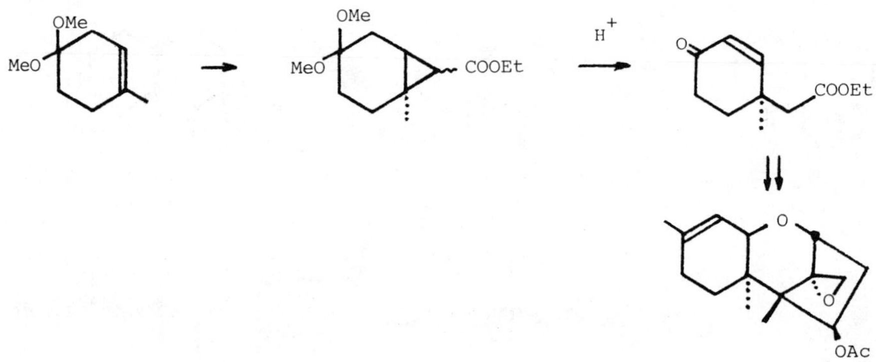

T24

12.1 CYCLOPROPANES

Fragmentation of two cyclopropane rings in one operation is featured in a synthesis of 9-pupukeanone (**P55**) (Schiehser, 1980).

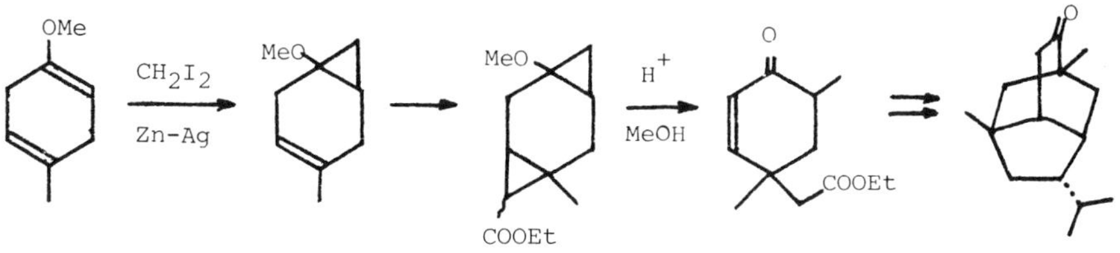

In a synthesis of pacifigorgiol (M. Martin, 1982) the isobutenyl chain is installed via solvolysis of a fused cyclopropylcarbinol which is derived from a dibromcarbene adduct.

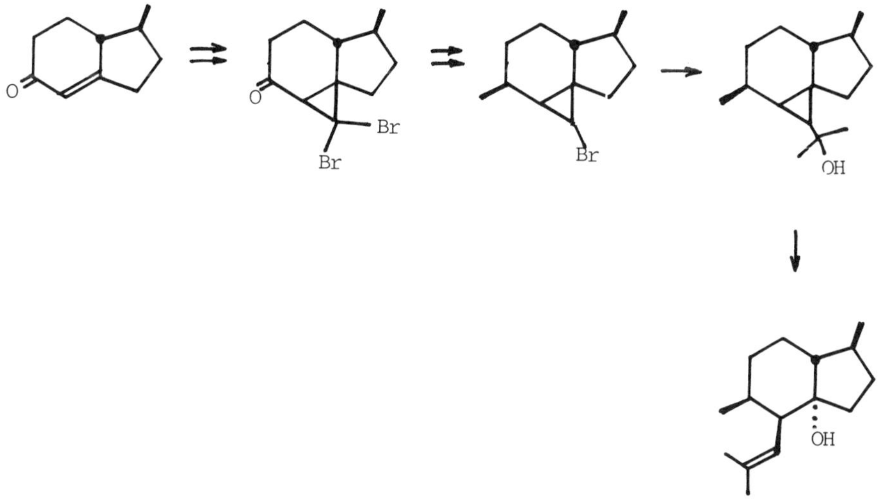

pacifigorgiol

Methylene transfer using diazomethane is now almost entirely superseded by the Simmons–Smith reaction. However, its appearance in a synthesis of patchouli alcohol (**P15**) (K. Yamada, 1979) is noted.

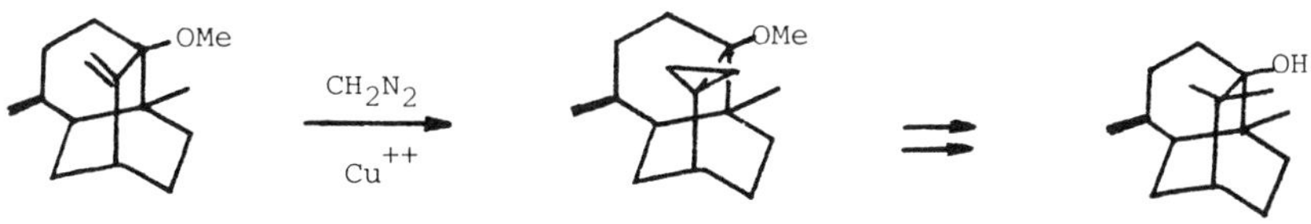

A synthesis of *threo*-juvabione (**J9**) (Morgans 1983) is based on a conjugate addition of a 2-methoxycyclopropylcopper reagent followed by ring cleavage. Diasteroselection to the degree of 5–6:1 may be achieved.

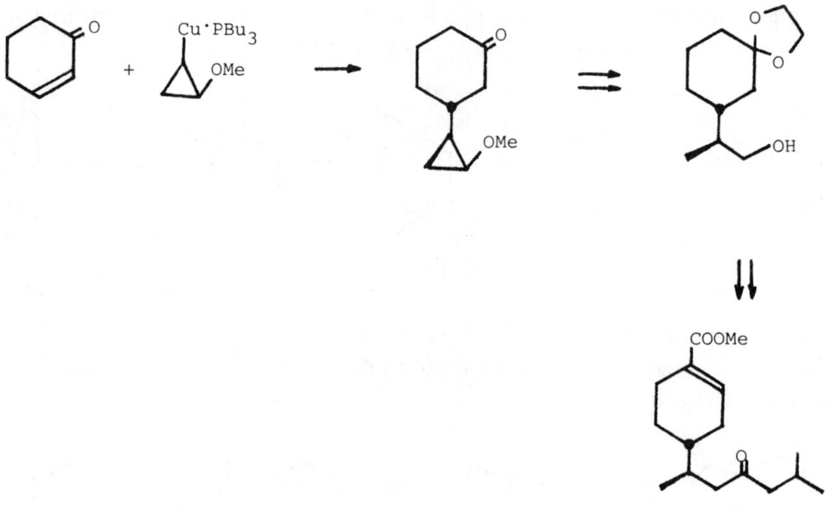

J9

The synthesis of α- and β-biotol (Grewal, 1987; Yates, 1987) patterns after a cedrol synthesis (Stork, 1961a) in the closure of the six-membered ring and sundry transformations.

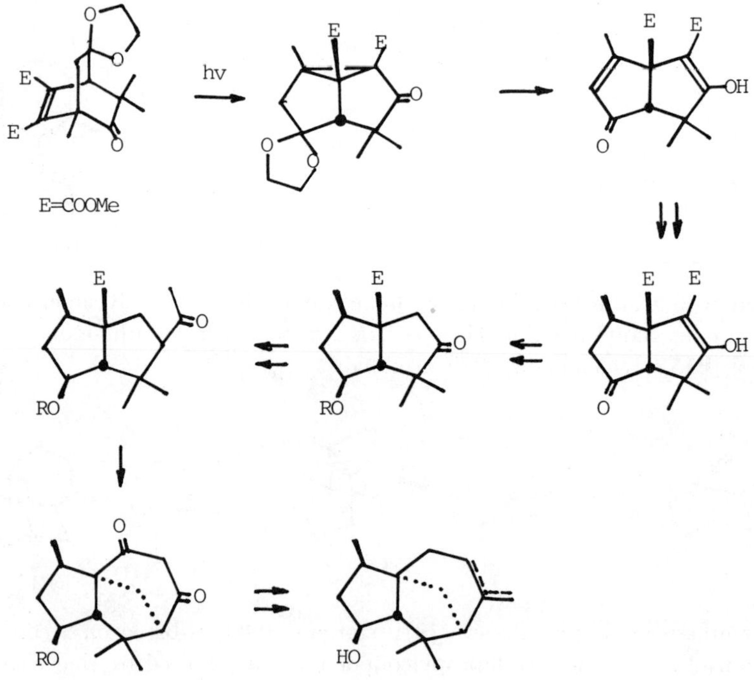

E=COOMe

biotol

12.1 CYCLOPROPANES

Reductive opening of a cyclopropanecarboxylate angularly fused to a diquinanone framework reveals an acetic chain for bridging the six-membered ring of quadrone (**Q1**). A ketone enolate is simultaneously created in a regioselective manner. Alkylation of the enolate provides the methylene group of the δ-lactone moiety (Iwata, 1985b).

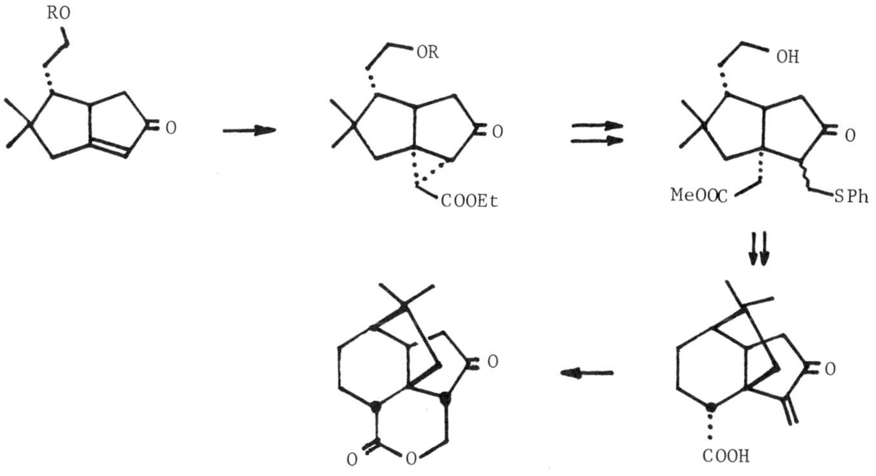

Q1

12.1.2 One-Carbon Insertion via Cyclopropanation; Ring Expansion

The methylene bridge of dihydrospiniferin I (**S46**) may be derived synthetically from a cyclopropane (Marshall, 1980). The structural characteristics of the molecule are such that a more accessible norcaradiene derivative is a surrogate of the cycloheptatriene tautomer.

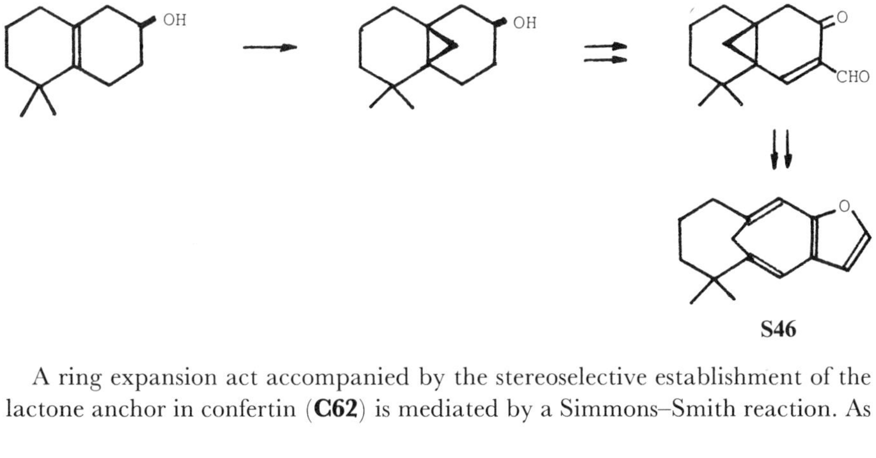

S46

A ring expansion act accompanied by the stereoselective establishment of the lactone anchor in confertin (**C62**) is mediated by a Simmons–Smith reaction. As a result of the concerted lactonization, the configuration of C_8 is actually determined by the hydroxyl group at the 5:6 ring juncture.

C62

One of the cedrol (**C38**) syntheses formally consists of inserting one- and two-carbon bridges in sequence to the α- and the β-carbon atoms of a cyclopentanone (Corey, 1973a). Thus, construction of the bicyclo[3.1.0]hexanone precursor with the aid of a Simmons–Smith reaction is followed by exposure of the compound to a powerful but nonnucleophilic *O*-acetylating agent (acetyl mesylate). That allows intramolecular interception of the tertiary carbenium by the double bond.

C38

Fused cyclopropanols with an angular hydroxyl group undergo ring expansion on exposure to bases to give the homologous ketones. This process has been used to advantage in converting 2-trimethylsiloxy-*exo*-dicyclopentadiene to hirsutene (**H17**) (Dawson, 1983). A skeletal rearrangement involving homoenolization is the crucial step of the synthesis.

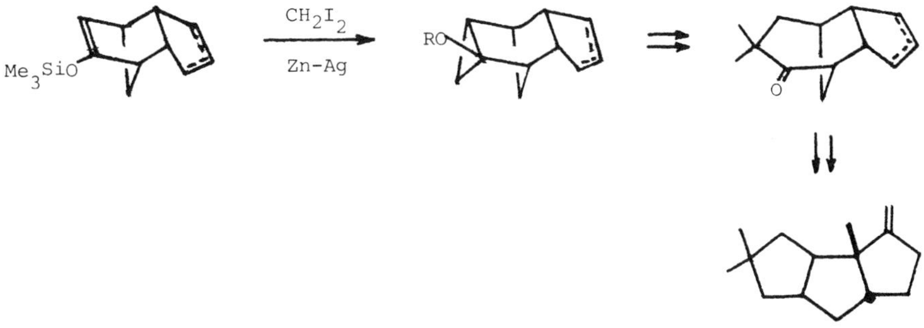

(A formal reductive homoenolization intervenes during treatment of 4-methylated Miescher–Wieland ketone with lithium in liquid ammonia. Here, the cyclopropanol cleavage actually accomplishes a ring contraction. The product is a useful precursor of β-vetivone (**V16**) (Subrahamanian, 1978).

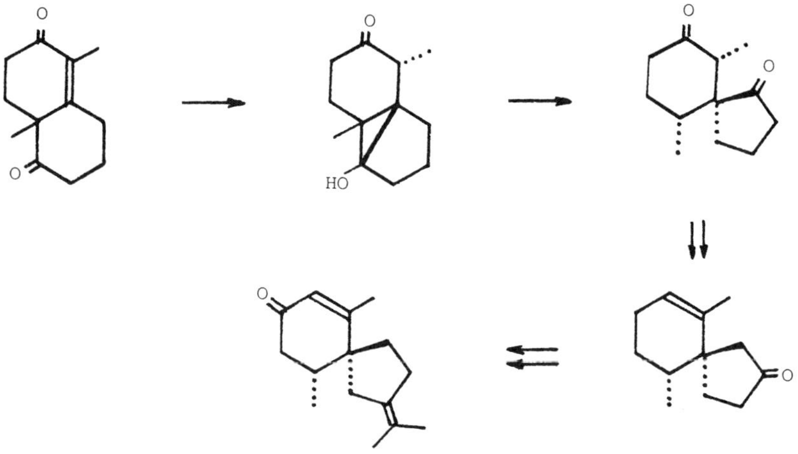

The methylchlorocarbene adducts of siloxyalkenes open readily to give β-alkyl methacroleins. The reaction sequence has found favorable employment in the synthesis of nuciferal (**N13**) (Blanco, 1981).

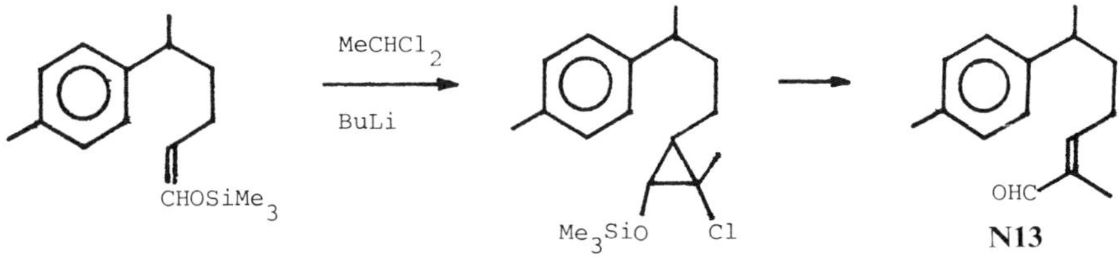

Based on the allene synthesis from *gem*-dihalocyclopropanes by treatment with carbon tetrabromide and methyllithium, a uniquely useful cyclononadienone for the synthesis of isocaryophyllene (**C33**) (Kumar, 1976) has been prepared from the abundant 1,5-cyclooctadiene.

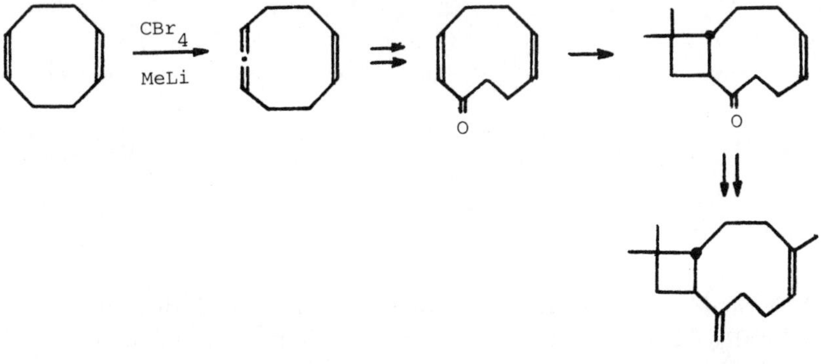

In a longifolene (**L24**) synthesis reported by McMurry (1972), the addition of dibromocarbene to a cyclohexene moiety precedes a mandatory expansion maneuver.

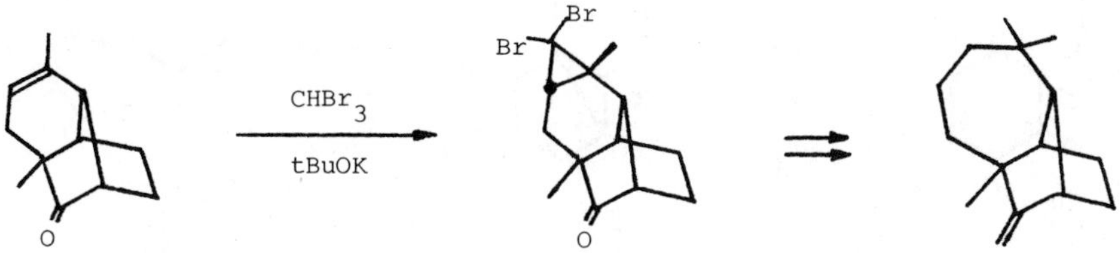

Perhaps the most unusual ring expansion to find synthetic applications is that involving breakage of the central bond of a bicyclo[1.1.0]butane. 1-(Arylsulfonyl)bicyclobutanes are available from γ,δ-epoxyalkyl aryl sulfones via a twofold intramolecular alkylation. They suffer ring cleavage on exposure to various organometallic reagents including LiAlH$_4$. Since the resulting cyclobutyl sulfones undergo alkylation and desulfurization readily, many cyclobutanes, including junionone (**J7**) and citrilol acetate (**C57**) may be synthesized (Gaoni, 1985).

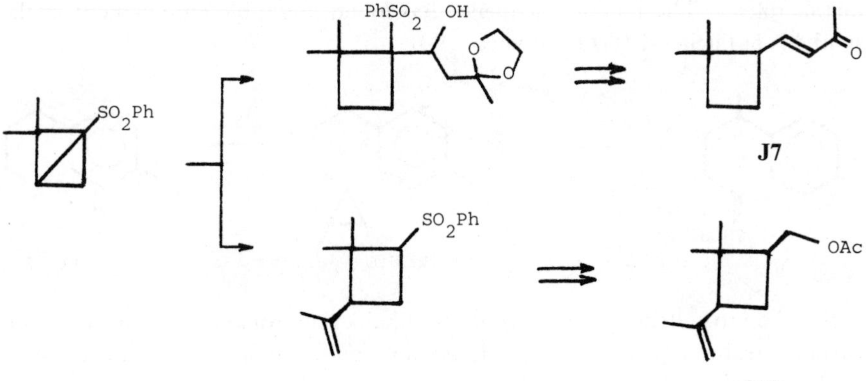

12.1.3 Indirect Alkylation Through Cyclopropanation

Although it seems uneconomical to achieve alkylation via formation and cleavage of cyclopropanes, the reaction sequence can become competitive when the individual processes are sufficiently efficient. More important, the cyclopropanation route may actually curtail synthetic operations by yielding products with polar substituents in the "wrong" positions. In other words, a desired umpolung in reactivity may be accomplished. (The following paragraphs offer examples in which compounds containing oxygen functions are separated by an even number of atoms.)

Spirocyclic systems may evolve from the regioselective cleavage of condensed rings interlocked by a cyclopropane nucleus. These tricyclic compounds are frequently encountered because they are easily assembled by (*a*) intramolecular carbenoid addition to alkenes, and (*b*) photochemically induced rearrangement of cross-conjugated cyclohexadienones.

The carbenoid addition–cleavage sequence has been employed in the syntheses of α-chamigrene (**C46**) (White, 1974; Iwata, 1979), (—)-acorenone B (**A10b**) (Ruppert, 1976), and agarospirol (**A17**) (Mongrain, 1970). Depending on the substitution pattern and the ultimate need of the synthesis, either reductive cleavage or acidolysis of the cyclopropyl ketone products may be enacted.

As expected, the cyclohexadienols derived from intramolecular reaction of acylcarbenoids and phenols are very unstable, they undergo cleavage *in situ* (see α-chamigrene synthesis, **C46**: Iwata, 1979).

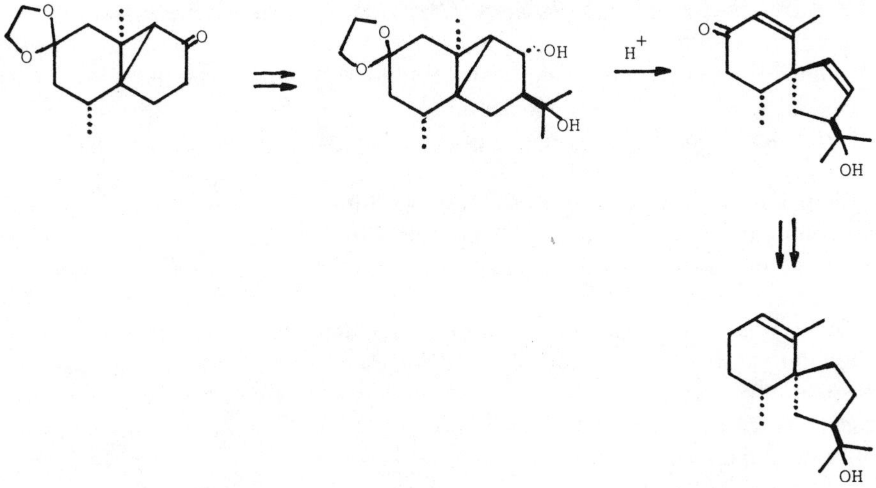

A17

In a synthesis of pentalenene (**P16**) (Imanishi, 1986b) the formation of a diquinane nucleus relies on the carbenoid insertion-reductive cleavage sequence. The third ring is constructed by intramolecular alkylation.

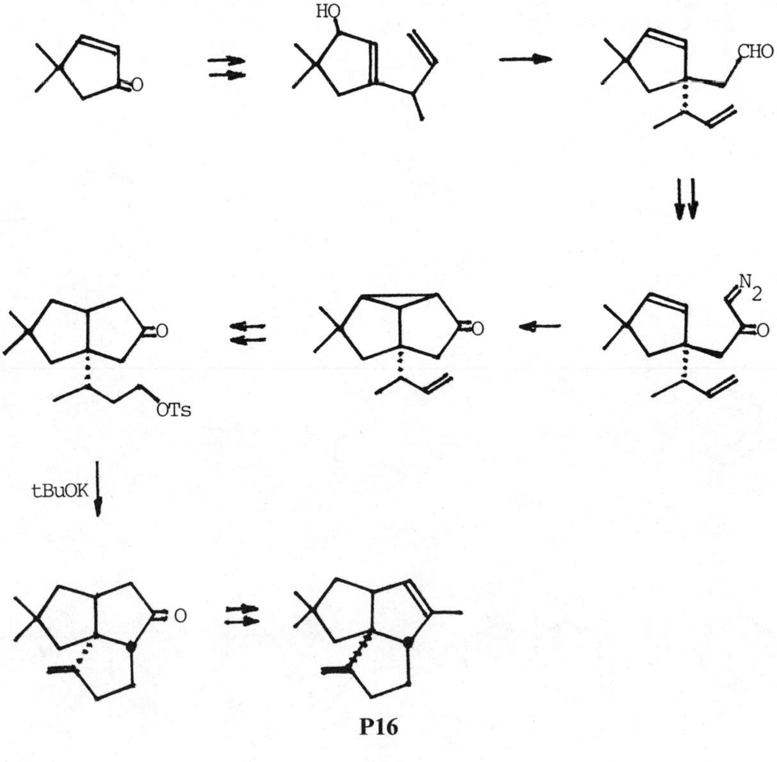

P16

12.1 CYCLOPROPANES

The synthesis of solavetivone (**S41**) (Iwata, 1981) from phenol alkylation is interesting in that reduction of the dienone proceeds stereoselectively with lithium in liquid ammonia. The hydroxyl group acts as an internal proton source.

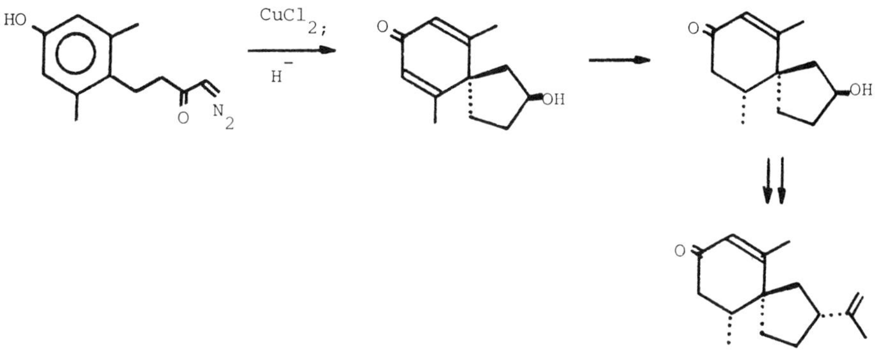

The same method has been used in a synthesis of the rearranged brominated chamigrane (Iwata, 1987) isolated from red algae.

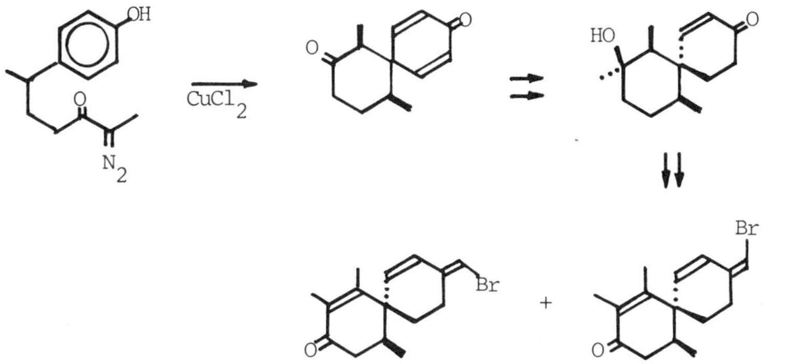

It is not surprising that a synthesis of spirolaurenone (**S47**) (Murai, 1982a) is based on an intramolecular carbenoid cycloaddition. It is important to note also that a potential complication accompanies this approach because the cyclohexene ring contains an asymmetric center. However, a high degree of stereoselectivity in the desired sense is attained thanks to the *gem*-dimethyl substituent. The transition state in which the diazoketone side chain interacts strongly with the axial methyl branch is destabilized.

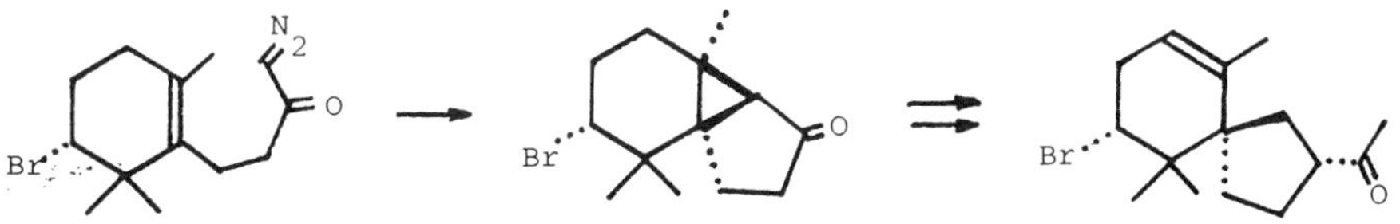

In the conversion of abietic acid into kaurene (**K4**) and phyllocladene (**P27**) (Tahara, 1973), the key steps are an intramolecular ketocarbene addition to an alkene linkage and the reductive cleavage (Li–NH$_3$) of the three-membered ring.

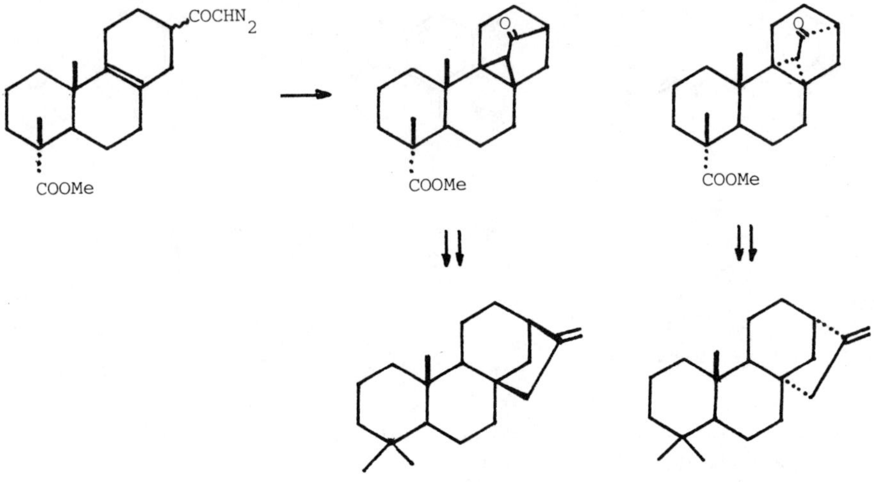

The conversion of gibberellin A$_7$ into antheridic acid (= antheridiogen An, **A35**) (Furber, 1987) involves a 1,2-bond shift via a cyclopropyl ketone intermediate.

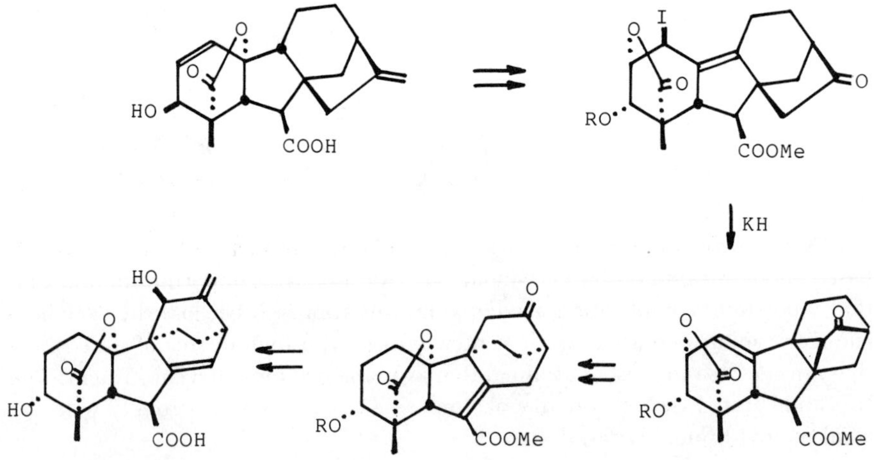

Cyclopropyl ketones derived from diazoketones and alkoxyalkenes are versatile intermediates for 1,4-dicarbonyl compounds and hence cyclopentenones. Synthons for α-cuparenone (**C85**) (Wenkert, 1978b) and acorone (**A11**) (Wenkert, 1978b) are readily available accordingly.

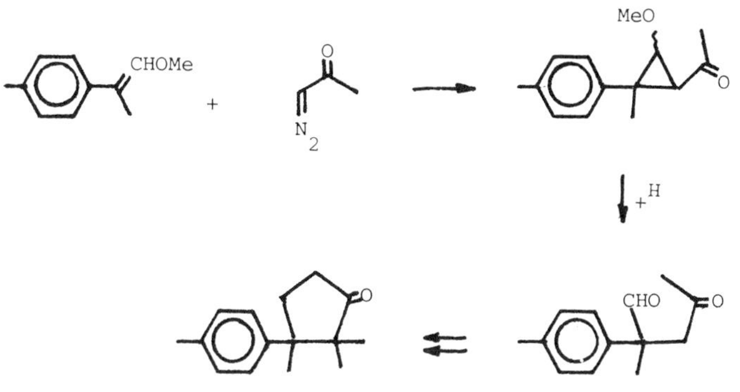

12.1.4 Preparation of γ-Functionalized Carbonyl Substances

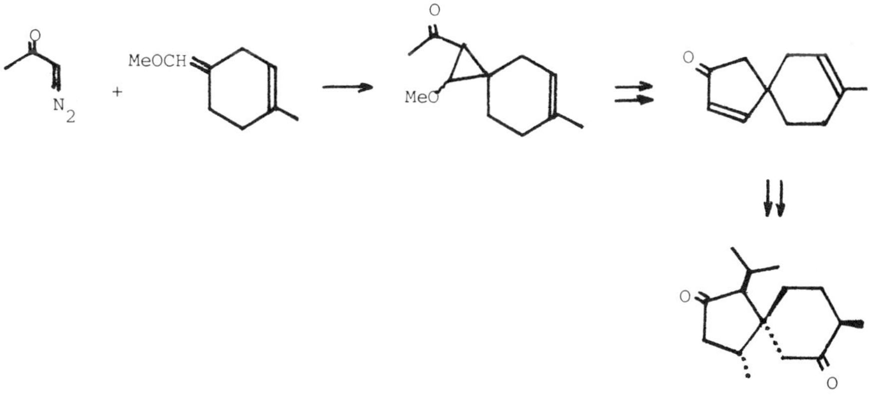

This section is an extension of the preceding one and it deals specifically with the introduction of an additional substituent to the γ position of the carbonyl group by virtue of the conjugated cyclopropane unit.

Tricyclo[3.3.0.0^{2,8}]octan-3-one derivatives are valuable building blocks for the synthesis of cyclopentanoids in view of the stereocontrol that is possible during maneuvers on the periphery. Furthermore, these substances are readily available from (2-cyclopenten-1-yl)methyl diazomethyl ketones.

The use of the tricyclic ketones in the syntheses of chrysomelidial (**C54**) (Kon, 1980), forsythide aglycone dimethyl ether (**F10**) (Takemoto, 1982), loganin (**L20**) (Kon, 1983), isoiridomyrmecin (**I9**) (Callant, 1981), and (+)-iridodial (**I5**) (Ritterskamp, 1984) is shown below.

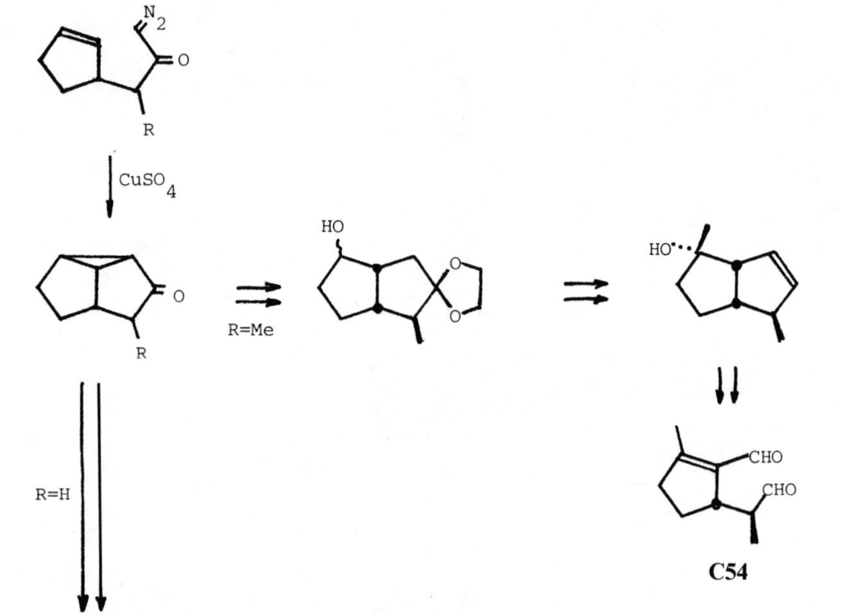

12 TRANSITORY ANNULATION

C54

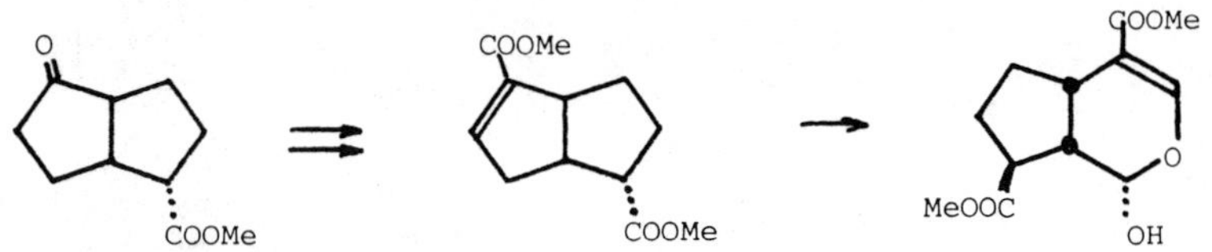

F10

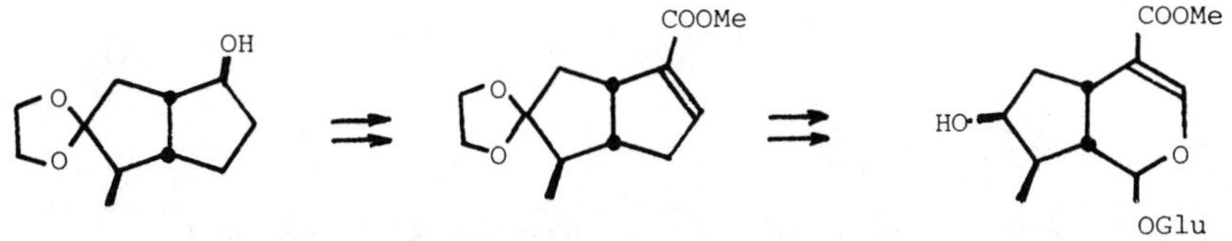

L20

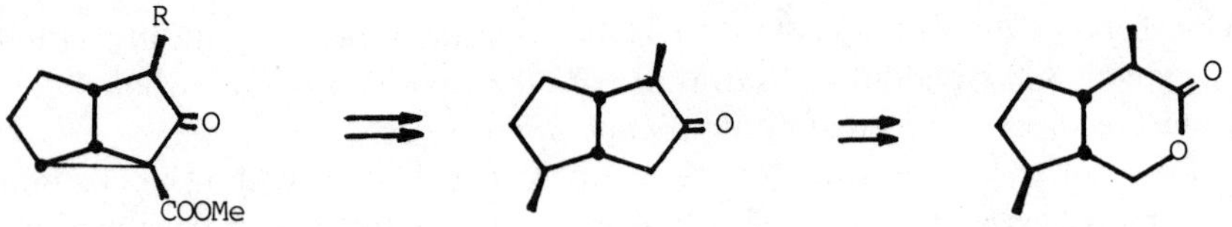

I9

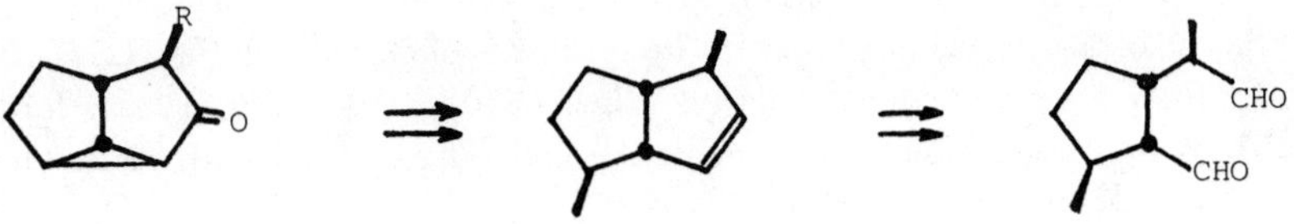

I5

12.1 CYCLOPROPANES

The versatility of these intermediates is further illustrated in an alternative synthetic application to the loganin-type substances such as verbenalol (**V9**) (Callant, 1981). Whereas in the loganin synthesis the preexisting five-membered ring eventually is cleaved, the verbenalol synthesis calls for a transitory annulation.

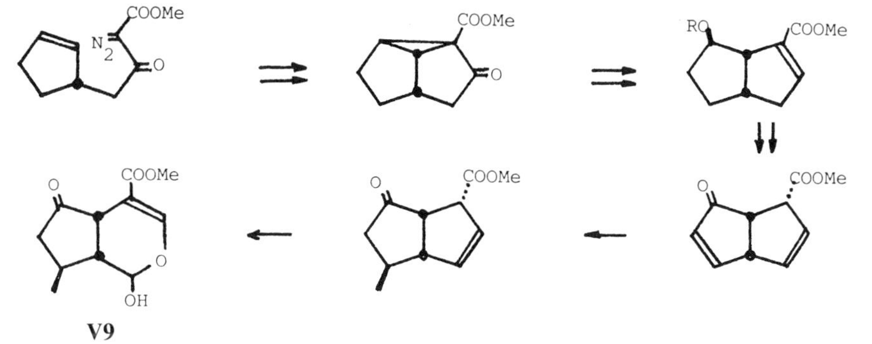

V9

The stereoselective cyanation of a bicyclo[3.1.0]hexanonecarboxylate is the key to establishing the correct stereorelationship of the two chiral centers of dehydroiridodiol (**I7**). The functional group distribution in the monocyclic product makes the completion of the synthesis rather straightforward (Nakayama, 1983).

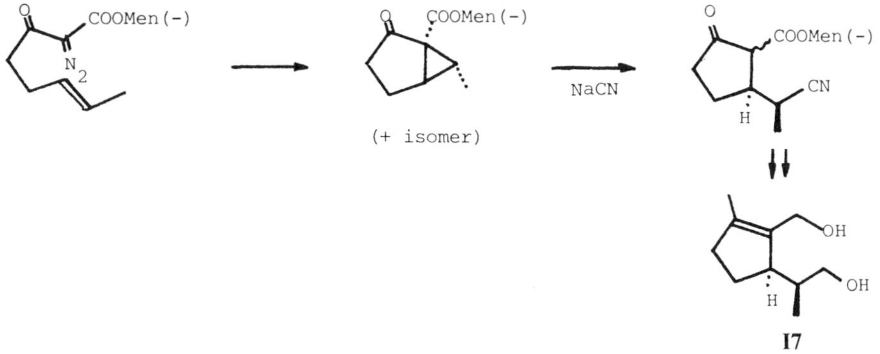

I7

The oxabridged tricyclic compound obtained from a diazoacetyldihydropyran undergoes ring opening together upon addition of a methyl group to give the oxo derivative of 1,8-cineole (Adams, 1986).

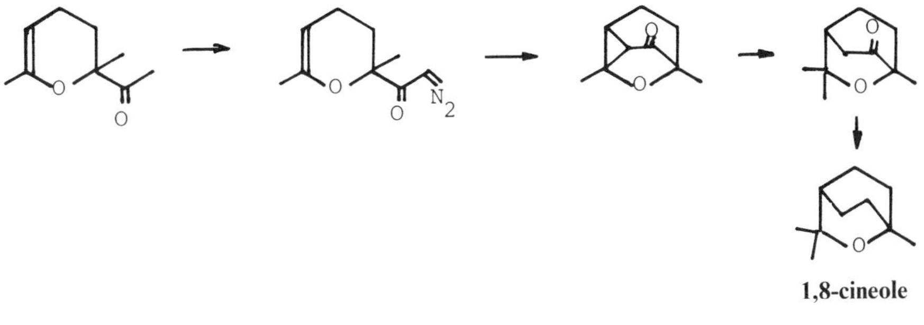

1,8-cineole

The seco-γ-functionalization sometimes may be performed in a stepwise manner. Thus, in a synthesis of hop ether (**H20**)(C. Johnson, 1982) a thermal cycloreversion of the bicyclic ketoester is achieved and a hydration of the isopropenyl group is effected later.

Solvolysis of cyclopropyl ketones generate 1,4-bifunctional compounds. This process has special ramifications in synthesis of guianolides such as grosshemin (Rigby, 1987) which establishes not only part of the oxygenation pattern, but also three contiguous chiral centers. Elaboration of the *cis*-perhydroazulene framework by this method from tropone paves an uneventful pathway toward the terpene molecule; the intermediate embodies well-defined functionalities that minimal number of steps suffice to convert it into the target.

grosshemin

In several vernolepin (**V10**) syntheses, the δ-lactone unit is constructed by linking the α carbon to an existing six-membered carbocycle. One of these (Zutterman, 1979) involves acylcarbene addition and benefits from the opportunity of functionalizing C_6 from the α side. In another approach (Isobe, 1978; Iio, 1979a, b), this peripheral cyclopropane bond spanning C_4 and C_6 is intentionally created via a bromination–dehydrobromination sequence and it is maintained over one-third of the synthetic course to ensure regioselective transformations. Its eventual cleavage serves to introduce an additional chiral center, which in turn makes possible an orderly attachment of the remaining substituents on the carbocycle.

A synthesis of descarboxyquadrone (Imanishi, 1986a) involves intramolecular cyclopropanation and acid-catalyzed ring opening of the cyclopropane. The regioselectivity of the latter reaction is higher when using a halide ion bulkier than formic acid, but conversion of the bicyclic halides into a carbonyl derivative has not been attempted. The unusual preference for nonperipheral bond scission is due to the presence of the *gem*-dimethyl group, which hinders the approach of the nucleophile to the alternative β-carbon atom.

This route is liable to modification such that quadrone (**Q1**) may be prepared. α-Carboalkoylation of the cyclopropyl ketone before the solvolytic ring cleavage is an obvious route.

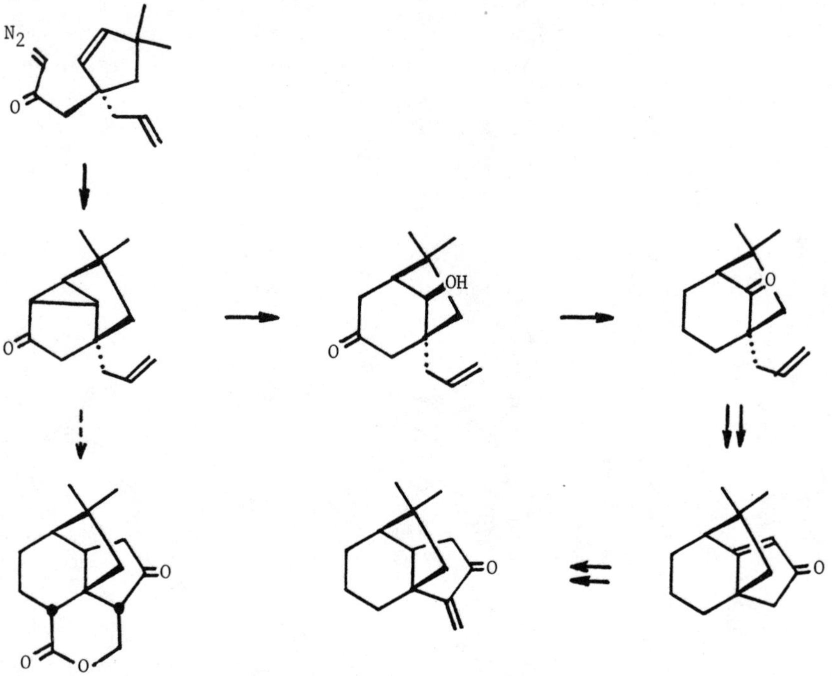

Q1

In a synthetic approach to modhephene (**M12**) reported by Wrobel (1983), a maneuver involving insertion of an α-ketocarbenoid into a remote C—H bond in the [3.3.3]propellane framework has a dual purpose. Cyclopropanation not only blocks one of the methylene groups from methylation, but also activates the γ-carbon atom so that the secondary methyl group may be added stereoselectively with reopening of the three-membered ring.

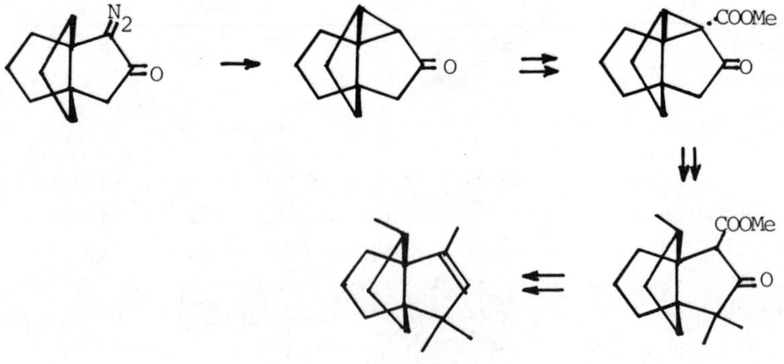

M12

12.1 CYCLOPROPANES

The enantiomer of (—)modhephene (**M12**) has been acquired via a Simmons–Smith reaction of a diquinane ketal followed by cleavage of the cyclopropane ring with iodotrimethylsilane, chain elongation at the angular substituent and further elaboration of this intermediate (Mash, 1987b). Consequently the previously assigned absolute configuration of this terpene must be revised.

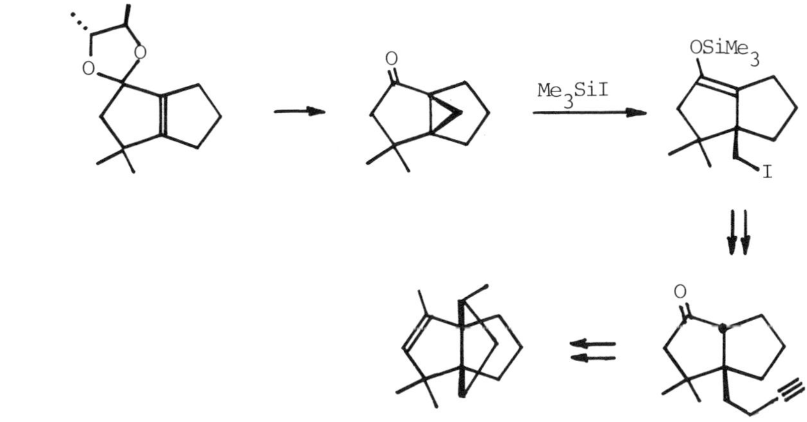

Ent-M12

A synthesis of (—)-chokol A (Mash, 1987a) is also based on similar operations. The remaining steps after bromolysis of the three-membered ring consist of mainly chain elongation.

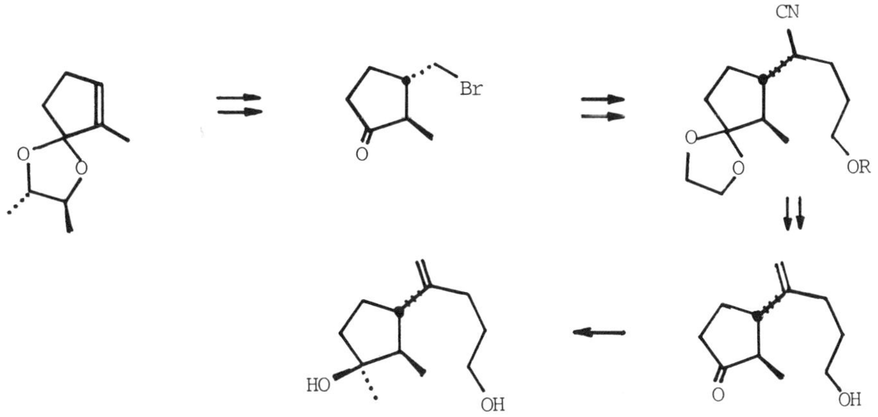

chokol A

12.1.5 Ring Contraction via Bicyclo[3.1.0]hexanones

A combination of carbocycle partition and allopatric cleavage of the bicyclic product constitutes a ring contraction operation. The usefulness of such an unorthodox protocol depends on the availability of efficient methods.

The fascinating photochemistry of cross-conjugated cyclohexadienones has its roots in the sesquiterpene α-santonin (**S9**). The structural elucidation of the various products and their correlation with reaction conditions have enabled synthetic planning for related systems. The transformation of the cyclohexadienones into cyclopropanated cyclopentenones is easily accomplished, and the latter substances can undergo cleavage at the cyclopropane periphery to yield cyclopentenones. Conceivably, spirocyclic terpenes are accessible based on these observations.

Sesquiterpenes that have yielded to synthesis featuring the cyclopropane fission of the "lumi" compounds include β-vetivone (**V16**) (Marshall, 1968c, 1970a,; Caine, 1976b), α-vetispirene (**V14**) (Caine, 1976a), and *ent*-axisonitrile-3(**ent-A55**) (Caine, 1978).

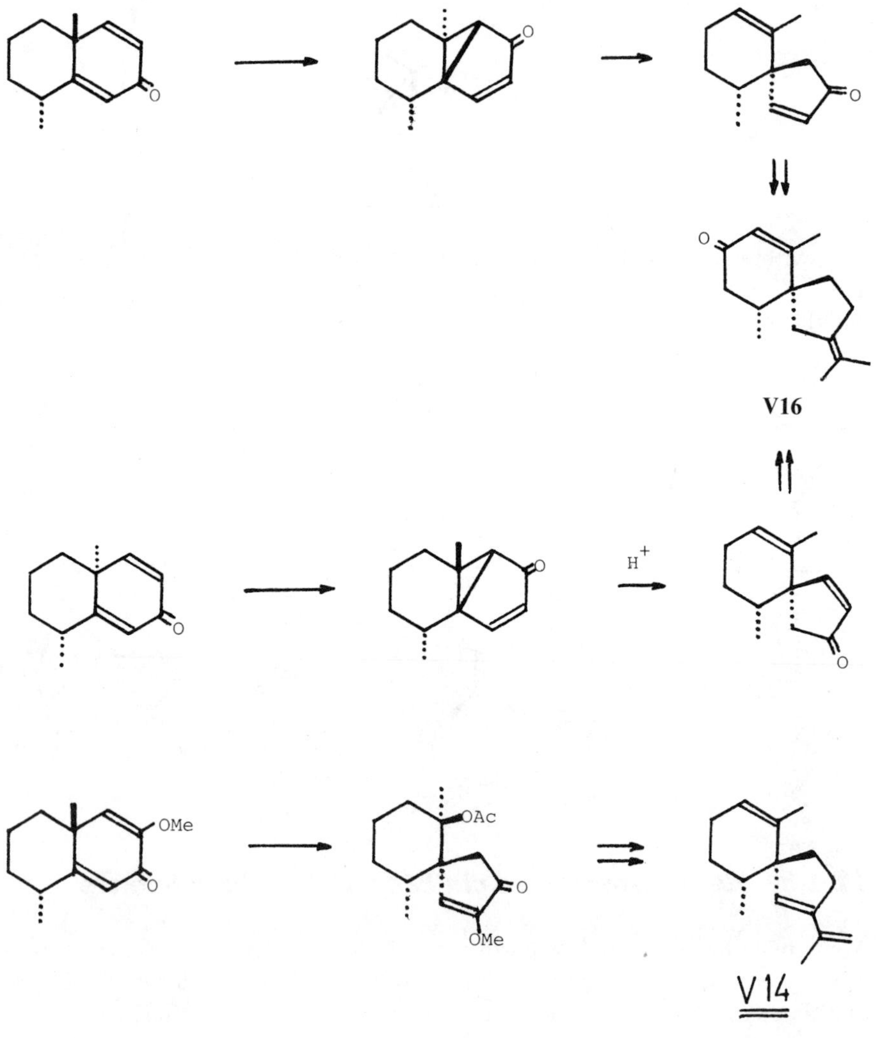

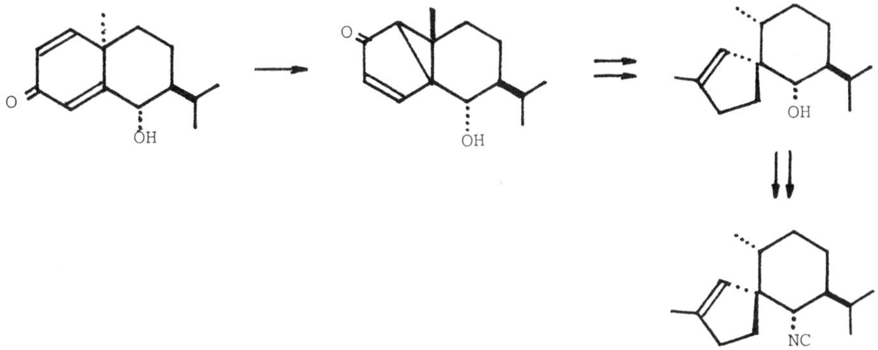

12.1 CYCLOPROPANES

ent-A55

A popular and useful photochemical process is the (oxa) di-π-methane rearrangement, whereby new carbon frameworks unfold. Its applications to syntheses of cedrene (**C37**) (K. Stevens, 1980), modhephene (**M12**) (Mehta, 1985b), and coriolin (**C70**) (Demuth, 1984) show that cyclopropyl ketone products can undergo selective alkylation and homoconjugate addition.

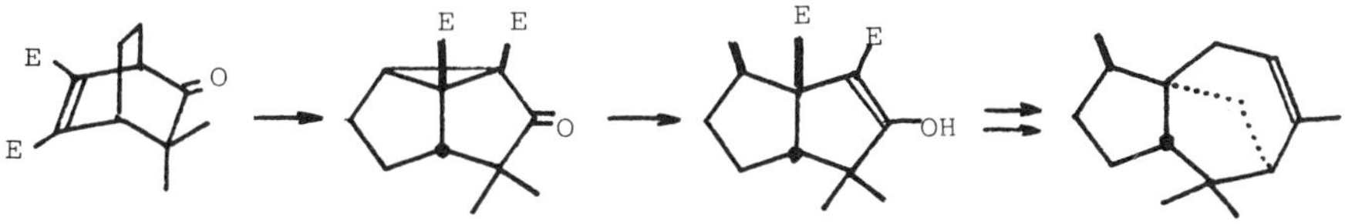

E = COOMe

C37

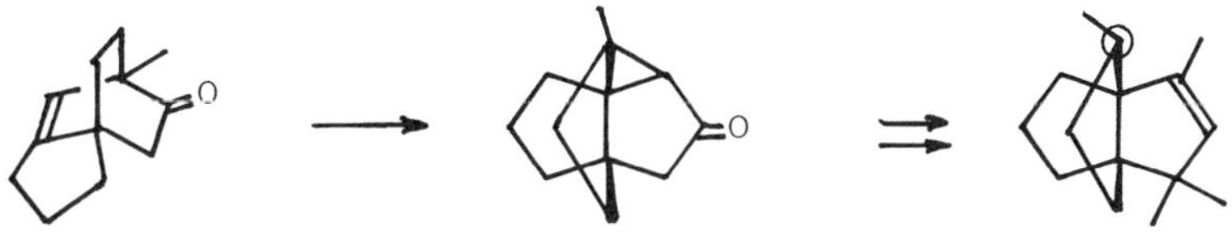

M12

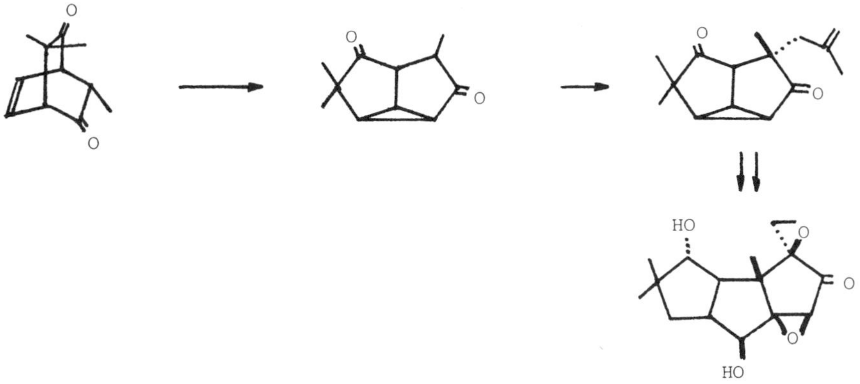

C70

A most interesting transformation during the synthesis of (+)-steviol (**S53**) (Mori, 1965, 1970b) is a Clemmensen reduction of a bridged 1,3-diketone. A cyclopropanediol is formed and then cleaved to give the norketone of steviol, accompanied by a smaller amount (41 % vs. 19 %) of the undesired tetracyclic product as a result of the alternative mode of cleavage. The preferential formation of the steviol skeleton reflects a lesser steric compression of the angular methyl group with the methano bridge than with the oxoethano bridge.

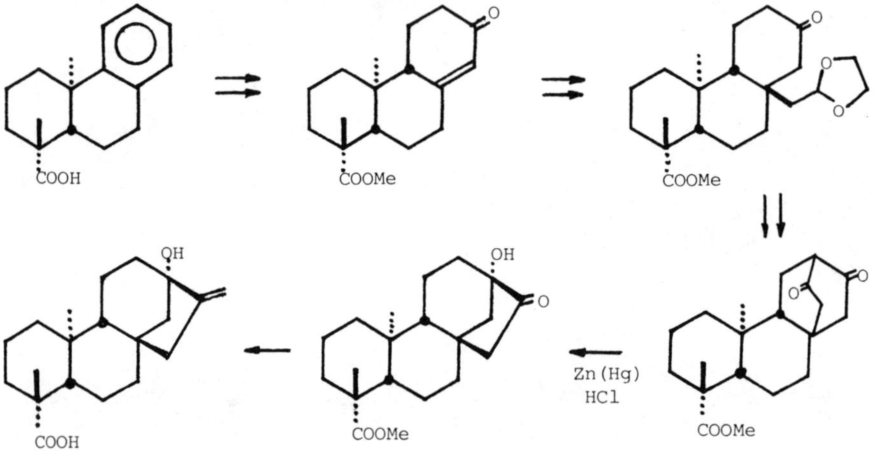

12.1.6 Cyclopropanocycloalkanone Synthons for Stereocontrol

In a synthesis of eremophilone (**E13**) reported by Ziegler (1977c) consisting of B-ring annulation, the stereochemical issue associated with the attachment of an isopropenyl pendant to a freely rotating side chain was addressed by a temporary asymmetrization of C_6 by the incorporation of this atom into a fused cyclopropane.

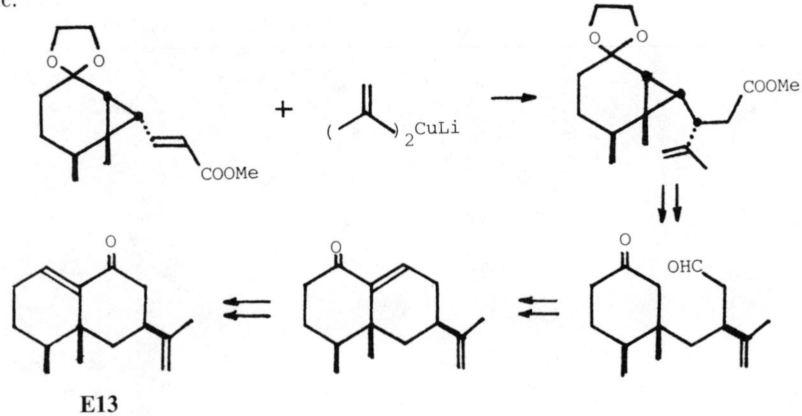

12.1 CYCLOPROPANES

Stereoselective isopropenylation by conjugate addition to an unsaturated ester appears to be dictated by the preferential adoption of a conformation in which nonbonding interactions are minimized while orbital overlap of the unsaturated system with the cyclopropane is maintained.

Strictly speaking, the use of natural bicyclic materials to synthesize bicyclic compounds of other types via tricyclic intermediates does not involve transitory annulation. However, inclusion of such work here is not only relevant, indeed, sometimes it might be more convenient to acquire these "natural" compounds by partial synthesis, as in the case of 2-caranone, which is readily prepared from dihydrocarvone.

The extra ring, often three- or four-membered, usually provides a steric shielding to one side of the main plane of the molecule, and external reagents are kept away from that side. The syntheses of (+)-β-cyperone (**C97**) (Kutney, 1980) and (—)-β-elemol (**E4**) (Kutney, 1982a) from (—)-thujone are two examples demonstrating the value of stereocontrol by a fused cyclopropane.

Analogously, *ent*-nootkatone (**ent-N11**) is reachable by processing (—)-sabina-ketone through the Robinson annulation/cyclopropane cleavage sequence (van der Gen, 1971b).

It is recognized that Robinson annulation of dihydrocarvone affords predominantly epi-α-cyperone on a stereoelectronic basis. In the case of 2-carone, the Michael acceptor can verge only from the side opposite the existing cyclopropane ring. Consequently, the proangular methyl substituent and the cyclopropane, (hence the isopropenyl group) in the adduct, are destined to be cis. α-Cyperone (**C95**) is ultimately produced (Caine, 1974). (Cf. a synthesis of dihydronootkatone from 2-carone reported by Caine, 1976c).

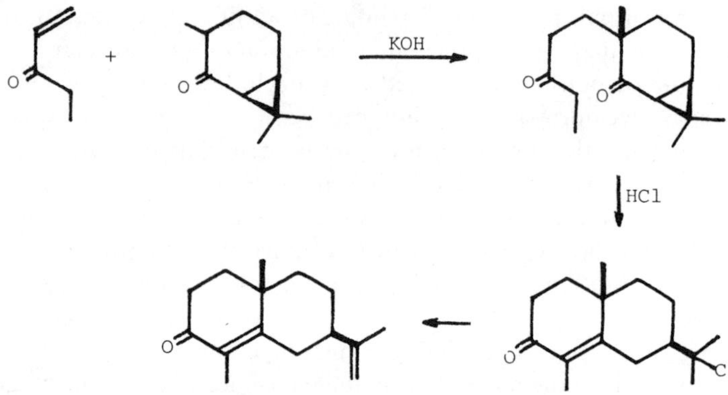

C95

Although phytuberin (**P28**) contains only one carbocycle, it proves profitable to use 2-carone as a building block (Caine, 1980).

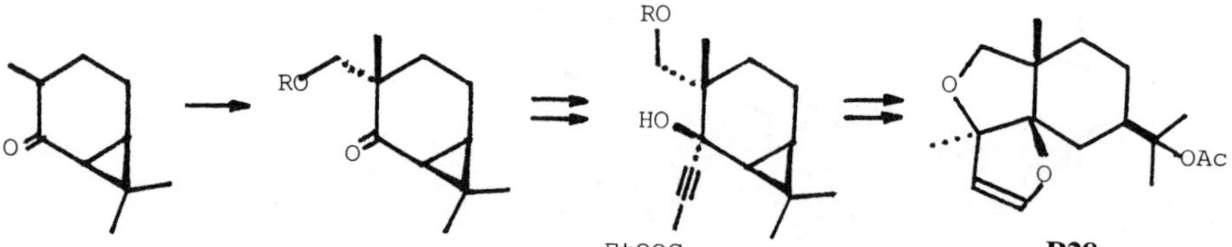

P28

12.1.7 Chain Building Based on Cyclopropanes

Chain lengthening by five carbon atoms to give a functionalized alkene homolog can be achieved using acetylcyclopropane as a synthon. This stereoselective reaction sequence is particularly useful in the synthesis of acyclic terpenes such as nerolidol (**N7**) (M. Julia, 1960).

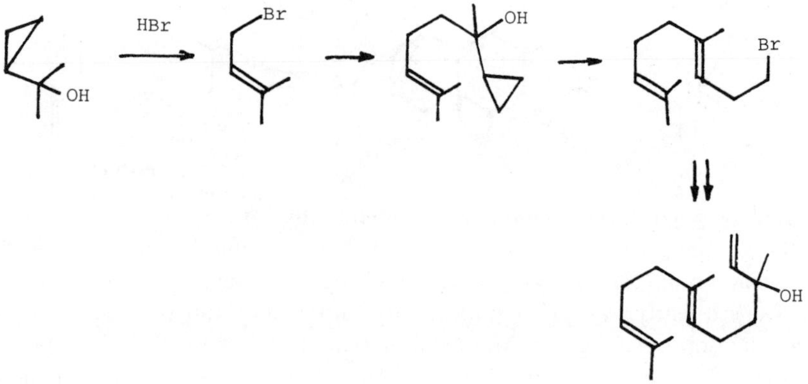

N7

Cyclopropenes are susceptible to attack by certain Grignard reagents; the resulting cyclopropylmagnesium halides may be trapped with ketones and aldehydes to furnish cyclopropylcarbinols. The use of the latter substances in the synthesis of alkenes including hotrienol (**H21**) and santolina alcohol (**S8**) (Moiseenkov, 1982) via solvolysis is easily conceived.

12.2 CYCLOBUTANES

The usefulness of cyclopropane-containing monoterpenes as synthons for sesquiterpenes was indicated above. Research has also been directed toward using the most abundant pinenes.

Nopinone is obtained from β-pinene by oxidative cleavage of the exocyclic methylene group. The presence of a bridged, asymmetric cyclobutane ring enables nopinone to undergo diastereofacially discriminative reactions. A synthesis of nootkatone (**N11**) (Yanami, 1979) has exploited this steric influence. The cyclobutane is eventually broken at the desired position by virtue of carbonyl activation on one side.

The C-alkylation of olivetol both with (—)-verbenol to produce (—)-Δ^6-tetrahydrocannabinol (**C17**) (Mechoulam, 1967, 1972) and with (+)-chrysanthenol to afford (—)-Δ^1-tetrahydrocannabinol (**C16**) (Razdan, 1975) is accompanied in each case by the formation of an O—C bond with opening of the four-membered ring in the substrates. [Cf. three-membered ring opening in the condensation of olivetol with (+)-3-carene oxide (Montero, 1976) and (+)-2-carene oxide (Razdan, 1970).]

A formal synthesis of β-elemenone (**E3**) (Ho, 1985) involving a conjugate vinylation of verbenone and fragmentation of the four-membered ring also takes advantage of the steric bias presented by the monoterpene.

Benzocyclobutenones have become very readily available in recent years, and an extensive exploitation of these substances as sophisticated synthons is continuing. An example in which a benzocyclobutenone acts as an acylating agent is found in a synthesis of taxodione (**T7**) reported by R. Stevens (1982).

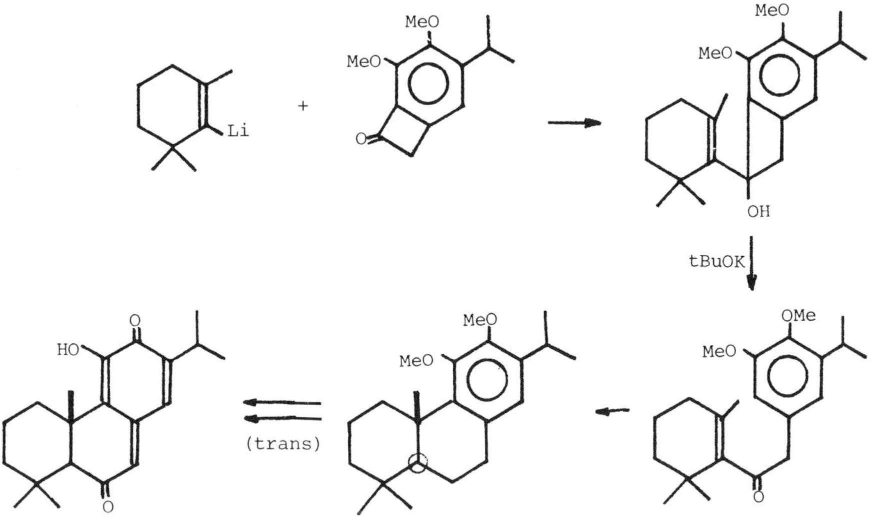

12.2 CYCLOBUTANES

T7

An interesting indirect intramolecular acylation intervenes during a synthesis of aphidicolin (**A36**) (Ireland, 1981, 1984). This important process, which establishes the complete skeleton of the diterpene (with functional group introduction and the A/B-ring juncture yet to be accomplished), involves a ketene cycloaddition and cleavage of the ensuing cyclobutanone in a desilylative reaction. The ketene is generated by a photolytic ring contraction of an α-diazoketone.

The unidirectional ring scission is gratifying in view of the local symmetry. It must be attributed to strain relief achieved upon removing the 1,3-diaxial interaction between the angular methyl group and the carbonyl bridge; the protonation of C_8 at the β side is in agreement with this rationale.

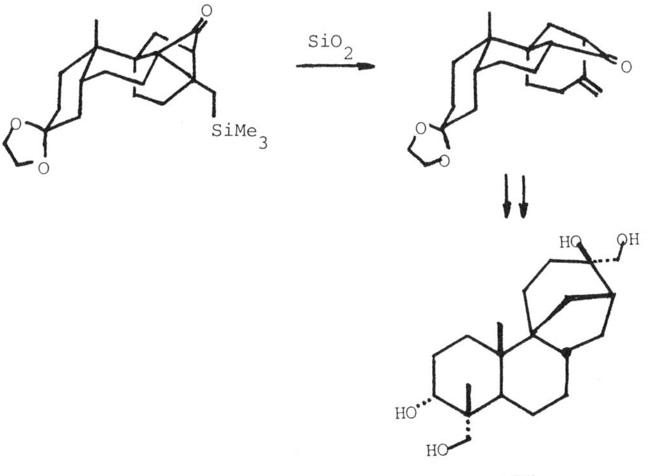

A36

The stereocontrolled synthesis of the very complex diterpene ginkgolide B (Corey, 1987b) represents a monumental advance in the terpene field. After introduction of the *t*-butyl group to a cyclopentane nucleus a latent spiroannulated γ-lactone is introduced. Next, a 3*Z*-pentenoic acid chain is annexed to set stage for a ketene/alkene cycloaddition that leads to a fused cyclobutanone and thence the second γ-lactone subunit. Oxygenation of the angular position adjacent to the lactone carbonyl and its bridging to the latent lactone establish a large portion of the target molecule. The cyclopentene now undergoes oxidation, chain grafting and participates in the formation of the linearly annulated lactone. Final unraveling of the dihydrofuran (latent lactone) culminates in the access of the terpene.

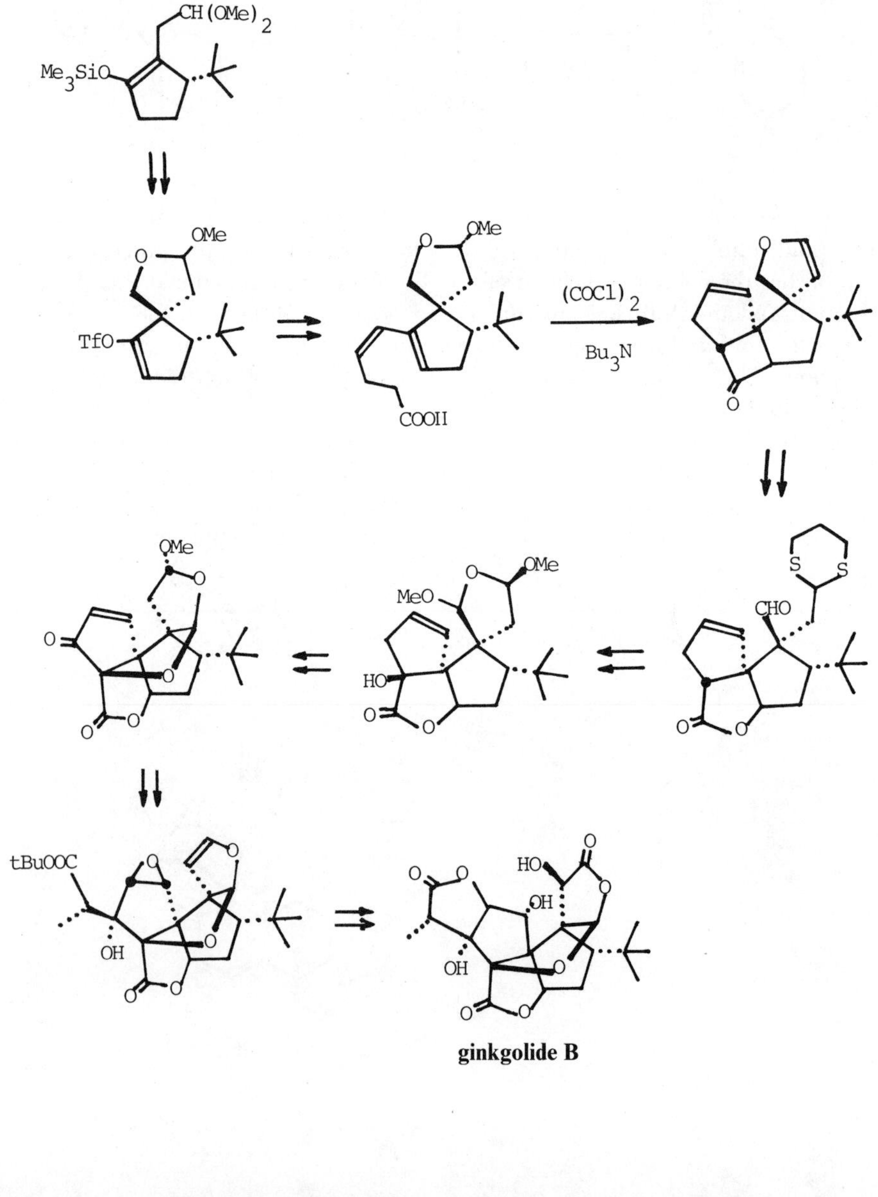

ginkgolide B

12.2 CYCLOBUTANES

A significant synthesis of the complex iridoids including plumericin (**P41**) and allamandin (Trost, 1986) from a readily available oxopentalene has been achieved along with development of several new methods. The stereocontrolled ring expansion of the cyclopropanol to a cyclobutanone with concurrent transposition of an allylic double bond, the hydroxyalkylation of γ-lactones via α,α-disulfenyl derivatives, and the indirect β-carboalkoxylation of dihydropyrans are highlights of this accomplishment.

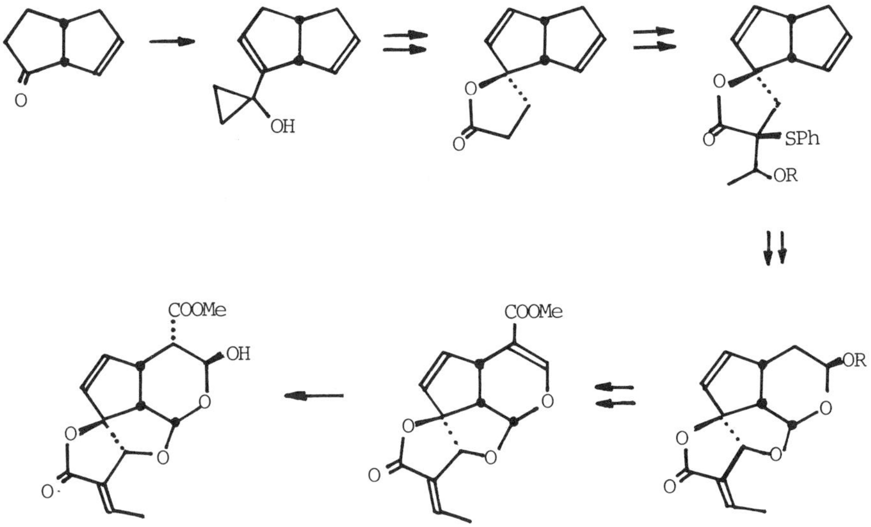

allamandin **P41**

The 1:1 photocycloadducts of piperitone and allene undergo isomerization on exposure to boron trifluoride etherate to give the 2-isopropenylcyclohexenone. This compound is a convenient precursor of shyobunone (**S32**), isoshyobunone (**S32i**) and their epimers (Weyerstahl, 1987).

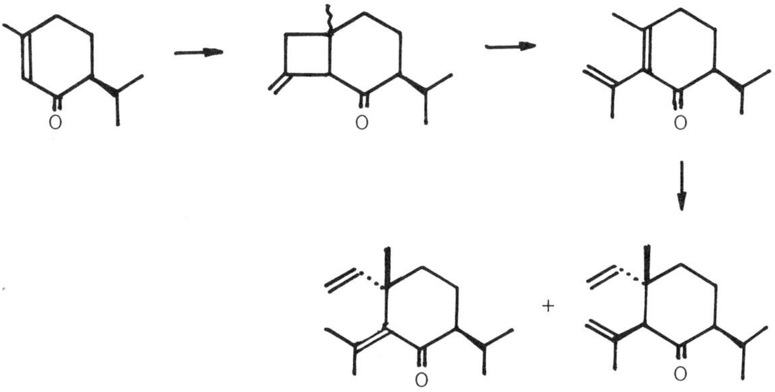

S32i **S32**

12.2.1 Reductive Alkylation of Enones via [2 + 2]Cycloaddition

Although the introduction of an alkyl group to the β position of an α,β-unsaturated ketone can be easily accomplished using organocopper reagents (Posner, 1972), alternative methods involving [2 + 2]cycloaddition are sometimes advantageous in terms of reagent availability and stereochemical consequences. Thus, a major difficulty pertaining to the synthesis of the insect juvenile hormones called juvabiones is in the stereoselective establishment of the side-chain asymmetric center(relative to the adjacent carbon in the cyclohexene ring). Of the several solutions, those using cyclic intermediates are most effective. For example, the thermal adduct of 2-cyclohexenone and 1-(N, N-dimethylamino)propyne is hydrolyzed to give a carboxylic acid whose configurations correspond to racemic *erythro*-juvabione (**J8**) (Ficini, 1974). Kinetic protonation of the bicyclic enamine, which occurs from the convex face (i.e., syn to the angular hydrogen atom), is followed by rapid ring opening.

Similarly, a facile synthesis of isodihydronepetalactone (**N6'**) (Ficini, 1976) has been accomplished.

Treatment of ω-methylenedihydro-β-ionone with tin(IV) chloride gives rise to a tricyclic ketone as the major product. The reaction course is undoubtedly stepwise. The methylcarbinol derived from this ketone undergoes fragmentation to furnish α- and β-chamigrenes (**C46** and **C47**) (Naegeli, 1981) in addition to three other tricyclic olefins.

C46, C47

Photocycloaddition of enones with unsaturated compounds forms four-membered carbocycles. When addends containing adequate latent functionalities are employed, cleavage of the four-membered ring in the adducts can be achieved. Since allene adds to photoactivated cyclic enones with a high degree of regioselectivity to give fused methylenecyclobutanocycloalkanones in which the two sp^2-hybridized carbon atoms are 1,3-related, a combined cycloaddition–oxidation maneuver becomes a process equivalent to a specific conjugate alkylation. The method is particularly valuable for erecting an acetic acid or acetaldehyde chain at an angular position.

Pursuant to this protocol, segments in a number of terpenes have been assembled. Examples are found in the syntheses of ishwarane (**I11**) (R. Kelly, 1970, 1971, 1972), 7,12-secoishwaran-12-ol (**I13**) (Funk, 1983), and a precursor of pentalenolactone E (**P19**) (Exon, 1981).

I11   **I13**

P19

Very frequently in the area of diterpene synthesis a bicyclo[2.2.2]octane system must be constructed. Since certain hydrophenathrenones [8(14)-en-13-one type in the diterpene numbering] are readily available, it cannot be more suitable to employ the reaction sequence under discussion to fulfill the need. In many cases the release of an acetaldehyde chain from the cyclobutane moiety is immediately followed by an aldol condensation at the α' position, and a synthetic step is saved.

Although the involvement of this indirect conjugate alkylation–aldolization sequence in a synthesis of trachylobane (**T22**) (R. Kelly, 1973a) may be surmised, the use of this sequence in the syntheses of napelline (**N3**) (Wiesner, 1974a), talatisamine (**T1**) (Wiesner, 1974c), and chasmanine (**C50**) (Tsai, 1977) is far from apparent. Such synthetic schemes could not have been developed without a thorough study of the rearrangement steps beyond.

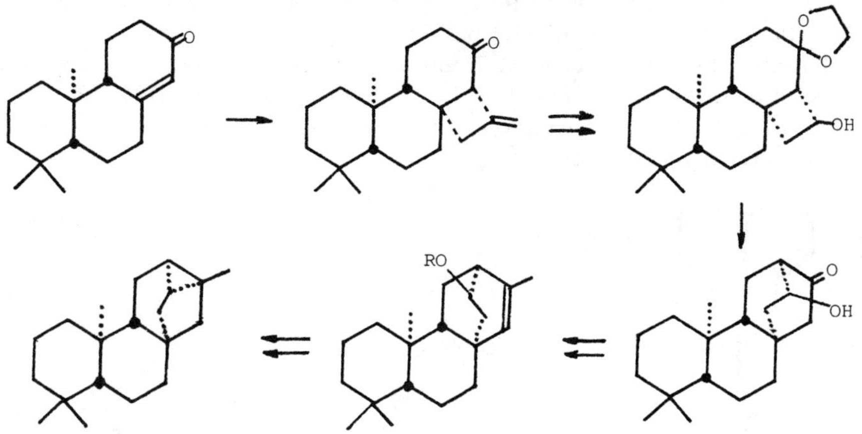

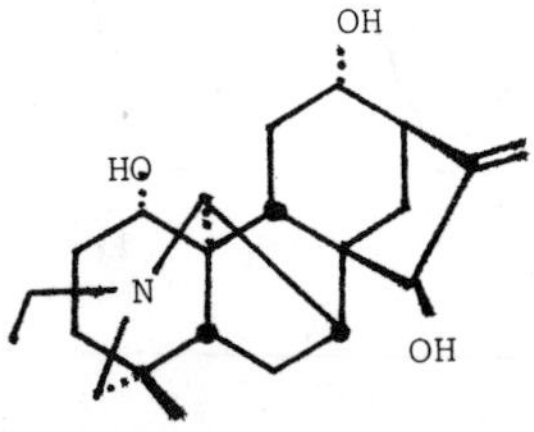

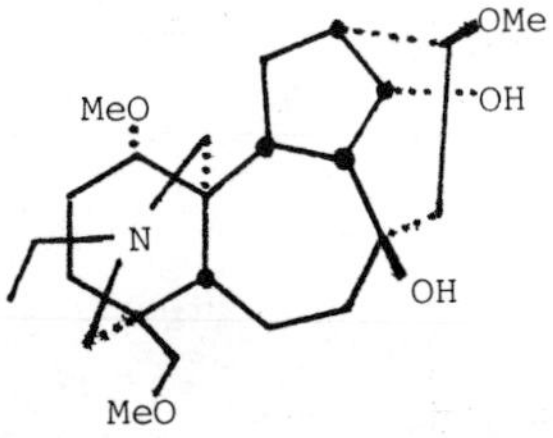

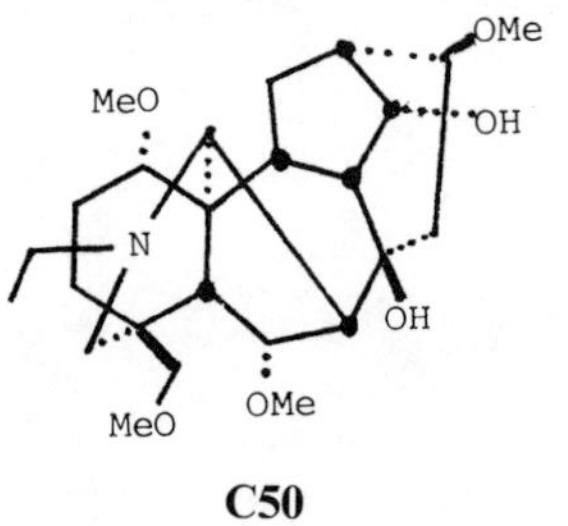

A remarkable construction of the skeleton of stemodin (**S50**) (Piers, 1982a) features a conversion of two diastereomeric β-diketones derived from an unexpectedly nonstereoselective cycloaddition of allene to a tricyclic enone and oxidative demethylenation to the same trans-syn-cis ketoester. This isomer is more stable on account of the B/C-cis fusion, and two pathways are open for its genesis from the trans-anti-trans diketone. Thus, methoxide ion can cleave either the four- or five-membered ring and effects reclosure of the ketoester to give the trans-syn-cis diketone and a bicyclo[2.2.1]heptanedione, respectively. The latter compound is a precursor of the more stable ketoester.

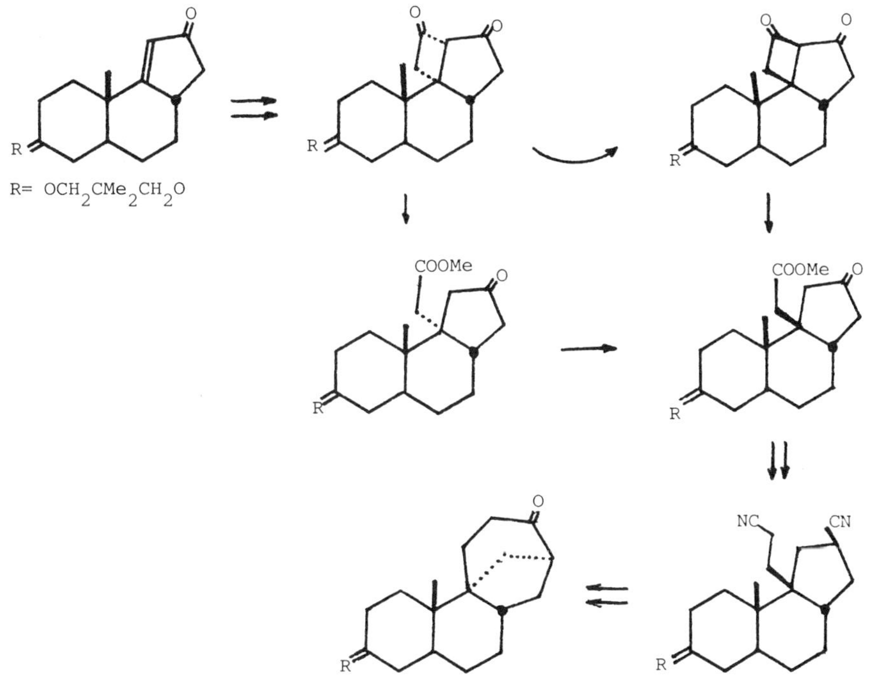

In the preparation of a cyathin A_3 (**C89**) precursor, the five-membered A ring is acquired via a photocycloaddition of an enedione and allene, disengagement of the cyclobutane ring, and aldolization (W. Ayer, 1981).

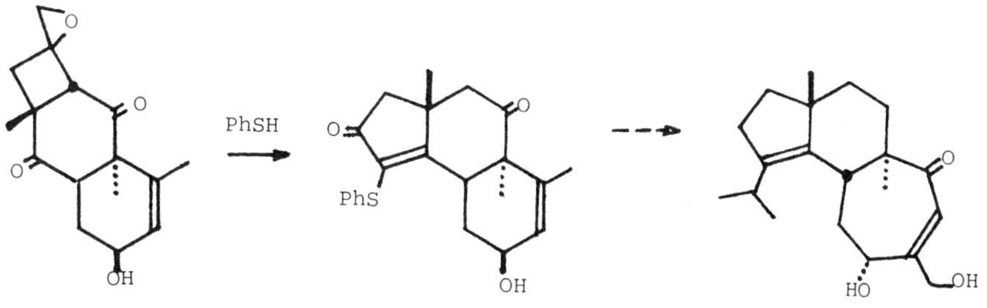

The problems inherent in the formation of a fully substituted C—C bond are compounded by a stereochemical issue when synthesis of a trichothecane by union of two carbocyclic synthons is contemplated. The photocycloaddition route is a definitely more attractive alternative, provided the addends are equipped with functionalities to unlink the cyclobutane ring subsequently. Two individual approaches to trichodiene (**T26**) (Yamakawa, 1976) and 12,13-epoxytrichothec-9-ene (**T27**) (Masuoka, 1976) are based on this concept.

A very interesting photocycloaddition–Norrish type II fragmentation sequence, when applied to 3-(6-methyl-5-hepten-2-yl)cyclohexenone, leads to a spirocyclic ketone. The discovery of a remarkable epimerization of the isopropenyl group with silver nitrate then allows a rapid unraveling of acorenone B (**A10b**) (Manh, 1981).

A rational approach to α-acoradiene (**A4**) (Oppolzer, 1983) is similar in design. However, a device for ionic fragmentation of the cyclobutane ring is supplied to ensure successful ring opening.

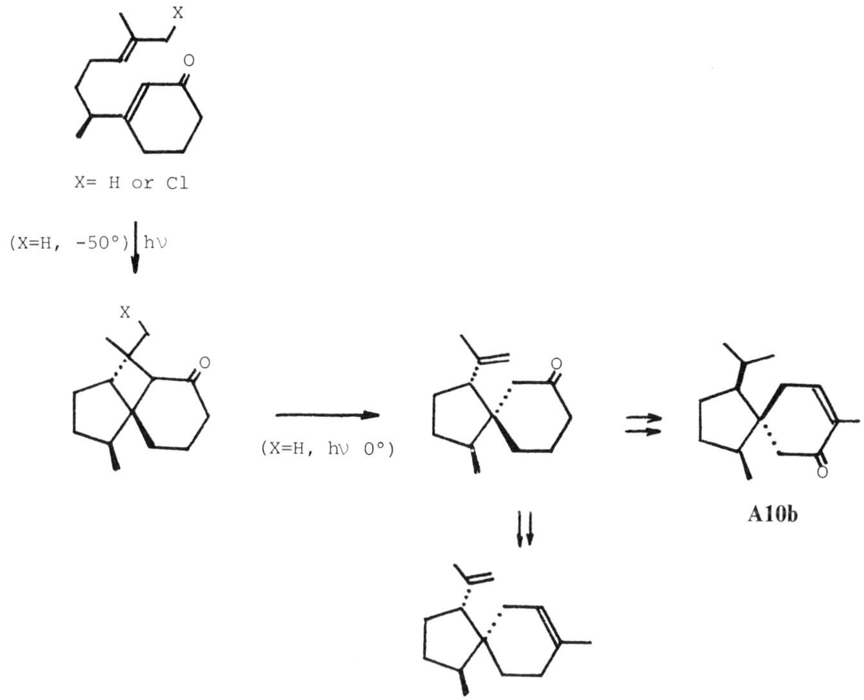

X= H or Cl

(X=H, −50°) | hν

(X=H, hν 0°)

A10b

A4

The production of a [4.5.5.5]fenestranone by an intramolecular photocycloaddition and its cleavage with iodotrimethylsilane constitute the key steps of an imaginative synthesis of silphinene (**S34**) (Crimmins, 1986). By the nature of the steric arrangements, the four chiral centers are already established at the bicyclic level. Since the exo preference of substituents on the diquinane periphery is well documented, the butenyl chain at the equilibratable γ position of the enone is expected to have the exo configuration.

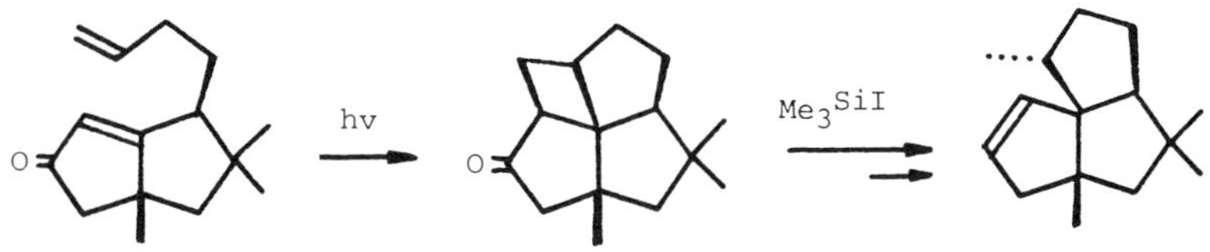

hν → Me₃SiI →

S34

The above reaction sequence has been further improved with respect to regioselective installation of the unsaturation in the final stage. Transformation of the tetracyclic alcohol to the iodide and treatment of the latter compound with hexabutylditin lead to ring opening. The resulting primary iodide is reducible to give only silphinene (**S34**) (Crimmins, 1987b).

The analogous stereocontrolled access of a tetraquinane skeleton via intramolecular photocycloaddition and cyclobutane cleavage protocol is apparently adaptable to synthesis of crinipellin A (Mehta, 1987b) by judicious modification of certain reactions.

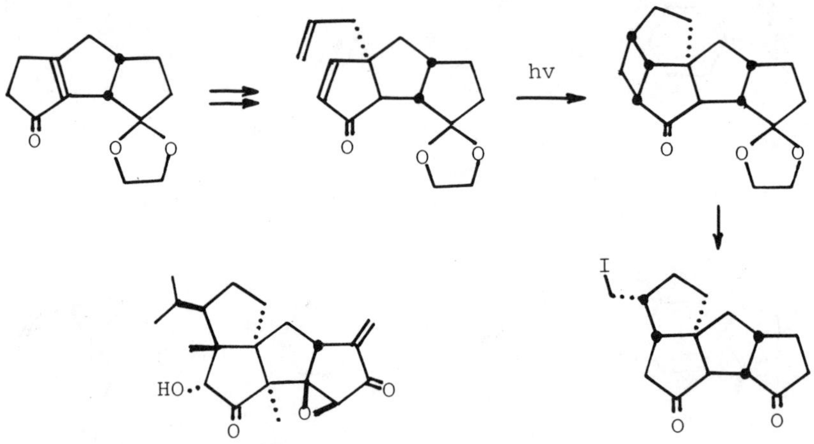

crinipellin A

The synthesis of laurenene (Crimmins, 1987a) is more complicated because the fenestrane skeleton must be locked and the secondary methyl group is in a more congested environment. The latter problem has been resolved by applying Still's method of transpositional homologation on the minor tetracyclic allylic alcohol followed by deoxygenation.

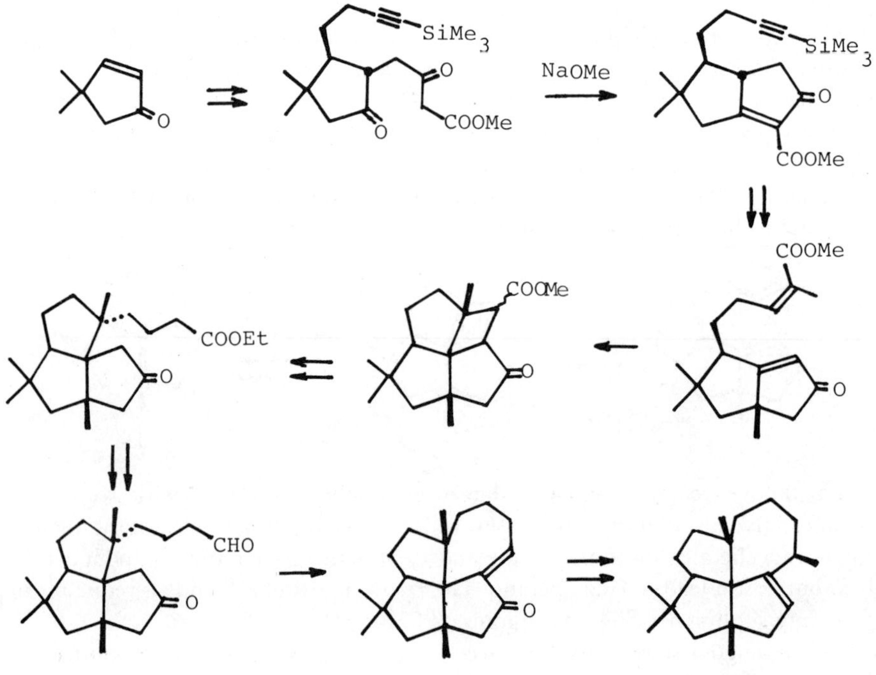

laurenene

The pentalenane-type molecules are commonly approached via the formation of a diquinane nucleus on which a chain is attached to an angle to initiate the closure of the third carbocycle. Thus, the route to pentalenic acid (**P17**) (Crimmins, 1984) involving an intramolecular photocycloaddition stands out from the conventional approach. Remarkably, three stereocenters are generated in the photochemical reaction in high selectivity. After reductive cleavage of the succinic ester a 13:1 epimer mixture about the secondary methyl group (in favor of the desired isomer) is obtained. Adequate functionalities are present in these intermediates for completing the carbon skeleton.

α-Alkylation of enones can be achieved using vinyl ethers and acetates as photoaddends. The oxygenated carbon atom in the cyclobutane moiety of such an adduct is now connected to the β position of the original enone, and the cleavage of this bond is most easily executed. In a synthesis of paniculide A (**P9**) (Smith, 1981d) the regioselectivity of the cycloaddition is fully exploited.

Cyclobutanones are obtainable from the reaction of ketenes (preferably chloroketenes) with alkenes. As expected, γ-lactones are only step(s) away from these intermediates. Application of the reaction sequence to the synthesis of lactonic terpenes include ivangulin (**I20**) (Grieco, 1977d) and eriolanin (**E14**) (Grieco, 1978).

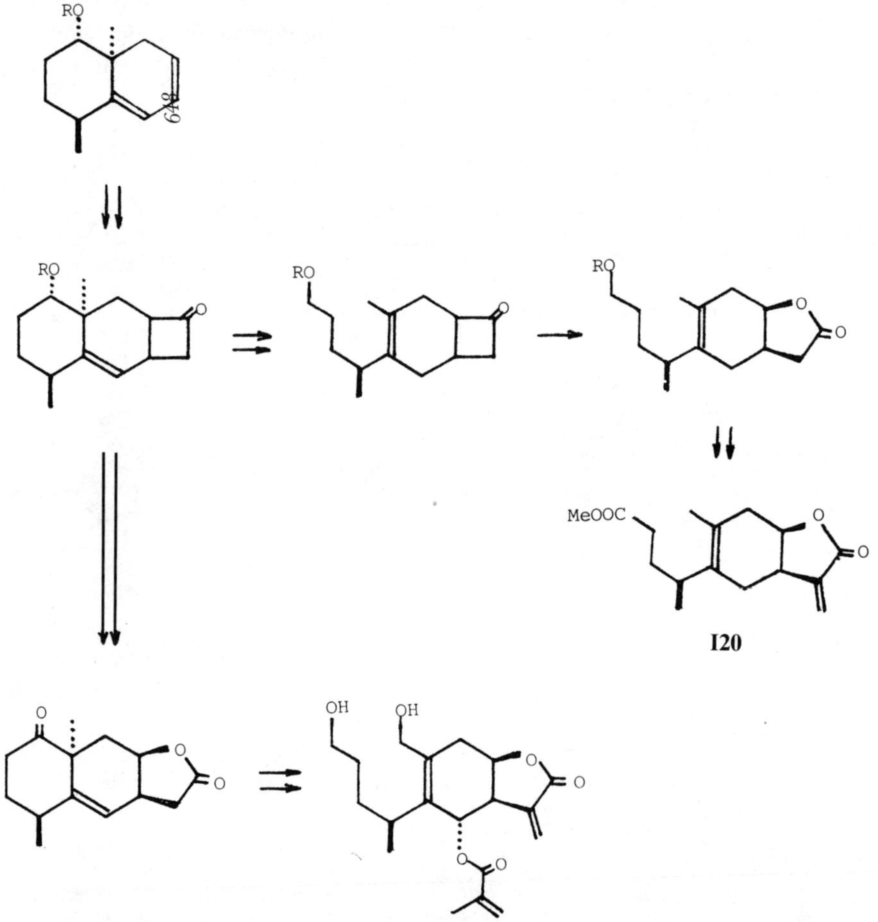

An elegant synthesis of the American cockroach sex pheromone periplanone B (**P22**) (Schreiber, 1984) starts from the head-to-head photoadducts of cryptone and allene. The medium ring containing the two double bonds and a ketone group is unfolded quickly via an oxy-Cope rearrangement and electrocycloreversion. The allene molecule is incorporated into the ring and the exocyclic methylene group. Its bonding to the original α carbon has remained unbroken, and the overall atomic redistribution indicates the result of a seco-α-alkylation process.

12.2.2 Additive 1,2-Dialkylation of Alkenes

The inherent strain of a cyclobutane is responsible for its facile scission at moderate temperatures. The direction of the cleavage is largely determined by the presence and location of substituents, whose stabilization of incipient radicals tends to lower the activation energies further. When a [2 + 2]cycloadduct undergoes a cyclobutane cleavage at an "aboriginal" single bond, the net effect on one addend is a dissociative 1,2-addition by the other addend. Thus, the adduct of (R)-piperitone and methylcyclobutene is readily converted into *ent*-shyobunone (**ent-S32**) (Williams, 1979) in a formal 1,2-dialkenylation.

An expedient route to hibiscone C (**H12**) (Koft, 1982b) focuses on the synthetic equivalence of furan and cyclobutene and the construction of a properly fused cyclobutene derivative by an intramolecular cycloaddition (enone + alkyne). An extra advantage of this approach is the facile resolution of its stereochemical problem during the assembly of the cyclohexenone precursor, which requires a trans relationship of the two side chains.

A convergent strategy for synthesis of bakkenolide A (**B1**) (Greene, 1985a) calls for the preparation of a *cis*-1, 2-bishalomethylcyclohexane and its use as alkylating agent to close the cyclopentane ring. The availability of a [2 + 2]cycloadduct from 1,6-dimethylcyclohexene and dichloroketene, and its facile dechloroacetylation (leading to a structure easily recognizable as a useful precursor of the required intermediate) are reassuring.

B1

The first synthesis of loganin (**L20**) (Büchi 1973) is distinguished by the masterful choice of a photocycloaddition route whereby the complete heterocyclic functionality is unraveled directly. The missing methyl group is then introduced via the cyclopentanone. [More deMayo reaction on p. 655].

A modified procedure using optically active (1*R*,2*R*)-2-methyl-3-cyclopenten-1-yl acetate as an addend leads to products containing four chiral centers, the major one being *O*-acetylloganin aglucone (Partridge, 1973).

R= H, R'= THP
R=Me, R'= Ac

L20

The loganin synthesis actually embodies a *cis*-alkylation/acylation of an alkene. By a similar process 1,5-dicarbonyl compounds are readily obtainable using (5*H*)-5,5-dimethylfuran-3-one, and appropriate precursors of (—)-acorenone (**A10**) (S. Baldwin, 1982a) and occidentalol (**O2**) (S. Baldwin, 1982b) have been synthesized.

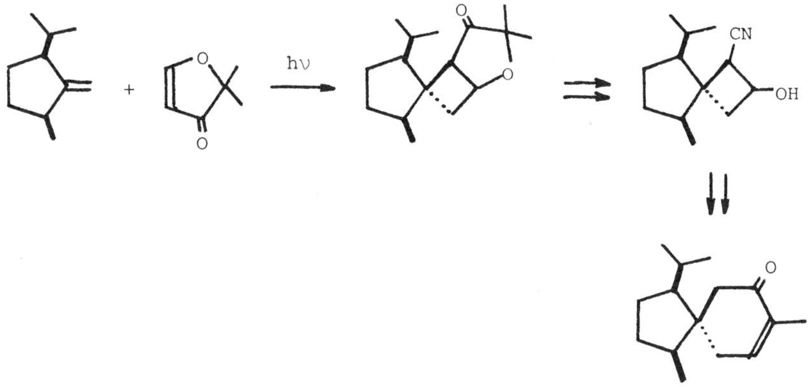

12.2 CYCLOBUTANES

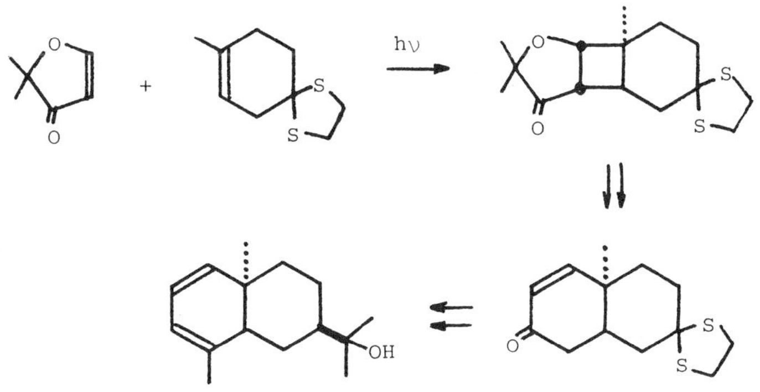

Since the photocyclization of α-terpineol with the homologous 2,2,6-trimethyl-4H-1,3-dioxin-4-one shows an essentially complete head-to-head regioselectivity and a good alkene face selectivity (3.35:1), the major adduct is suitable for elaboration of elemol (**E4**) (S. Baldwin, 1985). The employment of (—)-α-terpineol in this work is predicated on the emergence of unnatural (+)-elemol.

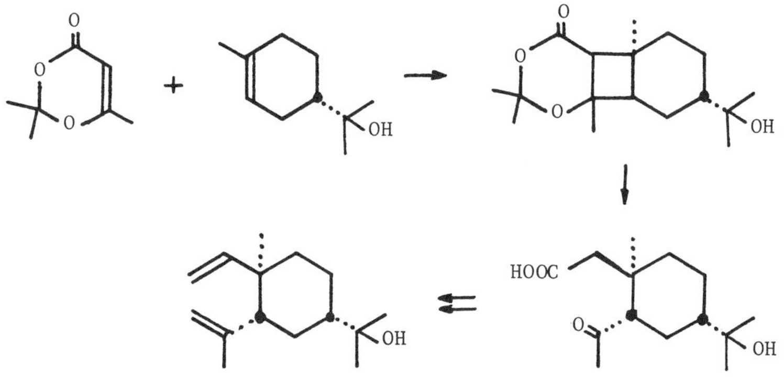

Any synthetic route to ingenol (**I4**) must address the inside-outside bridging characteristic of its carbon framework. An attractive solution to this intriguing problem is furnished by an intramolecular [2 + 2]photocycloaddition/fragmentation sequence. The pentacyclic system is formed from a pseudo-chair transition state in which the two reacting double bonds are aligned in parallel. This conformation also experiences the least nonbonding interactions.

The synthesis of a tricyclic intermediate of ingenol (Henegar, 1987, Winkler, 1987) starts from a hexahydropentalenol. Introduction of the side chain containing a chiral center to the angular position is accomplished via an ester-Claisen rearrangement.

The presence in a cyclic photoaddened of functional groups that can facilitate cyclobutane cleavage would enable the addition of a carbon chain to the double bond, effectively achieving a ring formation process. The adducts of 1,2-bis(trimethylsiloxy)cyclopentene with properly substituted cyclopentenones are perfect precursors of hydrazulenediones for the synthesis of damsin (**D5**)

(DeClercq, 1977), hysterin (**H25**) (Demuynck, 1979), carpesiolin (**C30**) (Kok, 1979), compressanolide (**C60**) (Devreese, 1980), and estafiatin (**E15**) (Demuynck, 1982).

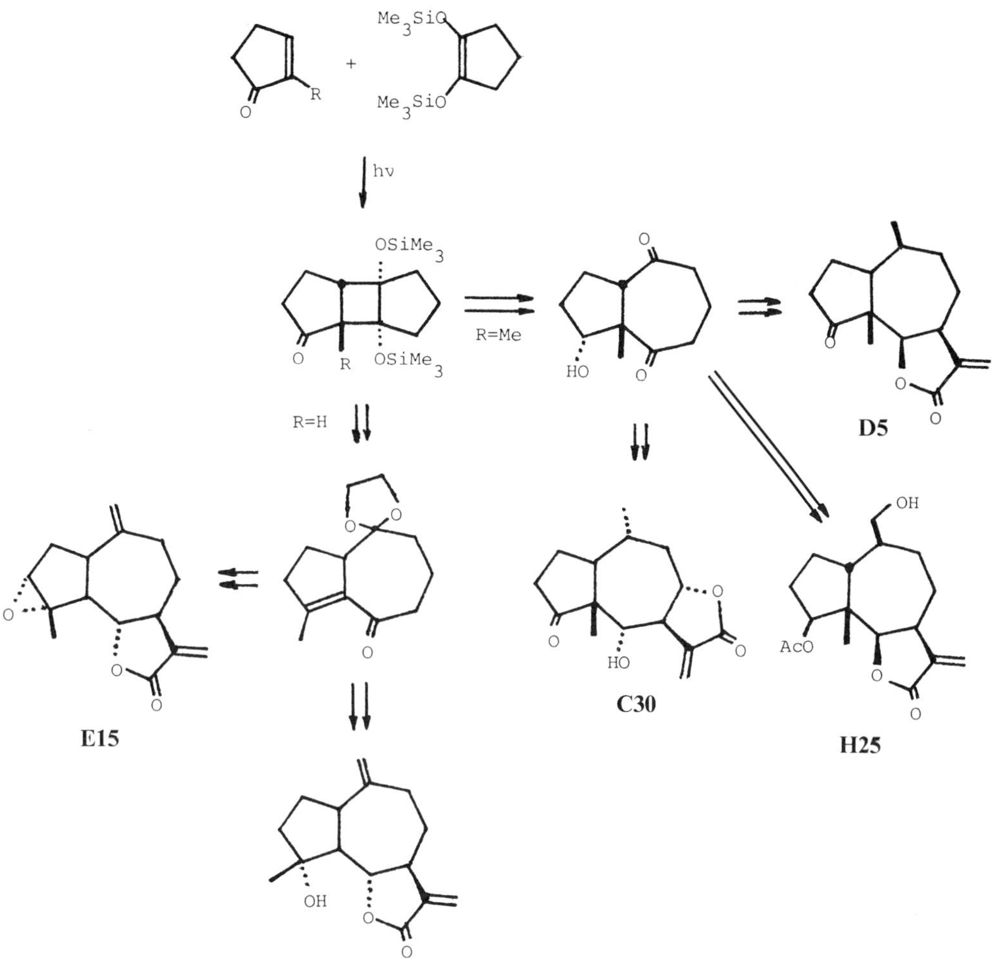

Decalin-1,4-diones are equally easy to make by this photocycloaddition/α-glycol cleavage sequence. As a result, 1-mono-oxygenated eudesmanes are directly available now, using suitable cyclohexenone derivatives as the other photoaddends. Moreover, the ketone group in the adducts can be reduced stereoselectively because external reagents must approach from the same side as the adjacent angular hydrogen atom. The relative configuration of the resulting hydroxyl group is the same as that usually found in eudesmanolides.

The cis ring juncture of the photoadducts causes no problems because liberation of the dione permits epimerization at that point. Further elaboration of such intermediates can lead to dihydroreynosin (**R2**), maritimin (**M8**), magnolialide (**M2**), α-santonin (**S9**) (van Hijfte, 1984), and 1-oxocostic acid (**C73**) (van Hijfte, 1982).

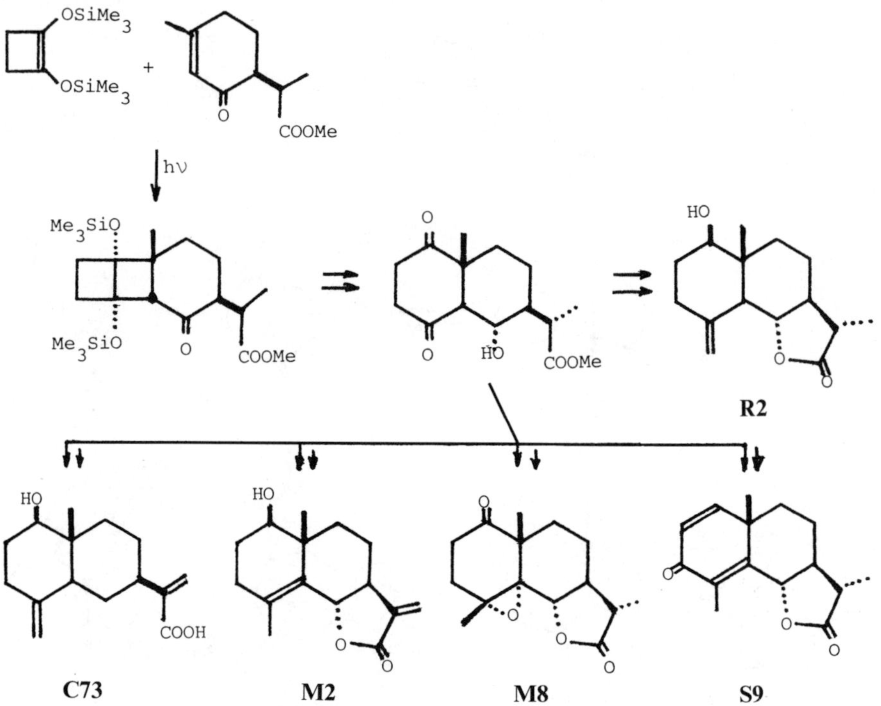

A less obvious way of controlling the fragmentation is illustrated in a synthesis of 10-epijunenol (**J6**) (Wender, 1978). The adduct from piperitone and methyl 1-cyclobutenecarboxylate can be transformed into a diol, and the reduction of the monophosphate causes a concurrent ring opening. The primary radical intermediate is expected to collapse exothermically to give the secondary radical before it encounters a hydrogen source.

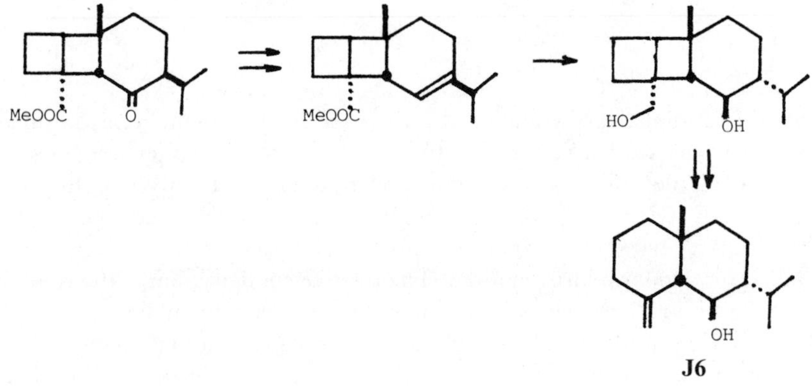

The deMayo reaction—that is, a combination of photocycloaddition of cycloalkenones with β-oxy-α,β-unsaturated compounds and fragmentation of the adducts by a retroaldolization or related process—has become a very popular method for natural product synthesis. Two of the examples are β-himachalene (**H14**) (Challand, 1967) and dehydrokessane (**K6**) (Liu, 1977). The photocycloaddition is subject to a distinct solvent effect. For example, the β-himachalene precursor predominates to the extent of 98:2 in cyclohexane, whereas in more polar solvents (acetonitrile or methanol) the same isomer is formed in slightly less amount (45%) than the head-to-tail isomer (Challand, 1968).

H14

K6

The deMayo reaction is unquestionably an excellent method for achieving a two-carbon expansion of cyclic β-diketones. In this context, approaches to hirsutene (**H17**) (Disanayaka. 1985) and a potential precursor of fusicoccin H (**F22**) (Grayson, 1984) should be mentioned. The former synthesis employs the free β-diketone, which allows the adducts to undergo a retroaldol reaction immediately. The unfavorable regioselectivity of the photocycloaddition is a serious drawback, however.

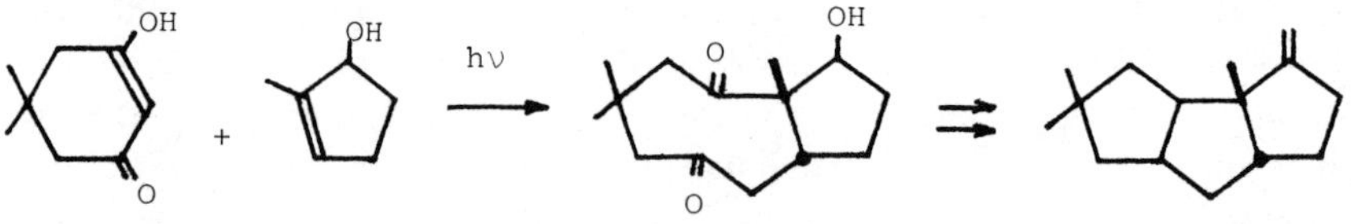

Particularly intriguing applications of the reaction sequence are those involving an intramolecular cycloaddition. The net result is an annulative ring expansion. The syntheses of longifolene (**L24**) (Oppolzer, 1978). (+)-sativene (**S11 +**) (Oppolzer, 1981), β-bulnesene (**B20**) and its 1-epimer (Oppolzer, 1980), zizaene (**Z1**) (Baker, 1981), and pentalenene (**P16**) (Pattenden, 1984) are all imaginative contributions to the current refinement of synthetic idioms.

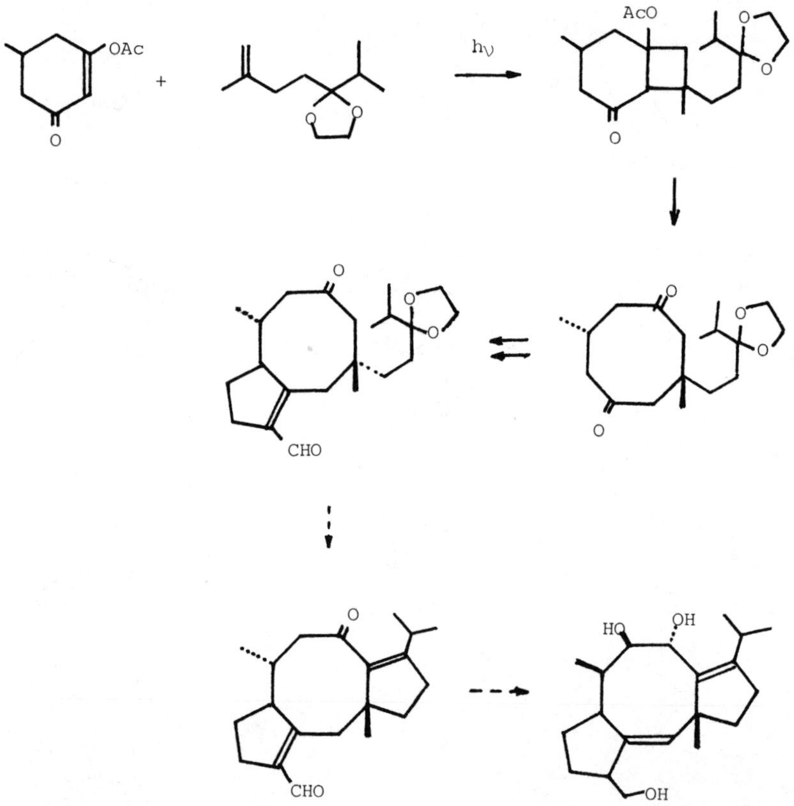

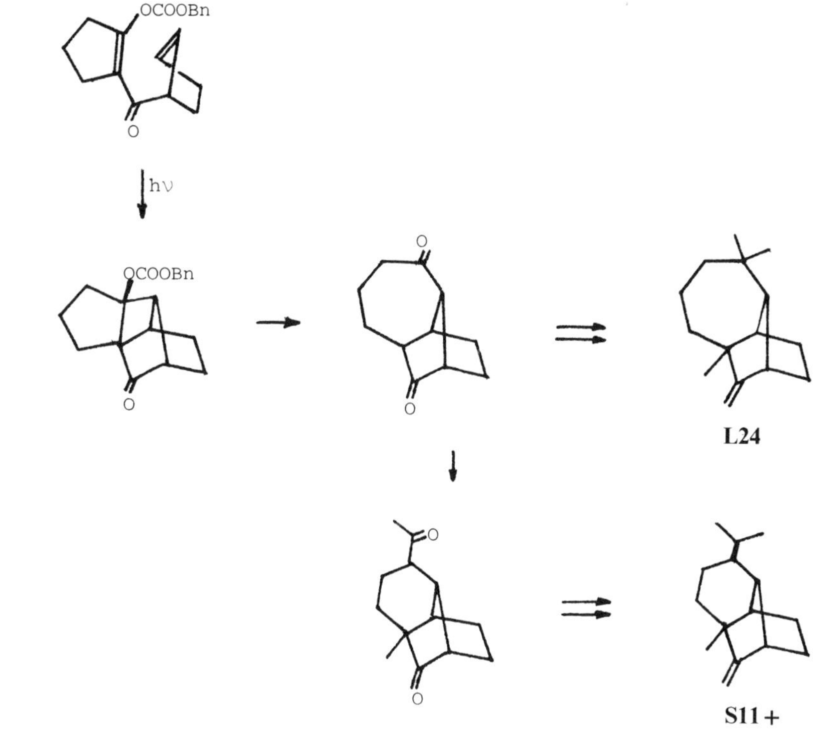

L24

S11+

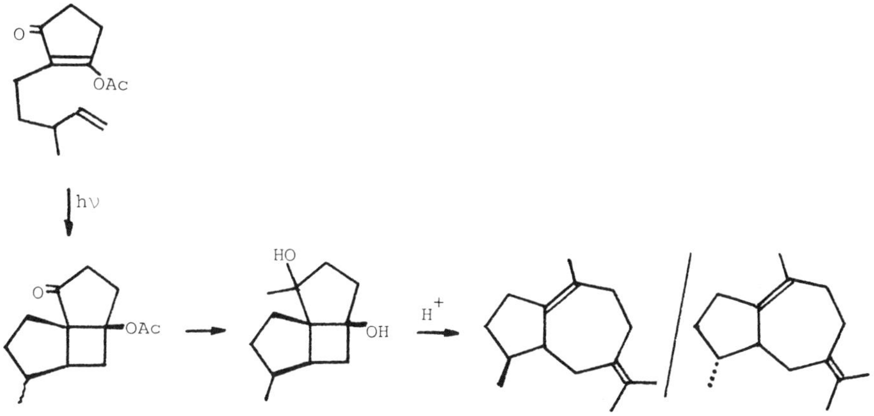

B20

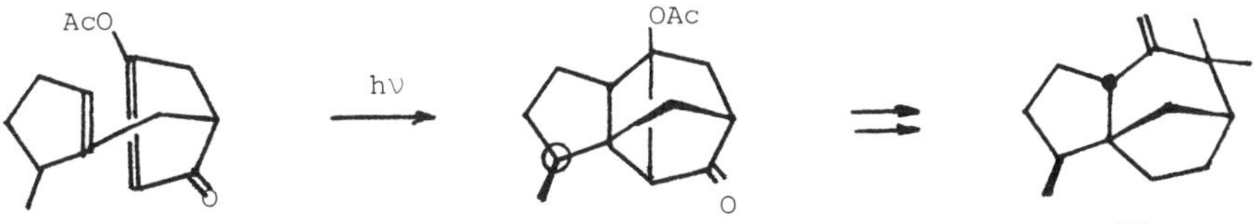

Z1

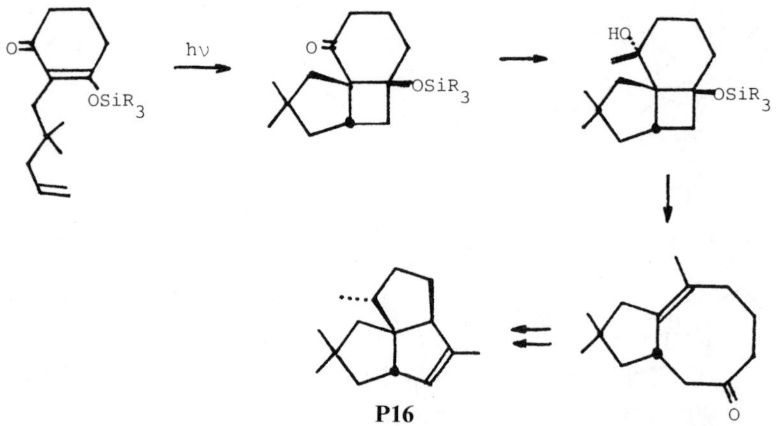

P16

Another recent example concerns a synthesis of daucene (**D8**) (Seto, 1985). The spontaneous aldolization of the diketone liberated from the photoisomer(s) provides a means of controlling the proper settlement of the double bond in the seven-membered ring. Independently, the desmethyl tricyclic ketol derivative has been prepared and further elaborated into isoamijiol (**A28**) (Pattenden, 1986).

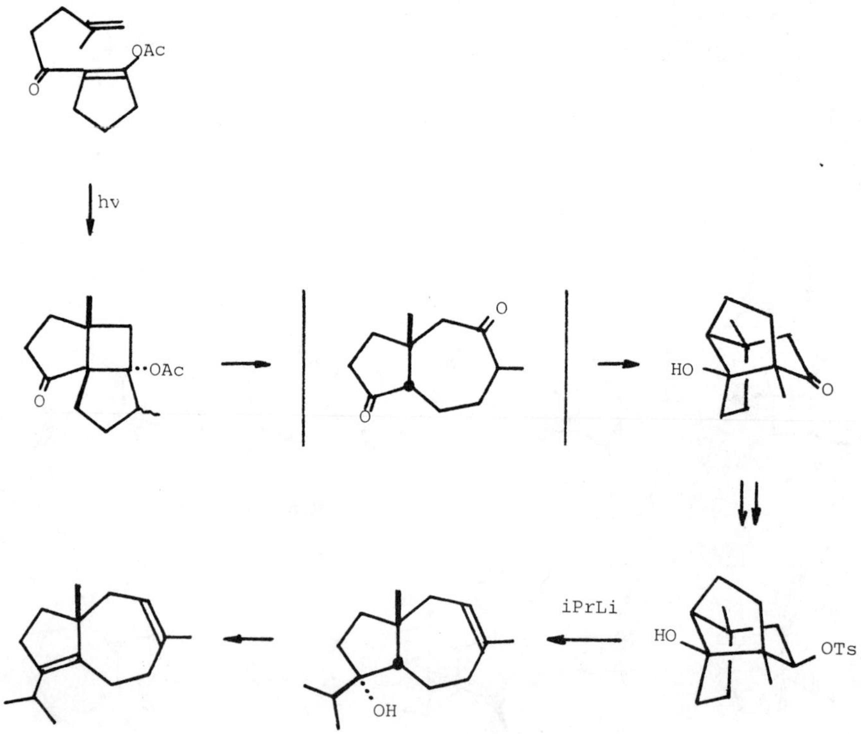

D8

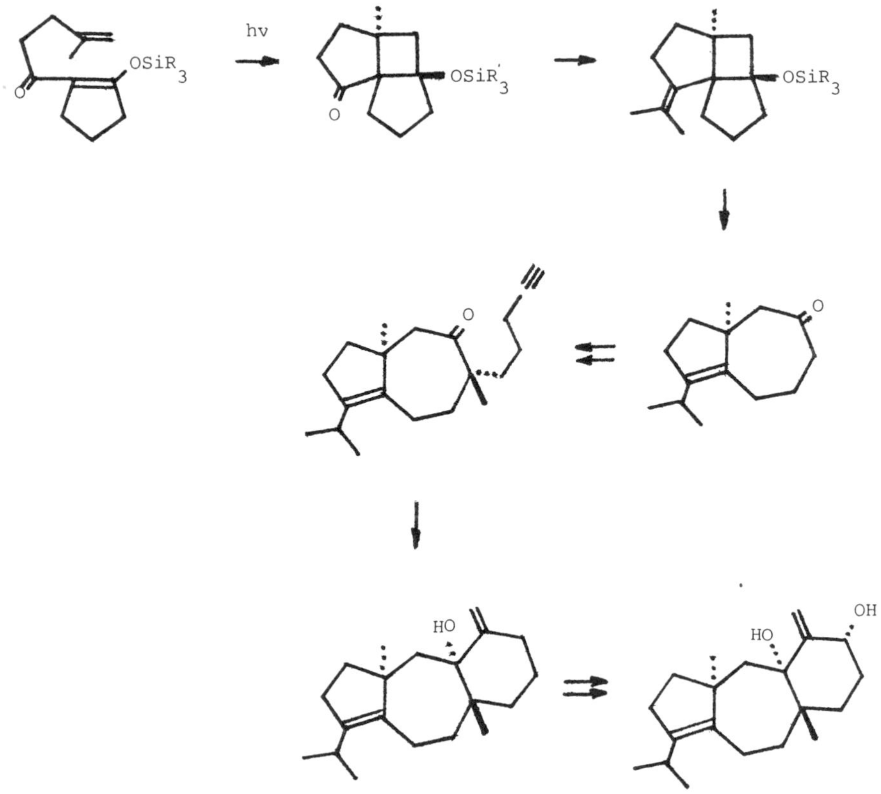

It must be emphasized that the method is sometimes limited by unfavorable steric influences. Thus, besides the example of β-bulnesene in which its 1-epimer is dominant (ratio, 3.3 : 1), an approach to precapnelladiene (A. M. Birch, 1980) has been thwarted by the formation exclusively of its methyl epimer (**P46e**).

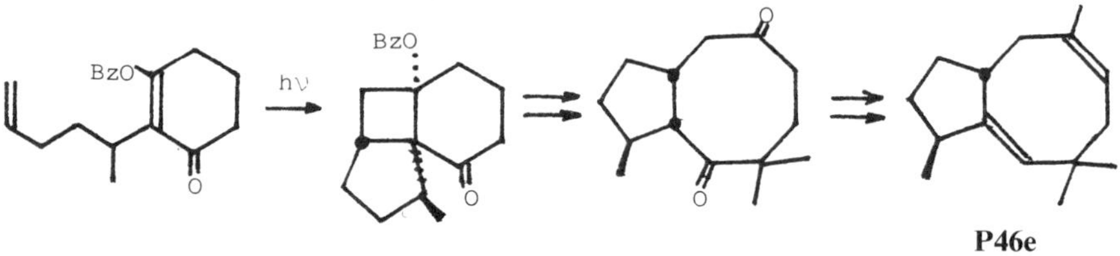

The model study toward the synthesis of the taxanes shown below (Kojima, 1985) made use of a modified deMayo sequence. However, there is a reservation with respect to its extrapolation to the natural products. In such a case a *gem*-dimethyl group on the methano bridge of the substrates would certainly hinder the approach of the cyclohexene moiety prior to the photocycloaddition. A trigonal carbon bridge should be installed.

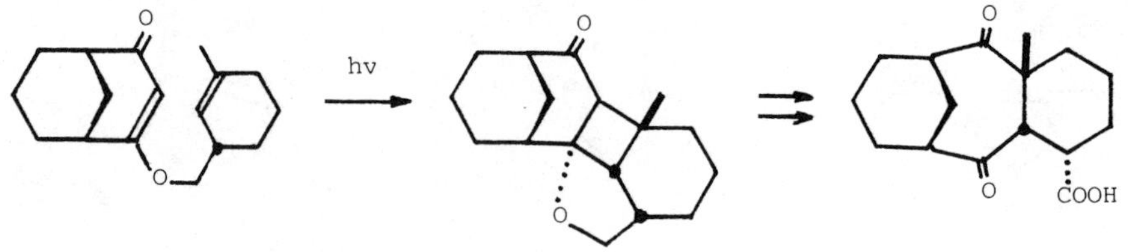

An intramolecular photocycloaddition-fragmentation pathway has been adopted to a synthesis of the BC ring system of taxanes (Swindell, 1987). The A ring can be closed by an intramolecular alkylation.

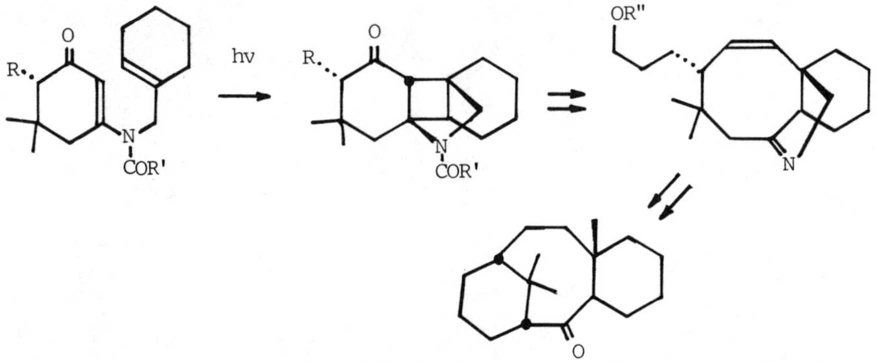

12.2.3 Alkene Metathesis by a Tandem Photothermal Process

Cage molecules are formed upon exposure of the Diels–Alder adducts of cyclopentadienes and benzoquinones. These compounds undergo thermolysis, which splits the cyclobutane ring in the orthogonal direction to give linear triquinanes. In this not only is the strain associated with the four-membred ring removed, but a product with two conjugate ketones is generated. Accordingly, sesquiterpenes such as hirsutene (**H17**) (Mehta, 1981) and $\Delta^{9(12)}$-capnellene (**C21**) (Mehta, 1983) have been synthesized from these "metathesis" products.

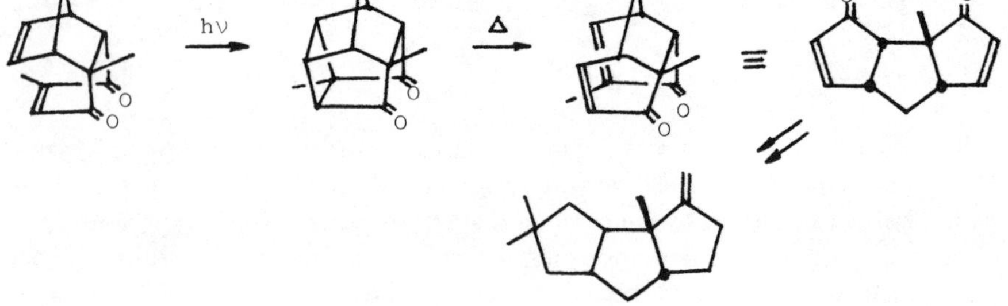

H17

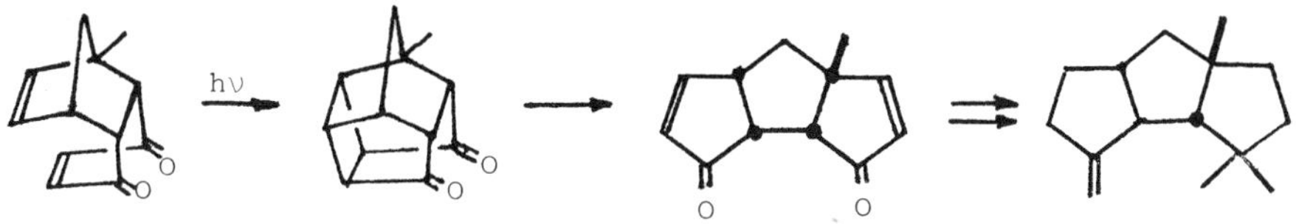

The involvement of a formal metathesis is less evident in a novel approach to β-eudesmol (**E18**) (Wender, 1980c) and warburganal (**W1**) (Wender, 1982b) because the mesocyclic intermediate undergoes a rapid transannular ene reaction to form a methyleneoctalin. When that secondary reaction is denied, the expected 1,5-cyclodecadiene can be isolated, and a synthesis of isabelin (**I10**) (Wender, 1980b) has been accomplished with the necessary structural adjustment.

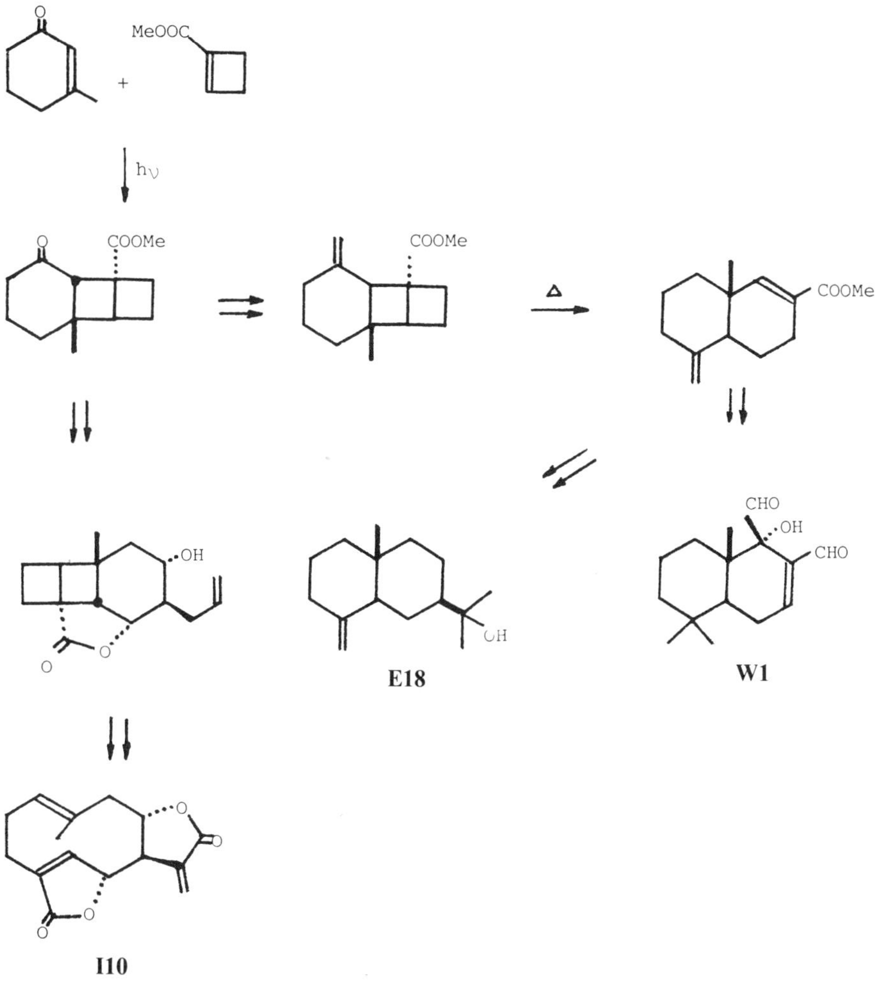

12.2.4 Reductive gem-Dialkylation of Carbonyl Compounds

A versatile method for cyclobutanone synthesis (Trost, 1974a) consists of reacting a carbonyl compound with α-sulfenylcyclopropyl carbanion or ylide to give an oxaspiro[2.2]pentane, and rearranging the latter species. Unsymmetrically substituted cyclobutanones so obtained may undergo ring cleavage, and their elements can be incorporated into other ring systems (cf. vetispiranes: Trost, 1975c; acorenone B: Trost, 1975a). A few examples describing the use of these cyclobutanones only as a source of *gem*-dialkyl substituents are given next.

In synthesis of the diterpene resin acids, a most critical aspect is probably the elaboration of the quaternary carbon atom C_4. The reductive *gem*-dialkylation procedure provides a solution to the podocarpic acid/callistrisic acid (**P43/C10**) series. As a result of the stereoselectivity of the reactions involved, the abietic acid series is not attainable by this method (Trost, 1973).

Application of the reductive dialkylation protocol to a synthesis of trihydroxy-decipiadiene (**D10**) (M. Greenlee, 1981) is even more germane because a wasteful chain degradation is not necessary. As expected, attack of the carbanion reagent comes from the exo face, and therefore a spirocyclic ketone with a functionalized exo branch is produced.

D10

Spiroannulation as a means of introducing the lactone moiety of plumericin (**P41**) is the key to a successful synthesis (Trost, 1983a). The stereoselectivity originates from an *exo*-selenylative migration of a cyclopropane bond. The only other crucial step remaining is carbomethoxylation of a dihydropyran, which has been achieved using trichloroacetyl chloride and methanolysis of the ensuing trichloromethyl ketone.

P41

12.3 CYCLOPENTANES

When cyclopentane derivatives are used as latent elements for other structural units the main purpose is stereocontrol. Sometimes the ready availability of the cyclopentanes and their suitability for the particular targets are too alluring to ignore.

12.3.1 Condensed Systems

Isolated cyclopentane rings are not ideal for exercising steric control, yet their presence in polycyclic environments, be they condensed or bridged, changes drastically their capability in directing stereoselective reactions. Thus, in an AC + B approach to the resin acid diterpenes such as podocarpic acid (**P43**) the latent B ring is incorporated in a hydrindenone nucleus such that the configurations of the pro-C_4 and pro-C_5 carbon atoms relative to pro-C_{10} may be regulated during the alkylation and reduction steps. The hydrindanone is then ruptured to release two carboxyl chains (Giarrusso, 1968).

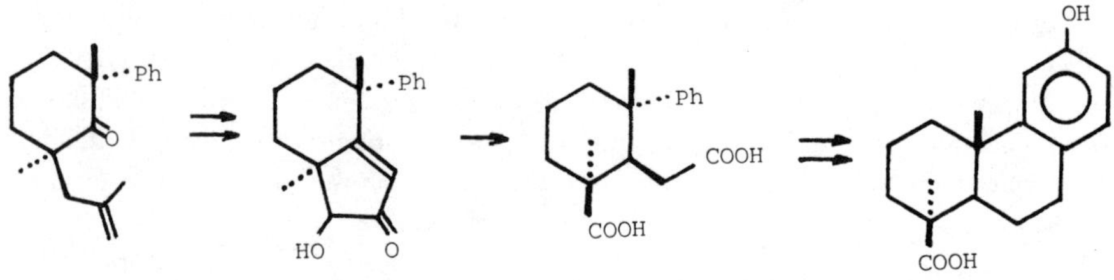

Control of the two chiral centers of chrysomelidial (**C54**) is facilitated by their involvement in a bicyclo[3.3.0]octane system. The determinant factor is the preference for substituents to exist in the exo orientation in its periphery. Thus a transitory annulation technique using an α-sulfenylvinylphosphonium salt to react with a 3-acetylcyclopentanone enolate, and regioselective reopening of the lately created carbocyle, accomplishes the difficult task (Hewson, 1985).

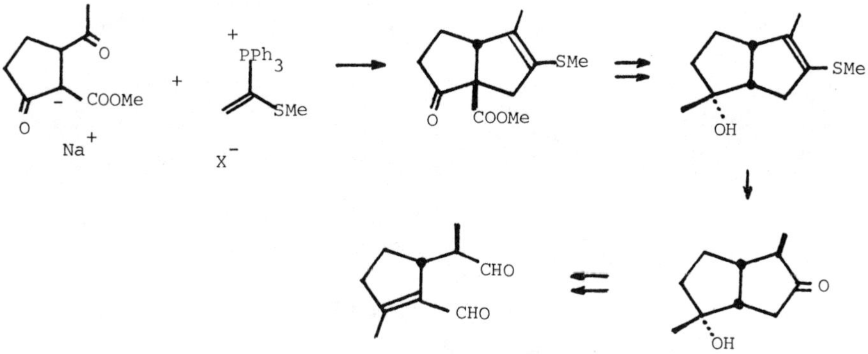

The symmetrical bicyclo[3.3.0]octan-3,7-dione is also useful for elaboration of loganin aglucone (**L20′**). The monoketal tosylhydrazone undergoes α-carbomethoxylation and Shapiro elimination to afford a β,γ-unsaturated ester. A methylation follows the recovery of the ketone (Caille, 1984).

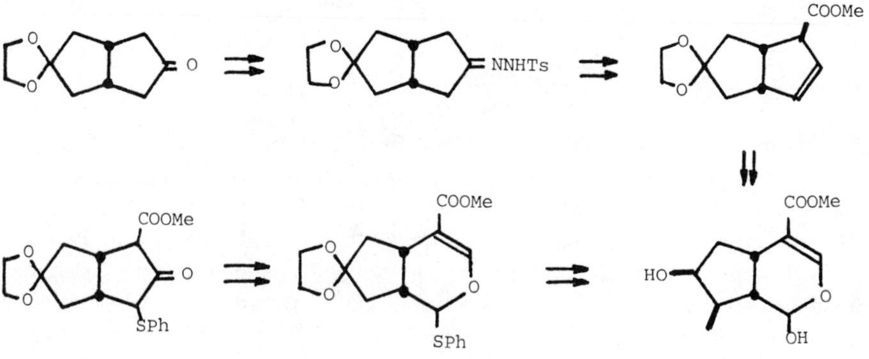

12.3 CYCLOPENTANES

Alternatively, the ketoester is sulfenylated using the dianion technique. Selective cleavage of the derived β-sulfenyl alcohol with lead(IV) acetate gives rise to the heterocycle (Hiroi, 1981).

Elemol has been partially degraded into a cyclopentenone, which is then systematically fashioned into the difuran segment of phytuberin (**P28**) (Kido, 1981a). In doing so, the five-membered carbocycle suffers cleavage. Its formation helps inversion of the ring juncture and ensures angular oxygenation from the β face.

P28

1,3-Diols are susceptible to fragmentation after activation at one site. In early 1960s when few methods were available for the selective formation of medium-sized rings, the synthesis of caryophyllene/isocaryophyllene (**P32/P33**) (Corey, 1963a, 1964b) was a monumental challenge. In this work the cyclononenes are created by fragmentation of a 5:6-fused system. The configuration of the double bond is determined by adjusting that of the tosylate group with respect to the adjacent angular methyl substituent.

P32

P33

12 TRANSITORY ANNULATION

Involuntary annulation sometimes intervenes in synthetic operations. For example, during a longifolene (**L24**) synthesis reported by McMurry (1972), conjugate methylation also causes an intramolecular aldol condensation. Fortunately the aldol cannot undergo dehydration, and the desired tricyclic skeleton can be regenerated by means of a fragmentation. (See also Section 12.2.2, daucene synthesis: Seto, 1985.)

L24

12.3.2 Bridged Systems

It is generally true that superior steric control is attainable within the frameworks of bridged carbocycles, exo/endo selectivity being more highly contrasting. Thus a stereocontrolled route to *threo*-juvabione (**J9**) (Larsen, 1979) involves formation of 4,7-dimethylbicyclo[3.2.1]oct-3-en-6-one and the kinetic protonation of its enolate. Hydride reduction of its hydrochloride is attended by fragmentation whereby the monocyclic intermediate is recovered. The exercise of forming and breaking the five-membered ring has for its sole purpose stereoregulation at the chiral centers.

The same bicyclic ketone is also available as one of the products from 1-menthene-8,9-diol on treatment with boron trifluoride etherate (I. Kitagawa, 1983).

J9

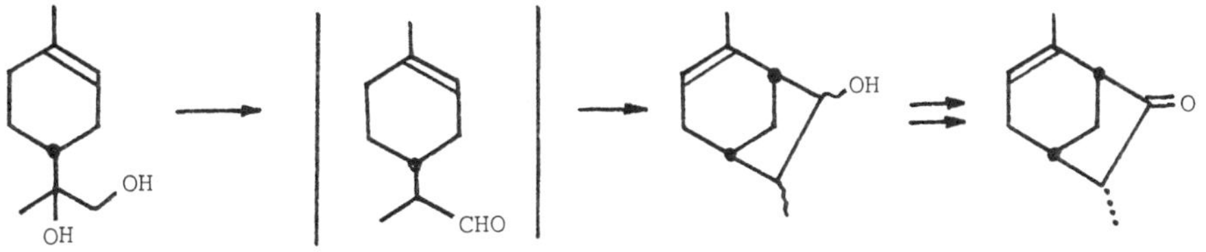

The fenchone system derived from rearrangement of a pinane derivative acts as a template for the generation of *ent*-hinesol (**ent-H15**) (Buddhasukh, 1975; Chass, 1978). Cleavage of the bicyclo[2.2.1]heptane skeleton is induced by a Dieckmann-type condensation, resulting in an anion of a 1, 3-diol monotosylate. The inherent strain of the cyclic array contributes to the ease of fragmentation.

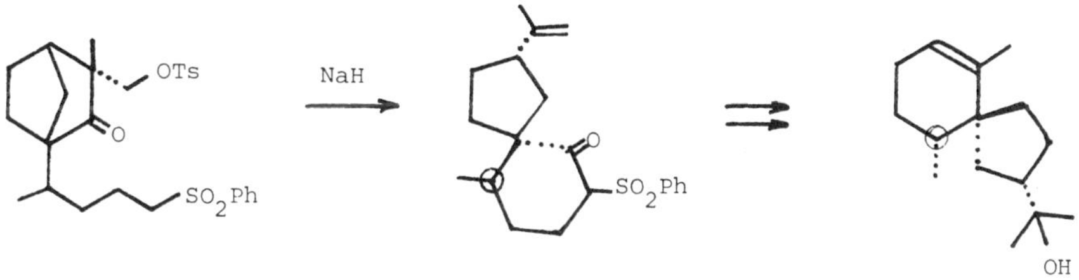

ent-H15

1,3-Dicarbonyl compounds interposed by a double bond are also liable to fragment. Thus the reaction has been applied to the assemblage of guaiol (**G20**) (Buchanan, 1971, 1973).

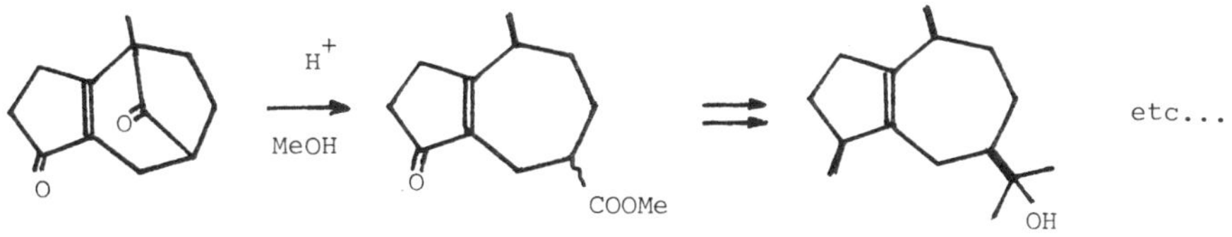

G20

(—)-Camphor-10-sulfonic acid is degradable to α-campholenic acid on treatment with alkali. This carboxylic acid is useful for the synthesis of (—)-khusimone (**K8**) (Liu, 1979) via cleavage of the cyclopentene and reclosure into a cyclohexenone.

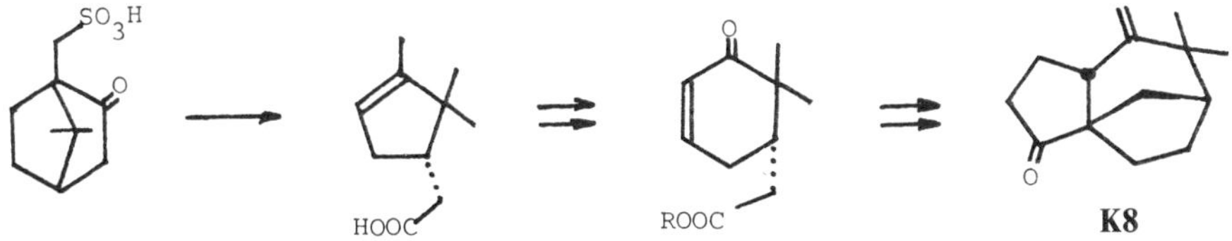

K8

Camphor can be functionalized regioselectively at the 7-*anti*-methyl group (e.g., by bromination) via a series of Wagner–Meerwein rearrangements. A synthon for balduilin (**B2**) (Lansbury, 1985) has been prepared by the Beckmann fragmentation of 9-acetoxycamphor oxime and elaboration of the released carbon chains by means of, among other reactions, a chelation-controlled homologation and aldol condensation.

Ring strain is always an important factor for instigating molecular disintegration. Thus a δ-bromoketone derived from longifolene is fragmented to a 6:7-condensed system, which is then elaborated into β-himachalene (**H14**) (Mehta, 1974).

Norbornadiene (bicyclo[2.2.1]hepta-2,5-diene) is an excellent building block for natural products containing a cyclopentane subunit. However, the mental association of pseudoguaianolides such as damsin (**D5**) (Grieco, 1977c) and helenalin (**H3**) (Ohfune, 1978) with the diene is remarkable. The rigid bicyclic skeleton furnishes superior regio- and stereocontrol, which permit the unambiguous establishment of three continuous asymmetric centers before the more flexible seven-membered ring is tied.

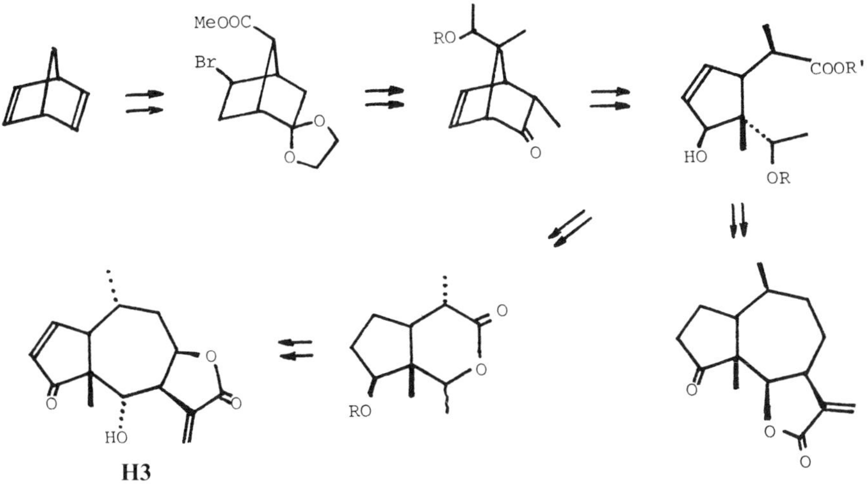

12.3 CYCLOPENTANES

H3

D5

A structural correlation of the iridoids with various bicyclo[2.2.1]octane derivatives has led to formulation of a new synthetic strategem (Callant, 1983) for boschnialactone (**B15**), teucriumlactone C (**T14**), loganin (**L20**), and more recently, specionin (**S44**) (Van der Eycken, 1985). Cleavage of the bicyclic system can be performed by photoinduced Norrish type I fission or by a Baeyer–Villiger reaction, depending on whether retention of an oxyfunction in the carbocycle is required.

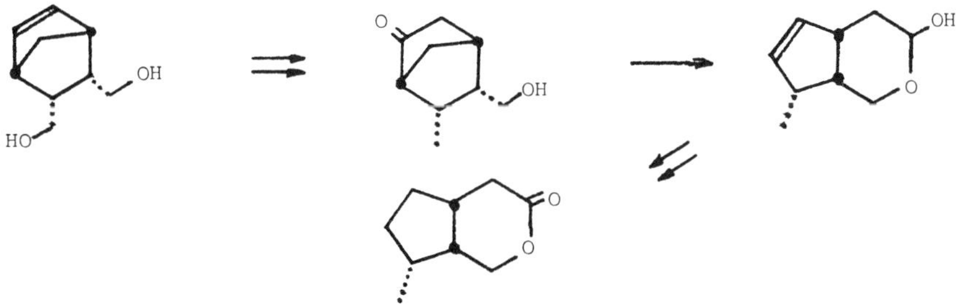

B15

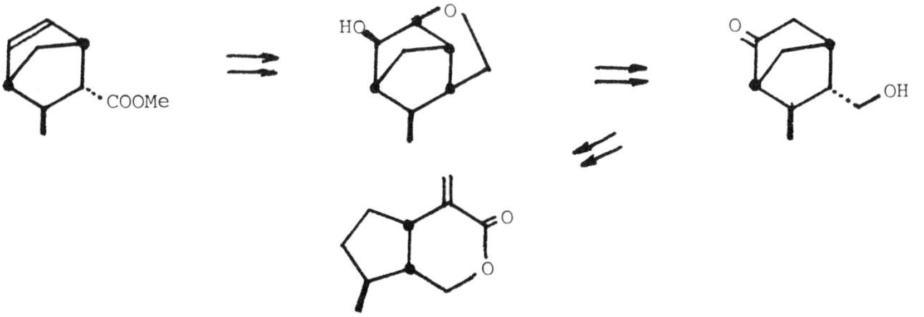

T14

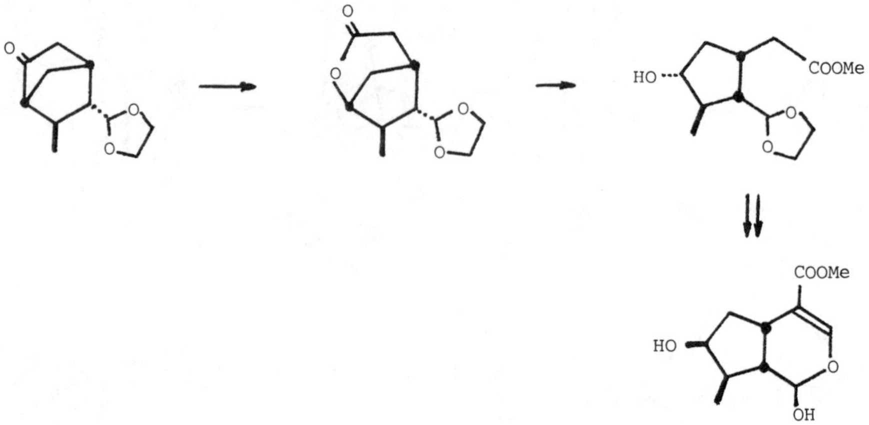

L20'

An approach to stoechospermol (**S54**) (Salomon, 1984) involves a photocycloaddition of a protected norbornenone to 2-cyclopentenone. Subsequently the norbornenone is ruptured by a Baeyer–Villiger reaction. Configurational correction of the released carbon chain is achieved, after its transformation into an alkylidenemalonate residue, by an intramolecular hydride reduction mediated by the hydroxyl group.

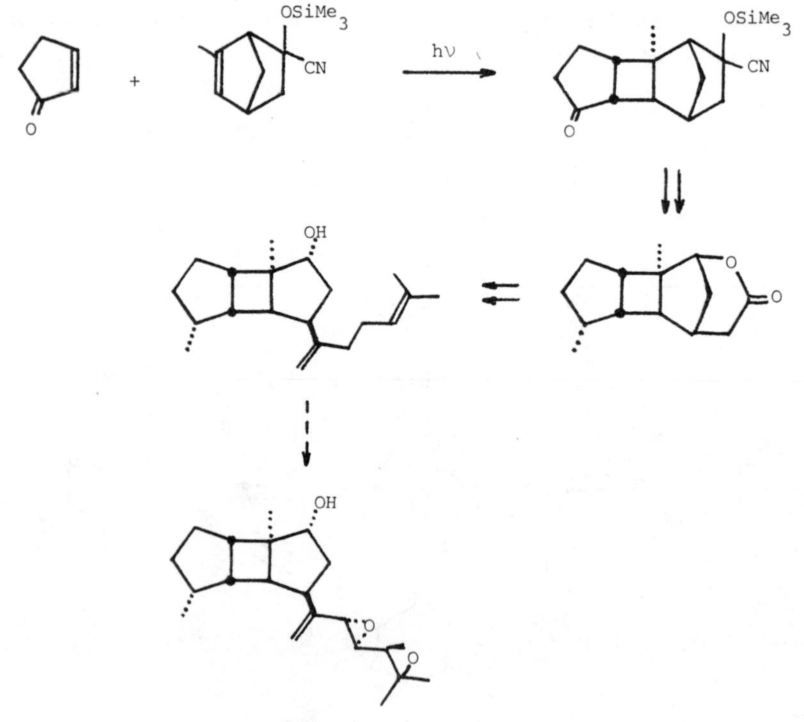

S54

12.3 CYCLOPENTANES

The versatility of superannulated norbornenes is related to the use of the double bond as latent carbonyl handles. A bridged-to-spiro cycloexchange process is the crux of an enlightening synthesis of silphinene (**S34**) (Sternbach, 1985).

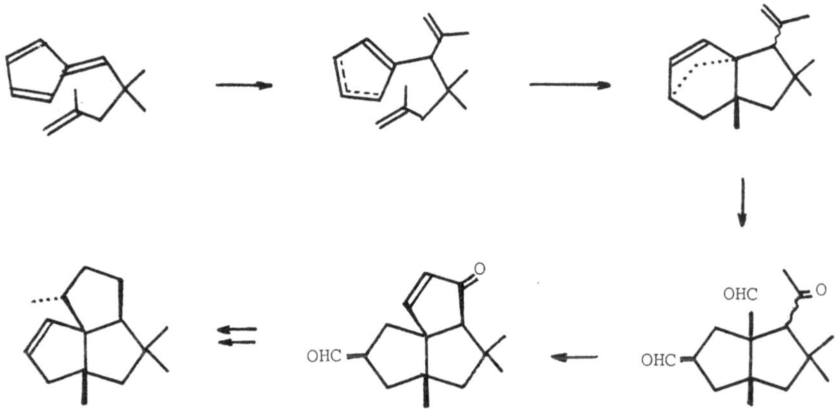

The concept involving intramolecular Diels–Alder reaction and cleavage of the double bond in the adduct to obtain a functionalized cyclopentane subunit has also been delineated for approaches to quadrone (**Q1**) (Sternbach, 1983) and retigeranic acid (Sternbach, 1984a).

A stereocontrolled approach to silphinene (**S34**) (Suri, 1984) also takes advantage of the norbornene unsaturation as latent methyl groups. Furthermore, the secondary aldehyde generated upon the ring fission plays a crucial role in the formation of the remaining five-membered ring of the terpene molecule. This synthesis has yet to be completed.

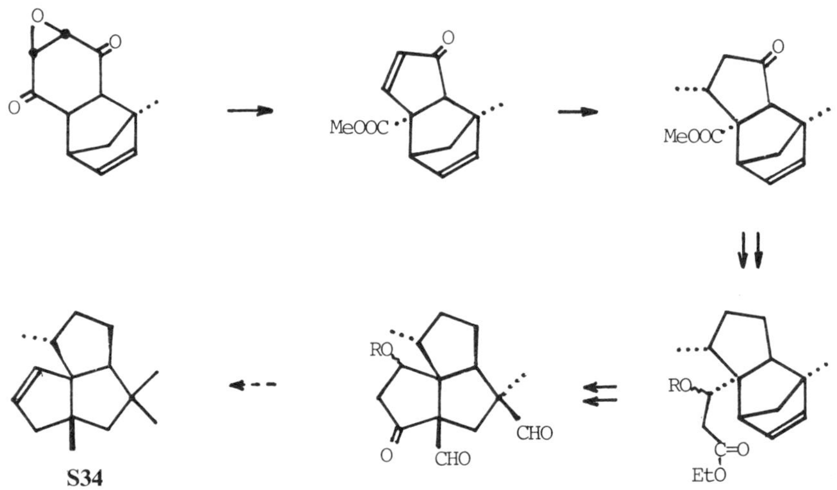

Verbenalol (**V9**), with an *exo*-methyl substituent at the β carbon of the ketone, is conceivably obtainable from the corresponding cyclopentenone via conjugate addition. The heterocyclic moiety is actually a dialdehyde ester and hence synthetically equivalent to a Δ^2-cyclopentenecarboxylate. In practice, dicyclopentadienone serves well as a starting material. It allows stereoselective introduction of the methyl group while presenting the etheno bridge for a systematic degradation into the hydroxydihydropyran (T. Sakan, 1968).

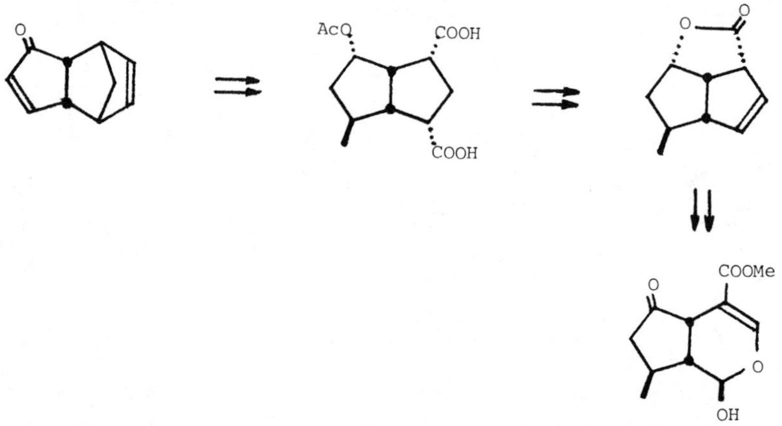

The same tricyclic dienone is also an excellent synthon for coriolin (**C70**) (Ito, 1982). The critical aspect of this synthesis is a configuration inversion at a neopentyl center: the desired alcohol is obtained via sulfonate displacement with potassium superoxide.

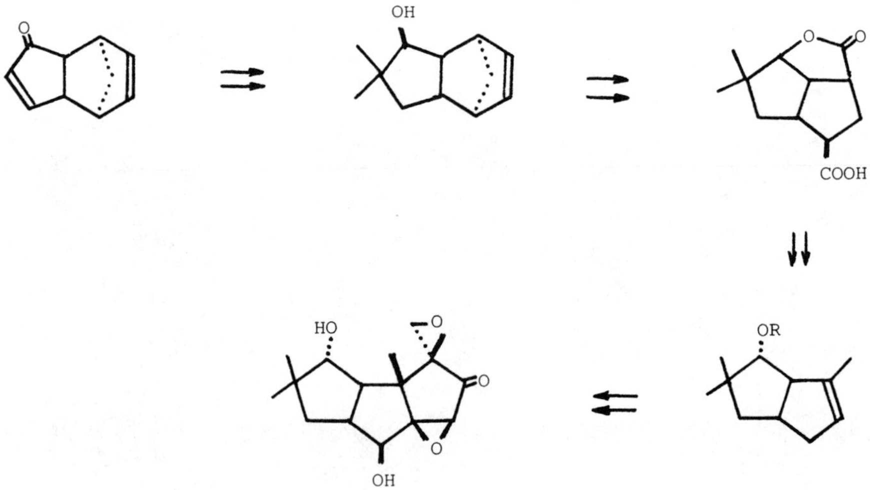

12.3 CYCLOPENTANES

Both the silphinene and the verbenalol syntheses involve the degradation of one of the olefinic carbon atoms in the norbornene framework, but the full utility of the carbon resources from the tricyclic intermediate of hinesol (**H15**) is in some way more satisfying (Helquist, 1985, Nystrom, 1985).

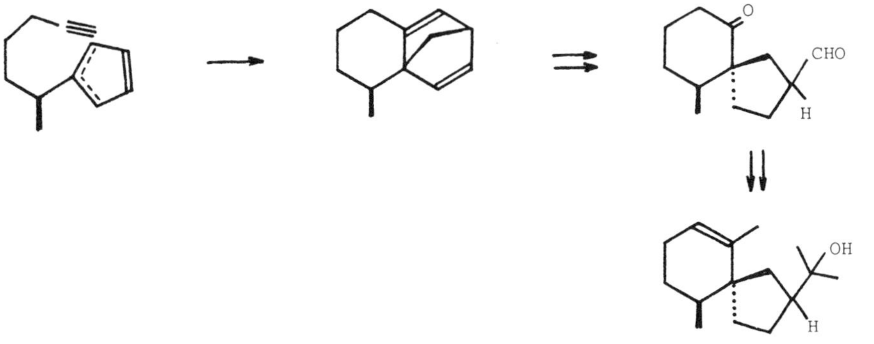

H15

The vastly different reactivities of the two double bonds of dicyclopentadiene make it easy to manipulate either site at will. Thus, partial hydrogenation and elaboration of the condensed cyclopentene ring readily accomplishes the syntheses of epi-β-santalene (**S5**) and sesquifenchene (**S28**) (Grieco, 1975b).

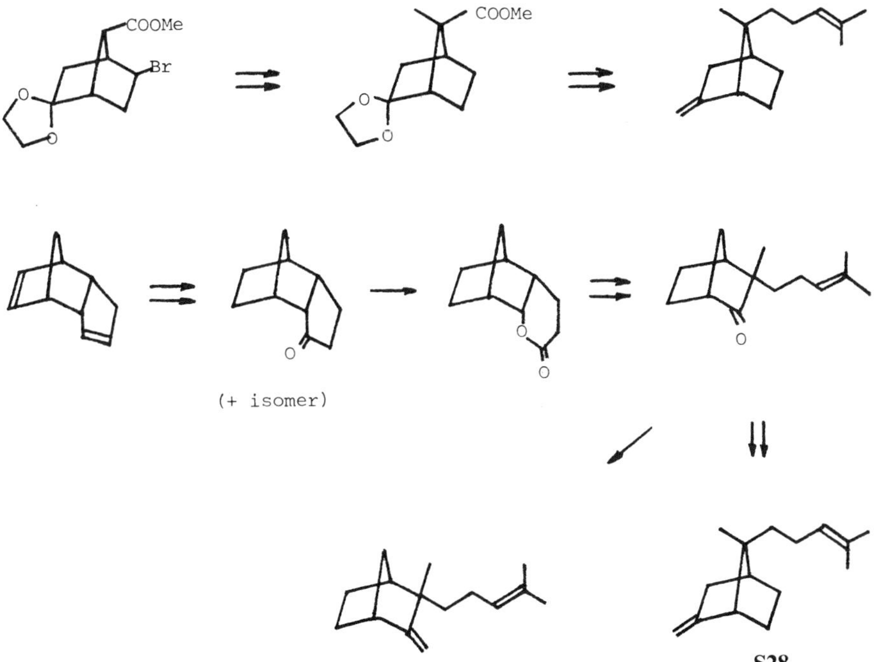

(+ isomer)

S5   **S28**

The intramolecular ene reaction that occurs after the Diel–Alder cycloaddition of a cyclopentane-1,3-dione acrylic ester and siloxy isoprene is a welcoming intervention during the synthesis of verrucarol (**V12**) (Trost, 1982a). One of the ketone groups is thereby masked as a homoallylic alcohol, leaving the other free to be modified. The alkene–ketone functions are recovered subsequently by pyrolysis.

V12

12.4 CYCLOHEXANES

As is well known, the Diels–Alder reaction effects a simultaneous formation of two bonds. Its efficiency and stereo- and regioselectivity advantages may still outweigh the merits of other reactions even when a synthetic operation calls for the formation of one bond only, which necessitates a ring cleavage of the cycloaddition product.

A relatively simple cyclohexenone synthesis may be illustrated with a precursor of trichodermin (**T24**) (Colvin, 1971, 1973), which is preferably unraveled from Diels–Alder adducts. An important element present in these adducts and in the diene addend is an alkoxy group, which is not only instrumental in unlocking the cyclohexenone regioselectively, but also is a determinant in the cycloaddition that leads predominantly to fragmentable intermediates. The methoxycyclohexadienes most frequently encountered are conveniently obtained by Birch reduction of the corresponding anisoles. These dienes undergo *in situ* conjugation (which can be catalyzed by dichloromaleic anhydride).

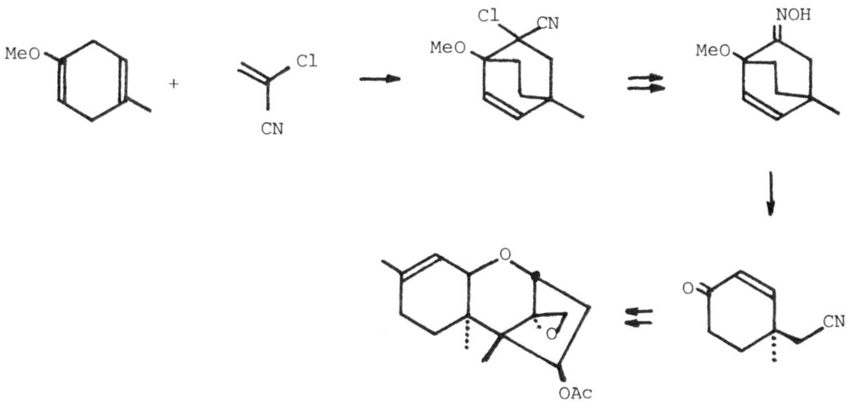

T24

The basic scheme was originally devised for a synthesis of *threo*-juvabione (**J9**) (A. J. Birch, 1969, 1970).

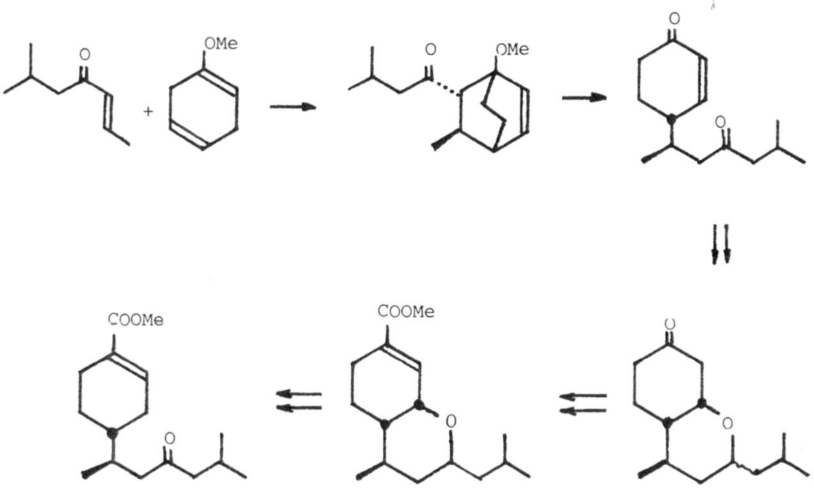

J9

An excellent route to nootkatone (**N11**) (Dastur, 1973, 1974) is based on the same strategy. The incorporation of an additional vinyl group conjugated to the masked enone, and the presence of an incipient trisubstituted double bond situated five carbon atoms away from the far end of the polyunsaturated system constitute an ideal environment in which the unfolding of the bridged rings is immediately followed by internal interception of the electron-deficient species to form a six-membered ring.

The configurations of the methyl substituents are secured at the Diels–Alder level on the ground of least hindered approach. The three-carbon chain is fixed at the equatorial orientation as the octalone is being created.

Glutinosone (**G15**) has been prepared in the same manner from the desmethyl analog. Only α-hydroxylation is needed after the octalone emerges (Murai, 1980).

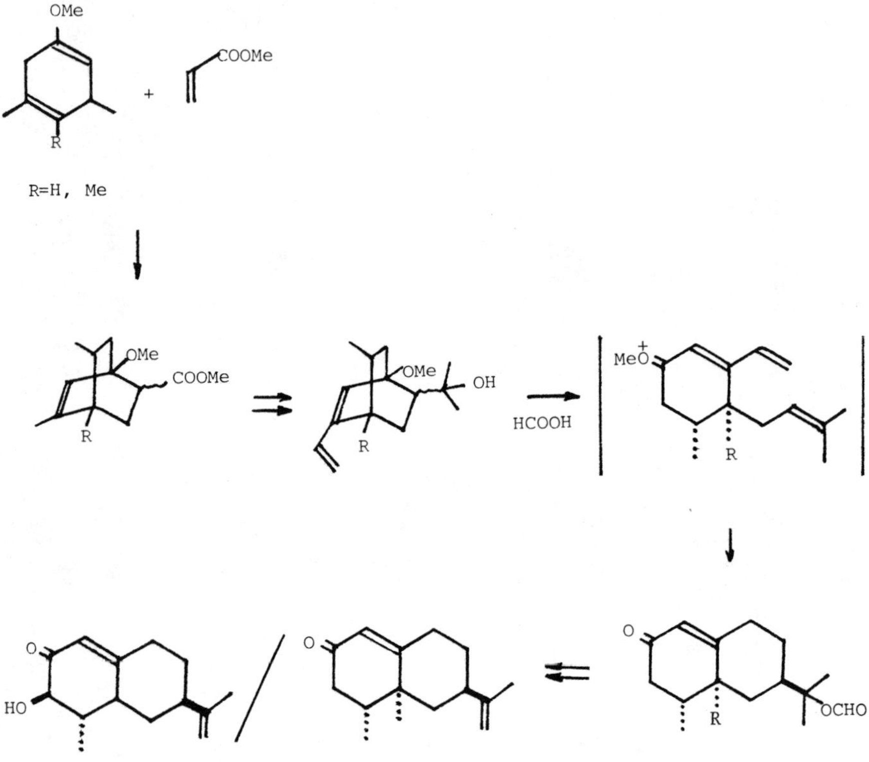

If a cyclohexenone so unveiled also carries a short chain with functionality at C_4, a spirocyclic system can be elaborated from it. Since both syn and anti isomers are formed (rather unexpectedly) from the Diels–Alder reaction of (2,6-dimethyl-4-methoxy-1,3-cyclohexadienyl)acetaldehyde ethyleneketal with methyl acrylate, β-vetivone (**V16**), hinesolone (**H16**), and solavetivone (**S41**) are obtainable (Murai, 1981), the first two terpenes from the syn adduct, the last one from the anti isomer.

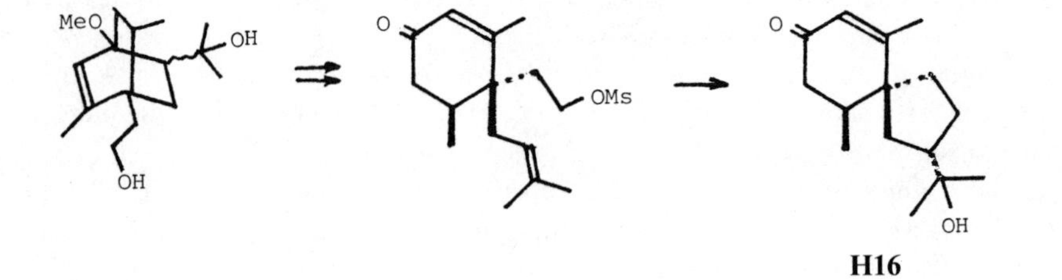

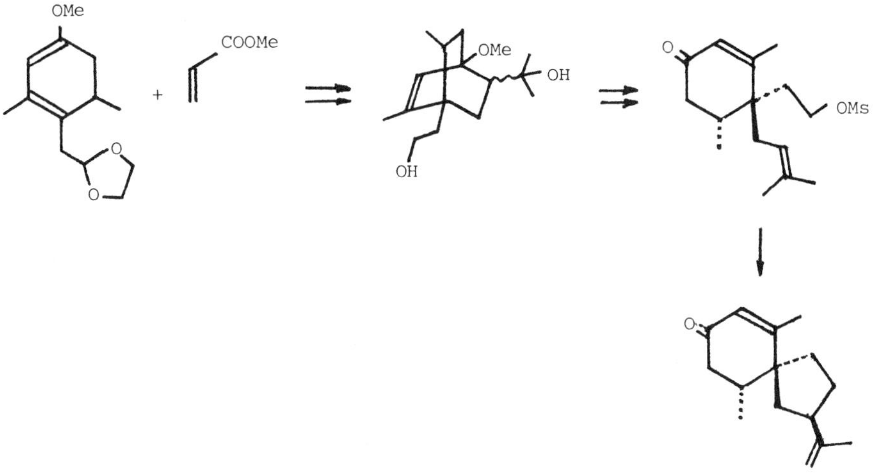

A variation of the cyclohexenone synthesis, with transposed functionalities, is via a reflexive Michael reaction of a monoprotected 1,3-cyclohexanedione with an unsaturated carbonyl substance, and transformation of the ketone group in the adduct into an alkene. The process appears in a synthesis of eriolanin (**E14**) (M. Roberts, 1981).

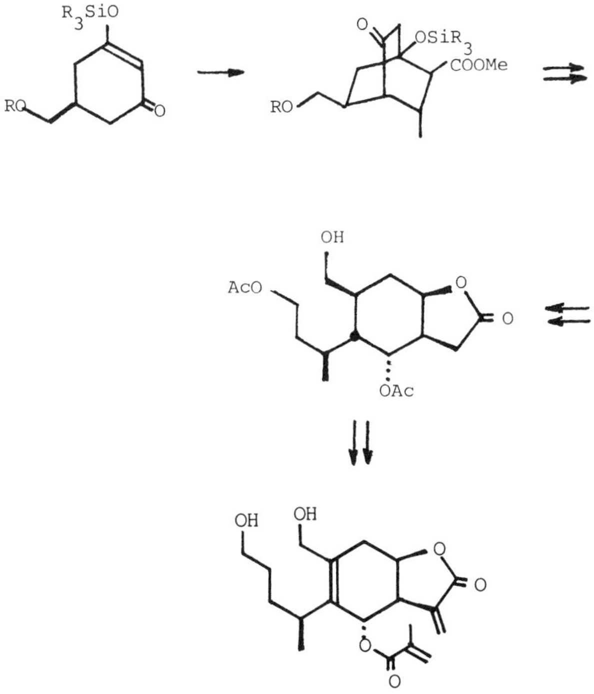

The Diels–Alder approach to trichodermol (**T25**) (Still, 1980) from benzoquinone is characterized by a high degree of regio- and stereoselectivity. All the structural modifications that are carried out on the enedione moiety of the adduct are controllable by virtue of its intimate association with the bicyclo[2.2.2]octane framework. Fragmentation is evoked when all the necessary substituents are properly marshalled. For example, the oxygen function on the five-membered ring exerts a directing effect on the process of epoxidation, which leads to the correct configuration at C_2 of the hydropyran ring.

T25

In a synthesis of pleuromutilin (**P39**) (Gibbons, 1982), cleavage of a C—C bond common to two cyclohexane rings to unveil the cyclooctane nucleus was planned. For reasons of synthetic expediency and steric control of the many substituents, an attempt to form the cyclooctane ring directly would not be promising. The transitory annulation tactic appears to be an excellent solution. Accordingly, an intermediate that allows a retroaldol fission was targeted, although slight modification in carrying out the ring cleavage was required.

P39

12.4 CYCLOHEXANES

For a study of taxane synthesis, a bicyclic precursor has been assembled (Nagaoka, 1984) with three key operations: a reflexive Michael addition to form a bicyclic template, an intramolecular alkylation to latch another bridged ring, and the removal of a C—C bond to unfold a cyclooctene unit. Another intramolecular Michael reaction is then used to introduce stereoselectivity a chain with which an eventual closure of the C ring is planned.

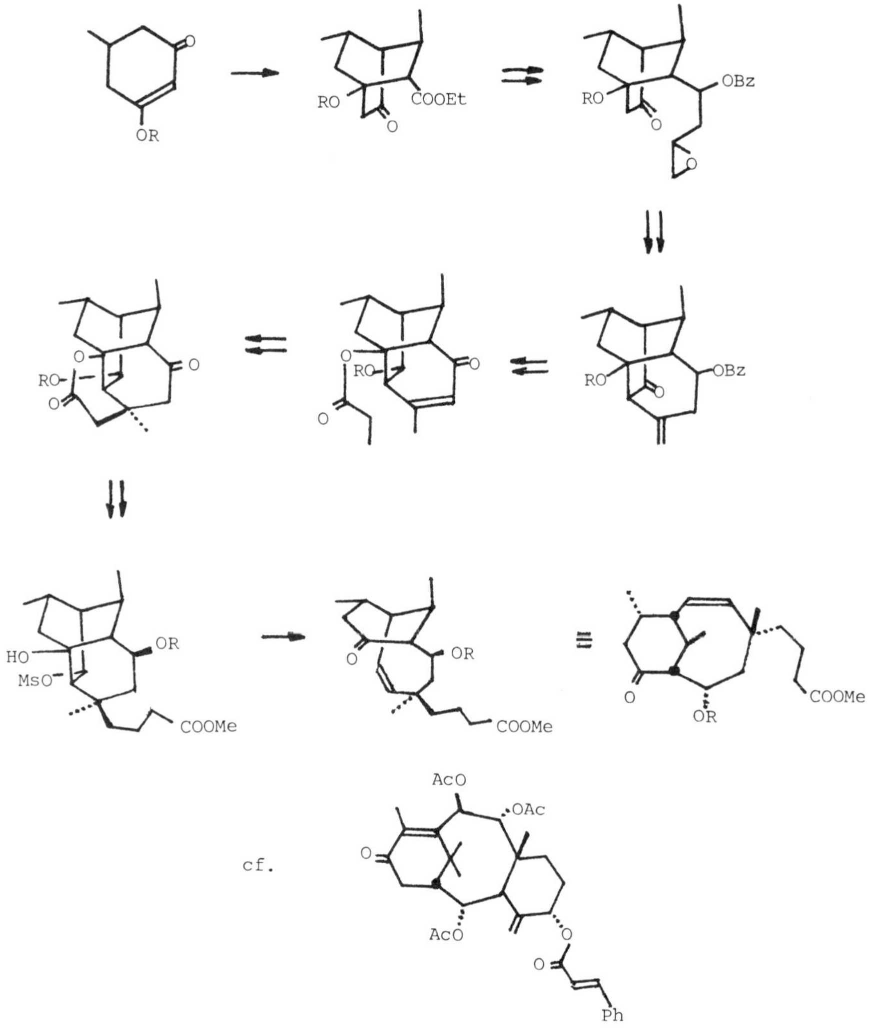

taxinine

In the construction of an A-secoquassin precursor (Stojanac, 1979) via a Diels–Alder reaction, the adduct is cleaved at the double bond to generate two carbonyl groups. These functions are to be incorporated into the A ring and the δ-lactone, respectively.

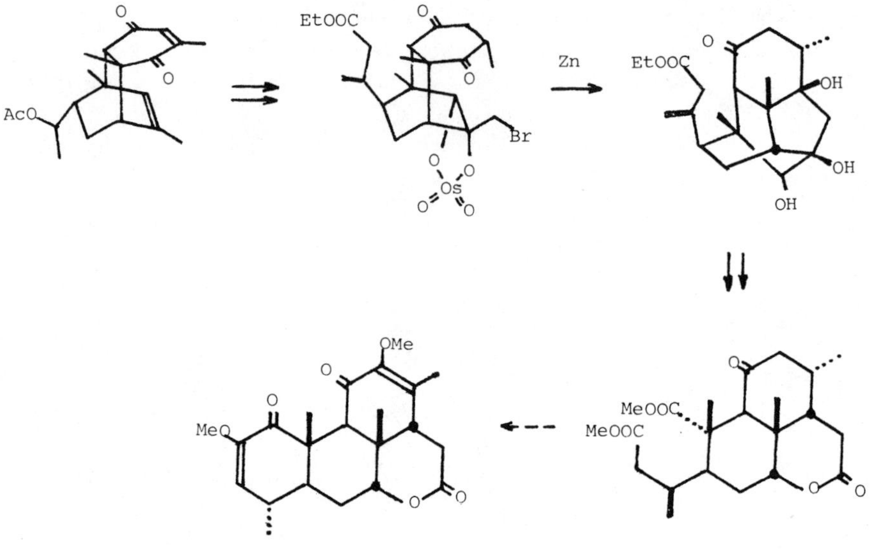

Decalins with stereodefined pendants are easily accessible. Consequently, many monocarbocyclic sesquiterpenes have been synthesized with preconceived ring fission of eudesmanes at a specific point. Furthermore, this consideration opens pathways wherein abundant compounds such as α-santonin may be used as starting material.

The elemanes are cyclohexanes with a *trans*-1,2-divinyl group. The crucial stereochemical relationship can be established at the bicyclic level by lithium/liquid ammonia reduction of octalones. Conversion of the resulting decalones into Δ^2-olefins sets the stage for the ring cleavage. This general strategy is applicable to the synthesis of γ-elemene (**E2**) (M. Kato, 1979), β-elemenone (**E3**) (Majetich, 1977), shyobunone (**S32**) (K. Kato, 1970, 1971), temsin (**T10**) (Nishizawa, 1978), saussurea lactone (**S14**) (Ando, 1978a, b), and costunolide (**C75**) (Grieco, 1977a).

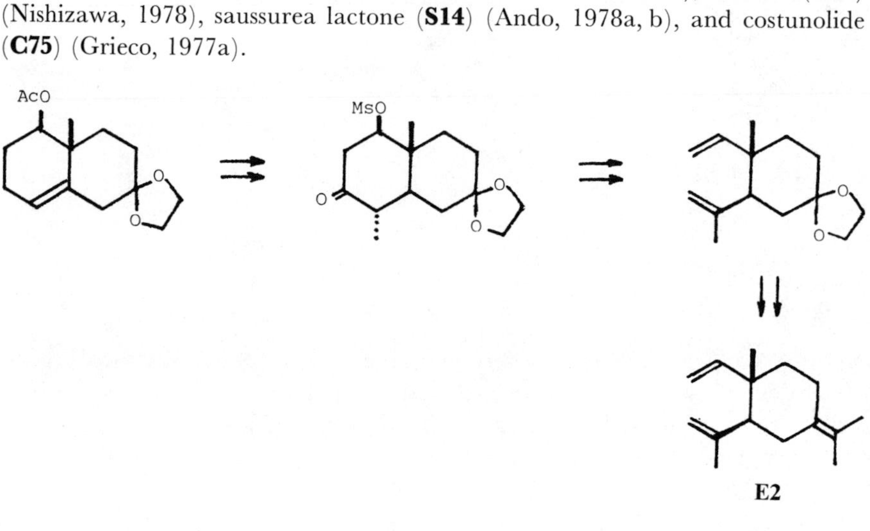

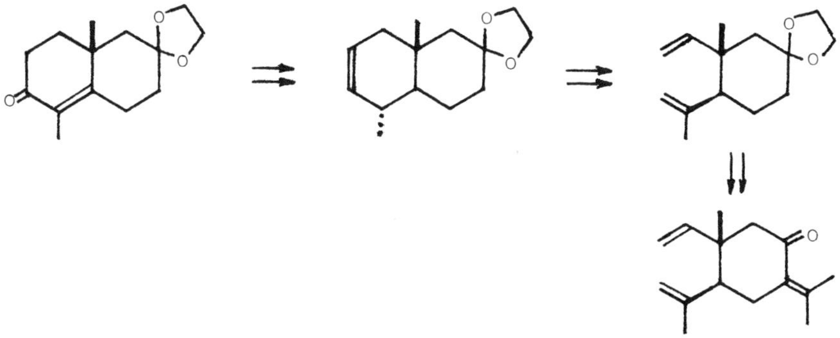

12.4 CYCLOHEXANES

E3

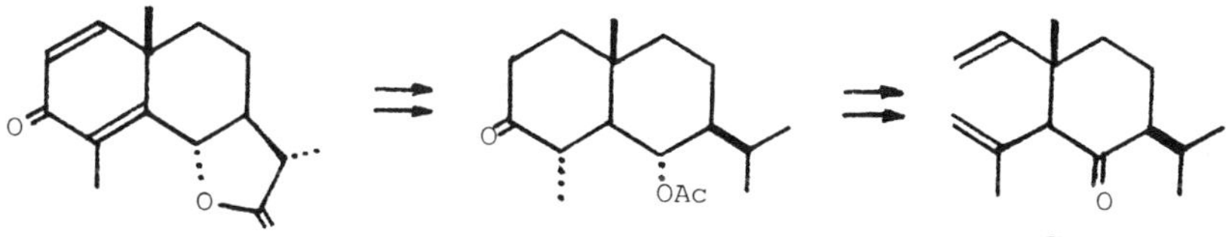

S32

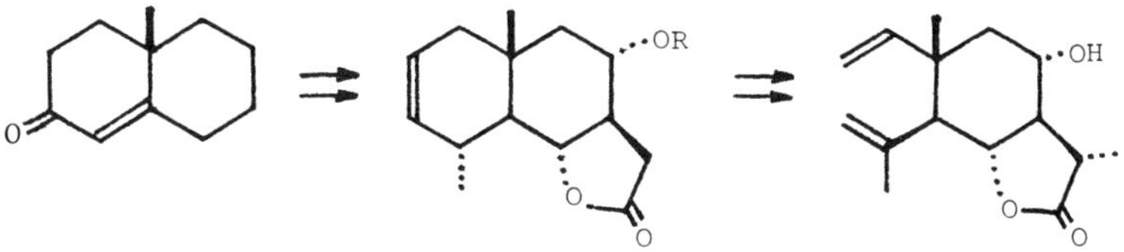

T10

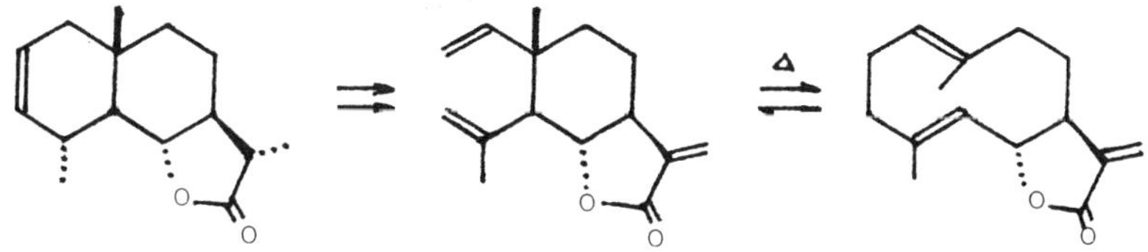

C75

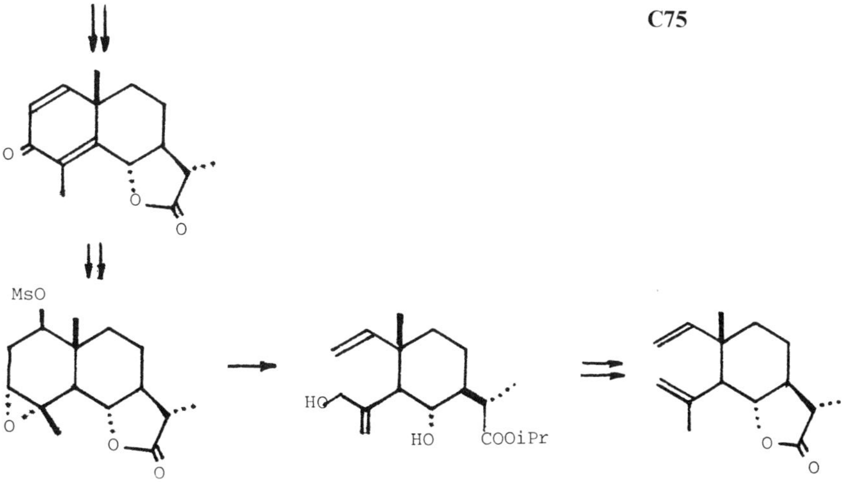

S14

Vernolepin (**V10**) and vernomenin (**V11**) are oxidized elemanes. The angular methyl group as well as the isopropenyl pendant are oxygenated and tied up in an α-methylene-δ-lactone unit. Consequently, the cyclohexane ring is fused to this lactone with a cis juncture. The "biogenetically patterned" approach to these sesquiterpenes accordingly passes through a *trans*-decalin in which the angle carries an alkoxymethyl substituent for eventual lactonization (Grieco, 1976a, 1977b).

V10

Three other classes of A-secoeudesmanes are typified by phytuberin, ivangulin and umbellifolide. A biomimetic route to phytuberin (**P28**) (Murai, 1978) calls for the preparation of a 1,2,5β-trihydroxy-α-eudesmol. When the α-glycol system is broken, and the functional groups are allowed to recombine, the difuran nucleus evolves.

P28

Chemists have taken advantage of this knowledge to extract the carbon chains of ivangulin (**I20**) (Grieco, 1977d) and eriolanin (**E14**) (Grieco, 1978) from the A ring of eudesmane-type intermediates. It is essential the configuration of the extracyclic tertiary carbon atom be set before a free-rotating chain is released. That the methyl group occupies an equatorial position in the decalin precursors simplifies much of the operation.

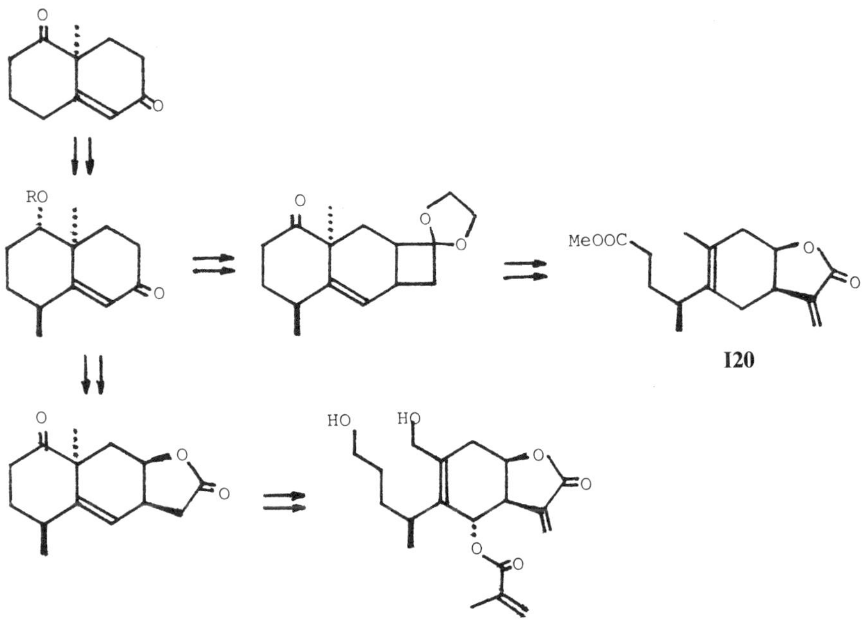

I20

E14

The conversion of (—)-artemisin into umbellifolide (Marco, 1987) centers on A ring modification to give a Δ^4-olefin and inversion of the lactone. Cleavage of the A ring and generation of the conjugated double bond have been accomplished in one step.

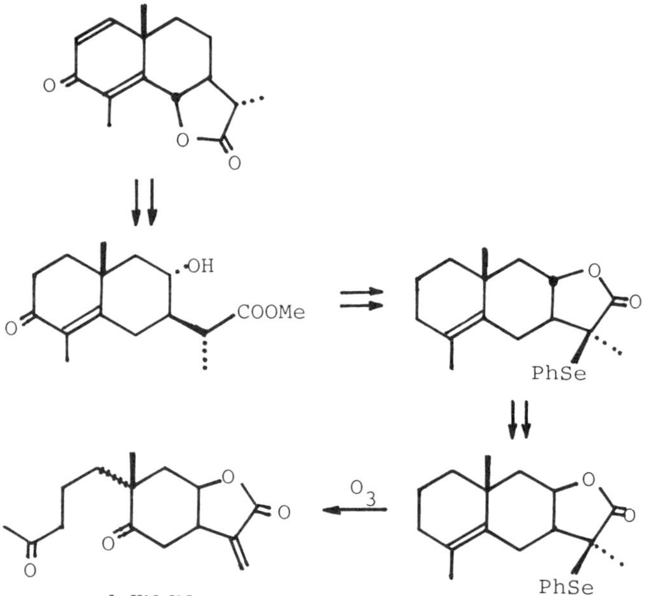

umbellifolide

A dimethyloctalone used in the synthesis of frullanolide (**F14**) (Semmelhack, 1981b) is cleaved and transformed into a properly protected monocyclic intermediate to which the elements for lactonization are added. The sesquiterpene is then assembled in one step.

F14

Dihydro-epi-α-cyperone supplies nearly all the carbon atoms of the bicyclo[3.2.1]octane segment of ylangocamphor (**Y3**) and sativene (**S11**) (Piers, 1975b). More important, the stereochemical relationship between the methyl and the isopropyl groups within the six-membered ring is established unambiguously. Direct alkylation of carvomenthone gives mainly the wrong isomer, which leads to copacamphor (Piers, 1971, 1975a).

S11 +   **Y3**

Formation of a bridged cyclopentane on the periphery of hydroxydecalone (actually via intramolecular alkylation of a hydroxytetralin) and subsequent dismantling the oxygenated ring form the crux of the first synthesis of hinesol (**H15**) (Marshall, 1969).

H15

The earliest approaches to germacranes involve contrathermodynamic rearrangement of elemanes. As a result, the mesocyclic products were isolated in small quantities and with difficulties due to their reversible transformations. By a fragmentative pathway hedycaryol (**H2**) (Wharton, 1972) has been prepared from an octalin derivative. The E,E-configuration of the 1,5-diene unit arises from 5α-H and 1β-OTs pattern.

H2

The cleavage of the common C—C bond of the two cyclohexane rings of a decalin is a crucial operation whereby a monocyclic diene (with one of the double bonds also conjugated to an exocyclic methylene residue) is to be made and recycled photochemically to a bridged precursor of longipinenes (**L25**, **L26**) (Miyashita, 1971, 1972, 1974).

L25, L26

The corresponding 1,6-dienes may undergo cyclization to a hydrazulene skeleton. Thus globulol (**G14**) (Marshall, 1974) owes its laboratory genesis to an octalindiol progenitor.

The *cis*-fused cyclohexano-δ-valerolactone feature of vernolepin (**V10**) probably supplied a major impetus to the development of the siloxydiene chemistry (Danishefsky, 1981). The Diels–Alder adducts of 1-methoxy-3-trimethylsiloxy-1,3-butadiene are synthetic equivalents of cyclohexenones, from which the removal of the α carbon is conceivably facile. This transformation is needed in the consolidation of the vernolepin skeleton (Danishefsky, 1976, 1977).

Since pentalenolactone (**P18**) (Danishefsky, 1978, 1979) also has a δ-lactone unit, now being condensed to a five-membered ring, formulation of a synthetic scheme based on the same theme as vernolepin is only natural. It turns out that a bicyclo[2.2.1]heptenedicarboxylic anhydride is a useful dienophile, and when the adduct is formed, three asymmetric centers are simultaneously established. The synthesis then requires the cleavage of the enone and the remote ethano bridge, re-formation of a cyclopentenone, and functional group adjustments.

P18

As for coriolin (**C70**) (Danishefsky, 1980c, d), indirect alkylation via the Diels–Alder adduct is perhaps a little circuitous and wasteful. However, it must be stressed that the regio- and stereoselective alkylation of the diquinane enedione is not simple (if one insists on using the enedione as the starting material).

C70

A most unusual synthetic design for hirsutic acid C (**H19**) (Trost, 1979) involves temporary connection of the two methyl groups in its precursors. A cyclohexene ring is the most useful latent form for these substituents. It is then possible to define other structural requirements for achieving the carbon framework of a diquinane, and forward elaboration has close analogies to this process.

An intramolecular Michael addition initiates construction from a substituted cyclohexanone. The resulting bridged system is further annulated using the existing carbon chain, while a double bond is also introduced into the six-membered ring at this phase of the synthesis. When the cyclohexene is opened, the molecule metamorphoses into a familiar diquinane structure whose similarity to the target becomes evident.

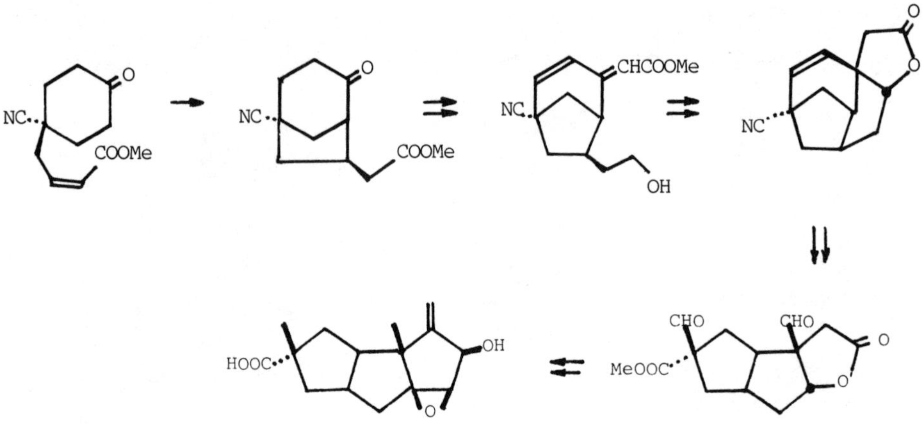

H19

An outstanding synthetic scheme for quadrone (**Q1**) (Dewanckele, 1983; Schlessinger, 1983) consists of an exo-selective, intramolecular Diels–Alder condensation and the subsequent degradation of the cyclohexene moiety of the adduct to provide a carboxyl (prolactone) group and an acetonyl chain. The latter is to be incorporated into the cyclopentanone ring by an aldol reaction.

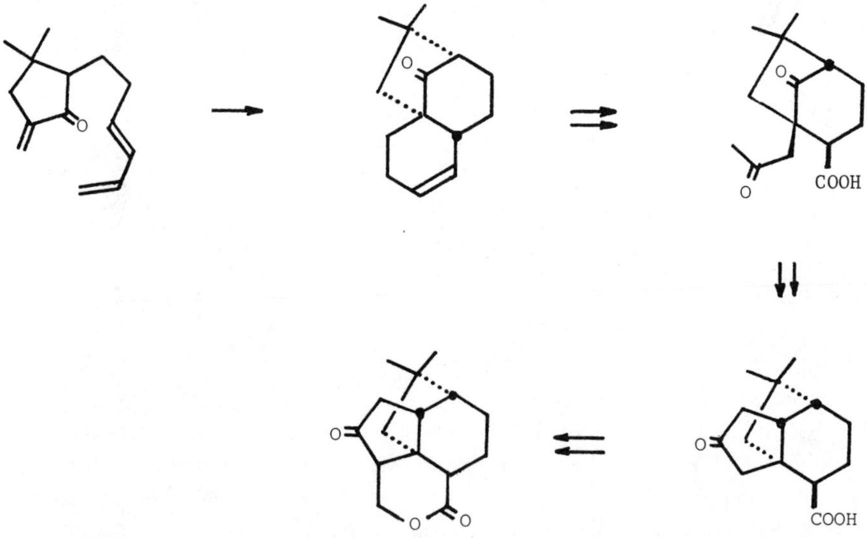

Q1

A formal synthesis of enmein (**E6**) (Fujita, 1972) involving the B-ring cleavage of a kaurene derivative may have a biogenetic parallel.

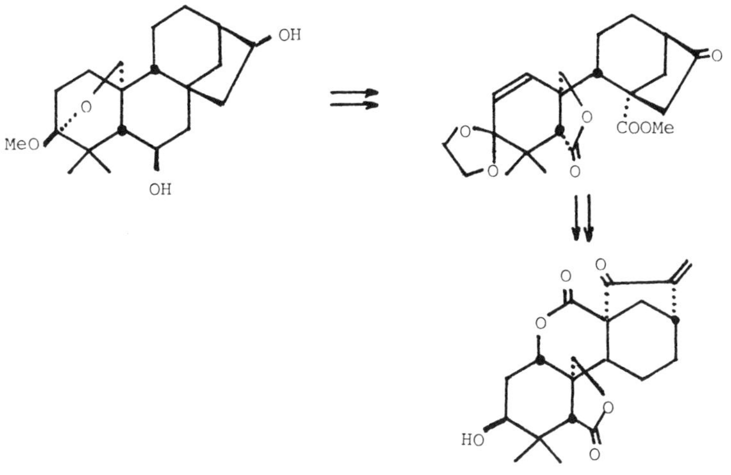

The most pleasing approach to the resin acids is probably the A + C + B scheme (Ireland, 1966). The potential carboxyl group and the ethano segment of the B ring are supplied from 3-buten-2-one. An exquisite stereocontrol at C_4 and C_5 is exercised by the order of alkylation and by lithium/liquid ammonia reduction, respectively. A synthesis of dehydroabietic acid (**A2**) is shown below.

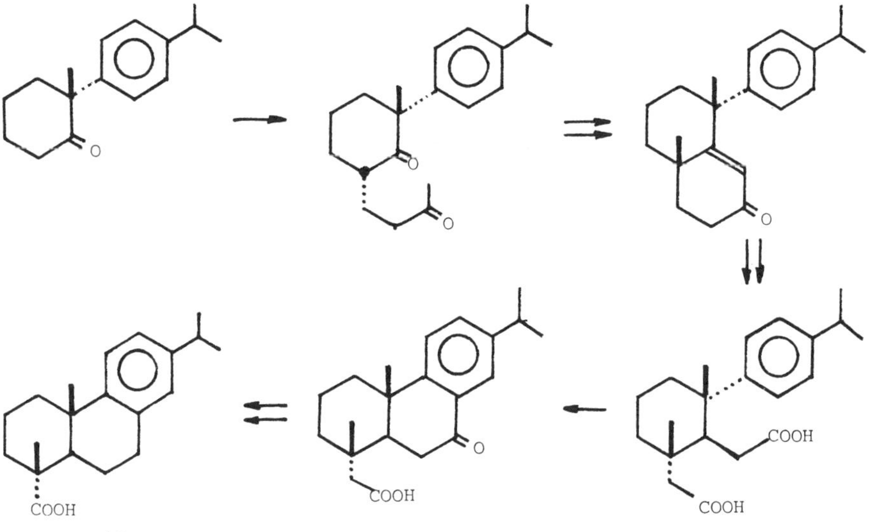

A synthesis of taxodione (**T7**) (Poirier, 1983) from dehydroabietane via a B-seco acid is also engaging. Benzylic oxidation of the hydrocarbon furnishes an α-tetralone which is then converted into a phenolic compound by a Baeyer–Villiger reaction and methanolysis. The veratryl nucleus can now be elaborated via an *o*-quinone. The B ring is reconstituted by a Friedel–Crafts acylation.

The important aspect of this route is the recognition of a latent symmetry in the aromatic nucleus on which a net meta-oxygenation may be achieved by the ring cleavage/reclosure operation.

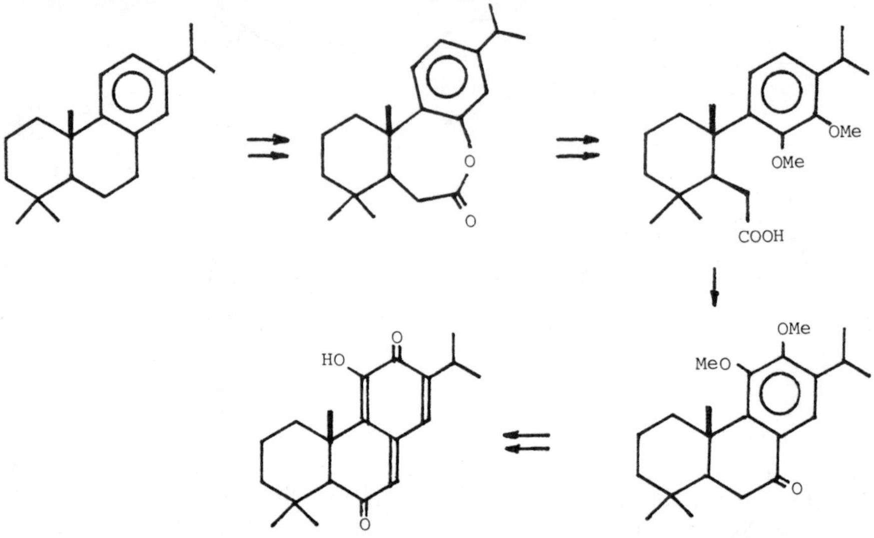

T7

An interesting transitory annulation sequence appears in a synthesis of portulal (**P45**) (Tokoroyama, 1974; Kanazawa, 1975), mainly for introduction of the angular aldehyde and the vinylic methyl group. Thus, a hydrazulenone undergoes a Michael–aldol reaction to give a bridged ketol; the β-configuration of the emerging formyl group is ensured.

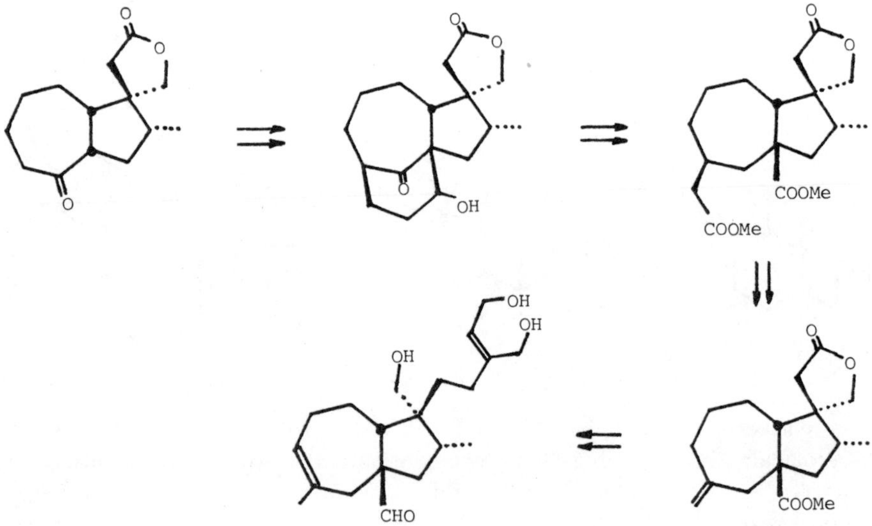

P45

12.4 CYCLOHEXANES

Aromatic compounds, especially phenolic derivatives, are latent dicarbonyl substances. In recognizing the facilitation of synthesis of zizanoic acid (**Z3**) (MacSweeney, 1971) by a 5:6 → 6:5 cycloexchange, a camphenilone derivative in which a γ-ketoester chain linked to the bridgehead is needed for preparation of the precursor. This ketoester is derivable from a *p*-cresyl ether via Birch reduction and conjugation of the methoxy-1,4-cyclohexadiene.

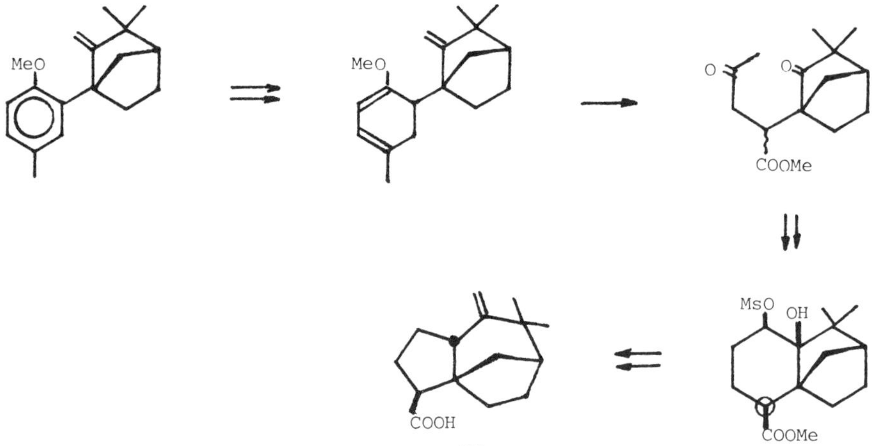

In one of the syntheses of juvenile hormone JH I (**C35**) of the cecropia moth (Corey, 1968), a keen correlation of the terminal epoxide geometry and hence that of the corresponding double bond with the Birch reduction product of *p*-cresyl methyl ether has simplified the preparation of the important synthon. The two double bonds in the dihydroaromatic compound are vastly different in reactivity toward ozone, so that selective cleavage of the ring is feasible.

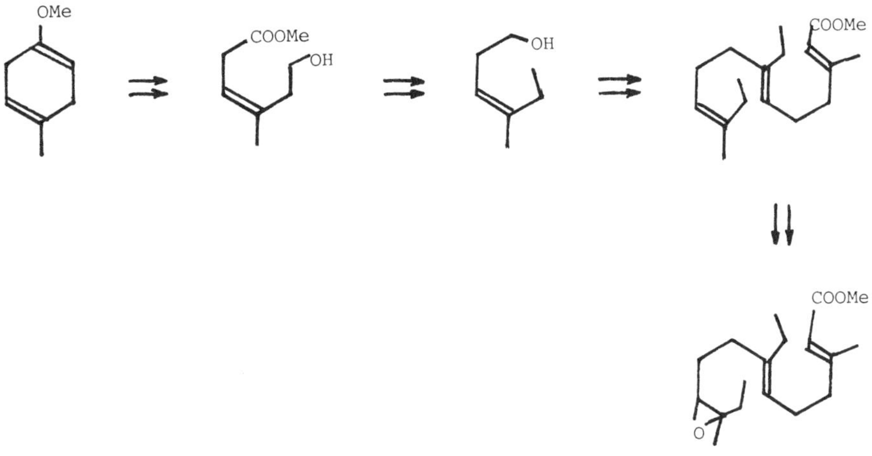

An interesting strategy in synthesis consists of annulation on a cyclic scaffold, which is dismantled after serving its designated role. In a synthesis of *erythro*-juvabione (**J8**) (Schultz, 1984) a kinetic intramolecular alkylation of a cyclohexenone leading to a bridged bicyclic framework permits stereoselective reduction of the double bond to establish the correct relative configuration of a secondary methyl substituent. Baeyer–Villiger oxidation of the saturated ketone, alcoholysis of the resulting lactone, and chain elongation give the known intermediate.

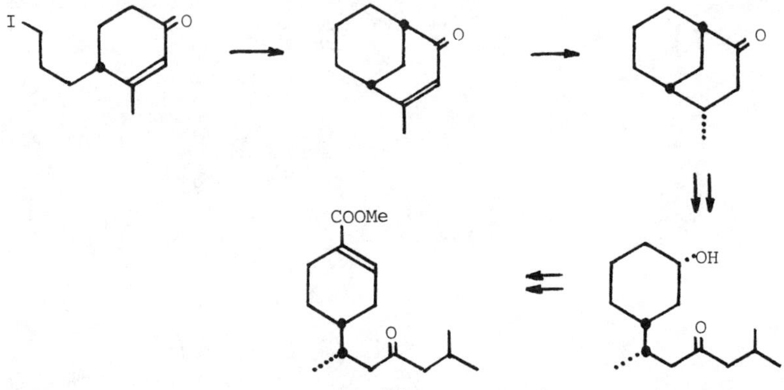

J8

The spiro[4.5]decane system unraveled from a bridged tricyclic precursor which is obtainable from a photochemical process can be exploited for synthesis of the acorane skeleton (SubbaRao, 1986).

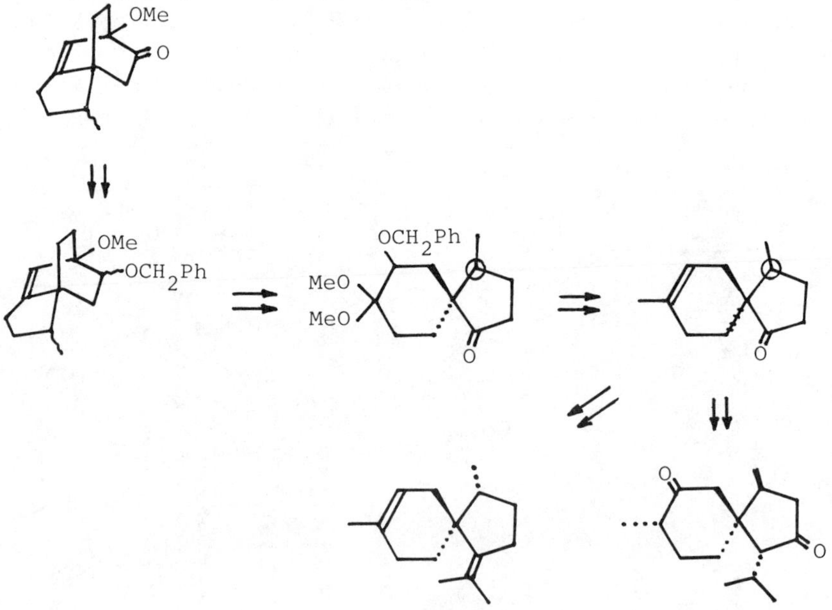

γ-**acoradiene**   **A11**

12.4 CYCLOHEXANES

Although the ophiobolins still await completion of synthesis, many model studies have succeeded in constructing the ring system. In one of these investigations (Boeckman, 1977) a full stereocontrol is attained at the level of a hydrindenone intermediate (two chiral centers), which then dictates the direction of conjugate addition with a cuprate reagent on an unsaturated lactone derived therefrom.

The eight-membered ring is created via an aldol protocol followed by oxidation and fragmentation. The third ring is then assembled by a Dieckmann condensation.

The flexibility of this approach is associated with the epimerizability of the hydrindenone ring juncture during an acid-catalyzed lactonization without affecting the stereochemical course of a subsequent cuprate addition. The atomic arrangement corresponding to the ceroplastols (e.g., ceroplastol II, (**C44**) is established.

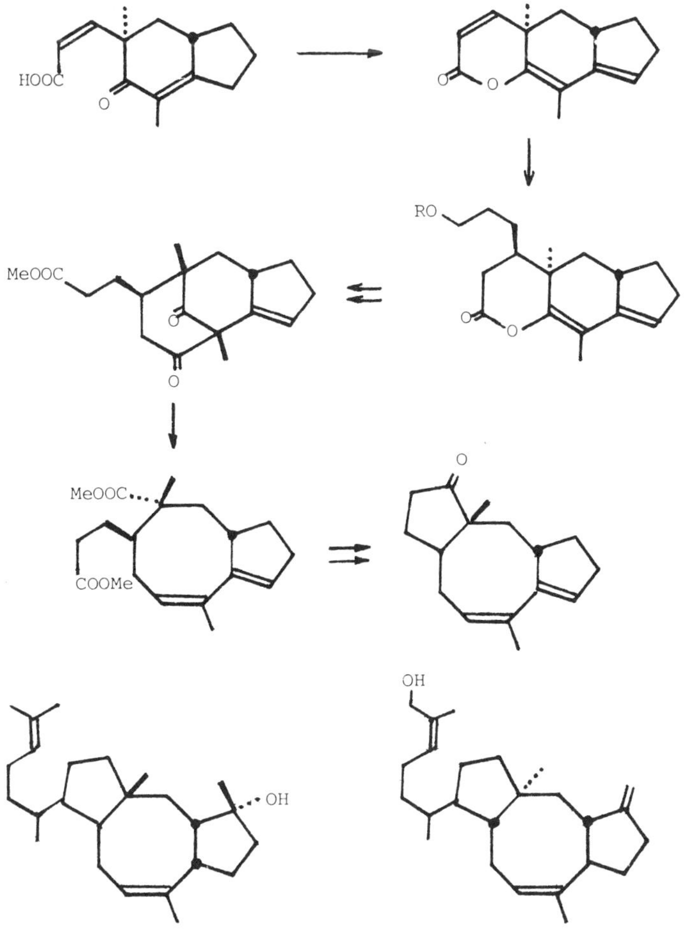

ophiobolin F

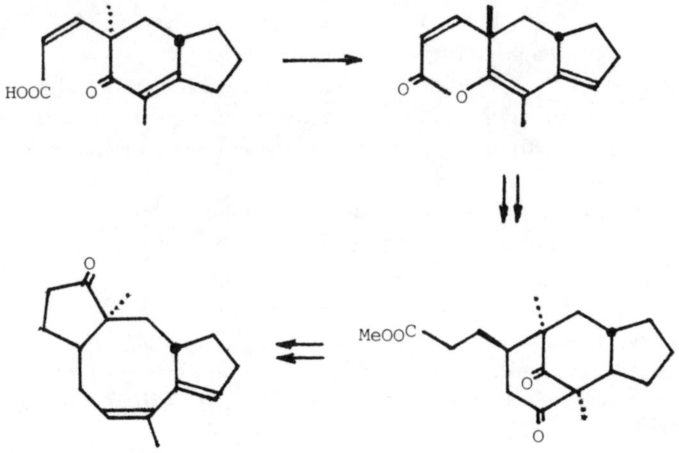

In a projected synthesis of vernolepin (**V10**) (Harding, 1981) a cycloheptene ring that constitutes a part of a bicyclo[4.3.1]decenone is to be severed; the bicyclic skeleton also is set up from the cycloheptenone. The ring cleavage should liberate two acetic acid chains correctly oriented for lactonization. Completion of the work is awaited.

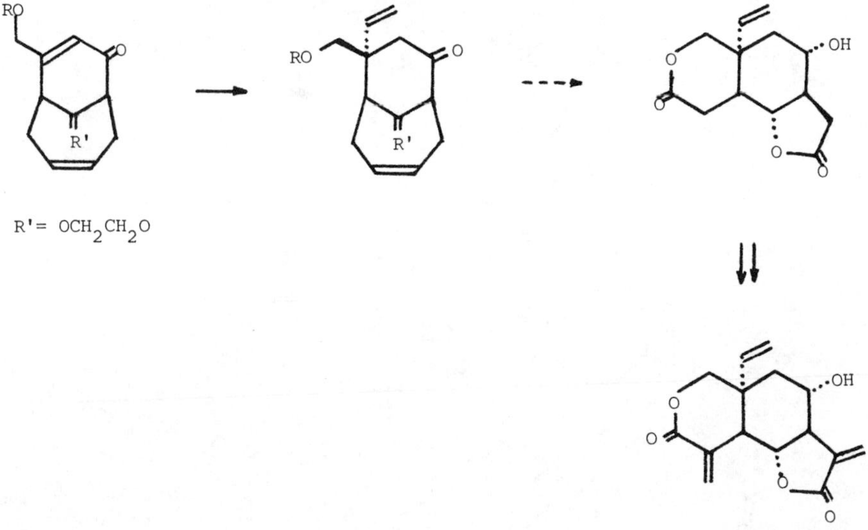

R' = OCH$_2$CH$_2$O

V10

An unusual synthesis of *ent*-copalic acid (**C68**) (Manh, 1975) involves a sensitized photochemical reaction of an allylic tertiary alcohol. The π electrons in the triplet state are avid hydrogen abstractors, and in the case of an allylic alcohol, the abstraction leads to a fragmentable species. The ketoolefin generated presently is two steps away from the target molecule.

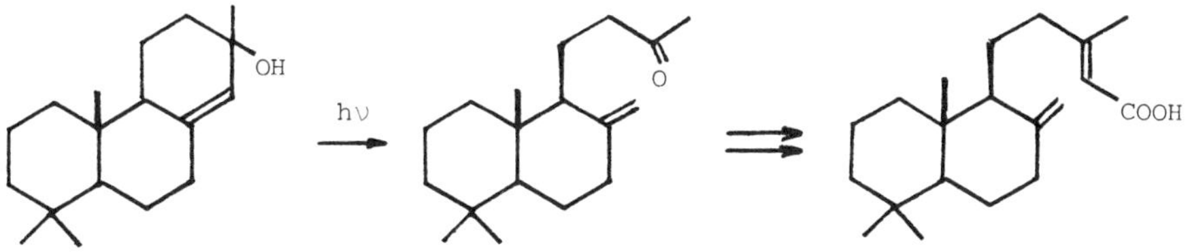

(+)-Grandisol (**G18**) is obtainable, albeit by a long sequence of reactions, from (—)-pinene via a photooximation of the *syn*-methyl group and homologation at that point. The six-membered ring is then degraded by its conversion into a saturated ketone and photolysis (Norrish Type I fragmentation) of the latter (Hobbs, 1974, 1976).

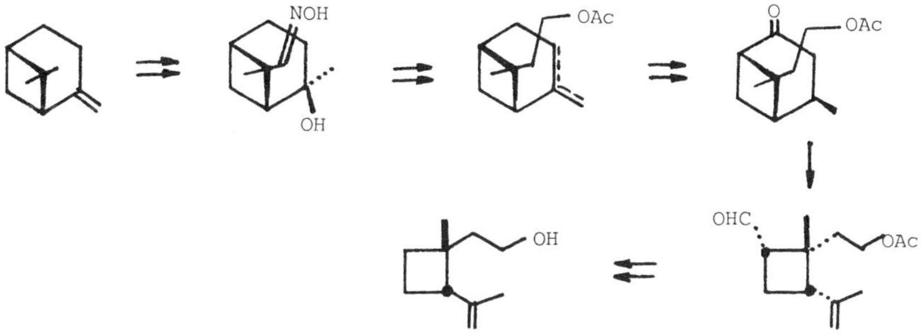

The inertness of furan toward dimethylmaleic anhydrides compels the adoption of more circuitous synthetic pathways for cantharidin (**C18**) (Stork, 1951, 1953), one of which involving the formation of an oxa-bridged octalin and a two-staged degradation of the cyclohexene moiety into a succinic anhydride.

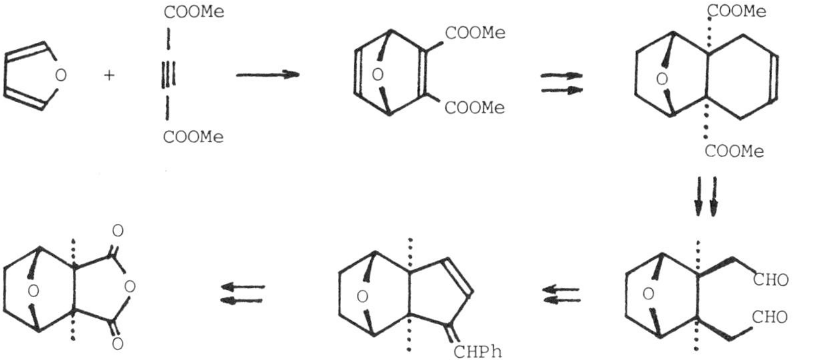

One of the most remarkable synthesis of JH1, the juvenile hormone of the cecropia moth (**C35**) (Zurflüh, 1968) sought to establish the geometry of the C_{15} dienone precursor by stereoselective construction of a hydrindane triol and a stepwise fragmentation.

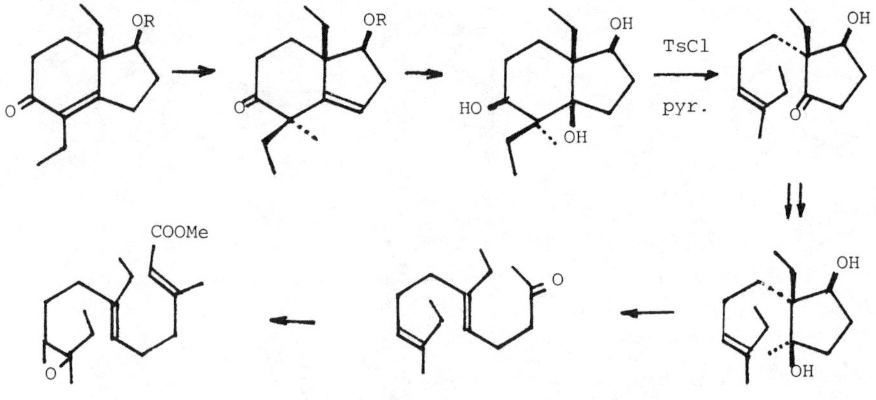

C35

An epimerization of the 19-methyl group is required during an impressive synthesis of butyrospermol (**B22**) (Kolaczkowski, 1985; Reusch, 1985) and it was planned into the reaction sequence. A Diels–Alder protocol for the formation of the B ring leads to a syn substitution pattern at C_9 and C_{10}, with C_9 being in the desired configuration. Epimerization of the 2,4-dien-1-one proceeds via the A-seco ketone by irradiation with 365-nm light in acetonitrile.

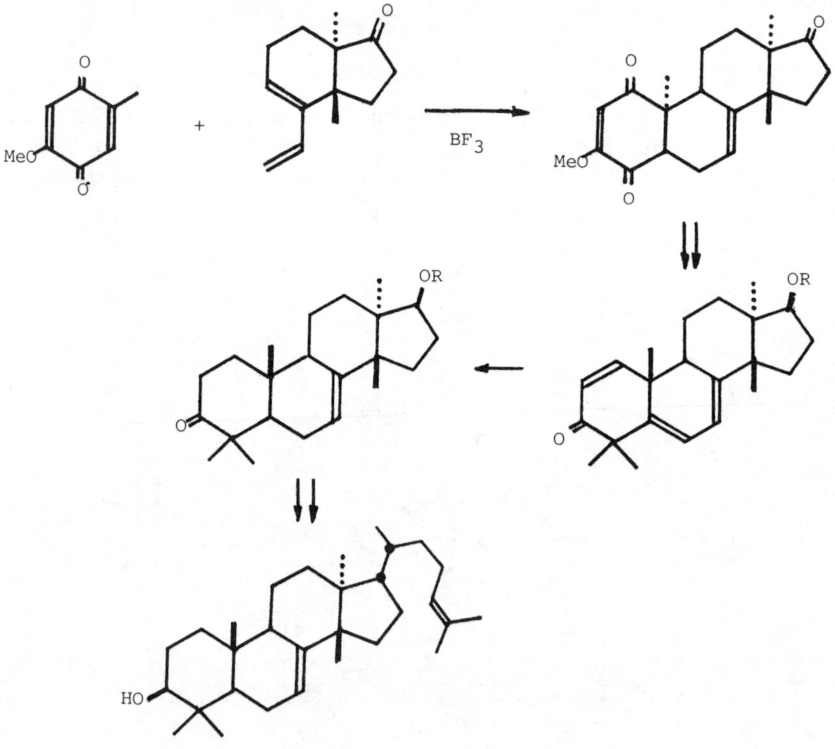

B22

12.5 HETEROCYCLES

Intramolecular formation of heterocycles occasionally serves the same end as the formation of carbocyclic counterparts in controlling stereochemistry of the products. Usually a heteroatom supplies a handle for ring cleavage after the desired transformations have been made.

One of the trichodiene (**T26**) syntheses uses a fused γ-butyrolactone as a platform for the construction of a cyclopentanone unit. The lactone is a V-shaped molecule, and it directs α-alkylation from the exo side. Accordingly, the two quaternary carbon centers are easily created stereoselectively (Welch, 1976).

T26

Calonectrin (**C12**), a more complex trichothecane, has been assembled (Kraus, 1982) from a pyranone via a heterocycloalkylation and aldolization of the acetaldehyde chain fashioned from the δ-lactone. The intramolecular alkylation ensures correct configuration at the bridging points of the ring system to be erected.

C12

12.6 RETROGRADE DIELS–ALDER REACTIONS

The retrograde Diels–Alder elimination is not very frequently employed in terpene synthesis. Extrusion of small molecules containing hetero-atoms is found in the syntheses of occidentalol (**O2**) (Watt, 1972) and copaene/ylangene

(**C66/Y1**) (Corey, 1973b), where carbon dioxide escapes from the Diels–Alder adduct of an α-pyrone.

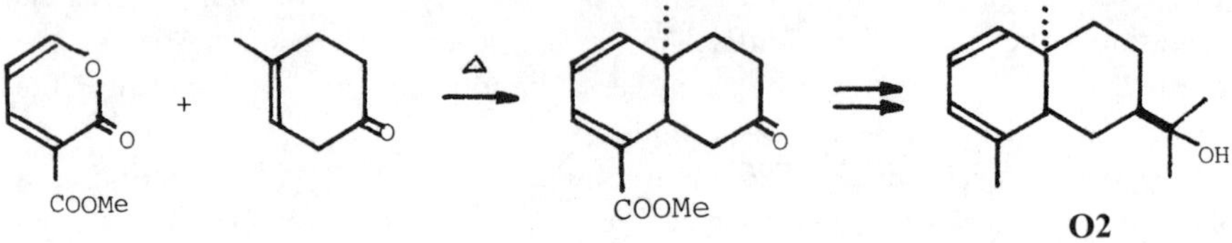

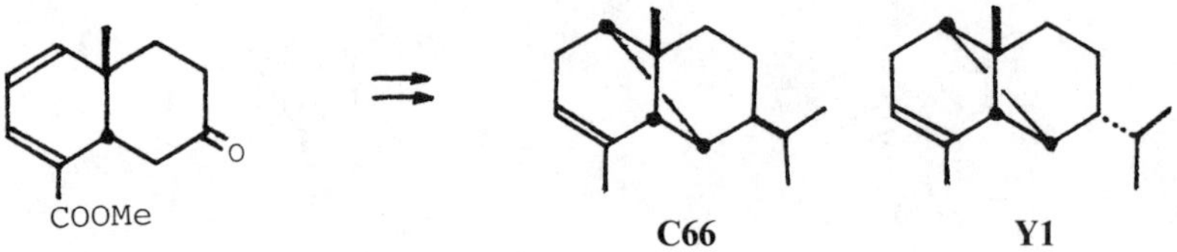

Alkyl cyanides are removed *in situ* when furanosesquiterpenes, including paniculide A (**P9**) (Jacobi, 1984a), gnididione (**G16**) (Jacobi, 1984c), and ligularone (**L13**) and petasalbine (**P23**) (Jacobi, 1984b) are constructed by an intramolecular reaction between an oxazole and an alkyne moiety.

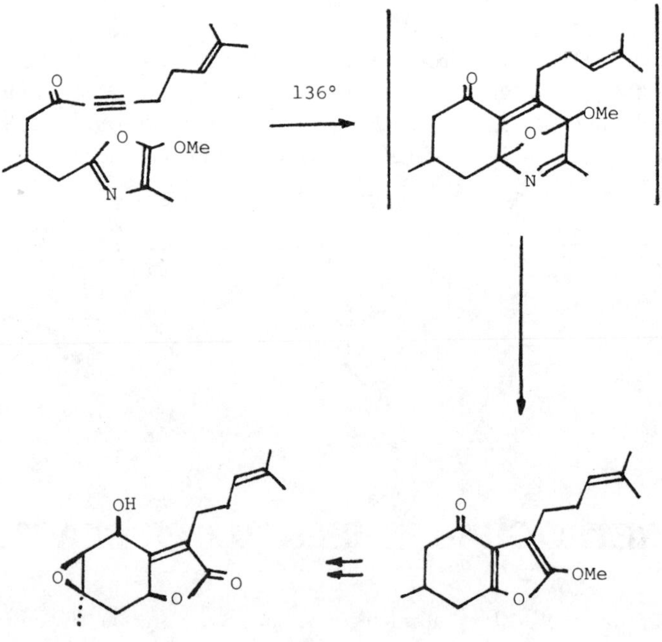

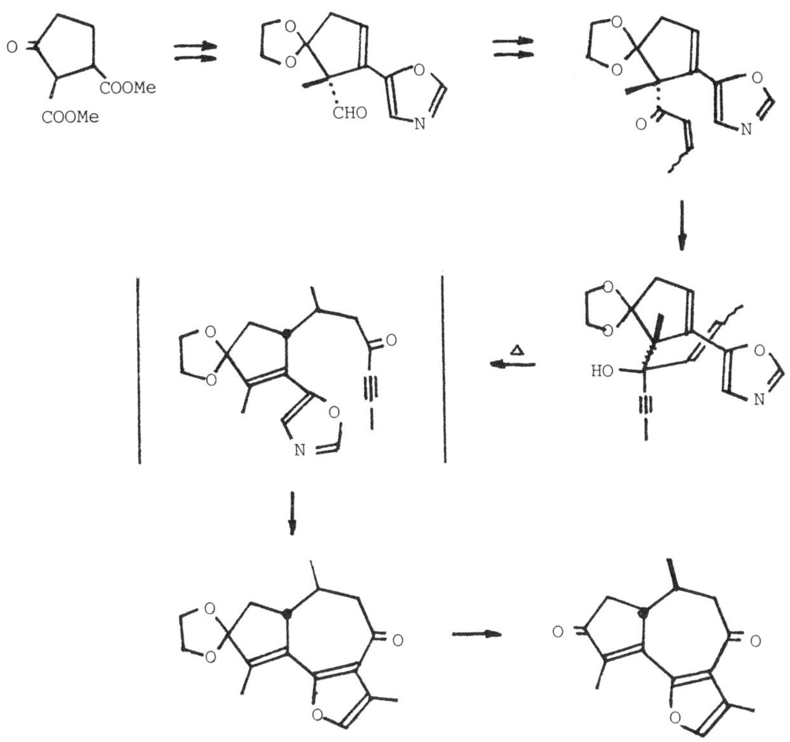

12.6 RETROGRADE DIELS–ALDER REACTIONS

G16

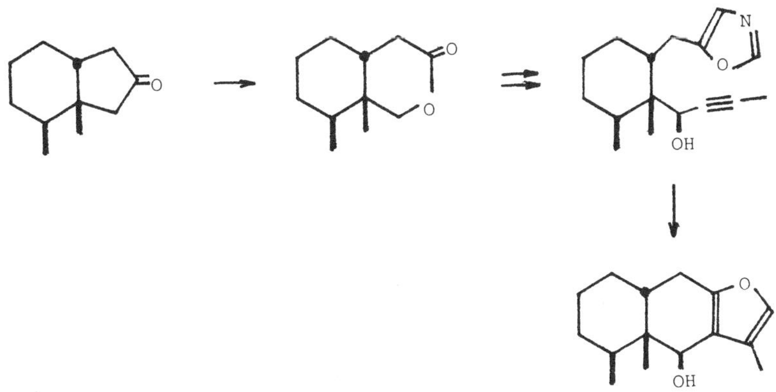

P23

Diels–Alder adducts of cyclopentadiene and furan are useful derivatives that may be elaborated until the "sensitive" double bond is needed. The syntheses of ar-turmerone (**T35**) (Ho, 1974), the *L. nidifica* alcohol (**N9**) (Bloch, 1983), and 8S,-14-cedranediol (**C36**) (Landry, 1983) incorporate this concept of protection.

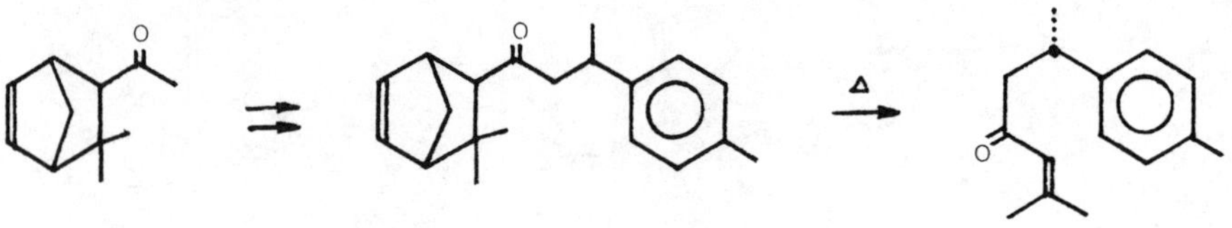

T35

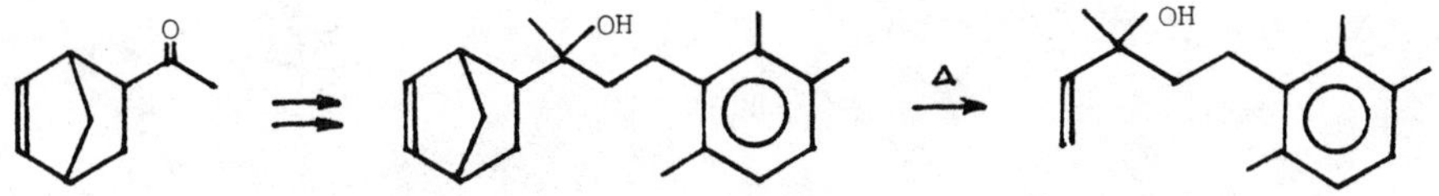

N9

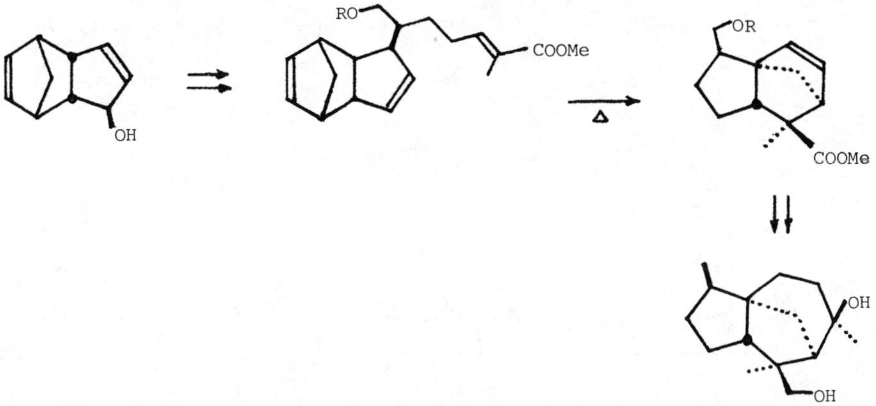

C36

References

Abe, Y., Harukawa, T., Ishiwaka, H., Miki, T., Sumi, M., Toga, T. **1953**, *J. Am. Chem. Soc. 75*, 2567.
Abe, Y., Harukawa, T., Ishikawa, H., Miki, T., Sumi, M., Toga, T. **1956**, *J. Am. Chem. Soc. 78*, 1422.
Abelman, M. M., Funk, R. L., Munger, J. D. **1982**, *J. Am. Chem. Soc. 104*, 4030.
Ackroyd, J., Karpf, M., Dreiding, A. **1985**, *Helv. Chim. Acta*, 68, 338.
Adams, D. R., Bhatnagar, S. P., Cookson, R. C., Tuddenham, R. M. **1975**, *J. Chem. Soc.. Perkin Trans. 1*, 1741.
Adams, J., Belley, M. **1986**, *Tetrahedron Lett. 27*, 2075.
Akita, H., Naito, T., Oishi, T., **1979**, *Chem. Lett.* 1365.
Akutagawa, S., Otsuka, S. **1975**, *J. Am. Chem. Soc. 97*, 6870.
Alder, K., Windmuth, E. **1939**, *Justus Liebigs Ann. Chem. 543*, 41.
Alexakis, A., Chapdelaine, M. J., Posner, G. H. **1978**, *Tetrahedron Lett.* 4209.
Alexander, J., Krishna Rao, G. S. **1973**, *Indian J. Chem.*, *11*, 859.
Alexandre, C., Rouessac, F. **1972**, *Bull. Chem. Soc. Jap. 45*, 2241.
Alexandre, C., Rouessac, F. **1975**, *Chem. Commun.* 275.
Alexandre, C., Rouessac, F. **1977**, *Chem. Commun.* 117.
Altenbach, H.-J. **1979**, *Angew. Chem. Int. Ed. 18*, 940.
Altman, L. J., Kowerski, R. C., Rilling, H. C. **1971**, *J. Am. Chem. Soc. 93*, 1782.
Amice, P., Conia, J.-M. **1970**, *Compt. Rend. 271C*, 948.
Anderson, N. H., Uh, H. **1973a**, *Synth. Commun. 3*, 115.
Anderson, N. H., Uh, H. **1973b**, *Tetrahedron Lett.* 2079.
Andersen, N. H., Golec, F. A., Jr. **1977**, *Tetrahedron Lett.* 3783.
Ando, M., Akahane, A., Takase, K. **1978a**, *Bull. Chem. Soc. Jap. 51*, 283.
Ando, M., Tajima, K., Takase, K. **1978b**, *Chem. Lett.* 617.
Ando, M., Wada, K., Takase, K. **1985**, *Tetrahedron Lett. 26*, 235.
Annis, G. D., Paquette, L. A. **1982**, *J. Am. Chem. Soc. 104*, 4504.
Antczak, K., Kingston, J. F., Fallis, A. G. **1985**, *Can. J. Chem. 63*, 993.
Aoki, M., Tooyama, Y., Uyehara, T., Kato, T. **1983**, *Tetrahedron Lett. 24*, 2267.
ApSimon, J. W., Yamasaki, K. **1977**, *Tetrahedron Lett.* 1453.
ApSimon, J. W., Badripersaud, S., Hooper, J. W., Pike, R., Birnbaum, G. I., Huber, C., Post, M. L. **1978**, *Can. J. Chem. 56*, 2139.
Arai, Y., Takeda, K., Masada, K., Koizumi, T. **1985**, *Chem. Lett.* 1531.
Aratani, T., Yoneyoshi, Y., Nagase, T. **1977**, *Tetrahedron Lett.* 2599.
Armstrong, R. J., Harris, F. L., Weiler, L. **1982**, *Can. J. Chem. 60*, 673.
Armstrong, R. J., Weiler, L. **1986**, *Can. J. Chem. 64*, 584.
Asaoka, M., Ishibashi, K., Yanagida, N., Takei, H. **1983**, *Tetrahedron Lett. 24*, 5127.
Asselin, A., Mongrain, M., Deslongchamps, P., **1968**, *Can. J. Chem. 46*, 2817.
Attah-Poku, S. K., Chau, F., Yadav, V. K., Fallis, A. G. **1985**, *J. Org. Chem. 50*, 3418.
Au-Yeung, B. W., Fleming, I. **1977**, *Chem. Commun.* 81.
Axon, B. W., Davis, B. R., Woodgate, P. D. **1981**, *J. Chem. Soc. Perkin Trans. 1*, 2956.
Ayer, S. W., Hellou, J., Tischer, M., Andersen, R. J. **1984**, *Tetrahedron Lett. 25*, 141.

Ayer, W. A., Browne, L. M. **1974**, *Can. J. Chem. 52*, 1352.
Ayer, W. A., Browne, L. M., Fung, S. **1976**, *Can. J. Chem. 54*, 3276.
Ayer, W. A., Ward, D. E., Browne, L. J., Delbaere, L.T.J., Hoyano, Y. **1981**, *Can. J. Chem. 59*, 2665.
Ayyar, K. S., Cookson, R. C., Kagi, D. A. **1973**, *Chem. Commun.* 161; **1975**, *J. Chem. Soc. Perkin Trans. I*, 1727.
Babin, D., Fourneron, J. D., Harwood, L. M., Julia, M. **1981**, *Tetrahedron, 37*, 325.
Babler, J. H., Olsen, D. O., Arnold, W. H. **1974**, *J. Org. Chem. 39*, 1656.
Babler, J. H. **1975**, *Tetrahedron Lett.* 2045.
Babler, J. H., Tortorello, A. J. **1976**, *J. Org. Chem. 41*, 885.
Babler, J. H., Invergo, B. J. **1981**, *Tetrahedron Lett. 22*, 2743.
Babler, J. H., Spina, K. P. **1985**, *Tetrahedron Lett. 26*, 1923.
Baker, R., Evans, D. A., McDowell, P. G. **1977**, *Chem. Commun.* 111.
Baker, R., Sims, R. J. **1981**, *J. Chem. Soc. Perkin Trans. I*, 3087.
Bakuzis, P., Campos, O.O.S., Bakuzis, M.L.F. **1976**, *J. Org. Chem. 41*, 3261.
Baldwin, J. E. **1976**, *Chem. Commun.* 734, 738.
Baldwin, J. E., Barden, T. C. **1981**, *J. Org. Chem.. 46*, 2442.
Baldwin, S. W., Tomesch, J. C. **1975**, *Tetrahedron Lett.* 1055.
Baldwin, S. W., Fredericks, J. E. **1982a**, *Tetrahedron Lett. 23*, 1235.
Baldwin, S. W., Landmesser, N. G. **1982b**, *Tetrahedron Lett. 23*, 4443.
Baldwin, S. W., Martin, G. F., Jr., Nunn, D. S. **1985**, *J. Org. Chem. 50*, 5720.
Balme, G. **1985**, *Tetrahedron Lett. 26*, 2309.
Banerjee, A. K., Hurtado, H. H., Carrasco, M. C. **1980**, *Synth. Commun. 10*, 261.
Banerjee, A. K., Carrasco, M. C. **1983**, *Synth. Commun. 13*, 281.
Banerjee, D. K., Angadi, V. B. **1973**, *Ind. J. Chem. 11*, 511.
Banerjee, U. K., Venkateswaran, R. V. **1983**, *Tetrahedron Lett. 24*. 423.
Baraldi, P. G., Barco, A., Benetti, S., Pollini, G. P., Polo, E., Simoni, D. **1986**, *Chem. Commun.* 757.
Barker, A. J., Pattenden, G. **1981**, *Tetrahedron Lett. 22*, 2599.
Barltrop, J. A., Littlehailes, J. D., Rushton, J. D., Rogers, N.A.J. **1962**, *Tetrahedron Lett.* 429.
Barltrop, J. A., Rogers, N.A.J. **1958**, *J. Chem. Soc.* 2566.
Barnier, J. P., Salaün, J. **1984**, *Tetrahedron Lett. 25*, 1273.
Barrett, H. C., Büchi, G. **1967**, *J. Am. Chem. Soc. 89*, 5665.
Bartlett, P. A. **1984**, in *Asymmetric Synthesis*, Vol. III, J. D. Morrison, Ed., Academic Press, New York, Chapter 5.
Barton, D.H.R., Kumari, D., Welzel, P., Danks, L. J., McGhie, J. F. **1969**, *J. Chem. Soc.* 332.
Bates, R. B., Büchi, G. Matsuura, T., Shaffer, R. R. **1960**, *J. Am. Chem. Soc. 82*, 2327.
Batt, D. G., Takamura, N., Ganem, B. **1984**, *J. Am. Chem. Soc. 106*, 3353.
Baudouy, R., Sartoretti, J., Choplin, F. **1983**, *Tetrahedron, 39*, 3293.
Baumann, M., Hoffmann, W. **1979**, *Justus Liebigs Ann. Chem.* 743.
Bedoukian, R. H., Wolinsky, J. **1975**, *J. Org. Chem. 40*, 2154.
Beereboom, J. J. **1965**, *J. Org. Chem. 30*, 4230.
Beereboom, J. J. **1966**, *J. Org. Chem. 31*, 2026.
Begley, M. J., Cameron, A. G., Knight, D. W. **1984**, *Chem. Commun.* 827.
Belanger, A., et al. **1979**, *Can. J. Chem. 57*, 3348.
Belavadi, V. K., Kulkarni, S. N. **1976**, *Indian J. Chem. 14B*, 901.
Bell, R. A., Ireland, R. E. Partyka, R. A. **1962**, *J. Org. Chem. 27*, 3741.
Bell, R. A., Ireland, R. E., Mander, L. N. **1966a**, *J. Org. Chem. 31*, 2536.
Bell, R. A., Ireland, R. E., Partyka, R. A. **1966b**, *J. Org. Chem. 31*, 2530.
Bell, V. L., Holmes, A. B., Hsu, S.-Y., Mock, G. A., Raphael, R. A. **1986**, *J. Chem. Soc. Perkin Trans. I*, 1507.
Bellesia, F., Ghelfi, F., Pagnoni, U. M., Pinetti, A. **1986**, *Tetrahedron Lett. 27*, 381.
Benayache, S., Fréjaville, C., Jullien, R., Wanat, M. **1978**, *Rev. Ital. Essenze, Profumi, Piante Off. 60*, 118.
Berger, C., Franck-Neumann, M., Ourisson, G. **1968**, *Tetrahedron Lett.* 3451.
Berlage, U., Schmidt, J., Peters, U., Welzel, P. **1987**, *Tetrahedron Lett. 28*, 3091.

Bernasconi, S., Gariboldi, P., Jommi, G., Montanari, S., Sisti, M. **1981**, *J. Chem. Soc. Perkin Trans. 1*, 2394.
Bertrand, M., Gras, J.-L. **1974**, *Tetrahedron*, **30**, 793.
Bertrand, M., Monti, H., Huong, K. C. **1979**, *Tetrahedron Lett.* 15.
Bertrand, M., Teisseire, P., Pelerin, G. **1980**, *Tetrahedron Lett.* 2055.
Bessière-Chrétien, Y., Grison, C. **1972**, *Compt. Rend. C275*, 503.
Best, W. M., Wege, D. **1981**, *Tetrahedron Lett. 22*, 4877.
Billups, W. E., Cross, J. H., Smith, C. V. **1973**, *J. Am. Chem. Soc. 95*, 3438.
Binger, P. **1983**, quoted in Trost, B. M. **1986**, *Angew. Chem. Int. Ed. 25*, 1.
Birch, A. J., Keeton, R. **1968**, *J. Chem. Soc. C*, 109.
Birch, A. J., Macdonald, P. L., Powell, V. H. **1969**, *Tetrahedron Lett.* 351; **1970**, *J. Chem. Soc. C*, 1469.
Birch, A. M., Pattenden, G. **1980**, *Chem. Commun.* 1195.
Bird, C. W., Yeong, Y. C., Hudec, J. **1974**, *Synthesis*, 27.
Blanco, L., Slougui, N., Rousseau, G., Conia, J. M. **1981**, *Tetrahedron Lett. 22*, 645.
Bloch, R., **1983**, *Tetrahedron, 39*, 639.
Bloomfield, J. J., Owsley, D. C., Nelke, J. M. **1976**, *Org. React. 23*, 259.
Boeckman, R. K., Jr., Silver, S. M. **1973**, *Tetrahedron Lett.* 3497; **1975**, *J. Org. Chem. 40*, 1755.
Boeckman, R. K., Jr., Bershas, J. P., Clardy, J., Solheim, B. **1977**, *J. Org. Chem. 42*, 3630.
Boeckman, R. K., Jr., Blum, D. M., Arthur, S. D. **1979**, *J. Am. Chem. Soc. 101*, 5060.
Boeckman, R. K., Jr., Ko, S. S. **1980**, *J. Am. Chem. Soc. 102*, 7146.
Boeckman, Jr., R. K. **1987**, Priv. Commun.
Bohlmann, F., Föster, H.-J., Fischer, C. H. **1976**, *Justus Liebigs Ann. Chem.* 1487.
Bohlmann, F. **1980a**, Lecture, 8th International Congress on Essential Oils, Cannes–Grasse.
Bohlmann, F., Eickeler, E. **1980b**, *Chem. Ber. 113*, 1189.
Bohlmann, F., Otto, W. 1982a, *Justus Liebigs Ann. Chem.* 186.
Bohlmann, F., Rotard, W. **1982b**, *Justus Liebigs Ann. Chem.* 1211.
Bold, G., Chao, S., Bhide, R., Wu, S.-H., Patel, D. V., Sih, C. J. **1987**, *Tetrahedron Lett. 28*, 1973.
Bonnert, R. V., Jenkins, P. R. **1987**, *Chem. Commun.*, 1540.
Bornack, W. K., Bhagwat, S. S., Ponton, J., Helquist, P. **1981**, *J. Am. Chem. Soc. 103*, 4467.
Bosch, M. P., Camps, F., Coll, J., Guerrero, A., Tatsuoka, T., Meinwald, J. **1986**, *J. Org. Chem. 51*, 773.
Boust, C., Leriverend, P. **1974**, *Bull. Soc. Chim. Fr.* 1201.
Bozzato, G., Bachmann, J.-P., Pesaro, M. **1974**, *Chem. Commun.* 1005.
Brady, W. T., Norton, S. J., Ko, J. **1983**, *Synthesis*, 1002.
Branca, S. J., Lock, R. L., Smith, A. B., III, **1977**, *J. Org. Chem. 42*, 3165.
Breitholle, E. G., Fallis, A. G., **1976**, *Can. J. Chem. 54*, 1991; **1978**, *J. Org. Chem. 43*, 1964.
Brieger, G. **1963a**, *Tetrahedron Lett.* 1949.
Brieger, G. **1963b**, *J. Am. Chem. Soc. 85*, 3783.
Brieger, G. **1965**, *Tetrahedron Lett.* 4429.
Brieger, G., Bennett, J. N. **1980**, *Chem. Rev. 80*, 63.
Brocard, J., Moinet, G., Conia, J. M. **1973**, *Bull. Soc. Chim. Fr.* 1711.
Bromidge, S. M., Sammes, P. G., Street, L. J. **1985**, *J. Chem. Soc. Perkin Trans. 1*, 1725.
Brooks, D. W., Grothaus, P. G. Mazdiyasni, H. **1983**, *J. Am. Chem. Soc. 105*, 4472.
Brooks, D. W., Bevinakatti, H. S., Kennedy, E., Hathaway, J. **1985**, *J. Org. Chem. 50*, 628.
Brown, M. **1968**, *J. Org. Chem. 33*, 162.
Bryce-Smith, D. Gilbert, A. **1976**, *Tetrahedron, 32*, 1309; **1977**, *33*, 2459.
Bryson, I., Mlotkiewicz, J. A., Roberts, J. S., **1979**, *Tetrahedron Lett.* 3891.
Bryson, T. A., McElligott, L. T. **1982**, *Synth. Commun. 12*, 307.
Buchanan, G. L., Young, G.A.R. **1971**, *Chem. Commun.* 643.
Buchanan, G. L., Young, G.A.R. **1973**, *J. Chem. Soc. Perkin Trans. 1*, 2404.
Büchi, G., MacLeod, W. D., Jr., Padilla, O. J. **1964a**, *J. Am. Chem. Soc. 86*, 4438.
Büchi, G., White, J. D. **1964b**, *J. Am. Chem. Soc. 86*, 2884.
Büchi, G., Hofheinz, W., Paukstelis, J. V. **1966**, *J. Am. Chem. Soc. 88*, 4113.
Büchi, G., Schneider, R. S., Wild, J. **1967**, *J. Am. Chem. Soc. 89*, 2776.

Büchi, G., Wüest, H. **1968**, *Tetrahedron*, 2049.
Büchi, G., Hofheinz, W., Paukstelis, J. V. **1969**, *J. Am. Chem. Soc. 91*, 6473.
Büchi, G., Wüest, H. **1971**, *Helv. Chim. Acta, 54*, 1767.
Büchi, G. Carlson, J. A., Powell, J. E., Jr., Tietze, L.-F. **1973**, *J. Am. Chem. Soc. 95*, 540.
Büchi, G., Berthet, D., Decorzant R., Grieder, A., Hauser, A. **1976**, J. Org. Chem. 41, 3208.
Büchi, G., Hauser, A., Limacher, J. **1977**, *J. Org. Chem. 42*, 3323.
Büchi, G., Chu, P. S. **1979a**, *J. Am. Chem. Soc. 101*, 6767.
Büchi, G., Wüest, H. **1979b**, *J. Org. Chem. 44*, 546.
Buckanin, R. S., Chen, S. J., Frieze, D. M., Sher, F. T., Berchtold, G. A. **1980**, *J. Am. Chem. Soc. 102*, 1200.
Buckwalter, B. L., Burfitt, I. R. Felkin, H., Joly-Goudket, M., Naemura, K., Salomon, M. F., Wenkert, E., Wovkulich, P. M. **1978**, *J. Am. Chem. Soc. 100*, 6445.
Buddhausukh, D., Magnus, P. **1975**, *Chem. Commun.* 952.
Buisson, D., Azerad, R., Revial, G., d'Angelo, J. **1984**, *Tetrahedron Lett. 25*, 6005.
Bullivant, M. J., Pattenden, G. **1976**, *J. Chem. Soc. Perkin Trans. I*, 256.
Burgstahler, A. W., Worden, L. R. **1961**, *J. Am. Chem. Soc. 83*, 2587.
Burk, L. A., Soffer, M. D. **1976**, *Tetrahedron, 32*, 2083.
Burke, S. D., Murtiashaw, C. W., Dike, M. S., Strickland, S.M.S., Saunders, J. O. **1981**, *J. Org. Chem. 46*, 2400.
Burke, S. D., Murtiashaw, C. W., Saunders, J. O., Dike, M. S. **1982**, *J. Am. Chem. Soc. 104*, 872.
Burke, S. D., Murtiashaw, C. W., Oplinger, J. A. **1983a**, *Tetrahedron Lett. 24*, 2949.
Burke, S. D., Powner, T. H., Kageyana, M. **1983b**, *Tetrahedron Lett. 24*, 4529.
Burke, S. D., Cobb, J. E., Takeuchi, T. **1985**, *J. Org. Chem. 50*, 3420.
Burke, S. D., Cobb, J. E. **1986**, *Tetrahedron Lett. 27*, 4237.
Burke, S. D.. **1987**, 194th ACS Nat. Meeting, CHED 48.
Caille, J. C., Bellamy, F., Guilard, R. **1984**, *Tetrahedron Lett. 25*, 2345.
Caine, D., Ingwalson, P. F. **1972**, *J. Org. Chem. 37*, 3751.
Caine, D., Tuller, F. N. **1973**, *J. Org. Chem. 38*, 3663.
Caine, D., Gupton, J. T., III, **1974**, *J. Org. Chem. 39*, 2654.
Caine, D., Hasenhuettl, G. **1975**, *Tetrahedron Lett.* 743.
Caine, D., Boucugnani, A. A., Chao, S. T., Dawson, J. B., Ingwalson, P. F. **1976a**, *J. Org. Chem. 41*, 1539.
Caine, D., Boucugnani, A. A., Pennington, W. R. **1976b**, *J. Org. Chem. 41*, 3632.
Caine, D., Graham, S. L. **1976c**, *Tetrahedron Lett.* 2521.
Caine, D., Frobese, A. A. **1977**, *Tetrahedron Lett.* 3107.
Caine, D., Deutsch, H. **1978**, *J. Am. Chem. Soc. 100*, 8030.
Caine, D., Smith, T. L., Jr. **1980**, *J. Am. Chem. Soc. 102*, 7568.
Cairns, P. M., Crombie, L., Patenden, G. **1982**, *Tetrahedron Lett. 23*, 1405.
Caliezi, A., Schinz, H. **1949**, *Helv. Chim. Acta 32*, 2556.
Callant, P., Ongena, R., Vandewalle, M. **1981**, *Tetrahedron*, 37, 2085.
Callant, P., Storme, P., Van der Eycken, E., Vandewalle, M. **1983**, *Tetrahedron Lett. 24*, 5797.
Campbell, I.G.M., Harper, S. H. **1945**, *J. Chen. Soc.* 283.
Campbell, R.V.M., Crombie, L., Pattenden, G. **1971**, *Chem. Commun.* 218.
Cane, D. E., Thomas, P. J. **1984**, *J. Am. Chem. Soc. 106*, 5295.
Cardwell, H.M.E., Cornforth, J. W., Duff, S. R., Holtermann, H., Robinson, R. **1951**, *Chem. Ind.* 389; *J. Chem. Soc.* 361.
Cargill, R. L., Wright, B. A. **1975**, *J. Org. Chem. 40*, 120.
Carlson, R. G., Zey, E. G. **1972**, *J. Org. Chem. 37*, 2468.
Casares, A., Maldonado, L. A. **1976**, *Synth. Commun. 6*, 11.
Cavill, G.W.K., Whitfield, F. B. **1964**, *Aust. J. Chem. 17*, 1260.
Cazes, B., Julia, S. **1978**, *Tetrahedron Lett.* 4065.
Challand, D. B., Kornis, G., Lange, G. L., deMayo P. **1967**, *Chem. Commun.* 704.
Challand, B. P., deMayo, P., **1968**, *Chem. Commun.* 982.
Chandrasekaran, S. Turner, J. V. **1982**, *Tetrahedron Lett. 23*, 3799.

Chass, D. A., Buddhasukh, D., Magnus, P. **1978**, *J. Org. Chem. 43*, 1750.
Chen, E. Y. **1982**, *Tetrahedron Lett. 23*, 4769.
Chetty, G. L., Krishna Rao, G. S., Dev, S., Banerjee, D. K. **1966**, *Tetrahedron*, 22, 2311.
Chidgey, R., Hoffmann, H.M.R. **1977**, Tetrahedron Lett. 2633.
Church, R. F., Ireland, R. E., Marshall, J. A. **1960**, *Tetrahedron Lett.* No. 17, 1.
Church, R. F., Ireland, R. E., Marshall, J. A. **1962**, *J. Org. Chem. 27*, 1118.
Church, R. F., Ireland, R. E., Marshall, J. A. **1966**, *J. Org. Chem. 31*, 2526.
Ciganek, E. **1984**, *Org. React. 32*, 1.
Clark, K. J., Fray, G. I., Jaeger, R. H., Robinson, R. **1959**, *Tetrahedron, 6*, 217.
Coates, R. M., Shaw, J. E. **1968**, *Chem. Commun.* 515.
Coates, R. M., Freidinger, R. M. **1969**, *Chem. Commun.* 871; **1970a**, *Tetrahedron, 26*, 3487.
Coates, R. M., Melvin, L. S., Jr. **1970b**, *J. Org. Chem. 35*, 865.
Coates, R. M., Shaw, J. E. **1970c**, *J. Am. Chem. Soc. 92*, 5657; **1970d**, *J. Org. Chem. 35*, 2597.
Coates, R. M., Robinson, W. H. **1971**, *J. Am. Chem. Soc. 93*, 1785.
Coates, R. M., Sowerby, R. L. **1972**, *J. Am. Chem. Soc. 94*, 5386.
Coates, R. M., Shah, S. K., Mason, R. W. **1979**, *J. Am. Chem. Soc. 101*, 6765.
Coates, R. M., Muskopf, J. W., Senter, P. A. **1985**, *J. Org. Chem. 50*, 3541.
Cocker, W., Geraghty, N.W.A., Grayson, D. H. **1978**, *J. Chem. Soc. Perkin Trans. I*, 1370.
Cohen, T., Bhupathy, M., Matz, J. R. **1983**, *J. Am. Chem. Soc. 105*, 520.
Collins, P. A., Wege, D. **1979**, *Aust. J. Chem. 32*, 1819.
Colvin, E. W., Raphael, R. A., Roberts, J. S. **1971**, *Chem. Commun.* 858.
Colvin, E. W., Malchenko, S., Raphael, R. A., Roberts, J. S. **1973**, *J. Chem. Soc. Perkin Trans. I*, 1989.
Colvin, E. W., Thom, I. G. **1986**, *Tetrahedron, 42*, 3137.
Conia, J.-M., Drouet, J. P., Gore, J. **1971**, *Tetrahedron, 27*, 2481.
Conia, J.-M., lePerchec, P., **1975**, *Synthesis*, 1.
Conia, J.-M., Lange, G. L. **1978**, *J. Org. Chem. 43*, 564.
Connolly, P. J., Heathcock, C. H. **1985**, *J. Org. Chem. 50*, 4135.
Cook, J. W., Raphael, R. A., Scott, A. I. **1951**, *J. Chem. Soc.* 695.
Corey, E. J., Chow, S. W., Scherrer, R. A. **1957**, *J. Am. Chem. Soc. 79*, 5773.
Corey, E. J., Ohno, M., Vatakencherry, P. A., Mitra, R. B. **1961**, *J. Am. Chem. Soc. 83*, 1251.
Corey, E. J., Hartmann, R., Vatakencherry, P. A. **1962**, *J. Am. Chem. Soc. 84*, 2611.
Corey, E. J., Mitra, R. B., Uda, H. **1963a**, *J. Am. Chem. Soc. 85*, 362.
Corey, E. J., Nozoe, S. **1963b**, *J. Am. Chem. Soc. 85*, 3527.
Corey, E. J., Hamanaka, E. **1964a**, *J. Am. Chem. Soc. 86*, 1641.
Corey, E. J., Mitra, R. B., Uda, H. **1964b**, *J. Am. Chem. Soc. 86*, 485.
Corey, E. J., Ohno, M., Mitra, R. B., Vatakencherry, P. A. **1964c**, *J. Am. Chem. Soc. 86*, 478.
Corey, E. J., Hortmann, A. G. **1965a**, *J. Am. Chem. Soc. 87*, 5736.
Corey, E. J., Nozoe, S. **1965b**, *J. Am. Chem. Soc. 87*, 5733.
Corey, E. J., Jautelat, M. **1967**, *J. Am. Chem. Soc. 89*, 3912.
Corey, E. J., Katzenellenbogen, J. A., Gilman, N. W., Roman, S. A., Erickson, B. W. **1968**, *J. Am. Chem. Soc. 90*, 5618.
Corey, E. J., Achiwa, K. **1969a**, *Tetrahedron Lett.* 1837; **1969b**, 3257.
Corey, E. J., Achiwa, K., Katzenellenbogen, J. A. **1969c**, *J. Am. Chem. Soc. 91*, 4318.
Corey, E. J., Broger, E. A. **1969**, *Tetrahedron Lett.* 1779.
Corey, E. J., Girotra, N. N., Mathew, C. T. **1969e**, *J. Am. Chem. Soc. 91*, 1557.
Corey, E. J., Achiwa, K. **1970**, *Tetrahedron Lett.* 2245.
Corey, E. J., Cane, D. E., Libit, L. **1971**, *J. Am. Chem. Soc. 93*, 7016.
Corey, E. J., Snider, B. B. **1972**, *J. Am. Chem. Soc. 94*, 2549.
Corey, E. J., Balanson, R. D. **1973a**, *Tetrahedron Lett.* 3153.
Corey, E. J., Watt, D. S. **1973b**, *J. Am. Chem. Soc. 95*, 2302.
Corey, E. J., Katzenellenbogen, J. A., Libit, L. **1976**, unpublished, from Katzenellenbogen, J. A. **1976**, *Science, 194*, 139.
Corey, E. J., Danheiser, R. L., Chandrasekaran, S., Keck, G. E., Gopalan, B., Larsen, S. D., Sinet, P., Gras, J.-L. **1978a**, *J. Am. Chem. Soc. 100*, 8034.

Corey, E. J., Danheiser, R. L., Chandrasekaran, S., Sinet, P., Keck, G. E., Gras, J.-L. **1978b**, *J. Am. Chem. Soc. 100*, 8031.
Corey, E. J., Behforous, M., Ishiguro, M. **1979a**, *J. Am. Chem. Soc. 101*, 1608.
Corey, E. J., Ishiguro, M. **1979b**, *Tetrahedron Lett.* 2745.
Corey, E. J., Pearce, H. L. **1979c**, *J. Am. Chem. Soc. 101*, 5841.
Corey, E. J., Smith, J. G. **1979d**, *J. Am. Chem. Soc. 101*, 1038.
Corey, E. J., Tius, M. A., Das, J. **1980a**, *J. Am. Chem. Soc. 102*, 1742: **1980b**, *102*, 7612.
Corey, E. J., Das, J. **1982a**, *J. Am. Chem. Soc. 104*, 5551.
Corey, E. J., Monroe, J. E. **1982b**, *J. Am. Chem. Soc. 104*, 6129.
Corey, E. J., Desai, M. C. **1985a**, *Tetrahedron Lett. 26*, 3535.
Corey, E. J., Desai, M. C., Engler, T. A. **1985b**, *J. Am. Chem. Soc. 107*, 4339.
Corey, E. J., Myers, A. G. **1985c**, *J. Am. Chem. Soc. 107*, 5574.
Corey, E. J., Seibel, W. L. **1986a**, *Tetrahedron Lett. 27*, 905; **1986b**, *27*, 909.
Corey, E. J., **1987a**, Priv. Commun.
Corey, E. J., Kang, M.-C., Desai, M. C., Ghosh, A. K., Houpis, I. N. **1987b**, *J. Am. Chem. Soc.* in press.
Corey, E. J., Magriotis, P. A. **1987c**, *J. Am. Chem. Soc. 109*, 287.
Corey, E. J., Reid, J. G., Myers, A. G., Hahl, R. W. **1987d**, *J. Am. Chem. Soc. 109*, 918.
Corey, E. J., Su, W.-g. **1987e**, *J. Am. Chem. Soc. 109*, 7534.
Corey, E. J., Wess, G., Xiang, Y. B., Singh, A. K. **1987f**, *J. Am. Chem. Soc. 109*, 4717.
Cory, R. M., McLaren, F. R. **1977**, *Chem. Commun.* 587.
Cory, R. M., Chan, D.M.T., McLaren, F. R., Rasmussen, M. H., Renneboog, R. M. **1979**, *Tetrahedron Lett.* 4133.
Cossy, J., Belotti, D., Pete, J. P. **1987**, *Tetrahedron Lett. 28*, 4547.
Crandall, T. G., Lawton, R. G. **1969**, *J. Am. Chem. Soc. 91*, 2127.
Crawford, R. J., Erman, W. F., Broaddus, C. D. **1972**, *J. Am. Chem. Soc. 94*, 4298.
Crimmins, M. T., DeLoach, J. A. **1984**, *J. Org. Chem. 49*, 2076.
Crimmins, M. T., Mascarella, S. W. **1986**, *J. Am. Chem. Soc., 108*, 3435.
Crimmins, M. T., Gould, L. D. **1987a**, *J. Am. Chem. Soc. 109*, 6199.
Crimmins, M. T., Mascarella, S. W. **1987b**, *Tetrahedron Lett. 28*, 5063.
Crisp, G. T., Scott, W. J., Stille, J. K. **1984**, *J. Am. Chem. Soc. 106*, 7500.
Crombie, L., Kneen, G., Pattenden, G. **1976**, *Chem. Commun.* 66.
Crowley, K. J. **1962**, *Proc. Chem. Soc.* 245, 334.
Curran, D. P., Chen, M.-H. **1985a**, *Tetrahedron Lett. 26*, 4991.
Curran, D. P., Jacobs, P. B. **1985b**, *Tetrahedron Lett. 26*, 2031.
Curran, D. P., Rakiewicz, D. M. **1985c**, *J. Am. Chem. Soc. 107*, 1448.
Curran, D. P., Kuo, S.-C. **1986**, *J. Am. Chem. Soc. 108*, 1106.
Curran, D. P., Chen, M.-H. **1987**, *J. Am. Chem. Soc. 109*, 6558.
d'Angelo, J., Revial, G. **1983**, *Tetrahedron Lett. 24*, 2103.
Danieli, N., Mazur, Y., Sondheimer, F. **1967**, *Tetrahedron, 23*, 509.
Danishefsky, S., Dumas, D. **1968**, *Chem. Commun.* 1287.
Danishefsky, S., Kitahara, T., Schuda, P. F., Etheredge, S. J. **1976**, *J. Am. Chem. Soc. 98*, 3028.
Danishefsky, S., Schuda, P. F., Kitahara, T., Etheredge, S. J. **1977**, *J. Am. Chem. Soc. 99*, 6066.
Danisheisky, S., Hirama, M., Gombatz, K., Harayama, T., Berman, E., Schuda, P. **1978**, *J. Am. Chem. Soc. 100*, 6536; **1979**, *J. Am. Chem. Soc. 101*, 7020.
Danishefsky, S., Tsuzuki, K. **1980a**, *J. Am. Chem. Soc. 102*, 6891.
Danishefsky, S., Vaughn, K., Gadwood, R., Tsuzuki, K. **1980b**, *J. Am. Chem. Soc. 102*, 4262.
Danishefsky, S., Zamboni, R. **1980c**, *Tetrahedron Lett.* 3439.
Danishefsky, S., Zamboni, R., Kahn, M., Etheredge, S. J. **1980d**, *J. Am. Chem. Soc. 102*, 2097.
Danishefsky, S., **1981**, *Acc. Chem. Res. 14*, 400.
Danishefsky, S., Harrison, P., Silvestri, M., Segmuller, B. **1984**, *J. Org. Chem. 49*, 1319.
Dastur, K. P. **1973**, *J. Am. Chem. Soc. 95*, 6509; **1974**, *96*, 2605.
Dauben, W. G., Ashcraft, A. C. **1963**, *J. Am. Chem. Soc. 85*, 3673.
Dauben, W. G., Ahlgren, G., Leitereg, T. J., Schwarzel, W. C., Yoshioko, M. **1972**, *J. Am. Chem. Soc. 94*, 8593.

Dauben, W. G., Hart, D. J., Ipaktschi, J., Kozikowski, A. P. **1973**, *Tetrahedron Lett.* 4425.
Dauben, W. G., Beasley, G. H., Broadhurst, M. D., Muller, B., Peppard, D. J., Pesnelle, P., Suter, C. **1974**, *J. Am. Chem. Soc. 96*, 4725; **1975a**, *97*, 4973.
Dauben, W. G., Hart, D. J. **1975b**, *J. Am. Chem. Soc. 97*, 1622.
Dauben, W. G., Kozikowski, A. P., Zimmerman, W. T. **1975c**, *Tetrahedron Lett.* 515.
Dauben, W. G., Hart, D. J. **1977a**, *J. Org. Chem. 42*, 922.
Dauben, W. G., Hart, D. J. **1977b**, *J. Am. Chem. Soc. 99*, 7307.
Dauben, W. G., Kessel, C. R., Takemura, K. H. **1980**, *J. Am. Chem. Soc. 102*, 6893.
Dauben, W. G., Walker, D. M. **1981**, *J. Org. Chem. 46*, 1103.
Dauben, W. G., Shapiro, G. **1984**, *J. Org. Chem. 49*, 4252.
Dauphin, G. **1979**, *Synthesis*, 799.
Davis, B. R., Johnson, S. J. **1978**, *Chem. Commun.* 614.
Dawson, B. A., Ghosh, A. K., Jurlina, J. L., Stothers, J. B. **1983**, *Chem. Commun.* 204.
deBroissia, H., Levisalles, J., Rudler, H. **1972a**, *Chem. Commun.* 855; **1972b**, *Bull. Soc. Chim. Fr.* 4314.
DeClercq, P., Vandewalle, M. **1977**, *J. Org. Chem. 42*, 3447.
Defauw, J., Majetich, G. **1987**, 194th ACS Nat. Meeting, ORGN 152.
deGroot, A., Broekhuysen, M. P., Doddema, L. L., Vollering, M. C., Westerbeek, J.M.M. **1982**, *Tetrahedron Lett. 23*, 4831.
Deighton, M., Hughes, C. R., Ramage, R. **1975**, *Chem. Commun.* 662.
Delay, F., Ohloff, G. **1979**, *Helv. Chim. Acta, 62*, 369.
Deljac, A., MacKay, W. D., Pan, C.S.J., Wiesner, K. J., Wiesner, K. **1972**, *Can. J. Chem. 50*, 726.
Dell, C. P., Knight, D. W. **1987**, *Chem. Commun.* 349.
Demole, E., Enggist, P. **1969**, *Chem. Commun.* 264.
Demole, E., Enggist, P. **1971a**, *Helv. Chim. Acta, 54*, 456.
Demole, E., Enggist, P., Borer, C. **1971b**, *Helv. Chim. Acta, 54*, 1845.
Demuth, M., **1984a**, *Chimia, 38*, 257.
Demuth, M., Ritterskamp, P., Schaffer, K. **1984b**, *Helv. Chim. Acta, 67*, 2023.
Demuth, M., Hinsken, W. **1985**, *Angew. Chem. Int. Ed. 24*, 973.
Demuynck, M., DeClercq, P. J., Vandewalle, M. **1979**, *J. Org. Chem. 44*, 4863.
Demuynck, M., Devreese, A. A., DeClercq, P. J., Vandewalle, M. **1982**, *Tetrahedron Lett. 23*, 2501.
Denis, J. M., Girard, C., Conia, J. M. **1972**, *Synthesis*, 549.
DeVos, M. J., Hevesi, L., Bayet, P., Krief, A. **1976**, *Tetrahedron Lett.* 3911.
DeVos, M. J., Krief, A. **1979**, *Tetrahedron Lett.* 1891.
DeVos, M. J., Krief, A. **1983**, *Tetrahedron Lett. 24*, 103.
Devreese, A. A., DeClercq, P. J., Vandewalle, M. **1980**, *Tetrahedron Lett. 21*, 4767.
Dewanckele, J. M., Zutterman, F., Vandewalle, M. E. **1983**, *Tetrahedron, 39*, 3235.
Dieter, R. K., Lin, Y. J. **1985**, *Tetrahedron Lett. 26*, 39.
Dilling, W. L. **1977**, *Photochem. Photobiol. 25*, 605.
Disanayaka, B. W., Weedon, A. C. **1985**, *Chem. Commun.* 1282.
Duc, D.K.M., Fetizon, M., Lazare, **1975**, *Chem. Commun.* 282; **1978**, *Tetrahedron, 34*, 1207.
Eck, C. R., Hodgson, G. L., MacSweeney, D. F., Mills, R. W., Money, T. **1974**, *J. Chem. Soc. Perkin Trans. I*, 1938.
Edwards, J. A., Schwarz, V., Fajkos, J., Maddox, M. L., Fried, J. H. **1971**, *Chem. Commun.* 292.
Eilbracht, P., Balas, E., Acker, M. **1984**, *Tetrahedron Lett. 25*, 1131.
Eilerman, R. G., Willis, B. J. **1981**, *Chem. Commun.* 30.
Ensley, H. E., Coppola, K., von Burg, G., Sastry, K. A. **1987**, 194th ACS Nat. Meeting, ORGN 50.
Enzell, C. **1962**, *Tetrahedron Lett.* 185.
Erman, W. F., Stone, L. C. **1971a**, *J. Am. Chem. Soc. 93*, 2821.
Erman, W. F., Treptow, R. S., Bakuzis, P., Wenkert, E. **1971b**, *J. Am. Chem. Soc. 93*, 657.
Escalone, H., Maldonado, L. A. **1980**, *Synth. Commun. 10*, 857.
Evans, D. A., Sims, C. L. **1973**, *Tetrahedron Lett.* 4691.
Evans, D. A., Sims, C. L., Andrews, G. C. **1977**, *J. Am. Chem. Soc. 99*, 5453.
Evans, D. A., Hart, D. J., Koelsch, P. M. **1978**, *J. Am. Chem. Soc. 100*, 4593.
Exon, C., Nobbs, M., Magnus, P. **1981**, *Tetrahedron, 37*, 4515.
Exon, C., Magnus, P. **1983**, *J. Am. Chem. Soc. 105*, 2477.

Fadel, A., Salaün, J. **1985**, *Tetrahedron, 41*, 413.
Fahrenholtz, K. E., Lurie, M., Kierstead, R. W. **1967**, *J. Am. Chem. Soc. 89*, 5934.
Fallis, A. G. **1984**, *Can. J. Chem. 62*, 183.
Fanta, W. I., Erman, W. F. **1968**, *J. Org. Chem. 33*, 1656.
Feutrill, G. I., Mirrington, R. N., Nichols, R. J. **1973**, *Aust. J. Chem. 26*, 345.
Fex, T., Froborg, J., Magnusson, G., Thorén, S. **1976**, *J. Org. Chem. 41*, 3518.
Ficini, J., d'Angelo, J., Noiré, J. **1974**, *J. Am. Chem. Soc. 96*, 1213.
Ficini, J., d'Angelo, J. **1976**, *Tetrahedron Lett.* 687.
Ficini, J., Touzin, A. M. **1977**, *Tetrahedron Lett.* 1081.
Ficini, J., Revial, G., Genet, J. P. **1981**, *Tetrahedron Lett. 22*, 629, 633.
Ficini, J., Falou, S., d'Angelo, J. **1983**, *Tetrahedron Lett. 24*, 375.
Fitzsimmons, B. J., Fraser-Reid, B. **1984**, *Tetrahedron, 40*, 1279.
Fleming, I., Terrett, N. K. **1984**, *Tetrahedron Lett. 25*, 5103.
Franck-Neumann, M., Lohmann, J. J. **1979**, *Tetrahedron Lett. 22*, 2075.
Franck-Neumann, M., Martina, D., Heitz, M. P. **1982**, *Tetrahedron Lett.. 23*, 3493.
Franck-Neumann, M., Sedrati, M., Vigneron, J.-P., Bloy, V. **1985**, *Angew. Chem. Int. Ed. 24*, 996.
Franke, L.R.R.A., Wolf, H., Wray, V. **1984**, *Tetrahedron, 40*, 3491.
Frater, G. **1974a**, *Helv. Chim. Acta, 57*, 172; **1974b**, *57*, 2446.
Frater, G. **1977**, *Helv. Chim. Acta, 60*, 515.
Frater, G. **1978**, *Helv. Chim. Acta, 61*, 2709.
Frater, G. **1982**, *Chem. Commun.* 521.
Frater, G., Wenger, J. **1984**, *Helv. Chim. Acta, 67*, 1702.
Fringuelli, F., Taticchi, A., Traverso, G. **1969**, *Gaz. Chim. Ital. 99*, 231.
Fringuelli, F., Pizzo, F., Taticchi, A., Ferreira, V. F., Michelotti, E. L., Porter, B., Wenkert, E. **1985**, *J. Org. Chem. 50*, 890.
Fringuelli, F., Pizzo, F., Taticchi, A., Wenkert, E. **1986**, *Synth. Commun. 16*, 245.
Froborg, J., Magnusson, G., Thorén, S. **1975**, *J. Org. Chem. 40*, 1595.
Froborg, J., Magnusson, G. **1978**, *J. Am. Chem. Soc. 100*, 6728.
Fuji, K., Node, M., Usami, Y., Kiryu, Y., **1987**, *Chem. Commun.* 449.
Fujimoto, Y., Yokura, S., Nakamura, T., Morikawa, T., Tatsuno, T. **1974**, *Tetrahedron Lett.* 2523.
Fujimoto, Y., Shimizu, T., Tatsuno, T. **1976**, *Tetrahedron Lett.* 2041.
Fujita, E., Shibuya, M., Nakamura, S., Okada, Y., Fujita, T. **1972**, *Chem. Commun.* 1107.
Fujiwara, S., Aoki, M., Uyehara, T., Kato, T. **1984**, *Tetrahedron Lett. 25*, 3003.
Fukamiya, N., Kato, M., Yoshikoshi, A. **1971**, *Chem. Commun.* 1120; **1973**, *J. Chem. Soc. Perkin Trans. I*, 1843.
Fukamiya, N., Oki, M., Okano, M., Aratani, T. **1981**, *Chem. Ind.* 96.
Fukuyama, Y., Tokoroyama, T., Kubota, T. **1972**, *Tetrahedron Lett.* 3401.
Fukuyama, Y., Tokoroyama, T., Kubota, T. **1973**, *Tetrahedron Lett.* 4869.
Funk, R. L., Horcher, L.H.M., III, Daggett, J. U., Hansen, M. M. **1983**, *J. Org. Chem. 48*, 2632.
Funk, R. L., Bolton, G. L. **1984**, *J. Org. Chem. 49*, 5021.
Funk, R. L., Abelman, M. M. **1985a**, 189th ACS Meeting, ORGN 35.
Funk, R. L., Munger, J. D., Jr. **1985b**, *J. Org. Chem. 50*, 707.
Funk, R. L., Abelman, M. M.**1986**, *J. Org. Chem. 51*, 3247.
Funk, R. L., **1987a**, Priv. Commun.
Funk, R. L., Bolton, G. L. **1987b**, *J. Org. Chem. 52*, 3173.
Furber, M., Mander, L. N. **1987**, *J. Am. Chem. Soc. 109*, 6389.
Furuichi, K., Miwa, T. **1974**, *Tetrahedron Lett.* 3689.
Fukukawa, J., Kawabata, N., Nishimura, J. **1966**, *Tetrahedron Lett.* 3353.
Furukawa, J., Morisaki, N., Kobayashi, H., Iwasaki, S., Nozoe, S., Okuda, S. **1985**, *Chem. Pharm. Bull. 33*, 440.
Gadwood, R. C. **1983**, *J. Org. Chem. 48*, 2098.
Gadwood, R. C., Lett, R. M., Wissinger, J. E. **1984**, *J. Am. Chem. Soc. 106*, 3869.
Gammill, R. B., Bryson, T. A. **1976**, *Synth. Commun. 6*, 209.
Gaoni, Y., Tomazic, A. **1985**, *J. Org. Chem. 50*, 2948.

Garbers, C. F., Steenkamp, J. A., Visagie, H. E. **1975**, *Tetrahedron Lett.* 3753.
Garbers, C. F., Beukes, M. S., Ehlers, C., McKenzie, M. J. **1978**, *Tetrahedron Lett.* 77.
Garratt, P. J., Porter, J. R. **1986**, *J. Org. Chem. 51*, 5450.
Garver, L. C., van Tamelen, E. E. **1982**, *J. Am. Chem. Soc. 104*, 867.
Gasa, S., Hamanaka, N., Matsunaga, S., Okuno, T., Takeda, N., Matsumoto, T. **1976**, *Tetrahedron Lett.* 553.
Gassman, P. G., Singleton, D. A.. **1986**, *J. Org. Chem. 51*, 3075.
Gawley, R. E. **1976**, *Synthesis*, 777.
Genêt, J. P., Piau, F., Ficini, J. **1980**, *Tetrahedron Lett. 21*, 3183.
Genêt, J. P., Piau, F. **1981**, *J. Org. Chem. 46*, 2414.
Gensler, W. J., Solomon, P. H. **1973**, *J. Org. Chem. 38*, 1726.
Georges, M., Fraser-Reid, B. **1985a**, 190th ACS Meeting, ORGN 64.
Georges, M., Fraser-Reid, B. **1985b**, *J. Org. Chem. 50*, 5754.
Georges, M., Tam, T.-F., Fraser-Reid, B. **1985c**, *J. Org. Chem. 50*, 5747.
Gerling, K.-G., Wolf, H. **1985**, *Tetrahedron Lett. 26*, 1293.
Gesson, J.-P., Jacquesy, J.-C., Renoux, B. **1986**, *Tetrahedron 27*, 4461.
Ghatak, U. R., Sanyal, B., Ghosh, S. **1976**, *J. Am. Chem. Soc. 98*, 3721.
Ghera, E., Sondheimer, F. **1964**, *Tetrahedron Lett.* 3887.
Ghera, E., Maurya, R., Ben-David, Y. **1986**, *Tetrahedron Lett. 27*, 3935.
Giarrusso, F., Ireland, R. E. **1968**, *J. Org. Chem. 33*, 3560.
Gibbons, E. G. **1982**, *J. Am. Chem. Soc. 104*, 1767.
Givaudi, E., Plattier, M., Teisseire, P. **1974**, *Recherches*, *19*, 205.
Givaudi, E., Ehret, C., Plattier, M., Achard, M., Corbier, B., Teissiere, P. **1980**, *Trans. 8th International Congress on Essential Oils*, Cannes–Grasse.
Godfrey, J. D., Schultz, A. G. **1979**, *Tetrahedron Lett.* 3241.
Goldberg, O., Deja, I., Rey, M., Dreiding, A. S. **1980**, *Helv. Chim. Acta, 63*, 2455.
Goldsmith, D. J., Sakano, I. **1976**, *J. Org. Chem. 41*, 2095.
Gonzalez, A. G., Martin, J. D., Rodriguez, M. L. **1973**, *Tetrahedron Lett.* 3657.
Gonzalez, A. G., Martin, J. D., Perez, C., Ramirez, M. A. **1976**, *Tetrahedron Lett.* 137.
Gonzalez, A. G., Darias, J., Martin, J. D., Melian, M. A. **1978**, *Tetrahedron Lett.* 481.
Gonzalez, A. G., Martin, J. D., Martin, V. S. , Norte, M., Perez, R. **1982**, *Tetrahedron Lett. 23*, 2395.
Gopalan, A., Magnus, P. **1980**, *J. Am. Chem. Soc. 102*, 1757.
Gopichand, Y., Chakravarti, K. K. **1974**, *Tetrahedron Lett.* 3851.
Grafen, P., Kabbe, H. J., Roos, O., Diana, G. D., Li, T., Turner, R. B. **1968**, *J. Am. Chem. Soc. 90*, 6131.
Grayson, D. H., Wilson, J.R.H. **1984**, *Chem. Commun.* 1695.
Greene, A. E. **1980**, *Tetrahedron Lett.* 3059.
Greene, A. E., Lansard, J.-P., Luche, J.-L., Petrier, C. **1984**, *J. Org. Chem. 49*, 931.
Greene, A. E., Deprés, J.-P., Coelho, F., Brocksom, T. J. **1985a**, *J. Org. Chem. 50*, 3943.
Greene, A. E., Luche, M.-J., Serra, A. A. **1985b**, *J. Org. Chem. 50*, 3957.
Greene, A. E., Coelho, F., Barreiro, E. J., Costa, P.R.R. **1986**, *J. Org. Chem. 51*, 4250.
Greene, A. E., Charbonnier, F., Luche, M.-J., Moyano, A. **1987a**, *J. Am. Chem. Soc. 109*, 4752.
Greene, A. E., Serra, A. A., Barreiro, E. J., Costa, P.R.R. **1987b**, *J. Org. Chem. 52*, 1169.
Greenlee, M. L. **1981**, *J. Am. Chem. Soc. 103*, 2425.
Greenlee, W. J., Woodward, R. B. **1976**, *J. Am. Chem. Soc. 98*, 6075.
Greenwood, J. M., Sutherland, J. K., Torre, A., *Chem. Commun.* 410.
Gregson, R. P., Mirrington, R. N. **1973**, *Chem. Commun.* 598.
Gregson, R. P., Mirrington, R. N. **1976**, *Aust. J. Chem. 29*, 2037.
Grewal, R. S., Hayes, P. C., Sawyer, J. F., Yates, P. **1987**, *Chem. Commun.* 1290.
Grieco, P. A. **1969**, *J. Am. Chem. Soc. 91*, 5660.
Grieco, P. A., Masaki, Y. **1975a**, *J. Org. Chem. 40*, 150.
Grieco, P. A., Reap, J. J. **1975b**, *Synth. Commun. 5*, 347.
Grieco, P. A., Nishizawa, M., Burke, S. D., Marinovic, N. **1976a**, *J. Am. Chem. Soc. 98*, 1612.
Grieco, P. A., Nishizawa, M. **1976b**, *Chem. Commun.* 582.

Grieco, P. A., Nishizawa, M. **1977a**, *J. Org. Chem. 42*, 1717.
Grieco, P. A., Nishizawa, M., Oguri, T., Burke, S. D., Marinovic, N. **1977b**, *J. Am. Chem. Soc. 99*, 5773.
Grieco, P. A., Ohfune, Y., Majetich, G. **1977c**, *J. Am. Chem. Soc. 99*, 7393.
Grieco, P. A., Oguri, T., Wang, C.L.J., Williams, E. **1977d**, *J. Org. Chem. 42*, 4113.
Grieco, P. A., Oguri, T., Gilman, S., DeTitta, G. T. **1978**, *J. Am. Chem. Soc. 100*, 1616.
Grieco, P. A., Ohfune, Y., Majetich, G. **1979a**, *J. Org. Chem. 44*, 3092.
Grieco, P. A., Ohfune, Y., Majetich, G. **1979b**, *Tetrahedron Lett.* 3265.
Grieco, P. A., Ferrino, S., Vidari, G. **1980a**, *J. Am. Chem. Soc. 102*, 7586.
Grieco, P. A., Vidari, G., Ferrino, S., Haltiwanger, R. C. **1980b**, *Tetrahedron Lett. 21*, 1619.
Grieco, P. A., Garner, P., He, Z.-m. **1983**, Tetrahedron Lett. 24, 1897.
Grieco, P. A., Nargund, R. P. **1986**, *Tetrahedron Lett. 27*, 4813.
Grootaert, W. M., DeClercq, P. J. **1982**, *Tetrahedron Lett. 23*, 3291.
Grootaert, W. M., De Clercq, P. J. **1986**, *Tetrahedron Lett. 27*, 1731.
Gueldner, R. C., Thompson, A. C., Hedin, P. A. **1972**, *J. Org. Chem. 37*, 1854.
Guest, I. G., Hughes, C. R., Ramage, R., Sattar, A. **1973**, *Chem. Commun.*, 526.
Guthrie, R. W., Henry, W. A., Immer, H., Wong, C. M., Valenta, Z., Wiesner, K. **1966**, *Collect. Czech. Chem. Commun. 31*, 602.
Gutsche, C. D., Redmore, D. **1968**, *Carbocyclic Ring Expansion Reactions*, Academic Press, New York.
Hackett, S., Livinghouse, T. **1986**, *J. Org. Chem. 51*, 879.
Hagiwara, H., Miyashita, M., Uda, H., Yoshikoshi, A. **1975**, *Bull. Chem. Soc. Jap. 48*, 3723.
Hagiwara, H. Uda, H., Kodama, T. **1979**, *Koen Yoshishu-Koryo, Terupen oyobi seiyu Kagaku ni Kansuru ToronKai* 23rd, 118.
Hagiwara, H., Okano, A., Uda, H. **1985**, *Chem. Commun.* 1047.
Hagiwara, H., Okano, A., Akama, T., Uda, H. **1987a**, *Chem. Commun.* 1333.
Hagiwara, H., Uda, H. **1987b**, *Chem. Commun.* 1351.
Hajos, Z. G., Micheli, R. A., Parrish, D. R., Oliveto, E. P. **1967**, *J. Org. Chem. 32*, 3008.
Halazy, S., Zutterman, F., Krief, A. **1982**, *Tetrahedron Lett. 23*, 4385.
Hamanaka, N., Matsumoto, T. **1972**, *Tetrahedron Lett.* 3087.
Han, Y.-K., Paquette, L. A. **1979**, *J. Org. Chem. 44*, 3731.
Harayama, T., Cho, H., Inubushi, Y. **1977**, *Tetrahedron Lett.* 3273.
Harayama, T., Shinkai, Y., Hashimoto, Y., Fukushi, H., Inubushi, Y. **1983**, *Tetrahedron Lett. 24*, 5241.
Harding, K. E., Clement, B. A., Moreno, L., Peter-Kalinic, J. **1981**, *J. Org. Chem. 46*, 940.
Harding, K. E., Clement, K. S. **1984**, *J. Org. Chem. 49*, 3870.
Hashimoto, H., Tsuzuki, K., Sakan, F., Shirahama, H., Matsumoto, T. **1974**, *Tetrahedron Lett.* 3745.
Hashimoto, S., Itoh, A., Kitagawa, Y., Yamamoto, H., Nozaki, H. **1977**, *J. Am. Chem. Soc. 99*, 4192.
Hashimoto, H., Furuichi, K., Miwa, T. **1987**, *Chem. Commun.* 1002.
Hauptmann, H., Mühlbauer, G., Walker, N.P.C. **1986**, *Tetrahedron Lett. 27*, 1315.
Hauser, C. R., Hudson, B. E. **1942**, *Org. React. 1*, 266.
Hayakawa, K., Ohsuki, S., Kanematsu, K. **1986**, *Tetrahedron Lett. 27*, 947.
Hayakawa, Y., Sakai, M., Noyori, R. **1975**, *Chem. Lett.* 509.
Hayakawa, Y., Shimizu, F., Noyori, R. **1978**, *Tetrahedron Lett.* 993.
Hayano, K., Ohfune, Y., Shirahama, H., Matsumoto, T. **1978**, *Tetrahedron Lett.* 1991.
Hayashi, K., Nakamura, H., Mitsuhashi, H. **1973**, *Chem. Pharm. Bull. 21*, 2806.
Hayashi, Y., Nishizawa, M., Harita, S., Sakan, T. **1972**, *Chem. Lett.* 375.
Hayashi, Y., Nishizawa, M., Sakan, T. **1975**, *Chem. Lett.* 387.
Heathcock, C. H. **1966**, *J. Am. Chem. Soc. 88*, 4110.
Heathcock, C. H., Badger, R. A., Patterson, J. W. **1967**, *J. Am. Chem. Soc. 89*, 4133.
Heathcock, C. H., Kelly, T. R. **1968**, *Tetrahedron, 24*, 1801.
Heathcock, C. H., Ratcliffe, R. **1971**, *J. Am. Chem. Soc. 93*, 1746.
Heathcock, C. H., Amano, Y. **1972**, *Can. J. Chem. 50*, 340.
Heathcock, C. H., Delmar, E. G., Graham, S. L. **1982a**, *J. Am. Chem. Soc. 104*, 1907.
Heathcock, C. H., Tice, C. M., Gemroth, T. C. **1982b**, *J. Am. Chem. Soc. 104*, 6081.
Heathcock, C. H., Mahaim, C., Schlecht, M. F., Utawanit, T. **1984**, *J. Org. Chem. 49*, 3264.

Heissler, D., Riehl, J.-J. **1980**, *Tetrahedron Lett. 21*, 4707.
Helquist, P. **1985**, private communication.
Henegar, K. E., Winkler, J. D. **1987**, 194th ACS Nat. Meeting, ORGN 117.
Herz, W., Mirrington, R. N., Young, H., Lin, Y. Y. **1968**, *J. Org. Chem. 33*, 4210.
Hewson, A. T., MacPherson, D. T. **1985**, *J. Chem. Soc. Perkin Trans. I*, 2625.
Higo, M., Toda, H., Suzuki, K., Nishida, Y. **1978**, German patent 2814558.
Hikino, H., Suzuki, N., Takemoto, T. **1966**, *Chem. Pharm. Bull. 14*, 1441.
Hiroi, K., Miura, H., Kotsuji, K., Sato, S. **1981**, *Chem. Lett.* 559.
Hiyama, T., Shinoda, M., Nozaki, H. **1979**, *Tetrahedron Lett.* 3529.
Ho, T.-L. **1971**, *Chem. Ind.* 487.
Ho, T.-L. **1972**, *Can. J. Chem. 50*, 1098.
Ho, T.-L. **1973**, *J. Chem. Soc. Perkin Trans. I*, 2579.
Ho, T.-L. **1974**, *Synth. Commun. 4*, 189.
Ho, T.-L., Liu, S. H. **1980**, *Synth. Commun. 10*, 603.
Ho, T.-L. **1982a**, *Synth. Commun. 12*, 633.
Ho, T.-L., Din, Z. U. **1982b**, *Synth. Commun. 12*, 257.
Ho, T.-L., **1984**, U.S. patent 4463194.
Ho, T.-L., **1985**, unpublished results.
Hobbs, P. D., Magnus, P. D. **1974**, *Chem. Commun.* 856; **1976**, *J. Am. Chem. Soc. 98*, 4594.
Hodgson, G. L., MacSweeney, D. F., Money, T. **1971**, *Chem. Commun.* 766.
Hodgson, G. L., MacSweeney, D. F., Money, T. **1972**, *Tetrahedron Lett.* 3683.
Hodgson, G. L., MacSweeney, D. F., Mills, R. W., Money, T. **1973**, *Chem. Commun.* 235.
Hoffmann, H.M.R. **1969**, *Angew. Chem. Int.* Ed. *8*, 556.
Holton, R. A., Kennedy, R. M., Kim, H.-B., Krafft, M. E. **1987**, *J. Am. Chem. Soc. 109*, 1597.
Honan, M. C. **1985**, *Tetrahedron Lett. 26*, 6393.
Hook, J. M., Mander, L. N., Urech, R. **1984**, *J. Org. Chem. 49*, 3250.
Hortmann, A. G., Daniel, D. S., Martinelli, J. E. **1973**, *J. Org. Chem. 38*, 728.
Horton, M., Pattenden, G. **1983**, *Tetrahedron Lett. 24*, 2125; **1984**, *J. Chem. Soc. Perkin Trans. I*, 811.
Hosomi, A., Iguchi, H., Sasaki, J., Sakurai, H. **1982**, *Tetrahedron Lett. 23*, 551.
Houghton, R. P., Humber, D. C., Pinder, A. R. **1966**, *Tetrahedron Lett.* 353.
Howe, R., McQuillin, F. J. **1955**, *J. Chem. So.* 2423.
Howell, S. C., Ley, S. C., Mahon, M., Worthington, P. A. **1983**, *J. Chem. Soc. Perkin Trans. I*, 1579.
Hoye, T. R., Kurth, M. J. **1979**, *J. Org. Chem. 44*, 3461.
Hua, D. H., Sinai-Zingde, G., Venkataraman, S. **1985**, *J. Am. Chem. Soc. 107*, 4088.
Hua, D. H. **1986**, *J. Am. Chem. Soc. 108*, 3835.
Huckenstein, M., Kreiser, W. Rüschenbaum, V. **1987**, *Helv. Chim. Acta 70*, 445.
Hudlicky, T., Koszyk, F. J., Kutchan, T. M., Sheth, J. P. **1980a**, *J. Org. Chem. 45*, 5020.
Hudlicky, T., Kutchan, T. **1980b**, *Tetrahedron Lett. 21*, 691.
Hudlicky, T., Kutchan, T. M., Wilson, S. R., Mao, D. T. **1980c**, *J. Am. Chem. Soc. 102*, 6353.
Hudlicky, T., Sarpeshkar, A. M., Hiranuma, S. **1983**, *186th ACS Nat. Meeting*, ORGN 207.
Hudlicky, T., Govindan, S. V., Frazier, J. O. **1985**, *J. Org. Chem. 50*, 4166.
Hudlicky, T. **1987a**, Private Commun.
Hudlicky, T., Natchusm, M. G., Sinai-Zingde, G. **1987b**, *J. Org. Chem. 52*, 4641.
Huffman, J. W., Mole, M. L. **1971**, *Tetrahedron Lett.* 501; *J. Org. Chem. 37*, 13.
Huffman, J. W., Pandian, R. **1979**, *J. Org. Chem. 44*, 1851.
Huffman, J. W., Potnis, S. M., Satish, A. V. **1985**, *J. Org. Chem. 50*, 4266.
Huguet, J., Karpf, M., Dreiding, A. S. **1982**, *Helv. Chim. Acta, 65*, 2413.
Huguet, J., Karpf, M., Dreiding, A. S. **1983**, *Tetrahedron Lett. 24*, 4177.
Humber, D. C., Pinder, A. R., Williams, R. A. **1967**, *J. Org. Chem. 32*, 2335.
Hutchins, R. O., Natale, N. R., Taffer, I. M., Zipkin, R. **1984**, *Synth. Commun. 14*, 445.
Ibuka, T., Hayashi, K., Minakata, H., Inubushi, Y. **1979**, *Tetrahedron Lett.* 159.
Ibuka, T., Hayashi, K., Minakata, H., Ito, Y., Inubushi, Y. **1979**, *Can. J. Chem. 57*, 1579.
Ichinose, I., Kato, T. **1979**, *Chem. Lett.* 61.
Ihara, M., Toyota, M., Fukumoto, K., Kametani, T. **1985**, *Tetrahedron Lett. 26*, 1537.

Ihara, M., Katogi, M., Fukumoto, K., Kametani, T. **1987a**, *Chem. Commun.* 721.
Ihara, M., Kawaguchi, A., Ueda, H., Chihiro, M., Fukumoto, K., Kametani, K. **1987b**, *J. Chem. Soc. Perkin Trans. I,* 1331.
Iio, H., Isobe, M., Kawai, T., Goto, T. **1979a**, *Tetrahedron, 35,* 941; **1979b**, *J. Am. Chem. Soc. 101,* 6076.
Iio, H., Monden, M., Okada. K., Tokoroyama, T. **1987**, *Chem. Commun.* 358.
Ikeda, T., Wakatsuki, K. **1936**, *J. Chem. Soc. Jap. 57,* 425.
Ikeda, T., Yue, S., Hutchinson, C. R. **1985**, *J. Org. Chem. 50,* 5193.
Imagawa, T., Murai, N., Akiyama, T., Kawanisi, M. **1979**, *Tetrahedron Lett.* 1691.
Imamura, P. M., Sierra, M. G., Ruveda, E. A. **1981**, *Chem. Commun.* 734.
Imanishi, T., Matsui, M., Yamashita, M., Iwata, C. **1986a**, *Tetrahedron Lett. 27,* 3161.
Imanishi, T., Ninbari, F., Yamashita, M., Iwata, C., **1986b**, *Chem. Pharm. Bull. 34,* 2268.
Inokuchi, T., Asanuma, G., Torii, S. **1982**, *J. Org. Chem. 47,* 4622.
Inouye, Y., Kakisawa, H. **1969**, *Bull. Chem. Soc. Jap. 42,* 3318.
Inouye, Y., Uchida, Y., Kakisawa, H. **1975**, *Chem. Litt.* 1317.
Inubushi, Y., Tsuda, Y., Sano, T., Konita, T., Suzuki, S., Ageta, H., Otake, Y. **1967**, *Chem. Pharm. Bull. 15,* 1153.
Inubushi, Y., Kikuchi, T., Ibuka, T., Tanaka, K., Saji, I., Tokane, K. **1972**, *Chem. Commun.* 1251; **1974**, *Chem. Pharm. Bull. 22,* 349.
Ireland, R. E., Mander, L. N. **1964**, *Tetrahedron Lett.* 3453.
Ireland, R. E., Mander, L. N. **1965**, *Tetrahedron Lett.* 2627.
Ireland, R. E., Kierstead, R. C. **1966**, *J. Org. Chem. 31,* 2543.
Ireland, R. E., Baldwin, S. W., Dawson, D. J., Dawson, M. I., Dolfini, J. E., Newbould, J., Johnson, W. S., Brown, M., Crawford, R. J., Hurdlik, P. F., Rasmussen, G. H., Schmiegel, K. K. **1970a**, *J. Am. Chem. Soc. 92,* 5743.
Ireland, R. E., Welch, S. C. **1970b**, *J. Am. Chem. Soc. 92,* 7232.
Ireland, R. E., Bey, P., Cheng, K. F., Czarney, R. J., Moser, J.-F., Trust, R. I. **1975a**, *J. Org. Chem. 40,* 1000.
Ireland, R. E., Dawson, M. I., Kowalski, C. J., Lipinski, C. A., Marshall, D. R., Tilley, J. W., Bordner, J., Trus, B. L. **1975b**, *J. Org. Chem. 40,* 973.
Ireland, R. E., Kowalski, C. J., Tilley, J. W., Walba, D. M. **1975c**, *J. Org. Chem. 40,* 990.
Ireland, R. E., McKenzie, T. C., Trust, R. I. **1975d**, *J. Org. Chem. 40,* 1007.
Ireland, R. E., Walba, D. M. **1976**, *Tetrahedron Lett.* 1071.
Ireland, R. E., Aristoff, P. A., Hoyng, C. F. **1979**, *J. Org. Chem. 44,* 4318.
Ireland, R. E., Godfrey, J. D., Thaisrivongs, S. **1981**, *J. Am. Chem. Soc. 103,* 2446.
Ireland, R. E., Dow, W. C., Godfrey, J. D., Thaisrivongs, S. **1984**, *J. Org. Chem. 49,* 1001.
Irie, H., Katakawa, J., Mizuno, Y., Udaka, S., Taga, T., Osaki, K. **1978**, *Chem. Commun.* 717.
Irie, H., Takeda, S., Yamamura, A., Mizuno, Y., Tomimasu, H., Ashizawa, K., Taga, T. **1984**, *Chem. Pharm. Bull. 32,* 2886.
Isobe, M., Iio, H., Kawai, T., Goto, T., **1978**, *J. Am. Chem. Soc. 100,* 1940.
Isoe, S., Hayase, Y., Sakan, T. **1971**, *Tetrahedron Lett.* 3691.
Ito, T., Tomiyoshi, N., Nakamura, K., Azuma, S., Izawa, M., Maruyama, F., Yanagiya, M., Shirahama, H., Matasumoto, T. **1982**, *Tetrahedron Lett. 23,* 1721.
Itoh, A., Nozaki, H., Yamamoto, H. **1978**, *Tetrahedron Lett.* 2903.
Iwashita, T., Kusumi, T., Kakisawa, H. **1979**, *Chem. Lett.* 947.
Iwata, C., Yamada, M., Shinoo, Y. **1979**, *Chem. Pharm. Bull. 27,* 274.
Iwata, C., Fusaka, T., Fujiwara, T., Tomita, K., Yamada, M. **1981**, *Chem. Commun.* 463.
Iwata, C., Nakamura, S., Shinoo, Y., Fusaka, T., Okada, H., Kishimoto, M., Uetsuji, H., Maezaki, N., Yamadada, M., Tanaka, T. **1985a**, *Chem. Pharm. Bull. 33,* 1961.
Iwata, C., Yamashita, M., Aoki, S., Suzuki, K., Takahashi, I., Arakawa, H., Imanishi, T., Tanaka, T. **1985b**, *Chem. Pharm. Bull. 33,* 436.
Iwata, C., Akiyama, T., Miyashita, K. **1987**, *Chem. Ind..* 294.
Iyoda, M., Kushida, T., Kitami, S., Oda, M. **1986**, *Chem. Commun.* 1049.
Iyoda, M., Kushida, T., Kitami, S., Oda, M. **1987b**, *Chem. Commun.* 1607.
Jacobi, P. A., Walker, D. G. **1981**, *J. Am. Chem. Soc. 103,* 4611.

Jacobi, P. A., Kaczmarek, C.S.R., Udodong, U. E. **1984a**, *Tetrahedron Lett. 25*, 4859.
Jacobi, P. A., Craig, T. A., Walker, D. G., Arrick, B. A., Frechette, R. F. **1984b**, *J. Am. Chem. Soc. 106*, 5585.
Jacobi, P. A., Selnick, H. G. **1984c**, *J. Am. Chem. Soc. 106*, 3041.
Jallali-Naini, M., Boussac, G., Lemaitre, P., Larcheveque, M., Guillerm, D., Lallemand, J. Y. **1981**, *Tetrahedron Lett. 22*, 2995.
Jalali-Naini, M., Guillerm, D., Lallemand, J.-Y. **1983**, *Tetrahedron, 39*, 749.
Janssen, C. G., Godefroi, E. F. **1984**, *J. Org. Chem. 49*, 3600.
Jen, T. Y., Hughes, G. A., Smith, H. **1967**, *J. Am. Chem. Soc. 89*, 4551.
Johnson, C. R., Meanwell, N. A., **1981**, *J. Am. Chem. Soc. 103*, 7667.
Johnson, C. R., Elliott, R. C., Meanwell, N. A. **1982**, *Tetrahedron Lett. 23*, 5005.
Johnson, C. R., Kadow, J. F. **1987**, *J. Org. Chem. 52*, 1493.
Johnson, W. S., Jensen, N. P., Hooz, J. **1966**, *J. Am. Chem. Soc. 88*, 3859.
Johnson, W. S., Shenvi, A. B., Boots, S. G. **1982**, *Tetrahedron, 38*, 1397.
Johnston, B. D., Slessor, K. N., Oehlschlager, A. C. **1985**, *J. Org. Chem. 50*, 114.
Joshi, G. D., Kulkarni, S. N. **1965**, *Indian J. Chem. 3*, 91; **1968**, *6*, 127.
Julia, M., Julia, S., Guégan, R. **1960**, *Bull. Soc. Chim. Fr.* 1072.
Julia, M., Julia, S., Langlois, M. **1964**, *Bull. Soc. Chim. Fr.* 1007.
Julia, M., Guy-Rouault, A. **1967**, *Bull. Soc. Chim. Fr.* 1411.
Julia, S., Julia, M., Linstrumelle, G. **1966**, *Bull. Soc. Chim. Fr.* 3499.
Jung, M. E., McCombs, C. A. **1978**, *J. Am. Chem. Soc. 100*, 5207.
Jung, M. E., Pan, Y. G. **1980a**, *Tetrahedron Lett.* 3127.
Jung, M. E., Radcliffe, C. D. **1980b**, *Tetrahedron Lett. 21*, 4397.
Jung, M. E., McCombs, C. A., Takeda, Y., Pan, Y.-G. **1981**, *J. Am. Chem. Soc. 103*, 6677.
Kagawa, S., Matsumoto, S., Nishida, S., Yü, S., Morita, J., Ichihara, A., Matsumoto, T. **1969**, *Tetrahedron Lett.* 3913.
Kakiuchi, K., Nakao, T., Takeda, M., Tobe, Y., Odaira, Y. **1984**, *Tetrahedron Lett. 25*, 557.
Kametani, T., Kato, Y., Honda, T., Fukumoto, K. **1976**, *J. Am. Chem. Soc. 98*, 8185.
Kametani, T., Hirai, Y., Satoh, F., Fukumoto, K. **1977**, *Chem. Commun.* 16.
Kametani, T., Hirai, Y., Shiratori, Y., Fukumoto, K., Satoh, S. **1978**, *J. Am. Chem. Soc. 100*, 554.
Kametani, T., Tsubuki, M., Nemoto, H. **1979**, *Heterocycles 12*, 791.
Kametani, T., Fukumoto, K., Kurobe, H., Nemoto, H. **1981**, *Tetrahedron Lett. 22*, 3653.
Kametani, T., Kawamura, K., Tsubuki, M., Honda, T. **1985a**, *Chem. Commun.* 1324.
Kametani, T., Kawamura, K., Tsubuki, M., Honda, T. **1985b**, *Chem. Pharm. Bull. 33*, 4821.
Kanazawa, R., Kotsuki, H., Tokoroyama, T. **1975**, *Tetrahedron Lett.* 3651.
Kandil, A. A., Slessor, K. N. **1985**, *J. Org. Chem. 50*, 5649.
Kaneko, T., Kawasaki, I., Okamoto, T. **1959**, *Chem. Ind.* 1191.
Kanjilal, P. R., Alam, S. K., Ghatak, U. R. **1981**, *Synth. Commun. 11*, 795.
Kanno, S., Kato, T., Kitahara, Y. **1967**, *Chem. Commun.* 1257.
Karpf, M., Dreiding, A. S. **1980**, *Tetrahedron Lett.* 4569; **1981**, *Helv. Chim. Acta, 64*, 1123.
Katayama, M., Marumo, S., Hattori, H. **1983**, *Tetrahedron Lett. 24*, 1703.
Kato, K., Hirata, Y., Yamamura, S. **1970**, *Chem. Commun.* 1324; **1971**, *Tetrahedron, 27*, 5987.
Kato, M., Kosugi, H., Yoshikoshi, A. **1970a**, *Chem. Commun.* 185; **1970b**, 934.
Kato, M., Kurihara, H., Yoshikoshi, A. **1979**, *J. Chem. Soc. Perkin Trans. I*, 2740.
Kato, M., Heima, K., Matsumura, Y., Yoshikoshi, A. **1981**, *J. Am. Chem. Soc. 103*, 2434.
Kato, T., Tanemura, M., Suzuki, T., Kitahara, Y. **1970**, *Chem. Commun.* 28.
Kato, T., Tanemura, M., Kano, S., Suzuki, T., Kitahara, Y. **1971**, *Bioorg. Chem. 1*, 84.
Kato, T., Ichinose, I., Kamoshida, A., Kitahara, Y. **1976**, *Chem. Commun.* 518.
Kato, T., Takayanagi, H., Suzuki, T., Uyehara, T. **1978**, *Tetrahedron Lett.* 1201.
Kato, T., Ishii, K., Ichinose, I., Nakai, Y., Kumagai, T. **1980**, *Chem. Commun.* 1106.
Kato, T., Mochizuki, M., Hirano, T., Fujiwara, S., Uyehara, T. **1984**, *Chem. Commun.* 1077.
Kato, T., Hirukawa, T., Uyehara, T., Yamamoto, Y. **1987**, *Tetrahedron Lett. 28*, 1439.
Katsui, N., Matsunaga, A., Imaizumi, K., Masamune, T., Tomiyama, K. **1971**, *Tetrahedron Lett.* 83; **1972**, *Bull. Chem. Soc. Jap. 45*, 2871.

Katsumura, S., Isoe, S. **1982**, *Chem. Lett.* 1689.
Katzenellenbogen, J. A., Crumrine, A. L. **1976**, *J. Am. Chem. Soc. 98*, 4925.
Kelly, R. B., Zamecnik, J. **1970**, *Chem. Commun.* 1102.
Kelly, R. B., Zamecnik, J., Beckett, B. A. **1971**, *Chem. Commun.* 479; **1972**, *Can. J. Chem. 50*, 3455.
Kelly, R. B., Eber, J., Hung, H.-K. **1973a**, *Can. J. Chem. 51*, 2534; **1973b**, *Chem. Commun.* 689.
Kelly, R. B., Alward, S. J., Murty, K. S., Stothers, J. B. **1978**, *Can. J. Chem. 56*, 2508.
Kelly, R. B., Harley, M. L., Alward, S. J. **1980**, *Can. J. Chem. 58*, 755.
Kelly, T. R. **1972**, *J. Org. Chem. 37*, 3393.
Kende, A. S., **1960**, *Org. React. 11*, 261.
Kende, A. S., Bentley, T. J., Mader, R. A., Ridge, D. **1974**, *J. Am. Chem. Soc. 96*, 4332.
Kende, A. S., Roth, B., Kubo, I. **1982a**, *Tetrahedron Lett. 23*, 1751.
Kende, A. S., Roth, B., Sanfilippo, P. J., Blacklock, T. J. **1982b**, *J. Am. Chem. Soc. 104*, 5808.
Kende, A. S., Chen, J. **1985**, *J. Am. Chem. Soc. 107*, 7184.
Kende, A. S., Johnson, S., Sanfilippo, P., Hodges, J. C., Jungheim, L. N. **1986**, *J. Am. Chem. Soc. 108*, 3513.
Kenny, M. J., Mander, L. N., Sethi, S. P. **1986**, *Tetrahedron Lett. 27*, 3923, 3927.
Kergomard, A., Veschambe, H. **1977**, *Tetrahedron, 33*, 2215.
Khand, I. U., Knox, G. R., Pauson, P. L., Watts, W. E., Foreman, M. I. **1973**, *J. Chem. Soc. Perkin Trans. I*, 977.
Kido, F., Uda, H., Yoshikoshi, A. **1969**, *Chem. Commun.* 1335.
Kido, F., Maruta, R., Tsutsumi, K., Yoshikoshi, A. **1979a**, *Chem. Lett.* 311.
Kido, F., Tsutsumi, K., Maruta, R., Yoshikoshi, A. **1979b**, *J. Am. Chem. Soc. 101*, 6420.
Kido, F., Kitahara, H., Yoshikoshi, A. **1981a**, *Chem. Commun.* 1236.
Kido, F., Noda, Y., Maruyama, T., Kabuto, C., Yoshikoshi, A. **1981b**, *J. Org. Chem. 46*, 4264.
Kido, F., Noda, Y., Yoshikoshi, A. **1982**, *Chem. Commun.* 1209.
Kido, F., Abe, T., Yoshikoshi, A. **1986**, *Chem. Commun.* 590.
Kieczykowski, G. R., Schlessinger, R. H. **1978**, *J. Am. Chem. Soc. 100*, 1938.
Kim, D., Kim, H. S. **1987**, *J. Org. Chem. 52*, 4633.
Kimura, H., Miyamoto, S., Shinkai, H., Kato, T. **1982**, *Chem. Pharm. Bull. 30*, 723.
King, F. E., King, T. J., Topliss, J. G. **1956**, *Chem. Ind.* 113; **1957**, *J. Chem. Soc.* 573.
Kinney, W. A., Coghlan, M. J., Paquette, L. A., **1984**, *J. Am. Chem. Soc. 106*, 6868.
Kirmse, W. **1971**, *Carbene Chemistry*, 2nd ed., Academic Press, New York.
Kirmse, W., Arend, G., **1972**, *Chem. Ber. 105*, 2746.
Kitagawa, I., Tsujii, S., Nishikawa, F., Shibuya, H. **1983**, *Chem. Pharm. Bull. 31*, 2639.
Kitagawa, Y., Itoh, A., Hashimoto, S., Yamamoto, H., Nozaki, H. **1977**, *J. Am. Chem. Soc. 99*, 3864.
Kitahara, T., Matsuoka, T., Katayama, M., Marumo, S., Mori, K. **1984**, *Tetrahedron Lett. 25*, 4685.
Kitahara, T., Mori, M., Koseki, K., Mori, K. **1986**, *Tetrahedron Lett. 27*, 1343.
Kitahara, T., Mori, M., Mori, K. **1987**, *Tetrahedron 43*, 2689.
Kitahara, Y., Yoshikoshi, A., Oida, S. **1964**, *Tetrahedron, 26*, 1763.
Kitahara, Y., Kato, T., Kobayashi, T., Moore, B. P. **1976**, *Tetrahedron Lett.* 219.
Kitatani, K., Hiyama, T., Nozaki, H. **1976**, *J. Am. Chem. Soc. 98*, 2362.
Klein, E., Rojahn, W. **1970**, *Tetrahedron Lett.* 279.
Klein, E., Rojahn, W. **1971**, *Dragoco Rep.* 239.
Knapp, S., Trope, A. F., Theodore, M. S., Hirata, N., Barchi, J. J. **1984**, *J. Org. Chem. 49*, 608.
Knapp, S., Sharma, S. **1985**, *J. Org. Chem. 50*, 4996.
Knöll, W., Tamm, C. **1975**, *Helv. Chim. Acta, 58*, 1162.
Knudsen, M. J., Shore, N. E. **1984**, *J. Org. Chem. 49*, 5025.
Kochi, H., Matsui, M. **1967**, *Agr. Biol. Chem. 31*, 625.
Kodama, M., Matsuki, Y., Ito, S. **1975**, *Tetrahedron Lett.* 3065.
Kodama, M., Matsuki, Y., Ito, S. **1976**, *Tetrahedron Lett.* 1121.
Kodama, M., Kurihara, T., Sasaki, J., Ito, S. **1979**, *Can. J. Chem. 57*, 3343.
Kodama, M., Takahashi, T., Kojima, T., Ito, S. **1982**, *Tetrahedron Lett. 23*, 3397.
Kodama, M., Okumura, K., Kobayashi, Y., Tsunoda, T., Ito, S. **1984**, *Tetrahedron Lett. 25*, 5781.
Kodama, M., Shiobara, Y., Sumitomo, H., Fukuzumi, K., Minami, H., Miyamoto, Y. **1986a**, *Tetrahedron Lett. 27*, 2157.

Kodama, M., Tambunan, U.S.F., Tsunoda, T., Ito, S. **1986b**, *Bull. Chem. Soc. Jap. 59*, 1897.
Koft, E. R., Smith, A. B., III, **1982a**, *J. Am. Chem. Soc. 104*, 2659; **1982b**, *104*, 5568.
Koft, E. R. **1987a**, *Tetrahedron 43*, 5775.
Kroft, E. R., Broadbent, T. A. **1987b**, *Org. Prep. Proc. Int.* in press.
Kojima, T., Inouye, Y., Kakisawa, H. **1985**, *Chem. Lett.* 323.
Kok, P., DeClercq, P., Vandewalle, M. **1979**, *J. Org. Chem. 44*, 4553.
Kolaczkowski, L., Reusch, W. **1985**, *J. Org. Chem. 50*, 4766.
Komppa, G. **1903**, *Berichte, 36*, 4332.
Kon, K., Isoe, S. **1980**, *Tetrahedron Lett. 21*, 3399.
Kon, K., Isoe, S. **1983**, *Helv. Chim. Acta, 66*, 755.
Kon, K., Ito, K., Isoe, S. **1984**, *Tetrahedron Lett. 25*, 3739.
Kondo, A., Ochi, T., Iio, H., Tokoroyama, T., Siro, M. **1987**, *Chem. Lett.* 1491.
Kondo, K., Matsui, K., Takahatake, Y. **1976**, *Tetrahedron Lett.* 4359.
Koreeda, M., Mislankar, S. G. **1983**, *J. Am. Chem. Soc. 105*, 7203.
Korte, F., Dlugosch, E., Claussen, U. **1966**, *Justus Liebigs Ann. Chem. 693*, 165.
Köster, F.-H., Wolf, H. **1981**, *Tetrahedron Lett. 22*, 3937.
Köster, F.-H., Wolf, H., Kluge, H. **1986**, *Justus Liebigs Ann. Chem.* 78.
Kotake, M., Kawasaki, I., Okamoto, T., Kusumoto, S., Kaneko, T. **1960**, *Justus Liebigs Ann. Chem. 636*, 158.
Kozikowski, A. P., Mugrage, B. B., Wang, B. C., Xu, Z.-b. **1983**, *Tetrahedron Lett. 24*, 3705.
Kraus, G. A., Taschner, M. J. **1980**, *J. Org. Chem. 45*, 1175.
Kraus, G. A., Roth, B., Frazier, K., Shimagaki, M. **1982**, *J. Am. Chem. Soc. 104*, 1114.
Kraus, G. A., Hon, Y.-S. **1986**, *J. Org. Chem. 51*, 116.
Kraus, G. A., Thomas, P. J. **1986**, *J. Org. Chem. 51*, 503.
Kreiser, W., Janitschke, L. **1979**, *Chem. Ber. 112*, 408.
Kretchmer, R. A., Thompson, W. J. **1976**, *J. Am. Chem. Soc. 98*, 3379.
Krief, A., Hevesi, L., Chaboteaux, G., Mathy, P., Sevrin, M., DeVos, M. J. **1985**, *Chem. Commun.* 1693.
Kuehne, M. E. **1970a**, *Synthesis*, 510.
Kuehne, M. E., Nelson, J. A. **1970b**, *J. Org. Chem. 35*, 161.
Kuhn, W., Schinz, H. **1953**, *Helv. Chim. Acta, 36*, 161.
Kulkarni, Y. S., Snider, B. B. **1985**, *J. Org. Chem., 50*, 2809.
Kumanireng, A., Kato, T., Kitahara, Y. **1973**, *Chem. Lett.* 1045
Kumar, A., Singh, A., Devaprabhakara, D. **1976**, *Tetrahedron Lett.* 2177.
Kuo, D. L., Money, T. **1986**, *Chem. Commun.* 1691.
Kuo, F., Fuchs, P. L. **1987**, *J. Am. Chem. Soc. 109*, 1122.
Kutney, J. P., Balsevich, J., Grice, P. **1980**, *Can. J. Chem. 58*, 2641.
Kutney, J. P., Singh, A. **1982a**, *Can. J. Chem. 60*, 1111; **1982b**, *Can. J. Chem. 60*, 1842.
Kutznetsov, K. V., Myrsina, R. A. **1969**, *Dopov. Akad. Nauk Ukr, RSR, Ser. B, 21*, 810.
Kwasigroch, C. **1987**, Ph.D. Dissertation, MIT.
LaBelle, B. E., Knudsen, M. J., Olmstead, M. M., Hope, H., Yanuck, M. D., Shore, N. E. **1985**, *J. Org. Chem. 50*, 5215.
Ladlow, M., Pattenden, G. **1985**, *Tetrahedron Lett. 26*, 4413.
Ladwa, P. D., Joshi, G. D., Kulkarni, S. N. **1968**, *Chem. Ind.* 1601.
Landry, D. W. **1983**, *Tetrahedron, 39*, 2761.
Lange, G. L., Orrom, W. J., Wallace, D. J. **1977**, *Tetrahedron Lett.* 4479.
Lange, G. L., Neidert, E. E., Orrom, W. J., Wallace, D. J. **1978**, *Can. J. Chem. 56*, 1628.
Lansbury, P. T., Hilfiker, F. R. **1969**, *Chem. Commun.* 619.
Lansbury, P. T., Wang, N. Y., Rhodes, J. E. **1971a**, *Tetrahedron Lett.* 1829.
Lansbury, P. T., Nazarenko, N. **1971b**, *Tetrahedron Lett.* 1833.
Lansbury, P. T., **1972**, *Acc. Chem. Res. 5*, 311.
Lansbury, P. T., Haddon, V. R., Stewart, R. C. **1974**, *J. Am. Chem. Soc. 96*, 896.
Lansbury, P. T., Serelis, A. K. **1978**, *Tetrahedron Lett.* 1909.
Lansbury, P. T., Hangauer, D. **1979**, *Tetrahedron Lett.* 3623.
Lansbury, P. T., Hangauer, D. G., Vacca, J. P. **1980**, *J. Am. Chem. Soc. 102*, 3964.

Lansbury, P. T., Mazur, D. J., Springer, J. P. **1985**, *J. Org. Chem. 50*, 1632.
Lansbury, P. T., Mojica, C. A. **1986**, *Tetrahedron Lett. 27*, 3967.
Larsen, S. D., Monti, S. A. **1977**, *J. Am. Chem. Soc. 99*, 8015.
Larsen, S. D., Monti, S. A. **1979**, *Synth. Commun. 9*, 141.
Lee, T. V., Toczek, J. **1985a**, *Tetrahedron Lett. 26*, 473.
Lee, T. V., Toczak, J., Roberts, S. M. **1985b**, *Chem. Commun.* 371.
Leimner, J., Marschall, H., Meier, N., Weyerstahl, P. **1984**, *Chem. Lett.* 1769.
Leone-Bay, A., Paquette, L. A. **1982**, *J. Org. Chem. 47*, 4173.
Leriverend, P., Conia, J.-M. **1970**, *Bull. Soc. Chim. Fr.* 1060.
Leriverend, P. **1973**, *Bull. Soc. Chim. Fr.* 3498.
Ley, S. V., Murray, P. J. **1982**, *Chem. Commun.* 1252.
Ley, S. V., Simpkins, N. S., Whittle, A. J. **1983**, *Chem. Commun.* 1378.
Liotta, D., Ott, W. **1987**, *Synth. Commun. 17*, 1655.
Little, R. D., Carroll, G. L. **1981a**, *Tetrahedron Lett. 22*, 4389.
Little, R. D., Muller, G. W. **1981b**, *J. Am. Chem. Soc. 103*, 2744.
Little, R. D., Wolin, R. L. **1987**, *193rd ACS Nat. Meeting*, ORGN 134.
Liu, H.-J., Lee, S. P. **1977**, *Tetrahedron Lett.* 3699.
Liu, H.-J., Chan, W. H. **1979**, *Can. J. Chem. 57*, 708.
Liu, H.-J., Browne, E.N.C. **1981**, *Can. J. Chem. 59*, 601.
Liu, H.-J., Chan, W. H. **1982**, *Can. J. Chem. 60*, 1081.
Liu, H.-J., Ngooi, T. K., **1984**, *Can. J. Chem. 62*, 2676.
Liu, H.-J., Kulkarni, M. G. **1985**, *Tetrahedron Lett. 26*, 4847.
Liu, H.-J., **1987a**, Private Commun.
Liu, H.-J., Llinas-Brunet, M. **1987b**, Unpublished Results.
Liu, H.-J., Ralitsch, M. **1987c**, Unpublished Results.
Lombardo, L., Mander, L. N., Turner, J. V. **1980**, *J. Am. Chem. Soc. 102*, 6626.
Lupi, A., Patamia, M., Grgurina, I., Marini-Bettolo, R., DiLeo, O., Gioia, P., Antonaroli, S. **1984**, *Helv. Chim. Acta, 67*, 2261.
MacAlpine, G. A., Raphael, R. A., Shaw, A., Taylor, A. W. **1974**, *Chem. Commun.* 834; **1976**, *J. Chem. Soc. Perkin Trans. I*, 410.
Macdonald, T. L. **1978**, *J. Org. Chem. 43*, 3621.
MacKenzie, B. D., Angelo, M. M., Wolinsky, J. **1979**, *J. Org. Chem. 44*, 4042.
MacSweeney, D. F., Ramage, R. **1971**, *Tetrahedron 27*, 1481.
Maercker, A. **1965**, *Org. Reac. 14*, 270.
Magari, H., Hirata, H., Takahashi, T., Matsuo, A., Uto, S., Nozaki, H., Nakayama, M., Hayashi, S. **1982**, *Chem. Lett.* 1143.
Magari, H., Hirota, H., Takahashi, T. **1987**, *Chem. Commun.* 1196.
Magnus, P., Quagliato, D. A. **1982**, *Organometallics, 1*, 1243.
Magnus, P., Exon, C., Albaugh-Robertson, P. **1985**, *Tetrahedron, 41*, 5861.
Magnus, P. D. **1987a**, *194th ACS Nat. Meeting*, INOR 356.
Magnus, P., Principe, L. M., Slater, M. J. **1987b**, *J. Org. Chem. 52*, 1483.
Majetich, G., Grieco, P. A., Nishizawa, M. **1977**, *J. Org. Chem. 42*, 2327.
Majetich, G., **1985a**, Private communication.
Majetich, G., Behnke, M., Hull, K. **1985b**, *J. Org. Chem. 50*, 3615.
Majetich, G., Hull, K., Desmond, R. **1985c**, *Tetrahedron Lett. 26*, 2751.
Majetich, G., Hull, K., Defauw, J., Desmond, R. **1985d**, *Tetrahedron Lett. 26*, 2747.
Majetich, G., Hull, K., Defauw, J., Shawe, T. **1985e**, *Tetrahedron Lett. 26*, 2755.
Majetich, G., Ringold, C. **1987**, *Heterocycles 25*, 271.
Majewski, M., Snieckus, V. **1984**, *J. Org. Chem. 49*, 2682.
Malanco, F. L., Maldonado, L. A. **1976**, *Synth. Commun. 6*, 515.
Mandai, T., Hara, K., Kawada, M., Nokami, J. **1983**, *Tetrahedron Lett. 24*, 1517.
Mane, R. B., Krishna Rao, G. S. **1973**, *J. Chem. Soc. Perkin Trans. I*, 1806.
Mangoni, L., Adinoff, M., Laonigro, G., Caputo, R. **1972**, *Tetrahedron, 28*, 611.
Manh, D.D.K., Fetizon, M., Kone, M. **1975**, *Tetrahedron, 31*, 1903.

Manh, D.D.K., Ecoto, J., Fetizon, M., Colin, H., Diez-Masa, J.-C. **1981**, *Chem. Commun.* 953.
Manzardo, G.G.G., Karpf, M., Dreiding, A. S. **1986**, *Helv. Chim. Acta 69*, 659.
Manjarrez, A., Rio, T., Guzman, A. **1964**, *Tetrahedron, 20*, 333.
Manjarrez, A., Guzman, A. **1966**, *J. Org. Chem. 31*, 348.
Manzardo, G.G.G., Karpf, M., Dreiding, A. **1983**, *Helv. Chim. Acta, 66*, 627.
Marco, J. A., Carda, M. **1987**, *Tetrahedron 43*, 2523.
Marini-Bettolo, R., Tagliatesta, P., Lupi, A.,Bravetti, D. **1983a**, *Helv. Chim. Acta, 66*, 760; **1983b**, *66*, 1922.
Marino, J. P., Silveira, C., Comasseto, J., Petragnani, N. **1987**, *J. Org. Chem. 52*, 4139.
Marshall, J. A., Cohen, N. **1965a**, *J. Am. Chem. Soc. 87*, 2773.
Marshall, J. A., Carroll, R. D. **1965b**, *Tetrahedron Lett.* 4223.
Marshall, J. A., Pike, M. T. **1965c**, *Tetrahedron Lett.* 3107.
Marshall, J. A., Cohen, N., Hochstetler, A. R. **1966a**, *J. Am. Chem. Soc. 88*, 3408.
Marshall, J. A., Pike, M. T. **1966b**, *Tetrahedron Lett.* 4989.
Marshall, J. A., Pike, M. T., Carroll, R. D. **1966c**, *J. Org. Chem. 31*, 2933.
Marshall, J. A., Faubl, H., Warne, T. M. **1967**, *Chem. Commun.* 753.
Marshall, J. A., Bundy, G. L., Fanta, W. I. **1968a**, *J. Org. Chem. 33*, 3913.
Marshall, J. A., Hochstetler, A. R. **1968b**, *J. Org. Chem. 33*, 2593.
Marshall, J. A., Johnson, P. C. **1968c**, *Chem. Commun.* 391.
Marshall, J. A., Partridge, J. J. **1968d**, *J. Am. Chem. Soc. 90*, 1090.
Marshall, J. A., Pike, M. T. **1968e**, *J. Org. Chem. 33*, 435.
Marshall, J. A., Brady, S. F. **1969a**, *Tetrahedron Lett.* 1387.
Marshall, J. A., Partridge, J. J. **1969b**, *Tetrahedron, 25*, 2159.
Marshall, J. A., Johnson, P. C. **1970a**, *J. Org. Chem. 35*, 192.
Marshall, J. A., Ruden, R. A. **1970b**, *Tetrahedron Lett.* 1239.
Marshall, J. A., Cohen, G. M. **1971a**, *J. Org. Chem. 36*, 877.
Marshall, J. A., Greene, A. E. **1971b**, *Tetrahedron Lett.* 859.
Marshall, J. A., Warne, T. M., Jr. **1971c**, *J. Org. Chem. 36*, 178.
Marshall, J. A., Greene, A. E. **1972**, *J. Org. Chem. 37*, 982.
Marshall, J. A., Ruth, J. A. **1974**, *J. Org. Chem. 39*, 1971.
Marshall, J. A., Ellison, R. H. **1976**, *J. Am. Chem. Soc. 98*, 4312.
Marshall, J. A., Wuts, P.G.M. **1977**, *J. Org. Chem. 42*, 1794.
Marshall, J. A., Wuts, P.G.M. **1978a**, *J. Org. Chem. 43*, 1086.
Marshall, J. A., Wuts, P.G.M. **1978b**, *J. Am. Chem. Soc. 100*, 1627.
Marshall, J. A., Conrow, R. E. **1980**, *J. Am. Chem. Soc. 102*, 4274.
Marshall, J. A., Conrow, R. E. **1983**, *J. Am. Chem. Soc. 105*, 5679.
Marshall, J. A., Andrews, R. C. **1986a**, *Tetrahedron Lett. 27*, 5197.
Marshall, J. A., Cleary, D. G. **1986b**, *J. Org. Chem. 51*, 858.
Marshall, J. A., DeHoff, B. S. **1986c**, *Tetrahedron Lett. 27*, 4873.
Marshall, J. A., Jensen, T. M., DeHoff, B. S. **1986d**, *J. Org. Chem. 51*, 4316.
Marshall, J. A., DeHoff, B. S., Crooks, S. L. **1987a**, *Tetrahedron Lett. 28*, 527, 5081.
Marshall, J. A., Lebreton, J. **1987b**, *Tetrahedron Lett. 28*, 3323.
Marshall, J. A., Lebreton, J., DeHoff, B. S., Jensen, T. M. **1987c**, *Tetrahedron Lett. 28*, 3323.
Martel, J., Huynh, C. **1967**, *Bull. Soc. Chim. Fr.* 985.
Martin, M., Clardy, J. **1982**, *Pure Appl. Chem. 54*, 1915.
Martin, S. F., Chou, T. **1978**, *J. Org. Chem. 43*, 1027.
Martin, S. F., Phillips, G. W., Puckette, T. A., Colapret, J. A. **1980**, *J. Am. Chem. Soc. 102*, 5866.
Martin, S. F., Dappen, M. S., Dupre, B., Murphy, C. J. **1987**, *J. Org. Chem. 52*, 3706.
Maruoka, K., Fukutani, Y., Yamamoto, H. **1985**, *J. Org. Chem. 50*, 4412.
Marx, J. N., Norman, L. R. **1973**, *Tetrahedron Lett.* 4375; **1975**, *J. Org. Chem. 40*, 1602.
Marx, J. N., Bih, Q.-R. **1987**, *J. Org. Chem. 52*, 336.
Masaki, Y., Hashimoto, K., Serizawa, Y., Kaji, K. **1982**, *Chem. Lett.* 1879.
Masamune, S. **1964**, *J. Am. Chem. Soc. 86*, 288, 289, 290.
Mase, T., Shibasaki, M. **1986**, *Tetrahedron Lett. 27*, 5245.

Mash, E. A. **1987a**, *J. Org. Chem. 52*, 4142.
Mash, E. A., Fryling, J. A. **1987a**, *J. Org. Chem. 52*, 3000.
Mash, E. A., Math, S. K. **1987b**, *194th ACS Nat. Meeting*, ORGN 154.
Mash, E. A., Heidt, P. C. **1987c**, *193rd ACS Nat. Meeting*, ORGN 19.
Masuoka, N., Kamikawa, T., Kubota, T. **1974**, *Chem. Lett.* 751.
Masuoka, N., Kamikawa, T. **1976**, *Tetrahedron Lett.* 1691.
Matsui, M., Yoshioka, H., Sakamoto, H., Yamada, Y., Kitahara, T. **1967**, *Agr. Biol. Chem. 31*, 33.
Matsuki, Y., Kodama, M., Ito, S. **1979**, *Tetrahedron Lett.* 2901.
Matsumoto, T., Miyano, K., Kagawa, S., Yü, S., Ogawa, J., Ichichara, A. **1971a**, *Tetrahedron Lett.* 3521.
Matsumoto, T., Shirahama, H., Ichihara, A., Shin, H., Kagawa, S., Sakan, F., Miyano, K. **1971b**, *Tetrahedron Lett.* 2049.
Matsumoto, T., Usui, S., Fukui, K. **1976**, *Chem. Lett.* 241.
Matsumoto, T., Ohmura, T. **1977a**, *Chem. Lett.* 335.
Matsumoto, T., Usui, S., Morimoto, T. **1977b**, *Bull. Chem. Soc. Jap. 50*, 1575.
Matsumoto, T., Usui, S. **1983**, *Bull. Chem. Soc. Jap. 56*, 491.
Matsumoto, T., Imai, S., Yoshinari, T., Matsuno, S. **1986**, *Bull. Chem. Soc. Jap. 59*, 3103.
Matsuo, T., Mori, K., Matsui, M. **1976**, *Tetrahedron Lett.* 1979.
Matthews, R. S., Whitesell, J. K. **1975**, *J. Org. Chem. 40*, 3312.
McCormick, J. P., Pachlatko, J. P., Schafer, T. R. **1978**, *Tetrahedron Lett.* 3993.
McCrae, D. A., Dolby, L. **1977**, *J. Org. Chem. 42*, 1607.
McCreadie, T., Overton, K. H., Allison, A. J. **1969**, *Chem. Commun.* 959.
McCurry, P. M., Jr., Singh, R. K. **1973a**, *Tetrahedron Lett.* 3325.
McCurry, P. M., Jr., Singh, R. K., Link, S. **1973b**, *Tetrahedron Lett.* 1155.
McElvain, S. M. **1948**, *Org. React. 4*, 256.
McGuire, H. M., Odom, H. C., Jr., Pinder, A. R. **1974**, *J. Chem. Soc. Perkin Trans. I*, 1879.
McKay, W. R., Ounsworth, J., Sum, P.-E., Weiler, L. **1982**, *Can. J. Chem. 60*, 872.
McMurry, J. E. **1968**, *J. Am. Chem. Soc. 90*, 6821.
McMurry, J. E. **1971**, *J. Org. Chem. 36*, 2826.
McMurry, J. E., Isser, S. J. **1972**, *J. Am. Chem. Soc. 94*, 7132.
McMurry, J. E., Blaszczak, L. C. **1974**, *J. Org. Chem. 39*, 2217.
McMurry, J. E., Musser, J. H., Ahmad, M. S., Blaszczak, L. C. **1975**, *J. Org. Chem. 40*, 1829.
McMurry, J. E., Silvestri, M. G. **1976**, *J. Org. Chem. 41*, 3953.
McMurry, J. E., Andrus, A., Ksander, G. M., Musser, J. H., Johnson, M. A. **1979**, *J. Am. Chem. Soc. 101*, 1330.
McMurry, J. E., Choy, W. **1980**, *Tetrahedron Lett. 21*, 2477.
McMurry, J. E., Matz, J. R. **1982a**, *Tetrahedron Lett. 23*, 2723.
McMurry, J. E., Matz, J. R., Kees, K. L., Bock, P. A. **1982b**, *Tetrahedron Lett. 23*, 1777.
McMurry, J. E. **1983a**, *Acc. Chem. Res. 16*, 405.
McMurry, J. E., Miller, D. D. **1983b**, *Tetrahedron Lett. 24*, 1885.
McMurry, J. E., Bosch, G. K. **1985a**, *Tetrahedron Lett. 26*, 2167.
McMurry, J. E., Erion, M. D. **1985**, *J. Am. Chem. Soc. 107*, 2712.
McMurry, J. E., Kocovsky, P. **1985c**, *Tetrahedron Lett. 26*, 2171.
McMurry, J. E., Bosch, G. K. **1987**, *J. Org. Chem. 52*, 4885.
Mechoulam, R., Braun, P., Gaoni, Y. **1967**, *J. Am. Chem. Soc. 89*, 4552; **1972**, *94*, 6159.
Mehta, G., Kapoor, S. K. **1974**, *J. Org. Chem. 39*, 2618.
Mehta, G., Singh, B. P. **1975**, *Tetrahedron Lett.* 4495.
Mehta, G., Reddy, A. V. **1979**, *Tetrahedron Lett.* 2625.
Mehta, G., Reddy, A. V. **1981**, *Chem. Commun.* 756.
Mehta, G., Reddy, A. V., Murthy, A. N., Reddy, D. S. **1982**, *Chem. Commun.* 540.
Mehta, G., Reddy, D. S., Murty, A. N **1983**, *Chem. Commun.* 824.
Mehta, G., Murty, A. N. **1984**, *Chem. Commun.* 1058.
Mehta, G., Rao, K. S. **1985a**, *Chem. Commun.* 1464.
Mehta, G., Subrahmanyam, D. **1985b**, *Chem. Commun.* 768.

Mehta, G., Rao, K. S. **1987b**, *Chem. Commun.* 1578.
Meyer, W. L., Clemans, G. B., Manning, R. A. **1975**, *J. Org. Chem. 40*, 3686.
Meyer, W. L., Manning, R. A., Schindler, E., Schroeder, R. S., Shew, D. C. **1976**, *J. Org. Chem. 41*, 1005.
Meyer, W. L., Sigel, C. W. **1977**, *J. Org. Chem. 42*, 2769.
Meyers, A. I., Fleming, S. A. **1986a**, *J. Am. Chem. Soc. 108*, 306.
Meyers, A. I., Lefker, B. A. **1986b**, *J. Org. Chem. 51*, 1541.
Miller, R. B., Behare, E. S. **1974**, *J. Am. Chem. Soc. 96*, 8102.
Mills, R. W., Murray, R.D.H., Raphael, R. A. **1971**, *Chem. Commun.* 555.
Minato, H., Nagasaki, T. **1966**, *Chem. Commun.* 347.
Minato, H., Horibe, I. **1967**, *Chem. Commun.* 358; **1968a**, *J. Chem. Soc., C*, 2131.
Minato, H., Nagasaki, T. **1968b**, *J. Chem. Soc., C*, 621.
Mirrington, R. N., Schmazl, K. J. **1972a**, *J. Org. Chem. 37*, 2871; **1972b**, *37*, 2877.
Mitra, R. B., Khanra, A. S. **1977**, *Synth. Commun. 7*, 245.
Mitra, R. B., Mahamulkar, B. G., Kulkarni, G. H. **1984**, *Synthesis*, 428.
Miyashita, M., Yoshikoshi, A. **1971**, *Chem. Commun.* 1091; **1972**, 1173; **1974**, *J. Am. Chem. Soc. 96*, 1917.
Miyashita, M., Kumazawa, T., Yoshikoshi, A. **1979**, *Chem. Lett.* 163; **1981**, 593.
Miyaura, N., Suginome, H., Suzuki, A. **1984**, *Tetrahedron Lett. 25*, 761.
Moens, L., Baizer, M. M., Little, R. D. **1986**, *J. Org. Chem. 51*, 4497.
Moiseenkov, A. M., Czeskis, B. A., Semenovsky, A. V. **1982**, *Chem. Commun.* 109.
Mongrain, M., Lafontaine, J., Belanger, A., Deslongchamps, P. **1970**, *Can. J. Chem. 48*, 3273.
Montero, J. L. **1976**, Thesis, Université de Lanquedoc, Montpellier, France.
Monti, H., Corriol, C., Bertrand, M. **1982**, *Tetrahedron Lett. 23*, 5539.
Monti, S. A., Larsen, S. D. **1978**, *J. Org. Chem. 43*, 2282.
Monti, S. A., Yang, Y.-L. **1979**, *J. Org. Chem. 44*, 897.
Monti, S. A., Dean, T. R. **1982**, *J. Org. Chem. 47*, 2679.
Moore, L., Gooding, D., Wolinsky, J. **1983**, *J. Org. Chem. 48*, 3750.
Morgans, D. J., Jr., Feigelson, G. B. **1983**, *J. Am. Chem. Soc. 105*, 5477.
Mori, K., Matsui, M. **1965**, *Tetrahedron Lett.* 2347.
Mori, K., Shiozaki, M., Itaya, N., Matsui, M., Sumiki, Y. **1969**, *Tetrahedron, 25*, 1293.
Mori, K., Matsui, M. **1970a**, *Tetrahedron, 26*, 2801.
Mori, K., Nakahara, Y., Matsui, M. **1970b**, *Tetrahedron Lett.* 2411.
Mori, K., Ohki, M., Kobayashi, A., Matsui, M. **1970c**, *Tetrahedron, 26*, 2815.
Mori, K., Ohki, M., Matsui, M. **1970d**, *Tetrahedron, 26*, 2821.
Mori, K., **1978**, *Tetrahedron, 34*, 915.
Mori, K., Sasaki, M. **1979**, *Tetrahedron Lett.* 1329.
Mori, K., Uematsu, T., Minobe, M., Yangi, K. **1983**, *Tetrahedron, 39*, 1735.
Mori, K., Watanabe, H. **1986b**, *Tetrahedron, 42*, 273.
Mukaiyama, T., Iwasawa, N. **1981**, *Chem. Lett.* 29.
Mukaiyama, T. **1982**, *Org. React. 28*, 203.
Mukherji, G., Ganguly, B. K., Banerjee, R C., Mukherji, D. **1963**, *J. Chem. Soc.* 2407.
Mulzer, J., Kappert, M. **1983**, *Angew. Chem. Int. Ed. 22*, 63.
Murai, A., Taketsuru, H., Fujisawa, K., Nakahara, Y., Takasugi, M., Masamune, T. **1977**, *Chem. Lett.* 665.
Murai, A., Ono, M., Abiko, A., Masamune, T. **1978**, *J. Am. Chem. Soc. 100*, 7751.
Murai, A., Takatsuru, H., Masamune, T. **1980**, *Bull. Chem. Soc. Jap. 53*, 1049.
Murai, A., Sato, S., Masamune, T. **1981**, *Tetrahedron Lett.* 1033.
Murai, A., Kato, K., Masamune, T. **1982a**, *Tetrahedron Lett. 23*, 2887.
Murai, A., Sato, S., Masamune, T. **1982b**, *Chem. Commun.* 511, 513.
Murali, D., KrishnaRao, G. S. **1987**, *Synthesis* 254.
Murata, Y., Ohtsuka, T., Shirahama, H., Matsumoto, T. **1981**, *Tetrahedron Lett. 22*, 4313.
Naegeli, P., Kaiser, R. **1972**, *Tetrahedron Lett.*, 2013.
Naegeli, P., Wetli, M. **1981**, *Tetrahedron, 37* Suppl. 1, 247.

Näf, F., Ohloff, G. **1974**, *Helv. Chim. Acta*, *57*, 1868.
Näf, F., Decorzant, R., Thommen, W. **1975**, *Helv. Chim. Acta*, *58*, 1808; **1979**, *62*, 114.
Näf, F., Decorzant, R., Giersch, W., Ohloff, G. **1981**, *Helv. Chim. Acta*, *64*, 1387.
Nagakura, I., Maeda, S., Ueno, M., Funamizu, M., Kitahara, Y. **1975**, *Chem. Lett.* 1143.
Nagao, K., Chiba, M., Yoshimura, I., Kim, S.-W. **1981**, *Chem. Pharm. Bull. 29*, 2733.
Nagaoka, H., Ohsawa, K., Takata, T., Yamada, Y. **1984**, *Tetrahedron Lett. 25*, 5389.
Nagaoka, H., Kobayashi, K., Matsui, T., Yamada, K. **1987**, *Tetrahedron Lett. 28*, 2021.
Nagata, W., Sugasawa, T., Narisada, M., Wakabayashi, T., Hayase, Y. **1963a**, *J. Am. Chem. Soc. 85*, 2341; **1963b**, *85*, 2342.
Nagata, W., Narisada, M., Wakabayashi, T., Sugasawa, T. **1967a**, *J. Am. Chem. Soc. 89*, 1499.
Nagata, W., Sugasawa, T., Narisada, M., Wakabayashi, T., Hayase, Y. **1967b**, *J. Am. Chem. Soc. 89*, 1483.
Nagata, W., Wakabayashi, T., Narisada, M., Hayase, Y., Kamata, S. **1971**, *J. Am. Chem. Soc. 93*, 5740.
Nakamura, T., Hirota, H., Kuroda, C., Takahashi, T. **1986**, *Chem. Lett.* 1879.
Nakano, T., Hernández, M. I. **1982**, *Tetrahedron Lett. 23*, 1423.
Nakatani, Y., Yamanishi, T. **1969**, *Agr. Biol. Chem. 33*, 1805.
Nakatani, Y., Kubota, K., Tahara, R., Shigematsu, Y. **1974**, *Agr. Biol. Chem. 38*, 1351.
Nakayama, M., Ohira, S., Shinke, S., Matsushita, Y., Matsuo, A., Hayashi, S. **1975**, *Chem. Lett.* 1249.
Nakayama, M., Ohira, S., Takata, S., Fukuda, K. **1983**, *Chem. Lett.* 147.
Nakazaki, M., Naemura, K. **1966**, *Tetrahedron Lett.* 2615.
Nambudiry, M.E.N., Krishna Rao, G. S. **1974**, *J. Chem. Soc. Perkin Trans. I*, 317.
Nasipuri, D., Das, G. **1979**, *J. Chem. Soc. Perkin Trans. I*, 2776.
Naves, Y.-R. **1942**, *Helv. Chim. Acta*, *25*, 732.
Nazarov, I. N. **1949**, *Usp. Khim. 18*, 377; **1951**, *20*, 71.
Negishi, E., Boardman, L. D., Tour, J. M., Sawada, H., Rand, C. L. **1983**, *J. Am. Chem. Soc. 105*, 6344.
Nemoto, H., Shitara, E., Fukumoto, K., Kametani, K. **1987**, *Heterocycles 25*, 51.
Nicolaou, K. C., Li, W. S. **1985**, *Chem. Commun.* 421.
Nielsen, A. T., Houlihan, W. J. **1968**, *Org. React. 16*, 1.
Nishimura, H., Takabatake, T., Kaku, K., Seo, A., Mizutani, J. **1981**, *Agr. Biol. Chem. 45*, 1861.
Nishizawa, M., Grieco, P. A., Burke, S. D., Metz, W. **1978**, *Chem. Commun.* 76.
Nishizawa, M., Takenaka, H., Hayashi, Y. **1983**, *Chem. Lett.* 1459.
Nishizawa, M., Nishide, H., Hayashi, Y. **1984a**, *Tetrahedron Lett. 25*, 5071.
Nishizawa, M., Nishide, H., Hayashi, Y. **1984b**, *Chem. Commun.* 467.
Nishizawa, M., Takenaka, H., Hayashi, Y. **1984c**, *Tetrahedron Lett. 25*, 437.
Nishizawa, M., Takenaka, H., Hayashi, Y. **1986a**, *J. Org. Chem. 51*, 806.
Nishizawa, M., Yamada, H., Hayashi, Y. **1986b**, *Tetrahedron Lett. 27*, 187,
Niwa, H., Ban, N., Yamada, K. **1983**, *Tetrahedron Lett. 24*, 937.
Niwa, H., Hasegawa, T., Ban, N., Yamada, K. **1984a**, *Tetrahedron Lett. 25*, 2797.
Niwa, H., Wakamatsu, Hida, T., Niiyama, K., Kigoshi, H., Yamada, M., Nagase, H., Suzuki, M., Yamada, K. **1984b**, *J. Am. Chem. Soc. 106*, 4547.
Noyori, R., Nishizawa, M., Shimizu, F., Hayakawa, Y., Maruoka, K., Hashimoto, S., Yamamoto, H., Nozaki, H. **1979**, *J. Am. Chem. Soc. 101*, 220.
Nozaki, K., Oshima, K., Utimoto, K. **1987**, *J. Am. Chem. Soc. 109*, 2547.
Nozoe, S., Goi, M., Morisaki, N. **1971**, *Tetrahedron Lett.* 3701.
Nozoe, S., Furukawa, J., Sankawa, U., Shibata, S. **1976**, *Tetrahedron Lett.* 195.
Nugent, W. A., Hobbs, Jr., F. W. **1986**, *J. Org. Chem. 51*, 3376.
Nystrom, J. E., McCanna, T. D., Helquist, P., Iyer, R. S. **1985**, *Tetrahedron Lett. 26*, 5393.
Odom, H. C., Jr., Pinder, A. R. **1972**, *J. Chem. Soc. Perkin Trans. I*, 2193.
Ohashi, M., Muruishi, T., Kakisawa, H. **1968**, *Tetrahedron Lett.* 719.
Ohashi, M. **1969**, *Chem. Commun.* 893.
Ohfune, Y., Shirahama, H., Matsumoto, T. **1975**, *Tetrahedron Lett.* 4377.
Ohfune, Y., Grieco, P. A., Wang, C.-L., Majetich, G. **1978**, *J. Am. Chem. Soc. 100*, 5946.

Ohtsuka, Y., Niitsuma, S., Tadokoro, H., Hayashi, T., Oishi, T. **1984**, *J. Org. Chem. 49*, 2326.
Ono, T., Tamaoka, T., Yuasa, Y., Matsuda, T., Nokami, J., Wakabayashi, S. **1984**, *J. Am. Chem. Soc. 106*, 7890.
Oplinger, J. A., Paquette, L. A. **1987**, *Tetrahedron Lett. 28*, 5441.
Oppolzer, W. **1973**, *Helv. Chim. Acta, 56*, 1812.
Oppolzer, W., Mahalanabis, K. K. **1975**, *Tetrahedron Lett.* 3411.
Oppolzer, W., Mahalanabis, K. K., Bättig, K. **1977**, *Helv. Chim. Acta, 60*, 2388.
Oppolzer, W., Godel, T. **1978a**, *J. Am. Chem. Soc. 100*, 2583.
Oppolzer, W., Snieckus, V. **1978b**, *Angew. Chem. Int. Ed. 17*, 476.
Oppolzer, W., Snowden, R. L. **1978c**, *Tetrahedron Lett.* 3505.
Oppolzer, W., Bättig, K., Hudlicky, T. **1979**, *Helv. Chim. Acta, 62*, 1493.
Oppolzer, W., Wylie, R. D. **1980**, *Helv. Chim. Acta, 63*, 1198.
Oppolzer, W. **1981a**, *Pure Appl. Chem. 532*, 1181.
Oppolzer, W., Bättig, K. **1981b**, *Helv. Chim. Acta, 64*, 2489.
Oppolzer, W., Marazza, F. **1981c**, *Helv. Chim. Acta, 64*, 1575.
Oppolzer, W., Snowden, R. L. **1981d**, *Helv. Chim. Acta, 64*, 2592.
Oppolzer, W. **1982a**, *Acc. Chem. Res. 15*, 135.
Oppolzer, W., Bättig, K. **1982b**, *Tetrahedron Lett. 23*, 4669.
Oppolzer, W., Pitteloud, R. **1982c**, *J. Am. Chem. Soc. 104*, 6478.
Oppolzer, W., Strauss, H. F., Simmons, D. P. **1982d**, *Tetrahedron Lett. 23*, 4673.
Oppolzer, W., Zutterman, F., Bättig, K. **1983**, *Helv. Chim. Acta, 66*, 522.
Oppolzer, W. **1984**, *Angew. Chem. Int. Ed. 23*, 876.
Oppolzer, W., Chapuis, C., Dupuis, D., Guo, M. **1985**, *Helv. Chim. Acta, 68*, 2100.
Oppolzer, W., Cunningham, A. F. **1986a**, *Tetrahedron Lett. 27*, 5467.
Oppolzer, W., Jacobsen, E. J. **1986b**, *Tetrahedron Lett. 27*, 1141.
Oppolzer, W., Nakao, A. **1986c**, *Tetrahedron Lett. 27*, 5471.
Oppolzer, W., Schneider, P. **1986d**, *Helv. Chim. Acta, 69*, 1817.
Orban, J., Turner, J. V. **1983**, *Tetrahedron Lett. 24*, 2697.
Paquette, L. A., Han, Y. K. **1979**, *J. Org. Chem. 44*, 4014.
Paquette, L. A., Fristad, W. E., Dime, D. S., Bailey, T. R. **1980**, *J. Org. Chem. 45*, 3017.
Paquette, L. A., Schostarez, H., Annis, G. D. **1981**, *J. Am. Chem. Soc. 103*, 6526.
Paquette, L. A., Andrews, D. R., Springer, J. P. **1983**, *J. Org. Chem. 48*, 1147.
Paquette, L. A., Roberts, R. A., Drtina, G. J. **1984**, *J. Am. Chem. Soc. 106*, 6690.
Paquette, L. A., Bulman-Page, P. C. **1985a**, *Tetrahedron Lett. 26*, 1607.
Paquette, L. A., Ham, W. H., Dime, D. S. **1985b**, *Tetrahedron Lett. 26*, 4983.
Paquette, L. A., Wiedeman, P. E. **1985c**, *Tetrahedron Lett. 26*, 1603.
Paquette, L. A., Ham, W. H. **1986a**, *Tetrahedron Lett. 27*, 2341.
Paquette, L. A., Sugimura, T. **1986b**, *J. Am. Chem. Soc. 108*, 3841.
Paquette, L. A., Lin, H.-S., Coghlan, M. J. **1987a**, *Tetrahedron Lett. 28*, 5017.
Paquette, L. A., Schaefer, A. G., Springer, J. P. **1987b**, *Tetrahedron 43*, 5567.
Paquette, L. A., Wright, J., Drtina, G. J., Roberts, R. A., **1987c**, *J. Org. Chem. 52*, 2960.
Paquette, L. A., Okazaki, M. E., Caille, J.-C. **1988a**, *J. Org. Chem. 53*, 477.
Park, O. S., Grillasca, Y., Garcia, G. A., Maldonado, L. A. **1977**, *Synth. Commun. 7*, 345.
Parker, K. A., Farmar, J. G. **1986**, *J. Org. Chem. 51*, 4023.
Parker, W., Ramage, R., Raphael, R. A. **1962**, *J. Chem. Soc.* 1558.
Parkes, K.E.B., Pattenden, G. **1986**, *Tetrahedron Lett. 27*, 1305.
Parsons, W. H., Schlessinger, R. H., Quesada, M. L. **1980**, *J. Am. Chem. Soc. 102*, 889.
Partridge, J. J., Chadha, N. K., Uskokovic, M. R. **1973**, *J. Am. Chem. Soc. 95*, 532.
Pattenden, G., Teague, S. J. **1982**, *Tetrahedron Lett. 23*, 5471; **1984**, *25*, 3021.
Pattenden, G., Robertson, G. M. **1985**, *Tetrahedron, 41*, 4001.
Pattenden, G., Robertson, G. M. **1986**, *Tetrahedron Lett. 27*, 399.
Pawson, B. A., Cheung, H. C., Gurbaxani, S., Saucy, G. **1968**, *Chem. Commun.* 1057; **1970**, *J. Am. Chem. Soc. 92*, 366.
Pelletier, S. W., Chappell, R. W., Prabhakar, S. **1966**, *Tetrahedron Lett.* 3489; **1968a**, *J. Am. Chem. Soc. 90*, 2889.

Pelletier, S. W., Prabhakar, S. **1968b**, *J. Am. Chem. Soc. 90*, 5318.
Pennanen, S. **1980**, , *Acta Chem. Scand. B34*, 261.
Perkin, W. H., Jr., Thorpe, J. F. **1904**, *J. Chem. Soc.* 146.
Pesaro, M., Bozzato, G., Schudel, P. **1968**, *Chem. Commun.* 1152.
Pesaro, M., Bachman, J.-P. **1978**, *Chem. Commun.* 203.
Piers, E., Britton, R. W., deWaal, W. **1969a**, *Can. J. Chem. 47*, 831; **1969b**, *Chem. Commun.* 1069; **1969c**, *Tetrahedron Lett.* 1251.
Piers, E., Smillie, D. R. **1970**, *J. Org. Chem. 35*, 3997.
Piers, E., Britton, R. W., Keziere, R. J., Smillie, R. D. **1971a**, *Can. J. Chem. 49*, 2620; **1971b**, *49*, 2623.
Piers, E., Geraghty, M. B., Soucy, M. **1973**, *Synth. Commun. 3*, 401.
Piers, E., Britton, R. W., Geraghty, M. R., Keziere, R. J., Smillie, R. D. **1975a**, *Can. J. Chem. 53*, 2827.
Piers, E., Britton, R. W., Geraghty, M. B., Keziere, R. J., Kido, F. **1975b**, *Can. J. Chem. 53*, 2838.
Piers, E., Geraghty, M. B., Smillie, R. D., Soucy, M. **1975c**, *Can. J. Chem. 53*, 2849.
Piers, E., Phillips-Johnson, W. M. **1975d**, *Can. J. Chem. 53*, 1281.
Piers, E., Isenring, H.-P. **1976**, *Synth. Commun. 6*, 221.
Piers, E., Hall, T.-W. **1977a**, *Chem. Commun.* 880.
Piers, E., Lau, C. K. **1977b**, *Synth. Commun. 7*, 495.
Piers, E., Isenring, H.-P. **1977c**, *Can. J. Chem. 55*, 1039.
Piers, E., Banville, J. **1979a**, *Chem. Commun.* 1138.
Piers, E., Ruediger, E. H. **1979b**, *Chem. Commun.* 166.
Piers, E., Hall, T.-W. **1980**, *Can. J. Chem. 58*, 2613.
Piers, E., Abeysekera, B. F., Herbert, D. J., Suckling, I. D. **1982a**, *Chem. Commun.* 404.
Piers, E., Winter, M. **1982b**, *Justus Liebigs Ann. Chem.* 973.
Piers, E., Karunaratne, V. **1984a**, *Chem. Commun.* 959; **1984b**, *Can. J. Chem. 62*, 629.
Piers, E., Abeysekera, B. F., Herbert, D. J., Suckling, I. D. **1985a**, *Can. J. Chem. 63*, 3418.
Piers, E., Jung, G. L. **1985b**, *Can. J. Chem. 63*, 996.
Piers, E., Moss, N. **1985c**, *Tetrahedron Lett. 26*, 2735.
Piers, E., Friesen, R. W. **1986a**, *J. Org. Chem. 51*, 3405.
Piers, E., Gavai, A. V. **1986b**, *Tetrahedron Lett. 27*, 313.
Piers, E., Yeung, B.W.A. **1986c**, *Can. J. Chem. 64*, 2475.
Piers, E., Jean, M., Marrs, P. S. **1987a**, *Tetrahedron Lett. 28*, 5075.
Piers, E., Wai, J.S.M. **1987b**, *Chem. Commun.* 1342.
Pinder, A. R., Williams, R. A. **1961**, *Chem. Ind.* 1714; **1963**, *J. Chem. Soc.* 2773.
Pirrung, M. C. **1979**, *J. Am. Chem. Soc. 101*, 7130; **1981**, *103*, 82.
Pirrung, M. C., Thomson, S. A. **1986**, *Tetrahedron Lett. 27*, 2703; **1987**, *J. Org. Chem. 53*, 227.
Plattier, M., Teisseire. P. **1974**, *Recherches, 19*, 153.
Plattner, J. J., Bhalero, U. T., Rapoport, H. **1969**, *J. Am. Chem. Soc. 91*, 4933.
Plattner, J. J., Rapoport, H. **1971**, *J. Am. Chem. Soc. 93*, 1758.
Poirier, D., Jean, M., Burnell, R. H. **1983**, *Synth. Commun. 13*, 201.
Posner, G. H. **1972**, *Org. React. 19*, 1.
Posner, G. H., Hamill, T. Unpublished Results.
Pramod, K., Subba Rao, G.S.R. **1982**, *Chem. Commun.* 762.
Prasad, C.V.C., Chan, T. H.. **1987**, *J. Org. Chem. 52*, 120.
Prestwich, G. D., Labovitz, J. N. **1974**, *J. Am. Chem. Soc. 96*, 7103.
Prisbylla, M. P., Takabe, K., White, J. D. **1979**, *J. Am. Chem. Soc. 101*, 762.
Quallich, G. J., Schlessinger, R. H. **1979**, *J. Am. Chem. Soc. 101*, 7627.
Quinkert, G., Schmalz, H. G., Walzer, E., Kowalczyk-Przewloka, T., Durner, G., Bats, J. W. **1987**, *Angew. Chem. Int. Ed. 26*, 61.
Rama Rao, A. V., Rao, M. N. Garyali, K. **1984**, *Synth. Commun. 14*, 557.
Ranu, B. C., Kavka, M., Higgs, L. A., Hudlicky, T. **1984**, *Tetrahedron Lett. 25*, 2447.
Raphael, R. A., Telfer, S. J. **1985**, *Tetrahedron Lett. 26*, 489.
Rapson, W. S., Robinson, R. **1935**, *J. Chem. Soc.* 1285.
Rathke, M. W. **1975**, *Org. React. 22*, 423.

Raucher, S., Chi, K.-W., Hwang, K.-J., Burks, Jr., J. E. **1986**, *J. Org. Chem. 51*, 5503.
Razdan, R. K., Handrick, G. R. **1970**, *J. Am. Chem. Soc. 92*, 6061.
Razdan, R. K., Handrick, G. R., Dalzell, H. C. **1975**, *Experientia, 31*, 16.
Reddy, P. A., Krishna Rao, G. S. **1979**, *J. Chem. Soc. Perkin Trans. I*, 237; **1980**, *Indian J. Chem. 19B*, 753.
Redmore, D., Gutsche, C. D. **1971**, *Adv. Alicyclic Chem. 3*, 1.
Reetz, M. T., Westermann, J., Steinbach, R. **1981**, *Chem. Commun.* 237.
Reich, H. J., Eisenhart, E. K. **1984**, *J. Org. Chem. 49*, 5282.
Reusch, W. **1985**, private communication.
Reuvers, J.T.A., deGroot, A. **1986**, *J. Org. Chem. 51*, 4594.
Rhoads, S. J., Raulins, N. R. **1975**, *Org. React.. 22*, 1.
Rigby, J. H., Wilson, J. Z. **1984**, *J. Am. Chem. Soc. 106*, 8217.
Rigby, J. H. **1985**, private communication.
Rigby, J. H., Moore, T. L., Rege, S. **1986**, *J. Org. Chem. 51*, 2398.
Rigby, J. H., Senanayake, C. **1987**, *J. Am. Chem. Soc. 109*, 3147.
Riss, B. P., Muckensturm, B. **1986**, *Tetrahedron Lett. 27*, 4979.
Ritterskamp, P., Demuth, M., Schaffner, K. **1984**, *J. Org. Chem. 49*, 1155.
Roberts, B. W., Poonian, M. S., Welch, S. C. **1969**, *Tetrahedron Lett.* 3400.
Roberts, M. R., Schlessinger, R. H. **1979**, *J. Am. Chem. Soc. 101*, 7626; **1981**, *103*, 724.
Ronald, R. C. **1976**, *Tetrahedron Lett.* 4413.
Rosini, G., Salomoni, A., Squarcia, F. **1979**, *Synthesis*, 942.
Rosini, G., Marotta, E., Petrini, M., Ballini, R. **1985**, *Tetrahedron, 41*, 4633.
Rouessac, F., Zamarlik, H., Gnonlonfoun, N. **1983**, *Tetrahedron Lett. 24*, 2247.
Roush, W. R. **1978**, *J. Am. Chem. Soc. 100*, 3599.
Roush, W. R., D'Ambra, T. E. **1983**, *J. Am. Chem. Soc. 105*, 1059.
Roy, J. K. **1954**, *Chem. Ind.* 1393.
Ruegg, R., Pfiffner, A., Montavon, M. **1966**, *Recherches, 15*, 3.
Ruppert, J. F., Avery, M. A., White, J. D. **1976**, *Chem. Commun.* 978.
Ruzicka, L., Capato, E. **1925**, *Helv. Chim. Acta, 8*, 259.
Saha, M., Bagby, B., Nicholas, K. M. **1986**, *Tetrahedron Lett. 27*, 915.
Saito, A., Matushita, H., Kaneko, H. **1983**, *Chem. Lett.* 729.
Saito, T., Itoh, A., Oshima, K. **1979**, *Tetrahedron Lett.* 3519.
Sakan, F., Hashimoto, H., Ichihara, A., Shirahama, H., Matsumoto, T. **1971**, *Tetrahedron Lett.* 3703.
Sakan, K., Craven, B. M. **1983**, *J. Am. Chem. Soc. 105*, 3732.
Sakan, T., Fujino, A., Murai, F., Suzuki, A., Batsugan, Y. **1960**, *Bull. Chem. Soc. Jap. 33*, 1737.
Sakan, T., Abe, K. **1968**, *Tetrahedron Lett.* 2471.
Sakurai, H., Shirahata, A., Hosomi, A. **1979**, *Angew. Chem. Int. Ed. 18*, 163.
Salomon, R. G., Sachinvala, N. D., Raychaudhuri, S. R., Miller, D. B. **1984**, *J. Am. Chem. Soc. 106*, 2211.
Sammes, P. G. **1970**, *Synthesis*, 636.
Sammes, P. G., Street, L. J. **1983**, *Chem. Commun.* 666.
Sampath, V., Lund, E. C., Knudsen, M. J., Olmstead, M. M., Shore, N. E. **1987**, *J. Org. Chem. 52*, 3595.
Santelli-Rouvier, C., Santelli, M. **1983**, *Synthesis*, 429.
Sarma, A. S., Chattopadhyay, P. **1982**, *J. Org. Chem. 47*, 1727.
Sarma, A. S., Gayen, A. K. **1985**, *Tetrahedron, 41*, 4581.
Sato, K., Miyamoto, O., Inoue, S., Honda, K. **1981**, *Chem. Lett.* 1183.
Schaefer, J. P., Bloomfield, J. J. **1967**, *Org. React. 15*, 1.
Schenck, G. D., Wirtz, R. **1953**, *Naturwissenschaften, 40*, 581.
Schiehser, G. A., White, J. D. **1980**, *J. Org. Chem. 45*, 1864.
Schlessinger, R. H., Nugent, R. A. **1982**, *J. Am. Chem. Soc. 104*, 1116.
Schlessinger, R. H., Schultz, J. A. **1983a**, *J. Org. Chem. 48*, 407.
Schlessinger, R. H., Wood, J. L., Poss, A. J., Nugent, R. A., Parsons, W. H. **1983b**, *J. Org. Chem. 48*, 1147.

Schlessinger, R. H., Wong, J. W., Poss, M. A., Springer, J. P. **1985**, *J. Org. Chem. 50*, 3950.
Schmazl, K. J., Mirrington, R. N. **1970**, *Tetrahedron Lett.* 3219.
Schostarez, H., Paquette, L. A. **1981a**, *J. Am. Chem. Soc. 103*, 722; **1981b**, *Tetrahedron, 37*, 4431.
Schreiber, S. L., Santini, C. **1984**, *J. Am. Chem. Soc. 106*, 4038.
Schreiber, S. L., Hawley, R. C. **1985**, *Tetrahedron Lett. 26*, 5971.
Schreiber, S. L., Meyers, H. V., Wiberg, K. B. **1986**, *J. Am. Chem. Soc. 108*, 8274.
Schuda, P. F., Heimann, M. R. **1984**, *Tetrahedron, 40*, 2365.
Schuda, P. F., Phillips, J. L., Morgan, T. M., **1986**, *J. Org. Chem. 51*, 2742.
Schulte-Elte, K. H., Strickler, H., Gautschi, F., Pickenhagen, W., Gadola, M., Limacher, J., Müller, B. L., Wuffli, F., Ohloff, G. **1975**, *Justus Liebigs Ann. Chem.* 484.
Schultz, A. G., Godfrey, J. D. **1976**, *J. Org. Chem. 41*, 3494.
Schultz, A. G., Motyka, L. A. **1982**, *J. Am. Chem. Soc. 104*, 5800.
Schultz, A. G., Dittami, J. P. **1984**, *J. Org. Chem. 49*, 2615.
Schultz, A. G., Puig, S. **1985**, *J. Org. Chem. 50*, 915.
Schwartz, M. A., Crowell, J. D., Musser, J. H. **1972**, *J. Am. Chem. Soc. 94*, 4361.
Schwartz, M. A., Swanson, G. C. **1979**, *J. Org. Chem. 44*, 953.
Schwartz, M. A., Willbrand, A. M. **1985**, *J. Org. Chem. 50*, 1359.
Secrist, J. A., III, Hickey, C. J., Norris, R. E. **1977**, *J. Org. Chem. 42*, 525.
Seebach, D. **1969**, *Synthesis, 17.* **1979**, *Angew. Chem. Int. Ed. 18*, 239.
Semmelhack, M. F. **1972**, *Org. React. 19*, 115.
Semmelhack, M. F., Yamashita, A., Tomesch, J. C., Hirotsu, K. **1978**, *J. Am. Chem. Soc. 100*, 5565.
Semmelhack, M. F., Tomoda, S., Hurst, K. M. **1980a**, *J. Am. Chem. Soc. 102*, 7567.
Semmelhack, M. F., Yamashita, A. **1980b**, *J. Am. Chem. Soc. 102*, 5924.
Semmelhack, M. F., Brickner, S. J. **1981a**, *J. Am. Chem. Soc. 103*, 3945.
Semmelhack, M. F., Tomoda, S. **1981b**, *J. Am. Chem. Soc. 103*, 2427.
Seto, H., Fujimoto, Y., Tatsuno, T., Yoshioka, H. **1985**, *Synth. Commun. 15*, 1217.
Sevrin, M., Hevesi, L., Krief, A. **1976**, *Tetrahedron Lett.* 3915.
Sharpless, K. B. **1970**, *J. Am. Chem. Soc. 92*, 6999.
Shibasaki, M., Iseki, K., Ikegami, S. **1980**, *Tetrahedron Lett.* 3587.
Shibasaki, M., Yamazaki, M., Iseki, K., Ikegami, S. **1982**, *Tetrahedron Lett. 23*, 5311.
Shibasaki, M., Mase, T., Ikegami, S. **1986**, *J. Am. Chem. Soc. 108*, 2090.
Shibuya, H., Ohashi, K., Kawashima, K., Hori, K., Murakami, N., Kitagawa, I. **1986**, *Chem. Lett.* 85.
Shimada, J., Hashimoto, K., Kim, B. H., Nakamura, E., Kuwajima, I. **1984**, *J. Am. Chem. Soc. 106*, 1759.
Shimada, K., Kodama, M., Ito, S. **1981**, *Tetrahedron Lett. 22*, 4275.
Shimagaki, M., Tahara, A. **1975**, *Tetrahedron Lett.* 1715.
Shimizu, N., Tsuno, Y., **1979**, *Chem. Lett.* 103.
Shishido, K., Hiroya, K., Fukumoto, K., Kametani, T. **1987**, *Chem. Commun.* 1360.
Shizuri, Y., Suyama, K., Yamamura, S. **1986**, *Chem. Commun.* 63.
Sigrist, R., Rey, M., Dreiding, A. S. **1986**, *Chem. Commun.* 944.
Simmons, H. E., Cairns, T. L., Vladuchick, S. A., Hoiness, C. M. **1973**, *Org. React. 20*, 1.
Singh, A. K., Bakshi, R. K., Corey, E. J. **1987**, *J. Am. Chem. Soc. 109*, 6187.
Skattebol, L., Stenstrom, Y. **1983**, *Tetrahedron Lett. 24*, 3021.
Skeean, R. W., Trammell, G. L., White, J. D., **1976**, *Tetrahedron Lett.* 525.
Slessor, K. N., Oehlschlager, A. C., Johnston, B. D., Pierce, H. D., Jr., Grewal, S. K., Wickremesinghe, L.K.G. **1980**, *J. Org. Chem. 45*, 2290.
Smith, A. B., III, Dieter, R. K. **1981a**, *Tetrahedron, 37*, 2407.
Smith, A. B., III, Guaciaro, M. A., Schow, S. R., Wovkulich, P. M., Toder, B. H., Hall, T. W. **1981b**, *J. Am. Chem. Soc. 103*, 219.
Smith, A. B., III, Jerris, P. J. **1981c**, *J. Am. Chem. Soc. 103*, 194.
Smith, A. B., III, Richmond, R. E. **1981d**, *J. Org. Chem. 46*, 4814; **1983**, *J. Am. Chem. Soc. 105*, 575.
Smith, A. B., III, Fukui, M. **1984a**, *187th ACS Meeting*, ORGN 6.
Smith, A. B., III, Konopelski, J. P. **1984b**, *J. Org. Chem. 49*, 4094.
Smith, A. B., III, Mewshaw, R. **1984c**, *J. Org. Chem. 49*, 3685.

Smith, A. B., III, Liverton, N. J., Hrib, N. J., Sivaramakrishnan, H., Winzenberg, K. **1985**, *J. Org. Chem. 50*, 3239.
Smith, A. B., III, Dorsey, B. D., Visnick, M., Maeda, T., Malamas, M. S. **1986**, *J. Am. Chem. Soc. 108*, 3110.
Smith, A. B., III, Fukui, M. **1987**, *J. Am. Chem. Soc. 109*, 1269.
Snider, B. B., Rodini, D. J., van Straten, J. **1980**, *J. Am. Chem. Soc. 102*, 5872.
Snider, B. B., Kulkarni, Y. S. **1985a**, *Tetrahedron Lett. 26*, 5675.
Snider, B. B., Mohan, R., Kates, S. A. **1985b**, *J. Org. Chem. 50*, 3659.
Snitman, D. L., Himmelsbach, R. J., Watt, D. S. **1978**, *J. Org. Chem. 43*, 4758.
Snitman, D. L., Himmelsbach, R. J., Haltiwanger, R. C., Watt, D. S. **1979**, *Tetrahedron Lett.* 2477.
Snowden, R. L. **1981a**, *Tetrahedron Lett. 22*, 101.
Snowden, R. L., Sonnay, P., Ohloff, G. **1981b**, *Helv. Chim. Acta, 64*, 25.
Snowden, R. L., Sonnay, P. **1984**, *J. Org. Chem. 49*, 1464.
Soffer, M. D., Günay, G. E., Korman, O., Adams, M. B. **1963**, *Tetrahedron Lett.* 389.
Soffer, M. D., Günay, G. E. **1965**, *Tetrahedron Lett.* 1355.
Soffer, M. D., Burk, L. A. **1970**, *Tetrahedron Lett.* 211.
Solar, D., Wolinsky, J. **1983**, *J. Org. Chem. 48*, 670.
Sonawane, H. R., Nanjundiah, B. S., Kumar, M. U. **1984**, *Tetrahedron Lett. 25*, 2245.
Spencer, T. A., Weaver, T. D., Villarica, R. M., Friary, R. J., Posler, J., Schwartz, M. A. **1968**, *J. Org. Chem. 33*, 712.
Spencer, T. A., Smith, R.A.J., Storm, D. L., Villarica, R. M. **1971**, *J. Am. Chem. Soc. 93*, 4856.
Spitzner, D. **1978**, *Tetrahedron Lett.* 3349.
Stadler, H., Rey, M., Dreiding, A. S. **1984**, *Helv. Chim. Acta, 67*, 1854.
Sternbach, D. D., Hughes, J. W.. **1983**, *186th ACS Nat. Meeting*, ORGN 205.
Sternbach, D. D., Hobbs, S. H. **1984a**, *188th ACS Nat. Meeting*, ORGN 52.
Sternbach, D. D., Hughes, J. W. **1984b**, *188th ACS Nat. Meeting*, ORGN 53.
Sternbach, D. D., Hughes, J. W., Burdi, D. F., Banks, B. A. **1985**, *J. Am. Chem. Soc. 107*, 2149.
Sternbach, D. D. **1987**, Priv. Commun.
Stevens, K. E., Yates, P. **1980**, *Chem. Commun.* 990.
Stevens, K. E., Paquette, L. A. **1981**, *Tetrahedron Lett. 22*, 4393.
Stevens, R. V., Bisacchi, G. S. **1982**, *J. Org. Chem. 47*, 2396.
Still, W. C. **1977a**, *J. Am. Chem. Soc. 99*, 4186.
Still, W. C., Schneider, M. J. **1977b**, *J. Am. Chem. Soc. 99*, 948.
Still, W. C. **1979**, *J. Am. Chem. Soc. 101*, 2493.
Still, W. C., Tsai, M.-Y. **1980**, *J. Am. Chem. Soc. 102*, 3654.
Still, W. C., Mobilio, D. **1983a**, *J. Org. Chem. 48*, 4785.
Still, W. C., Murata, S., Revial, G., Yoshihara, K. **1983b**, *J. Am. Chem. Soc. 105*, 625.
Stille, J. R., Grubbs, R. H. **1986**, *J. Am. Chem. Soc. 108*, 855.
Stipanovic, R. D., Bell, A. A., Lukefahr, M. J. **1976**, *172nd ACS Meeting*, PEST 78.
Stipanovic, R. D., Bell, A. A., O'Brien, D. H., Lukefahr, M. J. **1977**, *Tetrahedron Lett.* 567.
Stipanovic, R. D., Bell, A. A., O'Brien, D. H. **1978**, *Phytochemistry, 17*, 151.
Stojanac, N., Strojanac, Z., White, P. S., Valenta, Z. **1979**, *Can. J. Chem. 57*, 3346.
Stork, G., van Tamelen, E. E., Friedman, L. J., Burgstahler, A. W. **1951**, *J. Am. Chem. Soc. 73*, 4505; **1953**, *75*, 384.
Stork, G., Clarke, F. H. **1955**, *J. Am. Chem. Soc. 77*, 1072; **1961a**, *83*, 3114.
Stork, G., Tsuji, J. **1961b**, *J. Am. Chem. Soc. 83*, 2783.
Stork, G., Schulenberg, J. W. **1962**, *J. Am. Chem. Soc. 84*, 284.
Stork, G., Brizzolara, A., Landesman, H., Szmuszkovicz, J., Terrell, R. **1963a**, *J. Am. Chem. Soc. 85*, 207.
Stork, G., Meisles, A., Davies, J. E. **1963b**, *J. Am. Chem. Soc. 85*, 3419.
Stork, G., Rosen, P., Goldman, N., Coombs, R. V., Tsuji, J. **1965**, *J. Am. Chem. Soc. 87*, 275.
Stork, G., Danishefsky, S., Ohashi, M. **1967**, *J. Am. Chem. Soc. 89*, 5459.
Stork, G., Gregson, M. **1969**, *J. Am. Chem. Soc. 99*, 2373.
Stork, G., Uyeo, S., Wakamatsu, T., Grieco, P., Labovitz, J. **1971**, *J. Am. Chem. Soc. 93*, 4945.

Stork, G., Danheiser, R. L., Ganem, B. **1973a**, *J. Am. Chem. Soc. 95*, 3414.
Stork, G., Gardner, J. O., Boeckman, R. K., Parker, K. A. **1973b**, *J. Am. Chem. Soc. 95*, 2014.
Stork, G., Cohen, J. F. **1974**, *J. Am. Chem. Soc. 96, 5270.*
Stork, G., Boeckman, R. K., Taber, D. F., Still, W. C., Singh, J. **1979a**, *J. Am. Chem. Soc. 101*, 7107.
Stork, G., Singh, J. **1979b**, *J. Am. Chem. Soc. 101*, 7109.
Stork, G., Baine, N. H. **1985**, *Tetrahedron Lett. 26*, 5927.
Strickland, J. B., Romo, D., Pommerville, J. C., Harding, K. E. **1987**, *194th ACS Nat. Meeting*, ORGN 199.
Strickler, H., Ohloff, G., Kovats, E. **1967**, *Helv. Chim. Acta, 50*, 759.
Subba Rao, G.S.R., Pramod, K. **1986**, *Indian J. Chem. B25*, 783.
Subrahamanian, K. P., Reusch, W. **1978**, *Tetrahedron Lett.* 3789.
Sum, F. W., Weiler, L. **1979**, *Tetrahedron Lett.* 707.
Suri, S. C., Marchand, A. P. **1984**, *188th ACS Nat. Meeting*, ORGN. 55.
Suryawanshi, S. N., Fuchs, P. L. **1986**, *J. Org. Chem. 51*, 902.
Suzuki, T., Tanemura, M., Kato, T., Kitahara, Y. **1970**, *Bull. Chem. Soc. Jap. 43*, 1268.
Swindell, C. S., Patel, B. P. **1987**, *193rd ACS Nat. Meeting*, ORGN 79.
Taber, D. F., Korsmeyer, R. W. **1978**, *J. Org. Chem. 43*, 4925.
Taber, D. F., Gunn, B. P. **1979**, *J. Am. Chem. Soc. 101*, 3992.
Taber, D. F., Saleh, S. A. **1980**, *J. Am. Chem. Soc. 102*, 5085; **1982**, *Tetrahedron Lett. 23*, 2361.
Taber, D. F., Petty, E. H., Raman, K. **1985a**, *J. Am. Chem. Soc. 107*, 196.
Taber, D. F., Schuchardt, J. L. **1985b**, *J. Am. Chem. Soc. 107*, 5290.
Tachibana, Y. **1975**, *Bull. Chem. Soc. Jap. 48*, 298.
Tada, M., Sugimoto, Y., Takahashi, T. **1980**, *Bull. Chem. Soc. Jap. 53*, 2966.
Tada, M. **1982**, *Chem. Lett.* 441.
Tahara, A., Shimagaki, M., Ohara, S., Nakata, T. **1973**, *Tetrahedron Lett.* 1701.
Takagi, Y., Nakahara, Y., Matsui, M. **1978**, *Tetrahedron, 34*, 517.
Takahashi, S., Kusumi, T., Kakisawa, H. **1979**, *Chem. Lett.* 515.
Takahashi, T., Nemoto, H., Tsuji, J., Miura, I. **1983a**, *Tetrahedron Lett. 24*, 3485.
Takahashi, T., Kitamura, K., Nemoto, H., Tsuji, J., Miura, I. **1983b**, *Tetrahedron Lett. 24*, 3489.
Takahashi, T., Kanda, Y., Nemoto, H., Kitamura, K., Tsuji, J., Fukazawa, Y. **1986a**, *J. Org. Chem. 51*, 3393.
Takahashi, T., Nemoto, H., Kanda, Y., Tsuji, J., Fujise, Y. **1986b**, *J. Org. Chem. 51*, 4315; **1987**, *Heterocycles 25*, 139.
Takaki, K., Ohsugi, M., Okada, M., Yasumura, M., Negoro, K. **1984**, *J. Chem. Soc. Perkin Trans. I*, 741.
Takaya, H., Hayakawa, Y., Makino, S., Noyori, R. **1978**, *J. Am. Chem. Soc. 100*, 1778.
Takayanagi, H., Uyehara, T., Kato, T. **1978**, *Chem. Commun.* 359.
Takeda, A., Sakai, T., Shinohara, S., Tsuboi, S. **1977**, *Bull. Chem. Soc. Jap. 50*, 1133.
Takeda, K., Shimono, Y., Yoshii, E. **1983**, *J. Am. Chem. Soc. 105*, 563.
Takeda, K., Sato, M., Yoshii, E. **1986**, *Tetrahedron Lett. 27*, 3903.
Takemoto, T., Isoe, S. **1982**, *Chem. Lett.* **1931**.
Takeshita, H., Hatsui, T., Kato, N., Masuda, T., Tagoshi, H. **1982**, *Chem. Lett.* 1153.
Takeshita, H., Kato, N., Nakanishi, K., Tagoshi, H., Hatsui, T. **1984a**, *Chem. Lett.* 1495.
Takeshita, H., Mori, A., Nakamura, S. **1984b**, *Bull. Chem. Soc. Jap. 57*, 3152.
Tamai, Y., Hagiwara, H., Uda, H. **1986**, *J. Chem. Soc. Perkin Trans. I*, 1311.
Tamura, Y., Saito, T., Kiyokawa, H., Chen, L. C., Ishibashi, H. **1977**, *Tetrahedron Lett.* 4075.
Tamaru, Y., Harada, T., Yoshida, Z. **1980**, *J. Am. Chem. Soc. 102*, 2392.
Tanaka, A.,. Uda, H., Yoshikoshi, A. **1967**, *Chem. Commun.* 188; **1968**, 56, **1969**, 308.
Tanaka, A., Tanaka, R., Uda, H., Yoshikoshi, A. **1972**, *J. Chem. Soc.. Perkin Trans. I*, 1721.
Tanaka, K., Yoshikoshi, A. **1971**, *Tetrahedron, 27*, 4889.
Tanaka, K., Uchiyama, F., Sakamoto, K., Inubuishi, Y. **1982**, *J. Am. Chem. Soc. 104*, 4965.
Tanaka, M., Tomioka, K., Koga, K. **1985a**, *Tetrahedron Lett. 26*, 3035; **1985b**, *26*, 6109.
Tanis, S. P., Nakanishi, K. **1979**, *J. Am. Chem. Soc. 101*, 4398.
Tanis, S. P., Chuang, Y.-H., Head, D. B. **1985a**, *Tetrahedron Lett. 26*, 6147.
Tanis, S. P., Herrinton, P. M. **1985b**, *J. Org. Chem. 50*, 3988.

Taschner, M. J., Shahripour, A. **1985**, *J. Am. Chem. Soc. 107*, 5570.
Tateishi, M., Kusumi, T., Kakisawa, H. **1971**, *Tetrahedron, 27*, 237.
Taticchi, A., Fringuelli, F., Traverso, G. **1969**, *Gazz. Chim. Ital. 99*, 247.
Tatsuta, K., Akimoto, K., Kinoshita, M. **1979**, *J. Am. Chem. Soc. 101*, 6116.
Taylor, M. D., Smith, A. B., III, **1983**, *Tetrahedron Lett. 24*, 1867.
Teisseire, P., Pesnelle, P., Corbier, B., Plattier, M., Monpetit, P. **1974**, *Recherches, 19*, 69.
Thomas, A. F., Ozainne, M., Decorzant, R., Näf, F., Lukacs, G. **1976**, *Tetrahedron, 32*, 2261.
Thomas, M. T., Fallis, A. G. **1973**, *Tetrahedron Lett.* 4687.
Tius, M. A., Takaki, K. S. **1982**, *J. Org. Chem. 47*, 3166.
Tius, M. A., Fauq, A. H. **1986**, *J. Am. Chem. Soc. 108*, 1035.
Tobe, Y., Yamashita, S., Yamashita, T., Kakiuchi, K., Odaira, Y. **1984**, *Chem. Commun.* 1259.
Tobe, Y., Yamashita, T., Kakiuchi, K., Odaira, Y. **1985**, *Chem. Commun.* 898.
Tokoroyama, T., Matsuo, K., Kanazawa, R., Kotsuki, H., Kubota, T. **1974**, *Tetrahedron Lett.* 3093.
Tokoroyama, T., Fujimori, K., Shimizu, T., Yamagiwa, Y., Monden, M., Iio, H. **1983**, *Chem. Commun.* 1516.
Tokoroyama, T., Tsukamoto, M., Iio, H. **1984**, *Tetrahedron Lett. 25*, 5067.
Toma, K., Miyazaki, E., Murae, T., Takahashi, T. **1982**, *Chem. Lett.* 863.
Tomioka, K., Tanaka, M., Koga, K. **1982**, *Tetrahedron Lett. 23*, 3401.
Tomioka, K., Masumi, F., Yamashita, T., Koga, K. **1984**, *Tetrahedron Lett. 25*, 333.
Tomioka, K., Sugimori, M., Koga, K. **1987**, *Chem. Pharm. Bull. 35*, 906.
Torii, S., Oie, T., Tanaka, H., White, J. D., Furuta, T. **1973**, *Tetrahedron Lett.* 2471.
Torii, S., Okamoto, T. **1976a**, *Bull. Chem. Soc. Jap.. 49*, 771.
Torii, S., Uneyama, K., Kuyama, M. **1976b**, *Tetrahedron Lett.* 1513.
Torii, S., Uneyama, K., Kawahara, I., Kuyama, M. **1978a**, *Chem. Lett.* 455.
Torii, S., Uneyama, K., Okamoto, K. **1978b**, *Bull. Chem. Soc. Jap. 51*, 3590.
Torii, S., Inokuchi, T., Kawai, K. **1979a**, *Bull. Chem. Soc. Jap. 52*, 861.
Torii, S., Inokuchi, T., Yamafuji, T. **1979b**, *Bull. Chem. Soc. Jap. 52*, 2640.
Torii, S., Uneyama, K., Ichimura, H. **1979c**, *J. Org. Chem. 44*, 2292.
Torii, S., Inokuchi, T. **1980**, *Bull. Chem. Soc. Jap. 53*, 2642.
Torii, S., Inokuchi, T., Handa, K. **1982**, *Bull. Chem. Soc. Jap. 55*, 887.
Torii, S., Inokuchi, T. Oi, R. **1983**, *J. Org. Chem. 48*, 1944.
Torrence, A. K., Pinder, A. R. **1971**, *Tetrahedron Lett.* 745.
Trammell, G. L. **1978**, *Tetrahedron Lett.* 1525.
Trost, B. M., Preckel, M. **1973**, *J. Am. Chem. Soc. 95*, 7862.
Trost, B. M., **1974a**, *Acc. Chem. Res. 7*, 85.
Trost, B. M., Keely, D. E. **1974b**, *J. Am. Chem. Soc. 96*, 1252.
Trost, B. M., Hiroi, K., Holy, N. **1975a**, *J. Am. Chem. Soc. 97*, 5873.
Trost, B. M., Keely, D. E., **1975b**, *J. Org. Chem. 40*, 2013.
Trost, B. M., Preckel, M., Leichter, L. M. **1975c**, *J. Am. Chem. Soc. 97*, 2224.
Trost, B. M., Bridges, A. J. **1976**, *J. Am. Chem. Soc. 98*, 5017.
Trost, B. M., Nishimura, Y., Yamamoto, K. **1979a**, *J. Am. Chem. Soc. 101*, 1328.
Trost, B. M., Shuey, C. D., DiNinno, F., Jr. **1979b**, *J. Am. Chem. Soc. 101*, 1284.
Trost, B. M., Curran, D. P. **1980**, *J. Am. Chem. Soc. 102*, 5699.
Trost, B. M., Chan., D.M.T. **1981**, *J. Am. Chem. Soc. 103*, 5972.
Trost, B. M., McDougal, P. G. **1982a**, *J. Am. Chem. Soc. 104*, 6110.
Trost, B. M., Renaut, P. **1982b**, *J. Am. Chem. Soc. 104*, 6668.
Trost, B. M., Balkovec, J. M., Mao, M. K.-T. **1983a**, *J. Am. Chem. Soc. 105*, 6755.
Trost, B. M., Ornstein, P. L. **1983b**, *Tetrahedron Lett. 24*, 2833.
Trost, B. M., Chung, J.Y.L. **1985a**, *J. Am. Chem. Soc. 107*, 4586.
Trost, B. M., Nanninga, T. N. **1985b**, *J. Am. Chem. Soc. 107*, 1293.
Trost, B. M., Balkovec, J. M., Mao, M.K.-T. **1986**, *J. Am. Chem. Soc. 108*, 4974.
Trost, B. M., Jebaratnam, D. J. **1987**, *Tetrahedron Lett. 28*, 1611.
Tsai, T.Y.R., Tsai, C.S.J., Sy, W. W., Shanbhag, M. N., Liu, W. C., Lee, S. F., Wiesner, K. **1977**, *Heterocycles 7*, 217.
Tsang, R., Fraser-Reid, B. **1985**, *J. Org. Chem. 50*, 4659.

Tsizin, Y. S., Drabkina, A. A. **1972**, *Zh. Obshch. Khim. 42*, 1852.
Tsuda, Y., Kashiwaba, N., Kajitani, M., Yasui, J. **1981**, *Chem. Pharm. Bull. 29*, 3424.
Tsunoda, T., Kodama, M., Ito, S. **1983**, *Tetrahedron Lett. 24*, 83.
Tsunoda, T., Amaike, M., Tambunan, U.S.F., Fujise, Y., Ito, S., Kodama, M. **1987**, *Tetrahedron Lett. 28*, 2537.
Tumlinson, J. H., Gueldner, R. C., Hardee, D. D., Thompson, A. C., Hedin, P. A., Minyard, J. P. **1971**, *J. Org. Chem. 36*, 2616.
Turner, R. B., Gänshirt, K. H., Shaw, P. E., Tauber, J. D. **1966**, *J. Am. Chem. Soc. 88*, 1776.
Uneyama, K., Torii, S. **1976**, *Tetrahedron Lett.* 443.
Uneyama, K., Okamoto, K., Torii, S. **1977**, *Chem. Lett.* 493.
Uneyama, K., Masatsugu, Y., Ueda, T., Torii, S. **1984**, *Chem. Lett.* 529.
Uneyama, K., Date, T., Torii, S. **1985**, *J. Org. Chem. 50*, 3160.
Utake, M., Fujii, Y., Takeda, A. **1986**, *Chem. Lett.* 1103.
Uyehara, T., Ogata, K., Yamada, J., Kato, T. **1983**, *Chem. Commun.* 17.
Uyehara, T., Kabasawa, Y., Kato, T., Furuta, T. **1985a**, *Tetrahedron Lett. 26*, 2343.
Uyehara, T., Yamada, J., Kato, T., Bohlmann, F. **1985b**, *Bull. Chem. Soc. Jap. 58*, 861.
Uyehara, T., Yamada, J., Furuta, T., Kato, T. **1986**, *Chem. Lett.* 609.
Van der Eycken, E., Van der Eycken, J., Vandewalle, M. **1985**, *Chem. Commun.* 1719.
Van der Gen, A., van der Linde, L. M., Witteveen, J. G., Boelens, H. **1971a**, *Rec. Trav. Chim. Pays-Bas, 90*, 1034; **1971b**, *90*, 1045.
Van Hijfte, L., Vandewalle, M. **1982**, *Tetrahedron Lett. 23*, 2229; **1984**, *Tetrahedron, 40*, 4371.
Van Hijfte, L., Little, R. D. **1985**, *J. Org. Chem. 50*, 3940.
Van Middlesworth, F. L. **1986**, *J. Org. Chem. 51*, 5019.
van Tamelen, E. E., Storni, A., Hessler, E. J., Schwartz, M. **1963**, *J. Am. Chem. Soc. 85*, 3295.
van Tamelen, E. E., Coates, R. M., **1966**, *Chem. Commun.* 413.
van Tamelen, E. E., Anderson, R. J. **1972a**, *J. Am. Chem. Soc. 94*, 8225.
van Tamelen, E. E., Holton, R. A., Hopla, R. E., Konz, W. E. **1972b**, *J. Am. Chem. Soc. 94*, 8228.
van Tamelen, E. E., Seiler, M. P., Wierenga, W. **1972c**, *J. Am. Chem. Soc. 94*, 8229.
van Tamelen, E. E., Carlson, J. G., Russell, R. K., Zawacky, S. R. **1981**, *J. Am. Chem. Soc. 103*, 4615.
van Tamelen, E. E., Leiden, T. M. **1982**, *J. Am. Chem. Soc. 104*, 1785.
van Tamelen, E. E., Zawacky, S. R., Russell, R. K., Carlson, J. G. **1983**, *J. Am. Chem. Soc. 105*, 143.
Vaughan, W. R., Goetschel, C. T., Goodrow, M. H., Warren, C. L. **1963**, *J. Am. Chem. Soc. 85*, 2282.
Vettel, P. R., Coates, R. M. **1980**, *J. Org. Chem. 45*, 5430.
Vig, O. P., Matta, K. L., Singh, G., Raj, I. **1966a**, *J. Indian Chem. Soc. 43*, 27.
Vig, O. P., Sharma, S. D., Chandar, S., Raj, I. **1966b**, *Indian J. Chem. 4*, 275.
Vig, O. P., Salota, J. P., Vig, B., Ram, B. **1967**, *Indian J. Chem. Soc. 5*, 475.
Vig, O. P., Bhatia, M. S., Gupta, K. C., Matta, K. L. **1969a**, *J. Indian J. Chem. Soc. 46*, 991.
Vig, O. P., Raj, I., Salota, J. P., Matta, K. L. **1969b**, *J. Indian Chem. Soc. 46*, 205.
Vig, O. P., Sharma, S. D., Chugh, O. P., Vig, A. K. **1974**, *Indian J. Chem. 12*, 1050.
Vig, O. P., Sharma, S. D., Kad, G. L., Sharma, M. L. **1975**, *Indian J. Chem. 13*, 764.
Viswanatha, V., Krishna Rao, G. S. **1974a**, *Tetrahedron Lett.* 247; **1974b**, *J. Chem. Soc. Perkin Trans. I*, 450.
Volkmann, R. A., Andrews, G. C., Johnson, W. S. **1975**, *J. Am. Chem. Soc. 97*, 4777.
Voyle, M., Dunlap, N. K., Watt, D. S., Anderson, O. P. **1983a**, *J. Org. Chem. 48*, 3242.
Voyle, M., Kyler, K. S., Arseniyadis, S., Dunlap, N. K., Watt, D. S. **1983b**, *J. Org. Chem. 48*, 470.
Wakamatsu, T., Hara, H., Abe, K., Ban, Y. **1980**, 3rd International Conference on Organic Synthesis, Madison, WI, Abstract 58.
Wakamatsu, T., Hara, H., Ban, Y. **1985**, *J. Org. Chem. 50*, 108.
Walba, D. M. **1985**, *Tetrahedron, 41*, 3161.
Walker, J. **1935**, *J. Chem. Soc.* 1585.
Walls, F., Padilla, J., Joseph-Nathan, P., Giral, F., Romo, J. **1965**, *Tetrahedron Lett.* 1577.
Wang, D., Chan, T. H. **1984**, *Chem. Commun.* 1273.
Watt, D. S., Corey, E. J. **1972**, *Tetrahedron Lett.* 4651.
Welch, S. C., Walters, R. L. **1973a**, *Synth. Commun. 3*, 15; **1973b**, *3*, 419; **1974**, *J. Org. Chem. 39*, 2665.

Welch, S. C., Rao, A.S.C.P., Gibbs, C. G. **1976a**, *Synth. Commun. 6*, 485.
Welch, S. C., Rao, A.S.C.P., Wong, R. Y. **1976b**, *Synth. Commun. 6*, 443.
Welch, S. C., Hagan, C. P., Kim, J. H., Chu, P. S. **1977a**, *J. Org. Chem. 42*, 2879.
Welch, S. C., Hagan, C. P., White, D. H., Fleming, W. P., Trotter, J. W. **1977b**, *J. Am. Chem. Soc. 99*, 549.
Welch, S. C., Valdes, T. A. **1977c**, *J. Org. Chem. 42*, 2108.
Welch, S. C., Chayabunjonglerd, S. **1979**, *J. Am. Chem. Soc. 101*, 6768.
Welch, S. C., Gruber, J. M., Morrison, P. A. **1984**, *Tetrahedron Lett. 25*, 5497.
Wender, P. A., Filosa, M. P. **1976**, *J. Org. Chem. 41*, 3490.
Wender, P. A., Lechleiter, J. C. **1978**, *J. Am. Chem. Soc. 100*, 4321.
Wender, P. A., Eissenstat, M. A., Filosa, M. P. **1979**, *J. Am. Chem. Soc. 101*, 2196.
Wender, P. A., Hubbs, J. C. **1980a**, *J. Org. Chem. 45*, 365.
Wender, P. A., Lechleiter, J. C. **1980b**, *J. Am. Chem. Soc. 102*, 6340.
Wender, P. A., Letendre, L. J. **1980c**, *J. Org. Chem. 45*, 367.
Wender, P. A., Dreyer, G. B. **1981a**, *Tetrahedron, 37*, 4445.
Wender, P. A., Howbert, J. J. **1981b**, *J. Am. Chem. Soc. 103*, 688.
Wender, P. A., Dreyer, G. B. **1982a**, *J. Am. Chem. Soc. 104*, 5805.
Wender, P. A., Eck, S. L. **1982b**, *Tetrahedron Lett. 23*, 1871.
Wender, P. A., Howbert, J. J. **1982c**, *Tetrahedron Lett. 23*, 3985.
Wender, P. A., Dreyer, G. B. **1983a**, *Tetrahedron Lett. 24*, 4543.
Wender, P. A., Howbert, J. J. **1983b**, *Tetrahedron Lett. 24*, 5325.
Wender, P. A., **1984**, in *Selectivity—A Goal for Synthetic Efficiency*, W. Bartmann, and B. M. Trost, Ed., VCH, New York.
Wender, P. A., Holt, D. A. **1985a**, *J. Am. Chem. Soc. 107*, 7771.
Wender, P. A., Singh, S. K. **1985b**, *Tetrahedron Lett. 26*, 5987.
Wender, P. A., Ternansky, R. J. **1985c**, *Tetrahedron Lett. 26*, 2625.
Wender, P. A., Wolanin, D. J. **1985d**, *J. Org. Chem. 50*, 4418.
Wender, P. A., Fisher, K. **1986**, *Tetrahedron Lett. 27*, 1857.
Wender, P. A., **1987a**, *194th ACS Nat. Meeting*, ORGN 360.
Wender, P. A., Correia, C.R.D. **1987b**, *J. Am. Chem. Soc. 109*, 2523.
Wender, P. A., Keenan, R. M., Lee, H. Y. **1987c**, *J. Am. Chem. Soc. 109*, 4390.
Wenkert, E., Jackson, B. G. **1958**, *J. Am. Chem. Soc. 80*, 217; **1959**, *81*, 5601.
Wenkert, E., Tahara, A. **1960**, *J. Am. Chem. Soc. 82*, 3229.
Wenkert, E., Strike, D. P. **1964**, *J. Am. Chem. Soc. 86*, 2044.
Wenkert, E., Berges, D. A. **1967**, *J. Am. Chem. Soc. 89*, 2507.
Wenkert, E., Bindra, J. S., Mylari, B. L., Nussim, M., Wilson, N.D.V. **1973a**, *Synth. Commun. 3*, 431.
Wenkert, E., Naemura, K. **1973b**, *Synth. Commun. 3*, 45.
Wenkert, E., Berges, D. A., Golob, N. F. **1978a**, *J. Am. Chem. Soc. 100*, 1263.
Wenkert, E., Buckwalter, B. L., Craveiro, A. A., Sanchez, E. L., Sathe, S. S. **1978b**, *J. Am. Chem. Soc. 100*, 1267.
Wenkert, E., Arrhenius, T. S. **1983**, *J. Am. Chem. Soc. 105*, 2030.
Weyerstahl, P., Zombik, W., Gansau, C. **1986**, *Liebigs Ann. Chem.* 422.
Weyerstahl, P., Rilk, R., Marschall-Weyerstahl, H. **1987**, *Liebigs Ann. Chem.* 89.
Wharton, P. S., Sundin, C. E., Johnson, D. W., Kluender, H. C. **1972**, *J. Org. Chem. 37*, 34.
White, J. D., Gupta, D. N. **1966**, *J. Am. Chem. Soc. 88*, 5364; **1968**, *90*, 6171.
White, J. D., Torii, S., Nogami, J. **1974**, *Tetrahedron Lett.* 2879.
White, J. D., Matsui, T., Thomas, J. A. **1981**, *J. Org. Chem. 46*, 3376.
White, J. D., Avery, M. A., Carter, J. P. **1982**, *J. Am. Chem. Soc. 104*, 5486.
White, J. D., Skeean, R. W., Trammell, G. L. **1985**, *J. Org. Chem. 50*, 1939.
White, J. D., Somers, T. C. **1987**, *J. Am. Chem. Soc. 109*, 4424.
Whitesell, J. K., Fisher, M., Da Silva Jardine, P. **1983**, *J. Org. Chem. 48*, 1556.
Wichterle, O., Prochazka, J., Hofmann, J. **1948**, *Collect. Czech. Chem. Commun. 13*, 300.
Wiesner, K., Uyeo, S., Philipp, A., Valenta, Z. **1968**, *Tetrahedron Lett.* 6279.
Wiesner, K., Ho, P.-T., Tsai, C.S.J. **1974a**, *Can. J. Chem. 52*, 2353.

Wiesner, K., Ho, P.-T., Tsai, C.S.J., Lam, Y.-K. **1974b**, *Can. J. Chem. 52*, 2355.
Wiesner, K., Tsai, T.Y.R., Huber, K., Bolton, S. E., Vlahov, R. **1974c**, *J. Am. Chem. Soc. 96*, 4990.
Wiesner, K. **1977**, *Chem. Soc. Rev. 6*, 413.
Wilkening, D., Mundy, B. P. **1984**, *Tetrahedron Lett. 25*, 4619.
Williams, J. R., Callahan, J. F. **1979**, *Chem. Commun.* 404; **1980**, *J. Org. Chem. 45*, 4475, 4479.
Wilson, S. R., Mao, D. T. **1978**, *J. Am. Chem. Soc. 100*, 6289.
Wilson, S. R., Phillips, L. R., Natalie, K. J. **1979**, *J. Am. Chem. Soc. 101*, 3340.
Wilson, S. R., Misra, R. N. **1980**, *J. Org. Chem. 45*, 5079.
Winkler, J. D., Henegar, K. E. **1987**, *J. Am. Chem. Soc. 109*, 2850.
Wolf, H., Kolleck, M. **1975**, *Tetrahedron Lett.* 451.
Wolinsky, L. E., Faulkner, D. J. **1976**, *J. Org. Chem. 41*, 597.
Wollweber, H. **1972**, *Diels-Alder-Reaktion*, Thieme Verlag, Stuttgart.
Wong, D., Chan, T.-H. **1984**, *Chem. Commun.* 1273.
Woodward, R. B., Sondheimer, F., Taub, D., Heusler, K., McLamore, W. M. **1951**, *J. Am. Chem. Soc. 73*, 2403, 3547, 3548; **1952**, *74*, 4223.
Woodward, R. B., Patchett, A. A., Barton, D.H.R., Ives, D.A.J., Kelly, R. B. **1954**, *J. Am. Chem. Soc. 76*, 2852; **1957**, *J. Chem. Soc.* 1131.
Woodward, R. B., Hoffmann, R. **1971**, *The Conservation of Orbital Symmetry*, VC, New York.
Woodward, R. B., Hoye, T. R. **1977**, *J. Am. Chem. Soc. 99*, 8007.
Wrobel, J., Takahashi, K., Honkan, V., Lannoye, G., Cook, J. M., Bertz, S. H. **1983**, *J. Org. Chem. 48*, 139.
Wu, H.-J., Pan, K. **1987**, *Chem. Comm.* 898.
Wulff, W. D., Gilbertson, S. R. **1987**, *194th ACS Nat. Meeting*, ORGN 34.
Xu, X.-X., Zhu, J., Huang, D.-Z., Zhou, W.-S. **1986**, *Tetrahedron, 42*, 819.
Yamada, K., Suzuki, M., Hayakawa, Y., Aoki, K., Nakamura, H., Nagase, H., Hirata, Y. **1972**, *J. Am. Chem. Soc. 94*, 8278.
Yamada, K., Nagase, H., Hayakawa, Y., Aoki, K., Hirata, Y. **1973**, *Tetrahedron Lett.* 4963.
Yamada, K., Kyotani, Y., Manabe, S., Suzuki, M. **1979**, *Tetrahedron, 35*, 293.
Yamada, S., Takamura, N., Mizoguchi, T. **1975**, *Chem. Pharm. Bull. 23*, 2539.
Yamada, Y., Sanjoh, H., Iguchi, K. **1979**, *Tetrahedron Lett.* 1323.
Yamaguchi, M., Hasebe, K., Tanaka, S., Minami, T. **1986**, *Tetrahedron Lett. 27*, 959.
Yamaguchi, M., Hamada, M., Nakashima, H., Minami, T. **1987**, *Tetrahedron Lett. 28*, 1785.
Yamakawa, K., Izuta, I., Oka, H., Sakaguchi, R. **1974**, *Tetrahedron Lett.* 2187.
Yamakawa, K., Sakaguchi, R., Nakamura, T., Watanabe, K. **1976**, *Chem. Lett.* 991.
Yamakawa, K., Satoh, T. **1977a**, *Chem. Pharm. Bull. 25*, 2535; **1977b**, *Heterocycles, 8*, 221; **1978**, *Chem. Pharm. Bull. 26*, 3704.
Yamakawa, K., Izuta, I., Oka, H., Sakaguchi, R., Kobayashi, M., Satoh, T. **1979**, *Chem. Pharm. Bull. 27*, 331.
Yamamoto, H., Sham, H. L. **1979**, *J. Am. Chem. Soc. 101*, 1609.
Yamasaki, M. **1972**, *Chem. Commun.* 606.
Yamazaki, M., Shibasaki, M., Ikegami, S. **1981**, *Chem. Lett.* 1245.
Yan, T. H., Paquette, L. A. **1982**, *Tetrahedron Lett. 23*, 3227.
Yanagawa, H., Kato, T., Kitahara, Y. **1970**, *Synthesis*, 257.
Yanagiya, M., Kaneko, K., Kaji, T., Matsumoto, T. **1979**, *Tetrahedron Lett.* 1761.
Yanami, T., Miyashita, M., Yoshikoshi, A. **1979**, *Chem. Commun.* 525.
Yanami, T. **1980**, *J. Org. Chem. 45*, 607.
Yates, P., Burnell, D. J., Freer, V. J., Sawyer, J. F. **1987**, *Can. J. Chem. 65*, 69.
Yoshida, K., Grieco, P. A. **1984**, *J. Org. Chem. 49*, 5257.
Yoshikoshi, A., Miyashita, M. **1985**, *Acc. Chem. Res. 18*, 284.
Yuste, F., Walls, F. **1976**, *Aust. J. Chem. 29*, 2333.
Ziegler, F. E., Wender, P. A. **1974**, *Tetrahedron Lett.* 449.
Ziegler, F. E. **1977a**, *Acc. Chem. Res. 10*, 227.
Ziegler, F. E., Kloek, J. A. **1977b**, *Tetrahedron, 33*, 373.
Ziegler, F. E., Reid, G. R., Studt, W. L., Wender, P. A. **1977c**, *J. Org. Chem. 42*, 1991.

Ziegler, F. E., Piwinski, J. J. **1980**, *J. Am. Chem. Soc. 102*, 6577.
Ziegler, F. E., Fang, J.-M. **1981**, *J. Org. Chem. 46*, 825.
Ziegler, F. E., Hwang, K. J. **1983**, *J. Org. Chem. 48*, 3349.
Ziegler, F. E., Jaynes, B. H., Saindane, **1985a**, *Tetrahedron Lett. 26*, 3307.
Ziegler, F. E., Klein, S. I., Pati, U. K., Wang, T.-F. **1985b**, *J. Am. Chem. Soc. 107*, 2730.
Ziegler, F. E., **1987**, Priv. Commun. (*J. Am. Chem. Soc. 109*, 8115).
Zschiesche, R., Reissig, H.-U. **1987**, *Liebigs Ann. Chem.* 387.
Zurflüh, R., Wall, E. N., Siddall, J. B., Edwards, J. A. **1968**, *J. Am. Chem. Soc. 90*, 6224.
Zurflüh, R., Dunham, L. L., Spain, V. L., Siddall, J. B. **1970**, *J. Am. Chem. Soc. 92*, 425.
Zutterman, F., deWilde, H., Mijngheer, R., deClercq, P., Vandewalle, M. **1979**, *Tetrahedron, 35*, 2389.

Addendum

New syntheses reported since submission of the holographic manuscript are abstracted as follows. The sections to which each synthesis is pertinent are indicated at the beginning of the paragraphs.

2.2

The trimethyloctalone employed in a thujopsene synthesis (Dauben, 1963) gives a *trans*-decalone on reduction. This decalone has now been converted into polygodial (**P44**) and related drimane sesquiterpenes (Jansen, 1988). Muzigadial is similarly synthesized from the proper decalone.

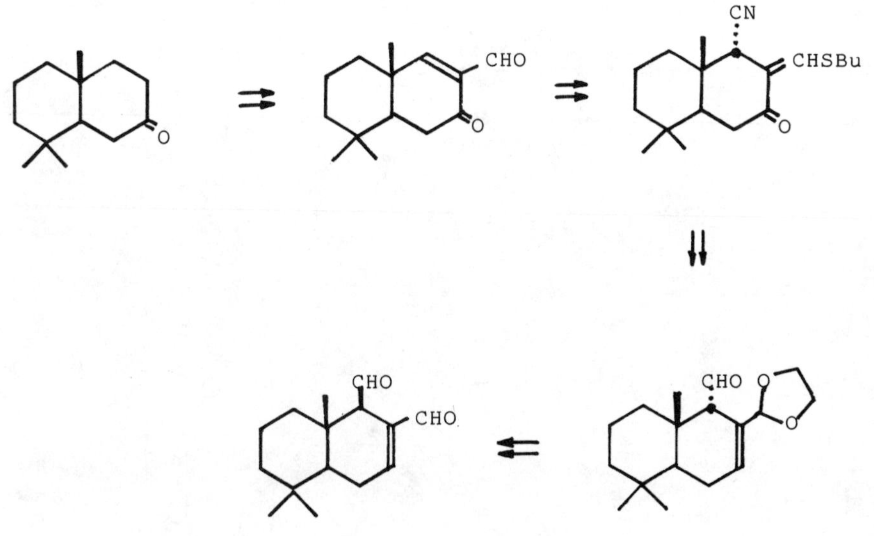

P44

2.4

4-Methylated W-M ketone serves as starting material in a synthesis of amarolide (Hirota, 1987b). The corresponding hydroxy enone undergoes another Robinson annulation to provide the hydrophenanthrene framework with adequate functional groups to introduce the missing substituents.

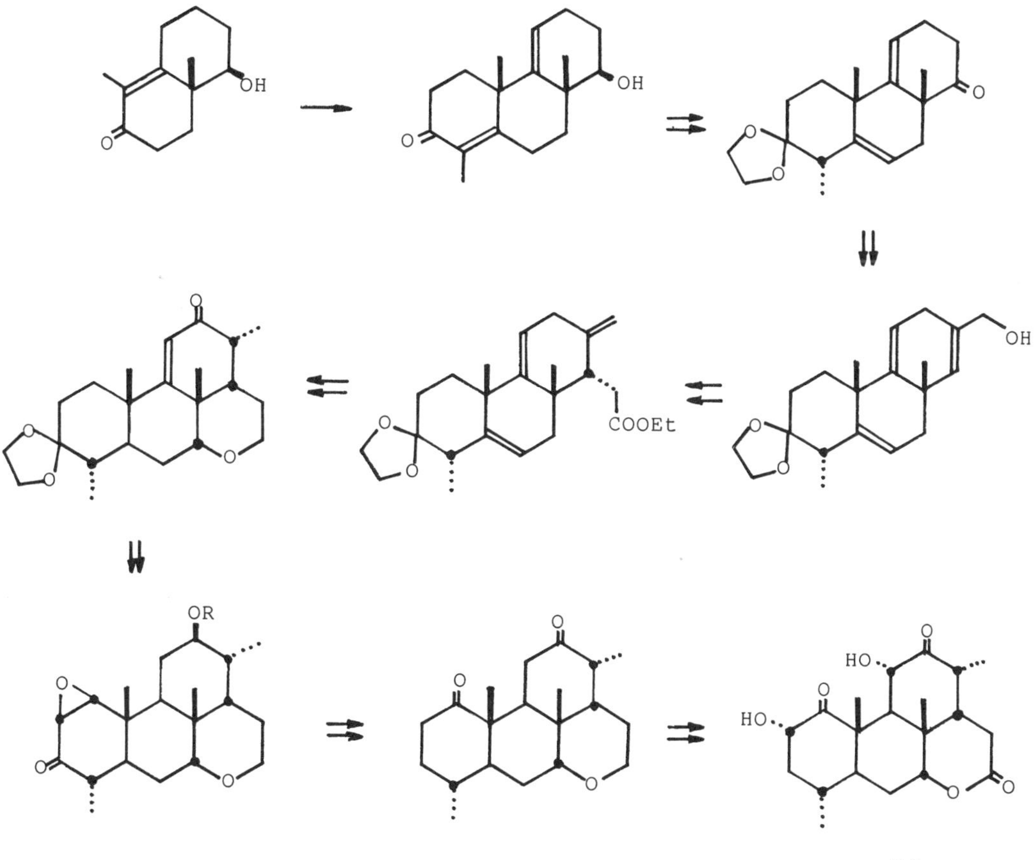

amarolide

2.7 and 4.8

Since the Robinson annulation product of 2-methyl-1,3-cyclopentanedione and 3-buten-2-one is readily obtained in high optical purity using proline as catalyst, the compound (in either enantiomer) has been exploited as a chiral building block in many syntheses of natural products. Its recent service is as a functionalized cyclopentanone subunit of the pseudoguanolides (Lansbury, 1988). The two carbonyl groups are respectively removed and homologated into an acetonitrile pendant for effecting a Dieckmann-type condensation with a four-carbon ester chain released from cleavage of the hydrindene double bond. The hydrazulenone intermediate is eminently suitable for elaboration into naturally occurring terpenes such as radiatin.

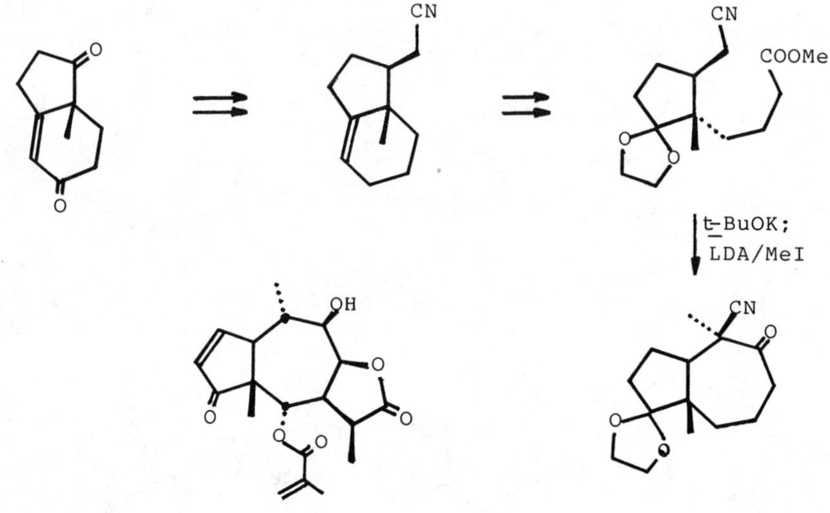

radiatin

3.1

Two iridoid building blocks derived from (+)-limonene have been joined, allosterically rearranged (Cope), and then sculptured into dictymal (N. Kato, 1987).

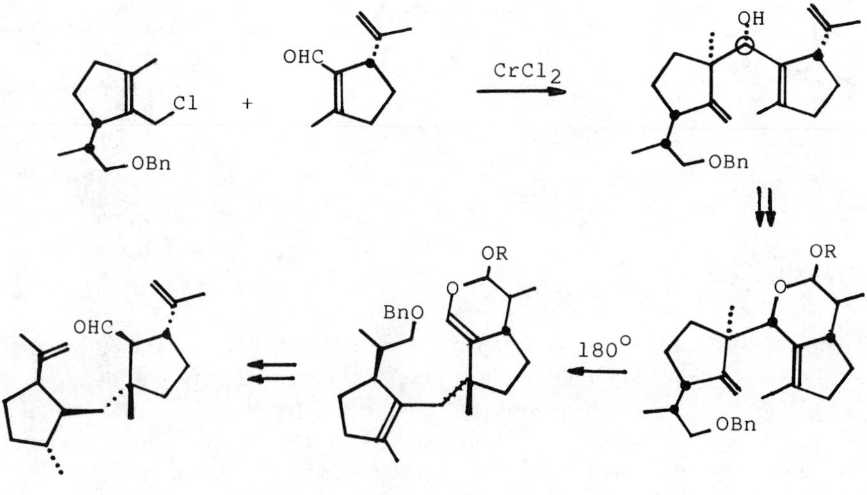

dictymal

3.4.1

Starting from β-ionone an intermediate for forskolin (**F9**) has been assembled (Hutchinson, 1987) by means of radical cyclization to form a lactol ether and a Mukaiyama aldolization to close the B-ring. The desired *cis*-addition to the double bond in the radical process is achievable using electrochemical reduction of the bromide in the presence of vitamin B_{12}. More conventional method of generating the radical by reaction with tin hydride leads to the trans isomer.

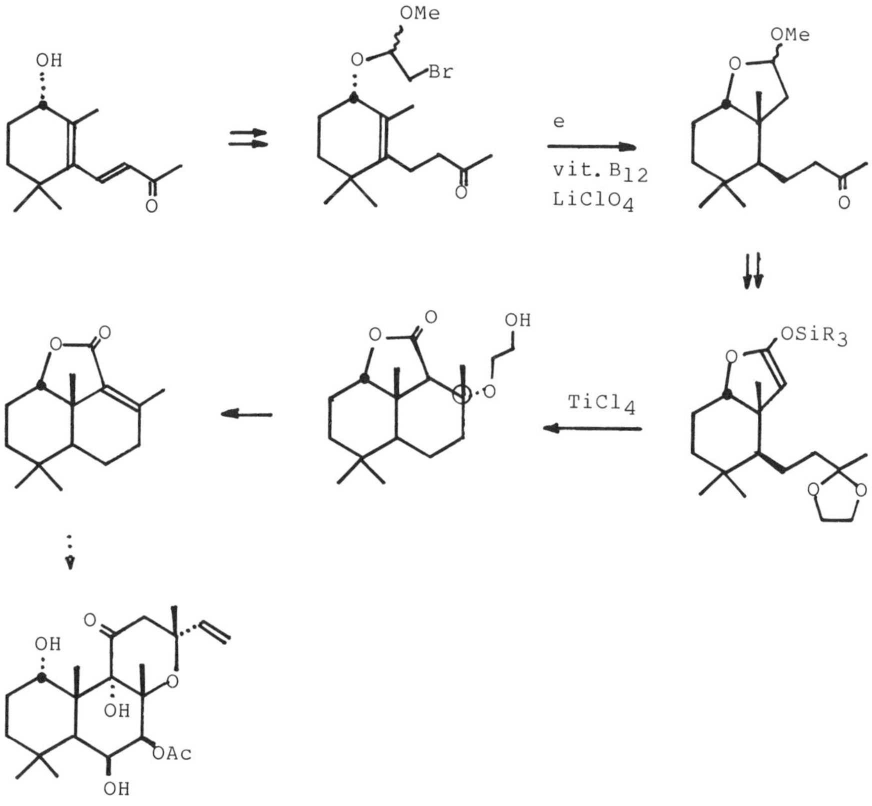

3.4.1

(+)-O-Methylpisiferic acid inhibits the hatching and feeding of the two-spotted spider mite. A synthesis of this valuable diterpene has been completed along a classical route that involves two Michael-aldol reaction sequences (Mori, 1986a). (S)-3-hydroxy-2,2-dimethylcyclohexanone, employed as starting material, is available from reduction of the diketone with baker's yeast. [This ketoalcohol has also been used in a synthesis of (+)-baiyunol (Mori, 1987), the diterpene aglycone of a sweet glycoside].

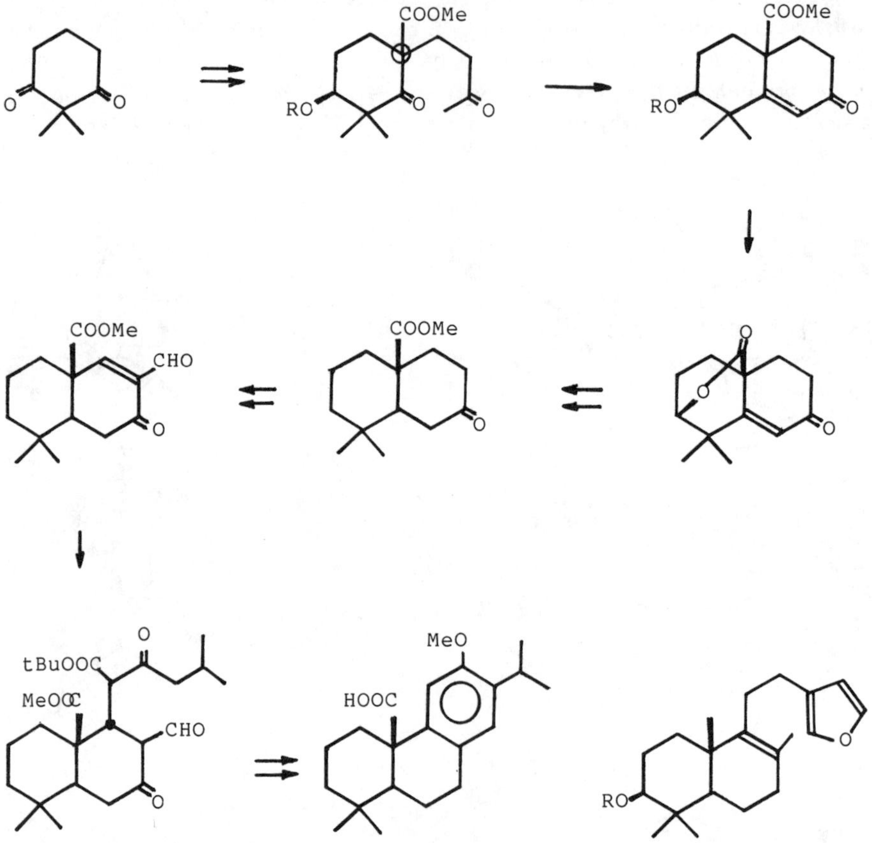

O-methylpisiferic acid

baiyunol

3.4.1 and 9.1

Closure of the third ring in a synthesis of silphinene (**S34**) (Y. K. Rao, 1988) has been delegated to an intramolecular radical addition to a diquinane enone. The Wittig–Horner-Emmons condensation is used to construct the diquinane framework. Without the phosphoryl activator, formation of a six-membered ring involving the other side chain is expected.

Stereospecific generation of the chiral center at the radical site is remarkable. Probably the angular methyl group inhibits cyclization leading to the epimer.

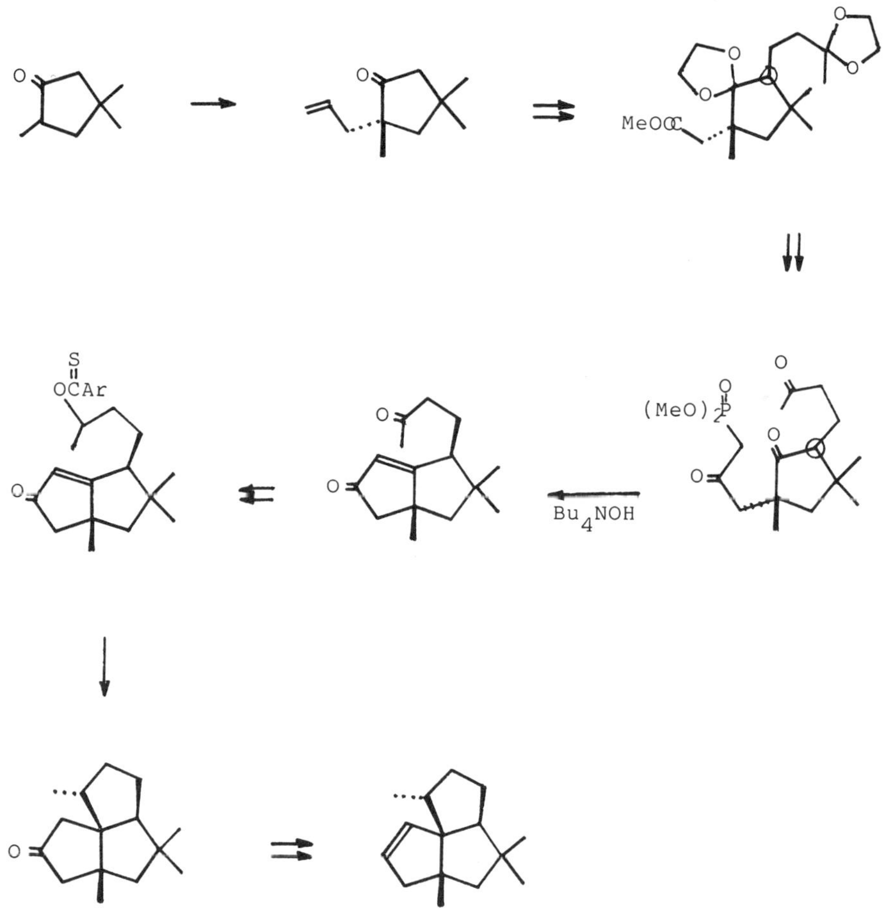

S34

3.4.2 and 7.4

In a synthesis of oppositol and its side chain prenolog, prepinnaterpene (Fukuzawa, 1987), a bromoperhydrindenecarboxaldehyde is the common precursor. This aldehyde has been obtained from the 1,3-butadiene-*p*-benzoquinone adduct via reduction, equilibration, cleavage of the cyclohexene ring, and reclosure to afford a hydrindene system by an intramolecular aldolization. This last reaction causes no regiochemical problems because the dialdehyde is symmetrical. Stereocontrol at various stages of the synthesis is exercised by neighboring group participation (e.g., γ-lactone formation) or equilibration (bromine atom and angular hydrogen).

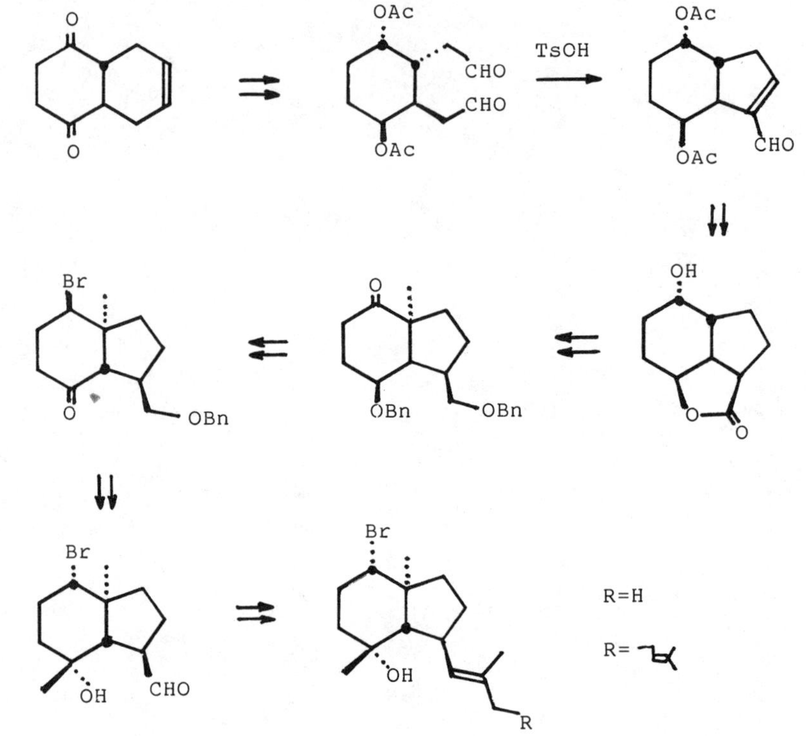

opositol
prepinnaterpene

3.4.2 and 8.8

A new approach to hirsutene (**H17**) (Levine, 1988) starts from neophyl chloride via 2,2-dimethylindane and its Birch reduction product. Cleavage of the perhydroindanediol followed by aldolization of the resulting dialdehyde leads to the diquinane. The third ring is closed via a Nazarov cylization.

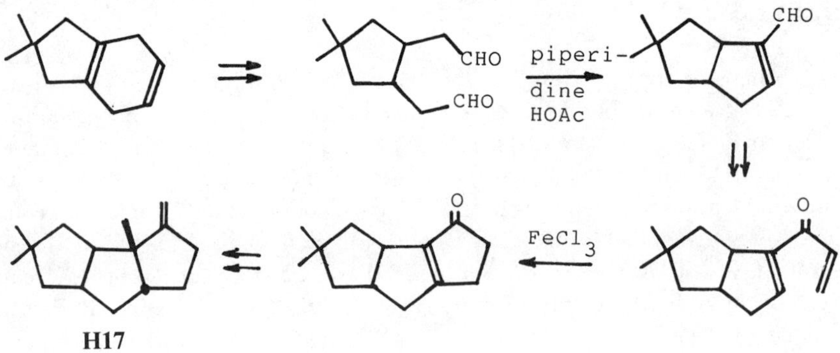

4.2

Bicyclo[3.3.0]octane-3,7-dione is a valuable building block for triquinane synthesis. This compound is readily prepared by reaction of glyoxal with acetonedicarboxylic ester. By means of microbial reduction of an (−)-enone derived from the monoketal of the diketone a chiral synthon for asymmetric synthesis of hirsutic acid (**H18**) (Xie, 1987) becomes available.

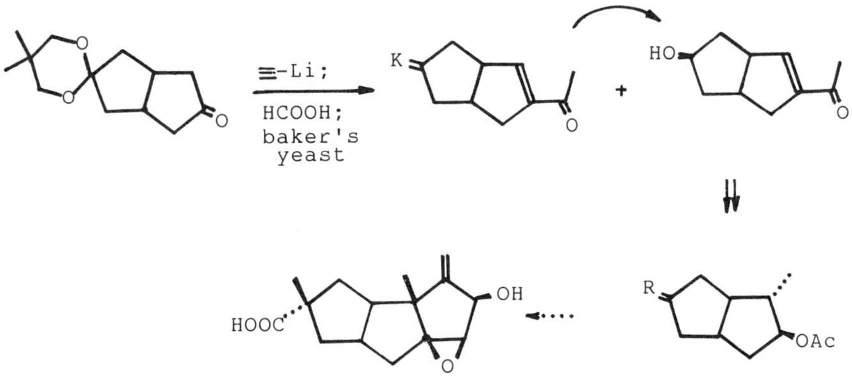

4.2

Intramolecular tandem Michael-aldol reactions provide a potential precursor of forskolin (**F9**) (Koft, 1988).

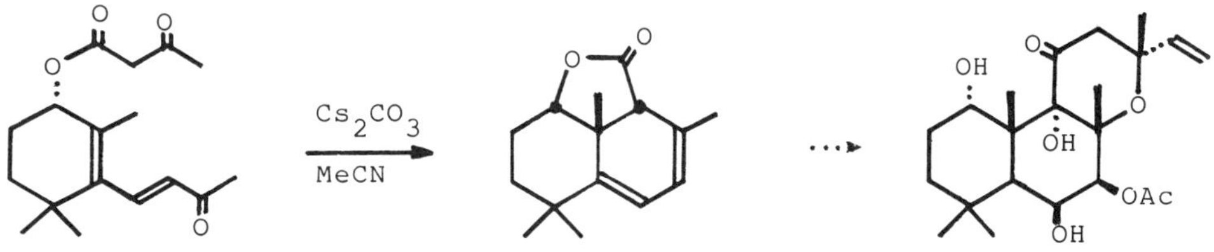

4.5

Cyclopropanation by the Michael-alkylation route in a synthesis of *trans*-chrysanthemic acid (**C51**) (Krief, 1988) has been applied to a diene diester derived from tartaric acid, resulting in a convenient access to the natural enantiomer.

C51

4.5

Reaction of 3-chloro-2-cyclopentenone with the dicuprate derived from 2,6-dihalo-2-heptene leads to the well-known spirocyclic ketone intermediate of the vetispirane sesquiterpenes (Cannone, 1988a).

4.5 and 4.7

The intricate ring skeleton of the alkaloid methyl homodaphniphyllate contains only two carbocycles. An excellent synthesis (Heathcock, 1986) of this compound starts from 2-oxocyclopentaneacetic acid which is easily converted into a tricyclic lactam. While the fourth ring is assembled from the dialkylated thiolactam derived therefrom, closure of the bridged system by an intramolecular Michael reaction requires transposition of the double bond. Deoxygenation after the isopropyl group is introduced via the bis(enol phosphate) is less than routine; the product retains a trisubstituted double bond which can only be saturated at high temperatures and under high pressure in the presence of a Pd(OH)$_2$/C catalyst. Moreover, an equimolar mixture of two epimers is obtained.

methyl homodaphniphyllate

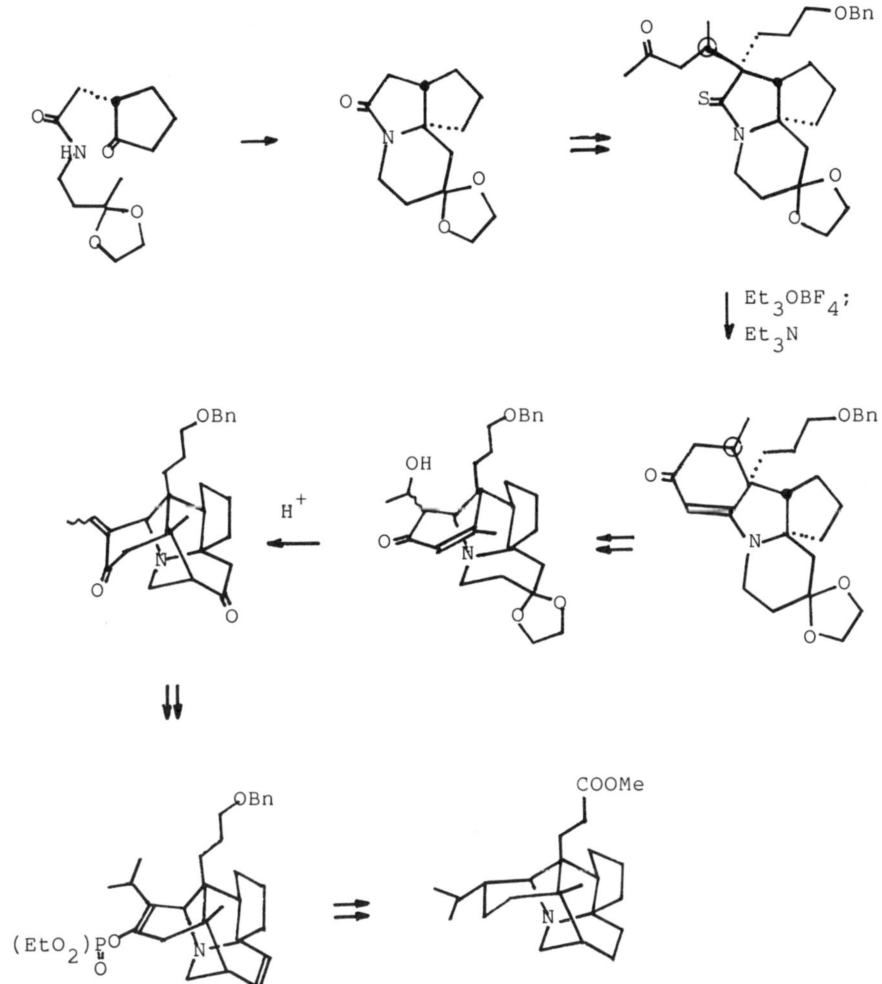

4.8

(+)-3-Trimethylsilyl-5-(*p*-toluenesulfenyl)cyclohexanone is accessible by conjugate addition of the toluenethiol to the silylated cyclohexenone in the presence of cinchonidine. Baeyer-Villiger reaction as controlled by the silyl substituent enables a parlay into an adipic diester and eventually, (+)-α-cuparenone (**C85**) via Dieckmann condensation and methylation (Asaoka, 1988).

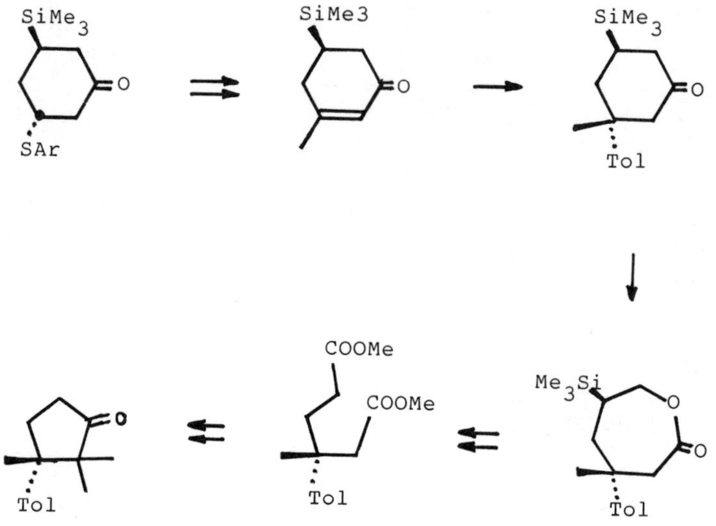

C85

5.1

An attempt at synthesizing parvifolin which has an eight-membered ring condensed to an aromatic nucleus by intramolecular phenolate alkylation failed. The reaction takes an SN2' course and thereby producing 7-hydroxycalamenene (**C8**) upon hydrogenation of the product (Krause, 1987).

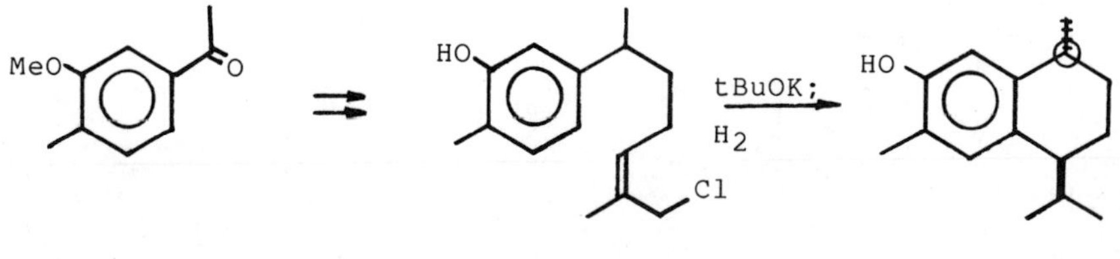

5.2

Spiroannulation by alkylation of a chiral γ-lactone with a dibromide derived from malic acid establishes all the chiral centers of $(-)$-solavetivone (**S41**) (Cannone, 1988b).

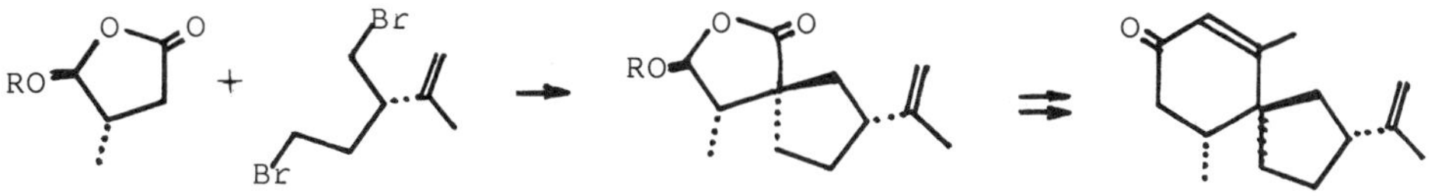

5.4

From L-serine a synthesis of (−)-nephthenol and (−)-cembrene A (**C40**) (Schwabe, 1988) has been developed. The macrocyclization is consigned to the intramolecular epoxide opening with allylic sulfenyl carbanion.

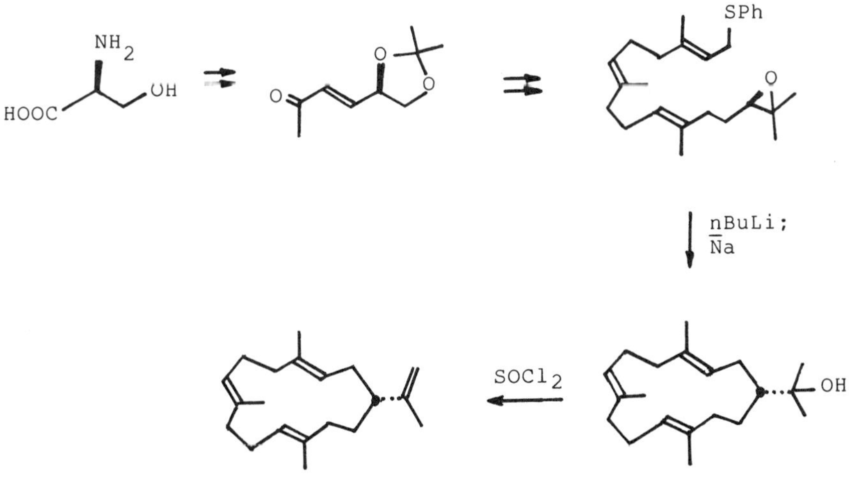

5.5

The Pd(II)-mediated cycloalkenylation of silyl enol ethers has been extended to the closure of the D-ring of phyllocladene (Kende, 1983).

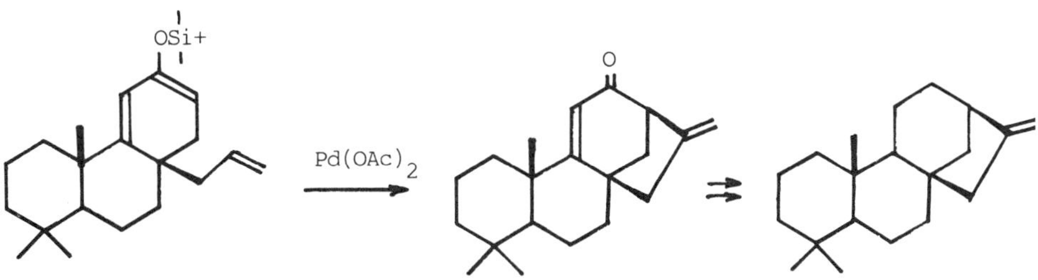

6.1.2

The norbornane lactone ester reported by Alder has been used in an approach to longifolene (**L24**) (T. L. Ho, unpublished). Conversion of the δ-lactone to a cyclohexenone allows establishment of the gem-dimethyl group which is followed by a ring expansion maneuver.

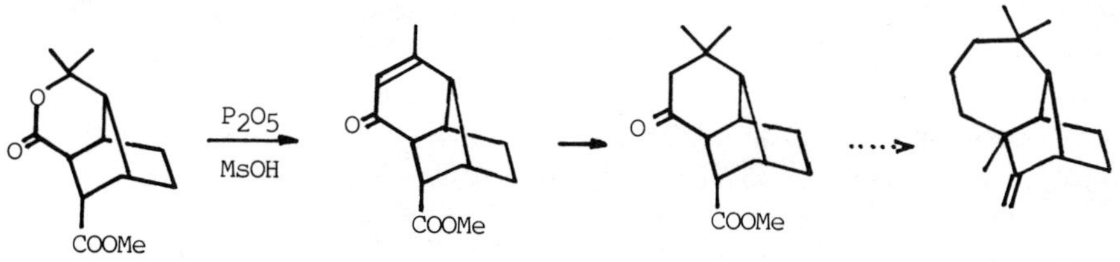

L24

6.2.5

When applied to synthesis of α-decalones the versatile cationic cyclization of alkenyl cyclohexenones possessing a silyl terminator proves to be an excellent method for the assembly of the carbocyclic skeleton of *cis*-clerodanes. It is possible not only to establish three contiguous chiral centers, but also to achieve a stereospecific angular alkylation in the same step. The resulting bicyclic ketone is well disposed to be elaborated into linaridial (Tokoroyama, 1987).

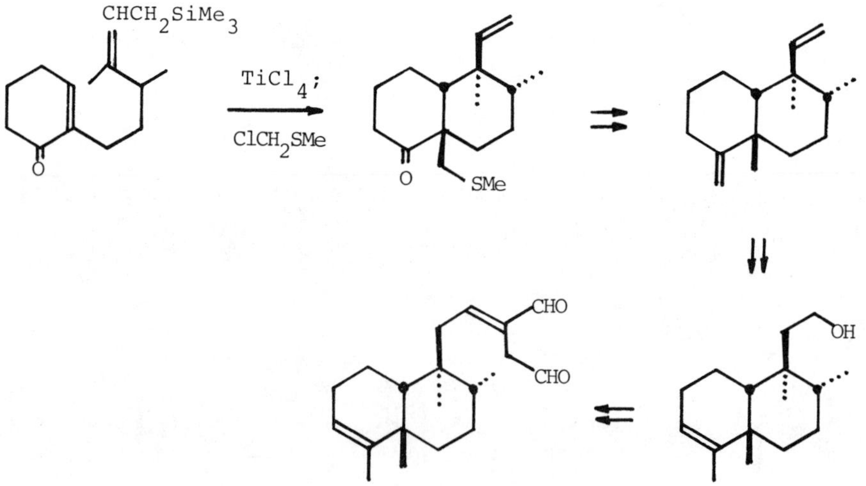

linaridial

6.2.5 and 9.1

In an approach to isoamijiol (Mehta, 1987a) the use of (+)-limonene of perforce leads to the enantiomer **ent-A28**. A 1,3-asymmetric induction during Claisen rearrangement of an iridenyl intermediate fixes the absolute configuration of the quaternary carbon atom common to the 5- and 7-membered ring.

Acid-catalyzed cyclization of an ene-enone furnishes a hydrazulenone in which the remote angular methyl group controls the stereochemistry in two consecutive alkylations. The penultimate step of the synthesis involves reductive cyclization of an alkynyl ketone with sodium naphthalenide, and the final allylic oxidation is slightly complicated by concurrent dehydrogenation of a portion of the substrate which results in a conjugate diene in the hydrazulene moeity.

ent-A28

7.5 and 8.6

The intricate array of four carbocycles in cerorubenic acid III is a formidable target for synthesis. One solution centers on a splendid evolvement of the tricyclo[5.4.1.0$^{2.11}$]-undec-7-ene network from a more innocent-looking bridged system via an oxyanionic Cope rearrangement (Paquette, 1988b). [A more conventional approach would involve intramolecular acylcarbenoid insertion of a cyclohexadiene derivative provided the proximate double is in a latent state.]

The six-membered ring which constitutes the bisabolane moiety of the sesterterpene molecular has been assembled using a Diels–Alder reaction under high pressure (98000 psi).

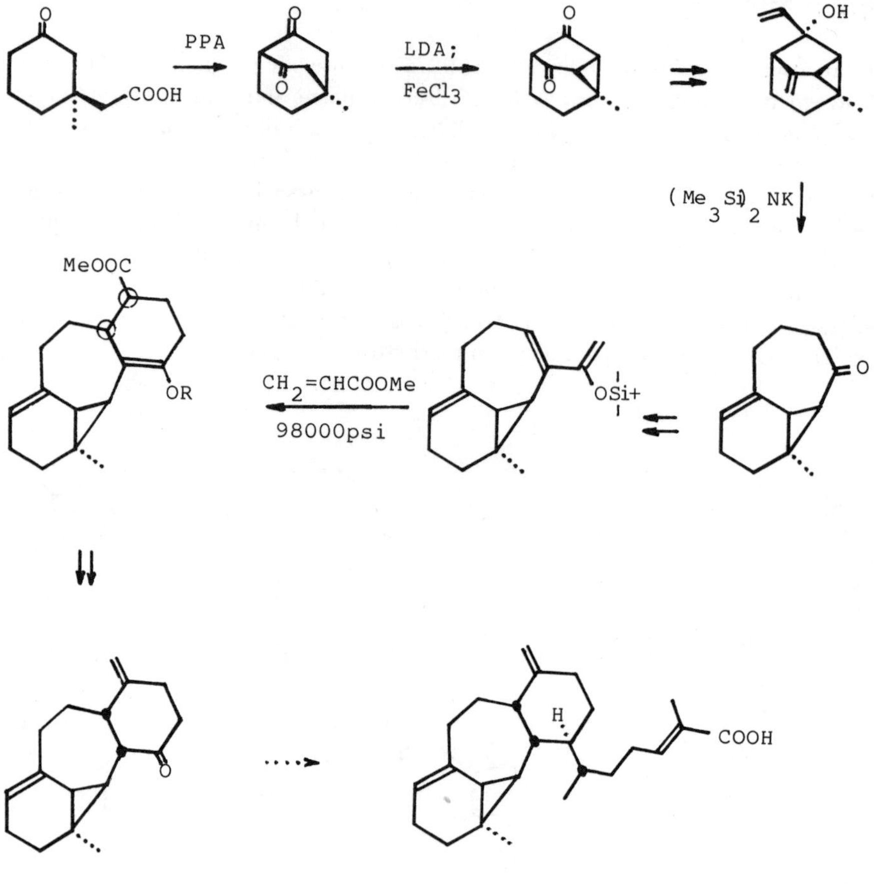

cerorubenic acid III

7.6

Frequently, enantioselective synthesis using chiral synthons (chirons) does not lead to the desired optical isomer, because chirons are available, in most cases, in only one form.

This issue sometimes can be circumvented. Thus, shrewd modifications of the Diels–Alder adduct of cyclopentadiene ans (4S)-hydroxymethylbutenolide into enantiomeric lactones has culminated in the synthesis of β-santalene (**S4**) (Takano, 1987a) of either optical series. The chiral butenolide is obtained from D-mannitol.

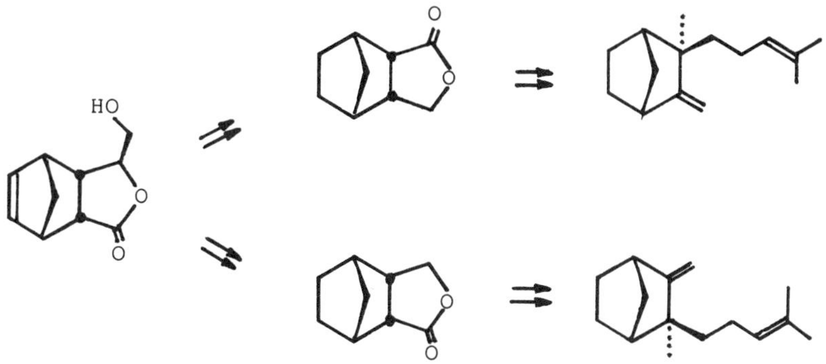

7.6

In conjunction with an asymmetric Diels–Alder reaction the flexible route to iridoids based on norbornene system has been applied to an enantioselective synthesis of loganin (**L20**) (Vandewalle, 1986). The substituted norbornene is available from condensation product of cyclopentadiene with a crotonamide which is linked to an auxiliary group prepared from camphor.

A more recent report describes a method using a diene diester derived from D-mannitol as the dienophile (Takano, 1987b).

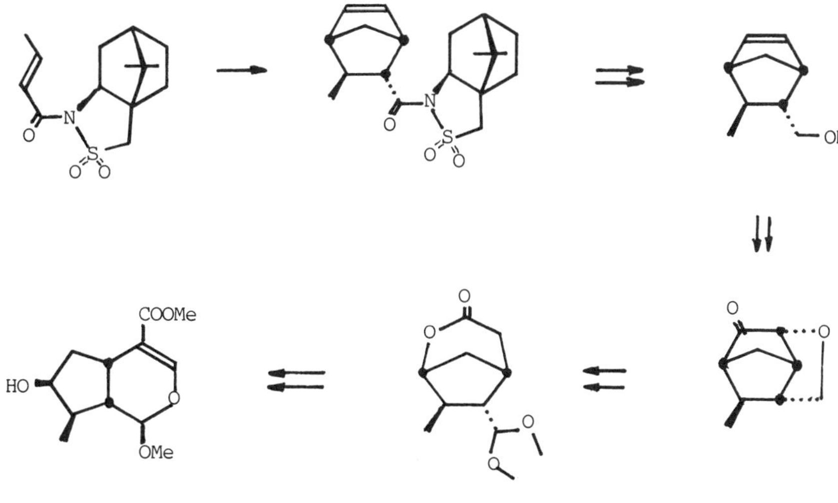

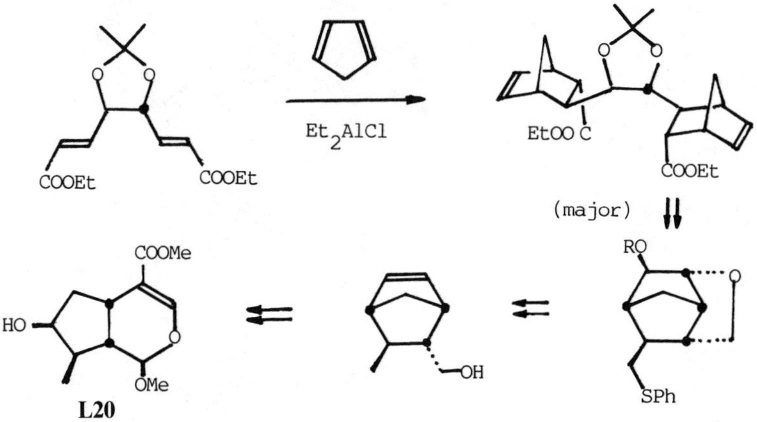

7.10 and 8.6

The presence of a *cis*-fused 6:6 ring system in furodysin and furodysinin, and particularly the location of the double bond in relation to the oxygen atom suggests the advantage of employing an oxy-Cope rearrangement to secure the carbocyclic skeleton of these marine sesquiterpenes (Hirota, 1987). The carbonyl group of the methyloctalone derived from a bicyclo[2.2.2]octeneone, itself obtained by a Diels–Alder route, becomes the pivot for dimethylation and the construction of the furan ring; the latter task via α-alkylation or α-hydroxylation as demanded.

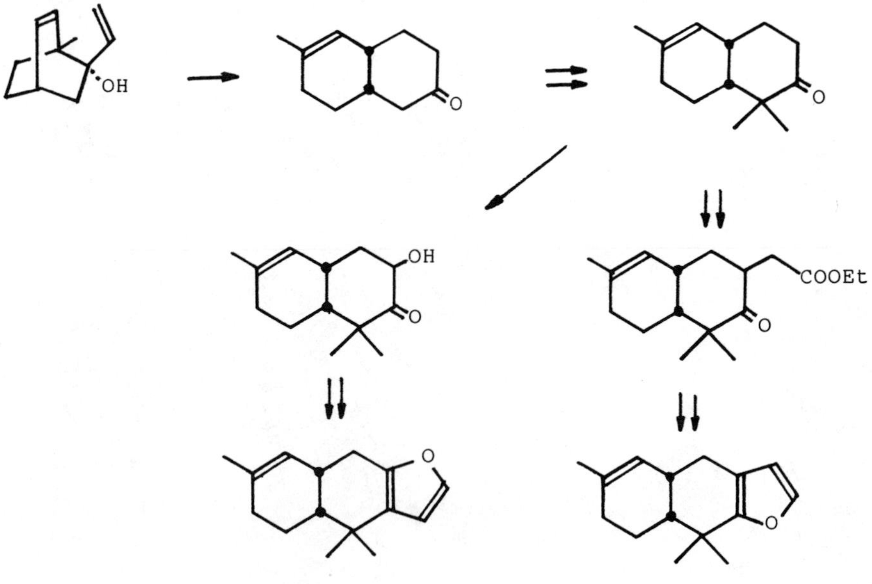

furodysinin

furodysin

7.11.1

Intramolecular Diels–Alder reaction of an acyclic precursor containing an *E*-diene subunit derived from (+)-carvone gives mainly a *trans*-octalone. On Wolff–Kishner reduction this ketone affords (−)-α-selinene (**S20**) (Caine, 1987). The unusual features of this route to the eudesmanes are the destruction of an existing carbocycle and reliance of conformational effects to determine the stereogenic centers during the cycloaddition process.

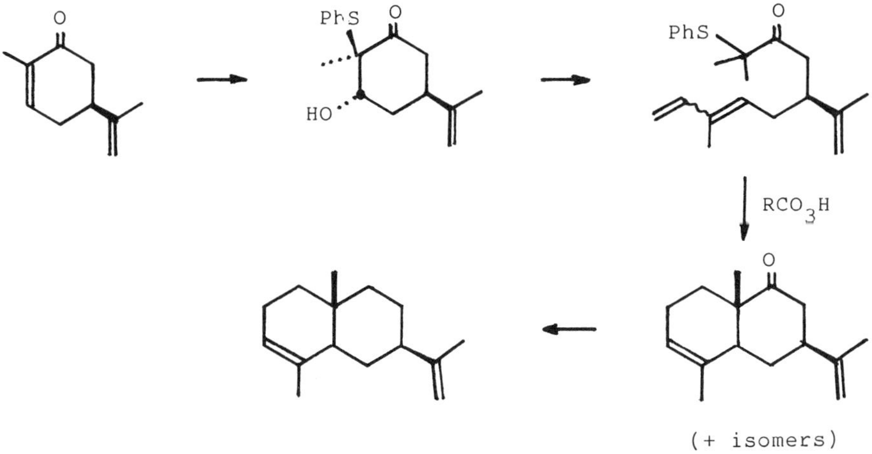

7.11.1

Another intramolecular Diels–Alder reaction conveniently elaborates the AB ring system of forskolin (**F9**) (Z.-Y. Liu, 1987).

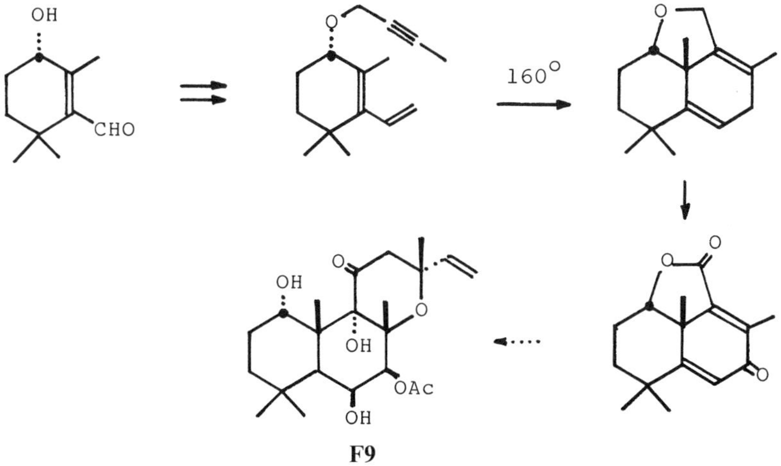

7.11.1

An asymmetric synthesis of (+)-4-oxo-5,6,9,10-tetradehydro-4,5-secofuranoeremophilane-5,1-carbolactone (K. Hayakawa, 1988) has been accomplished via a Diels–Alder route involving an allenyl ether which is generated in situ. The optically active acyclic alcohol is obtainable from a β-keto acid by reduction with baker's yeast.

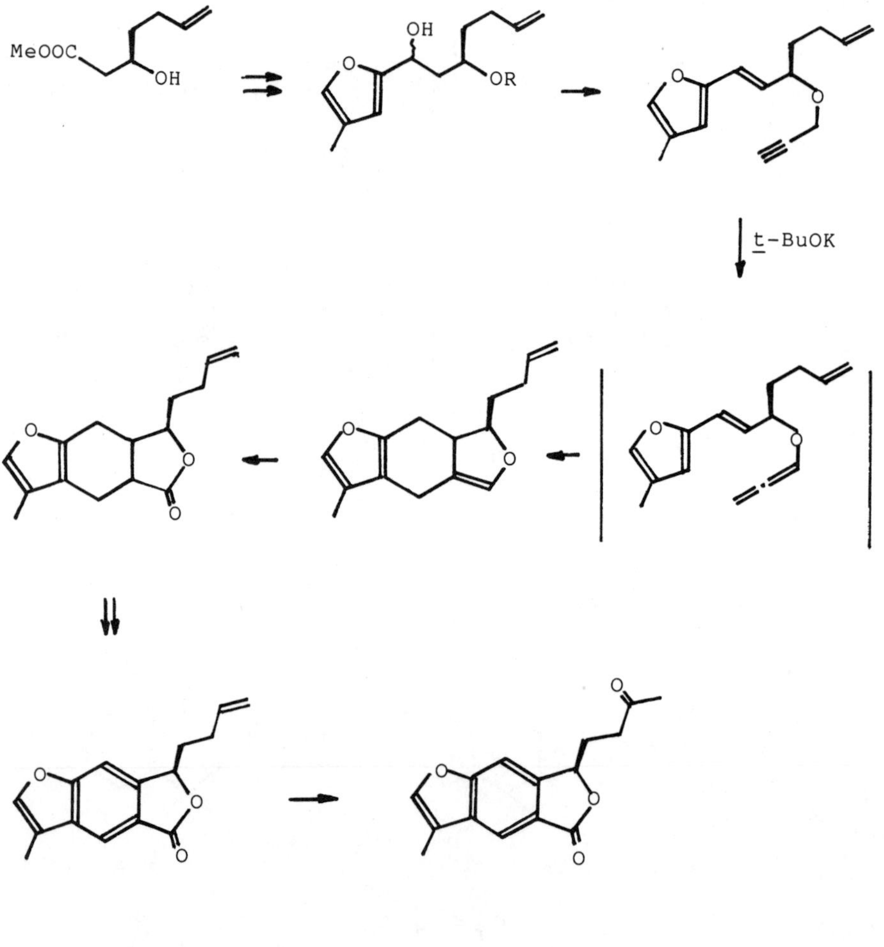

7.11.1

An excellent synthesis of verrucarol (**V12**) (Ricca, 1988) is based on an alumina-catalyzed intramolecular Diels–Alder reaction of a Z-dienol ether and a conformationally restricted dienophile.

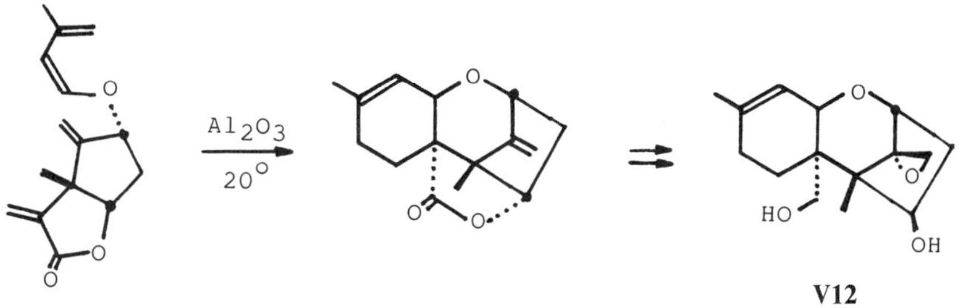

7.11.2

Intramolecular Diels–Alder reaction of a spirocyclopentadiene gives a potential precursor of longifolene (**L24**) (Bo, 1988).

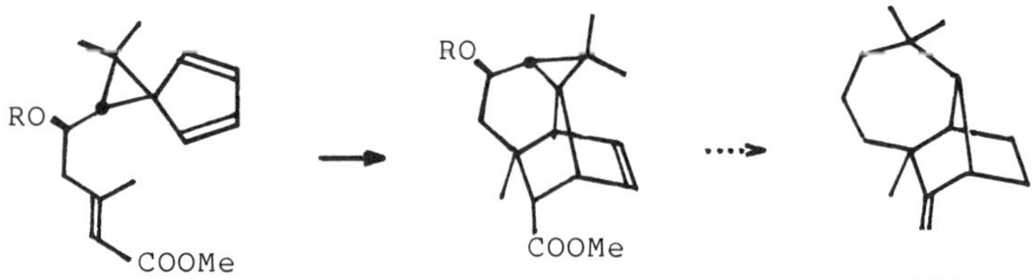

7.11.2 and 8.6

An exquisite design of gascardic acid (**G1**) synthesis (Bérubé, 1988) involves an intramolecular Diels–Alder reaction and oxy-Cope rearrangement of the bridged ring system after proper modification of the cycloadduct.

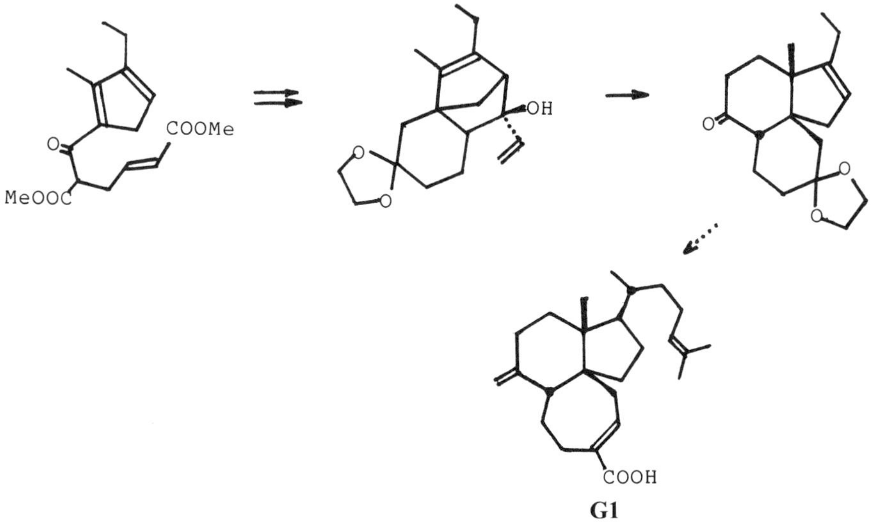

8.3

The key step of a synthesis of 8,14-cedrane oxide (Shizuri, 1987) involves intramolecular cycloaddition of a pentadienyl cation generated electrochemically from the appropriate phenol.

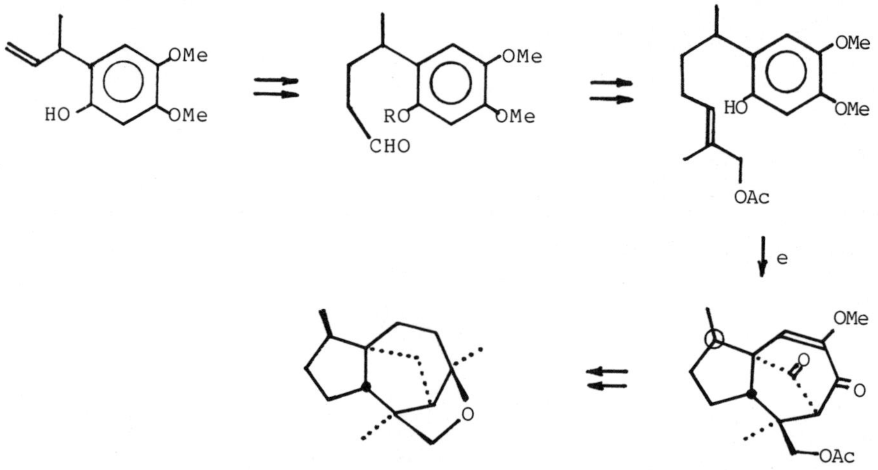

8,14-Cedrane oxide

8.3

An efficient route to calomelanolactone (Neeson, 1988) is based on rhodium-catalyzed intramolecular [2 + 2 + 2]cycloaddition of a triyne to form a pentasubstituted benzene nucleus with proper functionalities.

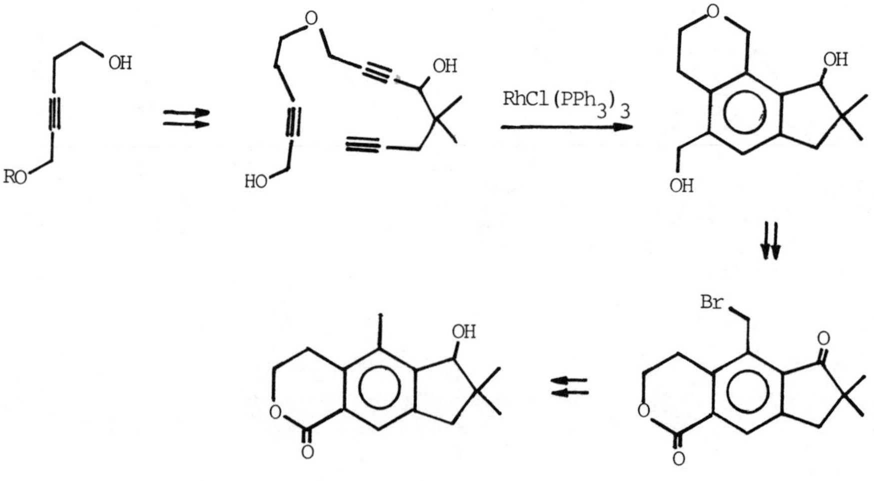

calomelanolactone

8.4

Diene stitching with an alkylborane followed by deboronative carbonylation constitutes a useful annulation tool. Recent applications of this method include assembly of two pseudoguianolide synthons which are epimeric at C_{10} from 2-methylcyclopentenone via a trans-2,3-dialkenylcyclopentanol derivative (M. Welch, 1988). These perhydrazulenones are known precursors of confertin (**C62**) and helenalin (**H3**), respectively.

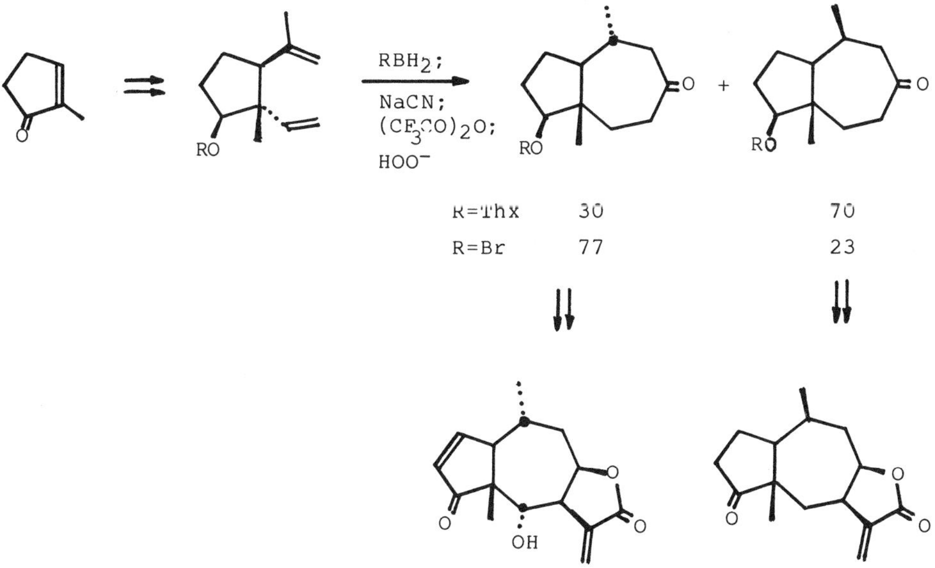

8.5

Construction of the A ring of anguidine (**A32**) has been accomplished by an intramolecular ene reaction (Ziegler, 1988).

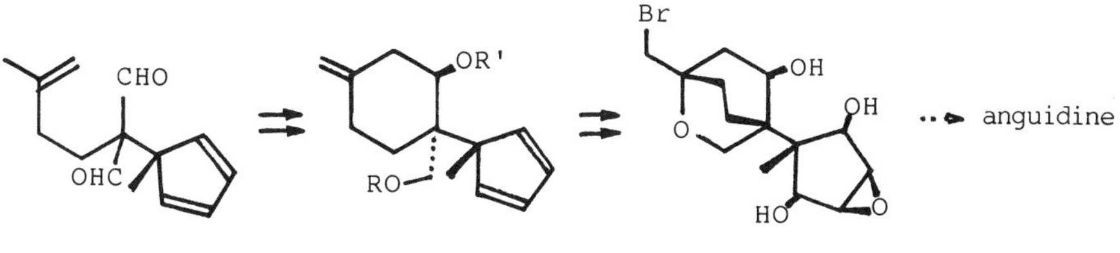

8.5

Palladium-catalyzed reductive enyne cyclization achieves a 1,3-diastereoselectivity which is very beneficial to the synthesis of β-necrodol (Trost, 1988).

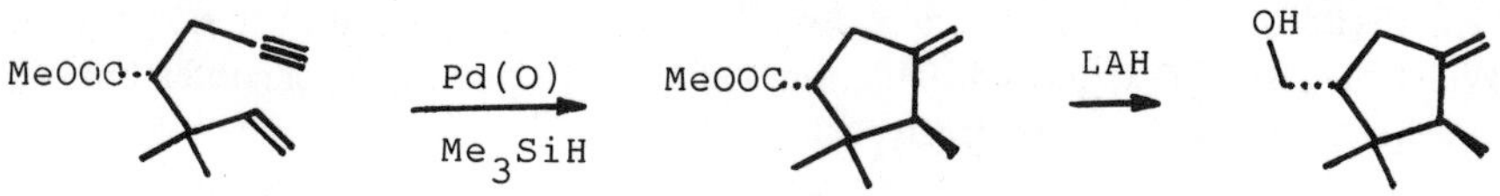

8.6

Homo–Cope rearrangement leading to 1,4-cycloheptadienes is useful for procuring intermediates in nezukone (**N8**) synthesis (Wenkert, 1987). The required precursor is readily available from the reaction of 1,3-butadiene with ethyl diazopyruvate.

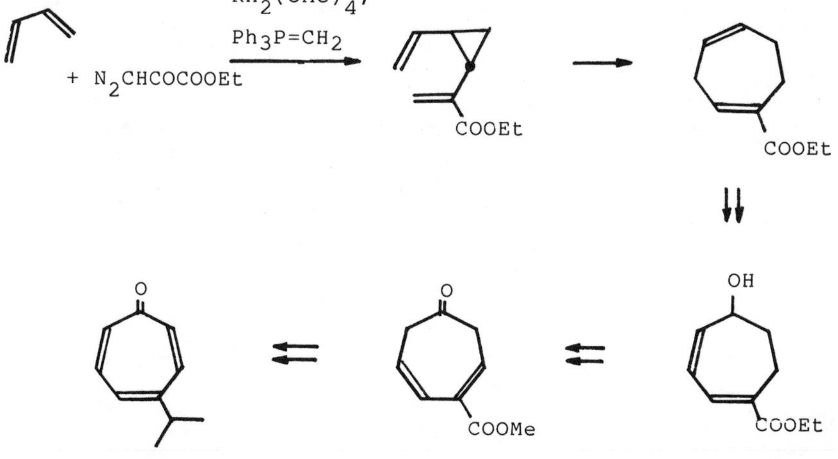

8.7

The [2.3]Wittig ring contraction has found another use in the preparation of a precursor of kallolide A (Marshall, 1988). It is interesting to note that, contrary to the rearrangement of acyclic *E*-crotyl propargylic ethers, a syn homoallylic alcohol is generated. Probably conformational strain of the macrocyclic system directs rearrangement along a different pathway.

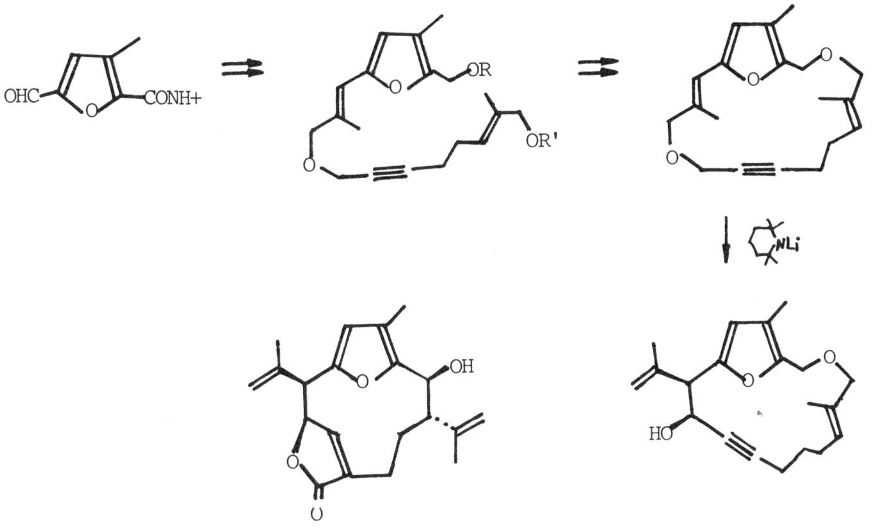

kallolide A

8.8

An electrocyclization following the generation of an oxyallene is the key step in the formation of the carbon framework of sterpurene (**S52**) (Gibbs, 1988) from a cyclobutene precursor.

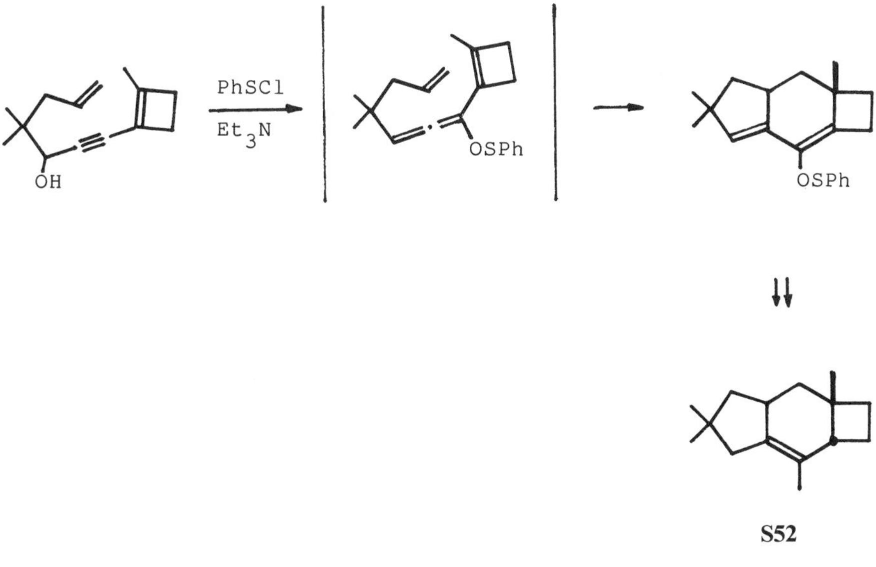

S52

8.9

An interesting synthesis of (−)-verbenalol (**V9**) (Laabassi, 1988) is based on an optically active, functionalized butadiene-iron tricarbonyl complex which is converted into a Meldrum acid derivative. The $Fe(CO)_3$ moiety provides stereocontrol over the delivery of the methyl group from a Grignard reagent. The diazoketone derived from the demetalated diene undergoes intramolecular carbenoid insertion to give mainly two diastereomers (1:1). Flash-vacuum thermolysis of the exo, exo isomer leads to a hydropentalenone and eventually, (−)-verbenalol.

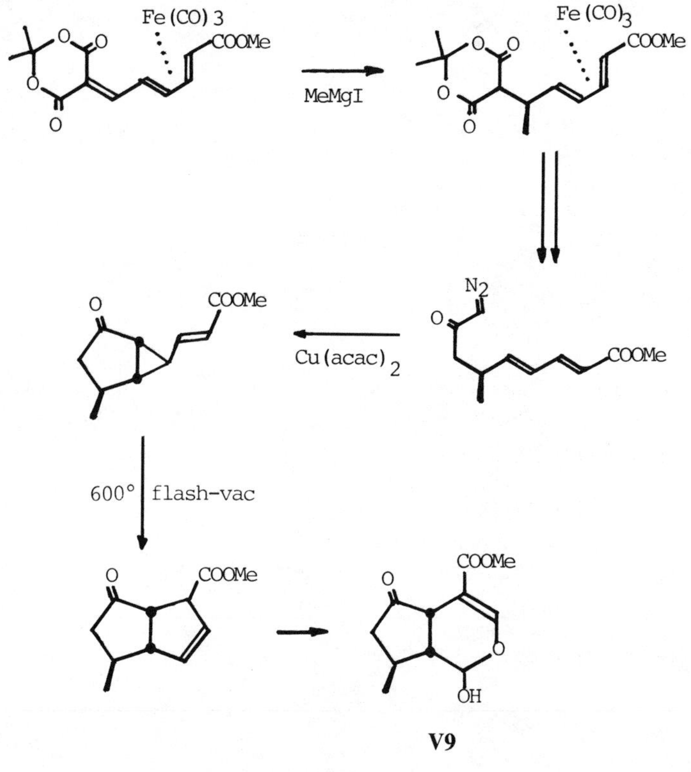

9.1

Radical cyclization using cyclic thiocarbonate-allylic thiocarbamate as initiator/terminator proves to be an excellent method for constructing isopropenylcyclopentanes. This process constitutes a novel access to a spirocyclic precursor of cedrol (**C38**) (Ziegler, 1987). Stereoselectivity of the cyclization is 12:1 in favor of the trans 1,3-dialkyl isomer.

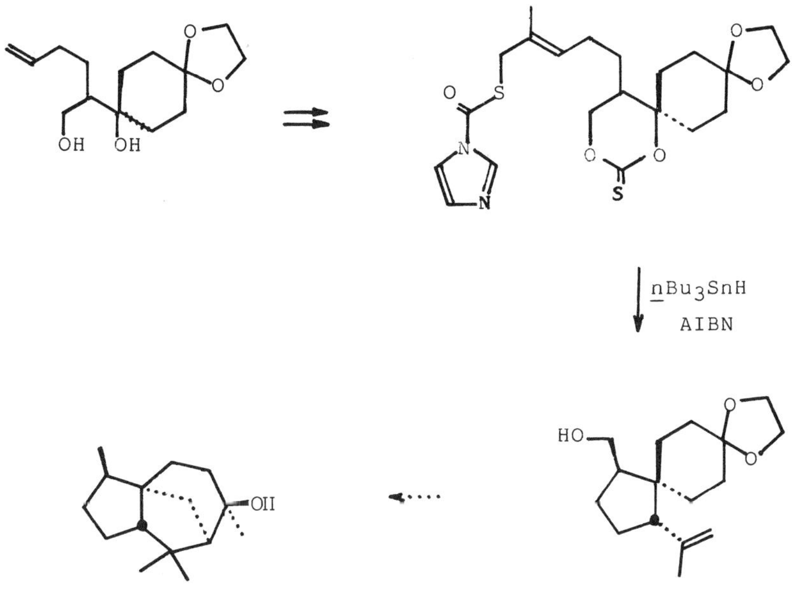

C38

10.2.2

Linear cis-anti-cis triquinanes have been secured by intramolecular ketene-alkene cycloaddition followed by ring expansion (DeMesmaeker, 1988). The applicability of this reaction sequence to the synthesis of certain sesquiterpenes such as hirsutene (**H17**) is beyond doubt.

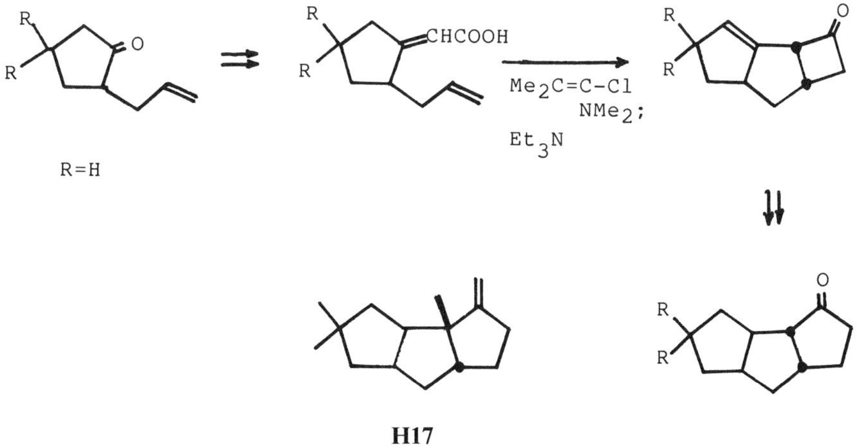

H17

11.1.1

Studies on a trisnor model has shown the feasibility of synthesizing modhephene (**M12**) by a fascinating cascade rearrangement (Fitjer, 1988).

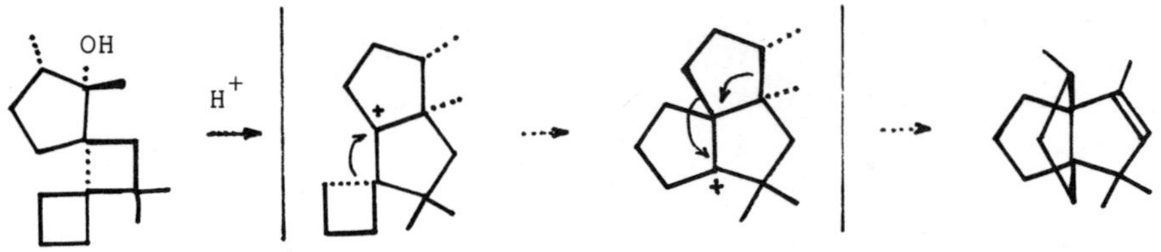

12.2.3

The critical aspect in synthesis of (+)-balanitol (Anglea, 1987) is the introduction of a β-hydroxyl group at C_1. The availability of various 1-menthene derivatives and facility of photocycloaddition of alkoxycyclobutenes with cyclohexenones indicate an annulation via a tricyclic intermediate is most profitable. The need for a secondary methyl group at C_4, a 5α stereochemistry, and abrogation of functionality at C_6 further points to a 1,2-dialkoxycyclobutene as the proper photoaddend because cleavage of the intracyclic bond of the bicyclo[2.2.0]hexane moiety of the photoadduct is facilitated.

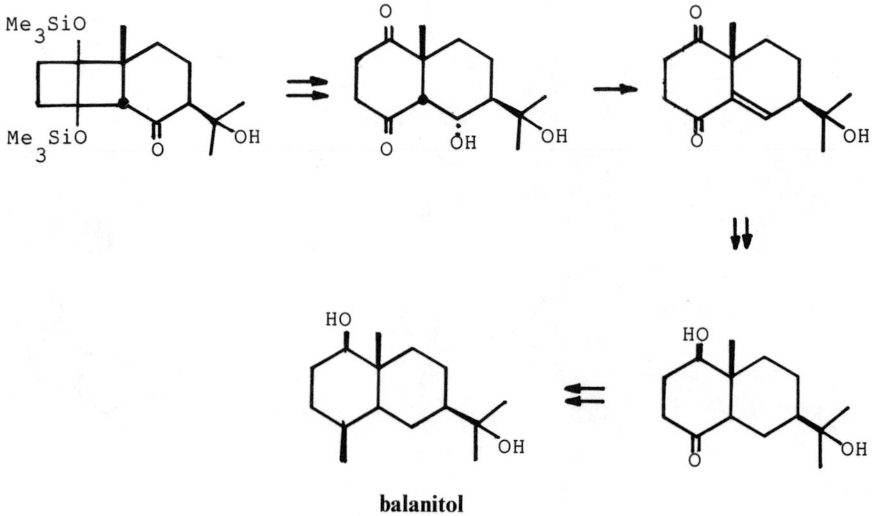

balanitol

12.2.3

A new method for hydrazulenone formation is based on thermolysis of a linear 4:4:6-ring system which is accessible from photocycloaddition. A synthesis of (+)-daucene (**D8**) (Audenaert, 1987) from (−)-piperitone is illuminative. Metathetical cleavage of the photoadduct gives a cyclodecadienone which is followed by trans-silylation (ene-type reaction) with attendant C—C bond formation.

D8

REFERENCES

Anglea, T. A., Pinder, A. R. **1987**, *Tetrahedron 43*, 5537.
Asaoka, M., Takenouchi, K., Takei, H. **1988**, *Tetrahedron Lett. 29*, 325.
Audenaert, F., DeKeukeleire, D., Vandewalle, M. **1987**, *Tetrahedron 43*, 5593.
Bérubé, G., Fallis, A. G. **1988**, *3rd Chem. Congr. N. America*, ORGN 25.
Bo, L., Fallis, A. G. **1988**, *3rd Chem. Congr. N. America*, ORGN 24.
Caine, D., Stanhope, B. **1987**, *Tetrahedron* 43, 5545.
Cannone, P., Boulanger, R. **1988a**, *3rd Chem. Congr. N. America*, ORGN 134.
Cannone, P., Plamondon, J. **1988b**, *3rd Chem. Congr. N. America*, ORGN 133.
DeMesmaeker, A., Veenstra, S. J., Ernst, B. **1988**, *Tetrahedron Lett. 29*, 459.
Fitjer, L., Majewski, M., Kanschik, A. **1988**, *Tetrahedron Lett. 29*, 1263.
Gibbs, R. A., Bartels, K., Okamura, W. H. **1988**, *3rd Chem. Congr. N. America*, ORGN 26.
Hayakawa, K., Nagatsugi, F., Kanematsu, K. **1988**, *J. Org. Chem. 53*, 860.
Heathcock, C. H., Davidsen, S. K., Mills, S., Sanner, M. A. **1986**, *J. Am. Chem. Soc. 108*, 5650.

Hirota, H., Kitano, M., Komatsubara, K., Takahashi, T. **1987a**, *Chem. Lett.* 2079.
Hirota, H., Yokoyama, A., Miyagi, K., Nakamura, T., Takahashi, T., **1987b**, *Tetrahedron Lett. 28*, 435.
Hutchinson, J. H., Pattenden, G., Myers, P. L. **1987**, *Tetrahedron Lett. 28*, 1313.
Jansen, B. J. M., Sengers, H. H. W. J. M., Bos, H. J. T., deGroot, A. **1988**, *J. Org. Chem. 53*, 855.
Kato, N., Tanaka, S., Kataoka, H., Takeshita, H. **1987**, *Chem Lett.* 2295.
Kende, A. S., Sanfilippo, P. J. **1983**, *Synth. Commun. 13*, 715.
Koft, E. R., Kotnis, A. S., Broadbent, T. A., **1987**, *Tetrahedron Lett. 28*, 2799.
Krause, W., Bohlmann, F. **1987**, *Tetrahedron Lett. 28*, 2575.
Krief, A., Dumont, W., Pasau, P. **1988**, *Tetrahedron Lett. 29*, 1079.
Laabassi, M., Gree, R. **1988**, *Tetrahedron Lett. 29*, 611.
Lansbury, P. T., Galbo, J. P., Springer, J. P. **1988**, *Tetrahedron Lett.* 29, 147.
Levine, S. G., Heard, N. E. **1988**, *3rd Chem. Congr. N. America*, ORGN 167.
Liu, Z.-Y., Zhou, X.-R., Wu, Z.-M. **1987**, *Chem. Commun.* 1868.
Marshall, J. A., Nelson, D. J. **1988**, *Tetrahedron Lett. 29*, 741.
Mehta, M., Krishnamurthy, N. **1987**, *Tetrahedron Lett. 28*, 5945.
Mori, K., Mori, H. **1986**, *Tetrahedron* 42, 5531.
Mori, K., Komatsu, M. **1987**, *Tetrahedron 43*, 3409.
Neeson, S. J., Stevenson, P. J. **1988**, *Tetrahedron Lett. 29*, 813.
Paquette, L. A., Poupart, M.-A. **1988**, *Tetrahedron Lett. 29*, 273.
Rao, Y. K., Nagarajan, M. **1988**, *Tetrahedron Lett. 29*, 107.
Ricca, D. J., Luengo, J. I., Koreeda, M. **1988**, *3rd Chem. Congr. N. America*, ORGN 170.
Schwabe, R., Farkas, I., Pfander, H. **1988**, *Helv. Chim. Acta 71*, 292.
Shizuri, Y., Okuno, Y., Shigemori, H., Yamamura, S. **1987**, *Tetrahedron Lett. 28*, 6661.
Takano, S., Inomata, K., Kurotaki, A., Ohkawa, T., Ogasawara, K. **1987a**, *Chem. Commun.* 1720.
Takano, S., Kurotaki, A., Ogasawara, K. **1987b**, *Tetrahedron Lett. 28*, 3991.
Tokoroyama, T., Tsukamoto, M., Asada, T., Iio, H. **1987**, *Tetrahedron Lett. 28*, 6645.
Trost, B. M., Braslau, R. **1988**, *Tetrahedron Lett. 29*, 1231.
Vandewalle, M., Van der Eycken, J., Oppolzer, W., Vullioud, C. **1986**, *Tetrahedron 42*, 4035.
Welch, M. C., Bryson, T. A. **1988**, *Tetrahedron Lett. 29*, 521.
Wenkert, E., Greenberg, R. S., Kim, H.-S. **1987**, *Helv. Chim. Acta 70*, 2159.
Xie, Z. F., Suemume, H., Nakamura, I., Sakai, K. **1987**, *Chem. Pharm. Bull. 35*, 4454.
Ziegler, F. E., Zheng, Z.-1. **1987**, *Tetrahedron Lett. 28*, 5973.
Ziegler, F. E., Sobolov, S. B. **1988**, *3rd Chem. Congr. N. America*, ORGN 174.

Compound Index

Compounds indicated by asterisks have not been synthesized

Abietic acid (A1), 49, 50, 408
3α-acetoxy-15β-hydroxy-7,16-seco-trinervita-7,11-diene (S17), 195
Acolamone (A3), 16
α-Acoradiene (A4), 181, 644
β-Acoradiene (A5), 217, 451, 504
γ-Acoradiene, 692
Acoradiene III (A6), 173
Acoragermacrone (G10), 243, 574
Acoratiene (A7), 275, 277, 450, 582
α-Acorenol (A8), 577
β-Acorenol (A9), 217, 577
Acorenone (A10), 62, 73, 174, 451, 478, 650
Acorenone B (A10b), 73, 174, 277, 451, 550, 619, 644
Acorone (A11), 73, 178, 331, 337, 457, 499, 622, 692
Aflavinine *, 169
Africanol, 263
Africanolone (A14), 328
α-Agarofuran (A15), 7
β-Agarofuran (A16), 15, 489
Agarospirol (A17), 75, 195, 515, 619
Ajugarin I, 134
Ajugarin IV (A19), 40
Alantolactone (A20), 279
Albene (A22), 103, 281, 376, 439, 492, 499
Albicanyl acetate (A23), 293
Allamcin, 229
Allamandin, 639
Alliacol B (A24), 129
Alliacolide (A25), 492
Alnusenone (A26), 53, 290, 418, 608
Amarolide, 733
Ambliol A (A27), 305
α-Amorphene (A29), 392, 462
ε-Amorphene, 318

β-Amyrin (A30), 280
δ-Amyrin (A31), 309
Anguidine (A32), 152, 753
Anhydro-β-rotunol (R8), 178, 216
Anisomelic acid, 118
Annonene (A34), 40, 156, 322
Antheridiogen An (A35), 384, 546, 622
Aphidicolin (A36), 38, 81, 87, 100, 144, 153, 219, 253, 303, 311, 312, 365, 370, 388, 469, 486, 585, 637
Aplysin (A37), 479
Arctiol (A38), 355
Aristolactone, 474
Aristolene (A39), 526
Aristolone (A40), 22, 175, 521, 525
Aromadendrene (A41), 239, 349, 592
Aromaticin (A42), 285
Aromatin (A43), 113, 466
Arteannuin (artemisinin, quinghaosu), 108
Arteannuin B (A44), 208, 252
Artecalin (A45), 201
Artemisin (A46), 48
Asperdiol, 116, 243, 251, 268
Aspterric acid (A47), 61, 585
Asteriscanolide, 443
Atisine (A48), 214, 417, 563
Atisirene (A49), 86
E-α-atlantone (A51), 331, 332, 335
Z-α-atlantone (A52), 331, 335
Atractyligenin, 321
Atractylon, 259
Aubergenone *, 48
Avarol (A54), 39
Axamide-1, 208
Axisonitrile (A55), 12, 630
Baiyunol, 736
Bakkenolide A (B1), 99, 228, 579, 650

761

Balanitol, 758
Balduilin * (B2), 113, 668
α-Barbatene (B3), 387
β-Barbatene (B4), 387
Barbatusol, 135, 483
α-trans-Bergamotene (B5), 231, 268, 540, 563
β-trans-Bergamotene (B6), 542
β-cis-Bergamotene (B7), 542
Bertyadionol, 116
Bicyclogermacrene (G7), 510, 541, 580
Biflora-4,10(19),15-triene (B8), 403, 404
Bilobalide, 196
Bilobanone, 175
α-Biotol, 614
β-Biotol, 614
Bisabolangelone, 437
α-Bisabolene (B9), 331
β-Bisabolene (B10), 331, 333, 335
γ-Bisabolene, 278
α-Bisabolol (B12), 277, 331, 333, 433
α-Bisabololone (B13), 331, 333
Boll weevil pheromones (BW1, BW2), 293
Bonandiol, 611
Boonein (B14), 555
Boschnialacetone (B15), 669
α-Bourbonene (B16), 102, 536, 537
β-Bourbonene (B17), 103, 536
Brasilenol, 132, 260
3β-Bromocaparrapi oxide (C20), 300, 302
10-Bromo-α-chamigrene (C48), 300
Bruceantin * (B18), 56, 370, 452
ε-Bulgarene, 318
α-Bulnesene (B19), 31, 487, 592
β-Bulnesene (B20), 443, 552, 656
Bulnesol (B21), 32, 161, 271, 583, 592
Butyrospermol (B22), 351, 696
Cacalol (C1), 256
β-Cadinene (C2), 339, 402
γ$_2$-Cadinene (C3), 339, 360
δ-Cadinene (C4), 142
ε-Cadinene (C5), 167, 318, 339, 349, 360
Cadinene hydrochloride, 142
α-Cadinol (C6), 59, 105, 591
Cafestol, 320
Calamenene (C7), 151, 284
Calameon (C9), 562
Callistrisic acid (C10), 51, 662
Calomelanolactone, 752
Calonectrin (C12), 78, 362, 697
Camphene (C13), 375
Campherenone (C14), 281, 439
Camphor (C15), 191, 192, 221, 375, 439, 456
Cantharidin (C18), 360, 396, 695
Caparrapi oxide (C19), 307

$\Delta^{9(12)}$-Capnellene (C21), 103, 120, 227, 323, 336, 378, 425, 426, 427, 454, 479, 480, 489, 499, 502, 511, 557, 604, 660
Capnellenediol (C22), 494
Capnellenetriol, 93, 255
Carabrone (C23), 336, 520
2-Carene (C25), 518, 521, 526
3-Carene (C26), 335, 518
Carissone (C27), 4
Carnosic acid (C28), 155
Carotol (C29), 9, 602
Carpesiolin (C30), 566, 653
Carvone (C31), 70, 371
Caryophyllene (C32), 189, 193, 327, 510, 665
Casbene (C34), 247, 509, 525
Cecropia juvenile hormone (C35), 691, 695
8S,14-Cedranediol (C36), 422, 569, 699
8,14-Cedrane oxide, 752
α-Cedrene (C37), 163, 216, 281, 385, 421, 422, 503, 563, 595, 604, 631
Cedrol (C38), 91, 183, 190, 216, 281, 320, 616, 756
Cembranolide, L, michaele, 246
Cembrene (C39), 247
Cembrene A (C40), 241, 743
3Z-Cembrene A (C41), 241, 576
Ceriferic acid, 117
Ceriferol (C43), 268
Ceroplastol II * (C44), 570, 602, 693
Cerorubenic acid III *, 745
Chamaecynone (C45), 369
α-Chamigrene (C46), 318, 337, 475, 509, 619, 640
β-Chamigrene (C47), 277, 318, 432, 469, 640
Chaminic acid (C49), 233
Chasmanic (C50), 389, 642
Chiloscyphone, 318
Chokol A, 455, 629
Cholesterol, 358
trans-Chrysanthemic acid (C51), 170, 171, 172, 236, 520, 525, 526, 531, 580, 739
cis-Chrysanthemic acid (C51c), 236, 238, 470, 520, 521, 526
Chryanthemol (C52), 237
Chrysanthemum dicarboxylic acid (C53), 530
Chrysomelidial (C54), 165, 176, 440, 512, 623, 664
Cinabicol (C55), 341
1,8-Cineole, 625
Cinnamolide (C56), 293
Cinncassiol D1, 416
Citrilol acetate (C57), 618
Colorata-4(13), 8-dienolide (C59), 44
Compresssanolide (C60), 653

Confertifolin (C61), 180, 293
Confertin (C62), 32, 60, 112, 114, 202, 250, 460, 567, 592, 615
Copacamphene (C63), 219, 489, 561
Copacamphor (C65), 182, 207, 218
α-Copaene (C66), 36, 231, 432, 697
β-Copaene (C67), 231, 491
ent-Copalic acid (C68), 694
Coriamyrtin (C69), 147, 188
Coriolin (C70), 95, 127, 158, 229, 251, 328, 377, 398, 447, 502, 507, 554, 631, 672, 687
Corymbol, 373
β-Costal (C71), 14, 355
β-Costic acid (C72), 14
β-Costol (C74), 14, 355
Costunolide (C75), 473, 573, 680
Crinipellin A, 646
Cryptofauronol (F3), 594
Cryptojaponol (C78), 25
Cryptone (C79), 141, 181
Cryptotanshinone (T3), 348
Cuauhtemone (C80), 47
β-Cubebene (C81), 522
Cubebol (C81a), 522
Cubitene (C82), 241
Culmorin (C83), 182
Cuparene (C84), 66, 190, 479, 497, 512, 552, 579
α-Cuparenone (C85), 67, 438, 445, 448, 497, 542, 551, 622, 741
β-Cuparenone (C86), 281, 381, 479, 541, 542, 551, 604
α-Curcumene (C87), 141
Curcuphenol acetate, 262
Curzerenone (C88), 179
Cyathin A3 ∗ (C89), 370, 643
Cybullol (C90), 19
Cycloartenol (C91), 233
Cyclocolorenone (C92), 101, 589
Cyclocopacamphene (C64), 527
Cycloeudesmol (E20), 19, 236, 522
Cyclonerodiol (C93), 306, 449
Cyclosativene (S13), 9, 80, 201, 326, 375, 527
Cycloseychellene (S30), 146, 161, 527
Cyperolone (C94), 5, 579
α-Cyperone (C95), 4, 5, 150, 633
epi-α-Cyperone (C96), 9
β-Cyperone (C97), 10, 633
Dactylol (D1), 263
β-Damascenone (D2), 176, 293, 341,
α-Damascone (D3), 293
δ-Damascone (D4), 341,
Damsin (D5), 112, 136, 207, 445, 603, 652, 668
Damsinic acid (D6), 214, 285, 460

Daucene (D8), 68, 285, 658, 759
Daucol (D9), 9
Debromolaurinterol acetate (L7), 517
Dehydroabietic acid (A2), 50, 154, 260, 689
Dehydrocostus lactone (C77), 273
Dehydrofukinone (F16), 110, 356, 411, 610
Dehydrofuropelargones (F19/20), 67
Dehydroiridodial (I6), 165, 512
Dehydroiridodiol (I7), 68, 197, 493, 625
Dehydrokessane (K6), 655
Dehydrosaussurea lactrone (S15), 573
Dendrobine (D11), 59, 120, 147, 358, 412, 600
Deodarone (D12), 331,
8-Deoxyanisatin (A13), 58
Deoxynupharidine (N14), 188
2-Deoxystemodinone, 452
Deoxytrisporone (T33), 293
Deoxynivalenol, 366,
Dictymal, 734
Dictyol C ∗ (D13), 467
Dictyolene (D14), 274
Dihydroactinidiolide (A13), 293
Dihydrocallistrisin (C11), 159, 184
Dihydrocostunolide (C76), 251, 466, 477
Dihydronepetalactone (N6), 463, 470
Dihydroreynosin (R2), 654
Dihydrospiniferin I (S46), 615
7,20-Diisocyanoadociane, 400
Dispermol (D15), 314
Dolabradiene (D16), 39, 112
β-Dolabrin, 569
14S-Dolasta-1(15),7,9,trien-14-ol (D17), 253
Dolichodials (D18), 164
Drimenin (D20), 180, 347
Drimenol (D22), 180, 293, 311
Effusin ∗, 217
β-Elemene (E1), 510, 586
γ-Elemene (E2), 33, 680
β-Elemenone (E3), 47, 636, 680
Elemol (E4), 459, 586, 633, 651
Emmotin H (E5), 256
Enmein (E6), 688
Epimukulol, 473
12,13-Epoxytrichothec-9-ene (T27), 78, 644
Eremolactone (E7), 383
Eremoligenol (E8), 45
Eremophilene (E11), 45
Eremophilenolide (E12), 279, 357,
Eremophilone (E13), 23, 132, 404, 632
Eriolanin (E14), 37, 189, 541, 648, 677, 682
Estafiatin (E15), 273, 653
Eucannabinolide (E16), 574
α-Eudesmol (E17), 5, 402, 434

β-Eudesmol (E18), 10, 13, 14, 15, 256, 266, 434, 562, 611, 661
γ-Eudesmol (E19), 4, 279
Evuncifer ether (E21), 8
Farnesiferol A (F1), 311
Farnesiferol C (F2), 429
Ferruginol (F4), 12, 109, 154, 287, 314
Fichtelite (F5), 290, 405
Filifolone (F6), 271, 542, 543, 599
Flexibilene (F7), 509
Fomannosin (F8), 373,
Forskolin (F9), 365, 410, 436, 465, 735, 739, 749
Forsythide (F10), 221, 623
Fragranol (F11), 544, 578
Fraxinellone (F12), 346
Friedelin (F13), 134, 284, 418
Frullanolide (F14), 19, 159, 250, 684
Fukinone (F15), 110, 175, 279
Fumagillin (F17), 344
Furanether B (F18), 445
Furanoeremophilane (E9), 168
Furanoeremophilan-14,6α-olide (E10), 185
Furodysin, 748
Furodysinin, 748
Furoventalene (F21), 159
Fusicoccin * (F22), 513, 655
Fusidic acid * (F23), 111
Gascardic acid (G1), 57, 199, 751
Geijerone (G2), 211, 354,
Genipin (G3), 476
Geosmin (G4), 19
Germacrene A (G5), 244
Germacrene D (G6), 574
Germacrone (G9), 243
Germanicol (G11), 275
Gibberellic acid (G12), 84, 89, 119, 163, 212, 225, 252, 269, 413, 429, 463, 511, 578
Gibberellin A5 (GA5), 430
Gibberellin A15 (G13), 122, 222, 269
Ginkgolide B, 638
Globulol (G14), 32, 285, 518, 685
Glutinosone (G15), 26, 274, 392, 676
Gnididione (G16), 199, 430, 698
β-Gorgonene, 15
Grandisol (G18), 232, 476, 532, 533, 538, 544, 545, 578, 598, 599, 695
Grayanotoxin II (G19), 83, 589
Grosshemin, 626
Guaiol (G20), 28, 121, 161, 271, 566, 667
Gymnomitrol (G21), 78, 94, 182, 387, 441
Haagenolide, 473
Hanegokedial (H1), 114
Hedycaryol (H2), 241, 350, 685

Helenalin (H3), 112, 668
Heliangolide, 465
Heliocides (H4-7), 344,
Helominthogermacrene (H8), 510
Helminthosporal (H9), 77, 161, 442, 601
Herbertene (H10), 582, 604
Hibaene (H11), 85, 187, 288, 498, 559
Hibiscone C (H12), 649
α-Himachalene (H13), 333, 337, 399, 563
β-Himachachalene (H14), 333, 337, 399, 460, 563, 604, 655, 668
Hinesol (H15), 139, 178, 215, 364, 425, 515, 550, 667, 673, 684
Hinesolone (H16), 676
Hirsutene (H17), 176, 224, 227, 254, 265, 280, 323, 378, 385, 435, 440, 486, 489, 495, 502, 507, 511, 541, 554, 581, 587, 592, 595, 596, 599, 616, 655, 660, 738, 757
Hirsutic acid (H18), 94, 739
Hirsutic acid C (H19), 95, 164, 447, 555, 606, 687
Homodaphniphyllic acid, Me ester, 740
Hop ether (H20), 585, 626
Hotrienol (H21), 635
Humulene (H22), 244, 246, 509, 583
7-Hydroxycalamenene (C8), 12, 742
3-Hydroxylabdadienoic acid, 313
3β-Hydroxynagilactone F, 43
Hypacrone (H23), 262
Hypolepin B (H24), 256
Hysterin (H25), 653
Illudin S (I1), 129
Illudimine (I2), 483
Illudol (I3), 105, 372, 535
Ingenol * (I4), 102, 444, 472, 652
Ircinianin, 408
Iridodial (I5), 164, 623
Iridomyrmecin (I8), 327, 454
Isabelin (I10), 562, 661
Ishwarane (I11), 85, 169, 233, 499, 519, 641
Ishwarone (I12), 207, 215, 356, 519
Isogatholactone (A18), 296, 306
Isoalantolactone (A21), 184
Isoamijiol (A28), 494, 658, 745
Isoaplysin, 20, 306
Isoatisirene (A50), 166
Isobicyclogermacrene (G8), 510
Isobicylogermacrenal (G8a), 459
Isobisabolene (B11), 194
Isocaryophyllene (C33), 189, 193, 510, 532, 541, 607, 618, 665
Isocomene (I14), 57, 94, 138, 185, 271, 443, 451, 487, 501, 505, 553, 576, 595
2-Isocyanopupukeanane (P53), 85, 420

9-Isocyanopupukeanane (P54), 210, 261, 420
Isodehydroiridodiol (17'), 197
Isodihydronepetalactone (N6'), 470, 640
Isodrimenin (D21), 180, 293
Isoiresin diacetate (I15), 41
Isoiridomyrmecin (I9), 508, 623
Isoligularone (L14), 366,
Isolinderalactone (L17), 573
Isolobophytolide, 245
Isonootkatone (N12), 27, 193
Isopetasol (P25), 49, 150, 356, 610
Isosesquicarene (S27), 335, 590
Isoshyobunone (S32i), 66, 69, 639
Isotelekin (T9), 47
Isotirucallol (T19), 309
α-Italicene (I17), 539
β-Italicene (I18), 539
Ivalin (I19), 273
Ivangulin (I20), 36, 541, 607, 648, 682
Jatropholone A, B (J1, 2), 397
Jatrophone (J3), 115
Jolkinodide E (J4), 154
Junenol (J5), 16, 271, 293
10-epi-Junenol (J6), 654
Junionone (J7), 618
erythro-Juvabione (J8), 333, 392, 640, 692
threo-Juvabione (J9), 268, 613, 666, 675
K76, 303, 608
Kallolide A *, 754
Karahanaenone (K1), 286, 313, 440, 460, 467, 565
Karahana ether (K2), 308, 313, 494
Karatavic acid (K3), 305
Kaurene (K4), 82, 217, 288, 622
Kessane (K5), 32, 592
Kessanol (K7), 271,
Khusimone (K8), 69, 215, 420, 454, 556, 595, 667
Khusitene (K9), 198, 345,
Kolavenic acid, methyl ester, 149
LL-Z1271, 43
Labda-7,14-dien-13-ol (L1), 297,
Lacinilene C, methyl ether (L2), 256
Lanceol (L3), 331, 332, 333
Lanosterol (L4), 358
Lansic acid (L5), 304
Latia luciferin (L6), 299
Laurenene, 130, 646
Lavender ketone (L8), 344,
Lemnalol (L9), 491
Lepidozene (L10), 510, 541, 580
α-Levantenolide (L11), 298
β-Levantenolide (L12), 298

Ligularone (L13), 99, 185, 362, 366, 430, 698
Linaridial, 744
Linderalactone (L16), 573
Lindestrene (L18), 30, 259
Lineatin (L19), 233, 534, 541, 543, 550
Loganin (L20), 176, 439, 491, 495, 556, 623, 650, 664, 669, 747
Loliolode (L21), 307
Longicamphor (L22), 79
Longicyclene (L23), 79
Longifolene (L24), 34, 35, 88, 98, 162, 219, 326, 423, 484, 527, 563, 582, 604, 618, 656, 666, 744, 751
α-Longipinene (L25), 36, 540, 685
β-Longipinene (L26), 36, 540, 685
Lophotoxian *, 72
Lubimin (L27), 392
Luperol (L28), 110, 224, 609
Maaliol (M1), 8, 239
Magnolialide (M2), 654
Malabaricanediol (M3), 310
Mansonone D (M4), 256
Mansonone E (M5), 428
Marasmic acid (M6), 234, 352, 412
Marine sponge metabolic (M7), 39
Maritimin (M8), 654
Maritimol (M9), 312, 388
Marrubiin (M10), 42
Maturone, 201
Menthone (M11), 333, 456
O-methylpisiferic acid, 735
Mitsugashiwalactone, 195
Modhephene (M12), 126, 138, 196, 224, 264, 385, 451, 457, 479, 499, 507, 549, 553, 601, 628, 629, 631, 758
γ-Muurolene (M13), 400
ε-Muurolene (M14), 339
Muzigadial, 29, 732
Mycorrhizin A (M15), 531
Nagilactone F, 314
Nakafuran 9 (N1), 317
Nanaimool (N2), 340,
Napellene (N3), 89, 642
β-Necrodol, 455, 754
Neocembrene (C42), 242, 268
Neolemnane * (N4), 325
Neolinderalactone, 573
Nepetalactone (N5), 143
Nephthenol, 743
Nerolidol (N7), 634
Nezukone (N8), 286, 441, 568, 754
Nidifica alcohol (N9), 699
Nimbiol (N10), 154

Nootkatone (N11), 22, 27, 48, 62, 108, 137, 194, 274, 323, 356, 361, 364, 392, 481, 566, 633, 635, 675
Norbisabolide, 361
Norpatchoulenol (P13), 69, 144, 146, 248, 381, 394, 418
Nuciferal (N13), 617
Obscuromatin (O1), 241
Occidentalol (O2), 7, 104, 274, 432, 478, 650, 697
Occidol (D3), 176, 256, 267, 353, 354,
18α-Olean-12-ene (O4), 280
αγ-Onoceradienodione (O5), 304
α-Onocerin (O6), 54, 610
Ophiobolins *, 178, 464, 693,
Oplopanone (O7), 106, 224, 226, 591
8-epi-Oplopanone (O8), 226, 317
Oppositol, 737
1-Oxocostic acid (C73), 654
3-Oxosilphinene, 414
Oxosilphiperfolene (S36), 385, 587
4-Oxo-5,6,9,10-tetradehydro-4,5-secofuranoeremophilane-5,1-carbolactone, 750
Pacifigorgiol, 613
Palauolide, 209
Pallescensin 1 (P1), 293
Pallescensin A (P2), 29, 293, 475
Pallescensin E (P3), 256
Pallescensin F (P4), 316
Pallescensin G (P5), 316
Palustric acid (P6), 50
α-Panasinsene (P7), 538
β-Panasinsene (P8), 538
Paniculide A (P9), 159, 430, 647, 698
Parkeol (P10), 309
Parthenin (P11), 124, 567
β-Patchoulene (P12), 136, 465
Patchoulenone (P14), 125, 270, 346,
Patchoulol (Patchouli alcohol) (P15), 146, 248, 267, 381, 418, 613
Pentalenene (P16), 101, 127, 226, 271, 319, 327, 541, 548, 554, 605, 620, 656
Pentalenic acid (P17), 166, 185, 647
Pentalenolactone (P18), 184, 263, 368, 686
Pentalenolactone E (P19), 92, 177, 497, 541, 577, 641
Pentalenolactone G, 539, 587
Pentalenolactone H, 446
Perforenone A (P20), 61, 323, 467
Perillaldehyde (P21), 340,
Periplanone B (P22), 242, 243, 364, 574, 648
Perrottetianal, 45
Petasalbine (P23), 430, 698

Petasitolone (P24), 362, 611
Phorbol, 406
Phyllanthocin (P26), 363, 364, 437; 450
Phyllocladene (P27), 51, 186, 192, 198, 288, 559, 622, 743
Phytuberia (P28), 6, 634, 665, 682
Picrotoxinin (P29), 80, 147
α-Pinene (P30), 231, 232, 598
β-Pinene (P31), 232, 598
Pinguisone (P32), 107, 385
Piperitenone (P33), 63, 262
Piperitone (P34), 63, 69
α-Pipitzol (P35), 427, 442
β-Pipitzol (P36), 427, 442
Pisiferin, 316
Pisiferol, 45
Pitoediol (P37), 561, 598
Platyphyllide, 410
Pleraphysillin I (P38), 308
Pleuromutilin (P39), 135, 165, 463, 481, 678
Plinols (P40), 69
Plumericin (P41), 639, 663
$\Delta^{8(14)}$-Podocarpen-13-one (P42), 156
Podocarpic acid (P43), 51, 55, 261, 287, 289, 493, 662, 663
Polygodial (P44), 293, 347, 732
Polypodatetraenes, 297
Portulol (P45), 348, 593, 690
Precapelladiene (P46), 469, 570, 659
Precarabrone (C24), 520
Preisocalamendiol (P47), 574
Prepinnaterpene, 737
Presqualene alcohol (P48), 172, 523
Prezizaene (Z2), 58, 461
6-Protoilludene, 106, 455
Pterosin E (P49), 256
Prychanolide * (P50), 501
Pulegone (P51), 262
Punctatin A, 496
9-Pupukeanone (P55), 613
Puupehenone (P56), 293
Pyroangolensolide (P57), 99
Pyrovellerolactone (V8), 212, 583
Quadrone (Q1), 91, 96, 97, 127, 138, 158, 210, 255, 266, 387, 421, 427, 446, 457, 461, 467, 507, 514, 548, 558, 597, 615, 628, 671, 688
Quassimarin * (Q2), 140, 343, 415
Quassin (Q3), 18, 38, 249, 342, 390
Radiatin *, 734
Retigeranic acid (R1), 63, 128, 319, 351, 406, 506, 549
Rhodolauradiol (R3), 313
Rimuene (R4), 52
Rishitinol (R5), 256

COMPOUND INDEX

Rosenonolactone (R6), 280
Rosmariquinone (R7), 348
Royleanone (R9), 52
Rudmollin (R10), 504
Ryanodol (R11), 122, 390
Sabinene (S1), 517, 521
Sabina ketone (S1'), 234
Sabinene hydrate (S2), 530
Sanadaol, 167
α-Santalene (S3), 268, 281, 528, 576
β-Santalene (S4), 379, 380, 595, 746
epi-β-Santalene (S5), 432, 673
β-Santalol (S6), 379
epi-β-Santalol (S6'), 432
Santolina alcohol (S8), 635
α-Santonin (S9), 21, 274, 630, 654
β-Santonin (S10), 21
Sativene (S11), 9, 35, 80, 218, 219, 424, 489, 561, 568, 582, 601, 656, 684
Sativenediol (S12), 239, 271, 484
Saussurea lactone (S14), 680
Sclerosporal (S16), 121, 400
Sclerosporin, 400
Secodaphniphyllene * (D7), 373
7,12-Secoishwaran-12-ol (I13), 435, 641
Selina-3,7(11)-diene (S18), 401
Selina-4(14),7(11)-dien-8-one (S19), 411
α-Selinene (S20), 5, 749
β-Selinene (S21), 14, 248, 266, 452
epi-γ-Selinene (S22)
δ-Selinene (S23), 133
Sempivirol (S24), 314
Serratenediol (S25), 293
Sesquicarene (S26), 398, 517, 524, 590
Sesquifenchene (S28), 374, 595, 673
Seychellene (S29), 33, 146, 161, 163, 169, 207, 210, 274, 381, 383, 418, 489
Shionone (S31), 284, 612
Shyobunone (S32), 639, 649, 680
Siccanin (S33), 49
Silphinene (S34), 104, 125, 377, 425, 505, 645, 671, 736
Silphiperfol-6-ene (S35), 126, 490, 505
Sinularene (S37), 207, 376, 423, 454, 461, 548, 607
Sirenin (S38), 239, 524, 581
α-Synderol (S39), 300
β-Snyderol (S40), 300
Solavetivone (S41), 394, 621, 676, 742
Spatol (S42), 537
Specionin (S44), 669
Spiniferin I (S45), 152, 240
Spirolaurenone (S47), 302, 621
Stachenone (S48), 55, 109, 584

Stemarin (S49), 86, 584
Stemodin (S50), 153, 203, 220, 643
Stemodinone, 585
Sterepolide (S51), 235, 255, 352, 481
Sterpurene (S52), 326, 515, 534, 755
Sterpuric acid, 534
Steviol (S53), 315, 560, 632
Stoechospermol (S54), 537, 670
Strigol (S55), 129, 191, 481
Sugiol (S56), 154
Talatisamine (T1), 88, 371, 594, 642
Tanshinone II (T2), 287, 348,
Taondiol (T4), 310
Taonianone, 68
Taxinine *, 471, 679
Taxodione (T7), 151, 292, 315, 636, 689
Taxusin * (T6), 408, 513,
Taylorione (T8), 66
Temsin (T10), 16, 680
δ-Terpineol (T11), 339,
Δ^1-Tetrahydrocannabinol (C16), 64, 200, 334, 636
$\Delta 6$-Tetrahydrocannabinol (C17), 64, 200, 334, 636
Tetrahymanol (T13), 309
Teucriumlactone C (T14), 669
α-Thujaplicin (T15), 563
β-Thujaplicin (T16), 443, 563, 570
γ-Thujaplicin (T17), 563, 568, 570
Thujopsene (T18), 16, 112, 516, 522, 523
Torrreyol (T20), 317, 400
Totarol (T21), 287
Trachylobane (T22), 528, 642
Trachylobanic acid (T23), 391, 529
Trichodermin (T24), 78, 394, 395, 612, 674
Trichodermol (T25), 579, 678
Trichodiene (T26), 192, 248, 332, 338, 398, 482, 644, 697
Trihydroxydecipiadiene (D10), 131, 539, 543, 662
Triptolide (T28), 123
Triptonide (T29), 299, 396
9Z-Trisporic acid B (T30), 70
9E-Trisporic acid B (T31), 70
Trisporol B (T32), 70
Tuberiferine (T34), 16, 201, 405
ar-Turmerone (T35), 699
Udoteatrial (U1), 221
Umbellifolide, 683
Umbellulone (U2), 104
Upial (U3), 77, 495
Valencene (V1), 45, 356,
Valeranone (V2), 11, 29, 279, 594, 607
Valerenal (V3), 308, 409

Valerianol (V4), 45
Veatchine (V5), 26, 82, 83, 186, 221, 565
Valleral (V6), 94, 569
Vellerolactone (V7), 569
Verbenalol (V9), 435, 625, 672, 756
Vernolepin (V10), 24, 173, 179, 342, 354, 357, 367, 626, 682, 686, 694
Vernomenin (V11), 24, 357, 682
Verrucarol (V12), 60, 223, 283, 368, 369, 380, 458, 560, 674, 750
Verticillol * (V13), 244
Vetiselinene, 196
α-Vetispirene (V14), 20, 119, 178, 484, 515, 630
β-Vetispirene (V15), 178
β-Vetivone (V16), 74, 139, 143, 178, 212, 213, 216, 230, 249, 265, 271, 277, 484, 485, 588, 617, 630, 676

Vinhaticoic acid, methyl ester (V17), 44
Vitrenal (V18), 119
Warburganal (W1), 347, 604, 661
Warburgiadione (W2), 150
Widdrol (W3), 16, 324, 467, 563, 596
Winterin (W4), 346,
α-Ylangene (Y1), 36, 231, 432, 497
β-Ylangene (Y2), 231, 491
Ylangocamphor (Y3), 9, 182, 207, 218, 684
Yomogin (Y4), 47, 193
Zizaene (Z1), 89, 148, 215, 388, 488, 595, 656
Zizanoic acid (Z3), 488, 592, 691
epiZizanoic acid (Z4), 107, 592
epiZonarene (Z5), 148, 402
Zonarol (Z6), 299